AF579156

MATHEMATICS FOR ELECTRICITY AND ELECTRONICS

MATHEMATICS FOR ELECTRICITY AND ELECTRONICS

GENE WARING

DELMAR PUBLISHERS INC.

COVER PHOTO: Courtesy of
Tottleben Micrographics
3526 Pestalozzi
St. Louis, Missouri 63118

Delmar Staff
Administrative Editor: Mark W. Huth
Production Editor: Eleanor Isenhart

For information address Delmar Publishers Inc.,
2 Computer Drive West, Box 15-015,
Albany, New York 12212

Printed in the United States of America
Published Simultaneously in Canada
by Nelson Canada,
A Division of International Thomson Limited

10 9 8 7 6 5 4 3 2 1

Library of Congress Cataloging in Publication Data

Waring, Gene
Mathematics for electricity and electronics.

1. Electric engineering–Mathematics. 2. Electronics–Mathematics.
I. Title
TK153.W28 1984 621.3'01'51 83-70906
ISBN 0-8273-1987-8

CONTENTS

Preface xii

SECTION 1 INTRODUCTION TO ELECTRICITY

1 INTRODUCTION AND CHOOSING AN ELECTRONIC CALCULATOR . . . 1

Mathematics and the Technician *1*
1-1 How to Study Electricity and Mathematics *2*
The Electronic Calculator *4*
1-2 Essential Functions *4*
1-3 Programmable Versus Nonprogrammable Calculators *4*
1-4 Choosing a Calculator *4*
Arithmetic of Whole Numbers *6*
1-5 Whole Numbers *6*
1-6 Place Value *6*
1-7 Addition *7*
1-8 Subtraction *8*
1-9 Multiplication *10*
1-10 Division *12*
1-11 Factoring Whole Numbers *13*
1-12 Mathematical Expressions *14*
1-13 Order of Operations *14*
1-14 Grouping Symbols *16*

2 BASIC ELECTRICAL CONCEPTS 19

Structure of Matter *20*
2-1 Subatomic Particles *21*
2-2 Planetary Model of the Atom *21*
2-3 Free Electrons *24*
2-4 Conductors, Semiconductors and Insulators *24*
The Electric Circuit and Basic Units *26*
2-5 Electric Current *28*
2-6 Electric Charge *29*
2-7 Resistance *29*
2-8 Electromotive Force and Potential Difference *30*
Electric Circuits and Measuring Instruments *32*
2-9 Circuit Symbols and Circuit Diagrams *32*
2-10 Series and Parallel Circuits *34*
2-11 Measuring Instruments *35*
2-12 Analog and Digital Readout *36*

3 THE SINGLE LOAD DC CIRCUIT. 38
Commmon Fractions 39
3-1 Mixed Numbers 39
3-2 Equivalent Fractions 40
3-3 Multiplying Common Fractions 42
3-4 Cancellation 42
3-5 Dividing Common Fractions 44
Decimals 47
3-6 Reading Decimal Fractions and Analog Scales 48
3-7 Changing Fractions to Decimals and Rounding Decimals 50
3-8 About the Calculator 52
3-9 Using the Calculator 52
3-10 Adding and Subtracting Decimals 53
3-11 Multiplying and Dividing Decimals 55
3-12 Multiple Operations 56
3-13 Reciprocals 59
3-14 Division Using the Reciprocal 60
3-15 Powers 61
3-16 Roots 63
Ohm's Law 65
3-17 The Single Load DC Circuit 66
3-18 Power Dissipation in Electric Circuits (DC) 68

4 GRAPHING ELECTRICAL QUANTITIES. 71
Introduction to Graphs 71
4-1 Variables 72
Constructing Graphs 74
4-2 Drawing the Curve 74
4-3 Graphing Rules 77
4-4 Electrical Graphs 80
4-5 Reading the Graph 82

5 NUMBER NOTATION FOR ELECTRICITY 86
Signed Numbers 87
5-1 Absolute Value of a Number 87
5-2 Adding and Subtracting Signed Numbers 88
5-3 Calculators and Signed Numbers 90
5-4 Vectors and Vector Addition 91
5-5 Rotation Operators 92
5-6 Vector Subtraction 92
5-7 Multiplying and Dividing Signed Numbers 93
Scientific Notation 94
5-8 Powers of Ten 95
5-9 Scientific Notation Defined 96
5-10 Significant Figures, Standard Position and Characteristic 96
5-11 Converting Between Decimal and Scientific Notation 99
5-12 Changing the Exponent 101
5-13 Color Codes 102
Arithmetic in Scientific Notation 105
5-14 Multiplying in Scientific Notation 106
5-15 Dividing in Scientific Notation 107
5-16 Powers, Roots and Reciprocals 109
5-17 Scientific Notation and the Calculator 111

6 MEASURING ELECTRICITY. 114
Reliability of Measurements 115
6-1 Uncertainty in Measurements 115
6-2 Maintaining Accuracy in Calculated Values 119

6 (Continued)
The SI Measuring System 120
6-3 Metric Units 121
6-4 Metric Prefixes 124
6-5 Rules for SI Usage 125
6-6 Metric to Metric Conversions 127
6-7 Common Units 129
6-8 More on Converting Units 130

7 USING ELECTRICAL FORMULAS 134
Formulas 135
7-1 Signs and Symbols 135
7-2 Multiplication and Division Axioms 138
7-3 Solving Equations 139
7-4 Translating Word Problems 141
Solving Electrical Problems 142
7-5 Formulas with Simple Products and Quotients 143
7-6 Problem Solving Procedure 145
7-7 Electrical Problems (DC) 146
Perimeter, Area and Volume 151
7-8 Plane Geometric Figures 151
7-9 Solid Geometric Figures 155
7-10 Wire Resistance 159

SECTION 2 DC CIRCUITS

8 SERIES CIRCUITS 165
Solving Equations 165
8-1 Terms 166
8-2 Like Terms 166
8-3 Addition and Subtraction Axioms 168
8-4 Solving Equations 168
8-5 Equations with Combined Operations 171
8-6 Formulas Containing Plus and Minus Signs 173
Series Circuits 174
8-7 Resistance in Series and Equivalent Circuits 175
8-8 Current in Series Circuits 176
8-9 Voltage in Series Circuits 178
8-10 Ohm's Law and Series Circuits 180
8-11 Internal Source Resistance 183
Power in Series Circuits 185
8-12 The Square and Square Root Axioms 185
8-13 Power 188

9 PARALLEL CIRCUITS 193
Addition and Subtraction of Common Fractions 194
9-1 The Least Common Denominator 194
9-2 Adding and Subtracting Fractions 196
Parallel Circuits 199
9-3 Resistance in Parallel Circuits and Equivalent Circuits 199
9-4 Voltage in Parallel Circuits 201
9-5 Division of Current in Parallel Circuits and Ohm's Law 202
9-6 Power Dissipation in Parallel Circuits 205
Simple Combination Circuits 206
9-7 Equivalent-circuit Method 207
9-8 Finding Total Resistance 207
9-9 Solving Combination Circuits 212

10 ADDITIONAL ELECTRICAL CONCEPTS 215
Variation of Electrical Quantities 216
10-1 Direct Variation 216
10-2 Inverse Variation 220
10-3 Joint Variation 224
Percents 228
10-4 Changing Fractions to Percents 228
10-5 Operations with Percents 230
10-6 Common Applications of Percents 230
10-7 Tolerance 232
Efficiency and Energy Consumption 234
10-8 Efficiency 236
10-9 Energy Consumption and Cost 242

11 SIMPLIFYING ALGEBRAIC EXPRESSIONS 246
Polynomials 246
11-1 Terms 247
11-2 Adding Monomials and Polynomials 248
11-3 Subtracting Monomials and Polynomials 250
11-4 Adding and Subtracting Horizontally 250
Multiplying and Dividing Polynomials 252
11-5 Exponents in Multiplication 252
11-6 Multiplying Monomials 253
11-7 Multiplying a Polynomial by a Monomial 254
11-8 Exponents in Division 255
11-9 Dividing a Monomial by a Monomial 257
11-10 Dividing a Polynomial by a Monomial 258
11-11 Powers and Roots of Monomials 259
11-12 Factoring Polynomials 261

12 PREPARING FOR CIRCUIT ANALYSIS 264
Solving Equations 264
12-1 Review of Simple Equations 265
12-2 Equations with Unknown in Several Terms 266
12-3 Equations with Unknown in Parentheses 268
12-4 Equations with Fractions 270
Simultaneous Equations 273
12-5 Graphing Equations 273
12-6 Graphical Method 276
12-7 Algebraic Methods 277
12-8 Addition-Subtraction Method 278
12-9 Substitution Method 281
12-10 Comparison Method 282
12-11 Word Problems 284
12-12 Deriving Formulas 287

13 CIRCUIT ANALYSIS 289
Kirchhoff's Laws 289
13-1 Three Equations with Three Unknowns 290
13-2 Kirchhoff's Loop Method 292
Other Methods 297
13-3 Voltage and Current Divider Rules 297
13-4 The Principle of Superposition 299
13-5 Thévenin's Theorem 305

SECTION 3 INTRODUCTION TO ac CIRCUITS

14 RATIO AND PROPORTION . . . 313
Ratio *313*
14-1 Simplifying Ratios *314*
14-2 Inverse, Decimal and Percent Ratios *315*
14-3 Electrical Applications of Ratios *316*
14-4 Efficiency *317*
14-5 Mechanical Advantage and Speed Ratio *318*
Proportion *321*
14-6 Cross-Multiplication *321*
14-7 Direct and Inverse Proportion *322*
14-8 Electrical Applications of Proportion *324*
14-9 Transformers *324*
14-10 Series and Parallel Circuits *328*
14-11 Ammeters and Voltmeters (DC) *329*
14-12 The Wheatstone Bridge *333*
14-13 Similar Triangles *334*

15 ANGLES AND RIGHT TRIANGLES . . . 339
Angles *339*
15-1 The Degree and Special Angles *340*
15-2 Generating Angles *341*
15-3 Angular Measurement (Degree and Radian) *343*
15-4 Other Angular Units *345*
The Right Triangle *346*
15-5 Angles of a Right Triangle *347*
15-6 Sides of a Right Triangle *348*
15-7 Drawing Right Triangles *352*

16 TRIGONOMETRY . . . 355
Trigonometric Functions *355*
16-1 Sine, Cosine and Tangent *356*
16-2 Functions of an Angle *357*
16-3 Estimating Sides and Angles *360*
16-4 Solving Side-Angle Right Triangles *362*
Inverse Trigonometric Functions *367*
16-5 Arc Sine, Arc Cosine and Arc Tangent *368*
16-6 Solving Side-Side Right Triangles *370*
Functions of Generated Right Triangles *375*
16-7 Generating Right Triangles *375*
16-8 Signs of the Functions *377*
16-9 Inverse Functions of Generated Triangles *380*

17 ALTERNATING CURRENT . . . 382
Voltage Generation *383*
17-1 Graphs of the Trigonometric Functions *383*
17-2 Magnetic Flux, Motion and Induced Current *386*
17-3 Mechanical Production of Alternating Voltage *390*
Fundamentals of the Periodic Cycle *394*
17-4 Period, Frequency and Angular Velocity *394*
17-5 Angular Frequency and the Voltage Equation *398*
17-6 Effective Alternating Voltage and Current in Resistive Circuits *401*
Reactance *406*
17-7 Inductance and Inductive Reactance *406*
17-8 The Inductive Circuit *408*

17 (Continued)
Quadratic Equations (Continued)
17-9 Capacitance and Capacitive Reactance *412*
17-10 The Capacitive Circuit *414*
17-11 Impedance *416*

18 THE j-OPERATOR 418
Binomial Products *419*
18-1 Multiplying Two Binomials *419*
18-2 Special Products *421*
18-3 Factoring Special Products *423*
Quadratic Equations *424*
18-4 Pure Quadratic Equations *424*
18-5 The Quadratic Formula *425*
18-6 j-Numbers *430*
Complex Numbers *431*
18-7 The j-Operator *432*
18-8 Adding and Subtracting Vectors *435*
18-9 Multiplying and Dividing Vectors *440*

SECTION 4 ANALYZING ac CIRCUITS

19 PHASORS 445
Phase Relationships in ac Circuits *446*
19-1 Circuits Containing Resistance and Reactance *448*
19-2 Phasors *450*
Transformation of Coordinates *453*
19-3 Impedance and Estimating Transformations *453*
19-4 Rectangular to Polar Transformation *457*
19-5 Polar to Rectangular Transformation *461*
Impedance in ac Circuits *464*
19-6 Phase and Power Factor *464*
19-7 Inductance and Capacitance *466*

20 ac SERIES CIRCUITS 469
Phasor Operations *470*
20-1 Adding Phasors *470*
20-2 Subtracting Phasors *473*
20-3 Multiplying Phasors *476*
20-4 Dividing Phasors *477*
Voltage and Current in ac Series Circuits *479*
20-5 Ohm's Law in ac Single-Element Circuits *480*
20-6 Two Element ac Series Circuits *482*
20-7 Series Circuits with Three or More Elements *485*
Power in ac Series Circuits *489*
20-8 Power in Single-Element ac Circuits *490*
20-9 Power in RL and RC Series Circuits *493*
20-10 Power in Multiple-Element Series Circuits *499*

21 ac PARALLEL CIRCUITS 503
Equivalent Circuits *503*
21-1 Two-branch Ideal Circuits *504*
21-2 Two-branch Practical Circuits *507*
21-3 The Admittance Method *509*
Voltage, Current and Power in Parallel Circuits *514*
21-4 Two-branch Ideal Circuits *514*
21-5 Practical Impedances in Parallel *520*

21 (Continued)

Series and Parallel Equivalent Circuits and Power Factor Correction *526*
21-6 Admittance, Conductance and Susceptance *526*
21-7 The Parallel Equivalent Circuit *530*
21-8 Power Factor Correction *532*

22 TWO-PHASE AND THREE-PHASE CIRCUITS **539**

Polyphase Generation *539*
22-1 Double-Subscript Notation and Two-Phase Alternators *540*
22-2 Current and Power in Two-Phase Circuits *544*
Three-Phase Generation *547*
22-3 The Wye-connected Alternator *547*
22-4 Current and Power in Four-Wire Wye Circuits *551*
22-5 The Wye-Delta System *556*
22-6 The Delta-Connected Alternator *563*

23 POWER LEVEL AND THE DECIBEL **569**

Logarithms *570*
23-1 Exponents and Logarithms *570*
23-2 Common Logarithms *572*
23-3 Antilogarithms *573*
23-4 Natural Logarithms *574*
23-5 Exponential Decay *575*
23-6 Exponential Growth *578*
Logarithmic Equations *580*
23-7 Properties of Logarithms *580*
23-8 Solving Logarithmic Equations *583*
Electrical Applications *585*
23-9 Power Ratio and Power Level *586*
23-10 Current and Voltage Ratios *587*
23-11 Approximating Decibels *590*
23-12 Transmission Lines *593*

Appendices **599**

A-1 American Wire Gage (AWG) Table *599*
A-2 Resistivity of Various Materials Table *600*
A-3 Temperature Coefficients of Resistance Table *600*
A-4 Standard Circuit Symbols *600*
A-5 The Greek Alphabet *601*
A-6 Conversion Factors *601*
B Trigonometric Table *602*
C Logarithmic Table *608*

Glossary **612**

Answers to Odd-Numbered Problems **623**

Index **666**

PREFACE

Mathematics for Electricity and Electronics is a specialized mathematics text. It is designed to provide the student with a knowledge of the mathematics required for the study of electricity. The mathematics chapters are interwoven with chapters on electrical theory. In this way, both subjects are developed in a logical and systematic sequence.

One of the primary goals of this text is to provide some of the basic electrical theory required by the electrical industry. Special emphasis is given to circuit analysis. A second goal is to provide the fundamentals of mathematics required for learning the electrical theory and for solving electrical problems at the technician level. The third primary goal is to develop each electrical and mathematical concept from basic principles and building in stages from simple to complex.

Some of the methods used to achieve these goals and to present the material in a consistent understandable way are described in the following:

- *Objectives* (goals) are given at the beginning of each chapter. The objectives indicate what student behavior should result after satisfactorily completing the chapter.
- *Pictures and Diagrams* are used extensively throughout the text to illustrate various electrical concepts and depict many problem situations.
- Numerous *solved examples* are given throughout the text. The examples are used to illustrate each new procedure or method and are typical of the problems that can be solved by that method. Checking answers is emphasized. Methods include quick mental checks as well as using alternate procedures for making the calculations.
- *Exercises* provide many practical problems of an electrical nature. Each exercise begins with the simple and advances to the more complex. An answer key is provided at the end of the text for the odd-numbered problems of each exercise.
- For easy reference, the *main topics* in each chapter are numbered in order by chapter and by topic. For example, Topic 5-9 refers to Chapter 5, Topic 9. In addition, figures and exercises have their own similar numbering system.
- A *unique glossary* containing both mathematical and electrical words and expressions is included in the Appendix.
- The use of the *electronic calculator* is encouraged in most chapters beginning in Chapter 3.

- *Scientific notation* is given an early high priority (Chapter 5).
- *Metric units* are explained in Chapter 6 and are used throughout the rest of the text.
- *Modern standards* are used. The Institute of Electrical and Electronic Engineers (IEEE) and the National Electrical Manufacturers Association (NEMA) are the references for most standard circuit symbols. The Systeme International d'Unités (SI) is the source for fundamental and related units, metric prefixes, abbreviations and symbols.

Many faculty and students have contributed their suggestions and comments. I am especially indebted to Reid Curtis whose influence can be found on many pages of the book; to Janalee Petsch who took many of the photographs and enlarged the prints; to A. J. Phillips, D. W. Pringle, T. Kucharski, and N. G. Karthas who extensively reviewed the chapters as the chapters were written; and to my wife, Nadine, who had the tedious job of preparing the manuscript. I am also very grateful to my wife and daughter, Nancy, for their support and encouragement.

SECTION 1 INTRODUCTION TO ELECTRICITY

INTRODUCTION AND CHOOSING AN ELECTRONIC CALCULATOR

OBJECTIVES

After satisfactorily completing this chapter, the student should be able to:

- Identify the primary goals of this text as the text relates to the electrical student.
- Choose an electronic calculator wisely.
- Define words and terms associated with the arithmetic of whole numbers.
- Perform basic operations (+, −, ×, ÷) on whole numbers including the solution of practical (word) problems.
- Find the various factors of a whole number including prime factors.
- Evaluate a mathematical expression including those containing two or more operations and grouping symbols.

Many fundamental ideas of electrical theory are based upon mathematics. Basic electrical principles are expressed in mathematical terms involving many mathematical operations. Electricians should be able to solve mathematics problems quickly and with confidence.

The mathematical content included in this text is the result of a national survey, an in-depth research of current electrical texts and years of experience. Only the mathematics that supports electrical curriculum is included.

MATHEMATICS AND THE TECHNICIAN

Sometimes a discouraged reader will ask, "Do I really need to know this much mathematics? Will I ever use it on the job?". The answer to

Figure 1-1 Both mathematical and electrical equipment are found in the toolkit of experienced electricians. (Courtesy of Southeast Nebraska Community College)

each question is "yes." The mathematical background requirements of electrical technicians depend on the type of jobs they hold.

Electricians can be found at residential, commercial or heavy industrial job sites. Electricians are found at power companies and telephone companies. Electricians are also found in electromechanical, fluid power, electronics and other fields. The electrician works in many fields doing many different jobs. Such activities as repair and maintenance, installation, fabrication, testing and design as well as the operation of electrical equipment are done by electricians. Most jobs overlap from one category to the next.

With such a wide variety of jobs in the electrical field, it is impossible to predict the kind of mathematics that will be needed. One can say for certain, however, that all electricians will use mathematics on the job. Without mathematics, analyzing electrical circuits, for example, would be difficult indeed.

A toolkit for an electrician consists of screwdrivers, soldering irons, wire strippers, conduit benders, voltmeters, a calculator and many other devices, Figure 1-1. Experienced electricians have found that a knowledge of mathematics is one of their most important tools.

1-1 HOW TO STUDY ELECTRICITY AND MATHEMATICS

The reader of this text is probably expecting to become a technician in an electrical field. The electrician's work will be much easier and more profitable with a good mathematical background. This text is intended for that purpose.

How students study has a strong influence on success. Learning is the responsibility of the student and the serious student will adopt tried and proven study methods and habits. The following suggestions work especially well in the study of mathematics and electricity.

- Choose a quiet place where noise and other distractions will be at a minimum.
- Study every day doing each assignment as it comes along. What you do in future weeks is based on what you study this week, Figure 1-2.
- In a particular study session:
  a. Read the text material in each section.
  b. Work each illustrative example problem using pencil, paper and a calculator.
  c. Study pictures and diagrams.
  d. Solve the problems in the exercises. Use the solved examples as guides, but don't be afraid to ask the instructor for help.

Figure 1-2 Electrical mathematics is learned the same way a transmission line is built: one step at a time. Many towers (mathematical concepts) with connecting lines form the whole. A distribution system (this textbook) is planned with a definite destination in mind. (Courtesy of Nebraska Public Power District)

Learning requires an adequate amount of study time. If students work few problems, the learning will be shallow and temporary, Figure 1-3. One cannot become skilled as an electrical technician with such temporary learning. Success comes by doing, not by watching someone else.

Figure 1-3 Mathematical concepts, like transmission line towers, are set on a firm foundation. In sandy areas, tower footings may be 9 feet across and 32 feet deep. (Courtesy of Nebraska Public Power District)

THE ELECTRONIC CALCULATOR

This text has been written to be used with an electronic pocket calculator. Trigonometric and logarithmic tables as well as the slide rule can still be used but a good calculator will save time and energy. The student should own a full function, scientific calculator. A name-brand calculator is recommended because an inadequate calculator will be disappointing.

1-2 ESSENTIAL FUNCTIONS

The mathematical functions and operations used in this text are listed in Figure 1-4. Any calculator without all of these is unsatisfactory for this text. Key numbers 15, 16, 28 and 29 are optional but very desirable. The leading manufacturers of calculators have models that have all the functions and operations shown. Many of these calculators are reasonably priced.

The many operations and functions listed in Figure 1-4 need not be frightening. There is a gradual progression in this text from the familiar to the unfamiliar. The calculator is used in most chapters. The student will be pleasantly surprised at the mathematics that can be learned while using the calculator.

The first sixteen functions and operations listed in Figure 1-4 are used in the first thirteen chapters. All others are introduced in later chapters. Any calculator having the first sixteen capabilities is adequate for the first thirteen chapters. A calculator with the other capabilities is required at the start of Chapter 15.

1-3 PROGRAMMABLE VERSUS NONPROGRAMMABLE CALCULATORS

Some calculators have most of the features described in Figure 1-4 and in addition, they have addressable memory stacks and storage registers. Programs that remember a series of keystrokes can be loaded into the calculator. The calculator then performs all the keystrokes automatically.

Programs can save time when certain calculations are done repetitively. In this text, a calculator with programming abilities is unnecessary. In fact, calculators with programming abilities may have some operations hidden as a second or third function of a given key. Thus, additional keystrokes are necessary to do some ordinary calculations. A calculator without programming abilities is probably best for this text. Of course, if a programmable calculator is available, use it.

1-4 CHOOSING A CALCULATOR

In choosing a personal calculator:

- Ask advanced students in the electrical department about their calculators.
- Shop around – look for name-brand calculators.
- Press the keys. Some have a different feel than others.

	KEY SYMBOL	ALTERNATE KEY SYMBOL	OPERATION OR FUNCTION	USAGE
1.	+		Add	Two number operations
2.	−		Subtract	
3.	×		Multiply	
4.	÷		Divide	
5.	y^x	x^y	Powers and roots	
6.	+/−	CHG	Change sign	Single number operations
7.	1/x		Reciprocal	
8.	$\sqrt{x}$		Square root	
9.	x^2		Square	
10.	FIX		Fixed point notation	Display and rounding
11.	SCI	ENG	Scientific notation	
12.	EE	EEX or EXP	Enter exponent	
13.	STO		Store	Memory
14.	RCL		Recall	
*15.	( )		Parentheses	
*16.	x⇄y	EXC	Exchange x and y	
17.	Sin		Sine	Trigonometric functions
18.	Cos		Cosine	
19.	Tan		Tangent	
20.	Sin^{-1}	INV + Sin	Arc sine	Inverse trig functions
21.	Cos^{-1}	INV + Cos	Arc cosine	
22.	Tan^{-1}	INV + Tan	Arc tangent	
23.	DEG		Degree	Angle selection and conversion
24.	Rad	DRG	Radian	
25.	π		3.14159 (rounded)	
26.	D→R	→DEG or →RAD	Degree to radians	
27.	D→DMS	→H.MS	Deg to deg-min-sec	
*28.	→R	P→R or →xy	Polar to rectangular	Coordinate transformation
*29.	→P	INV + P→R or →Rθ	Rectangular to polar	
30.	10^x	INV + LOG	Logarithms and antilogarithms	Log and exponent functions
31.	LOG	LOG x		
32.	e^x	INV + LN		
33.	LN			

*Optional

Figure 1-4 Essential calculator functions and operations

Figure 1-5A An LCD (liquid crystal diode) scientific calculator (Courtesy of Southeast Nebraska Community College)

Figure 1-5B An LED (light emitting diode) scientific calculator (Courtesy of Southeast Nebraska Community College)

- *LCD (liquid crystal diode).* These require little energy. The batteries will last a long time, Figure 1-5A.
- *LED (light emitting diode).* These require more energy. A battery charger for an LED calculator is recommended, Figure 1-5B.
- The calculator should be a scientific calculator which has the capabilities listed in Figure 1-4.

ARITHMETIC OF WHOLE NUMBERS

The electrician uses mathematics frequently. Solving problems is a daily experience. *Arithmetic of whole numbers* is the basis of all mathematics. The following is intended for those who require a brief review. It is also intended for use as a reference.

1-5 WHOLE NUMBERS

The *whole numbers* or natural numbers are the numbers used for counting. The counting numbers make up a number system using the *numerals:* 0, 1, 2, 3, 4, 5, 6, 7, 8, and 9. These numerals are known as *Arabic numerals.* With their use any whole number can be expressed. The system used to combine Arabic numerals to form various numbers is the number system used today. The system was originally devised in India. However, it did not use the Arabic numerals. Rightfully, the system should be called the Hindu system but it is commonly called the Arabic system.

1-6 PLACE VALUE

The Arabic system uses a system of *place value.* That is, the value of a numeral depends on its position or place. In the number 3333, for example,

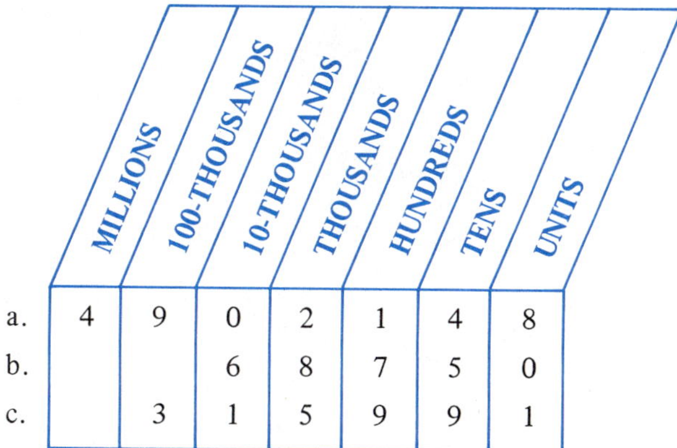

Figure 1-6 Place values

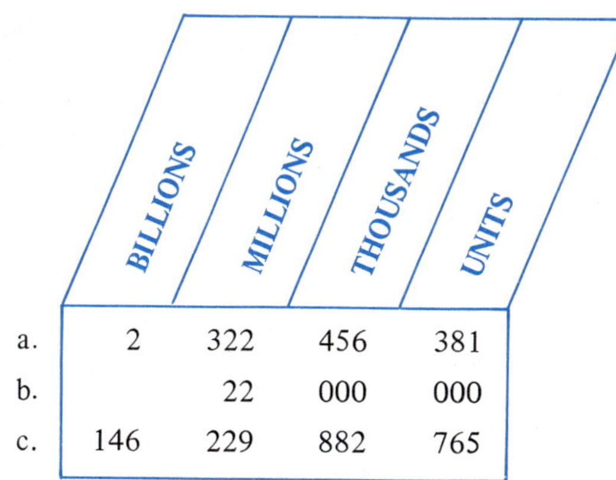

Figure 1-7 The digits of large numbers are separated in groups. Each group contains three digits.

the first numeral on the left represents 3000. The next numeral represents 300 and the third numeral represents 30. The last numeral has a value of 3. From left to right, each numeral represents the number of thousands, hundreds, tens and units, respectively. Figure 1-6 shows the place values of each numeral in the numbers (a) 4 902 148, (b) 68 750 and (c) 315 991.

In the decimal system these numerals are called *digits.* The digits of large numbers are separated in groups of three starting at the right side of whole numbers. Three examples are shown in Figure 1-7. Sometimes the separation is made with a comma. The SI (metric) measuring system separates the groups by a blank space. Since some countries use the comma to indicate the decimal point, the blank space has been adopted. This text will adopt the SI convention and omit the commas. No space will be used with four place numbers in this text.

EXERCISE 1-1

Write the following numbers in words.

1. 51 842
2. 8 560 000
3. 7 946 361 000
4. 2 316 000 000 000

Write the following numbers using numerals.

5. Thirty-nine thousand, four hundred eighty-five.
6. Seventeen million, seven hundred-thousand.
7. Four billion, two hundred eight million, eleven thousand.

Write the following numbers in words.

8. 32 617
9. 5 032 608
10. 14 561 000

1-7 ADDITION

The operation of *addition* is indicated by a plus sign (+). In the expression, 5 + 3 = 8, the numbers being added are called *addends.* The answer is called the *sum.* To add several numbers, list all the numbers. The units of each number must be in the same column. Likewise, tens, hundreds, etc. must be in their respective columns. List the numbers neatly in vertical columns.

Each column is now added separately starting with the units column. If the sum of any column is more than ten, write the last digit under the column being added. Add the other digit to the next column to the left. This process is called *carrying.* Once the answer is obtained it should be checked. One method of checking is simply adding in reverse order. Another method of checking is to add each column according to place value. Then total these values.

Example A Add: 5160 + 496 + 3603.

Solution:			**Check:**	
	⑴⑴			⑴⑴
	5160	(Addend)	(Reverse	3603
	496	(Addend)	order)	496
	+3603	(Addend)		+5160
	9259	(Sum)		9259

Example B Add: 8971 + 2698 + 904.

Solution:		**Check:**		
	⑵⑴⑴			
	8 971	(Sum of	13	Units
	2 698	columns)	160	Tens
	+ 904		2 400	Hundreds
	12 573		+10 000	Thousands
			12 573	

1-8 SUBTRACTION

Subtraction is indicated by the *minus* sign. In an expression such as 7 − 3 = 4, the 7 is the *minuend.* It is the number from which 3 is subtracted. The 3 is the *subtrahend.* It is the number being subtracted. The answer is the *remainder* or *difference.*

To subtract two whole numbers, place the smaller under the larger. Use care to place units under units, tens under tens, etc. Subtract each column starting with the units column. Check by adding the difference and the subtrahend. The result should equal the minuend.

Example A Subtract 521 from 652.

Solution:			**Check:**	
	652	(Minuend)		131
	−521	(Subtrahend)		+521
	131	(Difference)		652

If any digit in the subtrahend is larger than the corresponding digit in the minuend, it is necessary to *borrow.* The process of borrowing is shown in the following example.

Example B Subtract 586 from 723.

Solution:		**Check:**	
	⑪		137
	⑥ ~~2~~ ⑬		+586
	~~7~~ ~~2~~ ~~3~~		723
	−5 8 6		
	1 3 7		

Before 6 can be subtracted from 3, more units are needed. Borrow one of the tens and place it with the units to make 13. Now subtract the 6. Similarly, borrow one of the hundreds and place it with the tens.

EXERCISE 1-2

Add and check each of the following.

1. 6 + 7 + 4
2. 18 + 24 + 6
3. 319 + 817 + 701
4. 8 + 396 + 47
5. 8650 + 4534 + 6715 + 5847
6. 39 846 + 24 + 987 + 3791 + 22 222
7. 31 607 + 408 702 + 2 686 004

Subtract and check each of the following.

8. 14 − 3
9. 39 − 17
10. 50 − 14
11. 293 − 107
12. 2386 − 988
13. 4000 − 2641
14. Subtract 64 from 145.
15. Subtract 846 from 1299.
16. Take 521 from 1000.
17. From 4500 take 364.
18. Take 609 746 from 1 000 000.
19. Subtract 541 692 from 850 371.
20. The following pieces of wire are required for a job: 42 inches, 137 inches, 68 inches and 144 inches. Find, in inches, the total length of wire required.
21. Reels of cable and wire are shown in Figure 1-8. One of the reels of cable weighs 1285 lb, a block and tackle weighs 21 lb, a ladder weighs 68 lb, a transformer weighs 217 lb and a number of miscellaneous parts weighs 468 lb. If all of these are loaded onto a truck, find the total weight of the load.

Figure 1-8 Cable and wire reels vary from very small to very large.

22. Four wires are cut from a roll containing 1225 inches. How much wire is left on the roll if a total of 391 inches is required?
23. The total resistance in a series circuit is found by adding the individual resistances. The following resistances are connected in series: 46 000 ohms, 165 000 ohms, 242 000 ohms and 94 000 ohms. Find the total resistance.

24. A building for an electrical shop costs $228 625. The total cost of the building and lot is $256 375. Find the cost of the lot.
25. An electrician received a weekly wage of $394. After an increase, the electrician's weekly wage is $411. Find the weekly increase in wages.
26. The following mileages were recorded for a service truck: Monday 82 miles, Tuesday 35 miles, Wednesday 107 miles, Thursday 63 miles, Friday 128 miles and Saturday 28 miles. Find the total business mileage for the service truck.
27. The voltage drop in the line wire is the difference between input voltage and the voltage at the end of the line. A generator produces 238 volts. At the end of the line the voltage is 223 volts. What is the voltage drop in the line wires?

1-9 MULTIPLICATION

The process of *multiplication* is indicated by the times sign (X) or dot (•). Multiplication can also be indicated without either symbol by using parentheses. The expression 4 times 7 equals 28 can be written in these ways.

$$4 \times 7 = 28$$
$$4 \cdot 7 = 28$$
$$(4)(7) = 28$$

The number being multiplied is called the *multiplicand.* In the expression $4 \times 7 = 28$, the 4 is the multiplicand. The number by which the multiplicand is being multiplied is the *multiplier* (7). The answer is called the *product* (28).

Multiplication is a quick way to add. Adding seven fours is the same as multiplying four by seven:

$$4 + 4 + 4 + 4 + 4 + 4 + 4 = 28$$
$$4 \cdot 7 = 28$$

Multiplication can be performed in any order. The product 4•7 is the same as 7•4. If three or more numbers are to be multiplied, any order can be chosen:

$4 \cdot 7 \cdot 3 =$	$4 \cdot 3 \cdot 7 =$	$3 \cdot 7 \cdot 4 =$
$28 \cdot 3 =$	$12 \cdot 7 =$	$21 \cdot 4 =$
84	84	84

To check the answer in multiplication, simply multiply in a different order.

When multiplying by numbers with two or more digits, partial products are obtained, then added. When multiplying, write the numbers in columns with units over units, tens over tens, etc. The partial products are placed under each other. Multiply the multiplicand by the number in the units place of the multiplier. When the product of any two numbers is more than ten, record only the last digit. The other numbers are added to the next product. This is another example of *carrying.*

When multiplication by the units digit is complete, multiply by the tens digit, then the hundreds digit, etc. When all partial products are written down, they are added to obtain the overall product. The following examples indicate the process in more detail:

Example A Multiply 86 and 37.

Solution:		**Check:**
	86 (Multiplicand)	37
X	37 (Multiplier)	X 86
	602 (Partial	222
	258 products)	296
	3182 (product)	3182

Example B Multiple 4923 and 456.

Solution:		**Check:**
	4 923	456
X	456	X 4 923
	29 538	1 368
	246 15	9 12
	1 969 2	410 4
	2 244 888	1 824
		2 244 888

EXERCISE 1-3

Multiply and check each of the following.

1. Multiply 47 by 18.
2. 246 X 27
3. 328·92
4. (4437)(618)
5. (4)(7)(6)(8)(3)
6. 956 X 3 X 14
7. 42·652·12
8. 52 640 X 728 X 4
9. Number 4/0 copper stranded line weighs 653 pounds per thousand feet. Find, in pounds, the weight of 25 000 feet.
10. Number 4/0 aluminum wire weighs 195 pounds per thousand feet. Find, in pounds, the weight of 75 000 feet.
11. If 18 poles are required in each mile for a transmission line, how many poles are required for 25 miles?
12. Transformer oil weighs about 8 pounds per gallon. What is the weight of oil in a full 55-gallon oil drum?
13. Find the total wattage of six 75-watt lamps and nine 60-watt lamps.
14. A general purpose junction box, Figure 1-9, has the following inside dimensions: 14 inches (length) by 12 inches (height) by 4 inches (width). Find the internal volume in cubic inches. (Volume = length X width X height).
15. An electrician installing junction boxes worked an average of 3 hours on each junction box. If the electrician earns $9 per hour how much is earned while installing 14 junction boxes?
16. If 1 horsepower equals 746 watts, how many watts are in 15 horsepower?

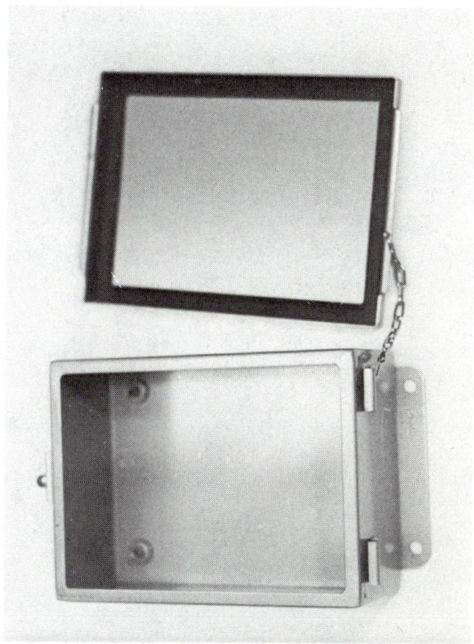

Figure 1-9 A general purpose junction box

1-10 DIVISION

The process of *division* is indicated by the symbols: ÷, $\overline{)\ \ }$, / (slash) or — (bar). Division is a method of finding how many times one number is contained in another. In the example: $12 \div 4 = 3$, the number 12 contains the number 4 three times. This is true because $3 \times 4 = 12$.

The number being divided is called the *dividend.* The number by which the dividend is divided is called the *divisor.* The answer in division is called the *quotient.* A *remainder* is left over when the division is not exact. In the expression $12 \div 4 = 3$, the number 12 is the dividend, 4 is the divisor and 3 is the quotient.

To divide whole numbers, use the symbol ($\overline{)\ \ }$). The divisor is placed to the left of the symbol and the dividend is inside the symbol. The quotient is placed above the symbol. Determine how many times the divisor is contained in the first digit of the dividend. If the divisor is larger than the first digit, use the first two digits. Use as many digits as necessary to get a one digit quotient.

Write the partial quotient directly above the last digit used. Multiply the divisor by the partial quotient. Subtract this product from the digit/digits that were used. Caution: keep all columns of numbers vertical. *Bring down* the next digit from the dividend. Annex it to the difference obtained above. Divide this new number by the divisor.

Repeat the above procedure over and over until all digits in the dividend are used. Check the answer by multiplying the quotient and the divisor. This product should equal the dividend. Study the following examples:

Example A Divide 581 by 7.

Solution:

```
   83
7)581
  56
   21
   21
```

Check: $83 \times 7 = 581$

Since 7 is larger than 5, 58 must be used. $58 \div 7 = 8$ with 2 left over. Write the 8 above the last digit used. Subtract 56 from 58. Bring down the 1. Now divide 21 by 7. $21 \div 7 = 3$. Write the 3 to the right of the 8 in the quotient.

Example B Divide 77 484 by 132.

Solution:

```
        587
132)77 484
    66 0
    11 48
    10 56
       924
       924
```

Check: $587 \times 132 = 77\ 484$

$5 \times 132 = 660$. The 5 is placed above the last digit used (4). The 660 is placed beneath 774. Subtract, bring down the 8. $8 \times 132 = 1056$, etc.

EXERCISE 1-4

Solve and check each of the following.

1. 231 ÷ 7
2. 168/3
3. 21)462
4. Divide 9198 by 73.
5. How much is 11 534 divided by 158?
6. How many times is 124 contained in 56 172?
7. An electrical shop uses 414 kilowatt-hours of electrical energy in a 23-day month. What is the average usage per day?
8. A certain length of copper line (00 gauge) weighs 50 316 kg. How many kilometres of line is there if each kilometre weighs 599 kg?
9. A power line is to be constructed with 14 towers per mile. How many miles of line can be constructed with 154 towers?
10. A reel of steel cable contains 432 metres. If this cable is to be cut into 9-metre lengths for guy wires, how many lengths will there be?

1-11 FACTORING WHOLE NUMBERS

A *factor* of a whole number is any number by which the whole number can be evenly divided. For example: the number 6 has factors of 2 and 3. The number 12 has factors of 2, 3, 4 and 6. Some numbers have many factors; some have very few. Some numbers are only divisible by 1 and by the number itself. The numbers 2, 3, 5 and 7 are examples. The set of whole numbers can be divided into two subsets. Those whole numbers that cannot be factored except by 1 and the number itself are called *prime numbers.* Numbers that can be factored are called *composite numbers.* Composite numbers can be expressed as a product of two or more smaller whole numbers (except 0 and 1).

The *prime factors* of a number are the prime numbers whose product is the original number. The numbers 3 and 5 are the prime factors of 15. Prime factors of a number are divisors of that number. The prime factors of a number can, therefore, be found by repeated division by prime numbers which exactly divides the given number. Divide even numbers by 2 as many times as necessary to obtain an odd number. Continue dividing by other primes.

Example A Find the prime factors of 168.

Solution:

2)168
2)84
2)42
3)21
7

The prime factors of 168 are 2, 2, 2, 3 and 7. Therefore: 2 × 2 × 2 × 3 × 7 = 168.

Example B Find the prime factors of 7854.

Solution:

$$\begin{array}{r} 2\overline{)7854} \\ 3\overline{)3927} \\ 7\overline{)1309} \\ 11\overline{)187} \\ 17 \end{array}$$

The prime factors of 7854 are 2, 3, 7, 11 and 17. Therefore: $2 \times 3 \times 7 \times 11 \times 17 = 7854$.

EXERCISE 1-5

1. List the first twelve prime numbers in order.

The number 12 has two pairs of factors: 3×4 and 6×2. For each of the following composite numbers, find as many such pairs of factors as possible.

2. 16
3. 24
4. 30
5. 32
6. 64
7. 18
8. 42
9. 100

Find all prime factors of each of the following.

10. 36
11. 60
12. 75
13. 275
14. 748
15. 3927

1-12 MATHEMATICAL EXPRESSIONS

A mathematical expression combines two or more numbers and one or more operations. Expressions range from simple to complex. The following are mathematical expressions:

a. $3 + 4$
b. $11 - 9$
c. 5×2
d. $28 \div 7$
e. $3 + 4 \times 2$
f. $5 - 8 \div 2$
g. $6 + 8 \times 4 - 3$
h. $7 - 12 \div 3 + 3 \times 2 + 4$

The *value* of an expression is the answer obtained when the numbers are combined according to the operations indicated. The value of an expression is the number represented by the expression. In the expressions listed, a and b represent the numbers 7 and 2, respectively.

A mathematical expression may consist of one or more terms. A *term* is an expression in which the parts are not separated by a plus (+) or a minus (–) sign. In the expressions listed, c and d are single term expressions. Single term expressions are sometimes called *monomials.*

In the expressions listed, a, b, e and f are two term expressions or *binomials.* Expression g has three terms and h has four terms. An expression with more than one term is called a *polynomial.*

1-13 ORDER OF OPERATIONS

A mathematical expression, however complex, represents a single number. For example, the expression $8 - 3 + 2$ represents the number 7. The expression is evaluated from left to right: $(8 - 3) + 2 = 5 + 2 = 7$. If it was evaluated from right to left, it would represent a different number: $8 - (3 + 2) = 8 - 5 = 3$. A mathematical expression must represent the same number for anyone who evaluates it. An expression may involve any or all of the four fundamental operations (+, –, ×, ÷).

The order in which the operations are done is important. A series of subtractions or divisions must be performed in the order given from left to right.

$$36 - 12 - 7 = 17$$
$$320 \div 16 \div 4 = 5$$

Suppose, in the subtraction problem, 7 is subtracted from 12 first, and then that answer is subtracted from 36. A different and wrong answer is obtained. Similarly, in the division problem, if $16 \div 4$ is evaluated first and then the 320 is divided by that answer, an incorrect answer will result. Such expressions must be evaluated from left to right.

A series of additions or multiplications may be performed in any order.

$$4 + 3 + 6 + 1 = 14$$
$$3 \times 2 \times 5 \times 4 = 120$$

The addition and multiplication problems can be evaluated from left to right, right to left or any other order. It is probably best, however, to perform all operations from left to right.

When two or more of the four operations are involved, the left to right order becomes critical. Use the following order of operations on all multiple operation problems.

ORDER OF OPERATIONS

- Perform all multiplications and divisions first in order from left to right. Each term must be evaluated before they can be added or subtracted.
- Combine terms by adding and subtracting in order from left to right. Terms are expressions or numbers separated by plus or minus signs.

To clarify the following examples, the operation performed is shown in the following line in parentheses at the right margin. In addition, partial answers are indicated in boldface print.

Example A Evaluate: $92 - 63 \div 9 \times 3 + 26$.

Solution: This expression has three terms.

$(92) - (63 \div 9 \times 3) + (26)$

The middle one must be evaluated first from left to right.

$= 92 - \mathbf{7} \times 3 + 26$ $\quad (63 \div 9)$

$= 92 - \mathbf{21} + 26$ $\quad (7 \times 3)$

Now terms can be combined from left to right.

$= \mathbf{71} + 26$ $\quad (92 - 21)$

$= 97$ $\quad (71 + 26)$

Example B Evaluate: $35 \div 7 + 2 \times 5 - 12 \div 3 \times 2 + 1$.

Solution: This expression has four terms.

$(35 \div 7) + (2 \times 5) - (12 \div 3 \times 2) + (1)$

$= \mathbf{5} + \mathbf{10} - \mathbf{8} + 1$ $\quad (\div, \times, \div \times)$

$= \mathbf{15} - 8 + 1$ $\quad (5 + 10)$

$= 7 + 1$ $\quad (15 - 8)$

$= 8$ $\quad (7 + 1)$

EXERCISE 1-6

How many terms are contained in each of the following expressions?

1. $4 + 2 + 3 - 5$
2. $286 \div 143$
3. $1 + 1$
4. $5 \times 4 - 3 + 32 \div 8$
5. $26 \times 2 \div 13 \times 4 \div 8$
6. $7 \div 7 + 42 + 820 \div 41 \div 5 - 1$
7. $4 \div 2 + 4 - 2 \times 2 + 6 \div 2 - 1$
8. $1 \times 1 + 1 \div 1 - 1 \times 1$
9. $6/2 + 48/4 - 64/8$ — Hint: Slash means divide.
10. $12 \div 3 + 8 \div 4 - 2 \times 2 + 3 \times 2 + 7 - 3 \times 2$

Each expression in problems 11-20 has been evaluated. The answers (some correct; some incorrect) are given in parentheses. For the answers that are correct write the word "correct" and for the answers that are incorrect write the correct answer.

11. $4 + 2 + 3 - 5$ (4)
12. $286 \div 143$ (2)
13. $1 + 1$ (1)
14. $5 \times 4 - 3 + 32 \div 8$ (9)
15. $26 \times 2 \div 13 \times 4 \div 8$ (2)
16. $7 \div 7 + 42 + 820 \div 41 \div 5 - 1$ (38)
17. $4 \div 2 + 4 - 2 \times 2 + 6 \div 2 - 1$ (4)
18. $1 \times 1 + 1 \div 1 - 1 \times 1$ (1)
19. $6/2 + 48/4 - 64/8$ (7)
20. $12 \div 3 + 8 \div 4 - 2 \times 2 + 3 \times 2 + 7 - 3 \times 2$ (8)

Evaluate these expressions.

21. $5 + 3 \times 2$
22. $5 \times 3 + 2$
23. $14 \times 3 \div 7 + 6$
24. $5 - 8 \div 2$
25. $14 - 12 \div 3 \times 2$
26. $16 - 12/4 + 2$
27. $29 - (4)(7) + 4$
28. $15 \div 3 - 8 \div 2$
29. $7 + 13 - 4 \times 2$
30. $27 \div 3 + 21 \div 7$

1-14 GROUPING SYMBOLS

In the expression $2 \times 3 + 4$, the multiplication must be done before the addition. Multiply first, then add. However, there are many times when it is necessary to do the addition before the multiplication. The expression $2 \times 3 + 4$ cannot be used if the addition must be performed first.

Whenever a certain operation needs to be performed first, *parentheses* ( ) or other *grouping symbols* are used. Grouping symbols are used to group terms together. This also establishes priority in the order of operations. The operations within such groups must be performed first. The expression $2(3 + 4)$ indicates addition before multiplication. Notice the times sign (X) is omitted. Compare the following expressions:

$$2 \times 3 + 4 = 6 + 4 = 10$$
$$2(3 + 4) = 2 \times 7 = 14$$

Notice the first expression has two terms but the second has only one. Even though there are two terms inside parentheses, overall, the second expression is a monomial.

Other grouping symbols are *brackets* [ ], *braces* { } and the *bar* or *vinculum* —. Brackets and braces are used to indicate groups within groups: $5 + \{3 + [4 + (3 - 2)]\}$. Always begin with the innermost group: parentheses first, brackets second and braces third.

Example A Evaluate: $5 + \{3 + [4 + (3 - 2)]\}$

Solution:

$$\begin{aligned} & 5 + \{3 + [4 + (3 - 2)]\} \\ &= 5 + \{3 + [4 + \mathbf{1}]\} && (3 - 2) \\ &= 5 + \{3 + \mathbf{5}\} && (4 + 1) \\ &= 5 + \mathbf{8} && (3 + 5) \\ &= 13 && (5 + 8) \end{aligned}$$

Parentheses, brackets and braces are used to group numbers together. They are also used to indicate multiplication if no operation symbol precedes them: $4(3 + 2)$ means $4 \times (3 + 2)$.

The bar is also a grouping symbol. It usually indicates division. The following expressions are exactly identical:

$$\frac{15}{3 + 2} = 15 \div (3 + 2)$$

The expression $3 + 2$ must be evaluated before dividing. In the electrical industry, the bar is used for division. One seldom sees the divide (÷) symbol in formulas.

In an expression consisting of two or more of the four operations (+, –, X, ÷) and grouping symbols, use the following order of operations.

ORDER OF OPERATIONS WITH GROUPING SYMBOLS

- Evaluate expressions in the innermost group first. Continue until all of the grouping symbols are eliminated.
- Perform all multiplications and divisions in order from left to right.
- Combine terms (addition and subtraction) in order from left to right.

Example B Evaluate: $2\{5 + 2[3(5 - 3) \div 2] - 7\}$

Solution:

$2\{5 + 2[3(5 - 3) \div 2] - 7\}$	
$= 2\{5 + 2[3 \times \mathbf{2} \div 2] - 7\}$	(5 − 3)
$= 2\{5 + 2[\mathbf{6} \div 2] - 7\}$	(3 × 2)
$= 2\{5 + 2 \times \mathbf{3} - 7\}$	(6 ÷ 2)
$= 2\{5 + \mathbf{6} - 7\}$	(2 × 3)
$= 2 \times \mathbf{4}$	(11 − 7)
$= 8$	(2 × 4)

Example C Evaluate: $\dfrac{27}{3 + 2(2 + 1)} + 4$

Note: This expression is equivalent to 27 ÷ [3 + 2(2 + 1)] + 4 with the bar replacing both the divide sign (÷) and the brackets.

Solution:

$\dfrac{27}{3 + 2(2 + 1)} + 4$	
$= \dfrac{27}{3 + 2 \times \mathbf{3}} + 4$	(2 + 1)
$= \dfrac{27}{3 + \mathbf{6}} + 4$	(2 × 3)
$= \dfrac{27}{\mathbf{9}} + 4$	(3 + 6)
$= \mathbf{3} + 4$	(27 ÷ 9)
$= 7$	(3 + 4)

EXERCISE 1-7

Find the numbers represented by each of these expressions.

1. $(5 - 3) \times 2$
2. $7(9 + 4)$
3. $6 + (8 - 5)$
4. $18 - (26 - 17)$
5. $(3 + 4)(11 - 4)$
6. $44 - 8(7 - 2)$
7. $12 + 3(7 + 8)$
8. $12(16 - 9) \div 14$
9. $5 - [(2 + 3) - (5 - 2)]$
10. $[(17 - 5) - 4] \div 2$
11. $296 - 12(8 + 7)$
12. $[5280 \div (740 - 80)] \times 4$
13. $5[26 + 2(28 - 21)]$
14. $3[14 - 5(31 - 29)] - (3 + 4)$
15. $28\{54 - 3[(7 + 5) + 2(71 - 68)]\} + 8(3 + 1)$
16. $\{[(17 + 16) \div 3] - 3(26 \div 13 + 1)\} \times 2 - 2$
17. $\dfrac{4(13 - 6)}{7} + 12$
18. $\dfrac{288}{5 + 3(16 - 9) - 2} - \dfrac{3(8 + 6 - 12)}{2}$
19. $\{3[17 - (14 - 11)] + 2[(16 - 12) + (17 - 9)]\} \div 2$
20. $\dfrac{(3 + 4)(12 - 5)}{8(2 + 3) + (8 - 5) \div 3}$

CHAPTER 2

BASIC ELECTRICAL CONCEPTS

OBJECTIVES

After satisfactorily completing this chapter, the student should be able to:

- Define various electrical words and terms introduced in this chapter.
- Describe the planetary model of the atom in terms of subatomic particles. Describe the action of free electrons.
- Name and/or describe electrical quantities and their units, various circuit elements and measuring instruments as they relate to the single load circuit and circuit diagrams.

Electricity is a form of energy which is produced in vast amounts at *power stations,* Figure 2-1. By transmitting electricity along power lines, the electricity is made available at places where it can be used conveniently.

Generators (electric machines) at the power station do not create the energy, Figure 2-2. The generators simply transform the energy from one form (coal, oil, water, wind, nuclear, etc.) to electrical energy. *Transmission lines* (electric circuits) provide a conduit by which electricity is transported to distant places. The electricity is controlled by other electric devices and finally transformed to other kinds of energy (light, heat, sound, etc.) or to useful work.

Electricity is the study of the electrical machines and devices and the electric circuits that make up an electrical distribution system. In this chapter, the basic ideas of electricity are presented beginning with the structure of matter.

Figure 2-1 Gerald Gentlemen Station, a coal fired electrical plant in Nebraska. Other plants use diesel fuel, moving water, wind, heat from the earth and nuclear power. (Courtesy of Nebraska Public Power District)

Figure 2-2 An alternating current generator driven by a diesel engine. The railing is about 3 1/2 feet high. This generator is typical of those used in small power plants.

STRUCTURE OF MATTER

Matter is defined as anything that has weight and occupies space. A few common examples are:

Solids: iron, copper, ice and salt
Liquids: water, gasoline, alcohol and mercury
Gases: air, hydrogen, natural gas and steam

Any sample of matter can be classified as either pure or impure. When a sample of a *pure substance* is broken down to the base particles, all of the particles will be alike. An *impure substance (mixture)* consists of two or more kinds of base particles.

In a pure substance, these small particles having all the physical properties of the substance are called *molecules.* It is entirely correct to speak of molecules of an element. The molecule is made up of atoms. An *atom* is the smallest particle that can take part in a chemical change. Atoms are the building blocks of molecules. Pure substances are subdivided into two kinds: elements and compounds. *Elements* have molecules that are made up of only one kind of atom. *Compounds* have molecules that are made up of more than one kind of atom.

There are 92 elements that occur in nature. About a dozen more have been produced in laboratories as a result of nuclear reactions. There are literally millions of compounds. Some are small ones with two-atom molecules. Others are made up of thousands of atoms.

PARTICLE	CHARGE	MASS	VOLUME
Electron	–	1	About
Proton	+	1836	the
Neutron	0	1837	same

Figure 2-3 Properties of subatomic particles

2-1 SUBATOMIC PARTICLES

The building blocks of atoms are *electrons, protons* and *neutrons.* Figure 2-3 shows the differences between the three particles. The differences in *electric charge* is probably the most commonly known feature of the three. Electrons have a negative charge, protons have a positive charge and neutrons are electrically neutral. Electric charge is a property of an electrified particle. Electrified particles obey the *first law of interaction:*

Like electric charges repel each other and unlike charges attract each other.

Two electrons repel each other. Two protons repel each other. An electron and a proton having unlike charges will attract each other. Uncharged particles such as the neutron are neither attracted to nor repelled by other neutrons or by electrons and protons.

A comparison of the mass or weight of the three subatomic particles is also given in Figure 2-3. The proton and neutron weigh about the same although the neutron is slightly heavier. In contrast, the electron is much, much lighter. The proton and neutron weigh over 1800 times more. The volume or size is about the same for all of the three subatomic particles.

2-2 PLANETARY MODEL OF THE ATOM

No one has ever seen the subatomic particles or even individual atoms. However, a mental picture or model of atoms has been devised that explains their behavior. The following numbered statements describe the basic model. This model is sometimes referred to as the *planetary model of the atom.*

1. The *nucleus* or center is made up of protons and neutrons.
2. Electrons revolve about the nucleus in definite *orbits* or *shells.*
3. A *normal* (electrically neutral) *atom* has the same number of electrons in orbit as there are protons in the nucleus.
4. Nearly all the mass of an atom is in its nucleus.
5. An atom is almost entirely empty space.

The nucleus consists of protons and neutrons. The protons give the nucleus a positive charge. The number of protons in the nucleus determines the kind of atom. The atoms of each element have different numbers of protons. The number of protons in the nucleus is called the *atomic number* of a given element. The atom described in Figure 2-4 is an atom of aluminum. Every aluminum atom has thirteen protons in the nucleus. If an atom's nucleus should gain or lose a proton or two, it becomes an atom of a different element. This would occur only in nuclear reactions.

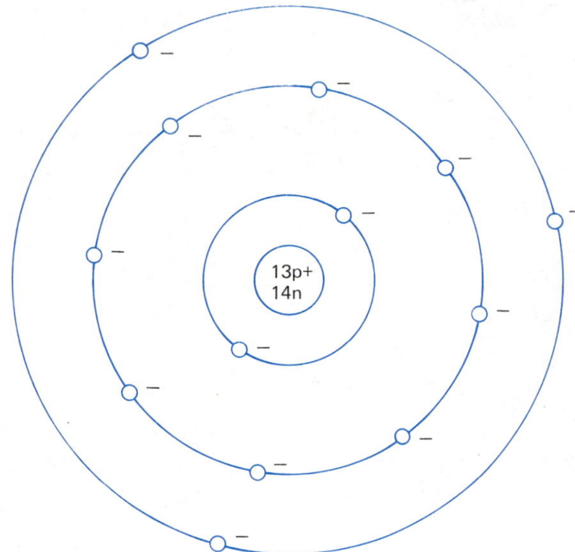

Figure 2-4 The planetary model of an aluminum atom

When protons are close together they tend to repel each other. However, when protons (and neutrons) are extremely close together, another stronger force takes over. It is called a *nuclear force* and is always an attractive force. Within the nucleus, the nuclear force of attraction may be 100 times greater than the repulsive electrical force. Nuclear forces bind protons and neutrons together to form the nucleus.

An electron must stay in an orbit or shell with a very definite radius. Figure 2-4 illustrates the planetary model of the atom. Two electrons (–) revolve about the nucleus in what is called the *innermost shell.* Only two electrons can occupy the innermost shell in any atom. Other electrons are found in shells with a larger radius.

The aluminum atom has thirteen electrons. It has two in the innermost shell. There are eight electrons in the second shell and three in the third. Other atoms may have electrons in a fourth, fifth or higher numbered orbit. More will be said about electron shells in the next topic.

The number of electrons in any normal atom is the same as the number of protons in the nucleus. One electron will balance one proton. A *normal* or *balanced* atom, therefore, is electrically neutral. An atom may gain or lose an electron or two. In so doing, they are still atoms of the same kind.

If an atom gains an electron, it becomes negatively charged. When an atom loses an electron, it will be positively charged. Charged atoms are called *ions.* An ion is an atom that has gained or lost some electrons. Atoms are usually ionized by a gain or loss of electrons from the *outermost occupied shell.* These outermost electrons are called *valence electrons.*

There are always matching numbers of protons and electrons in an uncharged atom. In most atoms there are more neutrons than protons. The massive protons and neutrons are in the nucleus so most of the mass is there. The tiny amount of mass not in the nucleus is possessed by the electrons.

Most of the space within an atom is empty. Electrons, protons and neutrons are all about the same size. However, these particles are very small compared to the whole atom. The diameter of most atoms is about 10 000 times larger than the nucleus. If an atom was a mile in diameter, the nucleus would range in size from a marble to the size of a tennis ball, depending on the kind of atom.

In summary, the nucleus containing protons and neutrons possesses most of the mass of the atom and a positive charge. Electrons orbit the nucleus in definite orbital shells in the same number as there are protons in the nucleus. The size of the electrons, protons and neutrons is very small compared to the size of the whole atom.

EXERCISE 2-1

Choose the one answer that best completes the statement.

1. The smallest particle that can take part in a chemical change is called (a) an atom, (b) a nucleus, (c) a molecule, (d) a crystal.
2. A material made up of only one kind of atom is (a) a compound, (b) a proton, (c) an element, (d) a solid.
3. The subatomic particle with a positive charge is (a) an electron, (b) a proton, (c) a neutron, (d) a positive ion.
4. The subatomic particle with a negative charge is (a) an electron, (b) a proton, (c) a neutron, (d) a negative ion.
5. The nucleus of an atom is composed of (a) protons and electrons, (b) electrons and neutrons, (c) electrons, protons and neutrons, (d) protons and neutrons.
6. The subatomic particle with the least mass (weight) is the (a) electron, (b) proton, (c) neutron, (d) nucleus.
7. An ion is an atom that has (a) lost or gained protons, (b) lost or gained neutrons, (c) lost or gained electrons, (d) more neutrons than protons.
8. The part of the atom in which electrons are found is called (a) the nucleus, (b) the orbits or shells, (c) the center, (d) the positively charged section.
9. The part of the atom which contains most of the mass is called the (a) orbits or shells, (b) negatively charged section, (c) ionic section, (d) nucleus.
10. A material having molecules made up of two or more kinds of atoms is (a) an element, (b) a compound, (c) a neutron, (d) an ion.
11. A normal or balanced atom is one that has equal numbers of (a) protons and neutrons, (b) electrons and neutrons, (c) electrons and protons, (d) valence electrons and planetary electrons.
12. Like (similar) charges tend to (a) repel each other, (b) attract each other, (c) orbit each other, (d) disintegrate when near each other.

2-3 FREE ELECTRONS

Figure 2-4 shows the planetary model of the aluminum atom. The first (innermost) orbit contains two electrons. The second orbit contains eight electrons. These ten electrons are very tightly bound to the nucleus. The force of attraction to the nucleus is very strong and they cannot escape.

The three valence electrons in the third (outermost) orbit are not so tightly bound. The third orbit in aluminum atoms is the *valence orbit*. Atoms with a higher atomic number have more protons in the nucleus. They can bind as many as eighteen electrons in the third orbit with more in a fourth orbit. Very large atoms have six orbits of very tightly bound electrons. In these, the seventh orbit contains the valence electrons.

The aluminum atom has only three occupied orbits. However, it has other unoccupied orbits beyond the valence (third) orbit. These orbits are called *conduction orbits,* Figure 2-5. Together, they make up a *conduction band* of orbits. Valence electrons can easily be bumped into the higher orbits. Heat and other kinds of energy can easily produce this effect.

A conduction orbit has a very large radius. There is almost no nuclear attraction acting on an electron in a conduction orbit. If an electron is lifted to a conduction orbit, it is essentially free to move from one atom to another. Electrons in the conduction orbits are often called *free electrons.*

2-4 CONDUCTORS, SEMICONDUCTORS AND INSULATORS

Figure 2-5(A) shows the aluminum atom. The third orbit is the valence orbit for aluminum atoms. Higher orbits are normally not occupied but are shown as a band of orbits. For convenience in drawing, the curved orbits are usually visualized as horizontal lines like those of Figure 2-5(B). Electrons in higher orbits have more energy than those in lower orbits. The various orbits are sometimes referred to as *energy levels.*

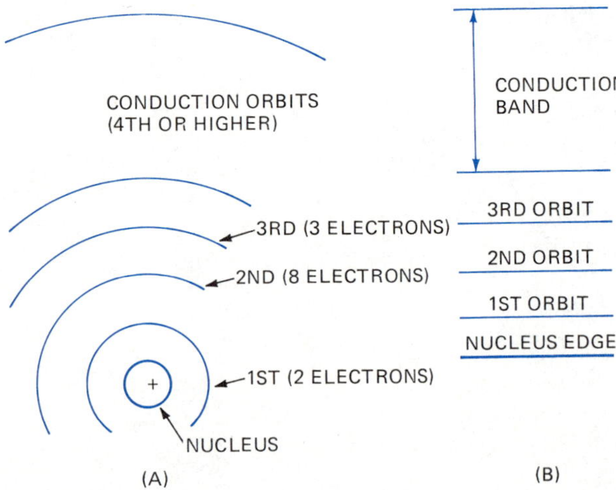

Figure 2-5 The aluminum atom showing normally occupied orbits and the conduction band. Electrons in the conduction band are free electrons.

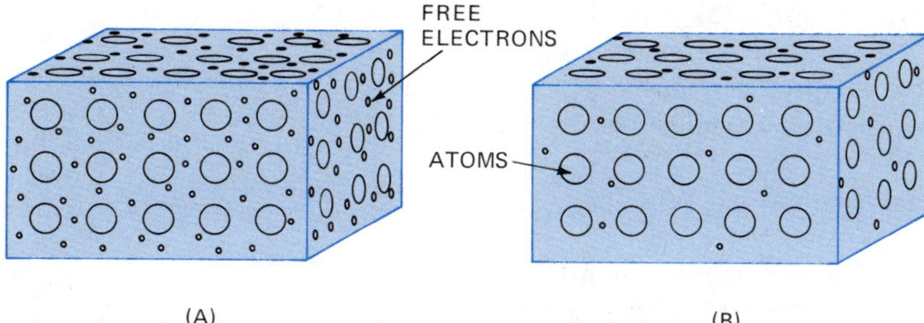

Figure 2-6 (A) Conductors have many free electrons (B) Insulators have very few free electrons

Matter is made up of many billions of atoms and molecules. In a small piece of copper wire, for instance, there are huge numbers of atoms. Each copper atom has one valence electron held very loosely. These electrons are constantly in the conduction band and drift from atom to atom. They move about in random directions.

Aluminum has large numbers of free electrons also but not as many as copper. Other metals have even fewer free electrons but still very large numbers. Such materials are said to be good *electrical conductors.*

A second group of materials is classified as *semiconductors*. These have smaller numbers of free electrons compared to conductors. Materials in a third group are called *insulators.* An insulator is a material that has very few free electrons for a given number of atoms, Figure 2-6. Some good conductors, semiconductors and insulators are listed:

CONDUCTORS	SEMICONDUCTORS	INSULATORS
Silver	Silicon	Mica
Copper	Germanium	Glass
Aluminum	Water (distilled)	Wood
(Metals)	Selenium	Ceramics

EXERCISE 2-2

Choose the one answer that best completes the statement.

1. Inner occupied shells are occupied by electrons that (a) can easily be lifted to the conduction orbit, (b) cannot easily leave their orbits, (c) are called valence electrons, (d) are called free electrons.
2. Valence electrons normally occupy (a) the innermost orbit, (b) a conduction orbit, (c) the outer portion of the nucleus, (d) the valence orbit.
3. A conduction orbit is (a) an orbit beyond the valence orbit, (b) the valence orbit, (c) the lowest energy level of all orbits, (d) the innermost orbit.
4. Free electrons can be found in (a) the valence orbit, (b) the innermost orbit, (c) conduction band, (d) the nucleus.
5. A normal atom contains electrons (a) only in their normal orbits, (b) only in the valence orbit, (c) only in the innermost orbit, (d) in a conduction orbit.

6. A free electron (a) can never escape from its orbit, (b) can easily fall into the nucleus, (c) can easily escape when attracted to other atoms, (d) is always attracted to other free electrons.
7. A good conductor has (a) many free electrons, (b) no free electrons, (c) no free protons, (d) many free protons.
8. A good class of conductors are (a) ceramics, (b) wood, (c) metals, (d) plastics.
9. A good insulator is (a) water, (b) glass, (c) copper, (d) aluminum.
10. The first law of interaction states partially that (a) like charges attract, (b) like charges repel, (c) unlike charges repel, (d) unlike charges neither attract nor repel.

THE ELECTRIC CIRCUIT AND BASIC UNITS

Electrons in the conduction band are free to migrate from atom to atom within a material, Figure 2-7. Such migrations occur in huge numbers in conductors such as a copper wire. However, there is no overall direction associated with the movement. Some free electrons move one way while others move in the opposite direction.

Free electrons can be forced to move in one direction. If one end of a copper wire is attached to a large negative charge, electrons will be repelled from that end. If the other end is attached to a large positive charge, the electrons are attracted to that end.

The negative terminal of a *battery* acts as a large negative charge which injects electrons into the copper wire. This pushes free electrons in the wire towards the other end. Electrons arriving at the positive side leave the wire and enter the battery through the positive terminal. Figure 2-8 shows the direction of the electron flow. The movement of electrons along a conductor is called *electron current* or *electron drift.*

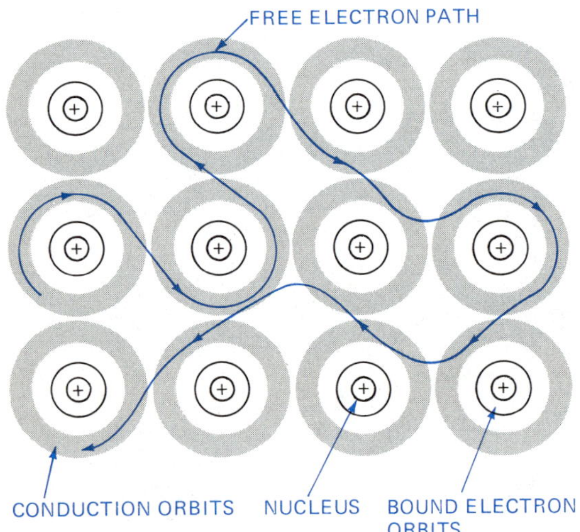

Figure 2-7 Free electrons can easily move from the conduction band of one atom to that of nearby atoms. Movement is random.

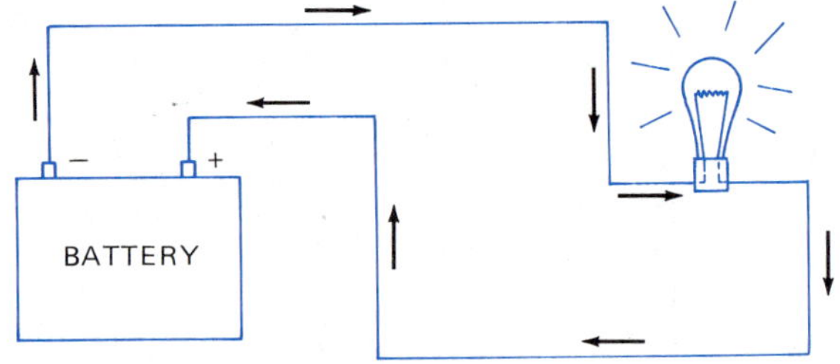

Figure 2-8 A battery causes movement of free electrons in the direction of the arrows in a complete circuit. The resulting flow is an electron current.

The battery is not a source of electrons. The battery supplies the energy that drives the free electrons through the wiring. If one electron is emitted by the negative terminal of the battery, another electron must enter the positive terminal. If ten electrons leave the battery, ten must enter. The number of free electrons at every point in the wiring, including the filament of the lamp, determines the number of electrons set into drift.

Figure 2-8 shows a simple complete circuit. A circuit is a complete path by which electrons can travel. Starting at any point, they travel along conductors, Figure 2-9, through the load (light bulb, heating element, motor windings, etc.), through the battery and back to the same point. A completed circuit is sometimes referred to as a *closed circuit.*

Loose connections, broken load windings, disconnected wires, etc. will stop the flow of electrons. The circuit is no longer a closed circuit. It would now be referred to as a *broken* or *open circuit.*

It is usually desirable to provide a means for connecting or disconnecting the circuit. An electrical *switch* is used for this purpose, Figures 2-10 and 2-11. A switch will stop the flow of electrons to the load and shut it off. A switch is used to shut off a light bulb, motor, heating element or whatever the load may be. Closing the switch turns the device back on.

Another piece of equipment found in an electrical circuit is a protective device. Such a device may be either a *fuse,* Figure 2-12, or a *circuit breaker.* A fuse is a piece of special metal that becomes a part of the circuit, Figure 2-13.

If the electron flow becomes too great, it produces heat. The fuse metal melts and opens the circuit. Thus, it is the fuse that melts and not the other (more expensive) electrical devices. Without the fuse, the other devices would burn up or a dangerous fire could be started.

A circuit breaker acts more like a switch. When the electron flow becomes excessive, heating occurs. A built-in thermal expansion apparatus flips the switch open when it gets hot. After a cooling period, the circuit breaker can be reset manually. Some circuit breakers such as those found on electric motors will reset themselves after cooling.

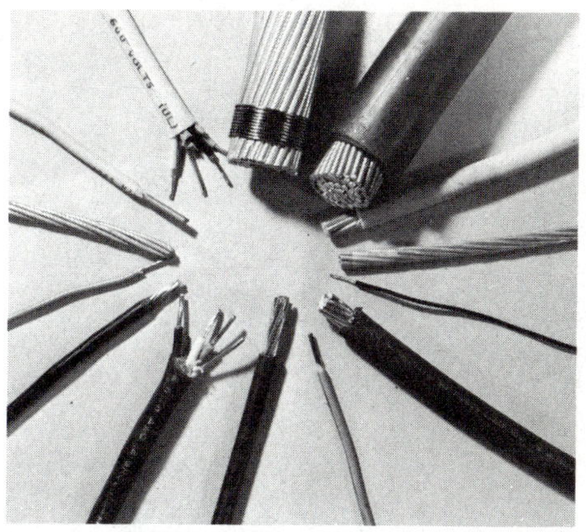

Figure 2-9 Various conductors: insulated and bare; single, multiple stranded and 3-wire with ground; and both copper and aluminum are included. (Courtesy of Southeast Nebraska Community College)

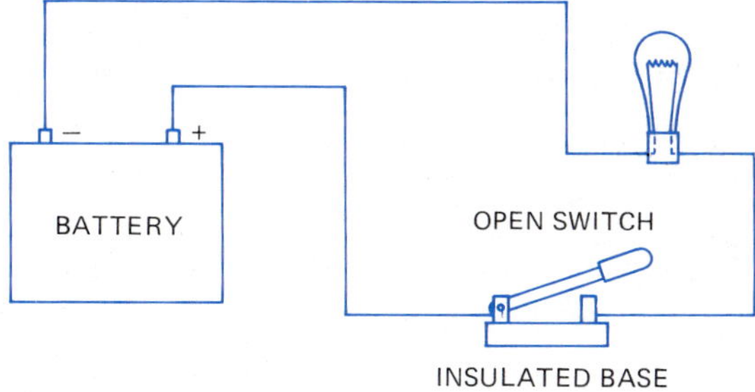

Figure 2-10 A circuit containing a battery, a lamp and an open switch connected by conductors. No current is flowing.

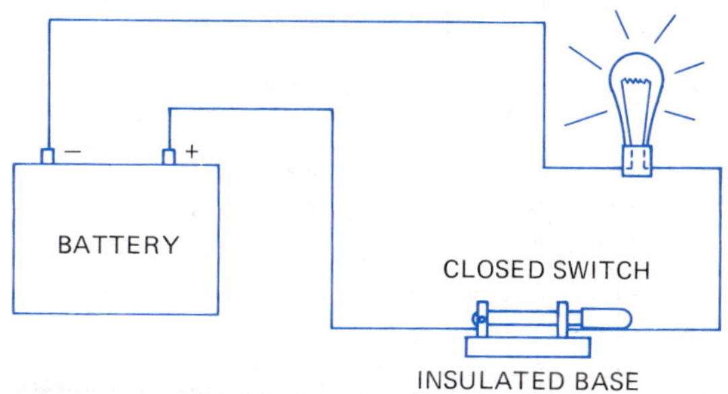

Figure 2-11 A circuit containing a closed switch. The current lights the lamp.

Figure 2-12 Various fuses. Large fuses have replaceable elements like the ones near the lower portion of the photograph. (Courtesy of Southeast Nebraska Community College)

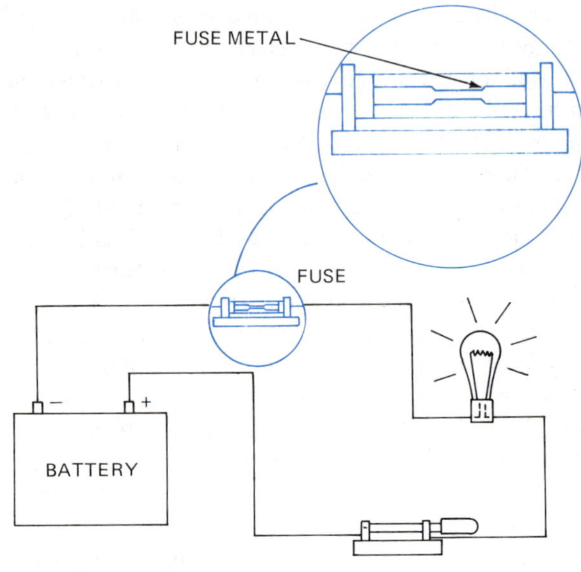

Figure 2-13 A circuit with a cartridge-type fuse. Magnified section shows details. Excessive current melts the fuse, opens the circuit and protects circuit elements.

In summary, an electric circuit consists of several parts. Every circuit has a source of energy (battery, generator, etc.), a load, a control (switch) and a protective device. These parts are interconnected by conductors (wires). The circuit is closed when the switch is closed.

2-5 ELECTRIC CURRENT

In a circuit, Figure 2-13, the conductor (wiring) will allow electrons to flow at almost any rate, including dangerously high rates. The rate of the flow (called *current*) is limited by the battery size and the load. The current flowing through the filament of a light bulb, for example, is restrained. The filament is made of a different material than the circuit conductor. It has fewer free electrons so the current will be much less than what is possible in the circuit conductors.

If a dozen electrons pass through the filament each second, only a dozen can flow past any other point in the circuit. The current is the same at every point in the circuit. Electrons cannot pile up anywhere without repelling each other. Thus, it is the load that determines the current in the circuit.

When only a few electrons pass by a point in the circuit, the current is too small to measure. The fundamental unit of current is called the ampere (A). In terms of electrons the ampere is very large. *One ampere* is the rate of flow when 6 240 000 000 000 000 000 electrons move past each point in the circuit in one second. The symbol for electric current is I.

When current flows continuously in one direction around a circuit, it is said to be a *direct current* (DC). Not all current is direct current. Certain generators or alternators will cause the current to flow in one direction, then the other. Such a current is called an *alternating current* (ac).

The reversal of flow may occur many times each second. Notice the letters (DC) are capitals while the letters (ac) are lowercase letters.

2-6 ELECTRIC CHARGE

The electron is the subatomic charged particle that transports energy in a completed circuit. However, the electric charge on the electron is much too small to be useful. The practical unit of charge is the coulomb (C). A *coulomb* is the amount of electric charge that moves past a point in a circuit in one second when the current is one ampere. The symbol for electric charge is Q.

A coulomb of charge is made up of 6 240 000 000 000 000 000 electrons. The definitions of the ampere and the coulomb confirm this. **A rate of flow of one ampere is equivalent to a flow of one coulomb per second.**

The flow of electric current in a conductor is similar to the flow of liquid in a pipe or stream. The amount of charge in coulombs is compared with the amount of water in gallons. The electric current (rate of flow) in amperes is compared with the rate of flow (movement) of the liquid in gallons per minute. At the same time, electrons might be compared with tiny droplets of the liquid. This analogy is usually helpful in studying electrical units.

2-7 RESISTANCE

Conductors are materials having many free electrons. Conductors offer only a slight opposition to the flow of electrons. Nonconductors (insulators) are materials with few free electrons. Insulators present a much greater opposition. The amount of opposition to the flow of electric current is called *resistance.* The symbol for resistance is R. For measuring resistance, the practical unit is the *ohm.* The abbreviation for the ohm is the Greek letter, capital omega (Ω).

The resistance of a wire depends on four things. First it depends on the material as discussed in the previous paragraph. It also depends on the cross-sectional area of the wire. A wire of large diameter has more free electrons than a smaller wire. The larger the diameter of the conductor, the lower the resistance.

The length of a conductor also affects the resistance. A 100-foot wire has only half the resistance of a 200-foot wire of the same size and material.

Fourth, the resistance of most materials increases when the temperature increases. A given change in temperature will affect the resistance much more in some materials than in others.

In summary, resistance of a wire depends on these four things:

- The material
- The cross-sectional area (size)
- The length
- The temperature

In an electrical circuit, resistance is encountered in the conductors. It is also found within the battery or generator. In general, however, most of the resistance of a circuit is located in the load or loads.

2-8 ELECTROMOTIVE FORCE AND POTENTIAL DIFFERENCE

In order to set electrons into drift action, an "electric pressure" must be applied. In the liquid pipe analogy, pressure causes the liquid to flow. Liquid pressure is applied by a pump. Fluid friction (resistance) is overcome by the pump pressure.

Free electrons are pushed away from the negative terminal of a battery and are attracted toward the positive terminal. The "electric pressure" provided by the battery is sometimes called *electromotive force (emf), potential difference* or *voltage.* All are indicated with the letter E. Each term, though, has a special meaning.

If two objects have different amounts of charge on them, there is a *potential difference* between them. If the two objects are connected with a conductor, charge will flow from one to the other. The flow of current stops when the charge on the two objects is equalized. There is no longer a potential difference between the two objects.

Some devices have the ability to maintain a potential difference between two points. They can maintain a difference in the charge on the two points or objects. This is true even when a current is flowing between them. Generators, thermocouples, photoelectric (solar) cells and batteries are examples of such devices. When operating, these devices are said to develop an *electromotive force (emf).*

A potential difference causes current to flow. An emf maintains the potential difference. The term *voltage* is used to mean a measure of either potential difference or emf. These three terms (potential difference, emf and voltage) do not mean the same thing but are usually used interchangeably. The word "voltage" will be used most frequently in this book. The term "electric pressure" will have no further use.

Actually, voltage is neither a pressure nor a force. **An emf is the work done on the free electrons by an energy source per unit charge.** Whenever work is done on the free electrons, a force acts on them and causes them to move. However, voltage is defined in terms of work and energy rather than the forces acting on the electrons.

An emf is a *voltage rise* and describes the energy given to the free electrons. This develops and maintains the current. The energy is dissipated in the circuit (mostly in the load). The electrical energy drawn from the emf source is converted to other forms of energy (heat, light, motion, etc.) in the load. This results in a *voltage drop* across the load.

In the circuit of Figure 2-13, a voltage rise occurs across the battery. A voltage drop occurs across the lamp. Small voltage drops also exist along the conductor, across the fuse, etc. A source of emf, such as a battery, is a source of energy. The energy is dissipated usually as heat when current flows through the load. The amount of energy used to push electrons against the resistance of the load is always equal to the energy drawn from the source of the energy. That is; in any electrical circuit:

voltage drops = voltage rises

Voltages are measured in units of *volts* (V). A voltage that can maintain a current of one ampere in a one ohm circuit is one volt. Typical circuit voltages range from small fractions of a volt to very high voltages. Dry cell batteries produce 1.5 V. Automobiles use a 12-V battery. Transmission lines use voltages as high as 345 000 V. Household circuits are mostly 110 V–120 V. Doorbell circuits (12 V), thermostat circuits (24 V) and air conditioners (220 V) are some exceptions.

EXERCISE 2-3

Choose the one answer that best completes the statement.

1. An electric current in a conductor is the movement of (a) protons, (b) electrons, (c) ions, (d) whole atoms.
2. A closed circuit is (a) a complete circuit, (b) an incomplete circuit, (c) electrically neutral, (d) a broken circuit.
3. An open circuit (a) is a complete circuit, (b) is an incomplete circuit, (c) has an unbroken path for current to flow, (d) is electrically neutral.
4. The unit of electric current is the (a) coulomb, (b) volt, (c) ampere, (d) ohm.
5. The unit of electromotive force is the (a) coulomb, (b) volt, (c) ampere, (d) ohm.
6. The unit of resistance is the (a) coulomb, (b) volt, (c) ampere, (d) ohm.
7. The unit of electric charge is the (a) coulomb, (b) volt, (c) ampere, (d) ohm.
8. A fuse is used to (a) open and close the circuit, (b) prevent shock, (c) protect the circuit from overheating, (d) provide a path from one circuit element to another.
9. A conductor is used to (a) open and close the circuit, (b) protect the circuit from overheating, (c) provide a source of voltage, (d) provide a path from one circuit element to another.
10. A switch is used to (a) open and close the circuit, (b) prevent shock, (c) protect the circuit from overheating, (d) provide a path from one circuit element to another.
11. A current that periodically reverses direction is called (a) a looping current, (b) a doubling current, (c) a direct current, (d) an alternating current.
12. A current that flows in only one direction is called (a) a straight current, (b) a regulated current, (c) a direct current, (d) an alternating current.

Match the following numbered circuit quantities with the correct lettered symbol.

13. Electric current (a) R
14. Electric charge (b) Q
15. Resistance (c) V
16. Voltage (d) Ω
(e) E
(f) A
(g) C
(h) I

Match the following numbered unit with the correct lettered symbol.

17. Ampere (a) R
18. Coulomb (b) Q
19. Ohm (c) V
20. Volt (d) Ω
(e) E
(f) A
(g) C
(h) I

Match the following numbered circuit quantities with the correct lettered unit.

21. Electric current (a) Coulomb
22. Electric charge (b) Watt
23. Resistance (c) DC
24. Voltage (d) Volt
(e) Ohm
(f) emf
(g) Ampere
(h) ac

ELECTRIC CIRCUITS AND MEASURING INSTRUMENTS

Every electric circuit contains the following parts:

- A source of energy or emf.
- A load or resistance.
- A continuous conductor connecting the various parts.
- A switch to control the flow of current.
- A fuse to protect the other parts.

2-9 CIRCUIT SYMBOLS AND CIRCUIT DIAGRAMS

Up to this point, the circuit diagrams have been drawn by picturing the various electric devices. It is easier, however, to use symbols that represent the various parts. The standard electrical symbol for many circuit elements are given in Figure 2-14. All symbols listed are in common usage. Most have been approved by various electrical associations. For various appliances (coffee pot, toaster, refrigerator, etc.) use the resistor symbol. The name of the appliances can be written above it.

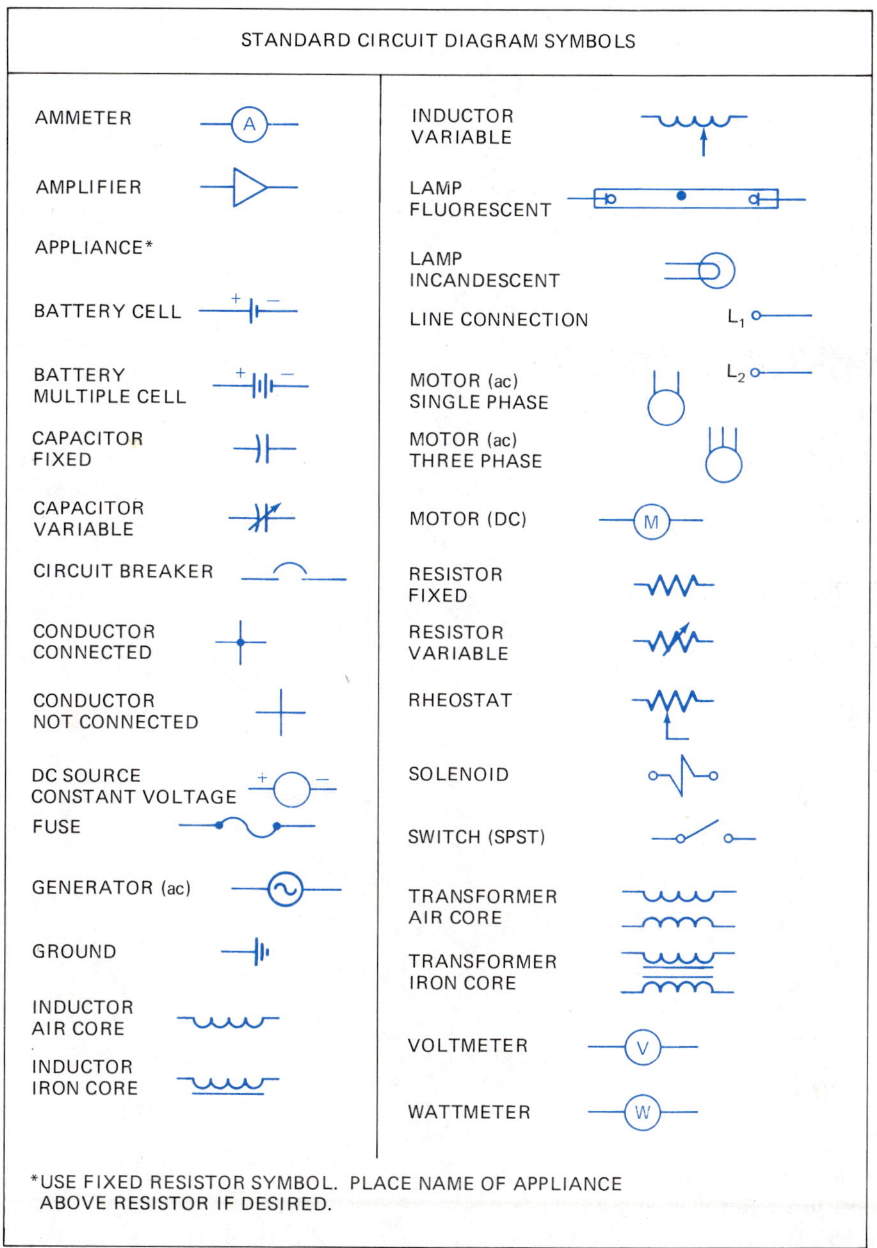

Figure 2-14 Standard circuit symbols

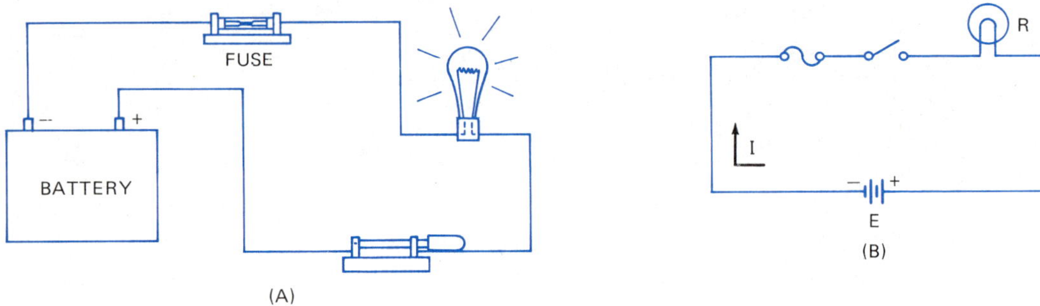

Figure 2-15 (A) Circuit with various elements (B) Circuit diagram for the circuit in A using symbols for the various elements

A diagram used previously is shown in Figure 2-15(A). The simplified version of the same circuit using standard symbols is shown in Figure 2-15(B).

EXERCISE 2-4

1. Name each of the following circuit symbols.
   - a.
   - b.
   - c.
   - d.
   - e.
   - f. M
   - g. V
   - h.
   - i.
   - j.
2. Draw the circuit symbol for each of the following.
   - a. Fluorescent lamp
   - b. Single phase ac motor
   - c. Three cell battery
   - d. Variable resistor
   - e. Capacitor
   - f. Ammeter
   - g. Single pole single throw (SPST) switch
   - h. Transformer

2-10 SERIES AND PARALLEL CIRCUITS

When more than one load is placed in a circuit, the loads can be connected in different ways. If all of the current flows through every load the circuit is called a *series circuit*. A resistor, a hotplate and a lamp are shown connected in series in Figure 2-16. The current passing through each circuit element is the same.

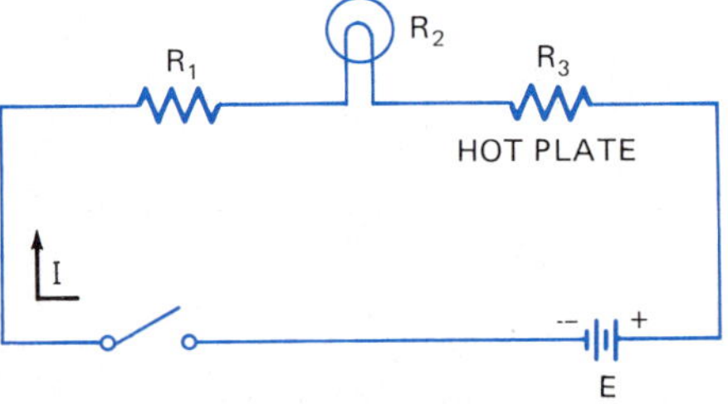

Figure 2-16 A series circuit with three loads. The current is the same in every circuit element.

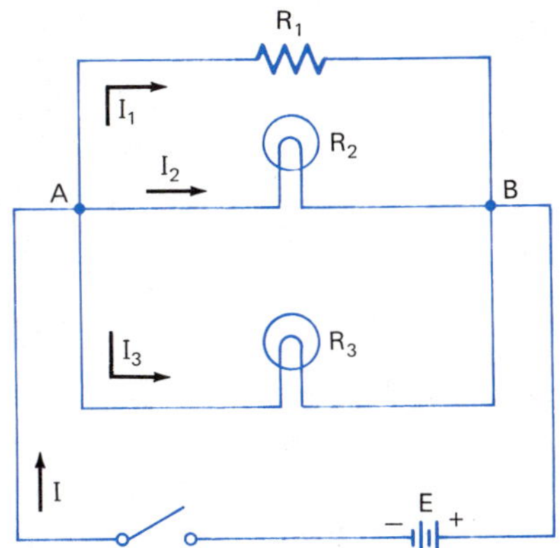

Figure 2-17 A parallel circuit with three loads. Current divides at point A, travels different paths and recombines at point B.

Figure 2-18 A typical panel ammeter (Courtesy of Triplett Corporation)

Circuit elements (loads) can be connected so the current in the main part of the circuit divides and travels different paths. Such a circuit is called a *parallel circuit.* A resistor and two lamps are shown connected in parallel in Figure 2-17. The electrons coming from the battery split at point A, travel separate routes, combine at point B and return to the battery together.

2-11 MEASURING INSTRUMENTS

Ammeter. The quantities: current, voltage and resistance are circuit quantities that are frequently measured. Electric current is measured with an ammeter. An ammeter measures the number of amperes (A) flowing in a circuit. A typical ammeter is shown in Figure 2-18.

To measure the current in a circuit, it must be connected as shown in Figure 2-19. The ammeter is inserted into the circuit. **The ammeter must be connected in series with the other circuit elements.** Notice that the electron leaving the generator must enter the ammeter through the negative (–) terminal. They leave the ammeter through the positive (+) terminal.

The ammeter has a very low resistance. In order to measure the current accurately, it must not affect the flow of current. Remember, an ammeter is always connected directly into a circuit.

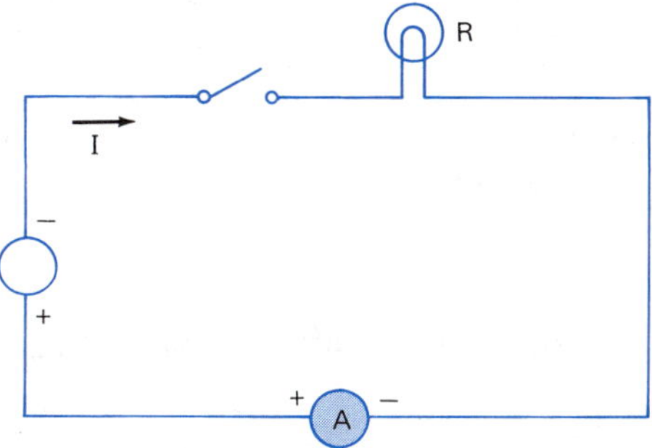

Figure 2-19 A circuit showing the proper way to connect an ammeter in series

Voltmeter. A voltmeter can be used to measure the emf of a battery or generator or the voltage drop across a load. A voltmeter measures the voltage in units of volts (V). A typical voltmeter is shown in Figure 2-20.

The method of connecting a voltmeter across a load is shown in Figure 2-21. It is connected in parallel with the load being measured. A voltmeter has a very high resistance. To measure voltage accurately, the voltmeter must not affect the flow of current from the source.

Notice the difference in the way ammeters and voltmeters are used. Compare Figures 2-19 and 2-21. An ammeter is connected or inserted into the circuit (in series with the load). **A voltmeter is connected across the load (in parallel).** Never connect an ammeter across the load. It has a low resistance and the huge current flow that would result would burn it up.

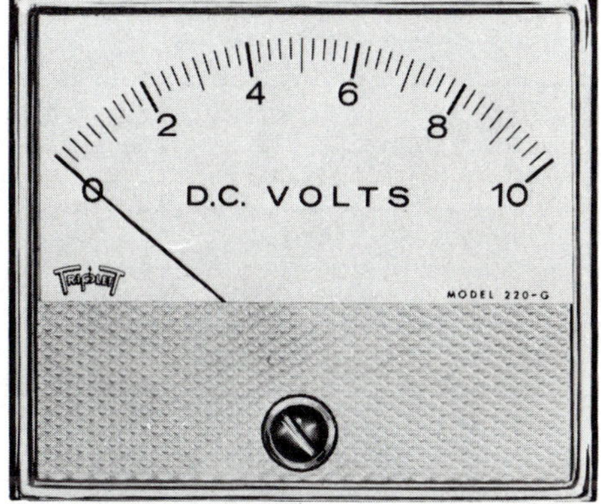

Figure 2-20 A typical panel voltmeter (Courtesy of Triplett Corporation)

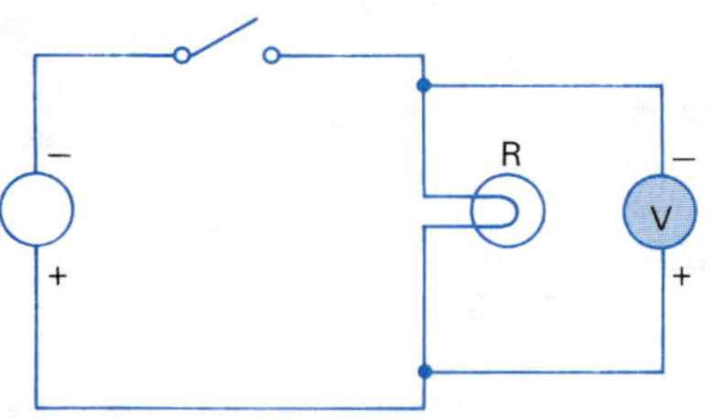

Figure 2-21 A circuit showing the proper way to connect a voltmeter to measure the voltage drop across a load. It is connected in parallel with the load.

Ohmmeter. An ohmmeter is a measuring instrument that measures the resistance (R) of various circuit elements (loads). It measures the resistance in units of ohms (Ω). Ohmmeters contain their own battery cells and must never be connected into a live circuit. The power must be turned off before connecting the ohmmeter. **To connect the ohmmeter, the leads of the ohmmeter are connected across the individual circuit devices being measured.**

Multimeter. In addition to individual meters, there are also multimeters, Figure 2-22. Multimeters come in various combinations. Some are volt-ammeters. Some are volt-ohmmeters. Others measure volts, amperes and ohms.

Figure 2-23 summarizes the various circuit quantities, their units, the measuring instruments and how they are used.

Figure 2-22 One of many kinds of multimeters (Courtesy of Triplett Corporation.)

2-12 ANALOG AND DIGITAL READOUT

The measuring devices pictured in Figures 2-18, 2-20 and 2-22 have analog scales. A needle points to the correct reading. The user determines the size of the smallest scale divisions by inspection. By counting the divisions from zero to the needle position, the reading is obtained. Usually, an estimate between the smallest scale divisions is also made.

Some measuring instruments have digital readouts. These are electronic devices in which the reading is displayed numerically. The calculator display is an example of a digital readout.

INSTRUMENT	CIRCUIT USE	QUANTITY MEASURED	UNITS
Ammeter	In series	Current (I)	Amperes (A)
Voltmeter	In parallel	Voltage (E)	Volts (V)
Ohmmeter	Power Off	Resistance (R)	Ohms (Ω)

Figure 2-23 Circuit quantities, units and measuring instruments

EXERCISE 2-5

Refer to the circuit symbols in Figure 2-14. Construct a circuit diagram for each of the following problems.

1. A two-cell battery, two light bulbs in series, a fuse and a switch.
2. A DC generator, a DC motor, a fuse and a switch.
3. An ac generator; a coffee pot and a toaster in parallel; one circuit breaker in the main line and a switch for each load.
4. A three-cell battery, two resistors and a light bulb in series, a fuse, a switch and an ammeter.
5. A two-cell battery, a DC motor and a heating element in parallel, a switch for each load, an ammeter to measure current through the heating element and another ammeter to measure circuit (total) current.
6. A DC generator, an air core inductor and a lamp in series, a switch, a fuse and a voltmeter used to measure voltage drop across the lamp.
7. An ac generator, two resistors in parallel, a switch, an ammeter to measure circuit (total) current and a voltmeter used to measure voltage drop across the bank of resistors.

CHAPTER 3

THE SINGLE LOAD DC CIRCUIT

OBJECTIVES

After satisfactorily completing this chapter, the student should be able to:

- Define the various electrical and mathematical words and terms introduced in this chapter.
- Perform manual arithmetic operations on common fractions.
- Evaluate arithmetic expressions involving decimals, including powers, roots and reciprocals using a calculator.
- Apply Ohm's Law and the definition of power using a consistent procedure to calculate the various circuit quantities in a single load circuit.

An electric circuit contains a source of energy (emf). Current flows in a closed circuit as a result of the emf but the current flow is limited by the load. The circuit itself is controlled by a switch and protected by a fuse. All circuits contain these elements.

The amount of *current* in a circuit is determined by the *emf* of the source and the *resistance* of the load. These three quantities are related to each other. The relationship is called Ohm's Law.

Chapter 2 described the electron flow, circuit elements and circuit quantities including measurement. Mastering more complex circuits requires experience with Ohm's Law. In addition, making and using graphs (Chapter 4) and the ability to convert units (Chapter 6) are necessary skills.

Ohm's Law will be studied along with power dissipation later in this chapter. Only one load circuits will be covered. The chapter begins with a review of the arithmetic of common fractions and decimal fractions. The use of the calculator begins with the review of decimals.

COMMON FRACTIONS

Objects or quantities are often divided into two or more equal parts. A *common fraction* is a way of expressing the number of the equal parts. The fraction symbol is the slash (/) or the bar (–). The fraction one-half can be expressed as:

$$1/2 \text{ or } \frac{1}{2}$$

The *denominator* is the divisor which indicates the number of equal parts into which the object or quantity has been divided. The denominator is the number below the bar or to the right of the slash in a fraction. The *numerator* indicates how many of the parts are taken. It is the number above the bar or to the left of the slash in a fraction:

$$\text{numerator/denominator or } \frac{\text{numerator}}{\text{denominator}}$$

The numerator and denominator are called the *terms* of a fraction. A fraction is sometimes called a *ratio*. The ratio of 5 to 8 is the fraction 5/8.

Figure 3-1 indicates the fractions: halves, quarters and eighths. The shaded area indicates the number of equal parts (numerator). The square shows the total number of equal parts (denominator). The fractions are shown with each figure.

3-1 *MIXED NUMBERS*

Mixed numbers are composed of a whole number and a fraction. An electrician might carry two new rolls of wire and one-third of another roll to the job site. This is indicated by the number 2 1/3 and means 2 + 1/3.

A *proper fraction* is one in which the numerator is smaller than the denominator. When the denominator is the smaller term, the fraction is called an *improper fraction*. A proper fraction has a value less than 1, while an improper fraction is greater than 1. Improper fractions can be written as mixed numbers.

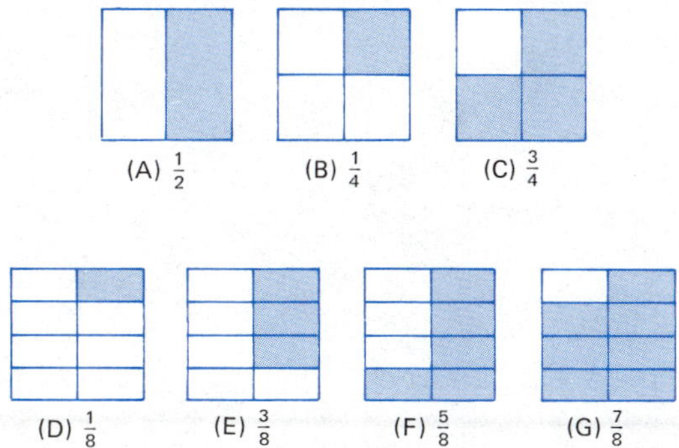

Figure 3-1 Fractional parts of a whole

Example A Express the fraction 41/16 as a mixed number.

Solution: Divide 41 by 16. This division results in 2 plus a remainder of 9/16.

$$\frac{41}{16} = 2\frac{9}{16}$$

Example B Express the mixed number 3 3/8 as an improper fraction.

(The number 3 is the same as $\frac{3 \times 8}{8} = \frac{24}{8}$)

Solution: $3\frac{3}{8} = 3 + \frac{3}{8} = \frac{24}{8} + \frac{3}{8} = \frac{27}{8}$

Note: Fractions with the same denominator are added by adding numerators. The sum has the same denominator as those being added.

3-2 EQUIVALENT FRACTIONS

A fraction is in *lowest terms* when the numerator and denominator have no common factors. A fraction can be *reduced to lowest terms* by dividing both terms by the largest common factor. *The largest common factor* is the largest number that evenly divides both terms of the fraction. The value of the fraction is not changed when both terms are divided by the same number. The new fraction is said to be an *equivalent* fraction.

Example A Reduce the fraction 16/24 to lowest terms.

Note: The largest common factor is 8. It is the largest number that evenly divides both 16 and 24.

Solution: Divide both terms by 8.

$$\frac{16 \div 8}{24 \div 8} = \frac{2}{3}$$

If the largest common factor cannot be determined easily, find the prime factors of both terms. The product of all common prime factors will be the largest common factor.

Example B Reduce the fraction 154/330 to lowest terms.

Solution: Find the prime factors.

$154 = 2 \times 7 \times 11$

$330 = 2 \times 3 \times 5 \times 11$

The factors 2 and 11 are common to both terms so divide both terms by the product ($2 \times 11 = 22$).

$$\frac{154 \div 22}{330 \div 22} = \frac{7}{15}$$

The terms of a fraction can be increased by multiplying both by the same number. Again, the value of the fraction remains unchanged. This process may be required when adding or subtracting fractions.

Example C Express the fraction 3/8 with a denominator of 48.

Note: The denominator (8) must be multiplied by 6 to obtain the new denominator. Thus both terms must be multiplied by 6:

Solution: $\frac{3}{8} = \frac{3 \times 6}{8 \times 6} = \frac{18}{48}$

EXERCISE 3-1

Express these improper fractions as mixed numbers.

1. 5/4
2. 5/3
3. 19/8
4. 19/12
5. 66/9
6. 501/4
7. 28/8
8. 99/7

Change these mixed numbers to improper fractions.

9. 3 1/2
10. 2 3/4
11. 5 2/3
12. 4 7/8
13. 7 5/16
14. 8 3/5

Reduce these fractions to lowest terms.

15. 9/12
16. 75/125
17. 6/18
18. 12/16
19. 126/198
20. 150/210

Change these fractions as indicated.

21. 1/2, 2/3 and 5/6 to twenty-fourths.
22. 1/2, 2/3 and 3/5 to fifteenths.
23. 1/3, 5/6 and 4/7 to forty-secondths.
24. A pole, Figure 3-2, is set in the ground.
    (a) What fractional part (reduced to lowest terms) is above ground?
    (b) What fractional part (reduced to lowest terms) is in the ground?
25. A 60-litre transformer housing contains 36 litres of oil. What fractional part (reduced to lowest terms) of the housing is filled with oil?
26. Twenty-five gallons of battery electrolyte is made up of 10 gallons of sulfuric acid and 15 gallons of water, Figure 3-3.
    (a) What fractional part (reduced to lowest terms) is sulfuric acid?
    (b) What fractional part (reduced to lowest terms) is water?
27. The end of a piece of #6 gage wire, Figure 3-4, has an area of about 26 000 units (circular mils). The area of the end of #12 gage wire is about 6500 units. What is the fractional part of the area of #12 wire compared to #6 wire? Reduce the answer to lowest terms.

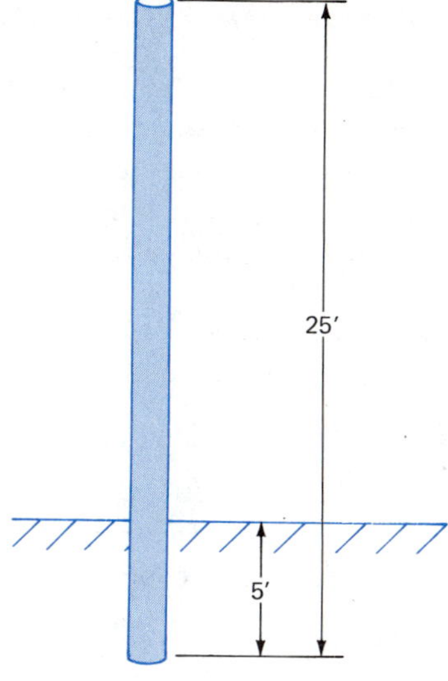

Figure 3-2

Figure 3-3

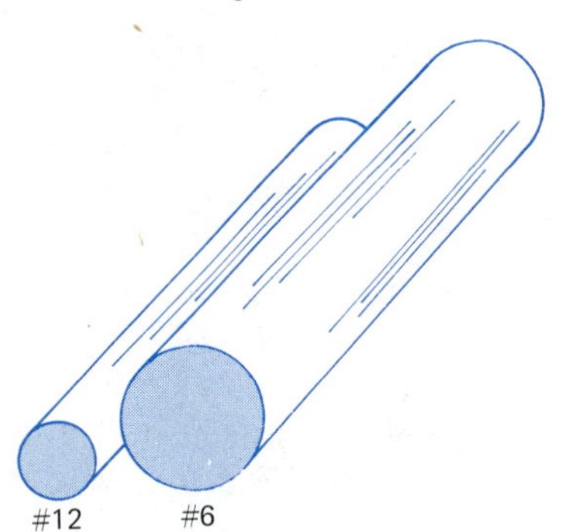

Figure 3-4

3-3 MULTIPLYING COMMON FRACTIONS

Most people agree it is easier to add or subtract whole numbers than it is to multiply or divide whole numbers. With common fractions, however, it is usually just the opposite. Addition and subtraction cannot proceed until all fractions have the same denominators. Fractions do not have to be changed before multiplying.

MULTIPLYING FRACTIONS

- Multiply the numerators to get the numerator of the product.
- Multiply the denominators to find the denominator of the product.
- Reduce the result to lowest terms if necessary.

Example A Find the product of 1/3, 7/4 and 2/5.

Solution: $\frac{1}{3} \times \frac{7}{4} \times \frac{2}{5} = \frac{1 \times 7 \times 2}{3 \times 4 \times 5} = \frac{14}{60} = \frac{7}{30}$

Notice the middle fraction is improper. The procedure for multiplying proper and improper fractions is the same.

When multiplying whole numbers and mixed numbers, the whole numbers are written as fractions with a denominator of 1 and the mixed numbers are changed to improper fractions. The fractions are then multiplied. If the product is improper, convert it to a mixed number.

Example B Evaluate: 3 2/5 × 4 2/3 × 1 1/2.

Solution: 3 2/5 = 15/5 + 2/5 = 17/5

4 2/3 = 12/3 + 2/3 = 14/3

1 1/2 = 2/2 + 1/2 = 3/2

$\frac{17}{5} \times \frac{14}{3} \times \frac{3}{2} = \frac{714}{30} = 23\frac{24}{30} = 23\frac{4}{5}$

3-4 CANCELLATION

When multiplying, the product should be reduced to lowest terms. This can sometimes be accomplished easier by *cancellation.* The value of a fraction is unchanged when both terms are divided by the same number. The order of multiplication makes no difference.

Example A Evaluate: 22/25 × 10/33.

Solution: $\frac{22}{25} \times \frac{10}{33} = \frac{22 \times 10}{25 \times 33} = \frac{220}{825} = \frac{4}{15}$

In this example, it is not easy to see that 220/825 reduces to 4/15. The following shows how the problem is solved by using cancellation.

$$\frac{\overset{2}{\cancel{22}}}{\underset{5}{\cancel{25}}} \times \frac{\overset{2}{\cancel{10}}}{\underset{3}{\cancel{33}}} = \frac{2 \times 2}{5 \times 3} = \frac{4}{15}$$

The terms 22 and 33 are divided by 11 and the terms 10 and 25 are divided by 5. The product is automatically reduced to lowest terms.

Caution: Use cancellation only in multiplication. If plus or minus signs occur in the numerator, denominator or both, or as an operation, do not cancel.

- $\frac{4}{3} + \frac{9}{16}$ (Do not cancel)
- $\frac{3-5}{15+9}$ (Do not cancel)

Do not cancel except in strict multiplication problems. Cancellation is especially convenient when multiplying several fractions.

Example B Evaluate: 4/5 × 7/12 × 15/32 × 8/9.

Solution: $\frac{\overset{1}{\cancel{4}}}{\underset{1}{\cancel{5}}} \times \frac{7}{\underset{\underset{1}{\cancel{3}}}{\cancel{12}}} \times \frac{\overset{\overset{1}{\cancel{3}}}{\cancel{15}}}{\underset{4}{\cancel{32}}} \times \frac{\overset{1}{\cancel{8}}}{9} = \frac{7}{4 \times 9} = \frac{7}{36}$

The terms 12 and 4 are divided by 4, 15 and 5 are divided by 5, and 32 and 8 are divided by 8. After the first cancellations, the terms 3 and 3 can be divided by 3.

EXERCISE 3-2

Evaluate the following products. Cancel when possible and reduce answers to lowest terms.

1. 4 × 2/3
2. 5 × 3/4
3. 4 1/3 × 1/2
4. 5/8 × 2/3
5. 5 1/3 × 3/8
6. 4 1/2 × 2 2/3
7. 5 × 1/2 × 3/5
8. 4 1/2 × 5/6 × 1 11/15
9. 12 1/4 × 8 2/3 × 6 2/5
10. 3 × 2 3/4 hours
11. 8 × 12 3/8 feet
12. 2 7/16 × 1 1/3 × 1 8/13

Solve the following:

13. A battery contains 2 7/8 litres of sulfuric acid, Figure 3-5. How many litres are needed to fill 12 batteries of this type?
14. A rod is a unit of measurement equal to 16 1/2 feet. If two poles are 22 1/2 rods apart, find the number of feet between poles.
15. Transformer oil weighs 15/16 kilograms per litre. A pole transformer contains 36 2/5 litres of oil. Find the weight of the oil contained in the transformer, in kilograms.
16. The weight of 0-gage copper wire is about 5/16 pounds per foot. Find the weight of 300 feet of this wire.

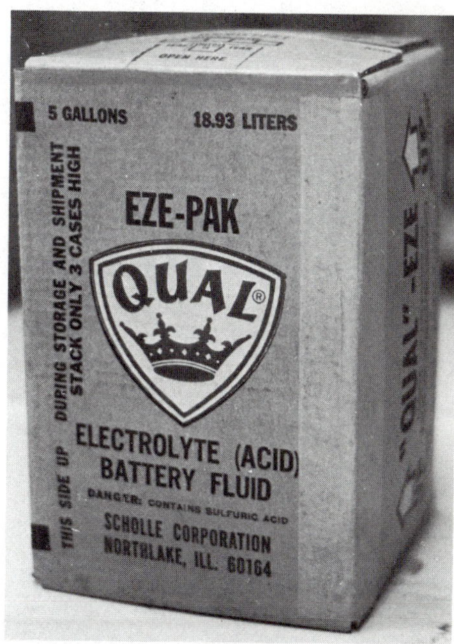

Figure 3-5 Sulfuric (battery) acid container

17. The distance around a circle (circumference) is about 3 1/7 times the distance across at the center (diameter). A coil of wire has 19 1/2 turns and is 2/3 of a metre across. How long is the coil in metres?
18. One kilowatt is equivalent to 1 1/3 horsepower. A coffee pot contains a 1 1/5 kW heating element. What is the equivalent horsepower rating of the heating element?
19. The volume of a rectangular solid is the length times width times height. Find the volume of 144 crossarms, Figure 3-6, each having the following dimensions in inches: 3 3/8 X 4 2/3 X 122 1/2.

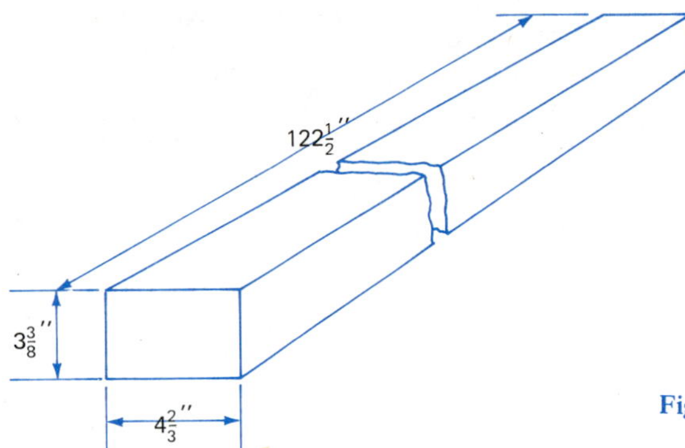

Figure 3-6 A typical wooden crossarm

20. Six conductors must be laid by an electrician. Each must be 5 3/8 thousand-feet long. The electrician decides to use three 2-wire cables. The cost of each 2-wire cable is $606 3/4 per thousand feet. Find the cost of the three cables.
21. An electrician has decided to use two 3-wire cables. The cost of each 3-wire cable is $880 2/3 per thousand feet. Each cable must be 5 3/8 thousand-feet long. Find the cost of the two cables.

3-5 *DIVIDING COMMON FRACTIONS*

Division follows easily from multiplication. As in multiplication of fractions, begin by changing whole numbers and mixed numbers to improper fractions. One initial step is required before division problems become multiplication problems. The divisor must be inverted. The inverted number is called the *reciprocal* of the divisor. Two numbers are said to be reciprocals if their product is 1. For example, 2/3 and 3/2 are reciprocals of each other. The reciprocal of 4 is 1/4.

DIVIDING FRACTIONS

- Multiply the dividend by the reciprocal of the divisor.

When dividing fractions, do not invert the wrong fraction. Always invert the divisor. In the example, $3/4 \div 2/5$, it is the second fraction (2/5) that must be inverted. When a division problem is written as a *complex fraction,*

$$\frac{\frac{3}{4}}{\frac{2}{5}}$$

it is the denominator (2/5) that must be inverted.

Example A Divide 25 by 2/3.

Solution: $\frac{25}{1} \div \frac{2}{3} = \frac{25}{1} \times \frac{3}{2} = \frac{75}{2} = 37\frac{1}{2}$

Note: When a number is divided by a proper fraction, the quotient is larger than the number (37 1/2 is larger than 25).

Example B Divide 2/3 by 8/5.

Solution: $\frac{2}{3} \div \frac{8}{5} = \frac{\overset{1}{\cancel{2}}}{3} \times \frac{5}{\underset{4}{\cancel{8}}} = \frac{5}{12}$

Note: When a number is divided by an improper fraction, the quotient is less than the number (5/12 is less than 2/3).

Example C Divide 4 1/2 by 5/6.

Solution: $4\frac{1}{2} \div \frac{5}{6} = \frac{9}{2} \times \frac{6}{5} = \frac{27}{5} = 5\frac{2}{5}$

Example D Divide 12 1/4 by 9 1/3.

Solution: $12\frac{1}{4} \div 9\frac{1}{3} = \frac{\overset{7}{\cancel{49}}}{4} \times \frac{3}{\underset{4}{\cancel{28}}} = \frac{21}{16} = 1\frac{5}{16}$

Example E Divide 4/9 by 5/6.

Solution: $\frac{\frac{4}{9}}{\frac{5}{6}} = \frac{4}{\underset{3}{\cancel{9}}} \times \frac{\overset{2}{\cancel{6}}}{5} = \frac{8}{15}$

EXERCISE 3-3

Find the reciprocals of these numbers.

1. 5
2. 1/3
3. 3/5
4. 27/32
5. 1
6. 3/27
7. 2 2/3
8. 5 4/5
9. 12 1/2
10. 64

Evaluate the following quotients. Cancel when possible and reduce answers to lowest terms.

11. 5 ÷ 3/4
12. 15/22 ÷ 5
13. 4 ÷ 2/3
14. 3/8 ÷ 3/4
15. $\frac{15/16}{5/8}$
16. $\frac{12\ 1/2}{1/2}$
17. 5/8 ÷ 15/16
18. 6 1/4 ÷ 1 3/5
19. 6/11 ÷ 12/33
20. 8 3/4 ÷ 2 1/2
21. $\frac{169/4}{13}$
22. 3 3/5 × 1/6 ÷ 3 1/5 Hint: Multiply first, then divide.

23. How many pieces of copper wire 3 1/2 centimetres long can be cut from a piece 48 centimetres long?
24. How many ohms of resistance is there in a soldering iron if the current is 7 2/3 amperes and the voltage is 115 volts?

    Note: resistance = voltage ÷ current.
25. Two line poles are 222 3/4 feet apart, Figure 3-7. The rod is a unit of linear measure which equals 16 1/2 feet. Express the distance between the poles in rods.
26. Good general-purpose solder, Figure 3-8, contains 3/5 tin and 2/5 lead.
    (a) How many kilograms of solder contains 2 3/4 kilograms of tin?
    (b) How many kilograms of solder contains 5 kilograms of lead?

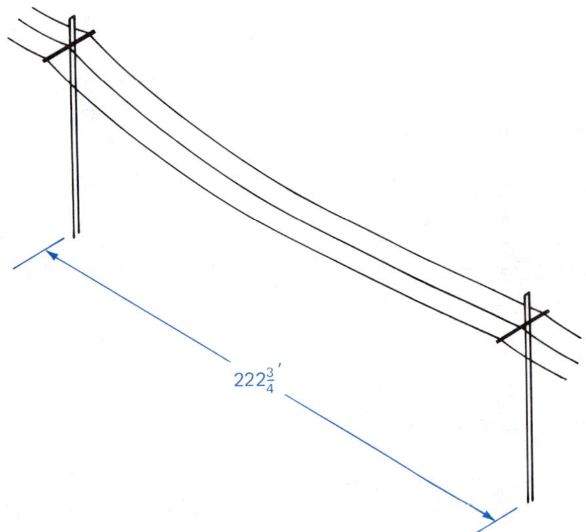

Figure 3-7

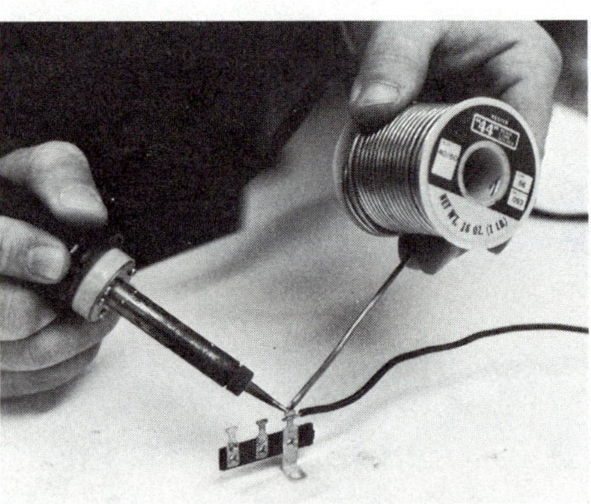

Figure 3-8 Soldering a connection with general purpose solder (Courtesy of Southeast Nebraska Community College)

27. The total weight of a reel of cable is 525 7/8 pounds. If 1 foot of this cable weighs 1 3/4 pounds, what is the length of the cable?
28. Copper splicing sleeves weigh 3/8 kilograms each. Find the number of sleeves in a shipment that weighs 150 kilograms.
29. Determine the rate of power consumption (kilowatt-hours per hour) if 8 3/4 kilowatt-hours are used in 5 1/3 hours.
30. At full load, a 1/3 horsepower (output) DC electric motor converts 3/4 of the electrical input power to mechanical power. What is the input power measured in horsepower?

DECIMALS

The ten Arabic numerals (0, 1, 2, 3, 4, 5, 6, 7, 8, 9,) have both face value and place value. The number 329, for example, means 3(hundreds) + 2(tens) + 9(units) = three hundred twenty-nine. The place value concept is extended to the right of the units digits. This allows fractional parts to be included in the system. A period, called a *decimal point,* separates the whole numbers from the fractional parts (some countries use a comma).

The digits to the right of the decimal make up the fractional part of the number and those to the left make up the whole number. A number such as 38.25 is a *mixed decimal fraction.*

The value of the digits varies according to their position. For every place a digit is moved to the left, its value increases by 10 times. If a digit is moved to the right, its value decreases by 10 times for each place. These relationships are true on either side of the decimal point. Figure 3-9 shows the place value for mixed decimal fractions. Fractions written in *decimal form* are called *decimal fractions* or simply *decimals.* They can be written as common fractions. Their denominators are 10, 100, 1000, 10 000, etc. Such numbers are called *powers of ten.*

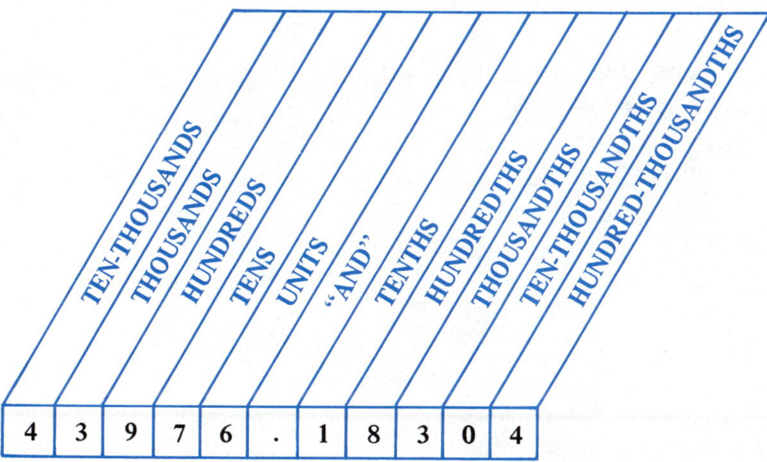

Figure 3-9 Place value for decimals

Example A Express the following fractions with a power of ten denominator. (1) 0.588 (2) 0.97 (3) 0.4 (4) 0.857 41

Solution: (1) $0.588 = \frac{588}{1000}$ (Three decimal places)

(2) $0.97 = \frac{97}{100}$ (Two decimal places)

(3) $0.4 = \frac{4}{10}$ (One decimal place)

(4) $0.857\,41 = \frac{85\,741}{100\,000}$ (Five decimal places)

Notice there are as many digits or places to the right of the decimal point as there are zeros in the denominator. If there are fewer digits in the numerator, zeros are added to the fraction to indicate place value.

Example B Express these fractions as decimals:

(1) $\frac{7}{10\,000}$ (2) $\frac{42}{1000}$

Solution: (1) $\frac{7}{10\,000} = 0.0007$ (2) $\frac{42}{1000} = 0.042$

Also, notice that a zero is placed to the left of the decimal point for *proper decimal fractions.*

3-6 READING DECIMAL FRACTIONS AND ANALOG SCALES

A whole number such as 15 283 is read, "fifteen thousand two hundred eighty-three." A proper fraction such as 0.3 is read three tenths. The number 0.005 is read "five thousandths." The denominator depends on the location of the decimal point, Figure 3-9. A fraction such as 0.4829 is read "four thousand eight hundred twenty-nine ten-thousandths."

A mixed decimal combines a whole number and a decimal fraction. They are read together with the word "and" separating the two parts.

Example A Read the number 12 386.429 16.

Solution: Twelve thousand three hundred eighty-six and forty-two thousand nine hundred sixteen hundred-thousandths.

Notice the word "and" is not used anywhere except to indicate the position of the decimal point.

Analog scales are read as decimals or mixed decimals. Determine the size of the smallest scale divisions by inspection. Count the divisions from zero to the needle position. Estimate the position between smallest divisions for the final digit.

EXERCISE 3-4

Write the following fractions in decimal form and in word form.

1. 7/10
2. 84/100
3. 987/1000
4. 43/100 000
5. $5\frac{4}{100}$
6. $87\frac{654}{10\,000}$
7. $\frac{3025}{100\,000}$
8. $\frac{90\,488}{1000}$

Write these decimals as common fractions. Reduce to lowest terms.

9. 0.75
10. 0.875
11. 0.1875
12. 0.983 25
13. 0.3005
14. 2125.984 375

Change the following word-forms to decimals and common fractions.

15. Nine-tenths
16. Three-hundredths
17. Three hundred seventy-nine thousandths
18. Six hundred and four tenths
19. The lines a, b, c and d (Figure 3-10) represent the needle (pointer) of an ammeter. Read the analog scale at each position. Estimate the digit between the smallest division. One answer is given as an example.

 a. ______________ c. ______________
 b. 1.19 A d. ______________

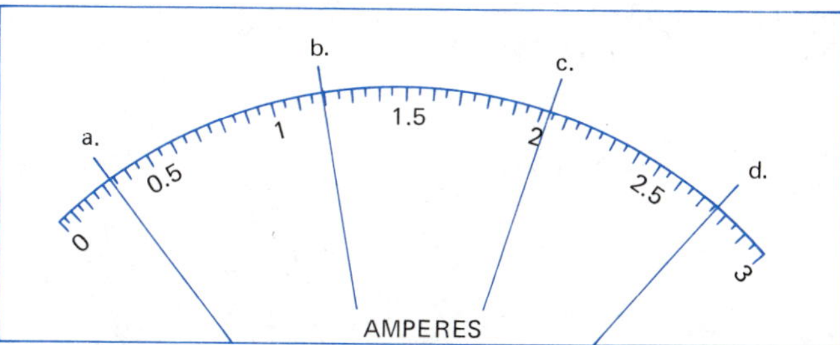

Figure 3-10

20. Give the ohmmeter readings a, b, c and d in Figure 3-11. Note: Read the scale from right to left. One example is given.

 a. 16.9 Ω c. ______________
 b. ______________ d. ______________

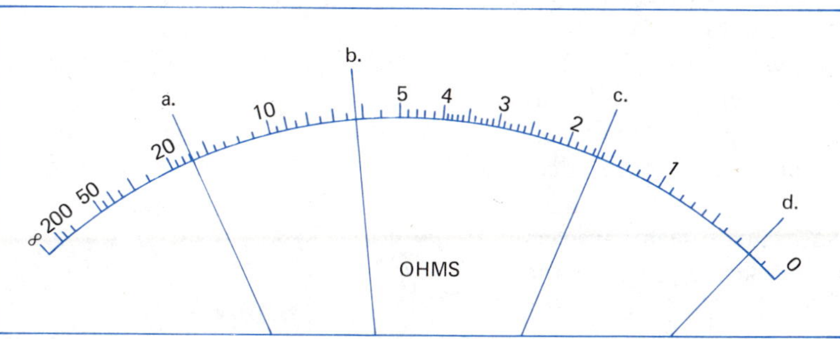

Figure 3-11

21. Give the voltmeter readings a, b, c and d in Figure 3-12 for each of the three scales. One example is given.

	10	50	250
a.	1.08 V	______	______
b.	______	______	______
c.	______	______	______
d.	______	______	______

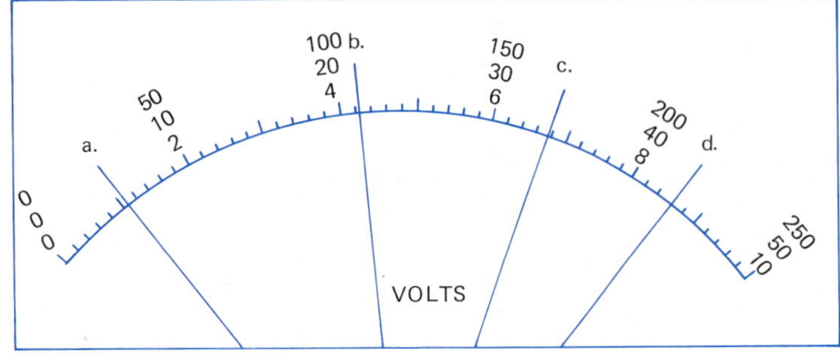

Figure 3-12

22. Arrange these decimals in order from smallest to largest:
(a) 0.020 56, (b) 0.2056, (c) 2.056, (d) 2.506,
(e) 0.025 06, (f) 0.2506, (g) 0.0205, (h) 2.05

3-7 CHANGING FRACTIONS TO DECIMALS AND ROUNDING DECIMALS

Common fractions can be changed to decimals by finding an equivalent fraction with a denominator of one of the powers of ten (10, 100, 1000, etc.).

Example A Convert the fraction 3/4 to a decimal fraction.

Note: If 4 is multiplied by 25, a power of ten (100) results. An equivalent fraction is found by multiplying both terms by 25.

Solution: $\frac{3}{4} \times \frac{25}{25} = \frac{75}{100} = 0.75$

The fraction 3/4 is the same as 0.75. This is an example of the many proper, improper or mixed fractions that can be converted to *exact decimals.* Others cannot because the denominator cannot be changed exactly to a power of 10. Some examples are: 1/3 = 0.333 333 . . . , 3/11 = 0.272 727 27 . . . . These are examples of *repeating decimals.* They must be rounded to meet the conditions of the problem. *Rounding* is the process of dropping digits after a certain significant place. Other fractions may not come out even until many decimal places have been calculated. These, too, must be rounded.

From this point on, students are expected to evaluate many answers using a calculator. The answers may show eight or more digits after the decimal point. Usually, there are more digits than are justified by the problem. Rounding, therefore, becomes a necessary part of solving problems.

Many calculators have rounding functions programmed into them. They use a simple procedure described by the conventions listed below. The symbol ≈ means "approximately equal to." It is used to indicate a number has been rounded.

ROUNDING NUMBERS

- If one digit is to be dropped, the digit preceding it is unchanged when the one being dropped is less than 5.
  Example: $5.293 \approx 5.29$.
- If one digit is to be dropped, the digit preceding it is increased by 1 when the one being dropped is 5 or greater.
  Example: $2.356 \approx 2.36$.
- If it is necessary to drop more than one digit, the rounding is done only once.
  Example: $5.2649 \approx 5.26$ but not $5.2649 \approx 5.265 \approx 5.27$.
- In cases where the rounding takes place to the left of the decimal point, the digits being dropped will be replaced with zeros.
  Example: $52\,947 \approx 53\,000$.

Example B Round these numbers to the nearest hundred.
(1) 5364 (2) 4921

Solution: (1) $5364 \approx 5400$
(2) $4921 \approx 4900$

Example C Round these numbers to two decimal places.
(1) 3.141 59 (2) 73.847 93 (3) 8.396 153

Solution: (1) $3.141\,59 \approx 3.14$
(2) $73.847\,93 \approx 73.85$
(3) $8.396\,153 \approx 8.40$

EXERCISE 3-5

Round these numbers to the nearest ten.

1. 53
2. 78
3. 5998
4. 6
5. 23.75
6. 94.465

Round these numbers to the nearest hundredths.

7. 3.1416
8. 2.7183
9. 0.070 92
10. 556.0369
11. 0.999 99
12. 0.0693

Round these numbers so there will be only two nonzero digits regardless of the decimal point location.

13. 144
14. 0.006 152
15. 1728
16. 3.141 59
17. 0.555 55
18. 4728

3-8 ABOUT THE CALCULATOR

By now you should own a scientific calculator. For the most part, it will be a valuable aid in learning mathematics. However, in learning some mathematical processes, the calculator should be set aside. In some sections of the book it will be of no help at all and may actually interfere with learning.

The calculator will do exactly what it is "told." If it isn't given correct instructions, it will give wrong answers. For example, an expression such as $4 - 3 + 7 - 2 - 4$ must be evaluated in the right order. If it isn't, the calculator will give the same wrong answer that is obtained manually with paper and pencil. The student must learn the correct procedures.

The evaluation of some mathematical expressions is easier and quicker when done manually. The expression in the previous paragraph is a good example. In the time it takes to do the problem manually, the problem wouldn't be half done with a calculator. In some portions of the text, the student is strongly urged not to use the calculator.

The calculator, however, was made to be used. For the most part, the student will be encouraged to take advantage of it. Specific calculator instructions will be given in many situations. Shortcuts will be suggested to save time.

3-9 USING THE CALCULATOR

There are many kinds of calculators. Some use different procedures than others for doing the same operation. Some have alternate symbols on the keyboard. For these reasons, definite instructions cannot be given for all calculators. The student is expected to use the *Owners Guide* for specific instructions.

To help clarify the many examples in this text, calculator operations will be boxed. The following symbols will be used: addition $\boxed{+}$, subtraction $\boxed{-}$, multiplication $\boxed{\times}$, division $\boxed{\div}$ and equal $\boxed{=}$. Manual operations will remain without a box.

Changing a common fraction to a decimal is easy with a calculator. Simply divide the numerator by the denominator. Mixed fractions can be converted by dividing the fractional part. Add the result to the whole number part manually. In all examples that follow, readers should evaluate the expression with their calculators. Use the procedure shown.

Example A Change 13/16 to decimal form.

Solution: 13 [÷] 16 [=] , 0.8125

Calculator note: The expression (13 [÷] 16 [=]) describes how the evaluation was made. (Enter 13, press [÷], enter 16, press [=]). The answer is now displayed. In the solution to the example, the answer is shown following a comma.

Example B Change 43 7/8 to decimal form.

43 7/8 = 43 + 7/8

Solution: 7 [÷] 8 [=] , 0.875

43 + 0.875 = 43.875

Note: The fraction is divided, then added manually.

EXERCISE 3-6

Convert these common fractions to decimals with the calculator and round them to three decimal places.

1. 1/8	5. 3/8	9. 15/16
2. 3/16	6. 5/32	10. 23/64
3. 5/16	7. 1/64	11. 3/4
4. 9/32	8. 7/8	12. 1/32

Manually convert these decimals to common or mixed fractions. Reduce to lowest terms. (The denominator is 10, 100, 1000, etc.).

13. 0.375 14. 14.75 15. 0.875 16. 4.85

Convert these mixed fractions to mixed decimal fractions. Round each answer to two decimal points.

17. 5 3/4	19. 17 9/32	21. 7 37/128
18. 24 7/16	20. 35 25/32	22. 12 55/64

3-10 ADDING AND SUBTRACTING DECIMALS

Adding and subtracting decimal fractions is similar to adding and subtracting whole numbers. In both operations, the various columns must be aligned properly. This is easy to do by simply aligning the decimal points. Carrying and borrowing procedures are the same.

When numbers are listed in a horizontal row, write them in a vertical column. Keep the decimal point aligned. The order of the list is not important when adding. When both operations are involved, remember to proceed from left to right.

With the calculator, much of the tedious work is eliminated. It automatically aligns the decimal point and performs the additions and subtractions in a fraction of a second. Of course, each entry must be made correctly with the right digits and the correct position of the decimal point.

Calculator notes: When several entries are to be made, it is not necessary to press the equal sign [=] after each entry. For example, 5 + 6 − 3 = ?. Enter 5, press [+], enter 6, press [-], enter 3, press [=]. The answer "8" will now be displayed. Notice that the equal sign was not pressed after 6 was entered. The intermediate answer (11) is displayed after the operation [-] is pressed. Press the equal sign only at the end of the evaluation.

Example A Combine: 8 + 3 − 7 + 2 − 1.
Solution: 8 [+] 3 [-] 7 [+] 2 [-] 1 [=], 5

EXERCISE 3-7

Write the following in a vertical column, then perform the indicated operations manually. Check with the calculator.

1. 32.04 + 8.3 + 9.125
2. $468.07 + $93.56 + $4.50
3. 0.0039 + 0.24 + 0.088 24
4. $0.17 + $5.08 + $3.40 − $2.16
5. $429.29 − $168.56
6. 2.625 + 1.25 − 3.5

Use the calculator to solve the following.

7. Add: 26.19, 8.391 and 12.03
8. Combine: $0.42 + $0.38 − $0.13 + $0.26 − $0.10
9. Subtract 5.3507 from 11.2.
10. Take $45.32 from $100.
11. Convert these to decimals, then add: 3 5/8, 2 1/4, 1/2 and 2 7/16.
12. Combine: 429 − 82.04 + 100.05 − 172.953
13. The following pieces of short wire are required for a job: 19.5 cm, 7.625 cm, 8.75 cm, 14.25 cm and 23 cm. How many centimetres of wire will be left from a piece 148 centimetres long?
14. A full 55-gallon oil drum was used to refill five pole transformers, Figure 3-13. How much oil was returned to the shop? The five transformers required the following amounts in gallons: 2.5, 5.25, 3.375, 4 and 6.625.
15. The following resistances in ohms are connected in series: 0.588, 1.33, 0.8 and 1.005. Find the total resistance by adding.
16. A repair truck carries several partially used coils of #12 three-wire W/G wiring of the following lengths expressed in metres: 24.333, 56.25, 35.875 and 82.5. If two pieces, 64.75 metres and 38.333 metres are used, find the total amount of wire remaining.
17. A repair truck travels the following distances in kilometres during one week: Monday (35.6 km), Tuesday (61.3 km), Wednesday (19.6 km), Thursday (45.2 km), Friday (106.9 km) and Saturday (8.4 km). What is the total distance traveled for the week?

Figure 3-13 A pole transformer (Courtesy of Westinghouse Electric Corporation)

18. The odometer in a truck reads 34 456.9 kilometres at the end of the week. What was the reading at the beginning of the week if the truck travels 277 kilometres during the week?
19. An order for electrical materials totaled $1755.89. If cash is paid within 10 days, a discount of $175.59 is offered. What is the discounted price?
20. A 500-foot span of copper power line undergoes thermal expansion. It increases by 0.302 foot in summer and decreases the same amount in winter. Give the summer and winter lengths of this span.

3-11 MULTIPLYING AND DIVIDING DECIMALS

As with addition and subtraction, multiplication and division of decimals is similar to working with whole numbers. It is a matter of locating the decimal point in the product or quotient. The procedure will not be described here. Multiplying and dividing decimals will be performed using a calculator.

The positioning of decimal points in multiplication and division is done automatically with a calculator. Each entry must be made with the correct digits. The correct location of the decimal point in each factor must be entered. In problems with multiple factors, the [=] is pressed only at the end.

Multiplication by 10, 100, 1000, etc. can be done by simply moving the decimal point. A calculator is not needed for this. Division is similar.

MULTIPLYING AND DIVIDING BY A POWER OF TEN

Count the zeros in the multiplier or divisor.

- When multiplying, move the decimal point that many places to the right. Add zeros if necessary.
- When dividing, move the decimal point that many places to the left. Affix zeros as necessary.

Example A Multiply each of these without the calculator:

(1) 5.4 × 1000
(2) 0.006 25 × 100
(3) 29.81 × 100 000

Solution:
(1) 5.4 × 1000 = 5400 (Three places right)
(2) 0.006 25 × 100 = 0.625 (Two places right)
(3) 29.81 × 100 000 = 2 981 000 (Five places right)

Example B Divide each of these without the calculator:

(1) 542 ÷ 1000 (2) 0.3 ÷ 10 000 (3) $\frac{576\ 000}{100}$

Solution:
(1) 542 ÷ 1000 = 0.542 (Three places left)
(2) 0.3 ÷ 10 000 = 0.000 03 (Four places left)
(3) $\frac{576\ 000}{100} = 5760$ (Two places left)

EXERCISE 3-8

Multiply these with the calculator.

1. 3.141 59 × 4
2. 0.035 × 9
3. 1.04 × 7.2
4. 2.006 × 3.01 × 0.009

Divide these with the calculator. Round to two decimal places.

5. 3.975 ÷ 3
6. 189.344 ÷ 3.14
7. Divide 65.435 by 0.075
8. $\frac{62\,458}{0.042}$

Multiply or divide these manually.

9. 4.53 × 10 000
10. 5280 × 100
11. 0.004 31 × 100 000
12. 86 400 ÷ 1000
13. $\frac{656.8}{100}$
14. 44.2 × 1000 × 10
15. 29.5 × 100 ÷ 10 000
16. 0.001 82 × 100
17. 0.001 82 ÷ 100
18. $\frac{0.009\,67 \times 100\,000}{1000}$

Use the calculator to solve these problems.

19. The distance around a circle is 3.1416 (π) times the distance across the center. A coil of wire has 25 turns and is 1.75 feet across. How many feet are in the coil? Use the [π] button on the calculator. Round the answer to two decimal places.
20. Copper wire (00 gage) weighs 0.4028 pound per foot. What is the weight of the three wires between towns 6.25 miles apart? (1 mi = 5280 ft).
21. Twenty-five light bulbs are connected in series. Each bulb has a resistance of 1.375 ohm. Find the total resistance of the circuit by multiplying.
22. A coil of wire 18.25 centimetres long is made by winding one layer of 10 gage wire on a pipe. If the wire has a diameter of 0.259 centimetres, how many turns are wound on the pipe?
23. The cost of making transformers increased $4.27 each. How much is the increase for 12 dozen transformers?
24. Ohm's Law states that voltage in volts divided by current in amperes equals resistance in ohms. What is the resistance of a coil if it draws 1.55 amperes from a 24-volt battery?
25. Eutectic solder is composed of 0.63 parts tin and 0.37 parts lead. How much of each metal is in a case containing 75 kilograms of solder?
26. Number 14 gage copper wire has 0.002 525 ohms of resistance per foot. What length of wire would have 1 ohm of resistance?

3-12 MULTIPLE OPERATIONS

The calculator may very well become a life-long companion to the reader. It is important to learn to use it efficiently. Some students tend to learn procedures that are very awkward. They use many more steps than necessary in evaluating relatively simple expressions.

This section is intended to show some short-cuts and alternative procedures before clumsy routines become habits. It is assumed the student has learned to use the [+], [−], [×] and [÷] operations involving two numbers.

Monomials. Monomials are single term expressions. They can, however, contain all four basic operations if grouping symbols are used. Thus, monomials can range from simple to complex. In the following discussion, a procedure is stated in the form of calculator notes and then an example is solved.

> *Calculator notes:* In problems in which only multiplication is involved, it is usually best to proceed from left to right. This avoids missing a factor. As with addition and subtraction, it is not necessary to press the equal sign until the end.

Example A Evaluate: 25 × 3.1416 × 16.2 × 8

Solution: 25 [×] 3.1416 [×] 16.2 [×] 8 [=], 10 180 (Rounded)

The division symbol (÷) is not used in electrical formulas. The bar is used instead. The symbol (÷) is used, however, on a calculator. Therefore, it is necessary to interpret between an expression using the bar and one using the (÷) symbol.

An expression such as:

$$\frac{48}{72.3 \times 0.001\,44}$$

can be written as 48 ÷ (72.3 × 0.001 44) or 48 ÷ 72.3 ÷ 0.001 44. It is the procedure shown in the latter expression that is simplest to use on a calculator. Notice that the numerator is simply divided by each factor in the denominator. It is unnecessary to translate the expression on paper before proceeding.

> *Calculator notes:* Multiply all factors in the numerator first proceeding from left to right. Now divide by each factor in the denominator from left to right. It is not necessary to press the equal sign until the end.

Example B Evaluate: $\dfrac{15 \times 7 \times 5}{22 \times 4 \times 9}$

Solution: 15 [×] 7 [×] 5 [÷] 22 [÷] 4 [÷] 9 [=], 0.663

The procedure just described avoids the necessity of writing intermediate answers on paper. Similarly, the storage [STO] and recall [RCL] buttons are not used. This procedure requires more steps: the denominator is evaluated first and stored. The numerator is then evaluated. Finally, the numerator is divided by the recalled denominator:

22 [×] 4 [×] 9 [=] [STO] 15 [×] 7 [×] 5 [÷] [RCL] [=], 0.663

Sometimes addition or subtraction operations show up in monomial expressions such as: $2\pi(28.6 - 17.4)$.

Calculator notes: The number π has a rounded value of 3.1416. It is stored in the calculator and can be entered by pressing [π]. In expressions of this type simply perform the operations in parentheses first. Press the [=] at the end. The problem then reduces to: $2\pi \times 11.2$ which is a simple multiplication problem. Since 11.2 is displayed, multiply it by 2π.

Example C Evaluate: $2\pi(28.6 - 17.4)$

Solution: 28.6 [-] 17.4 [=] [X] 2 [X] [π] [=] , 70.4

Some calculators have grouping symbol keys [(], [)] so the problem can be evaluated in order from left to right:
2 [X] [π] [X] [(] 28.6 [-] 17.4 [)] [=], 70.4

This method requires one additional step on the calculator.

Monomials with Grouping Symbols in the Numerator. The following expressions are mathematically the same:

$$\frac{17}{32}(62.4 + 249) = \frac{17(62.4 + 249)}{32}$$

Evaluation procedures are identical.

Calculator notes: Evaluate the group first, multiply by 17 and divide by 32.

Example D Evaluate: $\frac{17}{32}(62.4 + 249)$

Solution: 62.4 [+] 249 [=] [X] 17 [÷] 32 [=], 165.4
or using [(] [)]:
17 [X] [(] 62.4 [+] 24.9 [)] [÷] 32 [=], 165.4

EXERCISE 3-9

Multiply or divide the following expressions with a calculator. Do not use storage-recall or write down intermediate answers. Round to one decimal place.

1. $3 \times 1.14 \times 5.38$
2. $7 \times 2.99 \times 0.02 \times 1.41$
3. $56 \div 82 \div 0.53$
4. $6.28(597.3 - 409.7)$
5. $\frac{5}{8}(35.2 + 76.1 - 48.7)$
6. $\dfrac{46.7 \times 82\,400}{0.003\,16 \times 675}$
7. $\dfrac{28.2 \times \pi \times 8800}{1.38 \times 0.0075 \times 6}$
8. $\dfrac{35.7(3600 - 1728)}{52.8 \times 2}$
9. $\dfrac{\frac{15}{32}(44.6 + 15.2)}{3.14}$

Translate these expressions using the bar symbol for division to one using (÷). Do not evaluate.

10. $\frac{5 \times 8 \times 7}{3 \times 9}$

11. $\frac{5(17-8)}{3}$

12. $\frac{4}{5}(11-2)$

13. $\frac{\frac{5}{8}(256-129)}{17}$

Translate these expressions using the (÷) symbol to one using the bar. Evaluate each expression.

14. $56 \times 18 \div 54$

15. $244 \times 18 \div \pi \div 2 \times 4$

16. $5(8+5) \div 62.4$

17. $(5-2) \div (1728-1705)$

Solve these without using storage-recall or [(] [)] on the calculator.

18. $C = \frac{5}{9}(68 - 32)$ is an expression that converts 68°F to Celsius. Find the °C.
19. $E = 0.055(485 + 655 + 515)$ is an expression for emf in a three-load series circuit where the current is 0.055 A and the three resistances are 485 Ω, 655 Ω and 515 Ω. Find the voltage.
20. $X_L = 2\pi \times 600 \times 3$ is a formula for finding inductive reactance (X_L) in ohms when the frequency is 600 hertz and the inductance is 3 henrys. Find the reactance.
21. $X_C = \frac{1}{2\pi \times 600 \times 0.000\,015}$ is a formula for finding capacitive reactance (X_C) in ohms when the frequency is 600 hertz and the capacitance is 0.000 015 farads. Find the reactance.
22. $P = 0.055(26.7 + 36.0 + 28.3)$ is a formula for total power dissipation in watts in the series circuit of Problem 19. The voltage drops across the three resistors are 26.7 V, 36.0 V and 28.3 V, respectively. Find the power dissipation.

3-13 RECIPROCALS

Reciprocals were introduced in the explanation for dividing common fractions. Two numbers are said to be reciprocals if their product is 1. Every number except zero has a reciprocal. The reciprocal of any number is 1 divided by the number.

Calculator notes: Finding the reciprocal of a number with a calculator is easy. It is a one-number calculator function. Simply enter the number, then press the reciprocal button [1/x]. The reciprocal of the number is now displayed. In the solution to the following problem, the reciprocal of the given number is shown after a comma. The equal sign does not have to be pressed.

Example A Find the reciprocals of (1) 8, (2) 0.25, (3) 1, (4) 20 and (5) 3/4.

Solution: (1) 8 [1/x] , 0.125
(2) 0.25 [1/x] , 4
(3) 1 [1/x] , 1
(4) 20 [1/x] , 0.05
(5) 3 [÷] 4 [=] [1/x] , 1.33 (Rounded)

Note: The reciprocal of 3/4 is 4/3. Just inverting a fraction gives the reciprocal.

Example B Find the reciprocal of 16.28. Round the answer to four decimal places.
Solution: 16.28 [1/x] , 0.061 425 061 $\approx$ 0.0614

3-14 DIVISION USING THE RECIPROCAL

The rule for dividing complex fractions is to invert the denominator, then multiply. In other words, multiply the numerator by the reciprocal of the denominator (3/4 ÷ 5/8 = 3/4 × 8/5).

Whole numbers can be written as a fraction with a denominator of one (1). Thus, a fraction such as 3/4 can be written as a complex fraction: 3/1 ÷ 4/1. To evaluate, invert the denominator and multiply: 3/1 × 1/4 or simply 3 × 1/4. Any division problem can be changed to a multiplication problem (3/4 = 3 × 1/4).

Calculator notes: To divide one number by another, multiply the reciprocal of the denominator by the numerator.

Example A Divide 16 by 25 using the reciprocal method.
Written in symbols, the problem is:

$$16 \div 25 = 16 \times \frac{1}{25}$$

Solution: 25 [1/x] [×] 16 [=] , 0.64

Normally, one will use division in simple problems like the previous example. In evaluating expressions with combined operations, however, the [1/x] is a handy tool. This is especially true if the denominator has separate grouping symbols.

Monomials with Grouping Symbols in the Denominator. Whenever an expression contains grouping symbols in the denominator, it usually has to be evaluated first. Either it must be recorded on paper and later re-entered or stored and recalled. A few steps can be saved by using the reciprocal method.

Calculator notes: When there are grouping symbols in the denominator, evaluate it first. Find its reciprocal using [1/x] , then multiply by the numerator.

Example A Evaluate this expression using the reciprocal method:

$$\frac{28}{4(2.65 + 3.15)}$$

Solution: 2.65 [+] 3.15 [=] [×] 4 [=] [1/x] [×] 28 [=], 1.21

Notice that no storage or recall is required. No intermediate answers have to be written on separate paper.

EXERCISE 3-10

Write the reciprocal of these numbers manually. Express answers in decimal form rounded to no more than two decimal points.

1. 2
2. 3
3. 4
4. 5
5. 10
6. 20
7. 100
8. 0.25
9. 0.5
10. 0.001

Use [1/x] to find the reciprocal of these. Round the answers to three decimal points when necessary.

11. 6
12. 7
13. 8
14. 9
15. 1 000 000
16. 84.3
17. 0.789
18. 1.629
19. 12.5
20. 0.0054

Evaluate these expressions using the reciprocal method. Round to three places.

21. $\dfrac{25}{42}$
22. $\dfrac{60}{2\pi}$
23. $\dfrac{56}{9.1 \times 0.62}$
24. $\dfrac{75 \times 2.38}{0.048 \times 9.36}$
25. $\dfrac{5280}{2(44.3 + 62.9)}$
26. $\dfrac{1.07}{5(0.073 + 0.029) + 0.75}$

Evaluate these using the reciprocal method.

27. The following is an expression for the total resistance in a two-branch parallel circuit. One resistance is 500 Ω and the other is 800 Ω. Find the total resistance.

$$R_t = \frac{500 \times 800}{500 + 800}$$

28. The following expression is for the current in a three-load series circuit. The resistances are 155 Ω, 220 Ω and 335 Ω. The voltage is 220 V. Find the current.

$$I = \frac{220}{155 + 220 + 335}$$

3-15 POWERS

Many times a number is multiplied by itself. The number is occasionally used several times as a factor. When this occurs, the product is called a *power*. Whenever the factors are the same, the number (factor) is called the *base*. The number of factors is called an *exponent*. The exponent is indicated as a *superscript*. A superscript is a number placed above and to the right of the base number.

$$(\text{base})^{\text{exponent}} = \text{power}$$

The expression: 3 × 3 is written 3^2. It is read: "three to the second power" or "three *squared*." The base is 3, the exponent is 2 and the answer ($3^2 = 9$) is the second power of three. Some other examples are:

- $4 \times 4 = 4^2 = 16$ (Second power of four or four squared)
- $5 \times 5 \times 5 = 5^3 = 125$ (Third power of five or five cubed)
- $3^4 = 81$ (Fourth power of three)
- $2^5 = 32$ (Fifth power of two)

In electrical work, most problems involving powers of numbers are usually squares or cubes unless the base is ten. Powers of ten are discussed in Chapter 5. Most calculators have keys for finding powers. The button [x^2] is used for squaring a number and [y^x] is used for finding higher powers.

Calculator notes: The function [x^2] is a one number function. To square any number (base), enter it into the calculator. Press [x^2]. The square will now be displayed.

Example A Square the number 52.73.

Solution: 52.73 [x^2], 2780.45

Calculator notes: The function [y^x] is a two number function. To use the [y^x] function, enter the base number (y), press [y^x] then enter the exponent (x). Press the equal sign [=]. The power will now be displayed.

Example B Find the cube of 2.66.

Solution: 2.66 [y^x] 3 [=], 18.82

Monomials with Powers. Many electrical formulas have squared or cubed factors. Expressions such as $3(4)^2$, $(3 \times 4)^2$, $(3/4)^3$ and $4\pi(2)^3(5)^2(3)^2$ must be evaluated with care. Notice, especially, the difference between the first two.

Calculator notes: Expressions of $(3 \times 4)^2$ or $(3/4)^2$ are easy. Evaluate the expression inside parentheses first. Be sure to press [=] before finding the power.

Example C Evaluate: (1) $(3 \times 4)^2$ and (2) $(3/4)^2$.

Solution: (1) 3 [×] 4 [=] [x^2], 144

(2) 3 [÷] 4 [=] [x^2], 0.5625

Calculator notes: The expressions $3(4)^2$ and $(3 \times 4)^2$ are different. The first expression is easily confused with the second. The expression $3(4)^2$ means "three times the square of four" while $(3 \times 4)^2$ means "the square of the product of three and four." In evaluating these expressions, the [=] and [x^2] are interchanged. Multiplication expressions with powers can be evaluated in any order. It is probably best to handle the factors with powers first.

Example D Evaluate: (1) $3(4)^2$ (2) $4\pi(2)^3(5)^2(3)^2$

Solution: (1) 4 [x^2] [×] 3 [=], 48

(2) 2 [y^x] 3 [=] [×] 5 [x^2] [×] 3 [x^2] [×] 4 [×] [π] [=], 22 619.5

EXERCISE 3-11

Find the squares of the following numbers manually.

1. 1	4. 4	7. 7	10. 10	13. 15
2. 2	5. 5	8. 8	11. 11	14. 20
3. 3	6. 6	9. 9	12. 12	15. 25

Use the calculator to evaluate these expressions.

16. 14^2
17. 18^2
18. 2^3
19. 3^4
20. 217^2
21. $\pi 7^3$
22. 3^{12}
23. $(5/8)^2$
24. $(5/8)^3$
25. $(5 \times 144)^2$
26. $(1/2)^2$
27. $3(1.24)^2$
28. $4\pi(3.75)^2$
29. $4.3(7)^2(2)^3$
30. $\frac{4}{3}\pi(6)^3$
31. $\pi(5)^2 \times 2(3)^3(4)^2$
32. $2(72 \times 0.0045 \times 21.3)^2$
33. $\dfrac{17.5^2}{4(8.5 - 3.75)^2}$

Evaluate the following.

34. The expression $4\pi(6)^2$ will give the amount of surface area in square inches on a 12 inch spherical light globe. Evaluate the surface area.
35. The internal volume in cubic inches of the globe in Problem 34 is $\frac{4}{3}\pi(6)^3$. Calculate the internal volume.
36. The power in watts dissipated by a 3300-Ω resistor with an applied voltage of 110 V is $110^2/3300$. Calculate the power.

3-16 ROOTS

Finding the root or extracting the *root* of a number is the inverse of raising to a power. The *square root* of a given number is a number which when squared results in the given number. The symbol for indicating the square root is called the *radical sign* ($\sqrt{\ }$).

A small number above and to the left of the radical sign is the *index*. The symbol ($\sqrt[3]{\ }$) means the *cube root* since the index is 3. If the cube root of a number is cubed, the result is the same number. Whenever no index is shown, the sign is understood to mean square root. In electrical formulas, square roots are commonly used, cube roots are rarely used and other roots are never used.

To find the square root of 36, find the factor which, when squared, equals 36. It is the number 6 since $6^2 = 36$. Thus $\sqrt{36} = 6$. Similarly:

- $\sqrt{25} = 5$ since $5^2 = 25$
- $\sqrt[3]{27} = 3$ since $3^3 = 27$
- $\sqrt{144} = 12$ since $12^2 = 144$.

Roots can also be designated by exponents. The square root of a number is the number "raised" to the one-half power. Cube roots are expressed as a one-third power, etc.:

- $\sqrt{16} = (16)^{1/2} = 4$
- $\sqrt{25} = (25)^{1/2} = 5$
- $\sqrt[3]{8} = (8)^{1/3} = (8)^{0.333333} = 2$

There is a special calculator key for finding square root. It is the $\boxed{\sqrt{x}}$ button. For cube roots, use the $\boxed{y^x}$ button.

Calculator notes: The $\boxed{\sqrt{x}}$ button is a one number function. Simply enter the number, press $\boxed{\sqrt{x}}$ and the square root will be displayed. For cube roots, use the $\boxed{y^x}$ button, a two number function. The exponent for cube roots is 0.33333333. Use as many threes as the calculator will display. Enter the number, press $\boxed{y^x}$, enter 3, press $\boxed{1/x}$, press $\boxed{=}$ and the cube root will be displayed. The horizontal bar in the radical sign is a grouping symbol. Evaluate any expression "inside" before taking the root.

Example A Find the square root of 1000: $(\sqrt{1000})$

Solution: 1 000 $\boxed{\sqrt{x}}$,31.62

Example B Find the cube root of 729: $(729)^{0.3333333}$

Solution: 729 $\boxed{y^x}$ 3 $\boxed{1/x}$ $\boxed{=}$,9

Monomials Containing Radicals. As with monomials containing powers, it is best to evaluate the radicals first. Then multiply by other factors.

Example C Evaluate this expression: $4\pi^2\sqrt{\frac{0.0055 \times 3}{32}}$

Solution: 0.0055 $\boxed{\times}$ 3 $\boxed{\div}$ 32 $\boxed{=}$ $\boxed{\sqrt{x}}$ $\boxed{\times}$ $\boxed{\pi}$ $\boxed{x^2}$ $\boxed{\times}$ 4 $\boxed{=}$,0.896

EXERCISE 3-12

Manually find the square roots of these numbers.

1. 9	4. 64	7. 16	10. 4	13. 625
2. 25	5. 1	8. 81	11. 36	14. 49
3. 144	6. 225	9. 100	12. 121	15. 400

Find the following roots using the calculator.

16. $\sqrt{456}$	19. $18^{1/3}$	22. $\sqrt[3]{3/4}$
17. $\sqrt[3]{3.175}$	20. $23.75^{1/3}$	23. $(15\ 7/8)^{1/3}$
18. $0.0529^{1/2}$	21. $\sqrt{3/4}$	24. $(10\ 000\ 000)^{0.5}$

Evaluate these expressions with combined operations.

25. $10\sqrt{15 \times 27}$
26. $2\pi(5/8)^{0.5}$
27. $1.7\sqrt[3]{50 \times 7}$
28. $(\sqrt{169})^2$
29. $\sqrt{(14)^2}$
30. $\frac{8\pi}{2\sqrt{3.79 - 2.33}}$

Evaluate the following.

31. The diameter in inches of a lamp globe with a volume of one cubic foot is $2\sqrt[3]{3(1728)/4\pi}$. Find the diameter.
32. The voltage in a circuit in which the load has a resistance of 550 Ω and the power dissipation is 100 watts is $\sqrt{550 \times 100}$. Find the voltage.

OHM'S LAW

Every electric circuit, Figure 3-14, involves three electrical quantities:

- *Electromotive force* or *voltage* (E) measured in volts (V).
- *Electric current* (I) measured in amperes (A).
- *Resistance* (R) measured in ohms (Ω).

Voltage causes current to flow; resistance opposes the current.

Georg Ohm, a German physicist, discovered how the three electrical quantities were related. He found that doubling the voltage across a resistor caused the current to double. This led to a law relating the three quantities. The law, named after its discoverer, is called *Ohm's Law.* It states that: voltage equals current times resistance. In symbols, the law is:

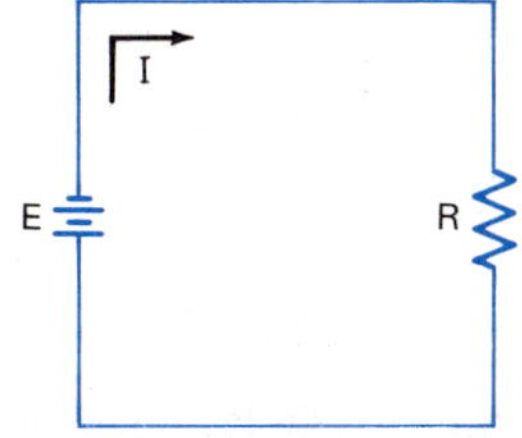

Figure 3-14 The three circuit quantities are voltage (E), current (I) and resistance (R).

$$E = IR$$

A *formula* is a rule written in algebraic symbols. A formula expresses the relationship between several quantities using symbols. When all the quantities on the right of the equal sign have known values, the quantity on the left of the equal sign can be calculated.

The equation E = IR can be solved for I or R, when required. To rearrange a formula, the unknown quantity (letter) must be isolated on one side of the equal sign. To do this, all other numbers, letters and other symbols must be removed. In the case of Ohm's Law, the standard form E = IR is often solved for R or I.

Ohm's Law can be solved for I by moving the letter R to the other side. This is done by dividing both sides of the equation by R. Start with the standard form:

$$E = IR$$

$$\frac{E}{R} = \frac{IR}{R}$$

The R's on the right cancel, so:

$$\frac{E}{R} = I$$

The two sides can now be interchanged to place I on the left:

$$I = \frac{E}{R}$$

Effectively the letter R was a multiplier on the right in the original form of Ohm's Law E = IR. In the form I = E/R, the letter R has become a divisor. This can be stated as a general rule.

REMOVING A FACTOR IN AN EQUATION BY DIVISION

- In a formula that does not contain plus or minus signs, a multiplier can be taken to the other side where it becomes a divisor.
- Interchange the two sides of the resulting formula so the unknown is isolated on the left.

Ohm's Law can be rearranged giving formulas for current and for resistance:

$$I = \frac{E}{R}$$

$$R = \frac{E}{I}$$

These three formulas are sometimes called the three forms of Ohm's Law. They are so important that all should be memorized. Notice that E is alone or always in the numerator. In all three forms of the law, current is measured in amperes (A), voltage in volts (V) and resistance in ohms (Ω).

3-17 THE SINGLE LOAD DC CIRCUIT

When a quantity is to be calculated using Ohm's Law, a step by step approach is best.

PROCEDURE FOR SOLVING PROBLEMS

- Always draw a circuit diagram showing the known quantities (data). Indicate the unknown with a question mark.
- Solve the equation for the indicated unknown.
- Substitute the known values of the other quantities into the formula in the exact order given by the formula.
- Evaluate the expression using a calculator. Round the answer to two places after the first nonzero digit (regardless of the decimal point location).
- Be sure to give the units of the answer as well as the numerical value.

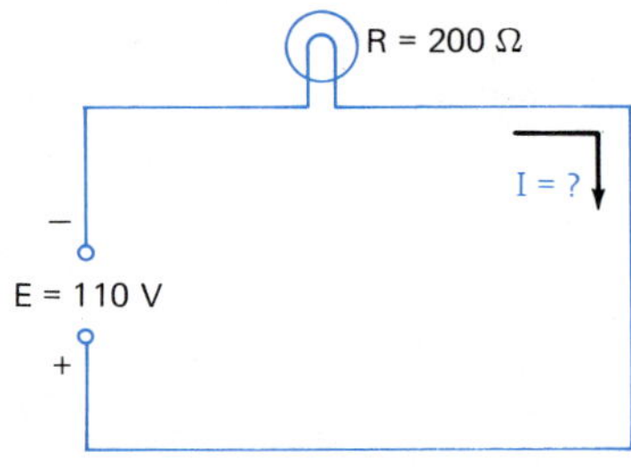

Figure 3-15

Example A A light bulb is plugged into a 110-V source. What is the current if the resistance of the filament is 200 Ω? Draw a circuit diagram.

Solution: Data is shown on the circuit diagram, Figure 3-15.

$E = IR$ (Ohm's Law)

$\frac{E}{R} = \frac{I\cancel{R}}{\cancel{R}}$ (Divide by R and cancel)

$I = \frac{E}{R}$ (Desired form)

$I = \frac{110\ V}{200\ \Omega}$ (Substitute known values)

$I = 0.550\ A$

Example B A headlamp on a car draws 4.07 A of current. What is the hot resistance of the bulb filament if the battery voltage is 12 V?

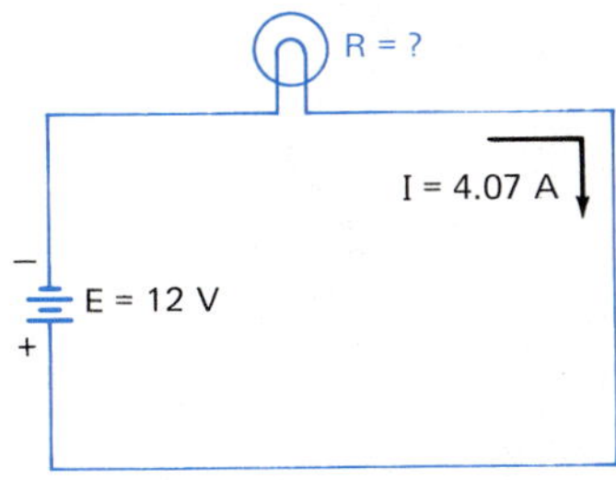

Figure 3-16

Solution: Data is shown on the circuit diagram, Figure 3-16.

$E = IR$ (Ohm's Law)

$\frac{E}{I} = \frac{\cancel{I}R}{\cancel{I}}$ (Divide by I and cancel)

$R = \frac{E}{I}$ (Desired form)

$R = \frac{12\ V}{4.07\ A}$ (Substitute known values)

$R = 2.95\ \Omega$

Example C The hot resistance of a pencil soldering iron is 960 Ω. What is the line voltage if the current is 0.125 A?

Solution: Draw diagram, Figure 3-17.

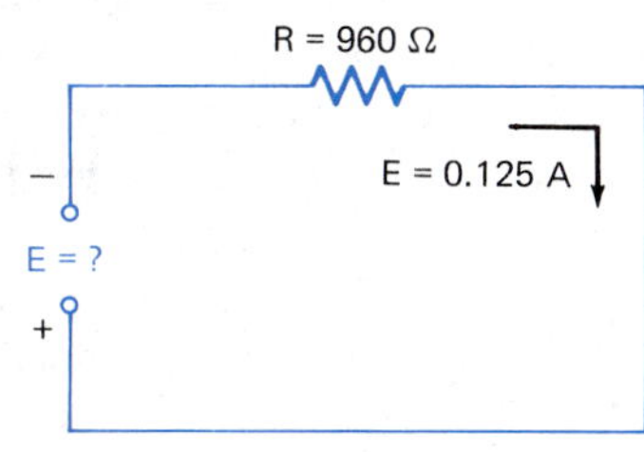

Figure 3-17

$E = IR$ (Choose formula)

$E = 0.125\ A \times 960\ \Omega$ (Substitute knowns)

$E = 120\ V$

In ac circuits, other quantities tend to oppose the flow of current. An ac circuit having only resistance is called a *resistive circuit.* Ohm's Law applies to all DC circuits and to resistive circuits.

EXERCISE 3-13

Draw circuit diagrams with data, write down the correct formula, substitute numbers and evaluate. Round to two places after the first nonzero digit. Give correct units with each answer.

1. A toaster operates on 110 V and has a resistance of 18.3 Ω. What current does it draw?
2. A coffee pot operates on 110 V and draws 10.9 A. Find the resistance of the heating element.
3. An electric clock draws 0.018 A. It has 6390 Ω of resistance. What voltage is required?
4. Find the current in the circuit shown in Figure 3-18.

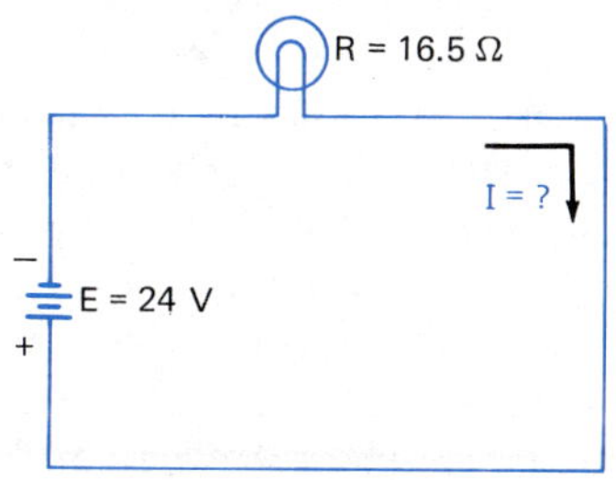

Figure 3-18

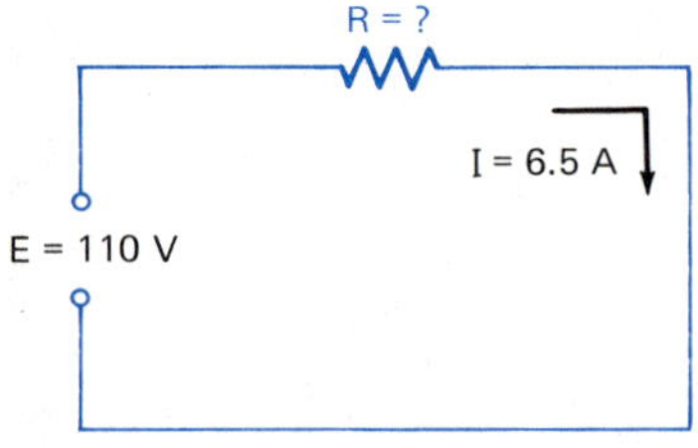

Figure 3-19

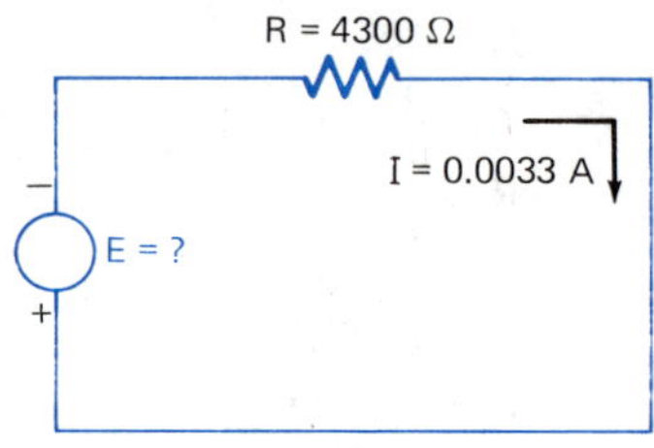

Figure 3-20

5. Find the resistance of the heater element in Figure 3-19.
6. Find the applied voltage in Figure 3-20.
7. A transmission line carries a current of 95.5 A. Find the voltage drop along the line if its resistance is 2.6 Ω.
8. A trouble lamp plugs into the cigarette lighter on a car. It has a hot resistance of 4.75 Ω and operates from a 12-V battery. What current does it draw?
9. A thermostat relay coil operates on 24 V. If it draws 0.135 A, what is its resistance?
10. An ammeter is placed in series with a load in a circuit. The resistance of the ammeter is 0.0048 Ω. If it carries a 5-A current, what is the voltage drop?
11. Find the resistance of a 12-V taillight on a car if it draws 0.45 A.
12. A 115-V electric heater element has a hot resistance of 8.8 Ω. Find the current through the element.
13. The heating element of an electric iron has a hot resistance of 25 Ω. What is the line voltage if it draws 4.4 A?
14. The starting motor (12 V) on a car draws 85 A. Find the resistance of the motor windings.
15. A 115-V furnace motor has a winding resistance of 41.3 Ω. What current does it draw?
16. An emf of 24 V is applied to a circuit. What is the least resistance of the circuit if the current must be limited to 1.5 A?
17. A fuse has a resistance of 0.025 Ω. If the voltage drop is 0.0005 V, what current is flowing?
18. A fuse has a resistance of 0.0195 Ω. What voltage drop will produce the 15 A required to melt the fuse?

3-18 POWER DISSIPATION IN ELECTRIC CIRCUITS (DC)

Power (P) can be described in terms of work or energy. *Power* is the rate at which work is done or energy is used (or produced). Power is measured in watts (W).

Work is related to force and movement. If work is done, forces are applied to an object and movement takes place. In electrical equipment, this occurs in devices such as motors, solenoids and vibrators. Work is measured in units of joules (pronounced jewels). The symbol for joules is (J).

Energy is the ability to do work. It is usually possessed by some object or material. It exists in several forms: mechanical (kinetic, gravitational and elastic), thermal, chemical, nuclear and electrical. The symbol for work and energy is the same. The italic letter (*W*) is used for both.

Work is required to move an electric charge around an electric circuit. Work usually manifests itself by the conversion of electric energy to heat as current flows through a load. Sound, light or mechanical (movement) energy may be produced also, depending on the type of load. The amount of energy conversion is usually referred to as energy consumption.

The rate at which energy is dissipated (consumed) is called *power loss.* The amount of energy consumed and the time involved will be discussed in

Chapter 7. For purposes at hand, power loss can be determined when the current in amperes and the voltage in volts are known. Power in watts is simply the product of these quantities in DC and resistive ac circuits:

$$P = EI$$

The other forms of this formula are:

$$E = \frac{P}{I}$$

$$I = \frac{P}{E}$$

Power in watts can be converted to horsepower by dividing by 746. (1 hp = 746 W). Notice that the power formula (P = EI) is similar to the formula used for Ohm's Law (E = IR). In the power formula, P is always in the numerator and E and I simply interchange positions.

Example A Find the input power to a 24-V DC electric motor if the current is 3.89 A. Express the answer in watts and horsepower.

Solution: Data is shown on the circuit diagram in Figure 3-21.

$P = EI$ (Choose formula)

$P = (24\ V)(3.89\ A)$ (Substitute)

$P = 93.4\ W$ (Evaluate)

$P = \dfrac{93.4\ W}{746\ W/hp} = 0.125\ hp$

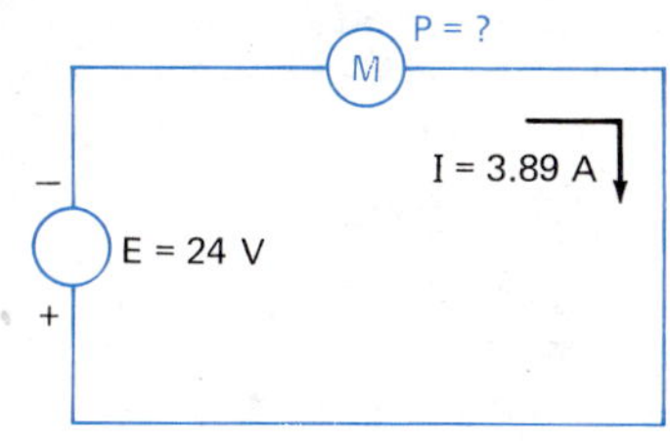

Figure 3-21

Example B Find the current through a 110-V motor if the power input is 310 W.

Solution: Data is entered on the circuit diagram in Figure 3-22.

$P = EI$ (Power formula)

$\dfrac{P}{E} = \dfrac{\not{E}I}{\not{E}}$ (Divide by E and cancel)

$I = \dfrac{P}{E}$ (Desired form)

$I = \dfrac{310\ W}{110\ V}$ (Substitute)

$I = 2.82\ A$

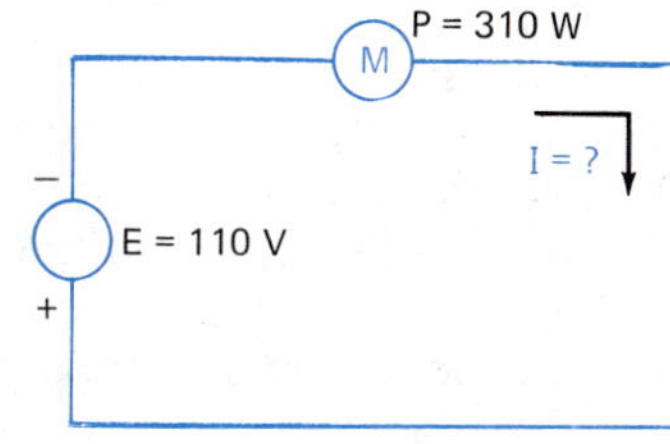

Figure 3-22

Example C Find the voltage across a heating element if the power consumption is 1600 W when the current is 6.96 A.

Solution: Data: Figure 3-23

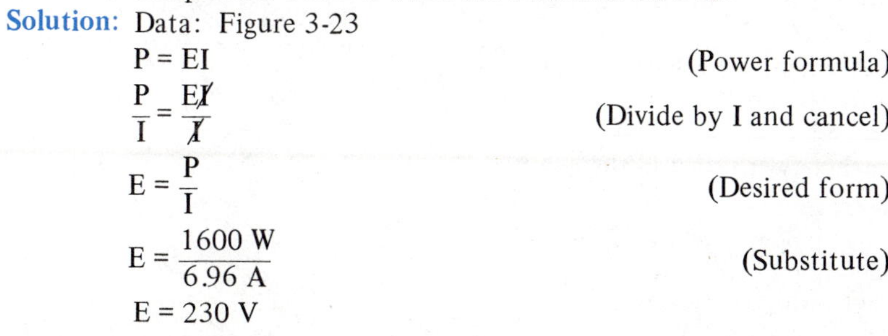

$P = EI$ (Power formula)

$\dfrac{P}{I} = \dfrac{E\not{I}}{\not{I}}$ (Divide by I and cancel)

$E = \dfrac{P}{I}$ (Desired form)

$E = \dfrac{1600\ W}{6.96\ A}$ (Substitute)

$E = 230\ V$

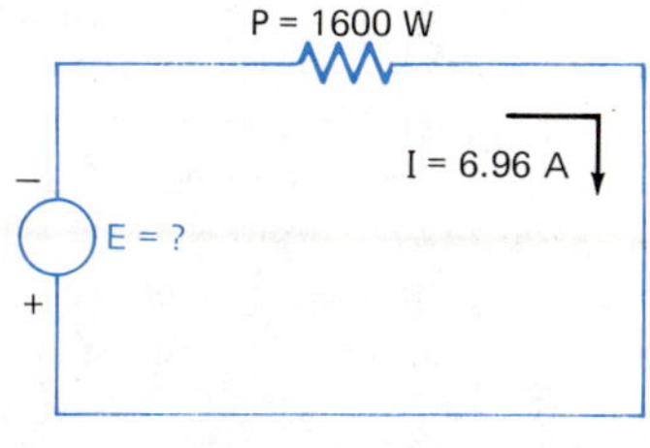

Figure 3-23

The *power rating* of electrical devices (other than resistors and motors) is simply the wattage (power) being drawn under normal conditions. These are operated at specified voltages. Resistors may be used at various voltages, however. Safe practices require that the power rating be specified as a fraction of that which will cause failure.

Light bulbs and heating elements are specified according to their power consumption (input). Electric motors are rated according to their mechanical output usually in horsepower (hp). In small fractional horsepower motors the mechanical output is specified in watts. In these devices, voltage is also specified.

EXERCISE 3-14

Solve the following formulas for the indicated unknown.

1. $C = \pi d$ for d
2. $A = LW$ for L
3. $F = PA$ for P
4. $W = Pt$ for t
5. $Q = CV$ for C
6. $Q = It$ for t
7. (a) Write the other two forms of the formula: $E = IR$.
   (b) Write the other two forms of the formula: $P = EI$.
8. Find the output power of a 1/4 hp electric motor expressed in watts.
9. What is the equivalent power of the 4500-W heating element in a water heater in horsepower?
10. An automobile starter motor requires 1.55 hp input. Give the equivalent input wattage.
11. A nuclear power plant has an output of 800 000 000 W. Express the output in horsepower.
12. Find the power rating of a soldering iron that draws 1.25 A when the voltage is 115 V.
13. A transmission line carries a current of 92.5 A. What is the power loss in the line if there is a voltage drop of 256 V along the line?
14. An electric clock draws 0.0175 A and has a power rating of 2 W. What voltage is used?
15. A thermostat relay coil operates on 24 V. If it has a power rating of 3.5 W, what current does it draw?
16. Find the power rating of the heating element in Figure 3-24.
17. What is the voltage in the automobile starter motor shown in Figure 3-25?
18. The service entrance to a home is rated at 11 500 W at 50 A. What voltage is this?
19. A furnace motor has an input power of 310 W. What is the current if the applied voltage is 115 V?
20. A 620-W street lamp draws a current of 5.67 A. Find the voltage.
21. A very small fractional horsepower motor has an input power of 7 W. What is the current if the voltage is 16 V?
22. A 440-V electric motor draws 12.4 A when operating at a normal load. What power is absorbed from the line?
23. A generator delivers 85 A of current and a voltage of 220 V. If the power input is 31.2 hp, how much power is lost in the generator in horsepower?
24. A resistor has a resistance of 8000 Ω. It is also rated at 20 W. If the current is 0.025 A, (a) find the applied voltage, and (b) is the resistor being used within its power rating?

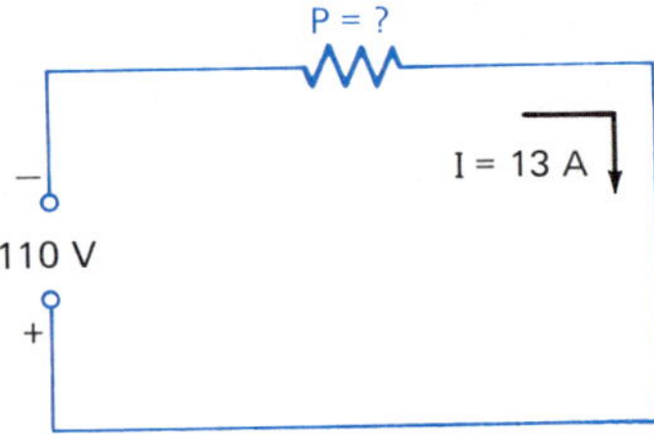

Figure 3-24

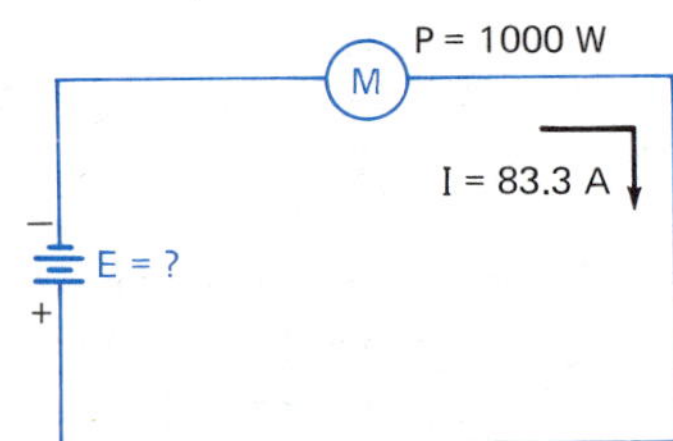

Figure 3-25

CHAPTER 4

GRAPHING ELECTRICAL QUANTITIES

OBJECTIVES

After satisfactorily completing this chapter, the student should be able to:

- Define various terms associated with graphs.
- Construct a graph correctly given a set of data.
- Find the values of one quantity from a graph given specific values of the related quantity.

Everyone has observed quantities changing. A *quantity* is a property, quality or dimension of an object or phenomenon that can be measured. Sound levels increase and decrease. Lighting levels (brightness) change. Colors fade. Automobile speeds increase and decrease. Thousands of examples can be given but it is the changes in electrical quantities that are important to the electrician.

A change in one quantity may cause a change in another quantity. In this case, there is a relationship between the two quantities. Such relationships can sometimes be described by a mathematical equation. Most relationships can be described by a set of data or by graphical means. This chapter describes how graphs are constructed and how they are used to describe various relationships.

INTRODUCTION TO GRAPHS

A *rectangular coordinate* system consists of two *axes* (lines), Figure 4-1. The two axes are at right angles to each other and divide the plane into four sections called *quadrants.* The four quadrants are numbered as shown. Most graphs in electricity use only the first quadrant.

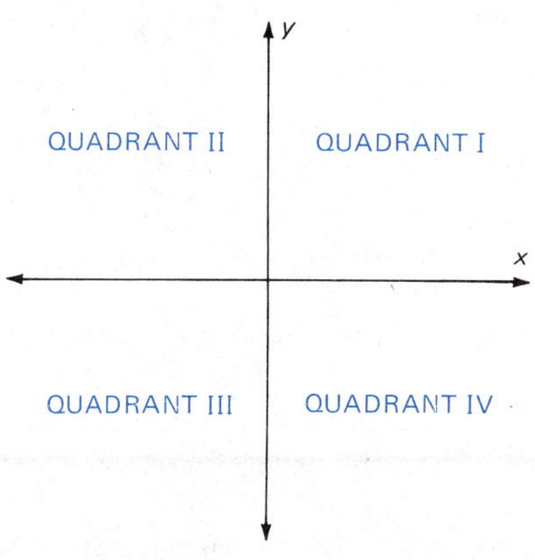

Figure 4-1 Rectangular coordinate system

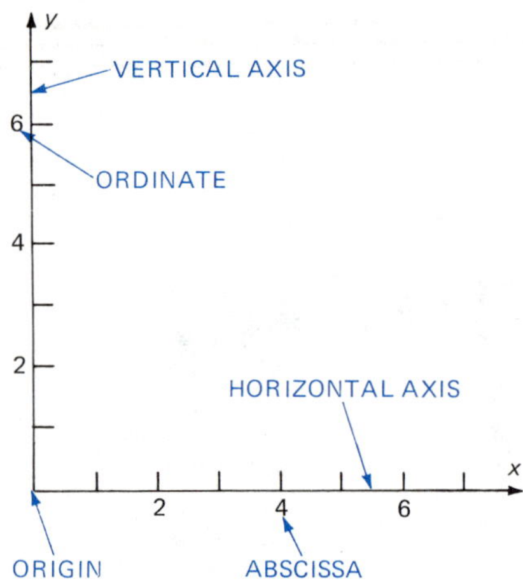

Figure 4-2 Graphical terms

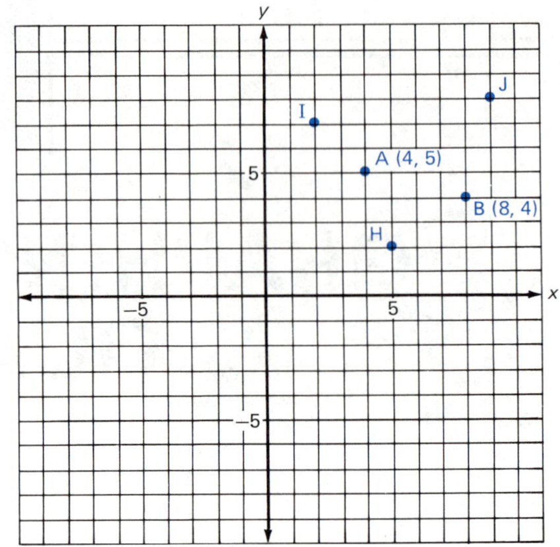

Figure 4-3 Plotting data points

In Figure 4-2, the intersection of the two axes is called the *origin.* The *horizontal axis* is sometimes called the *x-axis* or the *east-west axis.* The *vertical axis* is referred to as the *y-axis* or the *north-south* axis. The words horizontal and vertical are easy to distinguish. A horizontal line extends to the horizon.

The two axes are scaled off with one quantity plotted along each axis. The value of the quantity plotted along the horizontal axis is called the *x-coordinate* or *abscissa.* The value plotted on the vertical axis is called the *y-coordinate* or *ordinate.* The zeros of both quantities are located at the origin.

Each pair of numbers (abscissa, ordinate) represent a point in the plane. The abscissa is always given first, then the ordinate. They are separated by a comma and enclosed in parentheses. In Figure 4-3, point A has the coordinates (4,5). It is found by moving four spaces to the right of the origin then upward five spaces. Point B has the coordinates (8,4). Negative abscissas are to the left of the origin. Negative ordinates are below the origin. Negative numbers are discussed in Chapter 5.

4-1 VARIABLES

The two quantities that are plotted on such a graph can change values. Thus, they are called variable quantities or simply *variables.* One variable is plotted horizontally; one vertically. Some quantities do not change values. These are called *constants.* Constants, of course, would seldom be plotted on a graph.

Often, one quantity changes and this results in a change in a second quantity. The second quantity is said to be a *function* of the first quantity. For any given value of the first quantity, there can be only one value of the second. In the single load circuit, for example, a certain voltage will be applied. The resulting current will have one and only one definite value. If the voltage is changed, a new current will result having a new definite value.

In an experiment such as the one just described, the quantity that is controlled by the experimenter (voltage) is the *independent variable.* The second quantity is called the *dependent variable.* It is the one that is affected by changes in the independent variable. The values of the dependent variable *depend* on the values chosen for the independent variable.

It is usually easy to determine which quantity is independent. For example, consider the following experiments:

- An experimenter observes electrical energy consumption on the kilowatt-hour meter (Figure 4-4) on the side of a house. Time is recorded whenever the meter advances one kilowatt-hour. The experiment continues until ten kilowatt-hours are consumed. Time depends on how fast energy is being used. Thus, energy is the independent quantity and time is dependent.
- Another experimenter performs the same experiment but does it this way: the meter is read at the end of each hour and the experiment continues for ten hours. This experimenter doesn't have to stand continuously by the meter. Energy consumed depends on time. Time is the independent variable; just the reverse of the first experiment.

The quantity the experimenter controls is independent. Usually this quantity will be expressed in round numbers, but not always. The other quantity is dependent. Dependent values are recorded as observed.

Figure 4-4 A kilowatt-hour meter measures energy consumption in the home.

EXERCISE 4-1

1. On separate paper draw a figure similar to Figure 4-3. Number each of the four quadrants. Label the x-axis, y-axis and the origin. On which axis are the ordinates located? The abscissas?
2. On separate paper draw a figure similar to Figure 4-3. Locate each of the following points: C (2,5), D (4,4), E (2,4), F (3,1) and G (a point with an ordinate of 4 and an abscissa of 3). Give the coordinates of points H, I and J in Figure 4-3.

In each of the following experiments, (a) which variable is independent, and (b) which is dependent?

3. The voltage in a certain DC circuit is adjusted, Figure 4-5. The corresponding values of the electric current are recorded. The variables are voltage and current.
4. Ten various light bulbs are arranged according to their power ratings. The cold resistance is then measured with an ohmmeter. The variables are power rating and cold resistance.

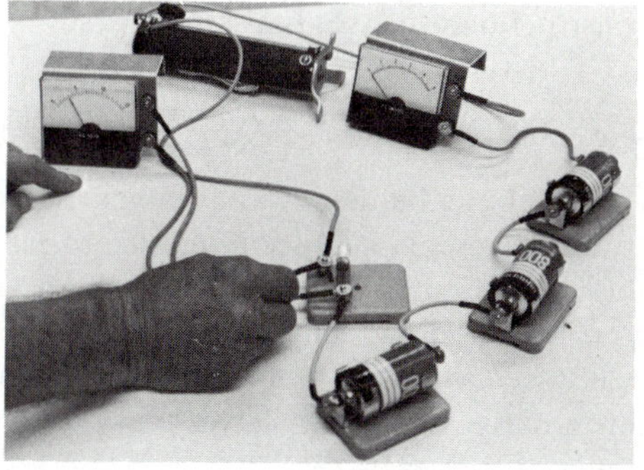

Figure 4-5 Making voltage and current measurements (Courtesy of Southeast Nebraska Community College)

5. Various loads are applied to a 1/4 hp electric motor. Angular speeds in revolutions per minute are recorded and the motor's efficiency (%) is determined at each load. The variables are angular speed and efficiency.
6. Several 110-V electric motors are accumulated and arranged according to the power rating. The motors are mounted on a device that measures power output *(dynamometer)*. The motors are loaded at their rated output. The full load current is measured. Data is shown in Figure 4-6. The variables are power (output) rating and full load current.
7. A sensor (a small wire coil) for a solar collector thermostat is heated to various temperatures. The resistance of the sensor is measured at each setting. Data is shown in Figure 4-7. The variables are resistance and temperature.

POWER RATING (hp)	FULL LOAD CURRENT (A)
$\frac{1}{4}$	5.7
$\frac{1}{2}$	9.8
1	16.1
2	23.9
3	34.0
5	56.0
$7\frac{1}{2}$	79.5
10	98.2

Figure 4-6 Full load current for various 110 V motors.

CONSTRUCTING GRAPHS

After the data are accumulated, the graph is then constructed. When making a graph, answers to numerous questions must be decided.

- How big should the graph be?
- Where do the axes go on the page?
- Which variable is plotted vertically?
- What scales should be used on each axis?
- How should the axes be captioned?
- What would be an acceptable title?

There are rules for answering these questions. Most graphing rules are flexible enough, however, that decisions must be made. Obviously, such choices must be settled first. After the axes are constructed, data points can be plotted. The *curve* for the plotted points is a smooth line drawn along the general lay of the data points. Graphing rules are given in Topic 4-3. The construction of curves is discussed first.

RESISTANCE (Ω)	TEMPERATURE (°C)
890	0
1035	27
1165	49
1290	71
1435	93
1580	116
1730	138
1890	160

Figure 4-7 Resistance of solar collector sensor at various temperatures

4-2 DRAWING THE CURVE

Curves can be described as either linear or nonlinear. *Linear curves* are straight lines. It may seem unusual to call a straight line a "curve" even if the word is qualified by the word linear. To justify the usage, consider a straight line drawn on level ground. Observed from outer space, it is actually seen as a section of the earth's circumference. Thus, calling a straight line a curve is proper. Curves that are clearly not straight lines are called *nonlinear curves.*

After the data points are plotted, use a pencil to draw a light, smooth curve through the points. Do not connect point-to-point with a series of straight lines. Make a smooth curve following the general drift or tendency of the points, Figure 4-8. The detailed procedure for making the curve is described in the following paragraphs.

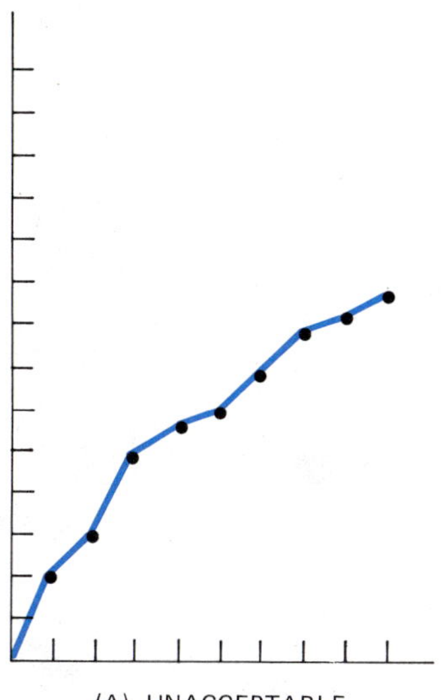

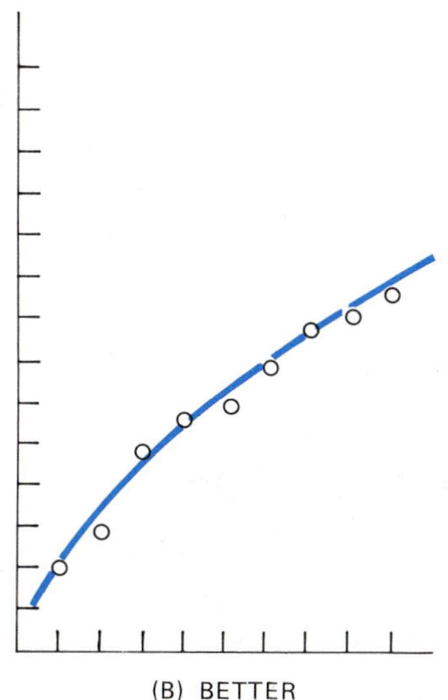

Figure 4-8 Do not connect points with a series of straight lines.

Linear Curves. When points are plotted, they should be enclosed in small, uniform-sized circles. Even with some scattering, the tendency will be along a straight line if it is linear. It may be helpful to observe the points from a distance.

The straight line to be drawn should be the one that fits the tendency of the points. The best fit or average fit curve will lay with about as many points above it as there are below.

To draw the curve, place a straight edge below the points. Adjust the straight edge so it lays parallel to the drift of the points. Move the straight edge closer to the points (keeping it parallel) until it just touches one of the circles. Draw a very light line along the straight edge. This line will be erased later. There will be less erasing if only the ends of the line are drawn.

Repeat the process on the upper side of the points. All the points should now be enclosed between two light lines, Figure 4-9. The best fit curve is a straight line that lays midway between the two outer lines. This line can now be drawn. Interrupt the line so it doesn't enter the circles. The curve will be completed after erasing the two outer lines.

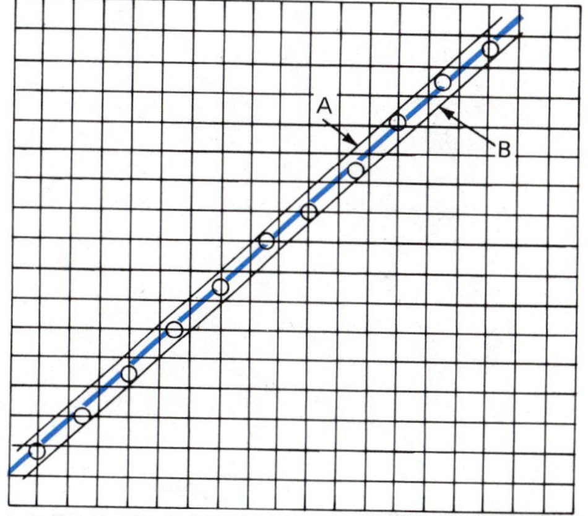

Figure 4-9 Finding the best-fit curve for linear data. Erase lines A and B to finish. Curve should not enter circles.

Nonlinear Curves. When nonlinear data are plotted, the points are also circled. Make the circles small and of uniform size. Again, do not connect point-to-point with a series of straight lines. A curve should never make sudden changes in direction.

When changes in one quantity affect another, the changes in the second quantity are always gradual. This is why the curve should follow a smoothly curved path. Figure 4-8(A) is unacceptable. Figure 4-8(B) is better.

Make the curve with a pencil. Begin by drawing a *very light,* smooth curve through the points. Interrupt the curve so it doesn't enter the circles. If the points are scattered, draw the curve so there are about equal numbers above and below the line. Sometimes, holding the paper at arm's length gives one a better feel for the shape of the curve.

After the curve is drawn very lightly, hold the paper horizontally in front of the face. Sight along the curve (like sighting along a curved railroad track), Figure 4-10. This will exaggerate any sideways jumps, humps, dips or jogs. Lightly straighten the curve at these places. Go over the curve several times. Gradually increase the intensity of the line but don't make it too heavy.

Use a lot of care in making the curve. Light erasing is acceptable but neatness must be emphasized. Freehand graphs are satisfactory for most laboratory work.

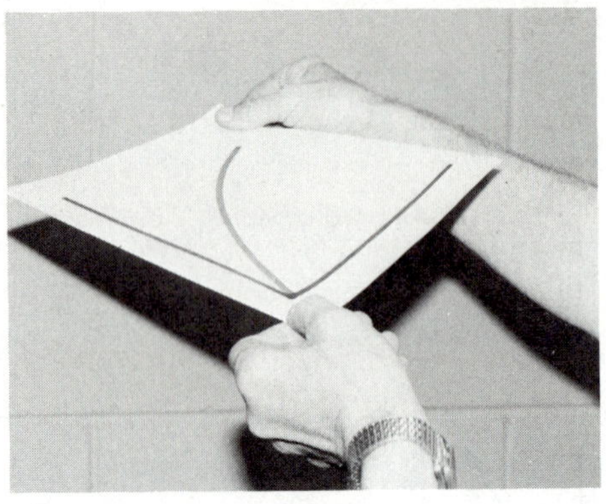

Figure 4-10 Sighting along the curve to find irregularities (Courtesy of Southeast Nebraska Community College)

EXERCISE 4-2

Place a sheet of paper over the figure for each problem and reproduce the plot of points. Make small, uniform circles around each point. Draw a curve through the points that best represents the plot. The curve should not enter the circles and will not pass through every point. Use the procedure described in the previous discussion about constructing linear curves. The scattering of points is exaggerated more than usual.

1. Draw the linear curve for the set of points in Figure 4-11.
2. Draw the nonlinear curve for the set of points in Figure 4-12.
3. Draw the nonlinear curve for the set of points in Figure 4-13.

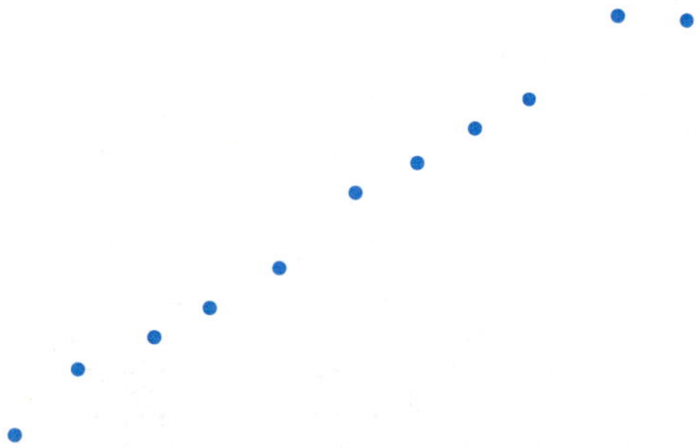

Figure 4-11

Figure 4-12

Figure 4-13

4-3 GRAPHING RULES

Most people spend many years in school. Much of the time is spent learning the meaning of words. Words must have the same meaning to everyone. Otherwise, people would be unable to talk or write to each other. Graphs are another means of displaying information and communicating with others. To be informative, graphs must also mean the same thing to everyone. To accomplish this, they must be constructed using a uniform procedure. The standard form is given in the following set of rules.

CONSTRUCTING GRAPHS

1. Make all graphs on graph paper using a pencil. Use a full page whenever possible. Smaller sized graphs can be made when only the shape of the curve is desired. The paper can be rotated 90° clockwise when necessary, Figure 4-14.
2. Do not write in the margins. If the graph paper doesn't have margins, leave 3/4 inch on each side and do not write in it. If the graph is a part of a bound report, leave a 1 1/2 inch margin on the left.
3. All lettering should be printed. Set all printed words so they can be read from left to right. The paper should be in the normal upright position or rotated 90° clockwise for reading. The paper should never have to be rotated counterclockwise to read printed matter, Figure 4-14.

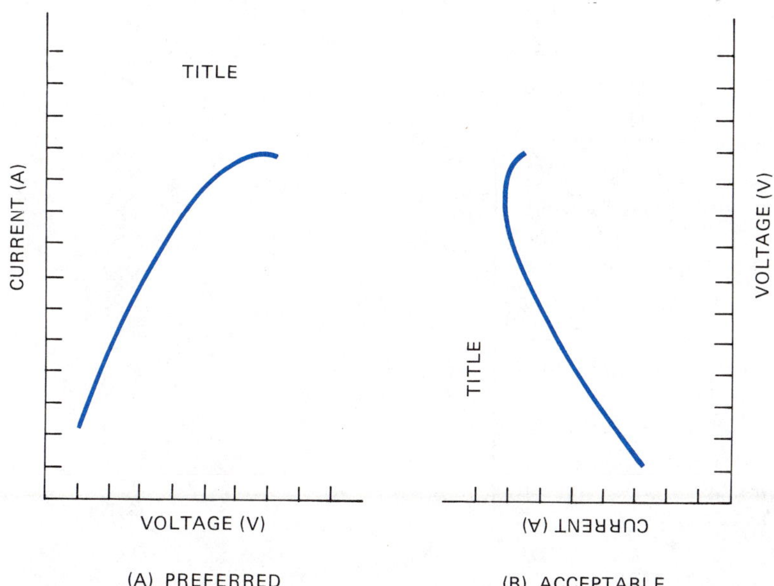

Figure 4-14 Orientation of graph on the page: (A) upright and (B) page rotated 90° clockwise.

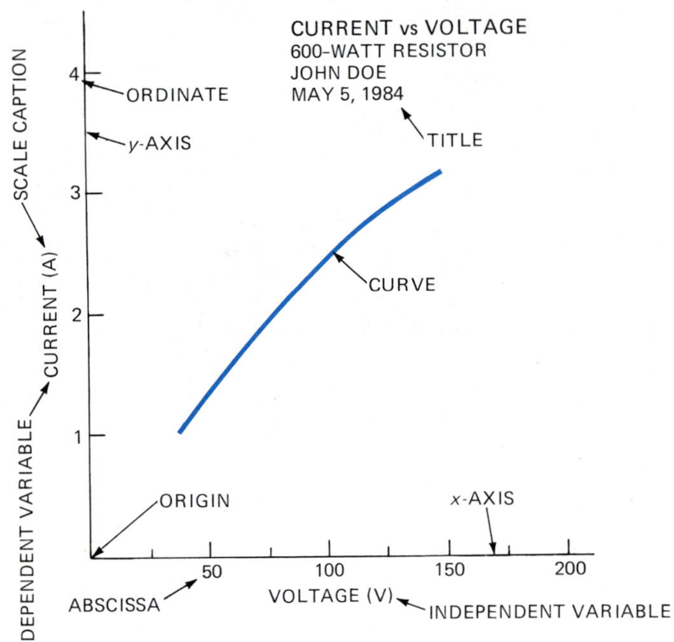

Figure 4-15 Graph nomenclature

4. Decide which variable is independent. It will be plotted along the horizontal axis, Figure 4-15. The dependent variable is plotted vertically. There are two exceptions. The variable time is almost always plotted horizontally, independent or not. Voltage-current graphs are exceptions, also. Current is usually plotted vertically. Use this latter rule unless lab instructions differ. There should be a compelling reason for doing otherwise.
5. Choose a scale for each axis that allows all the data to be placed on the graph. Keep the scale consistent (each division represents the same amount). Choose a scale that will be easy to read. Each division should be one, two or five units or power of ten multiples (10, 100, 1000, etc.) or submultiples (0.1, 0.01, 0.001, etc.) of those numbers. Angles (degrees) can be a simple submultiple of 90 or 360. It is not necessary to label every mark on the scale, Figure 4-15. Figures 4-16 and 4-17 show the effects of changing a scale.
6. Scale captions should be placed along each axis. Captions include only the names of the quantities and their units. Do not use the abbreviated symbols for the name of the quantity. Abbreviate the units and enclose them in parentheses, Figure 4-15.
7. The zero should be included on both axes. Sometimes a section of the scale can be omitted. The omission must be indicated by slash marks, Figure 4-17. Do not extend the curve through the omitted section.

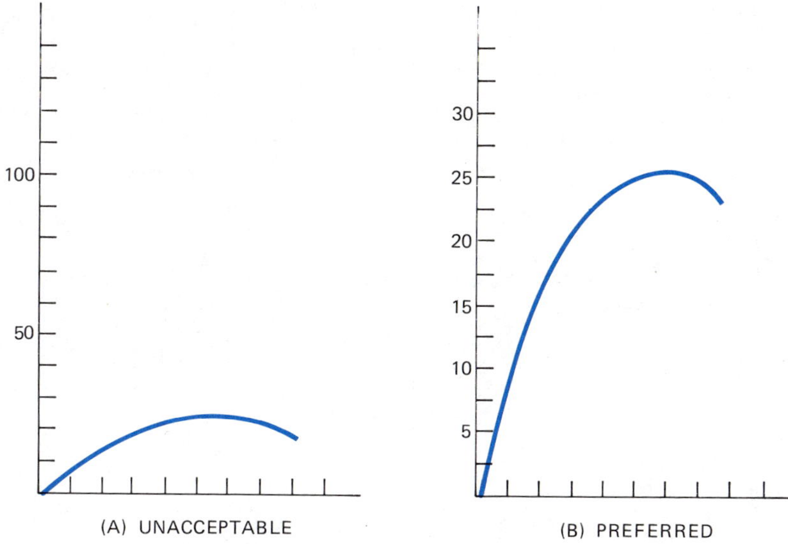

Figure 4-16 Effect of changing the scale

8. Plot the points represented by the data on the graph. Circle each point but keep the circles small, neat and of uniform size. Do not allow the circles to dominate the graph – keep them small.
9. Using a pencil, draw a very light, smooth curve through the points. Do not connect point-to-point with a series of straight lines. Make a smooth curve following the general drift or tendency of the points. The detailed procedure for making the curve was described previously.

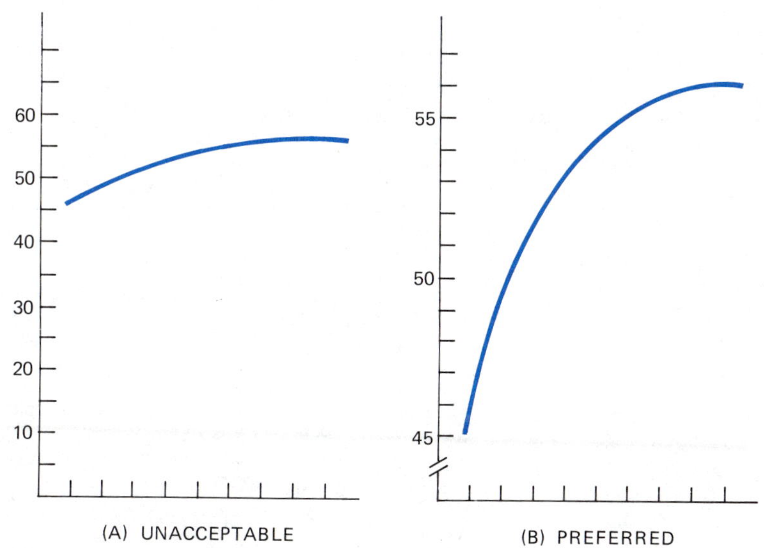

Figure 4-17 Effect of omitting part of the scale

10. Give the graph a title. Usually, the name of the dependent variable vs the name of the independent variable will do. For example: "Current vs Voltage" is satisfactory. Never abbreviate or use the names of the units. "I vs E" and "Amperes vs Volts" are unacceptable. If the equipment being used is of special significance, put the brand, model number, type of circuit, etc. under the title. Use smaller letters than those used in the main title. Finally, include your name and date. Place the entire title near the top of the graph if it doesn't interfere with the curve, Figure 4-15. Do not include lists such as data tables or calculations on the graph page.

4-4 ELECTRICAL GRAPHS

Example A An automobile radio is tuned by moving iron core rods into and out of the tuning coils. This movement was measured with a steel rule after every half-revolution of the tuning knob. The data is given in Figure 4-18. Plot a graph of these data.

Solution: Figure 4-19(A) is a graph of the iron core movement. It is very poorly constructed. Figure 4-19(B) is better. Compare the following list of errors found in Figure 4-19(A) with the rule that applies and with Figure 4-19(B). The violated rule is given in parentheses.

- Caption and title written (R3).
- Paper rotated counterclockwise to read vertical caption (R3).

KNOB ROTATION (r)	CORE MOVEMENT (cm)
1/2	0.3
1	0.55
1 1/2	0.8
2	1.1
2 1/2	1.4
3	1.7
3 1/2	2.0
4	2.2
4 1/2	2.5
5	2.8
5 1/2	3.1
6	3.3

Figure 4-18 Movement of iron core into and out of tuning coils as a function of the rotation of the tuning knob on a Ford car radio. Movement is measured in centimetres (cm) and rotation is in revolutions (r).

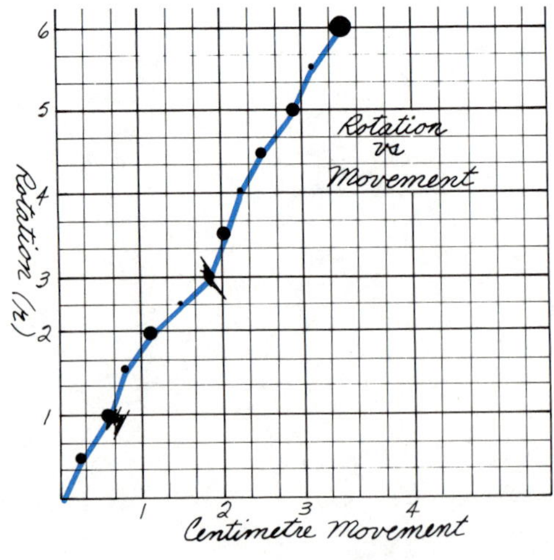

Figure 4-19(A) Poorly constructed graph: Axes scales inconsistent, scale captions crowded, caption and title should be printed, dependent variable should be vertical, scales not in 1, 2, 5 proportion, smeary, poor title, page rotated counterclockwise for vertical caption, curve not smooth, points not circled, and points and curve not uniform.

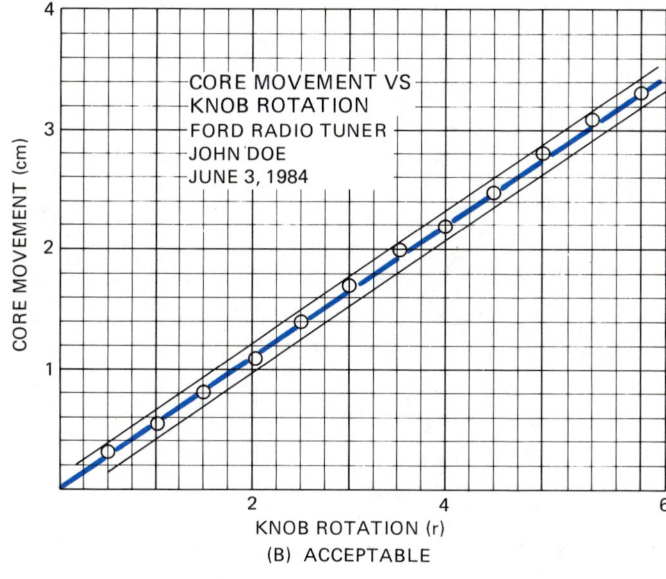

Figure 4-19(B) An acceptable graph. To finish, erase the two light lines.

- Independent variable plotted vertically (R4).
- Poor choice of vertical scale (R5).
- Scales inconsistent (R5).
- Units not clear (R6).
- Data points not precise (R8).
- Points connected by straight lines (R9).
- Smeary (R9).
- Very poor title (R10).
- Equipment not identified (R10).

Example B A prony brake dynamometer was used to measure the output power of a small electric motor designed for intermittent use. It operates on 110 V (ac). Gradually increasing the load resulted in the data in Figure 4-20. The angular speed in revolutions per second (r/s) was measured with a tachometer. The output power values were calculated. Make a graph of the data.

Solution: The graph is plotted in Figure 4-21. The reader should review the ten rules as the graph is studied.

ANGULAR SPEED (r/s)	POWER OUTPUT (W)
29.0	0
28.9	6.6
28.7	14.9
28.3	24.0
27.5	33.1
26.7	35.1
25.8	39.0
25.0	40.1
24.7	40.1
24.2	40.0
23.7	39.5
23.0	38.2

Figure 4-20

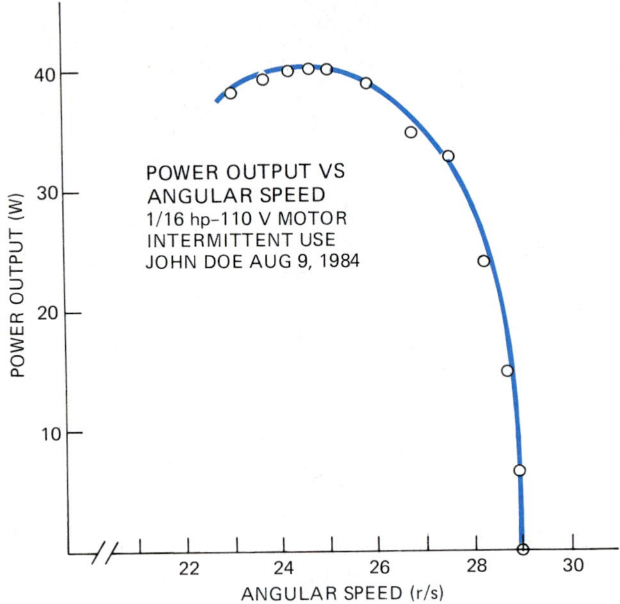

Figure 4-21

4-5 READING THE GRAPH

The following example illustrates how additional data can be obtained from a well constructed graph:

Example A Determine the power output from the graph, Figure 4-21 for each of the following angular speeds. Angular speeds are measured in revolutions per second (r/s) and power is measured in watts (W).

(1) 26 r/s, (2) 28 r/s. Find the angular speed at these power outputs: (3) 30 W, (4) 20 W.

Solution: (1) Find the angular speed of 26 r/s along the horizontal axis. Move upward until the curve is intersected. Now move left to the vertical axis and read the power output of 38 W. This is the answer.

(2) Similarly, the power at 28 r/s is 29 W.

(3) The angular speed is found by finding 30 W on the vertical axis. Move horizontally until the curve is intersected. Move downward and read an angular speed of 27.9 r/s on the horizontal axis.

(4) In a like manner, the angular speed is 28.7 r/s when the power output is 20 W.

EXERCISE 4-3

Use all the rules presented for constructing the graphs in each of the following problems.

1. Various sized circular objects are accumulated and measured, Figure 4-22. Make a graph of these data. Both circumference and diameter are given in centimetres. Use the graph to find the following:
   a. Find the circumference of a circle with a diameter of 45 cm.
   b. Find the diameter of a circle with a circumference of 50 cm.

OBJECT	DIAMETER (cm)	CIRCUMFERENCE (cm)
Scale weight	9.7	30.3
Pressure gauge	12.2	38.6
Clock	19.4	61.0
Vacuum plate	27.6	86.4
Waste basket	39.8	122.0
Radian circle	53.5	168.0

Figure 4-22

2. Two resistors are placed in series in a simple circuit. The source emf is adjusted until the current is 0.01 A, 0.02 A, 0.03 A, etc. At each setting, the voltage drop across each resistor is measured, Figure 4-23. Plot the data for both resistors on the same graph. Identify the two curves as R_1 and R_2, Figure 4-24. From the graph constructed, find the following:
   a. Find the voltage drop across each resistor when the current is 0.075 A.
   b. What is the current when the voltage drop across R_1 is 2.5 Ω?

CURRENT (A)	R_1 VOLTAGE DROP (V)	R_2 VOLTAGE DROP (V)
0.00	0	0
0.01	1.26	0.49
0.02	2.00	1.04
0.03	3.05	1.55
0.04	4.10	2.07
0.05	5.00	2.62
0.06	6.11	3.19
0.07	7.15	3.70
0.08	8.02	4.21
0.09	9.20	4.78
0.10	10.40	5.27

Figure 4-23

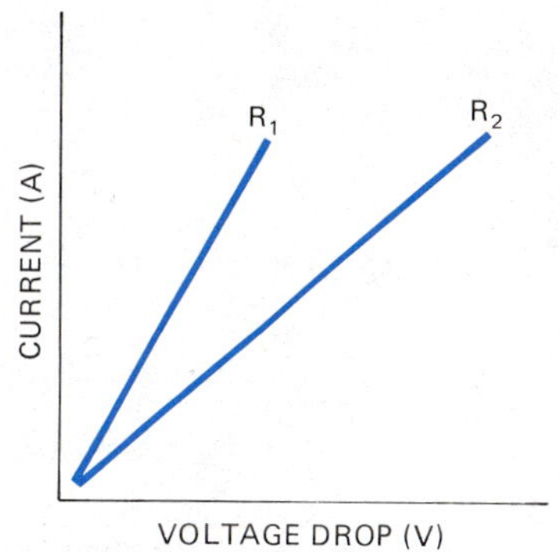

Figure 4-24 Identifying two or more curves on one graph

3. Various voltages are applied to a 75-W tungsten lamp. The filament has a cold resistance of 14 Ω. The resistance changes as the temperature changes and is calculated from current and voltage measurements. Construct a graph of the data given in Figure 4-25. Explain why this curve does or does not pass through the origin.
   a. Using the graph, what is the resistance of the lamp filament when the voltage is 75 V?
   b. Using the graph, what voltage produces a resistance of 100 Ω?

VOLTAGE (V)	RESISTANCE (Ω)
10	40
20	62
30	79
40	94
50	106
60	118
70	128
80	138
90	146
100	154

Figure 4-25

4. Various voltages are applied to a pilot lamp (an incandescent indicator lamp). The current is recorded at each setting, Figure 4-26. Construct a graph of these data. Use the graph to find the following:
   a. Find the current through the pilot lamp from the graph when the voltage is 7.5 V.
   b. What is the voltage when the current is 0.025 A?

VOLTAGE (V)	CURRENT (A)
5	0.014
10	0.022
15	0.028
20	0.033
25	0.037
30	0.040

Figure 4-26

5. The resistance of wire depends on the size of the wire and the material. Figure 4-27 lists the diameter in millimetres and resistance (per 1000 metres) in ohms for aluminum and copper wires of various gages. Make a graph showing the resistance of both aluminum (Al) and copper (Cu). Identify the two curves. Wire of 00 gage has a diameter of 9.27 mm. Find the resistance of 00 gage aluminum and copper wire per 1000 metres.

DIAMETER (mm)	COPPER RESISTANCE (Ω) PER 1000 METRES	ALUMINUM RESISTANCE (Ω) PER 1000 METRES
1.6	8.28	13.58
2.1	5.21	8.56
2.6	3.28	5.38
3.3	2.06	3.38
4.1	1.30	2.13
5.2	0.82	1.35
6.5	0.51	0.85
8.3	0.32	0.52
10.4	0.20	0.33

Figure 4-27

6. The luminous output power (luminous flux) of light bulbs decreases with the age of the bulb. The output in lumens is given in Figure 4-28 for a 450-W mercury street lamp. Time is measured in thousands of hours of usage. Plot a graph of these data. Use the graph to find the following:
   a. A 450-watt mercury lamp has a luminous flux of 8550 lumens. How long has it been in use?
   b. Find the luminous flux after the bulb has been used for 12 500 hours.

TIME (1000 h)	LUMINOUS FLUX (lm)
0	9500
2	9100
4	8900
6	8700
8	8600
10	8400
12	8300
14	8100
16	7900

Figure 4-28

7. Plot the data in Figure 4-29.

POWER RATING (hp)	FULL LOAD CURRENT (A)
$\frac{1}{4}$	5.7
$\frac{1}{2}$	9.8
1	16.1
2	23.9
3	34.0
5	56.0
$7\frac{1}{2}$	79.5
10	98.2

Figure 4-29 Full load current for various 110 V motors

8. Plot the data in Figure 4-30.

RESISTANCE (Ω)	TEMPERATURE (°C)
0	890
27	1035
49	1165
71	1290
93	1435
116	1580
138	1730
160	1890

Figure 4-30 Resistance of solar collector sensor at various temperatures

CHAPTER 5

NUMBER NOTATION FOR ELECTRICITY

OBJECTIVES

After satisfactorily completing this chapter, the student should be able to:

- Define various words and terms introduced in this chapter.
- Perform the four basic operations on signed numbers.
- Multiply, divide, square, cube, find roots and find reciprocals of powers of ten and other numbers expressed in scientific or engineering notation.
- Read electrical color codes for resistors and capacitors.

Whole numbers and fractions (common and decimal) have been discussed in the first four chapters. These numbers are all greater than zero and are called *positive numbers.* Another set of numbers are all less than zero. They are called *negative numbers.* Together, the positive and negative numbers are called *signed numbers.*

There are many applications of signed numbers in everyday life. Many additional uses are found in electrical fields.

- Temperature readings range from positive to negative on both Fahrenheit and Celsius scales.
- Bank balances undergo positive changes (deposits) and negative changes (withdrawals). Overdrawn accounts have negative balances.
- Electric charges can be positive (protons) or negative (electrons).
- Electric current in a conductor can flow one direction (positive) or in the opposite direction. This is determined by positive or negative voltage. Batteries undergo charging and discharging, Figure 5-1.

Figure 5-1 A battery charger

- In ac circuits, both current and voltage alternate between positive and negative directions.
- Some kinds of electrical equipment amplify power. Others lose power. They are rated as positive for gain; negative for loss.
- Rotating equipment can turn clockwise (negative) or counterclockwise (positive).

In addition to the applications of signed numbers just mentioned, there are many others. One that is very important is the usage in expressing numbers in scientific notation. Similarly, quantities are expressed using metric prefixes. Large or small amounts of current, voltage, resistance, capacitance, inductance and power are expressed this way. Metric prefixes indicate positive or negative shifts of the decimal point. Signed numbers and scientific notation are discussed in this chapter. The study of metric prefixes follows immediately in Chapter 6.

SIGNED NUMBERS

The numbers used for counting (1, 2, 3, 4 . . .) are called *natural numbers.* They can be represented geometrically on a straight line, Figure 5-2. Successive numbers are placed one unit apart. These numbers are the positive whole numbers. The position marked with the zero is called the origin. Proper fractions are located between 0 and 1. Improper fractions are located between other pairs of whole numbers.

If the number line is extended to the left of zero, Figure 5-3, numbers less than zero are required. A minus sign is used to indicate such numbers. They are called *negative numbers.* Positive numbers may be indicated with or without the plus sign. The number 3, for example, is understood to mean +3. Negative numbers (such as −3) are always preceded by the minus sign, Figure 5-3. Numbers with a + or − sign are called *signed numbers.*

5-1 ABSOLUTE VALUE OF A NUMBER

Signed numbers have two parts: the numerical part and the sign. The numerical part without the sign is the *absolute value* or *magnitude* of the quantity. The sign simply gives the *direction* (up or down, left or right, increase or decrease, etc.). The magnitudes of numbers such as +5, −3, +16 and −7 are 5, 3, 16 and 7 respectively.

The absolute values or magnitudes are sometimes indicated using the symbol $| \;\; |$. Thus: $|+4| = 4$, $|-12| = 12$ and $|-2| = 2$. Keep in mind that positive and negative numbers simply denote reference from the origin.

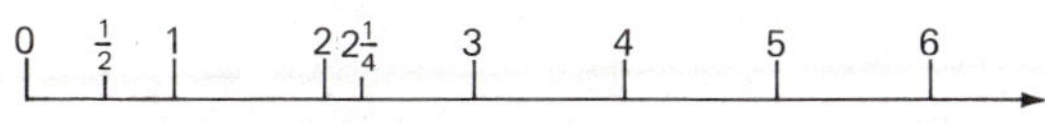

Figure 5-2 Scales, such as the number line, are often used as horizontal and vertical axes in graph making.

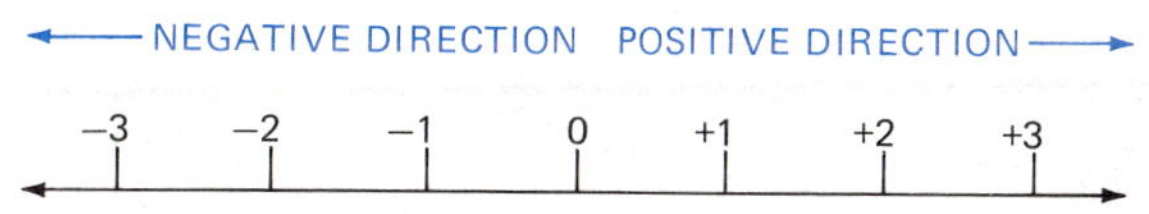

Figure 5-3 Number line including negative numbers

Negative numbers are no less meaningful than positive numbers. Seven miles south (−7 mi) requires as many steps to return to the origin as seven miles north (+7 mi). A negative electrical current can give the same shock as a positive current of the same magnitude.

The symbols (+) and (−) have a dual use. When they are attached to a number, they indicate direction. A +5 and a −5 are opposites, for example. The symbols also indicate the signs of operation (addition and subtraction). This occurs when the sign is located between pairs of numbers. The expression (−4) + (+6) contains both usages. The first (+) sign is an operation. It indicates that the numbers −4 and +6 are to be added.

5-2 ADDING AND SUBTRACTING SIGNED NUMBERS

After some practice with adding and subtracting signed numbers, any confusion in the dual usage will diminish. It is essential that the rules for adding and subtracting be memorized.

There are two rules governing the addition of a pair of signed numbers. One rule deals with addition of two numbers that have the same sign; the other with numbers having opposite signs.

ADDING SIGNED NUMBERS

Like Signs

- Add the absolute value of the numbers.
- Give the sum the same sign as the numbers.

Unlike signs

- Subtract the number with the smaller absolute value from that of the larger.
- Give the answer the same sign of that of the number with the largest absolute value.

Example A Evaluate: (The answer is given with each problem.)

Solution:

(Like Signs)	*(Unlike Signs)*
a. (+3) + (+6) = +9	c. (−2) + (+5) = +3
b. (−5) + (−7) = −12	d. (+11) + (−15) = −4

To combine three or more signed numbers, add all the positive numbers together. Separately, add all the negative numbers together. Now add the two resulting unlike sums.

Example B Evaluate: (−5) + (+4) + (−2) + (−8) + (+9) + (+1)

Solution: [(−5) + (−2) + (−8)] + [(+4) + (+9) + (+1)] =
(−15) + (+14) =
−1
Such combinations can also be added in sequence from left to right one pair at a time.

One interpretation of signed number problems is the following: the first number represents the starting value of a quantity. The second and other numbers represent changes. Positive numbers are increases; negative numbers are decreases. The sum is the value of the quantity after the change has been completed. Several such changes can be added in sequence:

Final value = starting value + changes

Example C The temperature one day at 6 AM was −14°F. By 10 AM the temperature had increased 11°F; by 2 PM, it had increased another 6°F. A drop of 4°F was noted by 6 PM and another drop of 8°F by 10 PM. What was the temperature at 10 PM?

Solution: Write the final temperature as a summation of signed numbers. Increases are positive; decreases are negative. Add left to right.

$T_f = (-14°F) + (+11°F) + (+6°F) + (-4°F) + (-8°F)$

$T_f = -9°F$

SUBTRACTING SIGNED NUMBERS

- Change the sign of the number being subtracted (subtrahend).
- Add the numbers using the rules for adding signed numbers.

Example D Evaluate the following expressions.

Note: The solutions are given with the problems.

Solutions: (1) $(+5) - (+6) = (+5) + (-6) = -1$

(2) $(-9) - (-7) = (-9) + (+7) = -2$

(3) $(-3) - (+6) = (-3) + (-6) = -9$

(4) $(+6) - (-8) = (+6) + (+8) = +14$

An interpretation of subtraction is given by the following formula:

Starting value = final value − changes

This equation has been solved for the starting value.

Example E An electrician's bank account indicated a balance of $549.50 on a Friday morning. During that same week, the following transactions had occurred: Monday, checks were written totalling $126.25. Tuesday, checks equalling $385.65 were written. Wednesday, a deposit of $436.75 was made and checks were written for $88.35. Thursday, no changes. What was the balance at the beginning of that same week?

Solution: ($549.50) − (−$126.25) − (−$385.65) − (+$436.75) − (−$88.35)

Change signs, then add:

= (+$549.50) + (+ $126.25) + (+$385.65) + (−$436.75) + (+$88.35)

= +$713.00

EXERCISE 5-1

Find the absolute value of each of the following signed numbers.

1. +7	3. +6	5. −16	7. −3/8
2. −2	4. −5	6. +14	8. +1/2

Combine the following by adding or subtracting. Do not use calculators.

9. (+5) + (+6)
10. (−2) + (+4)
11. (+2) + (−5)
12. (−6) + (−3)
13. (−1) + (−2)
14. (+6) + (−6)
15. (−4) + (+11)
16. (+5) + (+8)
17. (+5) − (+6)
18. (−17) − (−28)
19. (−2) − (+4)
20. (+2) − (−5)
21. (−6) − (−3)
22. (−1) − (−2)
23. (+6) − (−6)
24. (−4) − (−11)
25. (+5) − (+8)
26. (−3) + (−2) − (−4)
27. (−7) + (+3) − (+7) − (−1)
28. (+6) − (+1) + (−2) − (−5) + (+2)

5-3 CALCULATORS AND SIGNED NUMBERS

Scientific calculators have been programmed to perform arithmetic with signed numbers. The change sign key [+/−] or [CHG] is used to change the sign of the number displayed.

Use the Owners Manual and study the following example to learn how to work with signed numbers on the calculator.

Example A Evaluate this expression using the calculator.

(−5) + (+7) − (−8)

Solution: 5 [+/−] [+] 7 [−] 8 [+/−] [=] , +10

EXERCISE 5-2

Calculate each of the following expressions using a calculator.

1. (+4) + (−2)
2. (−3) + (+7)
3. (+16) − (−4) − (+26)
4. (+8) − (+1) + (−3) − (+11)
5. (+4) + (+2) + (−5) − (+4) − (−7)
6. (−3) − (−7) + (+1) + (−9) − (+6)

Rewrite the following into numerical expressions of signed numbers and evaluate.

7. In a power company's warehouse there were 214 reels of power line on the first of the month. During the following week, 64 reels were used. During the second week, 39 reels were used. In the third week, 200 reels were received and 46 were used. In the fourth week, 74 were used. How many reels are on hand at the end of the month?
8. An electrical shop buys a certain item for $38.75 wholesale. If the profit is $15.50, a $2.50 discount is allowed and sales tax is $2.07, find the price the customer must pay.

9. An electrician had a box of electrical nuts at a home he was wiring. He used 56, lost 5, broke 3 and 25 were stolen. He added another 100 from another box and an electrician friend borrowed 30. How many were originally in the box if there are 97 remaining?
10. On a three-wire line, the center or ground wire is neutral (zero voltage). One of the other wires is –110 V. The third is +110 V. Circuits can be connected across any two wires. What voltages are possible? In each case, subtract the lower voltage from the higher.

5-4 VECTORS AND VECTOR ADDITION

Signed numbers are sometimes called *directed numbers* or *vectors.* A vector is a quantity in which the direction is important. A vector can be represented by an arrow. The length of the arrow indicates the magnitude (numerical part) and the direction it points is given by the sign. Positive numbers are vectors pointing to the right while negative vectors point to the left. Examples are given on the number line of Figure 5-4. The vector +4 is indicated graphically by an arrow four units long. It points to the right. The vector –6 has a magnitude of 6 units. It points to the left.

Graphical addition of vectors is given by the following rules. It is sometimes called the *tail-to-tip method.*

VECTOR ADDITION (GRAPHICAL)

- On a number line, lay out the first vector in the direction of its sign. Place the tail of the arrow at the origin.
- Place the tail of the second vector at the tip of the first. Point it in the direction of its sign.
- Lay out the third vector similarly with its tail at the tip of the second vector. Place the fourth vector's tail at the tip of the third, etc.
- The *resultant* or *vector sum* is another vector. Locate its tail at the origin and its tip at the tip of the last vector plotted. Record its length and direction in the form of a signed number. The resultant is usually drawn thicker to distinguish it from the addends.

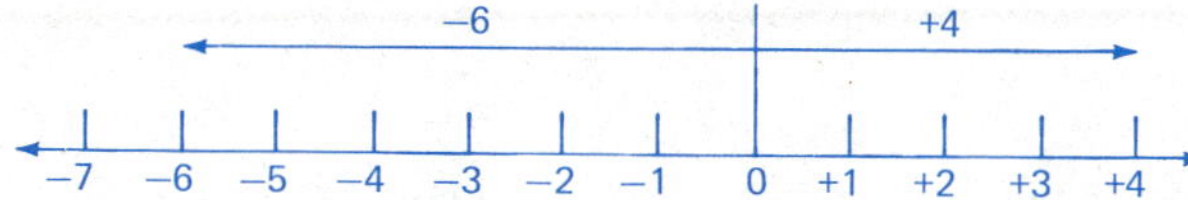

Figure 5-4 Positive vectors point to the right and negative vectors to the left as shown by the vectors for +4 and –6.

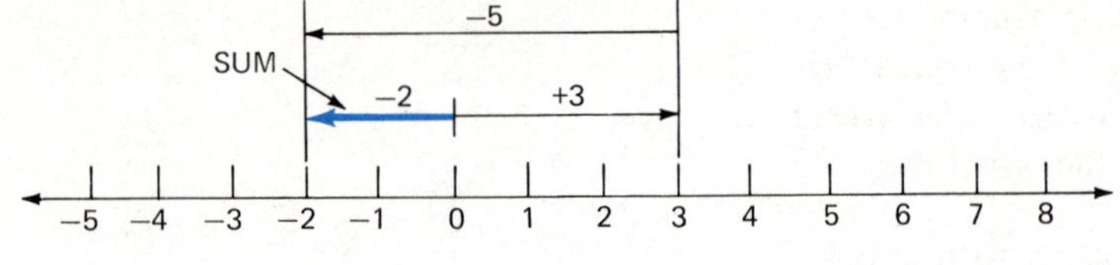

Figure 5-5 Addition of two vectors: (+3) + (−5) = −2.

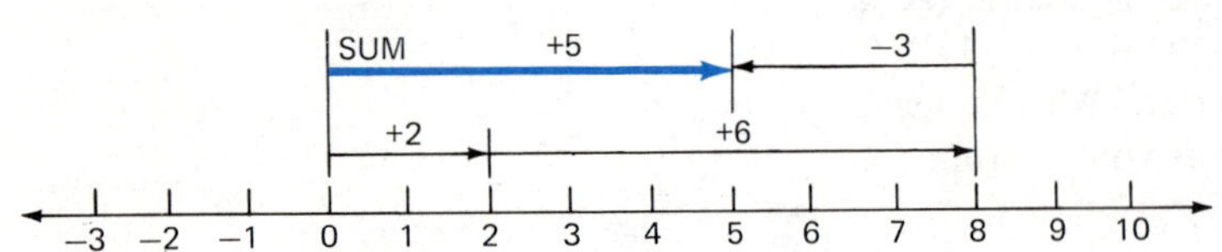

Figure 5-6 Vector sum: (+2) + (+6) + (−3) = (+5).

Example A Add the vectors (+3) and (−5) graphically.
Solution: The solution is shown in Figure 5-5.

Example B Add (+2), (+6) and (−3) graphically.
Solution: The solution is shown in Figure 5-6.

5-5 ROTATION OPERATORS

A negative number such as −6 can be thought of as the negative of a +6. That is, −6 = −(+6). Imagine the vector: +6. It points to the right and is six units long. Now imagine that it rotates 180° (one-half revolution). With the rotation, the vector +6 has changed to the vector −6.

Since the number −6 can be expressed as −(+6), the minus sign becomes a *rotation operator*. The expression −(+6) is the same as +(−6). Both the sign of operation and the number sign can be thought of as rotation operators. The vector −(−3) is rotated 180° twice. It is the same as though it had not been rotated at all: −(−3) = +(+3).

The sign of the subtrahend is changed by rotation and the operation changes from subtraction to addition. This result gives the subtraction rule for vectors.

5-6 VECTOR SUBTRACTION

VECTOR SUBTRACTION (GRAPHICAL)

- Rotate the vector being subtracted 180°.
- Now add the vectors according to the rules given previously.

Example A Solve graphically: (+4) − (+3)
Solution: The solution is shown in Figure 5-7.

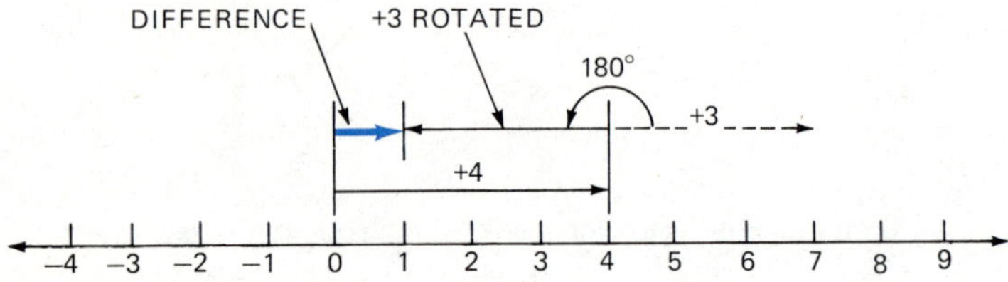

Figure 5-7 Subtracting vectors: (+4) − (+3).

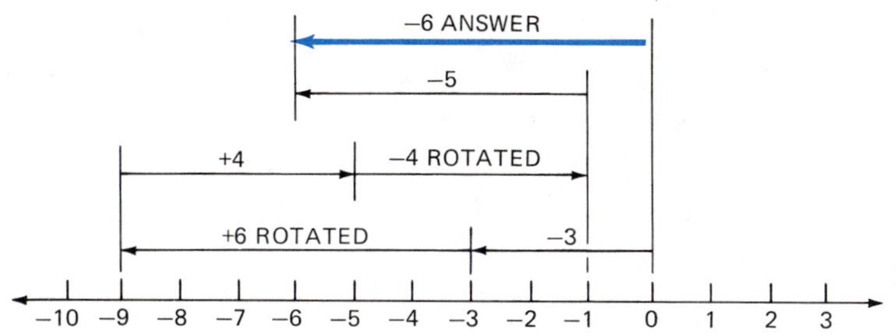

Figure 5-8 Combining vectors: (−3) − (+6) + (+4) − (−4) + (−5).

Example B Solve graphically: (−3) − (+6) + (+4) − (−4) + (−5)

Note: The second and fourth terms are to be subtracted.

Solution: The solution is shown in Figure 5-8.

The student should compare the rules for graphically adding and subtracting vectors with those for adding and subtracting signed numbers. These two sets of rules give graphical and analytical solutions to the same problems. In a later chapter, a 90° rotation operator is introduced. It allows expressions for vectors pointed in *any* direction; not just opposites. The graphical rules for adding and subtracting such vectors is exactly the same as those introduced here.

EXERCISE 5-3

Make a freehand number line and solve the following problems using the rules for vector addition and subtraction. Remember to put arrowheads on all arrows. Check all answers by calculating the answers using the rules for adding and subtracting signed numbers.

1. (−5) + (+6)
2. (−1) + (−3)
3. (+9) + (+2)
4. (+8) − (−4)
5. (−3) − (+2)
6. (−6) − (−9)
7. (+4) + (−2) + (+6)
8. (−2) − (+6) − (−8)
9. (+4) + (−5) − (−6) − (+9)
10. (−3) + (+7) − (+4) + (−4)
11. (−8) − (−6) + (+4) + (−3)
12. (+3) + (−5) − (+2) − (−6)

5-7 MULTIPLYING AND DIVIDING SIGNED NUMBERS

The rules for multiplication and division of signed numbers can be given together. Remember division is just the inverse of multiplication.

MULTIPLYING AND DIVIDING SIGNED NUMBERS

- The product or quotient of two signed numbers is found in the usual way using the absolute values of the numbers.
- *Like signs.* If the two numbers have the same sign (both positive or both negative), the product or quotient will be *positive.*
- *Unlike signs.* If the two numbers have unlike signs (one negative and one positive) the product or quotient will be *negative.*
- The product or quotient of three or more factors is positive if there is an even number of negative factors. It is negative if there is an odd number of negative factors.

Example A Evaluate the following expressions (solutions are given with the expression):

Solutions:
(1) $(+3)(+4) = +12$
(2) $(-3)(+4) = -12$
(3) $(-3)(-4) = +12$
(4) $\frac{+8}{+2} = +4$
(5) $\frac{-8}{+2} = -4$
(6) $\frac{-8}{-2} = +4$
(7) $\frac{(+4)(-3)(-2)(-6)}{(-2)(+12)} = +6$
(8) $\frac{(-8)(+4)(-3)(-7)(-2)}{(-12)(+14)(+2)} = -4$

Example (7) has an even number of negative factors; (8) has an odd number.

EXERCISE 5-4

Perform the indicated operations manually.

1. $(+3)(+4)$
2. $(+5)(-3)$
3. $(-4)(-6)$
4. $(-2)(+12)$
5. $(0)(-3)$
6. $(+4)(0)$
7. $(-1)(+6)$
8. $(+8)^2$
9. $(-6)^2$
10. $(+3)^3$
11. $(-5)^3$
12. $(-1)(-1)(-1)$
13. $(+4)(-5)(-2)$
14. $(-6)(-2)(+3)(-2)$
15. $(-72) \div (+12)$
16. $(+18) \div (-3)$
17. $(+6)(-4) \div (-8)$
18. $(-3) \div (-1.5)$
19. $\frac{1}{(-8)}$
20. $\frac{1}{(-1000)}$

Cancel wherever possible then solve manually and check using the calculator.

21. $\frac{(-2)(+4)(-5)}{(-10)(+4)}$
22. $\frac{(-12) + (+4)}{(-2)}$
23. $\frac{(-12)(+6)(-3)(-4)}{(-2)(+24)(-9)}$
24. $\frac{(-3)[(-7) - (+3) + (-5)]}{(-5)}$
25. $\frac{(+8) - (-10)}{(-3)}$
26. $\frac{(-3)(+6) - (+4)(+4)}{(-1) - (+1)}$

SCIENTIFIC NOTATION

There is a convenient notation for expressing very large and very small numbers. It is called scientific notation and makes use of powers of ten. *Powers of ten* are numbers such as $100 = 10^2$, $1000 = 10^3$ and $1\,000\,000 = 10^6$. These examples are some of the *positive powers of ten. Negative powers of ten* are numbers such as $0.01 = 1/100$, $0.001 = 1/1000$ and $0.000\,001 = 1/1\,000\,000$.

5-8 POWERS OF TEN

Positive powers were discussed briefly in Chapter 3. The following definition is repeated here for convenience.

$$(\text{base})^{\text{exponent}} = \text{power}$$

Negative powers are the reciprocals of the positive powers:

$$10^{-2} = \frac{1}{10^2} = \frac{1}{100} = 0.01$$

$$10^{-3} = \frac{1}{10^3} = \frac{1}{1000} = 0.001$$

The *zero power* of any number including ten is equal to one:

$$10^0 = 1$$

The *first power* of any number is simply equal to the number itself:

$$10^1 = 10$$

The following table collects all the previous definitions and indicates the general pattern.

EXPONENTIAL NOTATION		DECIMAL NOTATION
10^9	=	1 000 000 000
10^6	=	1 000 000
10^3	=	1 000
10^2	=	100
10^1	=	10
10^0	=	1
10^{-1}	=	0.1
10^{-2}	=	0.01
10^{-3}	=	0.001
10^{-6}	=	0.000 001
10^{-9}	=	0.000 000 001

From the position immediately following the one (1), count the number of places to the decimal place. That number is equal to the exponent. If the decimal is to the right, the exponent is positive. Whenever the decimal place is to the left of the one, the exponent is negative.

Example A (1) Convert 10^5 to decimal notation, (2) Convert 0.0001 to exponential notation.

Solution: (1) From the one, move the decimal to the right five places: 100 000.

(2) The decimal is four places to the left so the exponent is negative four: 10^{-4}.

Note: The number of zeros in decimal form equals the exponent.

Note: When the decimal is to the left, the number of zeros is one less than the exponent's absolute value.

EXERCISE 5-5

Convert the numbers in decimal notation to exponential notation and the numbers in exponential notation to decimal notation.

1. 10^{-5}	7. 0.01
2. 10^{8}	8. 0.000 000 000 000 01
3. 10^{-7}	9. 10 000 000 000
4. 10^{3}	10. 100 000
5. 10^{-11}	11. 0.000 000 01
6. 10^{10}	12. 10 000 000

13. The radius of most atoms is about 10^4 times larger than the nucleus. The volume is about 10^{12} larger than the volume of the nucleus. Express these numbers in decimal form.
14. The radius of electrons, protons and neutrons is approximately 0.000 000 000 000 1 centimetre. Most atoms have radii of about 0.000 000 01 centimetre. Express these measurements in exponential notation.

5-9 SCIENTIFIC NOTATION DEFINED

When a number is multiplied by a power of ten, the nonzero digits remain the same. Only the decimal point is shifted.

Example A Find the following values.

Solutions:
(1) $3.75 \times 1000 = 3750$
(2) $4.62 \times 0.001 = 0.004\ 62$
(3) $3.64 \times 10^9 = 3\ 640\ 000\ 000$
(4) $7.91 \times 10^{-8} = 0.000\ 000\ 079\ 1$

When the numbers are very large or very small such as the last two examples, it is easier to express them as the product on the left. A number such as 3.64×10^9 is said to be expressed in scientific notation. Its equivalent (3 640 000 000) is the *decimal form* or *decimal notation* of the same number. These are merely two different ways of writing the same number.

Scientific notation is widely used in the electrical industry. It is important for electricians to learn as much as possible about scientific notation. The arithmetic of numbers expressed in scientific notation will be discussed after the next topic.

5-10 SIGNIFICANT FIGURES, STANDARD POSITION AND CHARACTERISTIC

Numbers are ordinarily used to indicate various measurements. They are usually accompanied by units. In Figure 5-9 , the voltmeter reading is 7.53 V. The first two digits (7 and 5) have no uncertainty about them. Any electrician would surely agree that the first digit is 7 and the second is 5.

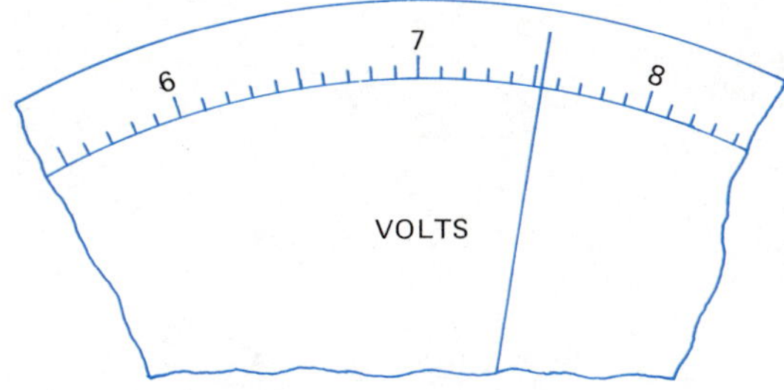

Figure 5-9 Voltmeter scale (7.53 V)

The third digit (3) is a different matter. It is the only digit of the three about which there may be some doubt. Some electricians might say it should be a 2, others might say 4. Recording it as a 3, however, is a reasonable estimate. Few would say it should be a 7 or 8.

In a measurement, the number of digits that are reliable and the first doubtful one are said to be *significant digits* or *significant figures*. The measurement 7.53 V taken from Figure 5-9, therefore, has three significant digits.

Since there is some doubt about the third digit, it is impossible to even guess the fourth digit. To determine the fourth digit requires a more accurate voltmeter. To obtain five or six significant digits in measurements requires very sophisticated (and expensive) equipment. More will be said about accuracy and uncertainty of measurements in the next chapter.

Not all digits in a recorded measurement are significant. By inspection, a technician should be able to state the number of significant digits in a measurement.

FINDING SIGNIFICANT DIGITS

The following digits in a measurement are significant:

- All nonzero digits.
  138.42 ft has five significant digits.
- All zeros that lie between significant digits.
  209.005 kg has six significant digits.
- All zeros to the right of a decimal point that are also to the right of the last nonzero digit.
  0.3600 gal has four significant digits.

The following digits in a measurement are not significant:

- All zeros at the beginning of a decimal fraction.
  0.000 32 s has two significant digits.
- The zeros at the end of a whole number unless their significance is indicated by some other means. Usually a bar is placed above the last zero that is significant.
  235 000 mm has three significant digits.
  3 400 $\bar{0}$00 lb has five significant digits.

If zeros only serve to locate the decimal point, they are not significant. Notice especially, that the location of the decimal point does not have any bearing on the significance of any of the digits.

When a number such as 3 640 000 is changed to scientific notation, it can be written with any power of ten desired. The decimal point and the exponent in the two parts are adjusted so the product yields the original number. These are all equivalent:

$$0.000\,364 \times 10^{10}$$
$$0.0364 \times 10^{8}$$
$$3.64 \times 10^{6}$$
$$36\,400 \times 10^{2}$$

Naturally, any number of combinations is possible.

Ordinarily, only a few of the possibilities are used. One is called *standard scientific notation* (SSN). A number expressed in SSN has the decimal point located in standard position. *Standard position* for any number is the position or place between the first and second significant digits. If there is only one significant digit, standard position is the place immediately following that digit. If a decimal point is located in standard position, the number has a value between one and ten.

Example A Locate standard position in each of the following numbers: (1) 4, (2) 55, (3) 278 000, (4) 0.000 314

Solution: An arrow is used to indicate the location:

(1) 4‸ (2) 5‸5 (3) 2‸78 000 (4) 0.000 3‸14
↑ ↑ ↑ ↑

Notice that standard position has nothing to do with the actual location of the decimal point.

Any number such as 438 000 can be converted to standard scientific notation (SSN). The decimal is moved five places left to standard position. In SSN, the number is 4.38×10^{5}. In decimal form the decimal point is located *five places* to the right of standard position. In SSN, the exponent is *five*. The five is called the characteristic of the number.

The *characteristic* of a number is the number of places from standard position to the deicmal point. The characteristic is positive when the decimal is to the right of standard position. It is negative when the decimal is to the left.

Example B Give the characteristic for each of the following numbers:

Solution:

Number	Characteristic
(1) 64	1
(2) 3894	3
(3) 0.0255	−2
(4) 246 084 000	8
(5) 4	0
(6) 0.000 000 661	−7

FINDING THE SIGN OF THE CHARACTERISTIC

- Any number equal to ten or greater: the characteristic is positive.
- Numbers between one and ten: the characteristic is zero.
- Any number less than one (proper fractions): the characteristic is negative.

EXERCISE 5-6

Rewrite the following numbers. Use an arrow to indicate the location of standard position. Give the characteristic and the number of significant digits for each number.

1. 640
2. 3.00
3. 0.4563
4. 85 600
5. 0.000 000 043 09
6. 905 $\bar{0}$00 000 000
7. 0.000 6
8. 1
9. 0.000 443 00
10. 0.000 456 780 043 21
11. 528$\bar{0}$
12. 0.004 00
13. 956 00$\bar{0}$ 000
14. 1728

Round each of these measurements to three significant figures.

15. Battery voltage: 12.0 V
16. Soldering iron tip: 565°C
17. Transmission line wave speed: 211 380 000 m/s
18. Time a wave travels one mile: 0.000 007 578 7 s
19. Time cutting tool falls 64 ft from pole: 2.00 s
20. Time between sparks, V-8 engine, 3000 r/min: 0.000 041 66 s
21. Electric meter reading: 2478 kW·h
22. Current in ac circuit: 0.1050 A
23. Span between poles: 45$\bar{0}$ ft
24. Router motor speed: 22 0$\bar{0}$0 r/min
25. Doorbell circuit resistance: 1184 Ω

5-11 CONVERTING BETWEEN DECIMAL AND SCIENTIFIC NOTATION

CHANGING FROM DECIMAL FORM TO STANDARD SCIENTIFIC NOTATION (SSN)

- Move the decimal point to standard position. This results in a number between one and ten.
- Give the power of ten an exponent equal to the characteristic of the original number.

Significant figures are given in the decimal factor *(mantissa)* of the expression. Rounding is done in the normal way to the mantissa. The power of ten factor merely indicates the location of the decimal point (characteristic).

Example A Change (1) 346 829 000 and (2) 0.000 089 557 to SSN. Round to three significant figures.

Solution: (1) 346 829 000 (Characteristic is +8)

$= 3.468\ 29 \times 10^8$

$= 3.47 \times 10^8$ (Rounded)

(2) 0.000 089 557 (Characteristic is −5)

$= 8.96 \times 10^{-5}$ (Rounded)

The procedure is simply reversed to convert a number in standard scientific notation (SSN) to its decimal form.

CHANGING FROM STANDARD SCIENTIFIC NOTATION (SSN) TO DECIMAL FORM

- Move the decimal point of the mantissa to the right if the characteristic is positive. Move it left for a negative characteristic.
- Move the decimal as many places as the absolute value of the characteristic.

Example B Convert (1) 7.84×10^{-4} and (2) 1.32×10^6 to decimal form.

Solution: (1) 7.84×10^{-4} (Characteristic is −4)

= 0.000 784

(2) 1.32×10^6 (Characteristic is +6)

= 1 320 000

EXERCISE 5-7

Express the following numbers in standard scientific notation (SSN). Round each answer to three significant figures.

1. 380 $\bar{0}$00
2. 0.000 771
3. 0.0314
4. 870.00
5. 44 305 000
6. 0.001 384 6
7. 824 706 000 000
8. 0.000 000 000 229 5
9. $0.000\ 552\ 1 \times 10^3$
10. 90.5×10^{-2}
11. 342.7×10^2
12. 7.000

Express the following numbers in decimal notation.

13. 5.92×10^2
14. 6.61×10^{-4}
15. 8.11×10^6
16. 1.64×10^0
17. 2.99×10^{-1}
18. 4.60×10^{12}
19. 3.00×10^{-12}
20. 7.94×10^{-7}
21. 3.34×10^4
22. 9.83×10^{-3}

23. Each cubic inch of copper contains 1.4×10^{24} free electrons. Write this number in decimal form.
24. The charge on an electron is 1.602×10^{-19} C. Rewrite this as a decimal.

25. One ampere is the current when 6 240 000 000 000 000 000 electrons pass by a point in the circuit each second. Express this number in standard scientific notation (SSN).
26. The mass of an electron in kilograms is 0.000 000 000 000 000 000 000 000 000 000 911. Convert to standard scientific notation (SSN).

5-12 CHANGING THE EXPONENT

Sometimes, it is necessary to write numbers in scientific notation when the decimal is not in standard position. In other cases, the number must be rewritten so the exponent has a certain value.

Another example where a shift in the exponent is required is in the product, quotient, square root and reciprocals of numbers written in standard scientific notation (SSN). Many times answers to such problems do not have the decimal point in standard position. A shift in the decimal point and an accompanying change in the exponent are required.

Such a procedure is easy to do. Keep in mind that a shift in the decimal by one place must be accompanied by a change in the value of the exponent by one. A shift of two places requires a change in the exponent of two.

The idea is not to change the value of the number. Whatever is done to the power of ten factor, the opposite must be done to the mantissa. If the power of ten is increased, the mantissa must be decreased by the same factor.

Example A Change (1) 492×10^{4}, (2) $0.006\ 84 \times 10^{8}$, (3) 328×10^{-4} and (4) $0.008\ 71 \times 10^{-2}$ to standard scientific notation (SSN).

Solution: (1) In the mantissa, the decimal point must be moved two places left. This is a decrease in the decimal factor. The power of ten must be increased by the same factor. The mantissa is divided by 100 so the power of ten must be multiplied by 100. This is accomplished by increasing the exponent to 6: Thus: $492 \times 10^{4} = 4.92 \times 10^{6}$

(2) The decimal is increased by 3 places so the power of ten must be decreased by 3: $0.006\ 84 \times 10^{8} = 6.84 \times 10^{5}$

(3) $328 \times 10^{-4} = 3.28 \times 10^{-2}$

(4) $0.008\ 71 \times 10^{-2} = 8.71 \times 10^{-5}$

Engineering Notation. The International Bureau of Weights and Measures adopted a number of rules for using the international measuring system. These will be discussed in the next chapter. One of their recommendations is to use only those metric prefixes that represent factors of 1000, 1 000 000, 1 000 000 000, 0.001, 0.000 001, etc. In other words, use only those prefixes that represent powers of ten in which the exponent is a multiple of three. These are called *common units.*

The electrical industry seldom uses other units anyway. Scientific notation in which the mantissa is a number between 1 and 1000 and the exponent is a multiple of 3 (3, 6, 9, 12, etc.) is called *engineering notation* (ENG). Engineering notation goes hand-in-hand with the use of common units.

Example B Convert the following decimals to engineering notation (ENG): (1) 14 600, (2) 288 000 000 and (3) 0.003 29.

Solution: (1) $14\ 600 = 14.6 \times 10^3$
(2) $288\ 000\ 000 = 288 \times 10^6$
(3) $0.003\ 29 = 3.29 \times 10^{-3}$

EXERCISE 5-8

Use these numbers for problems 1–3.

a. 0.004 49	e. 2	i. 6.61×10^{-2}
b. 348	f. 0.0998	j. 8.94×10^0
c. 794 000	g. 4.28×10^8	k. 1.77×10^{-5}
d. 0.000 114	h. 4.82×10^6	*l.* 2970×10^{-8}

1. Rewrite the numbers as a decimal times 10^2.
2. Rewrite the numbers as a decimal times 10^{-3}.
3. Rewrite the numbers in engineering notation (ENG).

5-13 COLOR CODES

Resistors are color coded to indicate the amount of resistance in ohms, Figure 5-10. Four-banded resistors give two significant figures; five-banded resistors give three, Figure 5-11. The bands showing the significant figures are followed by a band indicating the exponent in the power of ten factor. The last band indicates the tolerance. *Tolerance* indicates the amount by which a batch of resistors may vary. Resistors of 100 Ω and a tolerance

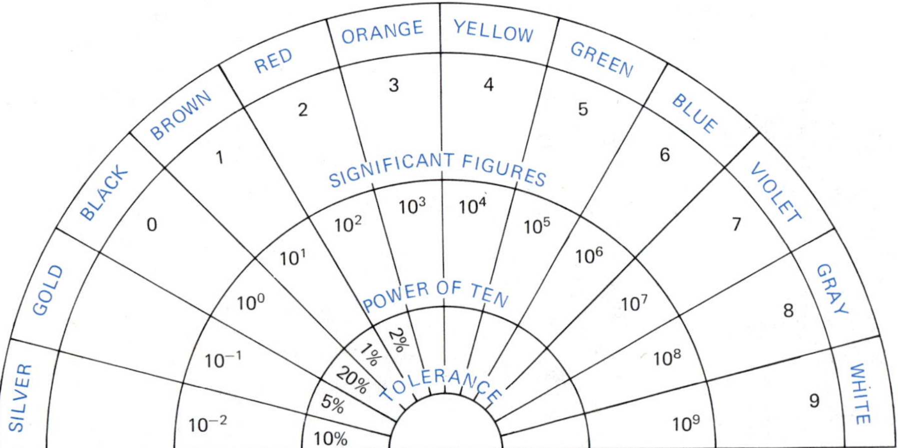

Figure 5-10 Color code for resistors and capacitors

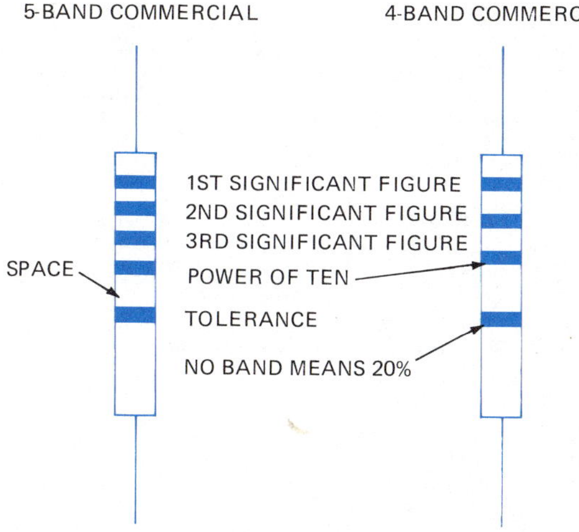

Figure 5-11 Color banding code for resistors

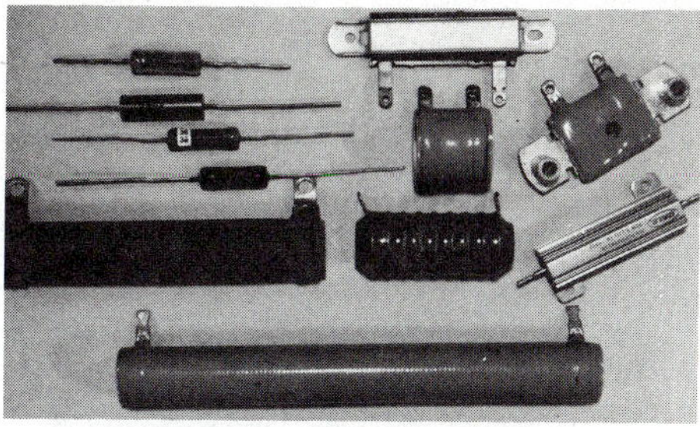

Figure 5-12 Various resistors (Courtesy of Southeast Nebraska Community College)

of ±10%, for example, may actually have a resistance anywhere between 90 Ω and 110 Ω. Tolerances are discussed in Chapter 10. Some resistors are stamped numerically, Figure 5-12.

Example A State the resistance of (1) the four-banded resistor and (2) the five-banded resistor shown in Figure 5-13.

Solution: (1) The blue and green bands indicate significant digits of 65. The third band (blue) means 10^6 and the fourth (silver) means a tolerance of ±10%. In standard scientific notation (SSN) the resistance is $6.5 \times 10^7\ \Omega \pm 10\%$.

(2) The color bands in this example are red, yellow and green (245), orange (10^3) and brown (±1%). The resistance is $245 \times 10^3\ \Omega$ or in standard scientific notation (SSN), $2.45 \times 10^5\ \Omega \pm 1\%$.

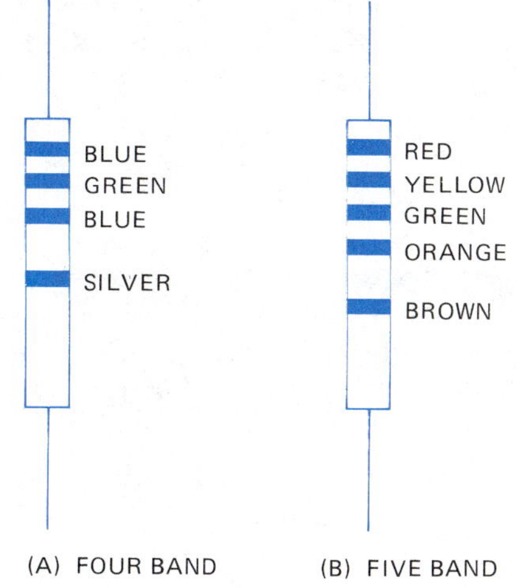

Figure 5-13

Capacitors are also coded in color for the amount of capacitance. A *capacitor* is an electrical device in which electric charges may be stored, Figure 5-14. Capacitors are usually called *condensers* in the automotive industry.

A capacitor is rated by its capacitance (C). *Capacitance* is a measure of the amount of charge a capacitor can store. Capacitors sometimes consist of two large metal (foil) plates separated by a thin insulating material such as waxed paper. They are then rolled into a small cylinder. Others are small square foil sheets separated by mica sheets. The latter kind are sealed within a ceramic coating. There are also other types.

Electrons flow from one plate to the other around the external circuit when attached across a battery or other voltage source. When charged, one plate of the capacitor is negative; the other positive. The amount of electric charge stored is the amount of charge on either plate.

Figure 5-14 Various capacitors (Courtesy of Southeast Nebraska Community College)

Capacitance is measured in farads (F). If one coulomb of electric charge is stored by the action of one volt potential difference, the capacitor has a rating (capacitance) of one farad. Most capacitors have a capacitance of only a few millionths of a farad or less.

Capacitors are color coded for the amount of capacitance in picofarads (pF = 10^{-12} F) and for other characteristics. Figure 5-15 indicates one of several systems of capacitor codes. The colors have the same meaning as before, Figure 5-10.

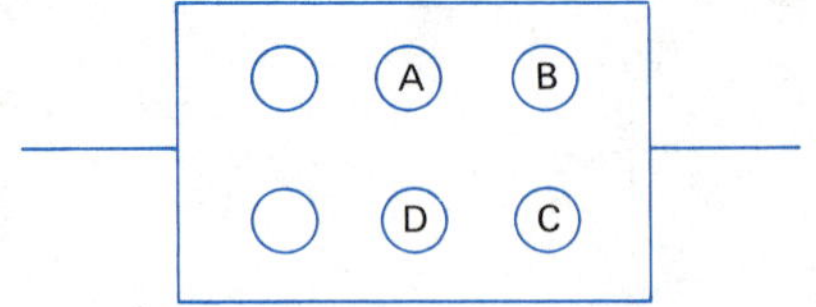

A – 1ST SIGNIFICANT FIGURE
B – 2ND SIGNIFICANT FIGURE
C – POWER-OF-TEN FACTOR
D – TOLERANCE

Figure 5-15 Color scheme for capacitors

Example B Find the capacitance of this capacitor in picofarads, Figure 5-16.

Solution: Red and green indicate significant figures of 25. Blue indicates 10^6. Black indicates a ±20% tolerance. In standard scientific notation (SSN) the capacitance is 2.5×10^7 pF ±20%.

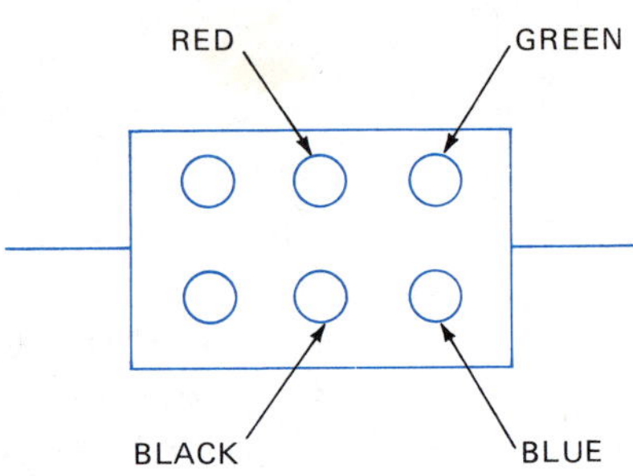

Figure 5-16

EXERCISE 5-9

What is the resistance of each of the following resistors? Express the resistance in engineering notation (ENG) and give the tolerance.

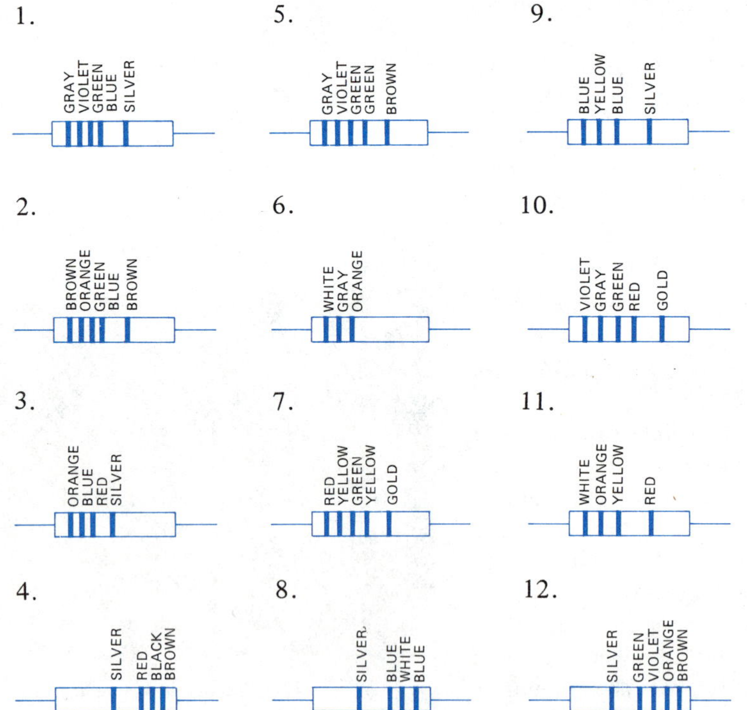

What is the capacitance in picofarads of each of these capacitors? Express answers in standard scientific notation (SSN) and give the tolerance.

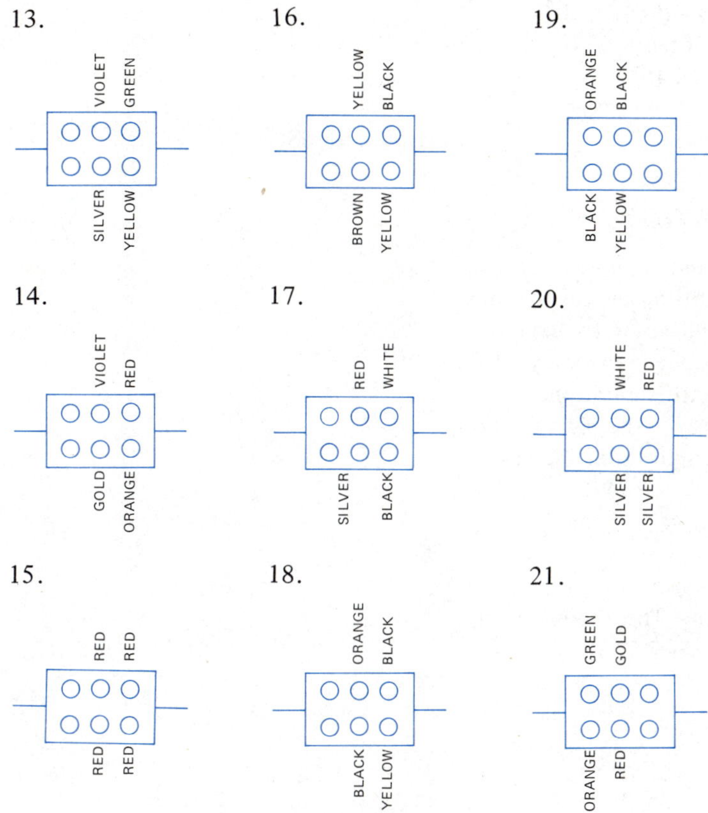

ARITHMETIC IN SCIENTIFIC NOTATION

Addition and Subtraction. When numbers written in scientific notation are to be added or subtracted, the numbers must all be expressed in the same power of ten. It does not matter what exponent is used as long as it is the same for all numbers being added or subtracted.

Example A Add: $(5 \times 10^2) + (4 \times 10^3)$

Solution: Convert to the same power of ten:
$= (0.5 \times 10^3) + (4 \times 10^3) = 4.5 \times 10^3$

An alternate solution is:
$= (5 \times 10^2) + (40 \times 10^2) = 45 \times 10^2$
$= 4.5 \times 10^3$

Since a step is saved if the exponent used is that of the larger addend, it makes sense to always use the larger exponent.

Example B Subtract: $(6.5 \times 10^{-3}) - (3 \times 10^{-4})$
Solution: Express both terms with the higher power of ten:
$= (6.5 \times 10^{-3}) - (0.3 \times 10^{-3}) = 6.2 \times 10^{-3}$

Example C Combine: $(33.4 \times 10^{3}) + (53.6 \times 10^{5}) - (98.2 \times 10^{4})$
Solution: Convert all terms to the highest power of ten:
$= (0.334 \times 10^{5}) + (53.6 \times 10^{5}) - (9.82 \times 10^{5})$
$= 44.114 \times 10^{5}$
$= 4.41 \times 10^{6}$ (Rounded, SSN)

5-14 MULTIPLYING IN SCIENTIFIC NOTATION

One advantage of scientific notation is that large and small numbers can be expressed in a simple way. Scientific notation also makes calculations involving such numbers easy to carry out. In multiplication, for example, the mantissas and the power of ten factors are multiplied separately. The product, therefore, is automatically expressed in scientific notation.

The decimal factors are combined in the usual way. The law of exponents is used to combine powers of ten. For multiplication, the law of exponents is:

$$\mathbf{A^{m} \times A^{n} = A^{m+n}}$$

The product of two powers of ten is a power of ten. The exponents are added.

Example A Multiply: $100 \times 10\,000$
Solution: Convert to powers of ten, then multiply
$10^{2} \times 10^{4} = 10^{2+4} = 10^{6}$

Example B Multiply: 100×0.0001
Solution: $10^{2} \times 10^{-4} = 10^{2+(-4)} = 10^{-2}$

Example C Multiply: $4560 \times 382\,000$. Round the product to three significant figures expressed in standard scientific notation (SSN).
Solution: $(4.56 \times 10^{3}) \times (3.82 \times 10^{5})$
$= (4.56 \times 3.82) \times (10^{3} \times 10^{5})$

Do not write out the first step. Do it mentally.
$= 17.4192 \times 10^{8}$
$= 1.74 \times 10^{9}$

Example D Multiply: $0.000\,082\,4 \times 144 \times 38\,000\,000$. Round to three significant figures expressed in engineering notation (ENG).
Solution: $8.24 \times 10^{-5} \times 1.44 \times 10^{2} \times 3.8 \times 10^{7}$
$= 45.089\,28 \times 10^{4}$
$= 451 \times 10^{3}$

Hint: multiply mantissas. Add exponents: 10^{-5+2+7}

EXERCISE 5-10

Add or subtract as indicated.

1. $3 \times 10^2 + 4 \times 10^3$
2. $2 \times 10^4 + 5 \times 10^6$
3. $7 \times 10^{-2} + 2 \times 10^{-3}$
4. $4 \times 10^{-5} + 5 \times 10^{-3}$
5. $2 \times 10^1 + 2 \times 10^{-1}$
6. $6.32 \times 10^2 + 5$
7. $4 \times 10^3 - 3 \times 10^2$
8. $5 \times 10^7 - 9 \times 10^4$
9. $3.2 \times 10^{-4} - 5 \times 10^{-5}$
10. $7 \times 10^4 - 2 \times 10^5$
11. $8 \times 10^5 - 4 \times 10^4 + 3 \times 10^5 - 12 \times 10^3$
12. $2 \times 10^{-3} + 4 \times 10^{-2} + 7 \times 10^{-3} - 1.1 \times 10^{-2} + 8.5 \times 10^{-3}$

Multiply the following expressions. Convert to power of ten notation before evaluating. Do not use calculators.

13. $10^3 \times 10^2$
14. $10^5 \times 10^{-3}$
15. $10^{-7} \times 10^4$
16. $10^{-2} \times 10^{-3}$
17. $10^4 \times 10^{-17} \times 10^{+17}$
18. $10^5 \times 10^{-6} \times 10^3 \times 10^{-3}$
19. $0.0001 \times 0.01 \times 1000$
20. $1\,000\,000 \times 10\,000 \times 0.000\,001$
21. $0.000\,000\,01 \times 0.000\,01 \times 1\,000\,000 \times 100 \times 0.0001$

Convert all numbers to scientific notation then evaluate. Round all answers to three significant digits. Solve without using calculators. Express answers in engineering notation (ENG).

22. $1.43 \times 10^2 \times 10^3$
23. $4 \times 10^5 \times 3$
24. $3 \times 10^{-2} \times 1000$
25. 500×300
26. $800 \times 38\,000$
27. $0.0004 \times 0.001 \times 8200$
28. $6\,000\,000 \times (4 \times 10^{-3})(8 \times 10^2)$
29. $300 \times 0.000\,000\,07 \times 2 \times 10^4$
30. $1000 \times 8 \times 10^{-6} \times 300\,000 \times 400$
31. $0.000\,002 \times 500 \times 7.5 \times 10^5 \times 40$
32. $500 \times 0.0005 \times 5 \times 50 \times 0.05$

5-15 *DIVIDING IN SCIENTIFIC NOTATION*

The law of combining exponents for division is:

$$\frac{A^m}{A^n} = A^{m-n}$$

In division, the exponents are subtracted to find the quotient. In scientific notation, the decimal factors are divided in the usual way. The power of tens are divided separately. The quotients are automatically given in scientific notation.

Example A Convert the following to powers of ten notation, then divide: 1 000 000 ÷ 10 000

Solution: $\frac{10^6}{10^4} = 10^{6-4} = 10^2$

Example B Divide 0.0001 by 0.000 01

Solution: $\frac{10^{-4}}{10^{-5}} = 10^{-4-(-5)} = 10^1$

Example C Divide 6480 by 2 710 000. Convert to standard scientific notation (SSN) before dividing. Round the quotient to three significant figures. Express quotient in SSN.

Solution: $\frac{6.48 \times 10^3}{2.71 \times 10^6}$ (Convert)

$= \frac{6.48}{2.71} \times 10^{3-6}$ (Do this step mentally)

$= 2.391\,143 \times 10^{-3} = 2.39 \times 10^{-3}$

Example D Evaluate: $\frac{1790}{0.046 \times 0.002\,94}$

Express quotient in engineering notation (ENG) rounded to three significant digits.

Solution: $\frac{1.79 \times 10^3}{(4.6 \times 10^{-2})(2.94 \times 10^{-3})}$

$= \frac{1.79}{4.6 \times 2.94} \times 10^{3-(-2)-(-3)}$ (Do mentally)

$= 0.132\,36 \times 10^8 = 13.2 \times 10^6$

EXERCISE 5-11

Divide the following expressions after converting decimals to powers of ten. Do not use calculators.

1. $10^{-8} \div 10^{-5}$
2. $10^3 \div 10^{-1}$
3. $10^4 \div 10^3$
4. $10^{-6} \div 10^2$
5. 1 000 000 ÷ 1000
6. 0.000 000 1 ÷ 1000
7. $\frac{100\,000 \times 100}{0.001}$
8. $\frac{10^5 \times 0.0001}{10^4 \times 10^{-6}}$
9. $\frac{10^{-5} \times 10\,000}{10^{-4} \times 10^6}$

Convert all factors in the following problems to standard scientific notation (SSN) and evaluate without using calculators. Express the answers in engineering notation (ENG). Cancel when possible.

10. $\frac{9 \times 10^6}{3 \times 10^4}$
11. $\frac{28 \times 10^2}{1.4 \times 10^4}$
12. $\frac{9\,500\,000}{50\,000}$
13. $\frac{125\,000}{250 \times 10^{-8}}$
14. $\frac{125 \times 10^{-4}}{25\,000\,000}$
15. $\frac{28\,000\,000}{7000}$

16. $\frac{0.000\,550}{0.0011}$

17. $\frac{(0.003)(80\,000)}{60}$

18. $\frac{(4 \times 10^{7})(6 \times 10^{-3})}{8 \times 10^{-2}}$

19. $\frac{(144 \times 10^{-1})(3.6 \times 10^{5})}{12}$

20. $\frac{0.522 \times 0.084}{261 \times 4}$

21. $\frac{(0.0003)(500)(0.4)}{(200)(30\,000)}$

22. $\frac{(80\,000)(0.6)}{(0.002)}$

23. $\frac{(600)(8000)}{(2000)(0.04)}$

24. $\frac{90\,000\,000 \times 0.000\,000\,000\,4 \times 140}{700 \times 0.8 \times 0.000\,000\,03}$

Make circuit diagrams for each of the following. Express all answers in standard scientific notation (SSN).

25. An electric clock draws 2×10^{-2} A and has a resistance of 6×10^{3} Ω. (a) What voltage does it operate on? (b) What power does it draw?
26. An ammeter has a resistance of 4.8×10^{-3} Ω. (a) If there is a voltage drop of 2.4×10^{-2} V across the ammeter, what current is flowing? (b) How much power is dissipated in the ammeter?
27. A 2.2×10^{2} V heater element draws a current of 10.0 A. Evaluate the (a) resistance and the (b) power.
28. A fractional horsepower motor operates on 1.6×10^{1} V. (a) If it draws 7 W, what is the current? (b) How much resistance does it have?
29. A resistor has a resistance of 8.25×10^{4} Ω and draws 2.53×10^{-3} A. Find (a) the voltage and (b) the power.
30. A 2.55×10^{4} Ω resistor draws 2.24×10^{-2} A. What voltage is used? If the current is maximum, what is the power rating of this resistor?

5-16 POWERS, ROOTS AND RECIPROCALS

The heading for this section could be expressed as: "Powers, Fractional Powers and Negative Powers." Roots and reciprocals can be expressed as powers so one exponent law explains all three:

$$(A^{m})^{n} = A^{m \times n}$$

For powers, the letter n is a positive whole number. For roots, it is a fraction. It is a −1 for reciprocals. If a power of ten is to be squared, just double the exponent. When a power of ten is cubed, triple the exponent.

Example A Evaluate: $(10^{3})^{2}$

Solution: $(10^{3})^{2} = 10^{3 \times 2} = 10^{6}$

Note: $1000^{2} = 1\,000\,000$

Example B Evaluate: $(10^{-4})^3$
Solution: $(10^{-4})^3 = 10^{-4 \times 3} = 10^{-12}$

When numbers are expressed in scientific notation, both the decimal and power of ten parts must be squared or cubed:

$$(a \times b)^n = a^n \times b^n$$

Example C Evaluate: $(5 \times 10^2)^3$
Solution: $(5 \times 10^2)^3$
$= 5^3 \times (10^2)^3 = 125 \times 10^6 = 1.25 \times 10^8$

To take a square root of a power of ten, write the exponent as an even number. Otherwise, the root is not a power of ten ($\sqrt{10^1} = 3.1623\ldots$, $\sqrt{10^3} = 31.623\ldots$, etc.). The number should be expressed in scientific notation so the exponent is an even number.

Example D Evaluate: $\sqrt{250\,000}$
Solution: $\sqrt{25 \times 10^4}$
$= \sqrt{25} \times \sqrt{10^4}$
$= 5 \times 10^2$

Alternate: $(25 \times 10^4)^{1/2}$
$= 25^{1/2} \times (10^4)^{1/2}$
$= 5 \times 10^2$

Example E Evaluate: $\sqrt{0.000\,016}$
Solution: $\sqrt{16 \times 10^{-6}} = \sqrt{16} \times \sqrt{10^{-6}} = 4 \times 10^{-3}$

The cube root of a power of ten will be a power of ten if the exponent is a multiple of three: $(10^6)^{1/3} = 10^{6 \times 1/3} = 10^2$

Example F Evaluate: $\sqrt[3]{27\,000\,000\,000}$
Solution: $\sqrt[3]{27 \times 10^9}$
$= \sqrt[3]{27} \times \sqrt[3]{10^9}$
$= 3 \times 10^3$

Alternate: $(27 \times 10^9)^{1/3}$
$= 27^{1/3} \times (10^9)^{1/3}$
$= 3 \times 10^3$

To take a reciprocal of a power of ten, simply change the sign of the exponent: $1/10^2 = (10^2)^{-1} = 10^{-2}$.

Example G Evaluate: $\dfrac{1}{25\,000}$

Solution: $\dfrac{1}{2.5 \times 10^4}$

$= \dfrac{1}{2.5} \times \dfrac{1}{10^4} = 0.4 \times 10^{-4} = 4 \times 10^{-5}$

Example H Evaluate: $\dfrac{1}{0.000\,125}$

Solution: $\dfrac{1}{1.25 \times 10^{-4}}$

$= \dfrac{1}{1.25} \times \dfrac{1}{10^{-4}} = 0.8 \times 10^4 = 8 \times 10^3$

Example I Evaluate: $(0.000\,05)^{-3}$

Note: the exponent (−3) can be written as 3(−1) so the expression can be rewritten in the form $[(0.000\,05)^3]^{-1}$.

Solution: $(5 \times 10^{-5})^{-3} = \dfrac{1}{(5 \times 10^{-5})^3}$

$= \dfrac{1}{5^3} \times \dfrac{1}{10^{-15}} = \dfrac{1}{125} \times 10^{15}$

$= 0.008 \times 10^{15} = 8 \times 10^{12}$

EXERCISE 5-12

Expesss the answer to each of the following problems in standard scientific notation (SSN). Solve without the aid of the calculator.

1. $(3 \times 10^2)^2$
2. $(5 \times 10^5)^2$
3. $(2 \times 10^3)^3$
4. $(4 \times 10^{-4})^2$
5. $(5 \times 10^{-5})^3$
6. $(2 \times 10^{-2})^3$
7. $(2 \times 10^3)^4$
8. $(3 \times 10^5)^3$

Express these powers in engineering notation (ENG). Do not use calculators.

9. $(2 \times 10^{-3})^2$
10. $(3 \times 10^2)^2$
11. $(8 \times 10^5)^2$
12. $(6 \times 10^{-5})^2$
13. $(7 \times 10^{-3})^3$
14. $(2 \times 10^8)^3$
15. $(3 \times 10^{-4})^4$
16. $(2 \times 10^2)^4$

Express these roots in standard scientific notation (SSN). Evaluate without calculators.

17. $\sqrt{4 \times 10^6}$
18. $\sqrt{40 \times 10^3}$
19. $\sqrt{9000 \times 10^5}$
20. $\sqrt{0.09 \times 10^{-4}}$
21. $\sqrt{1.6 \times 10^{-3}}$
22. $(2.5 \times 10^{-9})^{1/2}$

Without using a calculator, express the answers to these problems in engineering notation (ENG).

23. $(8.1 \times 10^3)^{1/2}$
24. $(2.7 \times 10^4)^{1/3}$
25. $\sqrt[3]{80 \times 10^{-7}}$
26. $\sqrt{1.44 \times 10^6}$
27. $(1.21 \times 10^{-4})^{1/2}$
28. $(6.4 \times 10^{10})^{1/3}$

Find these reciprocals and other negative powers without using a calculator. Express the answers for 29–33 in SSN and the answers for 34–38 in ENG.

29. $(10^5)^{-1}$
30. $(10^{-2})^{-1}$
31. $(2 \times 10^2)^{-1}$
32. $(4 \times 10^{-3})^{-1}$
33. $(5 \times 10^{-3})^{-2}$
34. $(4 \times 10^3)^{-2}$
35. $(2 \times 10^3)^{-3}$
36. $(3 \times 10^{-4})^{-3}$
37. $(7 \times 10^{-5})^{-2}$
38. $(1.6 \times 10^5)^{-1/2}$

5-17 SCIENTIFIC NOTATION AND THE CALCULATOR

The calculator operations of addition, subtraction, multiplication, division, powers, roots and reciprocals have been discussed previously. When numbers are expressed in scientific notation, these operations are the same. Only the method of entering the numbers and the mode of display are different.

Calculator Notes: To enter a number in scientific notation, use the [EE] key. Some calculators have the key labeled [EEX] or [EXP]. A number such as 4.25×10^5 is entered by performing the following steps:

4.25 [EE] 5, 4.25 5

The display follows the comma. Notice that the base 10 is not in the display. Only the mantissa and the exponent appear. The display (4.25 5) is read, "4.25 times ten to the fifth power."

If a number is negative (-3.28×10^2), it is entered this way:

3.28 [+/-] [EE] 2, −3.28 2

A negative exponent is entered by using the [+/-] in a different way. For a number such as 7.91×10^{-4}, press the [+/-] after the exponent is entered:

7.91 [EE] 4 [+/-], 7.91 −4

One of the most common mistakes is made when entering a simple power of ten such as 10^3. One has to remember that 10^3 equals 1×10^3. The error that is often made is to enter the number this way: 10 [EE] 3, 1.00 4, which is not correct. It should be entered this way: 1 [EE] 3, 1.00 3. Some calculators have the 1.00 programmed with the [EE] key. On these, the one (1) does not have to be keyed in: [EE] 3, 1.00 3. Read the Owners Manual for exact procedures.

Some calculators can be set to automatically display entries and answers in decimal form, standard scientific notation [SCI] or engineering notation [ENG]. The number of significant figures can be set for automatic display using [FIX]. The student should consult the Owners Manual for instructions. Some calculators have special keys that shift the exponent up or down one power at a time. The decimal point in the mantissa is shifted simultaneously.

EXERCISE 5-13

Enter these numbers in the calculator using [EE]. Use the rounding features of the calculator to round each number to three significant figures. Copy the display exactly as shown. If the calculator cannot perform this function, round manually.

1. 10^4
2. 10^{-3}
3. 6.2971×10^3
4. 9.3848×10^{-5}
5. -1.531×10^0
6. 5.293×10^{17}
7. -3.465×10^{-24}
8. 1.602×10^{-19}

Use the features available on your calculator to evaluate the following problems. Use [SCI] to obtain answers in standard scientific notation. Use [ENG] to get answers in engineering notation or else convert manually. Express all answers in the notation indicated, rounded to three significant figures.

9. $(9.85 \times 10^2)(6.04 \times 10^3)$ (SSN)
10. $(10^5)(10^{-3})$ (SSN)
11. $(4.25 \times 10^5)(10^{-3})$ (ENG)

12. $(7.62 \times 10^{-5})(2.91 \times 10^{10})^2$ (ENG)
13. $\dfrac{5.38 \times 10^{-1}}{6.72 \times 10^{7}}$ (SSN)
14. $\dfrac{10^4}{1.64 \times 10^2}$ (SSN)
15. $\dfrac{8.84 \times 10^{-3}}{3.76 \times 10^{-5}}$ (ENG)
16. $\dfrac{(6.95 \times 10^4)^2}{2.77 \times 10^{-2}}$ (ENG)
17. $(5.36 \times 10^{-2})(8.4 \times 10^2)(7.21 \times 10^1)$ (SSN)
18. $\dfrac{(7.27 \times 10^2)(4.23 \times 10^{-3})}{(6.72 \times 10^2)^{0.5}}$ (SSN)
19. $\dfrac{(0.000\ 354)(8.39 \times 10^6)^{1/2}}{(1.64 \times 10^2)}$ (ENG)
20. $\dfrac{(438\ 000)^2}{(6.21 \times 10^5)(9.62 \times 10^2)}$ (ENG)
21. $\dfrac{(0.003\ 41)(\pi)(5.41 \times 10^7)^2}{(84\ 700)^{1/2}(6.71 \times 10^{-3})(2.88 \times 10^{-4})}$ (SSN)
22. $\dfrac{2\pi^2(5.67 \times 10^5)(7.94 \times 10^{-2})}{\sqrt{5280}(3.66 \times 10^3)}$ (ENG)

Square.

23. 8.39×10^{-3} (SSN)
24. 3.71×10^2 (SSN)
25. 6.42×10^4 (ENG)
26. 2.88×10^{-7} (ENG)

Use $\boxed{y^x}$.

27. $(6.64 \times 10^2)^3$ (SSN)
28. $(1.71 \times 10^{-3})^5$ (SSN)
29. $(5.46 \times 10^{-5})^{-3}$ (ENG)
30. $(2.91 \times 10^4)^{-2}$ (ENG)

Find these square roots.

31. $\sqrt{3.64 \times 10^{-3}}$ (SSN)
32. $\sqrt{25.7 \times 10^4}$ (SSN)
33. $(9.11 \times 10^5)^{0.5}$ (ENG)
34. $(1.44 \times 10^6)^{1/2}$ (ENG)

Use $\boxed{y^x}$ to find these roots.

35. $\sqrt[3]{7.98 \times 10^{-4}}$ (SSN)
36. $\sqrt[5]{2.77 \times 10^{12}}$ (SSN)
37. $(8.92 \times 10^{-6})^{1/5}$ (ENG)
38. $(5.66 \times 10^{12})^{1/3}$ (ENG)

Find the reciprocal of these numbers.

39. 9×10^{-4} (SSN)
40. 1.21×10^5 (SSN)
41. 16.9×10^{-3} (ENG)
42. 1000×10^3 (ENG)

CHAPTER 6

MEASURING ELECTRICITY

OBJECTIVES

After satisfactorily completing this chapter, the student should be able to:

- Define various words and terms introduced in this chapter including those involving electrical quantity names and symbols, base unit names and abbreviations and metric prefix names and symbols.
- Convert a measurement from one unit to another similar unit.
- State the uncertainty in a measurement given its value and units and round calculated values to reflect the uncertainty in the components.

An understanding of the metric system is becoming more and more important in everyday usage. The litre, millimetre, degree Celsius and other units are becoming well known. The electrical industry has always required technicians to have a full understanding of the metric system. Electrical units such as the volt, ampere, ohm, farad and watt are metric units.

Electrical technicians use measuring instruments almost every working day. Voltage drops across electrical components must be within certain specifications. Resistance, current and power are only a few of the quantities future electricians will be expected to measure.

Accurate measuring instruments are essential. An ohmmeter that can only measure to the nearest tenth ohm is hardly adequate if the specified tolerance range is one hundredth ohm. The technician must be able to correctly mount the instrument on the object or circuit being measured and then read the device correctly.

This chapter begins with the discussion of precision measurements, the uncertainty in them and how the uncertainty carries through to calculated values. The SI (metric) measuring system is described in detail. The chapter concludes with conversion of units.

RELIABILITY OF MEASUREMENTS

Accuracy is a word that refers to the reliability of a measuring instrument. A micrometer, for example, is a more accurate device than a yardstick. A micrometer would be used to make measurements to the nearest thousandth inch. A yardstick cannot be used for measuring to the nearest thousandth inch. A sweep stop watch can measure a period of time to the nearest second. More accurate digital stop watches may gauge the same period to the nearest hundredth second.

Regardless of how accurate the measuring device, no matter how much skill the user has, all measurements are inexact. Every measurement contains uncertainty. *Uncertainty* refers to the inexactness of a measurement. It is expressed as a range of values within which the *theoretically true value* lies.

6-1 UNCERTAINTY IN MEASUREMENTS

A metric rule is used to measure the length of a rectangular object to the nearest millimetre, Figure 6-1. The measurement (49 mm) is actually just an approximation. The true value is probably somewhere between 48.5 mm and 49.5 mm. The measurement is, therefore: 49 mm ± 0.5 mm. The part expressing the range of values (± 0.5 mm) is the uncertainty. A tolerance, expressed similarly, is the range of acceptable values in a specification.

A micrometer (mike) is used to measure the width of the rectangular object, Figure 6-2. The micrometer, being more accurate, reads 25.39 mm.

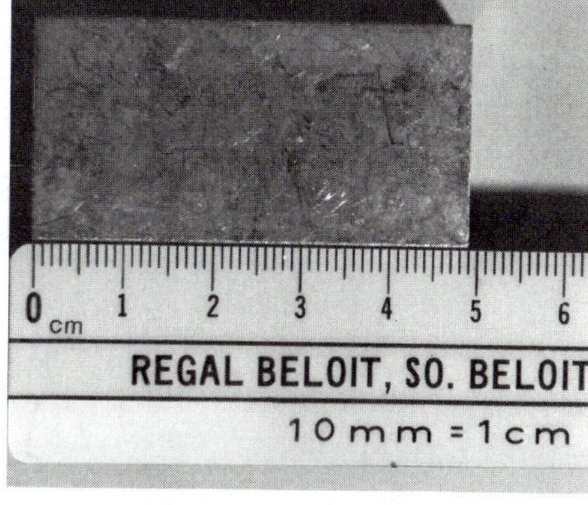

Figure 6-1 Measuring the length of an object with a metric rule. (Courtesy of Southeast Nebraska Community College)

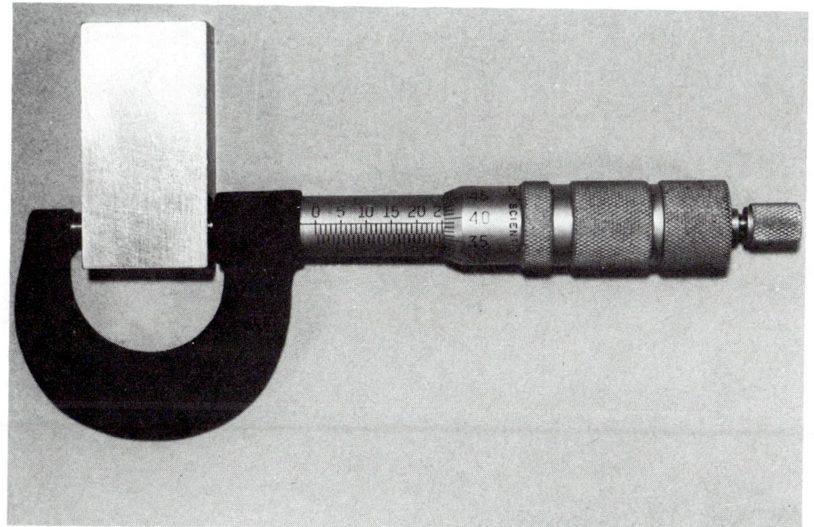

Figure 6-2 Measuring the width of an object with a metric micrometer caliper. Read 25 along the barrel and 0.39 on the thimble. The measurement is 25.39 millimetres. (Courtesy of Southeast Nebraska Community College)

The first two digits (25) are read on the barrel. The last two (0.39) are taken from the thimble. The true value of the width is between 25.385 mm and 25.395 mm. Thus the reading is 25.39 mm ± 0.005 mm.

Notice the uncertainty in the width (± 0.005 mm) using a micrometer is much smaller than the uncertainty in the length (± 0.5 mm) using a rule. The more accurate the measuring device, the smaller the uncertainty. Therefore, a micrometer is more accurate than a rule.

A measurement is normally given without specifying the uncertainty. Only the significant digits are given. The measurement 49 mm is understood to mean 49 mm ± 0.5 mm and 25.39 mm is understood to mean 25.39 mm ± 0.005 mm.

The uncertainty is understood to mean one-half of one unit in the last significant place value. For example, ±0.5 is one-half of one and ±0.005 is one-half of 0.01 in the previous measurements. A measurement of 500 V means 500 V ± 50 V. However, $5\bar{0}0$ V means $5\bar{0}0$ V ± 5 V and $50\bar{0}$ V means $50\bar{0}$ V ± 0.5 V.

EXERCISE 6-1

State the number of significant figures in each of the following measured quantities. Rewrite each measurement including the uncertainty using the (±) symbol.

1. 6 cm
2. 6.2 s
3. 24.0 V
4. 3.52×10^{-3} A
5. 1.61×10^{5} V
6. 4384 kW·h
7. 45.5×10^{6} Ω
8. 436×10^{-9} F
9. 0.3333 hp
10. 425×10^{6} V
11. 98.6°F
12. -1.602×10^{-19} C

Read the following meters and scales. State the number of significant digits in each measurement. Express the uncertainty with the (±) symbol.

13.

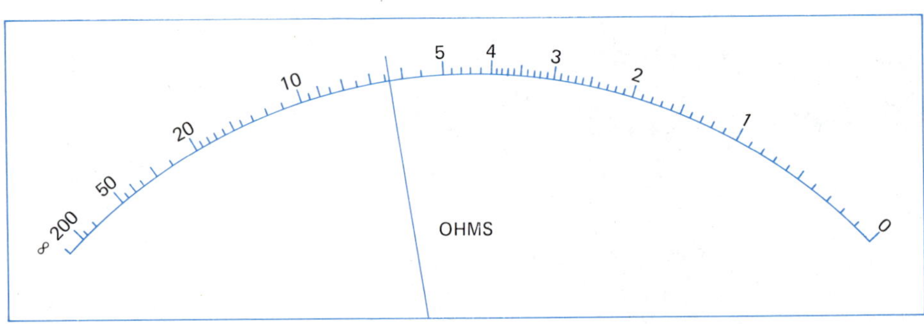

14.

0
50
100
150
200
250
VOLTS

15.

0
10
20
30
40
50
VOLTS

16.

0
2
4
6
8
10
VOLTS

17.

0
0.5
1
1.5
2
2.5
3
AMPERES

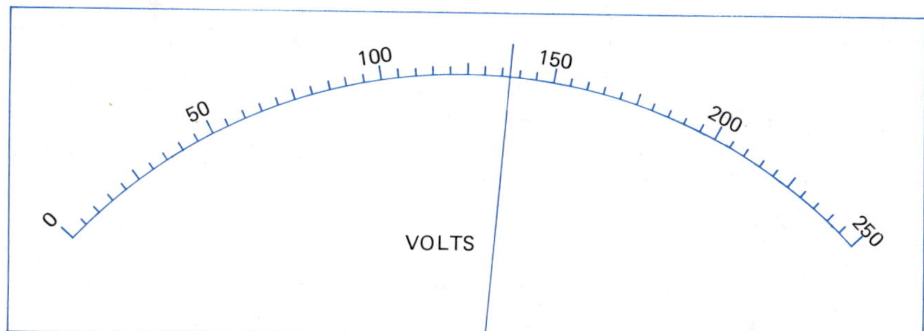

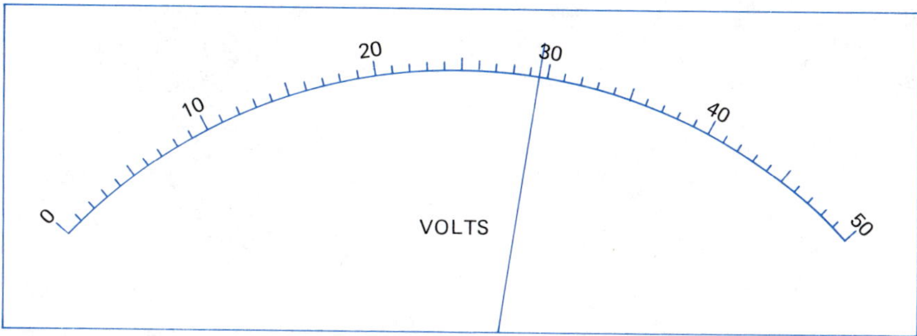

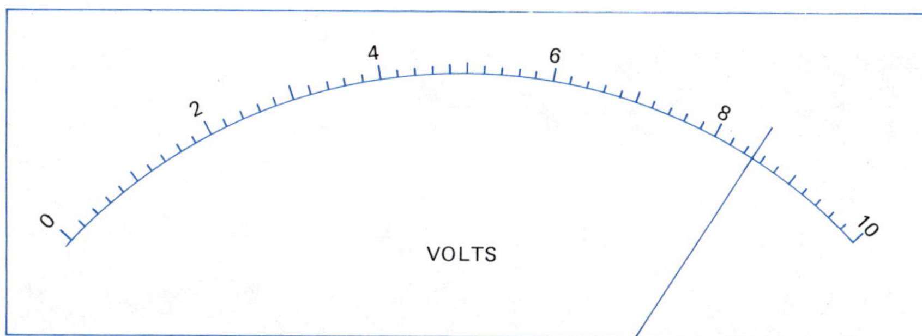

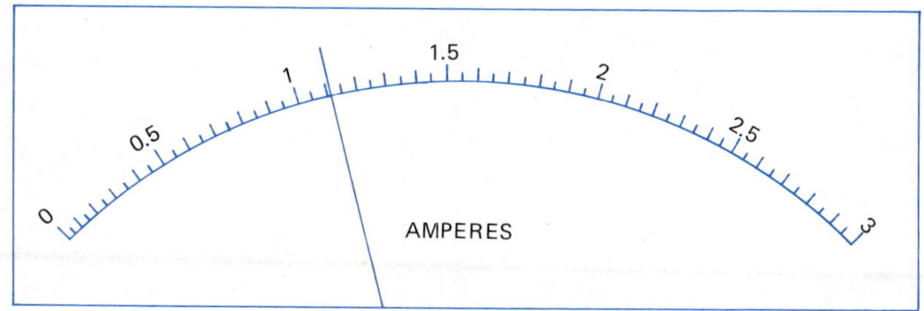

18.

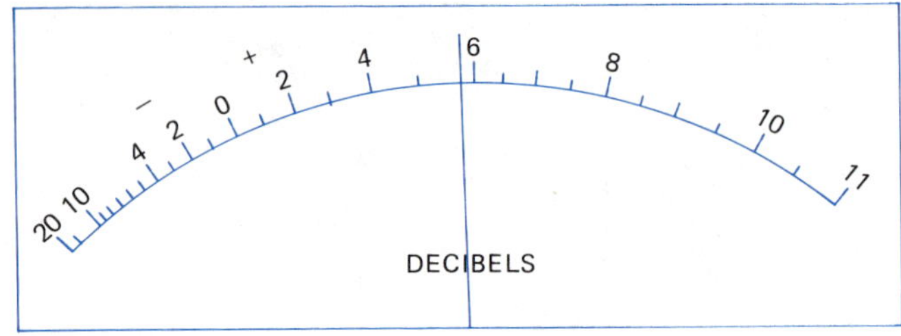

19.

NOTE: READ LEFT TO RIGHT. READ LARGEST DIGIT EACH POINTER EXCEEDS.

20.

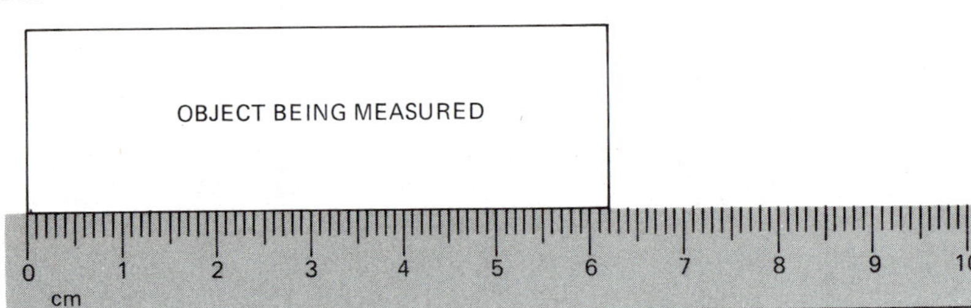

21.

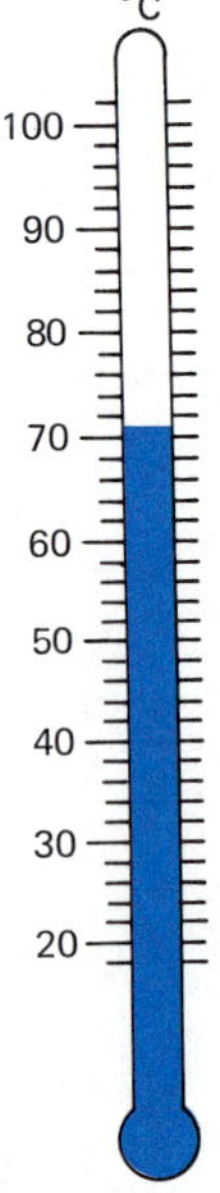

22.

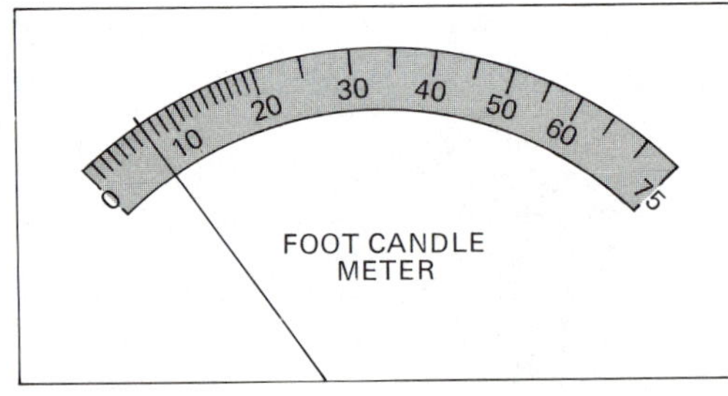

6-2 MAINTAINING ACCURACY IN CALCULATED VALUES

Often, measured values are used in calculations. The sum, difference, product or quotient will usually end up with many more digits than the original measured quantities. For example, the area of a circle with a radius of 1.12 centimetres will yield a calculated area of 3.940 935 04 square centimetres ($\pi = 3.1416$). Such an answer indicates an uncertainty of ± 0.000 000 005 square centimetres.

It is apparent that such a narrow uncertainty is unwarranted. The radius was measured with a much larger uncertainty (± 0.005 cm). Obviously, the answer should be rounded. Answers obtained with a calculator should almost always be rounded. The rules for rounding were given in Chapter 3.

There are two rules for determining the number of places to round the calculated answers. The rule depends on the type of calculation.

ROUNDING CALCULATED ANSWERS

- *Addition and Subtraction.* The answer is rounded to the place value of the last significant digit in the measurement with the greatest uncertainty.
- *Multiplication, Division, Powers, Roots and Reciprocals.* The answer is rounded to the number of significant digits contained in the factor having the least number of significant digits.

Example A Find the perimeter (distance around) of the rectangular object shown in Figure 6-1 and Figure 6-2. Use the measurements obtained by the rule and the micrometer caliper. The length is 49 mm and the width is 25.39 mm. The formula for perimeter is: $P = 2(L + w)$.

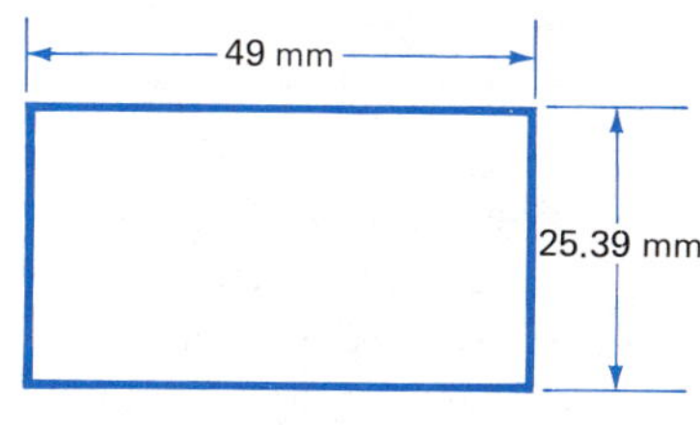

Figure 6-3

Solution: Diagram: See Figure 6-3.

$P = 2(L + w)$ (Formula)

$P = 2(49 \text{ mm} + 25.39 \text{ mm})$ (Substitute)

$P = 148.78 \text{ mm} \approx 149 \text{ mm}$

Note: The answer is rounded to the nearest millimetre because the length is known only to the nearest millimetre. The constant (2) is an exact number and has no uncertainty.

Example B Find the area of the rectangular object shown in Figure 6-1 and Figure 6-2. The length is 49 mm and the width is 25.39 mm. The formula for the area of a rectangle is: $A = Lw$.

Solution: See Figure 6-3.

$A = Lw$ (Formula)

$A = (49 \text{ mm})(25.39 \text{ mm})$ (Substitute)

$A = 1244.11 \text{ mm}^2 \approx 1200 \text{ mm}^2$

Note: The answer is rounded to two significant figures because the length has only two.

It is the *largest uncertainty* that carries through to sums and differences. With products, quotients, powers, roots and reciprocals it is the *smallest number* of significant figures.

EXERCISE 6-2

Perform the indicated operations with a calculator on the following measured quantities. Round all answers to the correct number of significant digits.

1. 649.0 V – 17.385 V
2. 44 Ω + 56.2 Ω + 10.05 Ω
3. 0.6253 A – 0.015 681 A
4. (72.91 m)(20.9 m)
5. (12.0 V)(0.002 A)
6. 2(3.141 59)(60.0 hertz)
7. $\dfrac{24.0\ \text{V}}{1084\ \Omega}$
8. $\dfrac{745.7\ \text{W}}{1\bar{1}0\ \text{V}}$
9. $\dfrac{(2\bar{4}0\ \text{V})^2}{1200\ \Omega}$
10. 35.6 ft + 28.00 ft + 8.753 ft + 100.00 ft

Draw diagrams if applicable, state the formula, substitute numbers and evaluate the following problems. Express the answer with the correct number of significant digits. Use the given formula.

11. The resistances in a series circuit are 645 Ω, 54.3 Ω, and 1600 Ω. Find the total resistance of this series circuit by adding the three resistances.
12. Find the current through a resistance of 432.8 Ω when the voltage is 12 V. ($I = E/R$).
13. Find the power dissipated in a circuit with a resistance of 3.4×10^6 Ω and a current of 7.39×10^{-4} A. ($P = I^2 R$).
14. A fluorescent lamp is 3.7 cm in diameter (d). The length (L) is 2.35×10^2 cm. Find the surface area (S) in square centimetres. ($S = \pi dL$).
15. In the triangle shown in Figure 6-4, $Z = \sqrt{R^2 + X^2}$. Find the length of Z if R is 2.8×10^2 metres and X is 3.915×10^2 metres.
16. A battery can move 1.435×10^2 coulombs of electric charge (Q) through a circuit in a time of 7.4 seconds. What is the current? ($I = Q/t$).
17. Copper has a resistivity coefficient (ρ) of 17.2×10^{-7} Ω•cm. Find the resistance (R) in ohms of a wire with a length (L) of 2.00×10^4 cm of 00 gage copper wire if the radius (r) is 0.4633 cm. The resistance is given by the formula: $R = \dfrac{\rho L}{\pi r^2}$.
18. A 45-foot southern yellow pine, Class 1 pole has a minimum radius of 14.1 inches at a point 6 feet from the butt end. Imagine a cut through the pole at that point. What would be the area of the cut? ($A = \pi r^2$).

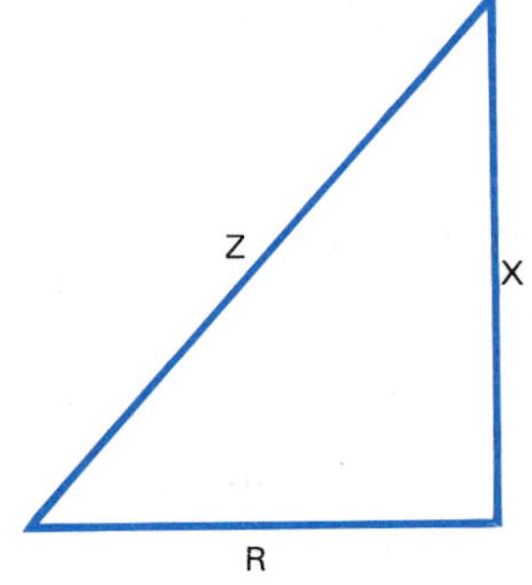

Figure 6-4

THE SI MEASURING SYSTEM

A consistent measuring system defines the measurement of a few *basic quantities* in fundamental units. *A fundamental unit* is defined so that no other units are involved. The metre is the unit of distance. Its value does not depend on time units, mass units, speed units or any others. The metre is independently defined.

The unit of time (second) and the unit of mass (kilogram) are also fundamental units. The metre, kilogram and second are the fundamental units for the study of mechanics. The units of all other mechanical quantities can be expressed in terms of those three. Nonfundamental units are called *derived units.*

Derived units are sometimes expressed as compounds of the metre, kilogram and second. Square metres (m^2), metres per second (m/s) and kilograms per cubic metre (kg/m^3) are a few examples. The more complex compound units are given a different name. The unit of force (pull of gravity), for example, has the compound ($kg \cdot m/s^2$). This complex compound is called the newton. Most such compounds are named after famous scientists.

Three fundamental units are required to derive all mechanical units. Additional fundamental units are needed in the study of heat, molecular physics, illumination and electricity. Altogether, there are seven fundamental units.

The system of measurement described in the previous paragraphs is called the *SI* (Système International) *measuring system.* With minor changes, it has replaced what was called the *mks* (metre, kilogram, second) *system* or the *practical system.* The latter name was derived from the practical size of most of the units.

English units are used occasionally in this book since they are widely used in the United States. The main emphasis, however, is on the SI system. Most countries of the world have now converted to the SI system.

6-3 METRIC UNITS

The metric units used most frequently are the electrical units. When those units were introduced in Chapter 2, several mechanical quantities were also discussed. Distance, mass, time, force, energy and power were all mentioned in that chapter.

Other quantities and units can be found in some of the practical problems used in the first five chapters. To mention a few: sound level (decibels), temperature (degrees Celsius), luminous flux (lumens), area (square metres), volume (litres), frequency (hertz) and density (kilograms per cubic metre).

A list of important quantity names and symbols and their SI unit names and symbols are given in Figure 6-5. Use care not to intermix the quantity symbols with the SI unit symbols. The two sets of symbols are exactly correct as shown. Do not substitute lowercase letters for capital letters or vice-versa. The SI system cannot be improved.

The quantity name symbols represent the quantity. They are used in formulas to show relationships between quantities. In contrast, unit symbols are used with numbers to represent specific measurements. Again, do not intermix the two sets of symbols.

QUANTITY NAME	QUANTITY SYMBOL	BASE SI UNIT	UNIT SYMBOL
Area	A	square metre	m^2
Capacitance	C	farad	F
Charge	Q	coulomb	C
Conductance	G	siemens	S
Current	I	ampere*	A
Distance	s	metre*	m
Force	F	newton	N
Frequency	f	hertz	Hz
Inductance	L	henry	H
Luminous Flux	F	lumen	lm
Luminous Intensity	I	candela*	cd
Mass	m	kilogram*	kg
Power	P	watt	W
Resistance	R	ohm	Ω
Temperature	T	kelvin*	K
Time	t	second*	s
Voltage	E	volt	V
Volume	V	cubic metre (litre)	m^3 (L)
Work-energy	*W*	joule	J
*Fundamental units			

Figure 6-5 Important quantities, SI units and symbols

To familiarize the reader with the metric system, the fundamental units are discussed briefly.

Distance. In the SI system, the *metre* (m) is a fundamental unit. It is defined in terms of the wavelength of light with a precision equivalent to two ten-millionths inch. The metre can be visualized as being about 3 1/3 inches longer than a yard.

Time. The fundamental unit of time is the familiar *second* (s). It is also defined with great precision in terms of light waves. The hour (h) and minute (min) are derived units.

Mass. Mass is a measure of the amount of matter an object contains. The fundamental unit is the *kilogram* (kg). It is the mass of the international kilogram (a platinum-iridium cylinder) kept at the International Bureau of Weights and Measures in Sevres, France. The *litre* (L) is a volume unit and is defined as the space occupied by one kilogram of water. A kilogram mass weighs about 2 1/5 pounds.

Kilogram Molecular Weight. The unit of molecular substance is the *kilomole* (kmol). It is one of the seven fundamental units but will not be used in this text.

Electrical Current. The unit of current is the *ampere* (A). The ampere is defined as the rate of electron flow in two long wires one metre apart that causes a magnetic force of 2×10^{-7} newtons for each metre of length. Although this is the official definition, it was described in Chapter 2 as a number of electrons flowing past a point each second.

Temperature. The fundamental unit is the *kelvin* (K). On this scale, the freezing point of water is about 273 K. The boiling point is about 373 K. The kelvin scale has the same size divisions as the Celsius scale. The degree mark (°) is not used with kelvins.

Luminous Intensity. The unit of luminous intensity is the *candela* (cd). This unit is defined by an international standard light source. It is approximately the same as the intensity of a burning candle.

The above mentioned units form the basis of the SI measuring system. All others are derived from them. Many important derived units are listed in Figure 6-5. Derived units can be found by two methods. The base unit of other quantities are worked out from the definition of the quantity. For example, velocity equals distance divided by time. Therefore, the units are: metres divided by seconds or metres per second (m/s). Such units are called compound units. Another way to derive units is to attach metric prefixes to the base unit. Prefixes are discussed next in this chapter.

EXERCISE 6-3

Match the following quantities with the correct metric unit.

1. Emf	a. ampere
2. Electric charge	b. kelvin
3. Distance	c. farad
4. Electric current	d. volt
5. Mass	e. litre
6. Resistance	f. coulomb
7. Time	g. ohm
8. Capacitance	h. metre
9. Temperature	i. kilogram
10. Volume	j. second

Match the following SI units with the correct unit symbols: Note: some are not used.

11. ampere	a. m	k. s
12. second	b. V	*l.* R
13. metre	c. p	m. W
14. coulomb	d. Ω	n. L
15. volt	e. I	o. C
16. watt	f. t	p. f
17. farad	g. Q	q. E
18. hertz	h. *W*	r. F
19. ohm	i. A	s. J
20. joule	j. Hz	t. N

Match the correct quantity symbols with the following quantity names. Some letters are not used.

21. Emf	a. I	k. J
22. Electric charge	b. e	*l.* R
23. Distance	c. *W*	m. A
24. Electric current	d. Q	n. q
25. Mass	e. Ω	o. N
26. Resistance	f. W	p. E
27. Time	g. s	q. V
28. Capacitance	h. t	r. L
29. Power	i. C	s. m
30. Energy	j. kg	t. P

6-4 METRIC PREFIXES

Some high voltage transmission lines have 161 000 V across the lines. A typical small capacitor is rated at 0.000 000 000 485 F. Coal fired power plants develop power in excess of 800 000 000 W. Such large and small measurements can be expressed in a less cumbersome way by using scientific notation. An even better way is to use larger and smaller units.

Larger and smaller units are formed in the metric system by attaching prefixes to the basic unit. Metric prefixes represent power of ten multiples or submultiples. A complete list of the prefixes, the power of ten multiples and the prefix symbols is given in Figure 6-6. As with quantity symbols and unit symbols, prefix symbols should be used exactly as shown. Do not interchange capital and lowercase letters.

Large and small units of any quantity such as those listed in Figure 6-5 can be formed by attaching prefixes. A centimetre (cm) is one-hundredth of a metre. A millivolt (mV) is one-thousandth of a volt. A kilowatt is one thousand watts. A megohm is one million ohms. Any prefix can be attached to any metric base unit to form new units.

MULTIPLICATION FACTOR	PREFIX	SYMBOL
10^{18}	exa	E
10^{15}	peta	P
10^{12}	tera	T
10^{9}	giga	G
10^{6}	mega	M
10^{3}	kilo	k
10^{2}	hecto	h*
10^{1}	deka	da*
10^{0} (base unit)		
10^{-1}	deci	d*
10^{-2}	centi	c*
10^{-3}	milli	m
10^{-6}	micro	μ
10^{-9}	nano	n
10^{-12}	pico	p
10^{-15}	femto	f
10^{-18}	atto	a

*Avoid these prefixes. The unit centimetre is acceptable.

Figure 6-6 Metric prefixes, multiplication factors and symbols

6-5 RULES FOR SI USAGE

The International Measuring System (SI) is constantly being improved by occasional conferences of the International Bureau of Weights and Measures. The following rules have recently been adopted. They will be strictly adhered to throughout the rest of this text.

RULES FOR SI USAGE

SI Prefixes and Units

- Prefixes and unit names (except Celsius) are never capitalized except at the beginning of a sentence.
- Unit symbols are capitalized if they are derived from a person's name, otherwise they are not capitalized. Only litre (L) is an exception. When in doubt, refer to Figure 6-5 and use them exactly as shown.
- Some prefix symbols are capitalized: exa (E), peta (P), tera (T), giga (G) and mega (M). None of the others are capitalized. Refer to Figure 6-6 and use them exactly as shown.
- The final vowel in a prefix is omitted in only two cases of interest: megohm and kilohm. Otherwise, both vowels are retained and pronounced. For example: milliampere.

- Unit symbols are never pluralized by adding an -s. The symbol is the same for one or for several units.
- A period is never placed after a unit symbol except at the end of a sentence.

Compound Units

- When expressing a compound unit such as metres per second in symbols, use the slash (m/s). When units combine as a product, use the multiplication dot (N•m).

Numbers

- When writing large numbers and small decimal fractions, separate the groups of three numerals with a blank space instead of a comma.
- When expressing a decimal fraction, always place a zero before the decimal point.

Measurements

- Leave a space between a number and the unit symbol.
- Use the unit symbol only when used with a number such as 5 MW. Write out the name of the unit in full when it is not used with a number.
- Never use a prefix alone to indicate the unit. For example: 5 k could mean 5 km, 5 kg, 5 kΩ, etc.; 5 K (capital K) means 5 kelvins (a temperature).

EXERCISE 6-4

Match the following metric prefixes with the correct prefix symbol.

1. tera	a. M	k. π
2. giga	b. n	*l.* N
3. mega	c. c	m. t
4. kilo	d. K	n. m
5. deci	e. g	o. C
6. centi	f. p	p. T
7. milli	g. d	q. P
8. micro	h. D	r. k
9. nano	i. μ	s. *W*
10. pico	j. G	t. Ω

What is wrong with each of the following sentences according to the rules for SI usage? Each has at least one error.

11. The abbreviation for Ampere is a.
12. A transformer has a mass of 985 KG.
13. An electric element has a resistance of 45 K.
14. An emf has a value of 220V.
15. The cost of 4 kW hours of electrical energy is 36¢.

16. A current of 7 amperes can also be written: seven A.
17. The reciprocal of 100,000,000 is .00000001.
18. Five L. of water has a mass of 5 kgs.

Write the name of the following units in full.

19.	a. μA	b. kV	c. MΩ	d. ms	e. pF
20.	a. MW	b. μF	c. km	d. kW•h	e. mg

Write the symbol for each of the following units.

21. a. millivolts
    b. kilohms
    c. microsecond
    d. kilogram
22. a. nanofarads
    b. centimetre
    c. megajoule
    d. microcoulomb

6-6 METRIC TO METRIC CONVERSIONS

Any metric prefix can be used with any metric base unit. Any metric unit of a given quantity is related to all other metric units of that quantity by a power of ten. It is easy, therefore, to convert from one metric unit to another. Simply move the decimal point or change the power of ten.

A conversion problem consists of converting a measurement expressed in certain units to different units. The original measurement is not changed in the conversion process. For example a 6.25-ft length of an object can be converted to 75.0 in. Either expression is a measure of the length of the object.

There are two simple rules which will help make a quick mental check of answers to conversion problems.

CHECKING ANSWERS TO CONVERSION PROBLEMS

- If a quantity is changed to smaller sized units, the numerical part must be increased.
  a. 3.5 ft = 42 in
  b. 25 m = 2500 cm
  c. 0.0003 kg = 0.3 g
- If a quantity is changed to larger units, the numerical part must be decreased.
  a. 150 min = 2.5 h
  b. 35.4 mm = 0.0354 m
  c. 400 000 μs = 400 ms

Notice the numerical part of these examples are changed by exactly the same factor by which the units are changed.

The number of places the decimal point must be moved is determined from a table of prefixes, Figure 6-6. Figure 6-7 may be a helpful aid in converting units. The following examples illustrate the process.

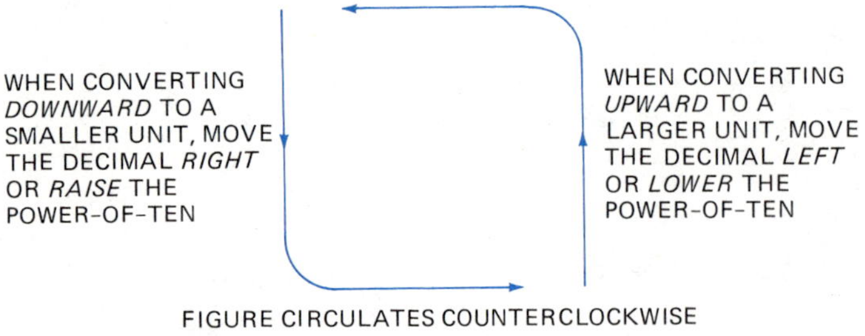

WHEN CONVERTING *DOWNWARD* TO A SMALLER UNIT, MOVE THE DECIMAL *RIGHT* OR *RAISE* THE POWER-OF-TEN

WHEN CONVERTING *UPWARD* TO A LARGER UNIT, MOVE THE DECIMAL *LEFT* OR *LOWER* THE POWER-OF-TEN

FIGURE CIRCULATES COUNTERCLOCKWISE

Figure 6-7 Converting aid for metric units

Example A Convert 35 $\bar{0}$00 to megawatts.

Solution: The conversion is to a larger sized unit so the decimal must be moved to the *left.* The prefixes kilo and mega mean 10^3 and 10^6, respectively. There is a three decimal place difference so the decimal must be moved *three places,* $(6 - 3 = 3)$. Therefore:

35 $\bar{0}$00 kW = 35.0 MW.

Example B Convert 2.74 mV to microvolts.

Solution: Conversion to a smaller sized unit means the decimal must be moved to the *right.* The prefixes milli and micro mean 10^{-3} and 10^{-6}, respectively. The decimal point must be moved *three places:* $-6 - (-3) = -3$.

2.74 mV = 2740 μV.

Example C Convert 1.61×10^{10} mΩ to megohms.

Solution: A change to larger units requires a *decrease* in the power of ten. From milli (10^{-3}) to mega (10^6) is nine places: $-3 - (+6) = -9$.

1.61×10^{10} mΩ = 1.61×10^1 MΩ.

Example D Change 1.7×10^{-4} A to microamperes.

Solution: *Increase* the power of ten by six places. The change is from the base unit (10^0) to micro (10^{-6}): $0 - (-6) = 6$.

1.7×10^{-4} A = 1.7×10^2 μA.

EXERCISE 6-5

Express the following measurements in the units indicated. Give the answer in decimal form.

Measurement	(a)	(b)
1. 435 000 Ω =	________ kΩ	________ MΩ
2. 0.008 58 V =	________ mV	________ μV
3. 38.4 A =	________ mA	________ μA
4. 395 mA =	________ A	________ μA
5. 75 μF =	________ pF	________ F
6. 11 650 MΩ =	________ kΩ	________ Ω
7. 0.0655×10^{-5} s =	________ ns	________ ms
8. 108.6 MHz =	________ kHz	________ GHz
9. 75 200 mH =	________ H	________ μH
10. 564×10^6 W =	________ kW	________ MW

Express the following in the units indicated. Give the answers in standard scientific notation (SSN) rounded to three significant figures.

	Measurement	(a)	(b)
11.	$4.1315 \times 10^5\ \Omega$ =	________ kΩ	________ MΩ
12.	0.000 540 0 μF =	________ pF	________ mF
13.	0.001 493 2 A =	________ mA	________ μA
14.	4.3864×10^2 mW =	________ kW	________ MW
15.	85.63×10^{-7} kA =	________ mA	________ μA
16.	154 300 kHz =	________ Hz	________ MHz
17.	$3.451 \times 10^3\ \mu$H =	________ mH	________ H
18.	0.1716×10^5 kW·h =	________ W·h	________ MW·h
19.	0.785 135 MΩ =	________ kΩ	________ Ω
20.	$0.005\ 555 \times 10^5$ V =	________ mV	________ kV

6-7 COMMON UNITS

Another way to express measurements is in common units. *Common units* for a measurement is formed by writing it as a number between 1 and 1000. The unit then uses a prefix for a multiple-of-three exponent. This is easily accomplished by changing to engineering notation using the calculator.

Example A Convert 64 200 pF to common units.

Solution: Change the existing prefix to a power of ten and combine with the numerical part: $64\ 200 \times 10^{-12}$ F. Rewrite the numerical part in engineering notation: 64.2×10^{-9} F. Change the power of ten to a metric prefix.

64 200 pF = 64.2 nF

Example B Change 204 000 000 ms to common units.

Solution: 204 000 000 ms

$= 204\ 000\ 000 \times 10^{-3}$ s

$= 204 \times 10^3$ s

= 204 ks

Example C Change 0.0724×10^{-4} GW to common units.

Solution: 0.0724×10^{-4} GW

$= 0.0724 \times 10^{-4} \times 10^9$ W

$= 7.24 \times 10^3$ W

= 7.24 kW

EXERCISE 6-6

Express the following measurements in common units.

1. 5200 V
2. 0.006 85 A
3. $7.76 \times 10^5\ \Omega$
4. $3.95 \times 10^7\ \Omega$
5. 99 600 kHz
6. 57 600 ms
7. 0.229 MΩ
8. 0.01×10^{-3} s
9. 5540 mH
10. $100 \times 10^5\ \Omega$

11. $46\,300 \times 10^1$ V
12. 0.004 575 W
13. 285×10^6 mW
14. 8500×10^0 Ω
15. 104.6 MHz
16. 0.004 75 kA
17. $34\,950 \times 10^{-10}$ F
18. 2.25×10^3 μH
19. 7.65×10^{-7} V
20. $49\,500 \times 10^1$ W

Express the following measurements in common units rounded to three significant figures.

21. 435 000 Ω
22. 0.008 58 V
23. 38.4 A
24. 395 mA
25. 75 μF
26. 11 650 MΩ
27. 0.0655×10^{-5} s
28. 108.6 mHz
29. 75 200 mH
30. 564×10^6 W
31. 4.1315×10^5 Ω
32. 0.000 540 0 μF
33. 0.001 493 2 A
34. 4.3864×10^2 mW
35. 85.63×10^{-7} kA
36. 154 300 kHz
37. 3.451×10^3 μH
38. 0.1716×10^5 kW·h
39. 0.785 135 MΩ
40. $0.005\,555 \times 10^5$ V

State the name of the quantity being measured for each of the following measurements. Rewrite each of them rounded to three significant digits in (a) the unit indicated in parentheses, (b) the base SI unit (no prefix except kg) and (c) the common unit.

41. 8.0951×10^{-4} MΩ (mΩ)
42. 1.921×10^3 mA (kA)
43. 1.902×10^5 mV (μV)
44. 0.002 506 mF (nF)
45. 0.001 34 kV (mV)
46. 108.4×10^6 Hz (GHz)
47. 2.038×10^{-4} ms (μs)
48. 506 400 μA (nA)
49. 0.000 000 403 7 GH (mH)
50. 3 834 000 mm (hm)
51. 32.85×10^{-4} mg (dg)
52. 0.000 400 00 mS (kS)
53. $200\bar{0}$ K (mK)
54. 3.156×10^7 s (ks)
55. 7.457 kW (mW)
56. 360.0×10^4 J (GJ)
57. 5.50×10^2 nm (mm)
58. $10\,00\bar{0}$ L (mL)
59. 1.602×10^{-16} mC (aC)
60. 101 325 Pa (MPa)
61. 3×10^8 m/s (km/s)
62. 161 000 V (MV)
63. 39 000 kW (MW)
64. 8×10^4 A (MA)
65. 34×10^8 mΩ (MΩ)

6-8 MORE ON CONVERTING UNITS

Converting from metric to metric units involves multiplication or division of multiples of ten. This is accomplished by simply moving the decimal point. Converting from British to British, British to metric or metric to British is a different matter. These involve conversion factors that are not multiples of ten.

Conversion factors have the form described in the following:

1 mi = 1760 yd
1 L = 0.264 gal
1 kg = 2.205 lb

The factor on either side of the equal sign can be divided by the other. The resulting ratio of the two equal quantities is equal to 1 (one). For example, the first conversion factor listed above becomes:

$$\frac{1 \text{ mi}}{1760 \text{ yd}} = 1 \text{ or } \frac{1760 \text{ yd}}{1 \text{ mi}} = 1$$

When a number is multiplied by the number one, the product is just the number itself. No change occurs. Multiplication of a quantity by a ratio equal to one also results in no change. A transformation of units is the only result. The following example illustrates this. This method should leave no doubt in one's mind about whether to multiply or divide by the conversion factor. A table of conversion factors is given in the Appendix.

Example A Convert a distance of 64.5 m to feet.

Solution: Find the conversion factor in the tables.
It is: 1 m = 3.281 ft.

Transform the conversion factor to a ratio and multiply it by the distance being converted.

$$64.5 \cancel{\text{m}} \left(\frac{3.281 \text{ ft}}{1 \cancel{\text{m}}}\right) = 212 \text{ ft}$$

Notice that metres was placed in the denominator so it would cancel the metres in the given quantity. The numbers are multiplied.

Another conversion factor (1 ft = 0.3048 m) is also given in the tables. An equivalent solution to the problem by dividing is:

$$64.5 \cancel{\text{m}} \left(\frac{1 \text{ ft}}{0.3048 \cancel{\text{m}}}\right) = 212 \text{ ft}$$

PROCEDURE FOR CONVERTING MEASUREMENTS

- Write down the measurement to be converted.
- Draw parentheses with a bar.
- Insert old units above or below the bar so they cancel with the given units. Insert the new units in the remaining space.
- Obtain conversion factor from table and insert in the parentheses.
- Evaluate with a calculator. A factor in the numerator is a multiplier. One in the denominator is a divisor.
- Write down the equivalent measurement with the new units.

Units cancel and obey other arithmetic procedures. They do so independently of the numerical part of the problem. Work with numbers and units separately.

Example B Find the number of seconds in three weeks.

Solution: Here, the conversion factor from weeks to seconds is not given in the table. A conversion to intermediate sized units is necessary.

$$3\ \text{wk}\left(\frac{7\ \text{da}}{1\ \text{wk}}\right)\left(\frac{24\ \text{h}}{1\ \text{da}}\right)\left(\frac{3600\ \text{s}}{1\ \text{h}}\right) = 1\ 810\ 000\ \text{s}$$

Notice the given quantity is multiplied by conversion ratios three times. The ratios are set up so all units cancel except that of the final answer.

Example C Convert a speed of 60 mi/h to feet per second.

Solution: Here, there is one unit (miles) in the numerator and one unit (hour) in the denominator. Conversion factor ratios must be set up so these units cancel.

$$60\frac{\text{mi}}{\text{h}}\left(\frac{5280\ \text{ft}}{1\ \text{mi}}\right)\left(\frac{1\ \text{h}}{3600\ \text{s}}\right) = 88\ \text{ft/s}$$

Notice that hours is placed in the numerator of the second conversion ratio in order to cancel hours in the original expression.

Always set up conversion problems (even simple ones) on paper in the manner illustrated. In the long run, time and effort will be saved.

EXERCISE 6-7

Use the conversion table in the Appendix. Convert the following measurements as indicated. Use the method described above. Round all answers according to the significant digits of the given measurements.

1. A 100 ft surveying tape standardized at 60°F increases in length by 0.2225 cm when used on a certain hot summer day. The tape increased by how many inches?
2. The span between poles on an electric transmission line is 350 ft. How many metres is this?
3. An electron weighs 2×10^{-30} lb. How much is this in kilograms?
4. The mass of an armature of a small electric motor is 2.35 kg. What is its weight in pounds? In newtons?
5. An electric motor that drives a press drill runs at 1750 r/min. It makes one revolution every 0.000 571 4 min. Express this time in decimal form in seconds and milliseconds.
6. A capacitor is made of two sheets of metal foil; each has an area of 600 in^2. Find the area of the two sheets in square centimetres.
7. A large transformer uses 245 gal of oil as a heat transfer medium. How many litres is this?
8. Sulfuric acid for batteries is transported in carboys that contain 18.93 L each. How many gallons does each carboy contain?

9. Which of the following is the largest volume?

a. 5200 mL	c. 0.004 15 m^3	e. 500 in^3
b. 18 pts	d. 0.25 ft^3	f. 1.5 gal

10. An electron in a hydrogen atom completes one orbit in 1.49×18^{-16} s. The distance traveled in one orbit is 3.32×10^{-10} m. Express these quantities in common units.

11. A certain brand of solder sells for 9.82 $/lb (dollars per pound). What is the cost in dollars per kilogram at the same rate?

12. The speed of a 60-Hz electric wave in a transmission line is 1.241×10^5 mi/s. What is the speed of the wave in kilometres per hour?

13. In the construction of a hydroelectric plant, water is to be fed to a turbine at the rate of 105 000 L/s. Express this rate in cubic feet per minute.

14. Speed is found by dividing distance by time. Find the speed of the hydrogen electron in Problem 10 in metres per second. Convert the answer to miles per hour.

15. A substation transformer, Figure 6-8, is cooled by two fans. Each fan moves 650 ft^3/min of air through the heat exchanger. How much air is moved by both fans in litres per second and cubic metres per hour?

Figure 6-8 A substation transformer: Two thermocontrolled fans circulate air through the heat exchanger to prevent overheating.

CHAPTER 7

USING ELECTRICAL FORMULAS

OBJECTIVES

After satisfactorily completing this chapter, the student should be able to:

- Define various words and terms introduced in this chapter.
- Solve simple equations, electrical formulas and geometrical formulas using the multiplication and division axioms.
- Apply the technique of problem solving to electric circuit problems, area-volume problems and wire sizing problems.

Imagine the following situation. A customer calls an electrician about a home electrical problem. The electrician must determine the cause of the complaint before attempting repair. Naturally, the customer will describe the problem. Additional facts can be acquired by asking the right questions: "Has there been lightning in the area?"; "Do the neighbors have electricity?"; "How long has it been this way?"; "Did you check the circuit breakers?", and "Was there any smoke or odors?".

If the trouble cannot be pin-pointed with the information now known, more will have to be obtained. Visual checks of the distribution center, outlet boxes, etc. and measurements with ohmmeters, voltmeters and other circuit testers may be necessary. At some point, the electrician recalls past experience with other home circuit problems. Sometimes, a circuit diagram is very helpful. With enough information, the cause of the complaint will be found.

Now assume the problem has been diagnosed. It is time to solve it. Solving electrical problems may require the replacement of a part, a modification of a circuit or an adjustment of some kind. Finally, a test run or some other check is made to make sure the problem has been corrected.

In various technologies or professions the terms "troubleshooting," "diagnosis" and "analysis" mean the same as "problem solving." All of these terms refer to a standard procedure, namely:

- **Gather** information and **data.**
- **Construct** or refer to **diagrams** and pictures.
- **Determine relationships** between different parts.
- **Solve the problem.**
- **Check** equipment or **solution** by conducting a test run.

The technique outlined above is used worldwide for solving every kind of problem imaginable. Policemen, lawyers and army commanders use it. Doctors, auto mechanics and engineers use it. It works! It works for solving mathematical problems, too. After a discussion about using formulas, the procedure will be used for solving many kinds of electrical problems.

FORMULAS

Many problems in electricity are mathematical in nature. Of these, most are solved by evaluating a quantity using a formula. A *formula* is a rule or law that describes the mathematical relationship between quantities. The most common electrical formula is Ohm's Law, Chapter 3.

In addition to Ohm's Law, there are many other electrical formulas. There are also many nonelectrical formulas that are important. Most electricians agree that the ability to solve formulas using algebra and arithmetic is essential. Algebra is an extension of arithmetic using letters to represent various values of quantities. Numbers are used to represent specific values. Letters are used to represent changing values.

7-1 SIGNS AND SYMBOLS

In algebra, letters or symbols represent the quantity. Such letters are called *literal numbers.* The quantity represented by the letter is called a *variable.* Actual values of the quantity can be substituted for the letter.

Symbols. Numbers and some letters may represent a *constant.* Pi (π) in the formula for circumference of a circle is a universal constant.

$$C = \pi d$$

The value of a constant never changes under any circumstances. Circumference (C) and diameter (d) are variables.

For a given electrical circuit, resistance is usually constant. The variable current will change if the voltage is changed. For that particular circuit, resistance will not change. Some letters in formulas may be constant for a given material. These are properties of matter. Such a constant is usually called a coefficient or a modulus.

GREEK LETTER	CAPITAL	LOWER CASE	QUANTITY
Alpha		α	Angles, Temperature coefficient
Beta		β	Angles
Gamma		γ	Angles
Delta	Δ		Change in a quantity
Theta		θ	Angles
Lamda		λ	Wavelength
Mu		μ	Micro
Pi		π	3.1416 (rounded)
Rho		ρ	Resistivity coefficient
Sigma	Σ		Summation
Tau		τ	Torque
Phi		ϕ	Angles
Omega	Ω	ω	Ohms – Angular velocity

Figure 7-1 Important Greek letter symbols

In algebra letters are used to represent numbers. Letters used in this way obey all the rules of arithmetic. The letters used for the quantity symbols are important in the study of electricity. Some of the Greek letters are commonly used in electricity as shown in Figure 7-1. The complete Greek alphabet is given in the Appendix.

Subscripts. Sometimes a given letter is used more than once in a formula. The three-resistor circuit, Figure 7-2, is an example. To distinguish the three, subscripts are used. A *subscript* is a number, letter or group of letters written in *smaller print* and placed to the right and below a literal number. In a given problem, letters with subscripts are usually constants. It is important to use smaller sized print for subscripts. Do not confuse the concepts of subscripts and superscripts. Superscripts are placed to the right and above a number. They represent powers of numbers. The symbols R_2, R^2 and 2R are not the same. Write them carefully.

The resistances in Figure 7-2 and other letters with subscripts are read in the following way:

R_1 is read "R one" or "R sub-one."
R_2 is read "R two" or "R sub-two."
R_3 is read "R three" or "R sub-three."
I_{min} is read "I minimum" (Minimum current)
E_{max} is read "E maximum" (Maximum voltage)
X_C is read "X sub C" (Capacitive reactance)
X_L is read "X sub L" (Inductive reactance)
d_{av} is read "d average" (Average diameter)
T_f is read "T final" (Final temperature)
T_i is read "T initial" (Initial temperature)

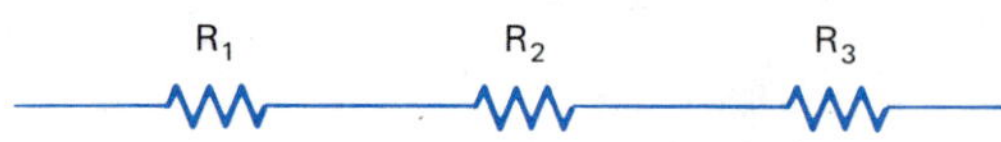

Figure 7-2 Different resistors are indicated by using subscripts.

If the letters denote an ac quantity, use lowercase letters for the subscript. For DC quantities, use capitals.

Delta. The capital Greek letter delta (Δ) is sometimes combined with the symbol for a quantity. The combination is then used to indicate the change in or a difference in that quantity. For instance, ΔT means: $\Delta T = T_f - T_i$ or $\Delta T = T_2 - T_1$. The symbol ΔT means a small change or difference in temperature and does not mean a quantity Δ times a quantity T. Other examples of the use of the letter Δ are: ΔL means a change in length, ΔP means a change in power and ΔE means a change in voltage. Be sure you understand the use of the Greek letter delta. It will not have any other meaning.

Signs of Operation and Grouping. An algebraic expression is like an arithmetic expression. It is a combination of literal numbers, numbers, exponents and radicals. The expression is tied together using grouping symbols and operation signs. Grouping symbols include parentheses, brackets, braces and the bar. Signs of operation are +, –, ×, ÷. Of the four multiplication representations: $a \times b$, $a \cdot b$, (a)(b) and ab, only the last two will be used. The first two are easily confused with the letter *x* and the decimal point. Parentheses are used for numbers or groups: (3.70)(6.31). No operational sign is used most often to express multiplication but only if there is no more than one number in the expression: $35a$, 6IR, xy, $2\pi f$.

The division symbol ÷ is not used. Both the bar and slash symbols are used for division but the bar is used most often. The following are examples of several algebraic expressions.

1. $a + 2$
2. $4P$
3. $\frac{1}{2}CE^2$
4. $\pi r^2 h$
5. a/c
6. $(4 - \theta)(4 + \theta)$
7. $(2)(78.2)Qt$
8. $\dfrac{\frac{1}{2}\theta R}{\phi}$
9. $\dfrac{R_1 R_2}{R_1 + R_2}$
10. $2\pi\sqrt{\dfrac{L}{g}}$

The symbols + and – are the only ones used for addition and subtraction. Expressions using either of these signs will not be discussed in this chapter. Only monomial expressions will be considered.

Factors. When a monomial product of two or more numbers or literal numbers is formed, each number is called a *factor*. The expression

$$5ab$$

has three factors (5, a and b). Any one of these is a factor of the product 5ab. Any one will evenly divide the expression. See Chapter 1 for a review of factoring and prime factors of whole numbers. Literal factors of a product are prime factors.

The numerical factor of an expression is called the *coefficient* of the product. The expression 60ϕ has a coefficient of 6 and $2\pi f$ has a coefficient of 2π. If an expression has no numerical part, the coefficient is understood to be 1. For example, EI = 1EI.

EXERCISE 7-1

1. Literal factors of a product are prime factors. Why?

Make a list of all the prime factors of each of the following expressions.

2. $2\pi r$
3. ωt
4. $\frac{1}{2}bh$
5. $30R^2t$
6. I^2R
7. $4E_1I_1$
8. $2\pi fL$
9. $\frac{1}{2}CE^2$
10. πr^2h
11. $2\pi\sqrt{L/g}$
12. $4\pi^2f^2R^2m$
13. E/R
14. P/I
15. $\frac{1}{2\pi fC}$
16. What is the coefficient of each expression in Problems 2–15?

7-2 MULTIPLICATION AND DIVISION AXIOMS

Two mathematical expressions set apart by an equal sign form an *equation.* An equation is a statement that the two expressions or quantities are equal. The equal sign means the expression on one side has the same value as that on the other side. Some examples are:

$$3x = 24 \qquad P = I^2R$$
$$E = IR \qquad \theta + \Phi = 90$$

Equations which have the same solution are called *equivalent equations.* When the terms and factors of an equation are rearranged, the new resulting equation must be equivalent to the original. When an equation is solved, methods used are derived from properties of numbers. These properties are stated as axioms. An *axiom* is a statement that is assumed to be true without proof.

Multiplication Axiom (MA). When equal numbers are multiplied by equal numbers, the products are equal. If a = b, then ac = bc.

Example A Demonstrate the multiplication axiom.

Solution:
(1) $x = 5$
(2) $7x = 7(5)$ (MA, × 7)
(3) $7x = 35$
(4) $7(5) = 35$

Note: The notation in line 2: (MA, × 7) means the multiplication axiom has been used to multiply both sides of the equation in the line above by 7. Also, notice that equations 1, 2 and 3 are equivalent. They have the same solution (5).

Division Axiom (DA). When equal numbers are divided by equal numbers the quotients are equal. If a = b, then a/c = b/c.

Example B Demonstrate the division axiom.

Solution:
(1) $x = 8$
(2) $x/2 = 8/2$ (DA, ÷ 2)
(3) $x/2 = 4$
(4) $(8)/2 = 4$

Note: Equations 1, 2 and 3 are equivalent equations. In line 2, (DA, ÷ 2) means both sides of line 1 have been divided by 2 using the division axiom.

7-3 SOLVING EQUATIONS

An equation or formula can be rearranged. The important thing to remember is that the two expressions must remain equal to each other. If one side is changed, the other side must be changed also. The change must be made in exactly the same amount and in exactly the same way. This process is used to derive other forms of the equation or to find the value of an unknown.

There are several procedures that are useful in rearranging an equation or formula. The purpose of rearranging an equation is to solve it. *Solving an equation* means to find the number which makes the equation true. This number is called the *solution of the equation.* Usually an equation is solved in this order.

- The terms and factors of an equation are rearranged so the unknown is isolated on one side of the equation using axioms.
- The expression on the other side is then evaluated.

If the unknown in an equation is divided by a certain number, multiply both sides by that number. This procedure will solve the equation.

Example A Solve and check: $\frac{x}{3} = 7$

Solution:
$\frac{x}{3} = 7$
$\frac{3x}{3} = 3(7)$ (MA, × 3)
$x = 21$ (Combine and cancel)
$\frac{21}{3} = 7$ (Check)

If the unknown in an equation is multiplied by a certain number, divide both sides by that number to solve for the unknown.

Example B Solve: $8x = 72$

Solution:
$\frac{8x}{8} = \frac{72}{8}$ (DA, ÷ 8)
$x = 9$ (Solution)
$8(9) = 72$ (Check)

Example C Solve: $-5 = 25R$

Solution: $\frac{-5}{25} = \frac{25R}{25}$ (DA, ÷ 25)

$-0.2 = R$

$-5 = 25(-0.2)$ (Check)

Occasionally, the unknown is found in the denominator in the original equation. If this occurs, use the multiplication axiom on the unknown. The effect is to put the unknown in the numerator on the other side of the equation. Now, solve the resulting equation.

Example D Solve: $\frac{24}{x} = 4$

Solution: $\frac{24x}{x} = 4x$ (MA, × x)

$24 = 4x$ (Cancel)

Now solve this equation:

$\frac{24}{4} = \frac{4x}{4}$ (DA, ÷ 4)

$6 = x$ or $x = 6$ (Combine terms)

$\frac{24}{6} = 4$ (Check)

EXERCISE 7-2

Solve and check (using MA):

1. $\frac{a}{5} = 3$
2. $\frac{x}{7} = 2$
3. $5 = R/4$
4. $-8 = \frac{I}{12}$
5. $\frac{E}{2.6} = 4.3$
6. $\frac{\phi}{56} = 129$
7. $0.073 = \theta/13.7$
8. $0.001\ 38 = \frac{\theta}{0.25}$
9. $\frac{144}{C} = 9$
10. $\frac{27}{x} = 9$

Solve and check (using DA):

11. $6a = 42$
12. $16R = 144$
13. $-121 = 11P$
14. $925 = -25I$
15. $0.006\ 31 = 0.006\ 31E$
16. $7.5\phi = 30$
17. $125\theta = -75$
18. $\frac{1}{4}A = 2$
19. $2.1X_C = 847$
20. $625 = R(125)$

Solve and check. Indicate which axiom is used:

21. $3.7R = 25.9$
22. $\frac{-I}{7.3} = 0.137$
23. $\frac{E}{17} = -29$
24. $624X_L = 4368$
25. $\frac{44}{X_C} = 11$
26. $\frac{350}{\phi} = 7$

27. $\frac{1}{4}P = 2\frac{1}{4}$

28. $-27Q = -297$

29. $\frac{B}{8.1} = 7.35$

30. $\frac{A}{624} = 0.009\ 62$

7-4 TRANSLATING WORD PROBLEMS

Many times an electrician will need to solve a problem stated in words. Translation from verbal expressions to algebraic is necessary. This is the chief difficulty in solving word problems. The equation must be formed from the information given. To learn this skill requires conscious effort and considerable amounts of study time.

Certain words indicate that two numbers should be multiplied. Other words indicate division. Some of these words are listed in the following table.

WORDS MEANING MULTIPLICATION	WORDS MEANING DIVISION
Multiplied by	Divided by
Times	Quotient
Product	Ratio
Twice or double	Half
Triple, etc.	Third, etc.
Of (One half of)	

Simple word problems can sometimes be solved by using only arithmetic. As problems get more complex, however, algebra is needed. In learning to use algebra, a start must be made with simple problems. Use algebra no matter how easy the problem is. Solving procedures cannot be learned for more difficult problems without practicing the simple ones first.

Put the calculator aside. It can be used to evaluate the final expression but it will not do algebra. To solve word problems, use the following steps. There is no other procedure that works.

SOLVING WORD PROBLEMS

- Read the problem carefully.
- Choose a symbol such as the letter x to represent the unknown. Write down exactly what the symbol represents.
- Sometimes drawing a diagram will help.
- Write an equation from the information given in the problem. Write in symbols exactly what the problem says in words.
- Solve the equation for the unknown.
- Check the answer to make sure all conditions of the problem are satisfied.

After reading the problem, choose a symbol for the unknown. Write down exactly what it stands for. It is easy to get mixed up. Take the time to write it down. This is a starting point. Many people will stare at a problem and puzzle over it for hours without starting to solve the problem. By beginning with simple problems and doing every step, these skills can be learned.

Example A A number times 25 is 175. Find the number.

Solution: Let x equal the unknown number.

$x(25) = 175$ (Equation)

$x = \frac{175}{25}$ (DA, ÷ 25)

$x = 7$ (Solution)

$(7)(25) = 175$ (Check)

Example B The ratio of a number and 4 is 36.

Solution: Let n be the unknown number.

$\frac{n}{4} = 36$ (Equation)

$n = 36(4)$ (MA, × 4)

$n = 144$ (Solution)

$\frac{144}{4} = 36$ (Check)

EXERCISE 7-3

Express each word problem algebraically and solve.

1. The product of a number and 5 is 65.
2. The ratio of 18 and a number is 9.
3. A number divided by 3 is equal to 17.
4. A frequency (f) multiplied by 6.28 is 314.
5. A voltage (E) when doubled equals 628 V.
6. A current (I) when halved is 0.0046 A.
7. The quotient of a current (I) and 10 is 0.146.
8. Two-thirds of a resistance (R) equals 3500 Ω.
9. A conductor is divided into four equal parts. Each part is 1.4 m long. Find the original length (L).
10. A voltage (E) after being tripled, then halved is 74 V. Find the original voltage.

SOLVING ELECTRICAL PROBLEMS

Formulas are rules or laws that show the relationship between two or more quantities. Such quantities are represented in the formula by letters. In the power formula (P = EI), for example, the letters E, I and P represent the quantities voltage, current and power respectively.

Some formulas may contain constants as well as variables. In the formula for the area of a circle ($A = \pi r^2$) for instance, the number π is a constant. In the formula $X_L = 2\pi fL$, there are three variables or quantities represented by letters: inductive reactance (X_L), frequency (f) and inductance (L). The product 2π is constant.

Solving a formula means transforming the stated formula into an equivalent (mathematically) formula with only the unknown on one side of the equal sign by itself. All other letters and numbers must be on the other side. This process is known as *solving a formula.* Once the formula has been solved, known values can be substituted for the quantities represented by letters. Now the unknown can be evaluated with a calculator and the *problem* will be solved. There are many formulas used in the electrical industry. An electrician should be able to rearrange them quickly and correctly.

7-5 FORMULAS WITH SIMPLE PRODUCTS AND QUOTIENTS

Simple formulas are formulas that have no grouping symbols of any kind. Rearrangement of the letters and constants in any formula must be in accordance with the rules of algebra. For the simple formulas in this chapter, only the multiplication axiom (MA) and the division axiom (DA) are employed. In addition, cancellation and other simplifying procedures are used.

SOLVING SIMPLE FORMULAS

- Remove coefficients and/or divisors of the unknown quantity using the multiplication axiom and/or the division axiom, respectively.
- Simplify.
- It makes no difference whether the unknown is on the right or left. However, most people prefer to place it on the left. This can be done by simply interchanging the two sides.

Example A Solve E = IR for I. Assume the values of E and R are given.

Solution: $E = IR$

$$\frac{E}{R} = \frac{I\cancel{R}}{\cancel{R}} \qquad \text{(DA, ÷ R and cancel)}$$

$$\frac{E}{R} = I$$

$$I = \frac{E}{R} \qquad \text{(Solution)}$$

Note: If the values of E and R were known, you could now substitute them in the solution and evaluate I.

Example B Solve $m = \dfrac{W}{9.8 \times 10^2}$ for W.

Solution:

$$m = \frac{W}{9.8 \times 10^2}$$

$$(9.8 \times 10^2)m = \frac{W\cancel{(9.8 \times 10^2)}}{\cancel{9.8 \times 10^2}} \qquad \text{(MA and cancel)}$$

$$(9.8 \times 10^2)m = W$$

$$W = 9.8 \times 10^2 m \qquad \text{(Solution)}$$

Example C Solve $P = W/t$ for t.

Solution: Note that t is in the denominator to start with. The simplest procedure is to multiply both sides by t, putting it in the numerator on the left side. Then solve the resulting equation.

$$P = \frac{W}{t}$$

$$Pt = \frac{W\not{t}}{\not{t}} \qquad \text{(MA, × t and cancel)}$$

$$Pt = W \qquad \text{(Now solve for t)}$$

$$\frac{Pt}{P} = \frac{W}{P} \qquad \text{(DA, ÷ P and cancel)}$$

$$t = \frac{W}{P} \qquad \text{(Solution)}$$

Example D Solve: $\theta = \frac{-2\pi\alpha^2\beta}{3\phi^2}$ for β

Solution: Remove the product $3\phi^2$ from the denominator using MA and the product $-2\pi\alpha^2$ from the numerator using DA. The minus sign acts as a factor of (−1) in the numerator.

$$\theta = \frac{-2\pi\alpha^2\beta}{3\phi^2}$$

$$\theta(3\phi^2) = \frac{-2\pi\alpha^2\beta\cancel{(3\phi^2)}}{\cancel{3\phi^2}} \qquad \text{(MA, × } 3\phi^2\text{)}$$

$$3\theta\phi^2 = -2\pi\alpha^2\beta \qquad \text{(Cancel)}$$

$$\frac{3\theta\phi^2}{-2\pi\alpha^2} = \frac{\cancel{-2\pi\alpha^2}\beta}{\cancel{-2\pi\alpha^2}} \qquad \text{(DA, ÷ } -2\pi\alpha^2\text{)}$$

$$\frac{-3\theta\phi^2}{2\pi\alpha^2} = \beta \qquad \text{(Cancel)}$$

$$\beta = -\frac{3\theta\phi^2}{2\pi\alpha^2} \qquad \text{(Solution)}$$

It should be noted that multipliers on one side become divisors on the other. Divisors on one side become multipliers when "taken to the other side." These observations can be used to do most of the formula solving manually. These are special formulas without grouping symbols and without addition and subtraction operators. The suggested shortcut applies only to such simple formulas.

EXERCISE 7-4

Solve these formulas for the letter indicated:

Given	*Solve for:*
1. $c = \lambda f$	f and λ
2. $C = 2\pi R$	R
3. $A = \frac{1}{2}bh$	h and b
4. $A = \pi R^2$	π
5. $V = \pi R^2 h$	h
6. $D = W/V$	W and V
7. $d = m/V$	V and m

	Given	*Solve for:*
8.	$P = I^2 Rt$	R and t
9.	$X_C = \dfrac{1}{2\pi fC}$	C and f
10.	$W = \dfrac{1}{2}CE^2$	C
11.	$Q = \dfrac{KA\Delta Tt}{L}$	ΔT and L
12.	$\dfrac{E_1}{I_1} = \dfrac{E_2}{I_2}$	I_2
13.	$X_L = 2\pi fL$	L and f
14.	$T = \dfrac{1}{f}$	f
15.	$\omega = 2\pi f$	f
16.	$R = \dfrac{\rho L}{A}$	A and L
17.	$E_1 I_1 = E_2 I_2$	I_2 and E_2
18.	$\theta = \omega t$	ω and t
19.	$C = \dfrac{Q}{E}$	E and Q
20.	$P = E^2/R$	R

Solve these circuit formulas for the letters indicated:

21.	$E = IR$	I and R
22.	$E = W/Q$	*W* and Q
23.	$Q = It$	I and t
24.	$P = W/t$	*W* and t
25.	$P = EI$	E and I

7-6 PROBLEM SOLVING PROCEDURE

Electricians will be required to solve many math problems as well as electrical ones. Learning to use an organized procedure will provide a skill that will be invaluable in the future. The general technique was briefly described at the beginning of this chapter. The step-by-step procedure provides a framework from which to view, understand and attack a problem.

1. **Gather information and data.** Read the problem carefully. Exactly what is known? What quantities are given? Make a list using the standard symbol for each quantity with the value and units of each. What is the unknown quantity? List it with the others followed by a question mark. At this point look at the units of all the quantities listed. Are they all British? All SI? If not, convert them. With practice, this can be done as the list of data is made.
2. **Construct a diagram.** Whenever possible, a freehand diagram or sketch of the situation (especially circuit diagrams) will help consolidate understanding of the problem and may even suggest how to solve the problem, Figure 7-3. Label the diagram care-

Figure 7-3 The use of circuit diagrams often suggests the solution to electrical problems. (Courtesy of Southeast Nebraska Community College)

fully. Sometimes, the data gathered in Step 1 can be written on the diagram instead of making a separate list.

3. **Determine the relationship between known and unknown quantities.** The relationship is usually a formula. From the list of known and unknown quantities with correct symbols (Step 1), there should be no trouble finding the required formula. Sometimes more than one are used.
4. **Solve the problem.** Solve the formula for the unknown if necessary by rearranging it using algebra. Substitute the known quantities (including units) into the formula. Evaluate the numerical part of the unknown using the calculator. Metric prefixes can be entered using [EE]. All electrical quantities must be entered in SI units. The quantity 39.4 mA, for example, must be entered as 0.0394 A or use the [EE] key: 39.4 [EE] 3 [+/-], 3.94 −2. When the answer is found, round it correctly. Record the answer with the correct units.
5. **Check the answer.** Are the units correct for the unknown quantity? Compare your answer to those given in the answer key in the Appendix. Determine the reasonableness of the answer in problems for which no answer is given.

7-7 ELECTRICAL PROBLEMS (DC)

Many students approach problem solving with a calculator in one hand. The index finger of the other hand is poised, ready to press the number keys. All this, while reading the problem for the first time. Such an approach is a mistake. Set the calculator aside until the latter part of the fourth step.

A shortened version may work on the simplest of problems. However, it is the procedure itself that is being emphasized here. In more complex problems, there is only one method that works. Concentrating on the step-by-step procedure now will pay rich dividends in later chapters. It will prevent many erratic and false starts and eliminate careless errors. Remember, the technique will carry over into the solution of electrical problems as well as mathematical ones.

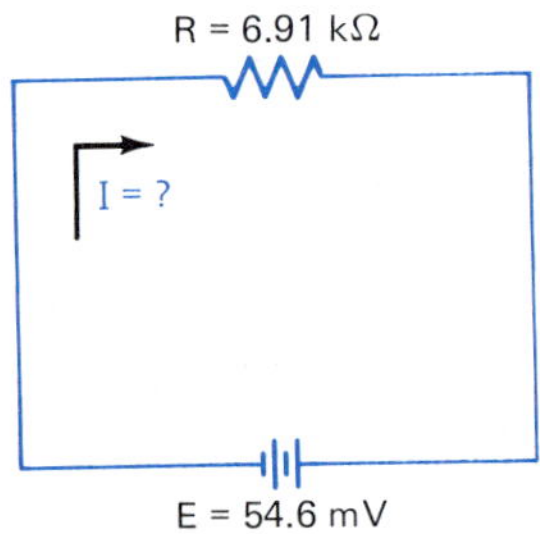

Figure 7-4

Example A An emf of 54.6 mV is applied to a load with a resistance of 6.91 kΩ. Find the current.

Solution:
Data: Data is given on the accompanying diagram, Figure 7-4.
Formula: $E = IR$
Solve: $I = E/R$ (DA, ÷ R)
Substitute: $I = \dfrac{54.6 \text{ mV}}{6.91 \text{ k}\Omega} = \dfrac{54.6 \times 10^{-3} \text{ V}}{6.91 \times 10^{3} \ \Omega}$
Evaluate: 54.6 [EE] 3 [+/-] [÷] 6.91 [EE] 3 [=], 7.90 −6
Answer: $I = 7.90 \times 10^{-6} \text{ A} = 7.90 \ \mu\text{A}$

Note: Three significant figures.

Check: Small voltage, large resistance = small current.

Example B A current of 34.8 mA is flowing in a circuit, Figure 7-5. How much time is required for 1.602 μC to flow past a given point?

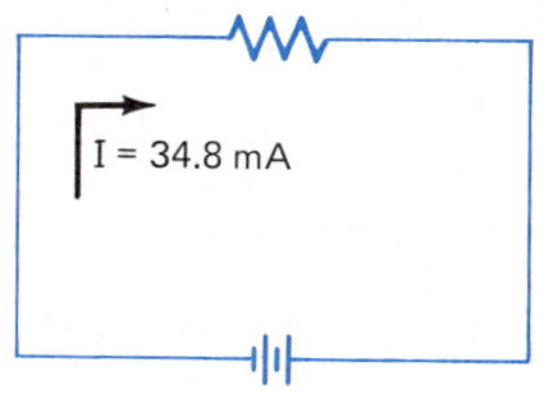

Figure 7-5

Solution: Data: $I = 34.8\ mA, Q = 1.602\ \mu C, t = ?$

Formula: $I = Q/t$

Solved F: $t = \dfrac{Q}{I}$ (MA, × t: DA, ÷ I)

Substitute: $t = \dfrac{1.602\ \mu C}{34.8\ mA}$

Evaluate: 1.602 [EE] 6 [+/-] [÷] 34.8 [EE] 3 [+/-] [=], 4.60 - 5

Answer: $4.60 \times 10^{-5}\ x = 46.0\ \mu s$

Check: coulomb = ampere-second, amperes cancel leaving seconds for the units of the answer (correct).

In future examples, the solved formula will show only the result. The method used will be shown only if that is important in the solution of the problem. The evaluation will not be shown either. The rearrangement of the formula is usually done on the side (scratch paper) and the evaluation is performed with a calculator.

Example C What length (L) is a 10 gage copper wire, Figure 7-6, having a resistance (R) of 2.00 Ω? The resistivity (ρ) of copper is $1.7 \times 10^{-8}\ \Omega \cdot m$ and the area is $5.27 \times 10^{-6}\ m^2$. The formula relating these quantities is $R = \rho L/A$.

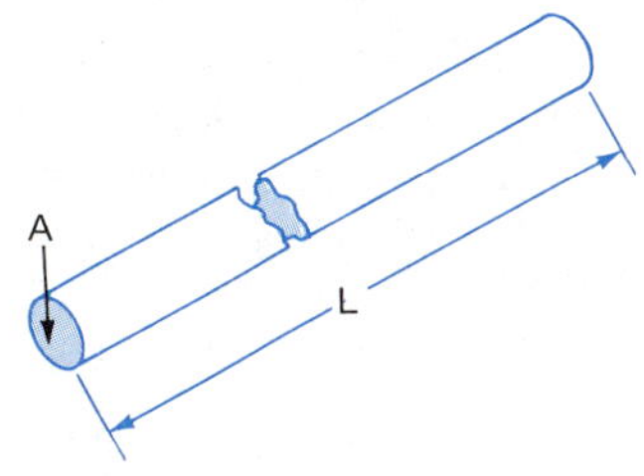

Figure 7-6

Solution: Data: $R = 2\ \Omega, \rho = 1.7 \times 10^{-8}\ \Omega \cdot m,$
$A = 5.27 \times 10^{-6}\ m^2, L = ?$

Formula: $R = \rho L/A$

$L = RA/\rho$ (DA, ÷ ρ; MA, × A)

Substitute: $L = \dfrac{(2\ \Omega)(5.27 \times 10^{-6}\ m^2)}{1.7 \times 10^{-8}\ \Omega \cdot m}$

$L = 620\ m$

Check: Units in the denominator cancel units in the numerator, leaving units of metres which is correct for length.

Several circuit quantities are listed in Figure 7-7. Included are the defining formulas for voltage, charge and power. Ohm's Law and the circuit power formula are added to complete the list for simple DC circuits. These should be memorized.

Together, the five formulas can be used to solve many kinds of circuit problems. Of the seven quantities, all except resistance appear in more than one formula. Often, one quantity can be calculated using one formula. Its value is then used in another formula to find the desired

QUANTITY	SI UNIT SYMBOL	DEFINING FORMULA	DERIVED FORMULA
Voltage (E)	V	$E = \frac{W}{Q}$	
Current (I)	A	Fundamental	
Resistance (R)	Ω	$R = \frac{E}{I}$	
Charge (Q)	C	Q = It	
Time (t)	s	Fundamental	
Work-energy (*W*)	J	Mechanical	
Power (P)	W	$P = \frac{W}{t}$	P = EI

Figure 7-7 Circuit quantities and formulas are listed in the table. All quantities except resistance appear in two of the five formulas.

unknown. Sometimes, three formulas are used before the unknown can be found.

Example D How much work is done in moving a charge of 5.85 mC through a resistance of 5.3 kΩ when the current is 45.3 mA?

Solution: Data: Figure 7-8, Q = 5.85 mC, *W* = ?

Formula: E = IR and E = *W*/Q

Note: Use Ohm's law to find the voltage, use the second formula to find the work done.

E = IR and *W* = EQ (MA, X Q)

Substitute: E = (43.3 mA)(5.3 kΩ)

E = 229 V

W = (229 V)(5.85 mC)

W = 1.3 J (Two significant digits)

Check: Sometimes the only check is to work the problem backwards:

$$E = \frac{W}{Q} = \frac{1.3\ \text{J}}{5.85\ \text{mC}} = 229\ \text{V}$$

$$I = \frac{E}{R} = \frac{229\ \text{V}}{5.3\ \text{k}\Omega} = 43.3\ \text{mA}$$

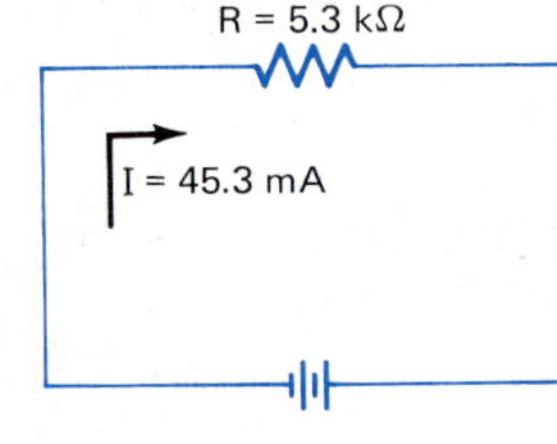

Figure 7-8

EXERCISE 7-5

Use the formulas in Figure 7-7 and the following problems to practice problem solving techniques. All exercises (problem solving type) in the rest of this text will include the following words: data, diagram, formula, substitute and check. These key words will help recall the step-by-step procedure for solving problems.

1. It takes 30 J of energy to move 20 C of charge through a battery. What is the emf of the battery?
2. How much energy is required to move 500 mC of charge through a battery with an emf of 12 V?

3. How much charge will move from the positive terminal to the negative terminal in a 6.0-V battery? Chemical energy of 25 mJ is converted.
4. In a battery, the charge moving from the positive terminal to the negative terminal is 4.2 mC. How many electrons are moving through the battery? $1\ e = 1.602 \times 10^{-19}$ C.
5. Find the current in a flashlight having an emf of 3.0 V if the lamp has a hot resistance of 12 Ω.
6. Find the hot resistance of a 240 V stove heating element if it draws a current of 2.5 A.
7. An electric clock motor, Figure 7-9, has a resistance of 5.77 kΩ. It draws 20.8 mA of current. Find the voltage.
8. A 12.0-V headlamp on a car draws 1.45 A. What power is dissipated by the lamp?
9. In an aluminum reduction plant, a current of 80 kA passes through each pot line. If the power dissipation is 68 MW, find the voltage across each line.
10. A $45\bar{0}$ W electric iron draws 3.84 A. Find the voltage.
11. An electric clock draws 20.5 mA of current. How much electric charge passes through the clock in 24.00 h? How many electrons is this? ($1\ e = 1.602 \times 10^{-19}$ C)
12. What is the current if 1 C of charge is moved through a switch each minute?
13. How much time is required for 2.07×10^{4} C to pass through the heating element on a toaster if the current is 5.75 A?
14. How much power is utilized by a house if 72.0×10^{2} kJ of energy are used in 4.00 h?
15. A flashlight bulb uses energy at a rate of 750 mW. How much time is required for it to dissipate $10\bar{0}$ J of energy?
16. How much energy is dissipated in a circuit in 10 min? The power is 24 mW.

Figure 7-9 An electric clock motor (2.5 W) (Courtesy of Southeast Nebraska Community College)

Solve the following problems. Each one requires at least one intermediate calculation:

17. What current is flowing in a circuit if the voltage drop across the load is 120 V? There is 1080 J of energy delivered to the load each minute.
18. How much energy is removed in 0.5 min from a battery if the current is 450 mA? The voltage is 1.5 V.
19. What is the voltage drop across a lamp if 3.12×10^{17} electrons pass through it? Heat and light in the amount of 6 J are produced.
20. A voltage of 24 V is applied across a resistance of 6 Ω for four minutes. How much charge passes through the load?
21. A trouble lamp has a hot resistance of 4.8 Ω. It is connected to a 12-V battery. How much time is required to pass 750 C of charge?

22. A transmission line with a resistance of 2.5 Ω has 200 V voltage drop along the line. Find the energy dissipated in one minute.
23. Find the energy dissipated by a 110-V toaster with a resistance of 18.3 Ω in one minute.
24. A current of 2 A is maintained in an electric circuit. The energy required to move the charge through the load each minute is 3.6 kJ. What is the voltage?
25. A current of 0.7 A is produced by 6 V across a load. How much energy is transferred to the load in 20 min?
26. A charge of 10 C passes through a resistor with a resistance of 500 Ω. How much time is required to dissipate 20.00 kJ of energy?
27. A charge of 25 C passes through a load in 1 min. If the resistance is 100 Ω, how much energy is dissipated?
28. For how much time will a steady current of 250 mA have to flow through a flashlight bulb if 1000 J of energy is drawn from the battery? There is a voltage drop of 3 V across the lamp. Also, find the resistance and the power rating of the bulb.
29. Suppose 8.9×10^{19} electrons pass through a $60\bar{0}$-Ω resistor. How much time is required if 120 J of heat is produced?
30. The $8\bar{0}$-kA current feeding the pot lines (Problem 9) passes through copper bus bars. If there is a 156 μΩ resistance in the bus bar, find the voltage drop, power dissipation and the heat produced in 24 hours in the bus bar.
31. What is the resistance of a voltmeter which reads 117 V? A charge of 4 mC passes through the meter in 1 min 40 s.
32. Find the resistance of an ammeter shunt. It is designed to have a 5-mV voltage drop when the power dissipation is 50 mW.
33. A fuse has a resistance of 0.021 Ω. What power does it dissipate if the voltage drop is 450 μV?
34. How much heat is produced in a toaster that draws 4.55 A from a 120-V source in 3 min?
35. A relay coil, Figure 7-10, is designed to energize with a minimum current of 22.5 mA through the coil. How long will it take to dissipate 5 J of heat if the resistance is 3.25 kΩ?

Figure 7-10 Various relays (Courtesy of Southeast Nebraska Community College)

36. An electric motor armature has a resistance of $20\bar{0}$ mΩ. If the armature current is 40.0 A, how much charge flows through the windings in the time $100\bar{0}$ J of heat is produced?

PERIMETER, AREA AND VOLUME

Electrical problems often require the calculation of perimeters, area and volume. The number of electrical outlets in a room depends on the room's *perimeter.* The resistance of a wire depends on the wire's cross-sectional *area.* The *volume* of a junction box determines the number of electrical components it can hold.

Geometrical concepts have special applications in the electrical fields. Many of these will be described in the example problems and exercises. At the same time, there will be more practice using problem solving techniques.

7-8 PLANE GEOMETRIC FIGURES

The *perimeter* (P) of an object is the distance around it. A large industrial building has a large perimeter. A small shed has a small one. Perimeter is a linear or distance measurement which is measured in units of metres, centimetres, feet, inches, etc. The distance around a circle is called *circumference* (C).

An *area* (A) is a quantity that describes the amount of surface an object has. A large workbench has more working area than a small one. A large showroom has more floor area (display area) than a small one. An area is measured by comparing the surface with "square units." The square centimetre (cm^2) and the square inch (in^2) are shown full scale in Figure 7-11. The square metre (m^2) is the SI unit for area.

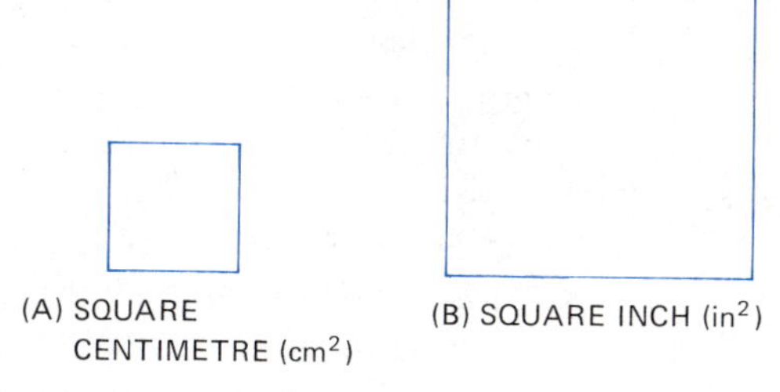

Figure 7-11 Area units

Perimeter and area are characteristics of *plane figures.* Plane figures are two-dimensional objects. They include rectangles, triangles and circles among the simple shapes having names.

A *rectangle* is formed by four sides, Figure 7-12(A). Adjacent sides are perpendicular (form right angles). A rectangle has a *length* (L) and *width* (w). If the length and width are equal, the rectangle is a *square,* Figure 7-12(B).

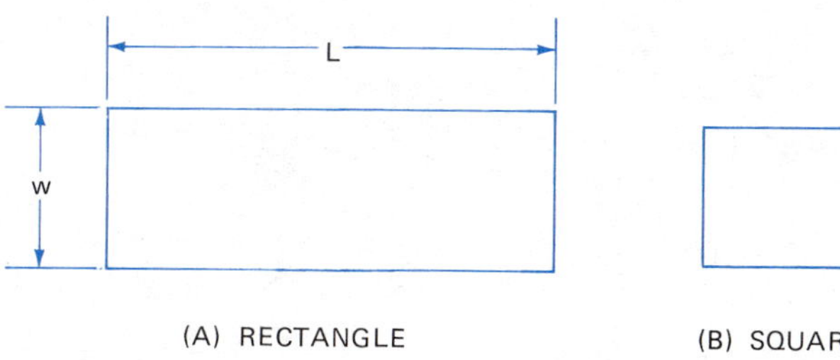

Figure 7-12 Rectangle and square: When length (L) equals width (w) the plane figure is a square.

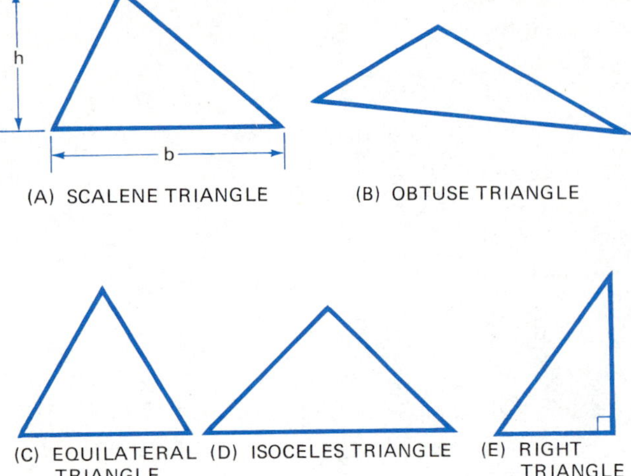

Figure 7-13 Various triangles

A *triangle* is a three-sided object, Figure 7-13(A). If none of the three sides are equal, it is a *scalene triangle*. The scalene triangle may have three acute angles (less than a right angle). If one angle is obtuse (more than a right angle) it is sometimes called an *obtuse triangle*, Figure 7-13(B). When all three are equal, it is an *equilateral triangle*, Figure 7-13(C). If two sides are equal, Figure 7-13(D), the triangle is *isosceles*. A triangle with a right angle is called a *right triangle*, Figure 7-13(E). The dimensions of the triangle, Figure 7-13(A), are *base* (b) and *altitude* (h).

A *circle* is a round plane object, Figure 7-14. The *diameter* (d) is the length of a straight line from edge to edge that passes through the center. The *radius* (r) is the distance from the center to the edge in any direction. The radius is one-half the diameter.

Formulas for finding perimeter or circumference and area of plane objects are given in Figure 7-15.

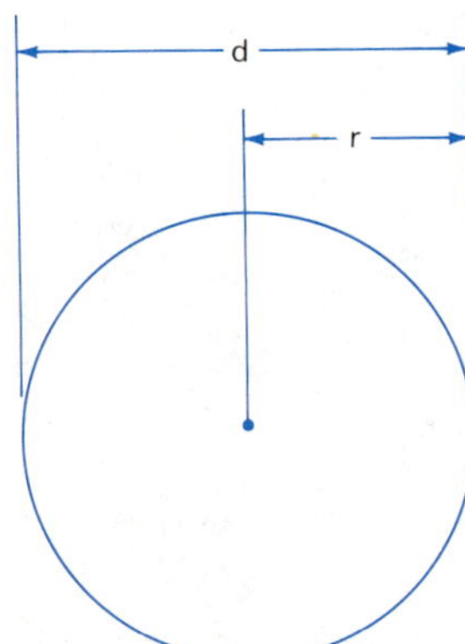

Figure 7-14 A circle

	PERIMETER	AREA
Rectangle	$P = 2(L + w)$	$A = Lw$
Triangle	$P = a + b + c$	$A = \frac{1}{2}bh$
Circle	$C = \pi d$	$A = \pi r^2$

Figure 7-15 Perimeter and area formulas for plane figures

Example A Find the circumference of a circle with a radius of 8.0 cm, Figure 7-16.

Solution: Data: $d = 2r = 2(8.0\text{ cm}) = 16.0\text{ cm}$

Formula: $C = \pi d$

Substitute: $C = \pi(16.0\text{ cm})$

$C = 50.3\text{ cm}$

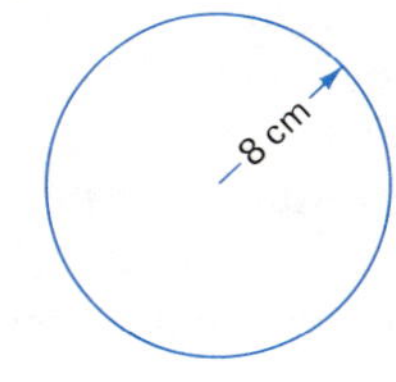

Figure 7-16

Example B Find the base side of a triangle with an area of 325 m². The altitude is 15.0 m, Figure 7-17.

Solution: Data: $h = 15.0\text{ m}, A = 325\text{ m}^2, b = ?$

Formula: $A = \frac{1}{2}bh$

$b = \frac{2A}{h}$ (MA, X 2; DA, ÷ h)

Substitute: $b = \frac{2(325\text{ m}^2)}{15.0\text{ m}} = 43.3\text{ m}$

Check: The unit (m) in the denominator cancels into the unit $m^2 = m \times m$ in the numerator leaving units of distance (m).

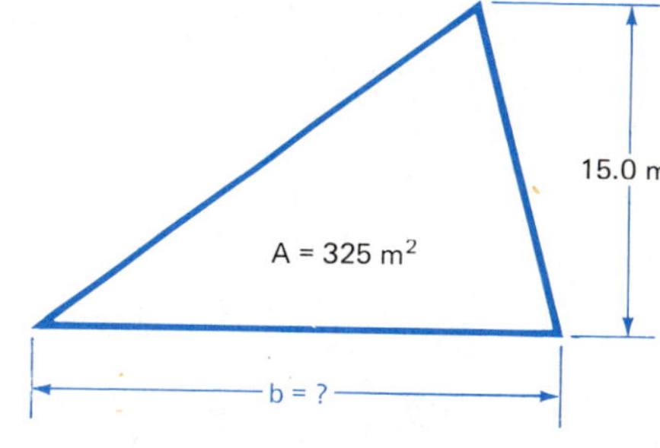

Figure 7-17

EXERCISE 7-6

Find the perimeter or circumference of each of the following objects. Round correctly according to convention.

1. A rectangle, sides are 48.1 cm and 36 cm.
2. A triangle, sides are 25.3 in, 38.25 in and 46 in.
3. A circle, diameter is 14.3 m.
4. A square, side is 2.45 ft.

Name the kind of triangle in each of the following examples. The length of the sides are:

5. 8.2 cm, 8.2 cm and 8.2 cm.
6. 7.50 in, 4.1 in and 7.50 in.
7. 3.58 m, 4.235 m and 6.43 m (two acute angles, one obtuse angle).
8. 8.1 ft, 4.44 ft and 6.97 ft (three acute angles).
9. 63 mm, 84 mm and 105 mm (one right angle).

Solve each of the following: (data, diagram, formula, substitute, check). Round as necessary.

10. Find the area of a rectangle; sides are 48.1 cm and 36 cm.
11. Find the area of a right traiangle; sides are 25.3 in, 38.25 in and 46 in.
12. Find the area of a circle; diameter is 14.3 m.
13. Find the area of a square; side is 2.45 ft.
14. A pole transformer has a diameter of 36 cm. Find its circumference.

15. A 3-hp motor has a circumference of 58.5 cm. What is its radius and diameter?
16. A substation, Figure 7-18, is located on a large triangular lot. If the area of the lot is 2820 m^2 and the triangle altitude is 55.0 m, find the length of the base.
17. The display room in an electrical shop has a length of 42.5 ft and a width of 28.25 ft. Find its perimeter and area.
18. Wood poles are classified by circumference at the top and at 2 m from the butt end. A class one, 12.2 m Douglas fir pole has a diameter of 21.8 cm at the top. Find the circumference.
19. A wood pole has a circumference of 107 cm at 2 m above the butt end. What is the diameter of the pole at that point?
20. Find the length of 25 turns of wire in a coil if the average diameter is $5\bar{0}$ cm.
21. A circular cover for a junction box has a diameter of 4.0 in. Find its area.
22. A capacitor is made up of two sheets of foil. Each is 5.0 cm by 750 cm. Find the area of the two sheets.
23. An electron in the innermost orbit of a hydrogen atom travels 3.32×10^{-10} m in one complete orbit. Find the diameter of the orbit.
24. List the following objects in order: largest area to smallest area.
    a. a square 12.5 cm on a side.
    b. a circle with a 7.11-cm radius.
    c. a right triangle with a 20.0 cm base, 15.8 cm altitude and 25.5 cm hypotenuse.
    d. a rectangle with 19.6 cm length and 8.01 cm width.
25. Notice that the objects in Problem 24 have about the same area. Now list them in order: largest perimeter to smallest perimeter. Compare this list with that of Problem 24. Can one conclude that the objects with the largest areas have the largest perimeters?
26. Find the length of thin-wall conduit in Figure 7-19. Note that the curved section is 1/4 of a circle.
27. A battery cell is made up of an element. An element consists of eleven negative plates and ten positive plates. If each plate is 15 cm by 15 cm, find the total area of the negative plates. Include both sides of each plate. Find the total area of the positive plates.
28. A 4.7-nF capacitor is constructed of foil squares separated by mica used as the dielectric. If there are 14 mica sheets each 15.43 mm square, what total area (one side) of mica is used? Express the answer in square centimetres.

Figure 7-18 Typical substation

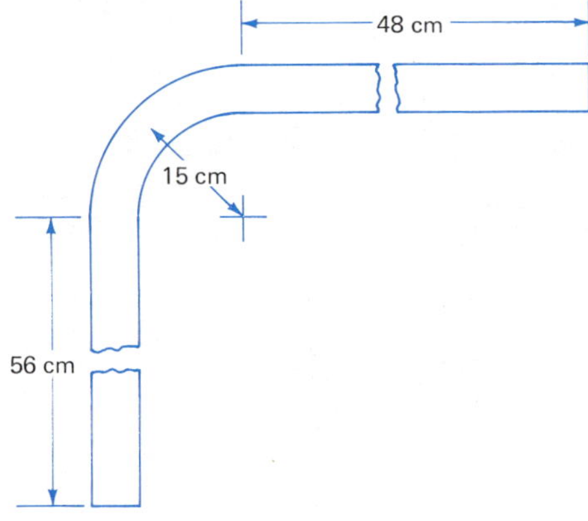

Figure 7-19

Note: The amount of area determines the amount of energy the battery can store.

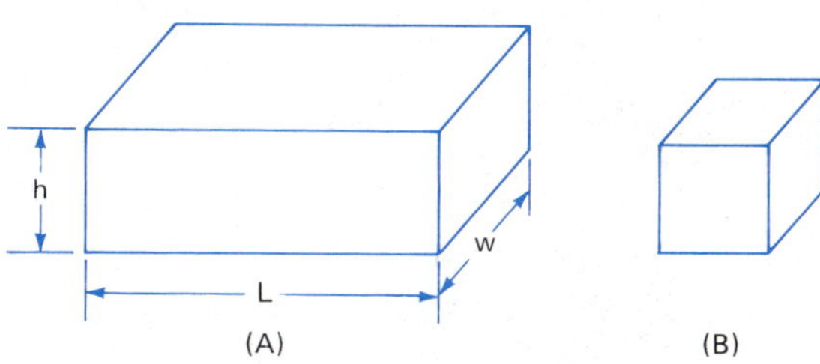

Figure 7-20 (A) Rectangular solid and (B) Cube

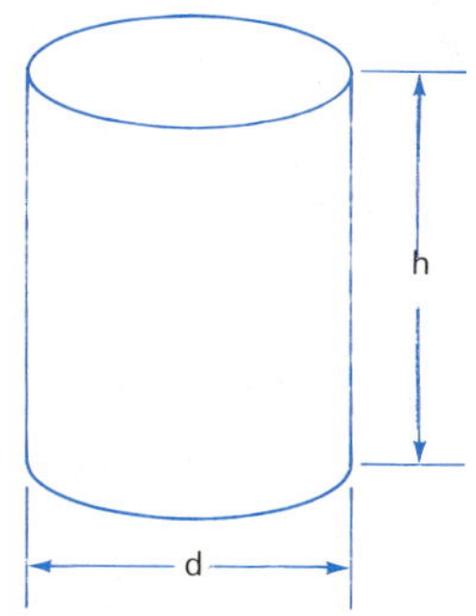

Figure 7-21 Right circular cylinder

7-9 SOLID GEOMETRIC FIGURES

Sometimes electricians need to calculate the surface area, cross-sectional area or the volume of solid objects and containers. The rectangular solid, cylinder and sphere are reviewed in the following paragraphs.

A *rectangular solid* has dimensions of length (L), width (w) and height (h), Figure 7-20(A). All adjacent sides are perpendicular to each other. When all three dimensions have the same length, the solid is a *cube,* Figure 7-20(B).

In this text, the use of the word *cylinder* will mean a *right circular cylinder.* It has a circular cross section. Its ends are cut perpendicular to a line passing through the centers of the two ends, Figure 7-21.

All points on the surface of a *sphere* are the same distance from the center, Figure 7-22. The distance from the center to the surface is the radius. The diameter is the length of a line from the surface to the opposite side and passing through the center. The diameter is twice the radius.

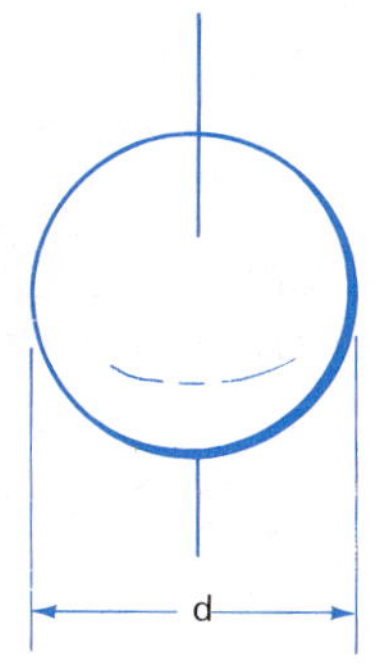

Figure 7-22 A sphere

Surface area of a solid object is a measure of the amount of surface. *Cross-sectional area* is the area of an imaginary surface. It usually applies to long, thin objects such as wire, conduit, poles, etc. Imagine that a wire or pipe is cut in two, Figures 7-23 A and 7-23 B . The cross-sectional area is the area found on the face of the perpendicular cut (shaded area). The area of a cross section of most objects is usually a simple plane figure.

Volume is a quantity that describes the amount of space an object occupies or contains. Large tanks hold more liquid than small ones so they contain more volume. Large rooms contain more space or volume than small ones.

The volume of an object or space is measured by comparing it with cubic volume units. Two common volume units are shown in Figure 7-24. These are the cubic centimetre (A) and the cubic inch (B). Other units are the cubic foot, cubic yard, cubic metre, etc. The volume of cylindrical, spherical and other objects can be determined as readily as rectangular solids.

When measuring a quantity of liquid, special units are used. Some examples are the gallon, pint, quart and litre. Other volume units are used to measure amounts of fruit, vegetables and grains. The bushel, peck, etc. are examples of volume units for dry measure. Dry volume units are not used in this text.

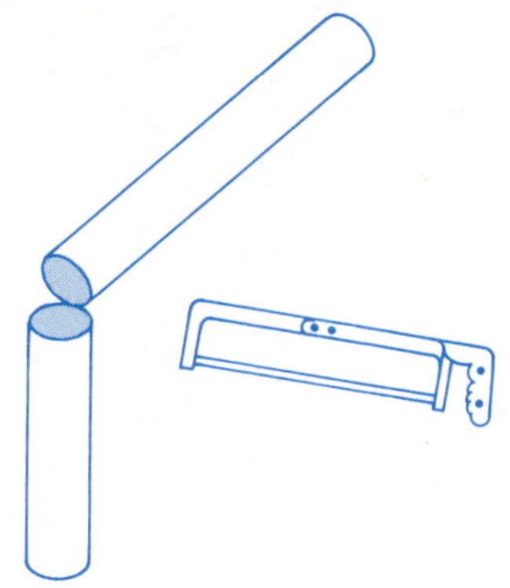

Figure 7-23A The cross-sectional area of an object is the area (shaded) on the face of an imaginary perpendicular cut.

Figure 7-23B Motor field windings cut to show the cross-sectional area (Courtesy of Southeast Nebraska Community College)

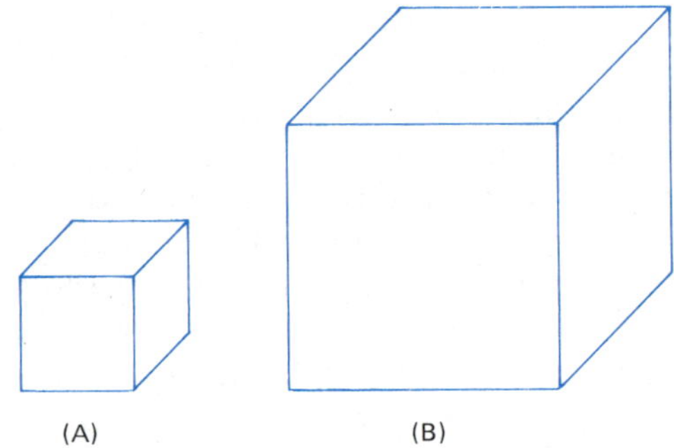

Figure 7-24 (A) Cubic centimetre and (B) cubic inch compared

The formulas for surface area and volume are given in Figure 7-25. Volume formulas never give liquid measure units. Once the number of cubic units are found, convert to liquid measure if desired.

SOLID FIGURE	SURFACE AREA	VOLUME
Rectangular solid	$A = 2Lw + 2Lh + 2wh$	$V = Lwh$
Cylinder	$A = \pi dh$ (lateral)	$V = \pi r^2 h$
Sphere	$A = 4\pi r^2$	$V = \frac{4}{3}\pi r^3$

Figure 7-25 Surface area and volume formulas for common solid figures

Example A Find the volume of a sphere if the radius is 5.00 ft, Figure 7-26.

Solution: Data: $r = 5.00$ ft, $V = ?$

Formula: $V = 4\pi r^3/3$

Substitute: $V = \dfrac{4\pi(5.00 \text{ ft})^3}{3}$

$V = 524 \text{ ft}^3$ (Cubic feet)

Check: A sphere has slightly more than half the volume of a cube whose sides equal the diameter: 524 is more than $\frac{1}{2}(10)^3 = 500$.

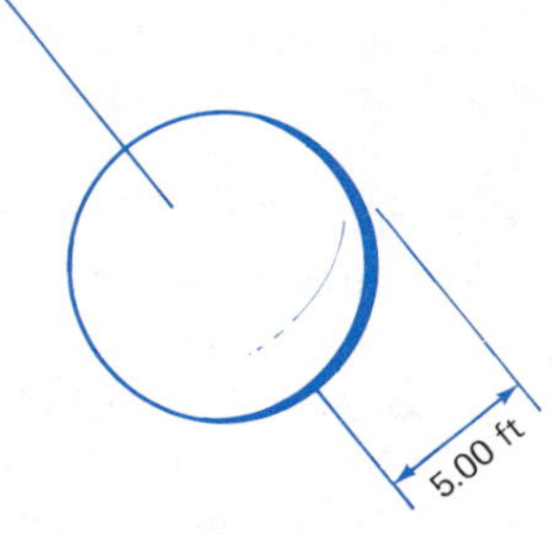

Figure 7-26

Example B Find the cross-sectional area of a pipe with an external diameter of 2.84 cm and an internal diameter of 2.60 cm. This can be accomplished by subtracting the inner circular area from the outer, Figure 7-27. This gives the area of the cross section (shaded).

Solution: Data: $r_o = 1.42$ cm, $r_i = 1.30$ cm, $A = ?$

Formula: $A = \pi r_o^2 - \pi r_i^2$

Substitute: $A = \pi(1.42)^2 - \pi(1.30)^2$

$A = 1.03 \text{ cm}^2$

Check: If the shaded part of Figure 7-27 could be cut and laid out lengthwise, it approximates a rectangle of length equal to average circumference ($\pi \times 2.7$). The width (thickness) is half the difference in diameters (about 1/8 cm). The area of such a rectangle is $\pi(2.7) \times 1/8 = 1.06$ cm. Thus, the answer (1.03 cm) is a reasonable value.

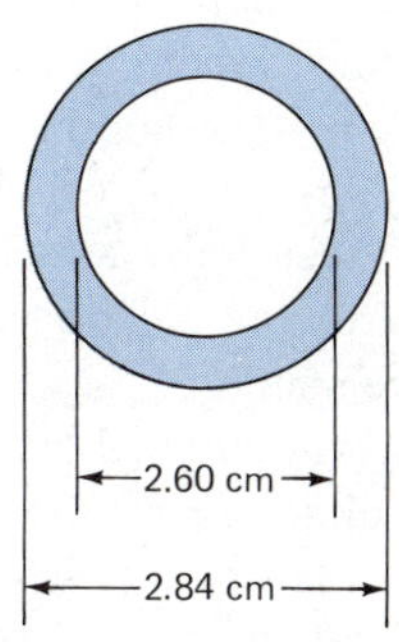

Figure 7-27

EXERCISE 7-7

Find the total surface area of the following objects. Round as necessary.

1. A sphere with a diameter of 3.00 in.
2. A cylinder with a length of 1.00 m and a diameter of 0.455 m.
3. A rectangular solid with dimensions: 44 cm, 35.35 cm and 10.0 cm.
4. A cube with 6.25-ft sides.

Which quantity (perimeter, area, volume) is described in each of the following?

5. The amount of wire connecting the outlet boxes in a room circuit.
6. The amount of space required to store 60 reels of wire cable.
7. The amount of sheet metal required to make a junction box.
8. The amount of work space on a bench.
9. The amount of strap iron holding a load of crossarms together.
10. The amount of oil in a transformer.

Solve the following. (data, diagram, formula, substitute, check). Round as necessary.

11. Find the volume of a sphere with a diameter of 3.00 in.
12. Find the volume of a cylinder with a length of 1.00 m and a diameter of 0.455 m.
13. Find the volume of a rectangular solid with dimensions: 44 cm, 35.35 cm and 10.0 cm.
14. Find the volume of a cube with 6.25-ft sides.
15. A Class 1, 13.7-m western red cedar pole has an average diameter of 31.85 cm. Find the volume of the pole in cubic metres. Assume it is a cylinder having the average diameter.
16. Crossarms measure 3.75 in x 4.75 in x $1\bar{0}$ ft. What volume of wood is in each crossarm?
17. Find the total surface area including top and bottom of a pole transformer housing. It has a diameter of 36 cm and a height of 75 cm.
18. A light fixture has a glass cover in the shape of a globe (sphere). If the globe has a diameter of 14 in, find its volume and surface area.
19. A transformer housing has a diameter of 36 cm and a height of 75 cm. The housing is 6/10 filled with oil. How many litres of oil does it contain? (1 L = 1000 cm^3).
20. Fourteen holes were dug for poles. Each hole has a 24-inch diameter and a depth of 6.0 feet. How many cubic yards of dirt were handled?
21. Find the internal volume of an electrical outlet box with internal dimensions of 35 x 93.5 x 96.5 mm. How many cubic centimetres is this? (1 cm^3 = 1000 mm^3).
22. A glass cover on a street lamp is in the shape of a hemisphere (one-half of a sphere). It has an internal radius of 6.25 in. What is the internal volume of the cover? What is the outer surface area if the outer radius is 6.75 in?
23. Find the surface area of a fluorescent lamp 3.7 cm in diameter and 235.0 cm long.
24. A trench for underground cable is 16 cm wide and $6\bar{0}$ cm deep. It is 110 m long. What volume of dirt was removed?
25. Find the cross-sectional area of a piece of rigid conduit (iron). The outside diameter is 1.625 inches and the inside diameter is 1.375 inches.
26. Find the cross-sectional area of each of the following wire sizes. Diameters are given in millimetres.
    a. 14 gage – 1.63 mm
    b. 10 gage – 2.59 mm
    c. 6 gage – 4.11 mm
    d. 2 gage – 6.543 mm
    e. 000 gage – 10.40 mm
27. For ordinary distribution work, the usual flat-strap brace is 0.25 in thick, 1.25 in wide and has a volume of 11.9 in^3. How long is it?

28. The gasoline tank (cylindrical) on a portable generator holds 45 L. It has a 15.0-cm radius. What is its length? (1 L = 1000 cm^3).
29. A car battery cell has an internal volume of 74 in^3. It is 6.5 in long and 1.75 in wide. How deep is it?
30. Electrons, protons and neutrons all have diameters of about 2×10^{-13} cm. Find the volume of these spherical subatomic particles.
31. The diameter of an aluminum nucleus is about 5.6×10^{-13} cm. Find its volume assuming it is a sphere.
32. A 10 gage wire has a diameter of 2.59 mm. (a) Find the volume of 1 km of wire in cubic centimetres. (b) If each cubic centimetre of copper has 8.5×10^{22} free electrons, how many free electrons does 1 km of 10 gage copper wire have?
33. A 4 x 4 x 1 1/2 in square outlet box has internal dimensions of 3.875 x 3.875 x 1.375 in. (a) What is its internal volume? (b) The free space allowed within the box for each No. 12 conductor is 2.25 in^3 (National Electric Code). How many conductors are allowed in this box?
34. A pole transformer has a circumference of 113 cm and a height of 76 cm. Find its volume.
35. A warehouse is 25 m long and its perimeter is 92.8 m. If the ceiling is 10.8 m high, what is the volume of the rectangular-shaped building? How much floor area does it have?

7-10 WIRE RESISTANCE

A simple electric circuit contains a battery and a load. The amount of current is determined by the emf and the resistance according to Ohm's Law. The resistance of the circuit stems mostly from the load. The battery also offers resistance called *internal resistance.* In addition, the conductors that connect the various circuit elements have some resistance.

To minimize the resistance, the wire must be large enough to carry the required current. Wire that is too small will increase the voltage drop in the circuit because of its higher resistance. Wire is available in different sizes (gages).

In the American Wire Gage (AWG) system, the heavier (larger diameter) wires have the smaller gage numbers. No 10 wire is larger than No. 14, for example. Larger or heavier wire has a larger cross-sectional area, Figure 7-28. Actual diameters and resistance for standard lengths of copper wire

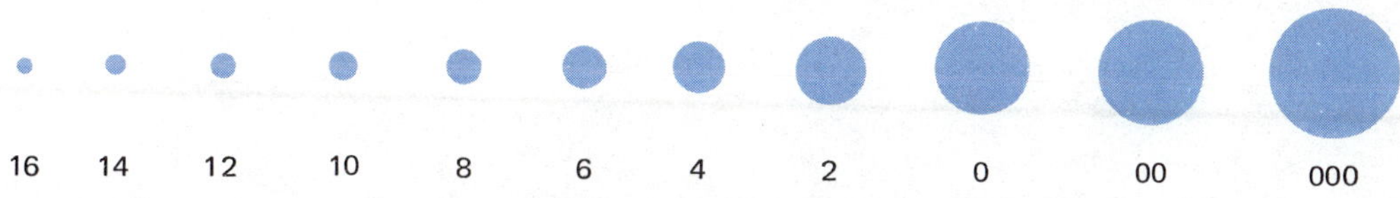

Figure 7-28 Actual cross section of various wire sizes (AWG)

GAGE	DIAMETER (mm)	Ω/km	DIAMETER (mil*)	Ω/1000 ft
36	0.1270	1361	5.000	414.8
32	0.2019	538.5	7.950	164.1
28	0.3211	213.0	12.64	64.90
24	0.5105	84.24	20.10	25.67
20	0.8118	33.31	31.96	10.15
16	1.291	13.17	50.82	4.016
14	1.628	8.285	64.08	2.525
12	2.053	5.211	80.81	1.588
10	2.588	3.277	101.9	0.9989
8	3.264	2.062	128.5	0.6282
6	4.115	1.297	162.0	0.3951
4	5.189	0.8155	204.3	0.2485
2	6.544	0.5129	257.6	0.1563
0	8.252	0.323	324.9	0.0983
00	9.266	0.256	364.8	0.0779
000	10.40	0.203	409.6	0.0618
0000	11.68	0.161	460.0	0.0490

*1 mil = 0.001 in

Figure 7-29 American Wire Gage (AWG) tables for solid annealed copper conductors (20°C)

are given for various gages in Figure 7-29. A more complete table is given in the Appendix.

The resistance of wires made of other metals are calculated by a formula that depends on the *resistivity* of the metal. Resistivity is indicated by the Greek letter rho (ρ). It is measured in ohm-metres ($\Omega \cdot m$). Resistance is given by the formula

$$R = \frac{\rho L}{A}$$

where L is the length of the wire in metres and A is the cross-sectional area in square metres. Some resistivities for several metals at 20°C are given in Figure 7-30 and in the Appendix.

SUBSTANCE	RESISTIVITIES AT 20°C OHM-METRES
Aluminum	28.3×10^{-9}
Carbon	$35\,000 \times 10^{-9}$
Copper	17.2×10^{-9}
Iron	103×10^{-9}
Manganin alloy	447×10^{-9}
Nichrome alloy	1000×10^{-9}
Silver	16.4×10^{-9}
Tungsten	55.2×10^{-9}

Figure 7-30 Resistivity of various materials

Example A Find the resistance at 20°C of a 20 gage Nichrome wire that is 3.00 m long.

Solution: Data: $L = 3.00\ m$, $\rho = 1000 \times 10^{-9}\ \Omega \cdot m$ (from table), $d = 0.812\ mm$, $A = \pi r^2 = \pi(0.000\ 406\ m)^2 = 5.18 \times 10^{-7}\ m^2$, $R = ?$

Formula: $R = \frac{\rho L}{A}$

Substitute: $R = \frac{(1000 \times 10^{-9}\ \Omega \cdot m)(3.00\ m)}{5.18 \times 10^{-7}\ m^2}$

$R = 5.79\ \Omega$

In the English system, resistivity is measured in ohms-circular mils per foot. This is usually written as ohms per mil-foot which is not strictly correct. The use of the *circular mil* (CM) as a unit of area has an advantage. If the diameter of the wire is expressed in *mils* (0.001 in), Figure 7-29, the area is d^2 CM. The circular mil is not a square area unit since π is not used. To obtain the resistivity in ohms per mil-foot, multiply the resistivity in ohm-metres by 6×10^8.

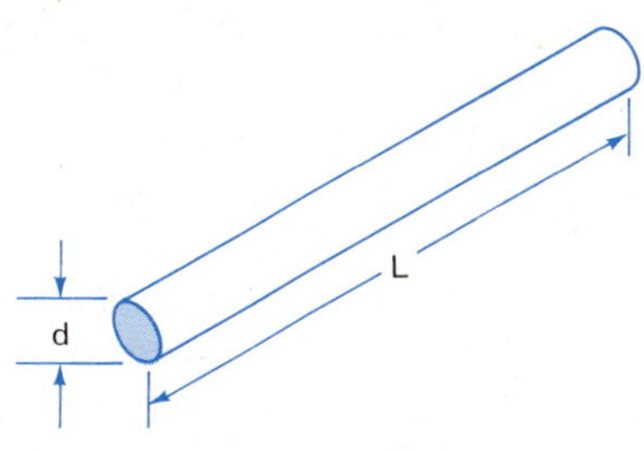

Figure 7-31

Example B Find the length of #10 copper wire having a resistance of 1.00 Ω at 20°C, Figure 7-31.

Solution: Data: $R = 1.00\ \Omega$, $\rho = 17.2 \times 10^{-9}(6 \times 10^8) = 10.3\ \Omega\cdot\text{CM/ft}$, $L = ?$, $A = d^2 = (101.9)^2 = 10\,380$ CM.

Formula: $R = \dfrac{\rho L}{A}$

$L = \dfrac{RA}{\rho}$

Substitute: $L = \dfrac{(1.00\ \Omega)(10\,380\ \text{CM})}{10.3\ \Omega\cdot\text{CM/ft}} = 1010\ \text{ft}$

Note: Don't confuse CM (circular mils) with cm (centimetres).

The resistance of most conductors increases when the temperature increases. The change in resistance (ΔR) is given by the formula:

$$\Delta R = \alpha R_o \Delta T$$

where R_o is the original resistance, ΔT is the change in temperature and α is the *temperature coefficient of resistance*. For most pure metals, the temperature coefficient has a value of about 0.004 per C°. For alloys it is much smaller, Figure 7-32.

SUBSTANCE	TEMPERATURE COEFFICIENT PER C°
Aluminum	0.0039
Carbon	−0.0005
Copper	0.003 93
Iron	0.006
Manganin alloy	0.000 01
Nichrome alloy	0.000 17
Silver	0.0038
Tungsten	0.0047

Figure 7-32 Temperature coefficients of resistance

Example C An aluminum wire has a resistance of 2.5 Ω at 20°C. What is its resistance at 80°C?

Solution: Data: $T_1 = 20°\text{C}$, $\Delta T = 60\ \text{C}°$, $R_o = 2.5\ \Omega$, $\alpha = 0.0040$ per C°, $\Delta R = ?$, $R = ?$

Formula: $\Delta R = \alpha R_o \Delta T$

Substitute: $\Delta R = (0.0040/\text{C}°)(2.5\ \Omega)(60\ \text{C}°) = 0.6\ \Omega$

Add the change in resistance to the original to find the new resistance.

$R = R_o + \Delta R = 2.5\ \Omega + 0.6\ \Omega = 3.1\ \Omega$

EXERCISE 7-8

(data, diagram, formula, substitute, check)

1. A copper bus bar is 0.80 cm thick, 1.8 cm wide and 25 cm long. Find the resistance along its length.
2. Find the length of a tungsten lamp filament. Its diameter is 0.025 mm. It has a resistance of 6.0 Ω at 20°C.

3. The copper field windings of an electric motor have a resistance of 35.2 Ω at 20°C. The windings are constructed of No. 32 gage wire. Find the length of wire used.
4. An electromagnet coil is wound with 114 m of No. 16 wire. Its resistance is 1.50 Ω. What kind of metal is the wire?
5. A transformer primary is wound with 175 m of No. 14 copper wire. Find its resistance.
6. The field coil of a motor has a resistance of 4.5 Ω at 20°C. The copper wires of the coil are No. 20. How long is the coil wire?
7. What is the resistance of two No. 8 service wires 150 ft long? The diameter is 128.5 mil. The wire is copper.
8. An 0000 gage aluminum wire at 20°C has a resistance of 132 Ω. How long is it?
9. Find the resistance of the aluminum wire in Problem 8 if the temperature is 38°C.
10. The resistance of the copper winding of a motor is 4.5 Ω at 0°C. When operating, it is 5.3 Ω. What is the operating temperature of the motor?
11. A tungsten lamp filament has a resistance of 8.5 Ω at 20°C. If the filament has an operating temperature of 2200°C, find its resistance. Assume the temperature coefficient does not change. What current does it draw from a 110-V line? What is the power rating of the lamp?
12. The resistance of an aluminum wire is 28.73 Ω at 0°C. What is its temperature if the resistance changes to 32.03 Ω?
13. The resistance of No. 0 wire at 20°C is 24.0 Ω. At 800°C the resistance is 27.2 Ω. Find the temperature coefficient of resistance.
14. A solenoid is wound with 1800 turns of No. 24 copper wire. The average diameter of each turn is 8.0 cm. Find the resistance at 60°C. Find the voltage if it draws 17 mA.
15. A motor's copper field coil operates at 115 V and draws 510 mA when it is started at 20°C. What current does it draw at its operating temperature of 60°C?
16. A telephone line is 10.0 km long and No. 20 copper wire is used. If the current is 185 mA, find the voltage drop across the line.
17. Find the current drawn from a 220-V line by a 3.0-m coil of No. 20 Nichrome wire used as a heating element operating at 900°C.
18. The resistance of a copper motor field winding is found to be 3.052 Ω after setting overnight at 20.0°C. Resistance is measured with a Wheatstone bridge. After several hours of operation, the resistance was 3.632 Ω. Find the average operating temperature. What is the "hot spot" temperature? The hottest part of the motor winding (hot spot) is assumed to be 15 C° above the average operating temperature. The hot spot temperature determines the expected life of motors.

19. The filament of a tungsten lamp is made of No. 50 wire. Its diameter is 0.002 54 cm. Its hot resistance is 12.1 Ω. What is its uncoiled length? If the voltage is 110 V, find the current and the power rating of the lamp.
20. Plot a graph of the resistance-diameter data in Figure 7-29. Use the SI units for gage No. 20 to No. 0000.

SECTION 1 REVIEW

Chapter 7 concludes Section 1: Introduction to Electricity. This review highlights the many topics in Section 1.

- *Number notation:* whole numbers, fractions, decimals, signed numbers and scientific notation (SSN and ENG).
- *Arithmetic:* addition, subtraction, multiplication, division, powers, roots and reciprocals (various notation).
- *Algebra:* solving simple equations and formulas using the multiplication and division axioms.
- *Structure of matter:* atomic structure, conduction band, free electrons, conductors, semiconductors, insulators.
- *Circuits:* circuit elements, circuit symbols, circuit diagrams, emf, current, charge, resistance, energy and power. Series and parallel circuits.
- *Measurements:* voltmeters, ammeters and ohmmeters. Electrical units and symbols, metric prefixes and symbols, common unit.
- *Accuracy and uncertainty:* significant digits, rounding. Accuracy of measuring instruments, uncertainty in measurements and tolerance (resistors and capacitors).
- *Problem solving:* method of viewing, understanding and attacking problems (data, diagram, formula, substitute, check). Solving circuit, wire size, area and volume problems.

SECTION 2
DC CIRCUITS

CHAPTER 8

SERIES CIRCUITS

OBJECTIVES

After satisfactorily completing this chapter, the student should be able to:

- Solve equations and formulas containing plus and minus signs using the addition and subtraction axioms.
- Describe and calculate current, total voltage and total resistance in series circuits and use Ohm's Law to solve series circuits.
- Solve formulas in which the unknown is taken to a power or root using the power and root axioms.
- Calculate energy and power dissipation in series circuits.

Most direct current (DC) electrical circuits consist of more than one load and sometimes more than one source of emf. There may be a large number of various elements interconnected in various ways. Instead of one current loop, there may be a whole network of loops.

Many times, a complex circuit can be replaced with a simpler equivalent circuit. After a few steps, a complex circuit reduces to a simple understandable one. Series circuits are studied in this chapter. Parallel circuits are studied in the next chapter and more complex circuits are studied in later chapters.

Series circuits were introduced in Chapter 2 to show how ammeters should be connected in a circuit. The study of series circuits is continued here in more detail. Before the discussion can begin, however, the student must be able to solve equations containing plus and minus signs.

SOLVING EQUATIONS

In the last chapter, the multiplication and division axioms were used to solve simple equations. The study of algebra is extended here to include the solving of equations with plus or minus signs. Equations with grouping symbols will be studied later.

Letters continue to be used for literal factors of various monomial expressions. Numerical factors are coefficients. Signs and symbols have the same meaning as before including subscripts and the delta symbol. Letters, used to represent numbers, obey all the rules of arithmetic.

8-1 TERMS

A term is a mathematical expression in which the various numbers are not separated by a plus or minus sign. In algebra, the definition is the same and includes literal factors as well as numerical ones. A term may be a single number, a single literal factor or a combination with several factors of both kinds.

A one term expression is called a monomial. Those with two terms are binomials. Polynomials are expressions with more than one term. Some examples are:

$2\pi fL$	(Monomial)
$2L + 2w$	(Binomial)
$R_1 + R_2 + R_3 + R_4$	(Polynomial with four terms)
$2\pi f - 2\pi f_o$	(Binomial)

Remember, terms are separated by plus or minus signs.

In the monomial $2\pi fL$, there are four factors. The numerical factor (2π) is called the coefficient. The expression 6Pt has a coefficient of 6. In an expression without a coefficient, the coefficient is understood to be 1. For example: $EI = 1EI$. An expression such as $-\omega t$ means $(-1)\omega t$. The coefficient is -1.

EXERCISE 8-1

How many terms are there in each of the following expressions? Indicate the type of expression for each (monomial, binomial or polynomial).

1. 256
2. $3R + 1 - 5Q$
3. $\sqrt{25\phi} - \theta$
4. $4\alpha\beta - 2\beta + 1$
5. $\dfrac{1}{2\pi fC}$
6. $7R - 3E/I$
7. $c^2 - b^2$
8. $4x^2 - 3x + 1$
9. $4\sqrt{PR} - I_o + E/R - 3$
10. $18D \div 4d \div \pi \div \dfrac{1}{2}$

Give the coefficient for each of these monomials.

11. $5ab$
12. $18Pt$
13. $-\frac{1}{2}b_oh$
14. $4\pi R^2$
15. $\alpha R\Delta T$
16. $-\theta\phi$
17. 26
18. $2\pi^2 f^2 R^2 m$
19. $-\pi R^2$
20. E/R

8-2 LIKE TERMS

If two or more terms have exactly identical literal factors, they are said to be *like terms.* The exponents or roots and the subscripts of each literal factor must be the same. Uppercase letters are not the same as lowercase letters. If there is any difference whatsoever in the literal factors, they are *unlike terms.*

The literal factors do not have to be given in the same order to be like terms. Some examples of like and unlike pairs of terms are:

LIKE TERMS	UNLIKE TERMS
$4R, 3R$	$7R, 8r$
$7a^2b, 8ba^2$	$5a^2b, 3ab^2$
$8EI_1, 4I_1E$	$7x\sqrt{y}, 5xy^2$
$11, -18$	$8I_1, 3I_2$
$2X_C, 9X_C$	$7X_C, 4X_L$

In arithmetic, the same kinds of quantities or objects can be added or subtracted. For example, 4 transformers and 3 transformers are 7 transformers. It makes no sense, though, to add 4 amperes and 3 ohms. In algebra, only like terms can be combined by addition or subtraction. Unlike terms can only be indicated.

To add like terms, combine the coefficients. The coefficient of the answer is the sum of the coefficients of the addends. The literal factors of the answer are the same as those of the addends. Subtraction is similar.

Example A Combine $4a + 3a - 2a$.

Solution: $4 + 3 - 2 = 5$ (Combine coefficients)

$4a + 3a - 2a = 5a$

Example B Combine $7ab^2 + 5b^2a - ab^2 - 7a^2b$.

Solution: Notice the last term is unlike the others but the first three terms can be combined.

$7 + 5 - 1 = 11$ (Combine coefficients)

$7ab^2 + 5b^2a - ab^2 - 7a^2b = 11ab^2 - 7a^2b$

EXERCISE 8-2

Combine like terms in each of these expressions.

1. $6R + 3R$
2. $5I^2 - 2I^2$
3. $4ac + ac - 9ac$
4. $4IR + 2IR - 3E$
5. $6P_1 + 3EI_1 - 4P_1 - 5EI_1$
6. $3 - 7C + 14 + 18C - 8$
7. $8a^2bc + 9ba^2c - 3cba^2$
8. $5E^2 - E + 4E$
9. $2\pi\sqrt{P/R} + \pi\sqrt{P/R} - 8\pi\sqrt{E^2/R^2}$
10. Find the perimeter of a triangle with these sides: 27D, 15D and 23D.
11. $8x - 3x^2 + 4y - 2 - 5x^2 + x - 18y + 1$
12. $7IR + 8P/I - 5 - IR + 2P/I + 4 - P/I$
13. $28R_1 + 33R_2 - R_1 + 17R_1 - 15R_1 + 9 - 9R_2$
14. $I_1{}^2R_1 + 5I_1{}^2R_2 - 4I_2{}^2R_1 + I_2{}^2R_2$
15. $5(2\pi fL) + 7(2\pi fL) - [2\pi fL + 4(2\pi fL)]$

8-3 ADDITION AND SUBTRACTION AXIOMS

An equation is a mathematical statement that indicates two expressions are equal to each other. If there is one literal factor in an equation, it is called the *unknown*. The multiplication and division axioms were used in the previous chapter to *solve* for the unknown. Solving for the unknown means finding a value of the unknown, which, when substituted into the equation, makes the two sides of the equation equal. Axioms are used for this purpose.

When both sides of an equation are multiplied by the same number, both sides are still equal (multiplication axiom). Similarly, if both are divided by the same number, both sides are still equal (division axiom). When these axioms are applied to an equation, the resulting equation is equivalent to the original. Axioms are used to isolate the unknown so its value can be determined. Two more axioms are the addition and subtraction axioms.

Addition Axiom (AA). When equal numbers are added to equal numbers, their sums are equal. If $a = b$, then $a + c = b + c$.

Example A Demonstrate the addition axiom using $x = 5$.

Solution:

$x = 5$

$x + 3 = 5 + 3$ (AA, + 3)

$x + 3 = 8$

$5 + 3 = 8$

Note: The abbreviation (AA, + 3) is used to indicate the addition axiom has been used to add 3 to both sides of the previous equation.

Subtraction Axiom (SA). When equal numbers are subtracted from equal numbers, the differences are equal. If $a = b$, then $a - c = b - c$.

Example B Demonstrate the subtraction axiom using $x = 7$.

Solution:

$x = 7$

$x - 4 = 7 - 4$ (SA, − 4)

$x - 4 = 3$

$7 - 4 = 3$

Note: (SA, − 4) means the subtraction axiom was used to subtract 4 from both sides of the equation on the line above.

8-4 SOLVING EQUATIONS

If the same number is added to both sides of an equation, the resulting expressions must also be equal. The new expressions have different values than the old. However, the new equation is equivalent to the old. This means it has the same solution.

The axioms are used to solve the equation. Whenever a certain number is subtracted from the unknown in the original equation, add that number to both sides. This process will isolate the unknown.

Example A Solve $x - 5 = 4$ for the unknown.

Solution:

$x - 5 = 4$	
$x - 5 + 5 = 4 + 5$	(AA, + 5)
$x + 0 = 9$	(Combine terms)
$x = 9$	(Answer)
$9 - 5 = 4$	(Check)

Note: If 5 is added to the left side, only the variable x will remain because $-5 + 5 = 0$.

If a certain number is added to the unknown in the original equation, subtract that number from both sides.

Example B Solve $x + 7 = 12$ for x.

Solution:

$x + 7 = 12$	
$x + 7 - 7 = 12 - 7$	(SA, – 7)
$x + 0 = 5$	(Combine terms)
$x = 5$	(Answer)
$5 + 7 = 12$	(Check)

Example C Solve $4 + R = 11$ for R.

Solution:

$4 + R = 11$	
$4 - 4 + R = 11 - 4$	(SA, – 4)
$0 + R = 7$	(Combine terms)
$R = 7$	(Answer)
$4 + 7 = 11$	(Check)

Example D Solve $7 = 13 - I$ for the unknown.

Solution:

$7 = 13 - I$	
$7 - 13 = 13 - 13 - I$	(SA, – 13)
$-6 = 0 - I$	(Combine terms)
$-6 = -I$	(Change signs to find I)
$6 = I$	(Answer)
$7 = 13 - 6$	(Check)

Note: The answer $6 = I$ can just as well be written $I = 6$. The equation is solved whether it is on the left or right side. Changing the signs of both sides is a correct procedure, (MA, $\times -1$).

In the process of using the addition and subtraction axioms, a term disappears from one side of the equation. It reappears on the other side with its sign changed. This fact leads to a short cut in applying these axioms. The process is called *transposing;* moving a term to the other side of the equal sign. A term can be transposed from one side to the other by changing the sign.

In written sentences, certain words indicate that two numbers are to be added. Others designate subtraction. Some examples are:

Words Meaning Addition	Words Meaning Subtraction
Sum	Difference
Plus	Minus
Increase	Decrease
More than	Less than
Greater than	Fewer than
Larger than	Smaller than
Rise	Drop or fall

In subtraction, the terms of an expression are not *commutative*. That means they may not be interchanged.

"The number R minus 8" is R − 8 not 8 − R.
"Seven decreased by *b*" is 7 − *b* not *b* − 7.
"Nine less than P" is P − 9 not 9 − P.

Notice the last example. Whenever expressions such as: "less than," "fewer than," or "smaller than" are used, be careful in translating. The expressions "8 minus 15" and "8 less than 15" are not the same.

Addition is commutative. The expression: "the sum of Q and 5" can be written either "Q + 5" or "5 + Q." The words, "is," "is equal to" and "equals" are replaced by the equal sign. A letter such as *x* should replace the unknown if another letter is not given.

Example E Express algebraically and solve: What number added to 7 is 19?

Solution: Let *x* be the number

$x + 7 = 19$ (Equation)

$x = 19 - 7$ (Transpose 7)

$x = 12$ (Answer)

Example F Find a weight (W) if 500 lb less than W is 950 lb.

Solution: W is the symbol given for the unknown weight.

$W - 500 = 950$ (Equation)

$W = 950 + 500$ (Transpose 500)

$W = 1450$ lb (Answer)

EXERCISE 8-3

Solve (AA) and check.

1. $x - 5 = 3$
2. $x - 9 = 8$
3. $R - 8 = -2$
4. $E - 17 = -29$
5. $i - 2.25 = 3.15$
6. $-9 - x = 4$
7. $5 = x - 11$
8. $-18 = \theta - 25$
9. $3\ 1/2 = \alpha - 6\ 1/2$
10. $-\frac{15}{8} - x = -\frac{3}{8}$

Solve (SA) and check.

11. $x + 7 = 8$
12. $x + 14 = 23$
13. $x + 7 = -9$
14. $E + 11 = 28$
15. $10 + \theta = 35$
16. $8 + \alpha = 17$
17. $4 = \beta + 7$
18. $-9 + R = 3$
19. $-26 = 13 + t$
20. $\frac{15}{8} + I = \frac{5}{8}$

Solve and check these. State which axiom is used.

21. $P - 14 = 34$
22. $15 - R = -5$
23. $-\theta + 7 = -90$
24. $4.25 = 4 + \alpha$
25. $I - 8\frac{1}{2} = 17$
26. $-E + 7 = -5$
27. $0 = P_1 - 7.2$
28. $956.58 = 530.23 + W$
29. $P - 3.56 \times 10^3 = 1.04 \times 10^3$
30. $y - \frac{2}{3} = 1$

31. $y + \frac{2}{3} = 1$
32. $7 - Z_2 = 2$
33. $X_C + 950 = -325$
34. $X_L - 475 = 640$
35. $C + 3 \times 10^{-7} = 7 \times 10^{-6}$
36. $2\frac{1}{2} + W = 13\frac{1}{2}$
37. $54.5 + \Delta T = 38.2$
38. $\Delta f + 1028 = 1092$

Express in algebraic form then solve for the unknown.

39. The sum of 5 and m equals 18.
40. B plus seven equals eleven.
41. A current (I) increased by 3 A is 5 A.
42. A current (I) decreased by 6 A is 7 A.
43. A voltage (E) added to 12 V is 28 V.
44. 25 C° less than temperature (T) equals 75°C.
45. 90° smaller than angle (θ) equals 47°.
46. 3×10^{-9} F less than capacitance (C) is 5×10^{-9} F.
47. 50 m less than span (D) is 85 m.
48. 3500 Ω greater than resistance (R) equals 7250 Ω.

8-5 EQUATIONS WITH COMBINED OPERATIONS

Most equations involve more than one operation. These are solved one step at a time using the four axioms discussed previously. The unknown is isolated on one side of the equation. Either side is all right.

To isolate the unknown, all numbers on that side of the equation must be removed. This is done by performing the *inverse* of the operations that tie the expression together.

An expression such as $x - 5 = 7$ indicates the subtraction operation. Addition is the inverse of subtraction. The five is removed by adding 5 to both sides. Similarly, multiplication is used when the equation indicates division. In the equation, $x/3 = 4$, multiply by 3. Always perform the inverse of the operation shown in the equation:

Subtraction is the inverse of addition
Addition is the inverse of subtraction
Multiplication is the inverse of division
Division is the inverse of multiplication

When more than one operation is indicated, the unknown must be isolated a step at a time. Just perform the inverse operations of those shown. Do them in the following sequence. Omit any step not required. Perform the steps in *exactly* the order shown:

SOLVING EQUATIONS

- Use the addition and/or subtraction axioms to place all *terms* containing the unknown on one side of the equation and all other *terms* on the opposite side.
- Combine like terms by adding or subtracting. There should now be only one term on each side.
- Use the multiplication and/or division axioms to remove coefficients of the unknown (including negative signs).

Example A Solve: $4x - 9 = 3$ for x

Solution:

$4x - 9 + 9 = 3 + 9$	(AA, + 9)
$4x = 12$	(Combine terms)
$\frac{4x}{4} = \frac{12}{4}$	(DA, ÷ 4)
$x = 3$	(Cancel and evaluate)
$4(3) - 9 = 3$	(Check)
$12 - 9 = 3$	

Example B Solve: $9x + 12 = 4x - 3$ for the unknown

Solution:

$9x + 12 - 12 = 4x - 3 - 12$	(SA, − 12)
$9x = 4x - 15$	(Combine terms)
$9x - 4x = 4x - 4x - 15$	(SA, − 4x)
$5x = -15$	(Combine terms)
$\frac{5x}{5} = -\frac{15}{5}$	(DA, ÷ 5)
$x = -3$	(Cancel and evaluate)
$9(-3) + 12 = 4(-3) - 3$	(Check)
$-27 + 12 = -12 - 3$	

When there is more than one operation in a word problem, translate to algebraic expressions carefully. The expression must say exactly the same thing that the word statement does.

Example C Express algebraically and solve: ten increased by twice a number is thirty-one minus the number.

Solution: Let n represent the number

$10 + 2n = 31 - n$	
$10 + 2n + n = 31 - n + n$	(AA, + n)
$10 + 3n = 31$	
$10 + 3n - 10 = 31 - 10$	(SA, − 10)
$3n = 21$	
$n = 7$	(DA, ÷ 3)
$10 + 14 = 31 - 7$	(Check)

EXERCISE 8-4

Solve and check. Write out each step and indicate the axioms used. Do not use a calculator.

1. $5Z - 14 = 6$
2. $1 - 12R = 13$
3. $8.3E + 4.6 = 54.4$
4. $9 - \frac{C}{11} = -6$
5. $2\theta = 51 - \theta$
6. $6t - 3 = 15$
7. $\frac{6N}{4} - 8 = 7$
8. $\frac{3L}{4} + 9 = 12$
9. $7 - 5P = -8$
10. $-1 = 1 - \frac{P}{2}$
11. $3D - 5 - D - 2 = 0$
12. $5G - 12 - 4 = 3G$
13. $-1.2 - 2.3s = 0.9s - 0.4$
14. $10^2 = \frac{R}{3} + 10$
15. $\frac{2X_C}{3} - 5 = 7$

16. $8 - \frac{I}{7} = 15$
17. $\frac{8}{X_C} + 16 = 32$
18. $8P = 32 + 4P$
19. $\frac{\phi}{2} + 3 = 7 - \frac{\phi}{2}$
20. $2\alpha - 5 = -\alpha + 4$
21. $25\beta - 25 = 600$
22. $8 = 10 + 3R$
23. $7W + 3 = 30 - 2W$
24. $7 - \frac{9}{I} = 10$
25. $8 + 9E = 4E + 53$
26. $\frac{1}{Q} + 0.0045 = 0.0085$
27. $14 - 5C = 3C + 6$
28. $4E - 6 + 34 = 28 - 4E$
29. $0 = 3I - 65 + 2I$
30. $\frac{5T}{8} - 4 = 6$
31. $3C - 2.75C + 4 = 8 - 0.25C + 6 - C$
32. $4\theta + 6 - \theta - 2 = 7\theta - 11 + 5\theta - 3$

Express these statements algebraically and solve.

33. Ten increased by twice a number is twenty-two.
34. One-half of a number plus seven is seven.
35. Twenty-five amperes less than the product of five and a current (I) is thirty-five amperes.
36. The quotient of a voltage (E) and five, added to four volts is one-half of a volt.
37. Twice a resistance (R) plus twenty ohms is fifty ohms minus the resistance.
38. Six times a voltage (E) minus eighty volts is twice the voltage.
39. The quotient of twice an area (A) and eight is the area less than four square metres.
40. A distance that is 60 m less than five times another distance (s) is one-half the second distance plus 105 m. Find the distance s.

8-6 FORMULAS CONTAINING PLUS AND MINUS SIGNS

When an equation has one unknown, it has only one literal factor. Most formulas have more than one. Any one of the literal factors can be designated as the unknown. Solving formulas is different, therefore, because there are few like terms that can be combined after axioms are used. Additions and subtractions of unlike literal terms can only be indicated.

When solving formulas with plus or minus signs use the following procedure. The steps must be performed in the exact order listed.

SOLVING FORMULAS

- Use the addition axiom (AA) or the subtraction axiom (SA) to remove terms.
- Use the multiplication axiom (MA) or the division axiom (DA) to remove divisors or factors of the term containing the unknown.
- Evaluate numerical parts after each step to simplify if this is possible.

Example A Solve $P = 2L + 2w$ for L.

Solution:

$P - 2w = 2L + 2w - 2w$ (SA, – 2w)

$P - 2w = 2L$

$\frac{P - 2w}{2} = \frac{2L}{2}$ (DA, ÷ 2)

$L = \frac{P - 2w}{2}$ (Cancel)

EXERCISE 8-5

Solve these formulas for the indicated unknown.

	Formula	Solve for
1.	$E_T = E_1 + E_2$	E_1
2.	$R_T = R_1 + R_2 + R_3$	R_3
3.	$P_T = P_1 + P_2 + P_3 + P_4$	P_2
4.	$R_2 = R_T - 3R_1$	R_1
5.	$e = E - IR$	R and I
6.	$E = IR_1 + IR_2$	R_2
7.	$P = E_1 I + E_2 I + E_3 I$	E_3
8.	$P = I^2 R_1 + I^2 R_2 + I^2 R_3$	R_1
9.	$I = 2I_1 + 4I_2$	I_1
10.	$5R_1 = R_T - 6R_2 - 8R_3 - 3R_4$	R_2
11.	$\theta + \phi = 90$	ϕ
12.	$\alpha + \beta + \gamma = 180$	α

SERIES CIRCUITS

Ohm's Law and other relationships between electrical quantities have been used to demonstrate the use of formulas. Only circuits with a single load have been described. Most electrical circuits contain several loads. One way they can be connected is in series.

A *series circuit* is one in which the current is the same at every point. There is only one current path, Figure 8-1. If an ammeter is installed first at one place in a series circuit, then at other points, the reading will always be the same. If several ammeters are added at the same time, Figure 8-2, they all read the same. The meters between the battery and R_1, between R_1 and R_2, between R_2 and R_3 or between R_3 and the battery all give identical readings.

It is important to have a full understanding of the concept of current. It is a measure of the rate of flow of electric charge. Current is the number of charges (electrons) moving past a point each second. **One ampere is one coulomb per second or 6.24×10^{18} electrons per second.**

Almost all circuits contain a switch for control and a fuse for safety. In a series circuit these elements are also connected in series. They will not be shown in circuit diagrams in this and the following chapters unless there is a need to show them. Keep in mind, however, that all circuits require them.

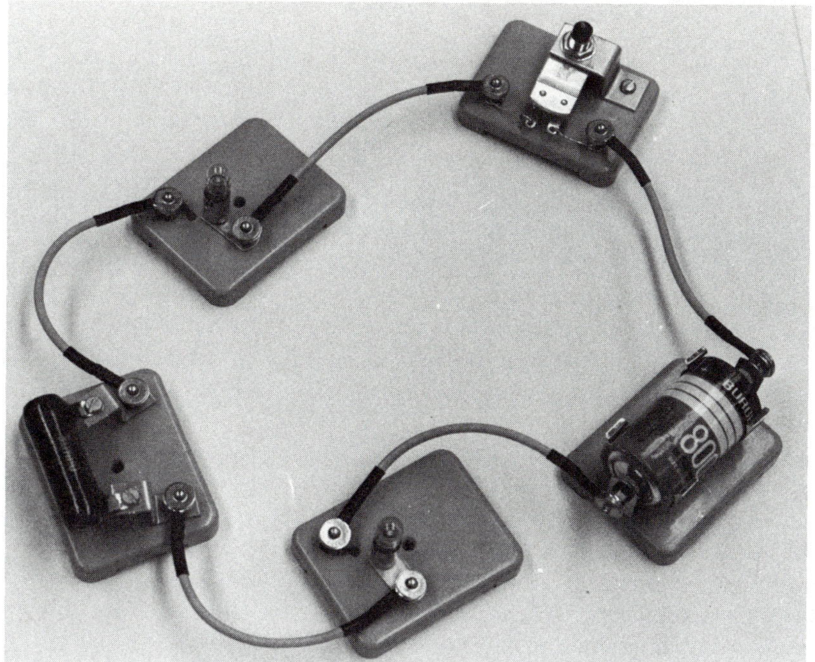

Figure 8-1 A battery, a lamp, a resistor, another lamp and a switch are connected in series (Courtesy of Southeast Nebraska Community College)

Figure 8-2 The four ammeters in the series circuit diagram all give the same reading.

8-7 RESISTANCE IN SERIES AND EQUIVALENT CIRCUITS

An *equivalent circuit* is a circuit containing a single resistance that is equal to the total resistance of a given circuit. Suppose an ohmmeter is used to measure the resistance of three resistors individually, Figure 8-3(A).

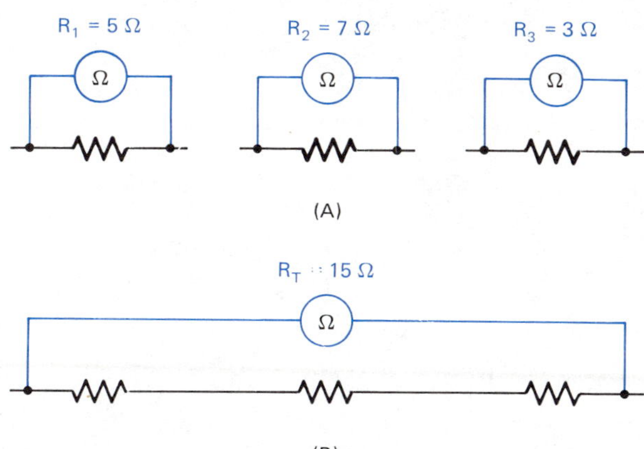

Figure 8-3 (A) Ohmmeters are used to measure individual resistances. (B) Total resistance equals the sum of the individual resistances. *Caution: Be sure all power sources are turned off when using an ohmmeter.*

The resistors are then connected together in series (end to end) and the resistance of the combination is measured. The combination resistance will be the sum of the three resistances, Figure 8-3(B).

In a series circuit, the value of the equivalent or total resistance (R_T) is the sum of all the resistances.

$$R_T = R_1 + R_2 + R_3 \ldots$$

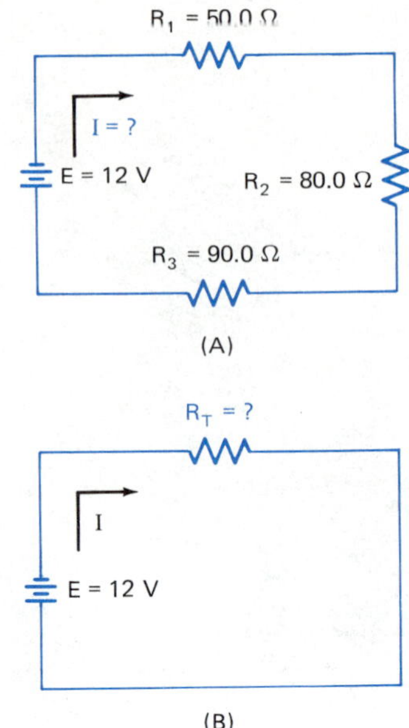

Figure 8-4

Example A Figure 8-4(A) shows a series circuit with three resistors. Make a diagram of the equivalent circuit and find the equivalent resistance.

Solution: Data: Given on circuit diagram. $R_T = ?$

Formula: $R_T = R_1 + R_2 + R_3$

Substitute: $R_T = 50.0\ \Omega + 80.0\ \Omega + 90.0\ \Omega$

$R_T = 220.0\ \Omega$

The equivalent circuit is shown in Figure 8-4(B).

No matter how many loads there are in series, the equivalent or total resistance is found by adding. Sometimes there are several loads in series having equal resistance. The equivalent resistance is found by multiplying the resistance by the number of loads. For example, six $5\bar{0}$-Ω resistors have an equivalent resistance of $30\bar{0}$ Ω.

8-8 CURRENT IN SERIES CIRCUITS

An equivalent circuit is one in which several resistances in combination are replaced by the equivalent resistance. The circuit voltage and current in the equivalent circuit are the same as in the original circuit. When the voltage and the equivalent resistance are known, it is easy to find the current. It is a simple Ohm's Law problem concerning a single load circuit.

Example A Find the current in the circuit shown in Figure 8-4.

Solution: Data: $R_T = 220.0\ \Omega, E = 12\ V, I = ?$

Formula: $I = \frac{E}{R}$ (Ohm's Law)

Substitute: $I = \frac{12\ V}{220.0\ \Omega}$

$I = 0.055\ A$

It has been stated that the current is the same at every point in a series circuit. This fact can be derived from Kirchhoff's Current Law:

> **Kirchhoff's Current Law:** The sum of all currents flowing into any point in a circuit is equal to the sum of all currents flowing out of that point.

This law simply states that current cannot pile up at any point in the circuit. In complex circuits, one usually applies this law at points where the current divides. In series circuits, choose any point on the circuit. The current flowing into the point equals the current flowing out. Traveling from point to point all the way around the circuit, one has to conclude that the current is the same at all points. In the circuit diagram, Figure 8-4(A), the current (0.055 A) is the same through each resistor as it is at all other points:

$$I = I_1 = I_2 = I_3$$

The current calculated using the equivalent resistance, Figure 8-4(B), is the same at every point in the original circuit.

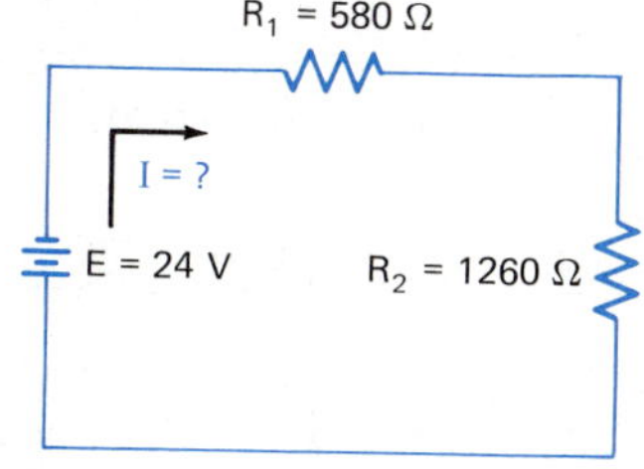

Figure 8-5

EXERCISE 8-6

Data, diagram, formula, substitute, check. Draw a circuit diagram and an equivalent diagram for each problem.

1. A circuit diagram is shown in Figure 8-5. (a) Determine the equivalent resistance of this circuit. (b) Find the current in the circuit.
2. A circuit diagram is shown in Figure 8-6. (a) Determine the equivalent resistance of this circuit. (b) Find the emf of the battery if the current is 47.5 mA.
3. A string of Christmas lights has 12 lamps connected in series. (a) Determine the total resistance if each lamp has a resistance of 9.83 Ω. (b) What is the current on a 118-V line?
4. An ammeter has a resistance of 0.005 Ω. It is connected in series with a 1-Ω coil. (a) Find the current when a 24.0-V battery is switched into the circuit. (b) What is the current if the ammeter is removed?
5. An electric clock has a small neon lamp and a resistor (39 kΩ) connected in series in a separate circuit. The total hot resistance is 80.6 kΩ when the applied voltage is 120 V. (a) Find the resistance of the lamp. (b) What current does this circuit draw from the 120-V line?
6. A 1.5-V dry cell is connected to one resistor. The current is 2.5 mA. What is the current if two more identical resistors are added in series with the first one?
7. Each of two field coils of a generator has a resistance of 65 Ω. The field coils are connected in series to a 115-V line. Find the current.
8. A 12-V car battery has a badly corroded terminal. There is a voltage drop of 3.5 V when the current is 45 A. (a) Find the resistance of the terminal. (b) If the starter motor has a resistance of 0.133 Ω, find the total resistance of the starter motor circuit including the terminal resistance. (c) Find the current in the circuit. (d) What is the current if the battery terminal is cleaned?

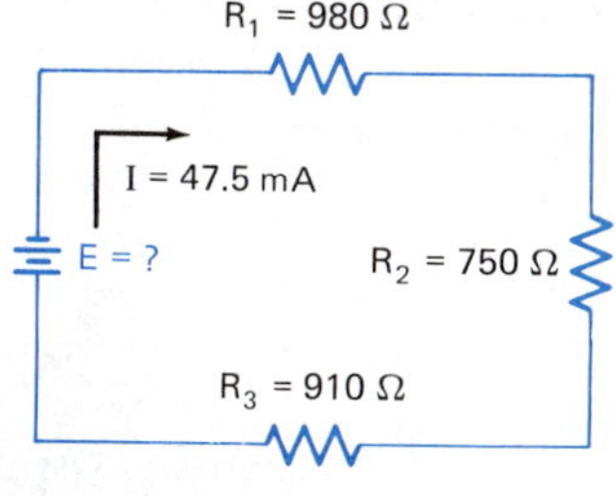

Figure 8-6

9. Four identical pilot (indicator) lamps are connected in series. The hot resistance of the total circuit is 2.8 kΩ. (a) Find the resistance of each lamp. If the circuit is connected across a 120-V source, find (b) the total current and (c) the current through each lamp.
10. Using No. 12 gage copper-conductor wire, street lights are connected in series along both sides of a street which is 2 km long. There is a total of 30 lamps on the street. The same current flows through each lamp so all have the same brightness. If the voltage across the circuit is 1135 V and the current is 6.60 A, find the resistance of each lamp. (Use the American Wire Gage Table found in the Appendix to find the resistance of 4 km of No. 12 copper-conductor wire.)

8-9 VOLTAGE IN SERIES CIRCUITS

In order to maintain a steady current in a circuit, an electromotive force (emf) is required. An emf can be produced by batteries, photoelectric or solar cells, Figure 8-7, thermocouples, generators and other devices. In each device, energy in some other form is converted into electrical energy. In batteries, chemical energy is converted to electrical energy. Solar cells use energy from the sun. Thermocouples use heat energy. Generators use the wind, water flow, coal, oil, gas and nuclear energy.

Sources. As charge moves through a source of emf, work is done by the source on the charge. A potential difference is created across the external circuit. It is the potential difference that causes the current flow in the

Figure 8-7 Solar cells compared to the size of a penny. Cells come in various sizes and shapes. All solar cells produce about 0.45 V. Cells are connected in series to produce the required voltage. Current and power depend on light intensity and total surface. (Courtesy of Southeast Nebraska Community College)

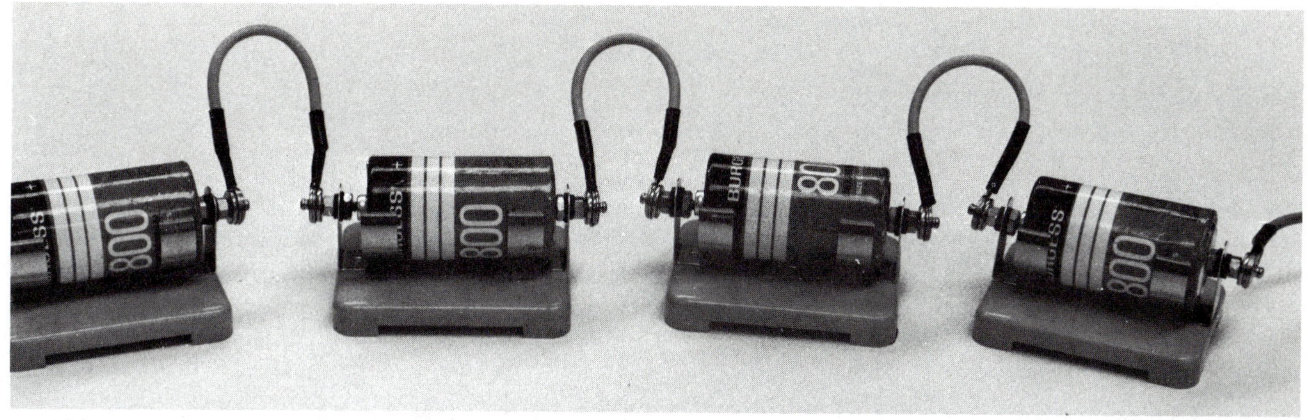

Figure 8-8 In the four cells connected in series, the total voltage is four times the voltage of each cell: 4(1.5 V) = 6 V. (Courtesy of Southeast Nebraska Community College)

circuit. Energy absorbed by the charge in the source is dissipated in the circuit, usually in the form of heat, light or useful work.

Sometimes cells are connected in series. If they are connected in such a way that all produce a current in the same direction, the emf is the sum of the emfs of all the cells. The emf of two cells connected so they produce current in opposite directions is the difference of the emfs of the two cells.

Figure 8-8 shows four identical dry cells in series. The emf of the combination (6.0 V) is four times the emf of a single cell (1.5 V). The type of cells used in an automobile battery have an emf of about 2 V. Three such cells in series produce 6 V. The more common 12-V battery has six cells connected in series.

Several cells in series form a battery of cells or simply, a battery. The entire combination acts as a single source. The free terminals of the first and last cell become the battery terminals; one positive, one negative.

Voltage Drop. Voltage sources are called *active* circuit elements. Internally, electrons flow from the positive terminal to the negative. In their passage through the battery, they gain energy. They undergo a *voltage gain* or *rise.*

In the external part of the circuit, electrons flow through devices called *passive* circuit elements. The loads, switches, fuses, voltmeters, ammeters and even the conductors that connect them are passive elements. Electrons flow from negative to positive in passive elements, Figure 8-9.

In active elements, there is a *voltage rise.* Electrons flow from (+) to (–) and gain energy as they are "pumped uphill." In passive elements, there are *voltage drops.* Electrons flow from (–) to (+) and dissipate (lose) energy as they "flow downhill." All the energy gained in the source is dissipated in the passive circuit elements. Voltage gains must be equal to the sum of all voltage drops.

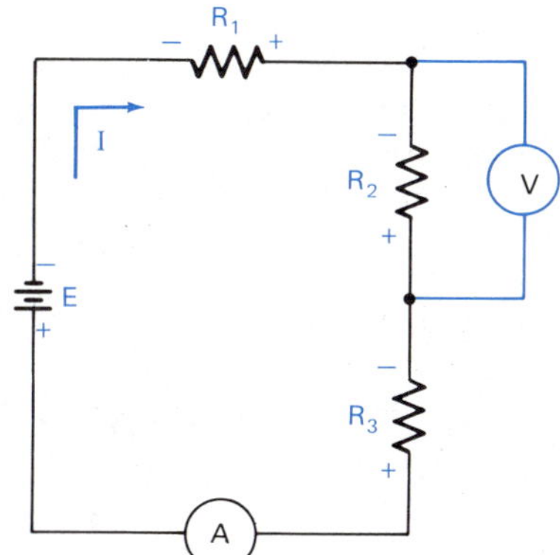

Figure 8-9 In the circuit diagram, the voltage meter shows the voltage drop across load (R_2).

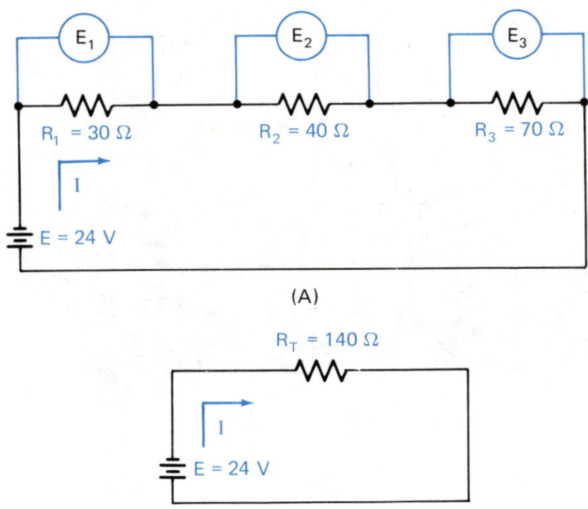

Figure 8-10 (A) Circuit diagram (B) Equivalent

8-10 OHM'S LAW AND SERIES CIRCUITS

Voltmeters can be used to find the voltage drops across the loads, Figure 8-10(A). Since voltage gains (emf of the source) must equal the sum of the voltage drops:

$$E_T = E_1 + E_2 + E_3 \ldots$$

where E_1, E_2 and E_3 are the voltage drops across the loads.

In the equivalent circuit, Figure 8-10(B), Ohm's Law can be used to find the current: $I = \frac{E_T}{R_T}$. When the current is determined, the voltage drop across each load can be calculated: $E_1 = IR_1$, $E_2 = IR_2$, $E_3 = IR_3$. Voltage drops are sometimes called *IR drops.*

Example A Find the voltage drops across each resistance in Figure 8-10 (A).

Solution: Data: $E = 24.0\ V$, $R_1 = 30\ \Omega$, $R_2 = 40\ \Omega$, $R_3 = 70\ \Omega$.

Note: Find the total resistance, find the current, then the voltage drops can be calculated.

Formula: $R_T = R_1 + R_2 + R_3$

Substitute: $R_T = 30\ \Omega + 40\ \Omega + 70\ \Omega = 140\ \Omega$

Formula: $I = \frac{E_T}{R_T}$ (Ohm's Law)

Substitute: $I = \frac{24\ V}{140\ \Omega} = 0.171\ A$

Formula: $E = IR$ (Ohm's Law)

Substitute: $E_1 = 0.171\ A \times 30\ \Omega = 5.13\ V$

$E_2 = 0.171\ A \times 40\ \Omega = 6.84\ V$

$E_3 = 0.171\ A \times 70\ \Omega = 12.0\ V$

Check: $E_T = 5.13\ V + 6.84\ V + 12.0\ V = 23.97\ V$

$E_T = 24.0\ V$ (Rounded)

The example verifies that the voltage gains equal the sum of the voltage drops in a series circuit. A more general statement is given by Kirchhoff's Voltage Law.

Kirchhoff's Voltage Law: In any complete conducting path, the sum of the voltage gains is equal to the sum of the voltage drops.

Kirchhoff's two laws can be used to study all types of electric circuits. They will be used in Chapter 13 as one method to analyze combination circuits.

The following statements give three important facts about series circuits:

- The voltage source equals the sum of the voltage (IR) drops.

$$E_T = IR_1 + IR_2 + IR_3 = IR_T$$

- The current is the same at every point in the circuit.

$$I = I_1 = I_2 = I_3$$

- The equivalent resistance equals the sum of the resistance of the different loads.

$$R_T = R_1 + R_2 + R_3$$

EXERCISE 8-7

Data, diagram, formula, substitute, check.

1. In the accompanying diagram, Figure 8-11, the two resistances are given. The current is 150 mA. Find (a) the equivalent resistance, (b) the voltage drop across R_1, (c) the voltage drop across R_2 and (d) the source voltage.

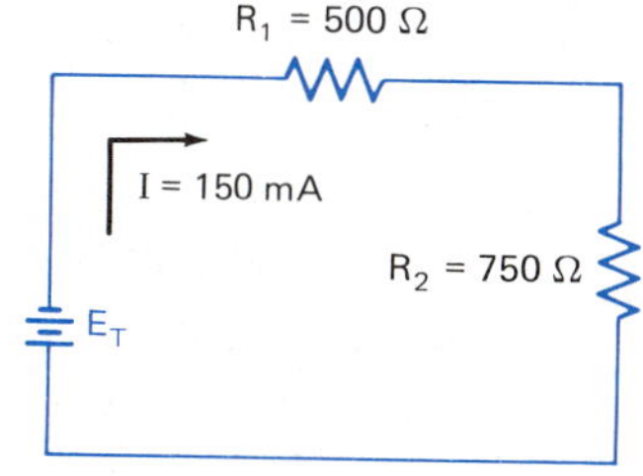

Figure 8-11

2. The voltage drop across R_1 is 50 V, Figure 8-12. The resistance R_1 is 550 Ω. Voltage drops E_2 and E_3 are 30 V and 40 V, respectively. Find (a) the current, (b) the resistance R_2, (c) the resistance R_3 and (d) the source voltage. (d) What is the voltage drop across the combination of R_1 and R_2?

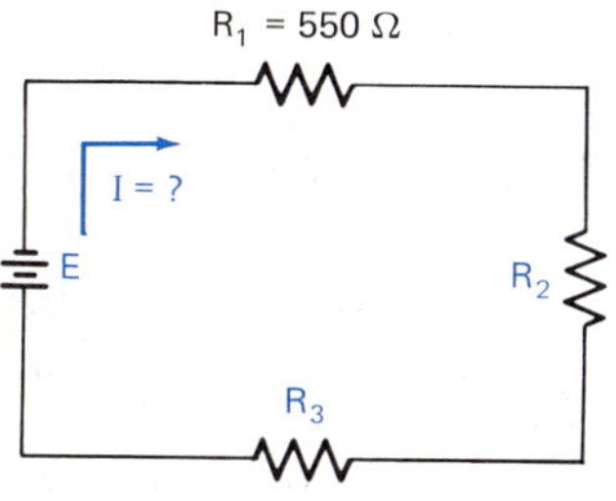

Figure 8-12

3. A circuit diagram is shown in Figure 8-13. (a) Find the current in this circuit. (b) What is the voltage drop across R_1? (c) What is the voltage drop across R_2? (d) What is the voltage drop across R_3? (e) What is the voltage drop across R_1 and R_2 combined? (f) What is the voltage drop across R_2 and R_3 combined?

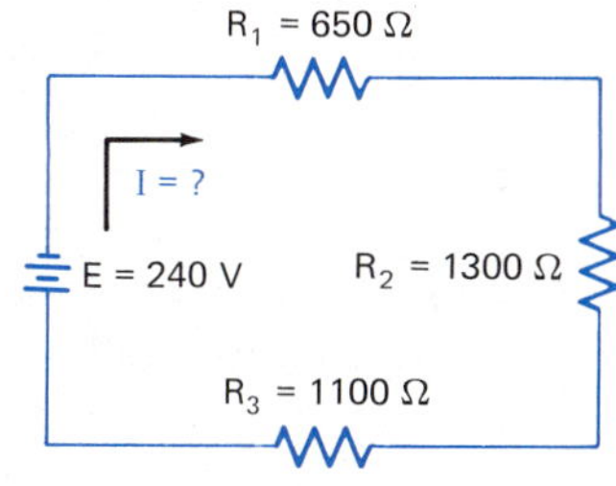

Figure 8-13

4. A voltmeter connected across R_1 reads 13 V, Figure 8-14. Across R_2, it reads 9.0 V. The source voltage is 28 V. R_3 is 75 Ω. Find (a) the current, (b) the voltage drop across R_3, (c) the resistance R_1 and (d) the resistance R_2.

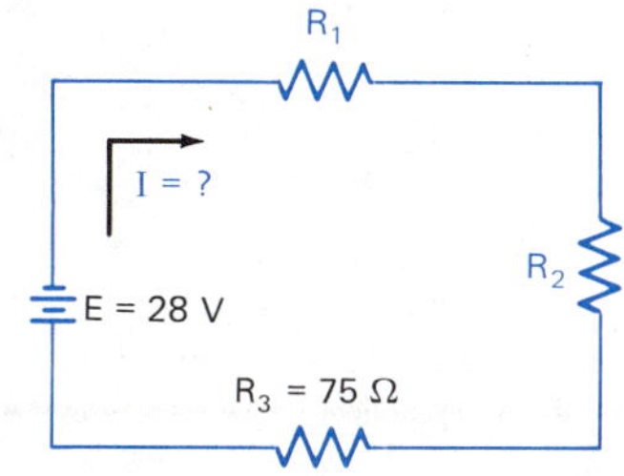

Figure 8-14

5. Four identical lamps are connected in series across a 120-V source, Figure 8-15. The hot resistance of each is 28 Ω. Find (a) the current and (b) the voltage drop across each lamp. (c) What happens if one bulb burns out?
6. An electrical cable has a resistance of 7.0×10^{-3} Ω/m of length. The cable is to carry 20 A at a voltage of 120 V. (a) Find the maximum length if the voltage drop of the cable cannot exceed 5 V. (b) What is the voltage drop per metre? (c) Find the resistance of the load.
7. A series street lighting system of 40 lamps draws a constant current of 6.60 A. (a) Find the voltage drop across each lamp if each lamp has 5.18 Ω of hot resistance. (b) Find the voltage drop along the line if the line has a resistance of 8.50 Ω. (c) What is the total resistance of the circuit? (d) What is the total voltage applied to the circuit?
8. A 1.5-V dry cell connected to one resistor draws 2.5 mA. Three identical resistors and two additional dry cells are added in series (aiding). Find (a) the total voltage, (b) the total resistance and (c) the current in the new circuit.
9. A two-wire power line delivers 30.3 A of current. It is ten kilometres long and delivers power to a factory. The resistance of each wire is 1.50 Ω/km. (a) If the voltage at the power plant is 33 $\bar{0}$00 V, what is the voltage at the factory? (b) What is the voltage drop per kilometre?
10. Three series resistors are connected to a 110-V source. R_1 has a current of 350 mA, R_2 has a voltage drop of 45 V and R_3 has a resistance of 120 Ω. Find (a) R_1, (b) R_2, (c) R_T, (d) E_1 and (e) E_3.
11. Three resistors are connected in a series circuit to a 115-V source. The total resistance is 460 Ω. The voltage drop across R_1 and R_2 combined is 75 V, Figure 8-16. Across R_2 and R_3 combined, the voltage drop is 80 V. Draw a circuit diagram showing the source and the three resistors. (a) Indicate the amount of resistance of and the voltage drop across R_1. (b) Indicate the amount of resistance of and the voltage drop across R_2. (c) Indicate the amount of resistance of and the voltage drop across R_3. (d) Indicate the current.
12. Three resistors are connected in a series circuit to a 22$\bar{0}$-V source. The voltage drop across R_1 and R_2 combined is 18$\bar{0}$ V. Resistors R_1 and R_3 have resistances of 82$\bar{0}$ Ω and 55$\bar{0}$ Ω, respectively. Find (a) I, (b) R_2, (c) R_T, (d) E_1 and (e) E_2.

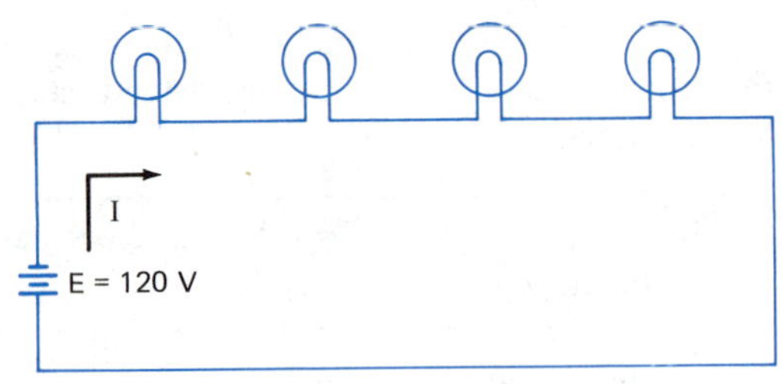

Figure 8-15

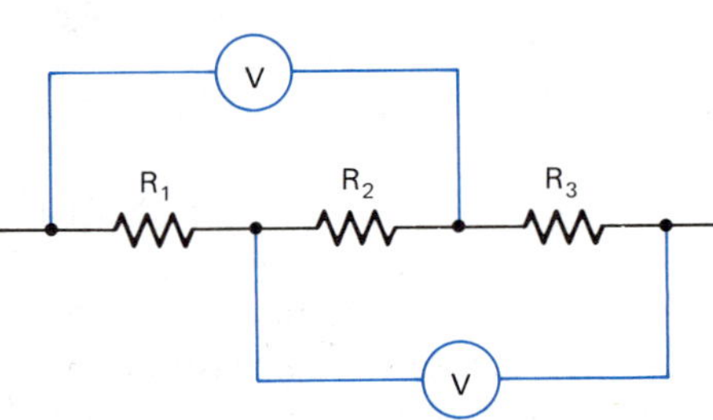

Figure 8-16

Sometimes resistors are placed in series with a load to limit the current through the load.

13. A 6.0-V car radio draws a current of 6.25 A. What resistance in series is necessary to operate the radio with a 12-V battery?

14. An electric motor is designed to limit the current through the coils when it first starts. A 50-hp, 230-V DC motor must be limited to an inrush current of 315 A. If the internal resistance of the motor is 0.063 Ω, what external series resistance is required at the instant of starting?
15. A rheostat is connected in series with a bank of stage lights. It is used to reduce the brightness of the lights by reducing the current. The 115-V stage lights have a total resistance of 4.6 Ω. The rheostat resistance varies from 0 to 25 Ω. Assume the resistance of the lights does not change. Find (a) the maximum voltage across the bank of lamps, (b) the minimum voltage across the bank of lamps, (c) the maximum current and (d) the minimum current.

8-11 INTERNAL SOURCE RESISTANCE

Batteries, generators and other sources of emf have some *internal resistance*. For practical purposes, a battery, for example, can be pictured as a pure emf and a series resistance. The internal resistance is represented by the lowercase (r).

The emf of a battery can be read with a high resistance voltmeter, Figure 8-17, connected across the open circuit terminals. There will be very little voltage drop across the internal resistance since the current is extremely small.

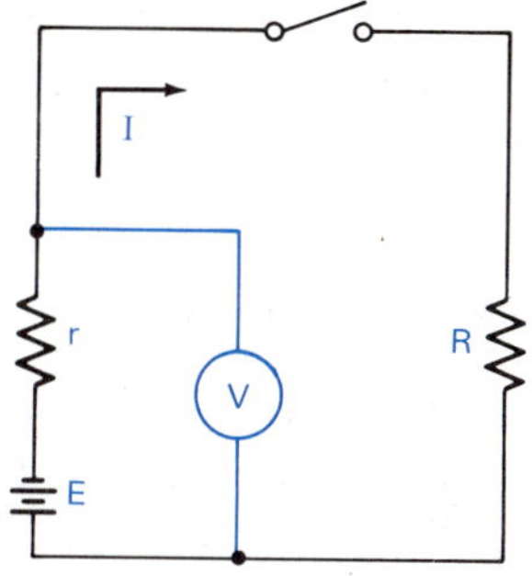

Figure 8-17 Circuit diagram showing internal resistance of source. In an open circuit, a high resistance voltmeter will indicate the emf.

When the circuit switch is closed, current flows through the external load and the internal resistance. If the load resistance is not too large, and a high current is flowing, the voltage drop across the internal resistance (Ir) may be significant. The *terminal voltage* E_T equals the emf minus the voltage drop across the internal resistance.

$$E_T = E - Ir$$

Example A A battery with a 12.0-V emf and an internal resistance of 0.80 Ω is connected in series with a 40.0-Ω load and a 19.2-Ω load, Figure 8-18. Find the current and the terminal voltage of the battery.

Solution: Data: $E = 12.0\text{ V}, r = 0.80\ \Omega, R_1 = 40.0\ \Omega, R_2 = 19.2\ \Omega$

Formula: $R_T = R_1 + R_2 + r$

Substitute: $R_T = 40.0\ \Omega + 19.2\ \Omega + 0.80\ \Omega = 60.0\ \Omega$

Formula: $I = \dfrac{E}{R_T}$

Substitute: $I = \dfrac{12.0\text{ V}}{60.0\ \Omega} = 0.20\text{ A}$

Formula: $E_T = E - Ir$

Substitute: $E_T = 12.0\text{ V} - (0.20\text{ A})(0.80\ \Omega)$

$E_T = 12.0\text{ V} - 0.16\text{ V} = 11.84\text{ V} \approx 11.8\text{ V}$

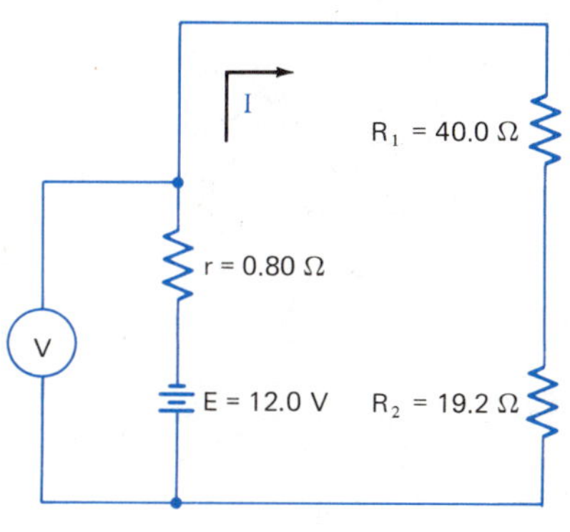

Figure 8-18

EXERCISE 8-8

Data, diagram, formula, substitute, check.

1. A battery has an open-circuit voltage of 12.0 V. When the current is 20 A, the terminal voltage drops to 10.2 V. (a) Find the internal resistance of the battery. (b) What is the resistance of the circuit load?
2. A battery has an open-circuit voltage of 24 V and an internal resistance of 190 mΩ. (a) How much current is flowing if the terminal voltage is 19.6 V? (b) Find the resistance of the circuit load.
3. An automobile battery has an emf of 12.0 V and an internal resistance of 8.5 mΩ. (a) If the starter motor draws 125 A, find the terminal voltage. (b) What is the starter-motor resistance?
4. The automobile in Problem 3 develops a corroded terminal connection with a voltage drop of 2.00 V when the starter is engaged. Find (a) the current in the starter circuit and (b) the terminal voltage of the battery. (c) What voltage is applied to the starter motor?

Use Figure 8-19 and the given data for problems 5–7. A certain battery has the characteristics shown in Figure 8-19. Its emf is given at the intersection of the curve and the vertical axis.

5. When the current is 10 A, find (a) the terminal voltage, (b) the internal resistance and (c) the resistance of the load.
6. When the current is 20 A, find (a) the terminal voltage, (b) the internal resistance and (c) the resistance of the load.
7. Find the resistance of the load when the terminal voltage is one-half the emf of the battery.
8. A 12.0-V automobile battery delivers a current of 15.0 A when the headlights are turned on. If the internal resistance of the battery is 0.015 Ω, find the voltage applied to the headlights.

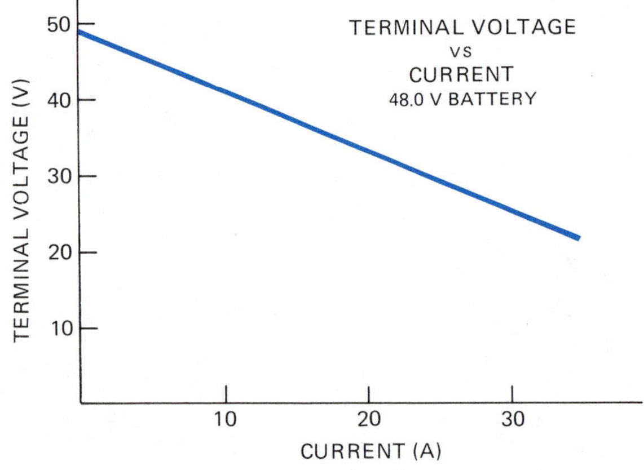

Figure 8-19

POWER IN SERIES CIRCUITS

The concept of power was introduced in Topic 3-18. At that time, single-load circuit applications were described. The student may wish to review that topic before proceeding.

Power can be defined in terms of work or energy. It is the rate at which work is done or the rate at which energy is generated or dissipated. Notice the words: generate and dissipate. They do not mean the same as create and destroy. Energy cannot be created. Electrical energy is generated at the expense of some other form of energy. It is dissipated (spread out) usually after conversion to heat energy. Energy cannot be destroyed.

Examples and applications of power dissipation in series circuits will be given immediately after some additional algebraic concepts are presented.

8-12 THE SQUARE AND SQUARE ROOT AXIOMS

Sometimes, the unknown in an equation is raised to a power or located within a radical sign. These can be solved using the *power axiom* and the *root axiom.* Only second and third powers and roots will be discussed.

Power Axiom (PA). When equal numbers are raised to the same power, the powers are equal. If $a = b$, then $a^n = b^n$.

Example A Demonstrate the power axiom using $x = 5$.

Solution:

$x = 5$

$x^3 = 5^3$ (PA, exp 3)

$x^3 = 125$

$5^3 = 125$

Note: (PA, exp 3) means the power axiom was used to take both sides of the previous equation to the 3rd power (exponent).

Root Axiom (RA). When the same root is taken of equal numbers, the roots are equal. If $a = b$, then $\sqrt[n]{a} = \sqrt[n]{b}$.

Example B Demonstrate the root axiom using $x = 16$.

Solution:

$x = 16$

$\sqrt{x} = \sqrt{16}$ (RA, square root)

$\sqrt{x} = 4$

$\sqrt{16} = 4$

Note: (RA, square root) means the square root of both sides of the equation in the line above was taken using the root axiom.

A radical equation contains the unknown under the radical sign. To solve radical equations, apply the power axiom. Keep in mind the following rules from Topic 5-16 regarding powers and roots.

Radical Form	*Exponential Form*
$(\sqrt{a})^2 = a$	$(a^{1/2})^2 = a$
$(\sqrt[3]{a})^3 = a$	$(a^{1/3})^3 = a$
$(\sqrt[n]{a})^n = a$	$(a^{1/n})^n = a$

Example C Solve for x. $\sqrt{x} = 4$

Solution:

$\sqrt{x} = 4$	
$(\sqrt{x})^2 = 4^2$	(PA, exp 2)
$x = 16$	(Answer)
$\sqrt{16} = 4$	(Check)

Example D Solve for x. $\sqrt{x - 5} = 4$

Solution:

$\sqrt{x - 5} = 4$	
$(\sqrt{x - 5})^2 = 4^2$	(PA, exp 2)
$x - 5 = 16$	
$x - 5 + 5 = 16 + 5$	(AA, + 5)
$x + 0 = 21$	
$x = 21$	(Answer)
$\sqrt{21 - 5} = 4$	(Check)

In an equation such as $x^3 = 27$, the unknown x is desired, not x^3. The unknown x is found by taking the cube root, ($x = 3$). Keep in mind the following rules concerning powers and roots. Compare these rules with those preceding Example C.

Radical Form	*Exponential Form*
$\sqrt{a^2} = a$	$(a^2)^{1/2} = a$
$\sqrt[3]{a^3} = a$	$(a^3)^{1/3} = a$
$\sqrt[n]{a^n} = a$	$(a^n)^{1/n} = a$

Example E Solve for a. $a^3 = 64$

Solution:

$a^3 = 64$	
$\sqrt[3]{a^3} = \sqrt[3]{64}$	(RA, cube root)
$a = 4$	(Answer)
$4^3 = 64$	(Check) (Use $\boxed{y^x}$)

Example F Solve for y. $(y - 2)^2 = 9$

Solution:

$(y - 2)^2 = 9$	
$\sqrt{(y - 2)^2} = \sqrt{9}$	(RA, square root)
$y - 2 = 3$	
$y - 2 + 2 = 3 + 2$	(AA, + 2)
$y = 5$	(Answer)
$(5 - 2)^2 = 9$	(Check)
$3^2 = 9$	

Formulas containing the unknown under a radical or exponent can be solved using the power or root axiom. Remove all terms and factors to the other side using the four basic axioms, then apply the power or root axiom. Remember, the bar in a radical sign is a grouping symbol.

Example G Solve for t. $s = \frac{1}{2}gt^2$

Solution:

$$s = \frac{1}{2}gt^2$$

$$2s = \left(\frac{2}{1}\right)\frac{1}{2}gt^2 \qquad \text{(MA, × 2 and cancel)}$$

$$2s = gt^2$$

$$\frac{2s}{g} = \frac{gt^2}{g} \qquad \text{(DA, ÷ g and cancel)}$$

$$\frac{2s}{g} = t^2$$

$$\sqrt{\frac{2s}{g}} = \sqrt{t^2} \qquad \text{(RA, } \sqrt{\ } \text{)}$$

$$\sqrt{\frac{2s}{g}} = t$$

$$t = \sqrt{\frac{2s}{g}}$$

Example H Solve for a. $c = \sqrt{a^2 + b^2}$

Solution:

$$c = \sqrt{a^2 + b^2}$$

$$c^2 = (\sqrt{a^2 + b^2})^2 \qquad \text{(PA, exp 2)}$$

$$c^2 = a^2 + b^2$$

$$c^2 - b^2 = a^2 + b^2 - b^2 \qquad \text{(SA, } -b^2\text{)}$$

$$c^2 - b^2 = a^2$$

$$\sqrt{c^2 - b^2} = \sqrt{a^2} \qquad \text{(RA, } \sqrt{\ } \text{)}$$

$$\sqrt{c^2 - b^2} = a$$

$$a = \sqrt{c^2 - b^2}$$

EXERCISE 8-9

Solve and check.

1. $\sqrt{x} = 5$
2. $\sqrt{E} = 8 \times 10^3$
3. $\sqrt{I} = 25$
4. $\sqrt[3]{\theta} = 5$
5. $\sqrt{R + 7} = 4$
6. $\sqrt{\phi + 9} = 3$
7. $\sqrt{\alpha - 5} = 15$
8. $\sqrt[3]{D + 172} = 7$
9. $\sqrt{3 - D} = 4$
10. $\sqrt[3]{X_L - 7} = 2$

Solve and check. Use calculators as necessary.

11. $R^2 = 36$
12. $P^2 = 576$
13. $\beta^3 = 27$
14. $\alpha^3 = 125$
15. $\omega^2 = 0.25 \times 10^{-4}$
16. $(I - 4)^2 = 16$
17. $(E + 5)^2 = 49$
18. $(C - 0.3)^3 = 125 \times 10^{-3}$
19. $R^3 = \frac{27}{64}$
20. $(X_L + 10)^2 = 1.6 \times 10^3$

Solve these formulas for the indicated unknown.

21. $A = \pi r^2$ (r)
22. $W = \frac{1}{2} I\omega^2$ (ω)
23. $Z^2 = R^2 + X^2$ (R)
24. $W = \frac{1}{2} mv^2$ (v)
25. $V = \frac{4}{3}\pi r^3$ (r)
26. $W = \frac{1}{2} CE^2$ (E)
27. $E = \sqrt{PR}$ (P)
28. $I = \sqrt{\frac{P}{R}}$ (R)
29. $W = I^2 Rt$ (I)
30. $f = \frac{1}{2\pi\sqrt{LC}}$ (C)
31. $X_C = \sqrt{Z^2 - R^2}$ (R)
32. $E = \sqrt{\frac{R}{t}}$ (t)
33. $I = \frac{E}{\sqrt{X^2 + R^2}}$ (X)

Solve these for the indicated unknown, then calculate its value using the given quantities. Round correctly.

Formula	Unknown	Knowns
34. $x^2 + y^2 = R^2$	(y)	$R = 1\bar{0}0$, $x = 5\bar{0}$
35. $P = I^2 Rt$	(I)	P = 12 W, $R = 4\bar{0}0$ Ω, t = 2.0 s
36. $r = \sqrt{\frac{A}{\pi}}$	(A)	r = 1 in
37. $P = \frac{E^2}{R}$	(E)	$P = 10\bar{0}$ W, R = 0.909 Ω
38. $P = I^2 R$	(I)	$P = 10\bar{0}$ W, R = 0.909 Ω
39. $r = \sqrt[3]{\frac{3A}{4\pi}}$	(A)	r = 2.0 ft
40. $E = \sqrt{\frac{2W}{C}}$	(C)	$W = 2.75 \times 10^{-4}$, $E = 10^2$

8-13 POWER

Power (P) does not describe the amount of energy generated or dissipated or the amount of work (W) accomplished. It simply tells how fast these things are happening:

$$P = \frac{W}{t}$$

If W is measured in joules, t in seconds, then P is in watts. The energy given up by the energy source to the electrons in a circuit and the energy dissipated in the load is very difficult to measure. The formula $P = W/t$ is not a very usable formula, therefore, for finding power in electric circuits.

In terms of circuit quantities, power is given by the formula:

$$P = EI$$

The power dissipated across a load is the product of the current and voltage. If the current is in amperes and the voltage in volts, the power unit is the volt-ampere (V·A). The compound unit (volt-ampere) is called the watt (1 V·A = 1 W).

If the right side of the formula $I = E/R$ (Ohm's Law) is substituted for current in the formula $P = EI$, another formula for power can be derived:

$$P = \frac{E^2}{R}$$

This formula is useful when the current is not known.

Similarly, if the right side of the formula $E = IR$ is substituted for E in the formula $P = EI$, a fourth formula for power is derived:

$$P = I^2R$$

This formula can be used to calculate power whenever the current and resistance are known. The power formulas and Ohm's Law apply to all DC circuits and to pure resistive ac circuits.

Many electrical circuits have more than one load. The loads may be connected in a series or parallel circuit or they may be arranged in a more complex network. Regardless of how they are coupled, the total power loss is equal to the sum of the power losses of all loads.

$$P_T = P_1 + P_2 + P_3 + \dots$$

The power loss in each load can be calculated. If voltage and current are known, use $P = EI$. When voltage and resistance are known, use $P = E^2/R$. Use $P = I^2R$ for circuit elements where the current and resistance are known. Keep in mind that the current is the same through every circuit element in a series circuit. Total power in a series circuit can be found with the formula $P_T = I^2R_T$.

Example A An 8.4-kΩ resistor has a 225-V voltage drop across it. If power is being dissipated at its rated amount for this resistor, what is the power rating?

Solution: Data: $R = 8.4\ k\Omega$, $E = 225\ V$, $P = ?$

Formula: $P = E^2/R$

Substitute: $P = \dfrac{(225\ V)^2}{8400\ \Omega} = 6.0\ W$

If a larger voltage was applied to this resistor, heat would be produced at a faster rate. Its temperature would increase to a point where insulation materials would be jeopardized. Since it cannot dissipate heat at a rate faster than 6 J/s, it would be destroyed.

Example B A heating element is rated at 1500 W. Find the current if the resistance is 8.10 Ω.

Solution: Data: $P = 15\bar{0}0\ W, R = 8.10\ \Omega, I = ?$

Formula: $P = I^2R$

$I = \sqrt{P/R}$ (DA, ÷ R; RA, $\sqrt{\ }$)

Substitute: $I = \sqrt{\dfrac{1500\ W}{8.10\ \Omega}} = 13.6\ A$

Check: Find voltage (E = IR). Find power (P = EI).

$E = 13.6\ A \times 8.10\ \Omega = 11\bar{0}\ V$

$P = 11\bar{0}\ V \times 13.6\ A = 15\bar{0}0\ W$

Example C Find the total power dissipation in a circuit having three resistors in series, Figure 8-20. One is being used at its power rating (1$\bar{0}$ kΩ, 2$\bar{0}$ W). The second has a resistance of 8$\bar{0}$ kΩ, and the third has a voltage drop of 135 V across it.

Solution: Data: $R_1 = 1\bar{0}\ k\Omega, P_1 = 2\bar{0}\ W, R_2 = 8\bar{0}\ k\Omega, E_3 = 135\ V, P_T = ?$

Formula: $P = I^2R$

$I = \sqrt{P/R}$ (DA, ÷ R; RA, $\sqrt{\ }$)

Substitute: $I = \sqrt{\dfrac{20\ W}{10\ 000\ \Omega}} = 45\ mA$

Formula: $P_2 = I^2R_2$

Substitute: $P_2 = (0.045\ A)^2(8\bar{0}\ 000\ \Omega)$

$P_2 = 160\ W$

Formula: $P_3 = E_3I$

Substitute: $P_3 = (135\ V)(0.045\ A) = 6\ W$

Formula: $P_T = P_1 + P_2 + P_3$

Substitute: $P_T = 20\ W + 160\ W + 6\ W = 186\ W$

Note: In this problem, the current can be found in the first resistor. It is the same in all parts of the circuit.

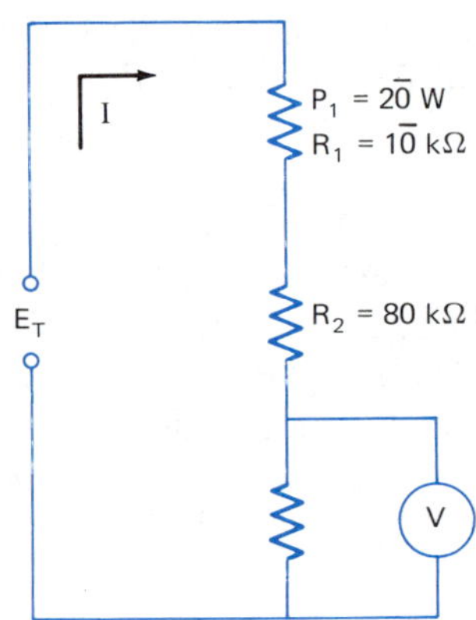

Figure 8-20

If the voltage across a load is changed, the current will also change. Unless the temperature changes drastically, the resistance will remain nearly the same. In situations like this, calculate the resistance to find the new power dissipation.

Example D A 22$\bar{0}$-V heating element is rated at 8.5 kW, Figure 8-21. If this element is connected across a 110-V line, what is the power dissipation?

Solution: Data: $E_1 = 22\bar{0}\ V, P_1 = 8.5\ kW, E_2 = 110\ V, P_2 = ?$

Formula: $P_1 = {E_1}^2/R$

$R = {E_1}^2/P_1$ (MA, × R; DA, ÷ P_1)

Substitute: $R = \dfrac{(220\ V)^2}{8500\ W} = 5.7\ \Omega$

Formula: $P_2 = \dfrac{{E_2}^2}{R}$

Substitute: $P_2 = \dfrac{(110\ V)^2}{5.7\ \Omega} = 2.1\ kW$

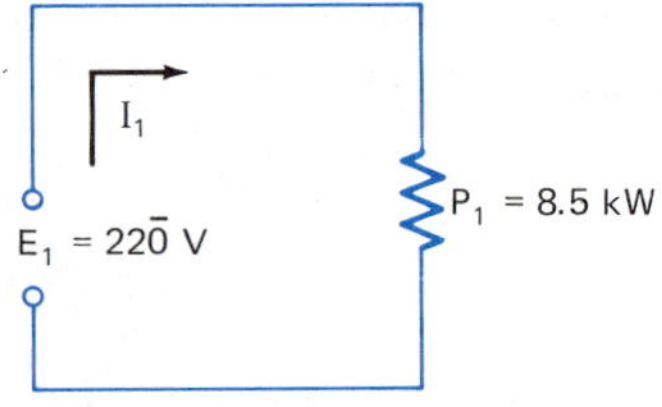

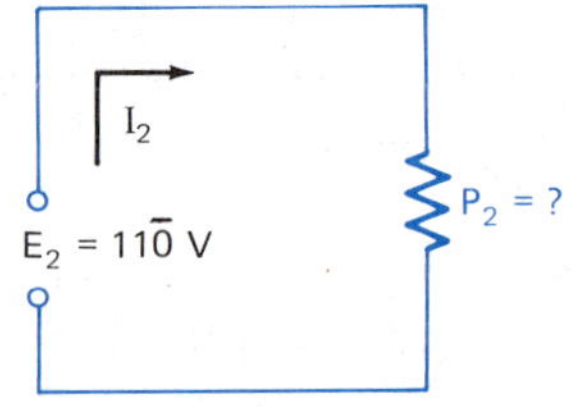

Figure 8-21

EXERCISE 8-10

Data, diagram, formula, substitute, check.

1. A room air conditioner has a resistance of 8.5 Ω. If it uses energy at a rate of 1560 W, what voltage is applied?
2. A line supplies 380 W of power to a hair dryer having a resistance of 34.8 Ω. What current is flowing?
3. Find the current through a corroded terminal if it uses 3.5 W of power and has a resistance of 42 Ω.
4. A washing machine operates on 115 V and has a resistance of 26.5 Ω. What power does it draw from the line?
5. A range with a self-cleaning oven draws 53 A. It has a resistance of 4.3 Ω. Find (a) the power and (b) the voltage.
6. A series circuit consists of R_1 = 35 kΩ, R_2 = 185 kΩ and R_3 = 9.4 kΩ. The current is 25 mA. Find the power dissipation in (a) R_1, (b) R_2 and (c) R_3. (d) Find the total power dissipated by the circuit. Check by using total resistance.
7. Two resistors, R_1 = 15 Ω and R_2 = $2\bar{0}$ Ω are connected in series to a 24-V source. Find (a) the voltage drop and (b) the power loss across each resistor.
8. Three lamps are connected in series in which the current is 1.2 A. One lamp is operating at its rated power of 60 W. The second lamp has a hot resistance of 192 Ω. The third has a voltage drop of 75 V across it. Find (a) the total power, (b) equivalent resistance and (c) the total voltage.
9. A 110-V Christmas tree light set has fifteen 6.0-W lamps in series. (a) What is the voltage drop across each lamp? (b) Find the current. (c) What is the hot resistance of each lamp? (d) Find the total power for the light set.
10. A 110-V Christmas tree light set has fifteen 6.0-W lamps in series. The light set has some frayed wiring. A do-it-yourself electrician removes three of the lamps and reconnects the remaining twelve lamps in series. Assume the hot resistance is the same. (a) What is the current? (b) Find the voltage drop across each lamp. (c) How much power is dissipated in each lamp? (d) What is the total power? (e) Why would it be dangerous to use this altered set on a dry Christmas tree?
11. Three resistors: 22 kΩ, 25 W; 58 kΩ, 100 W; and 90 kΩ, 15 W are connected in series. Find the maximum emf that can be applied without exceeding the power rating of any of them.

    Hint: The current will be the same through all of them.
12. Three resistors are connected in series: R_1 = 80 kΩ, R_2 = 200 kΩ and R_3 = 5 kΩ. Find the maximum voltages these resistors can be connected to without overheating any of them. All three are rated at 0.5 W.
13. A 12.0-V automobile battery delivers a current of 15.0 A when the headlights are on. (a) If the internal resistance of the battery is 0.015 Ω, find the power dissipated within the battery. (b) How much power is dissipated in the headlights?

14. A transmission line has a resistance of 2$\bar{0}$ Ω. The conductor resistance is in series with the primary of a transformer at the end of the system. The transformer draws 1.2 MW at a certain time. How much power is lost in the line if the voltage drop across the transformer is 13 800 V?
15. A stove element is rated at 600 W, 115 V. If two of these elements are connected in series to a 115-V line, what is the power dissipation? Assume no change in resistance.
16. (a) What resistance must be connected in series with a 6.0-V car radio if it is to be operated by a 12-V battery? It draws a current of 6.3 A. (b) What power rating must the resistor have?
17. Constant current street lamps each have a hot resistance of 5.18 Ω. Thirty lamps are connected by 4 km of No. 12 copper conductor. The current is 6.6 A. Find (a) the total voltage applied and (b) the total power dissipation of the circuit.
18. A 12-V battery connected to a resistor draws 25 mA. If four identical resistors and two additional batteries are added in series (batteries aiding), find the power dissipation in the circuit.
19. Three resistors are connected in series. The first has a resistance of 50 Ω. The second has a voltage drop of 60 V. The third passes a current of 450 mA and has a voltage drop of 32.5 V. Find the total power dissipation in the circuit.
20. A battery has an emf of 75 V and an internal resistance of 5 Ω. A variable series resistor is adjusted to 1 Ω, 2 Ω, 3 Ω, 4 Ω, 5 Ω, 6 Ω, 7 Ω, 8 Ω, 9 Ω and 10 Ω successively. (a) Find the power dissipated in the load at each setting. (b) Construct a graph of these results (power dissipation as a function of load resistance). (c) Is it true that maximum power is delivered to the load when its resistance matches the internal resistance of the source?
21. Figure 8-22 represents the lines from a 120-V generator to a load with a resistance R_3. A voltmeter across the load reads 108 V. R_1 and R_2 represent the line resistance between the generator and load and are equal. The current is 480 mA. From the data given, determine the value of (a) R_1, (b) R_2, (c) R_3, (d) P_1, (e) P_2 and (f) P_3.

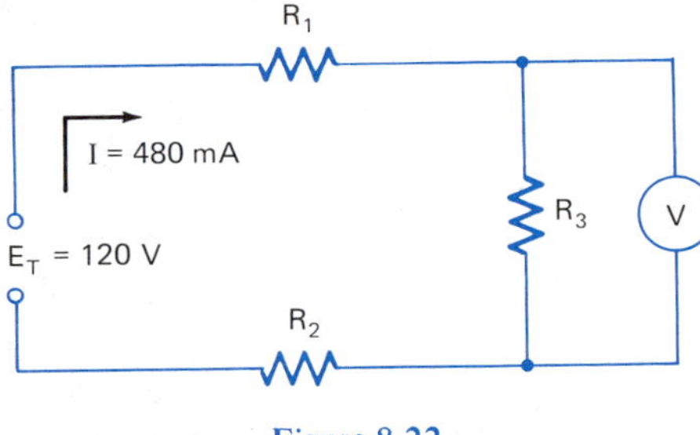

Figure 8-22

22. Three lamps (40 W, 60 W and 100 W) are connected in series. All are rated at 115 V. What is the maximum voltage that can be applied to the circuit without burning out any of the lamps?
23. Thirty-two solar cells are connected in series to form a solar panel. Each cell develops 0.45 V and the entire panel delivers 1.39 A in full sun. The panel is used to charge a 12-V battery having an internal resistance of 0.4 Ω. Find (a) the voltage produced by the solar panel and (b) the voltage drop across the internal resistance. (c) Find the power developed by the panel and (d) find the power dissipated by the internal battery resistance. (e) At what rate is the battery being charged with energy?
24. Repeat Problem 23 at one-fourth full sun. Voltages are the same but current is now 346 mA.

CHAPTER

9

PARALLEL CIRCUITS

OBJECTIVES

After satisfactorily completing this chapter, the student should be able to:

- Add and subtract common fractions including reciprocals.
- Describe and calculate voltage, total current and total resistance in parallel and combination circuits and analyze such circuits using Ohm's Law.
- Find the equivalent resistance in series-parallel combination circuits and find current and voltage in any branch.
- Calculate power dissipation in parallel and combination circuits.

Many multiple load electric circuits have the loads connected in parallel. The reason for this is that most electrical devices (light bulbs, motors, home appliances, etc.) have the power rating specified at a certain voltage. These electrical devices either do not work well or do not work at all at any voltage other than the specified voltage.

All 120-V circuits in a home (kitchen, utility room, living room, etc.) are connected in parallel to each other. Devices plugged into the outlets of any of these circuits are all automatically connected in parallel with each other. This ensures that the applied voltage will be 120 V. That is one of the characteristics of parallel circuits. The voltage is the same on all loads.

Parallel circuits will be covered in more detail later in this chapter. The addition and subtraction of common fractions is discussed first and is required for finding equivalent resistance in parallel circuits.

ADDITION AND SUBTRACTION OF COMMON FRACTIONS

Common fractions were introduced in the beginning topics of Chapter 3. The reader may wish to review those topics before continuing. Some terminology is briefly discussed in this chapter as a quick review.

The upper and lower terms of a common fraction are called the numerator and the denominator, respectively. A proper fraction is one in which the numerator is less than the denominator. The value of a proper fraction is less than one. Improper fractions have a larger numerator than denominator. The value of an improper fraction is greater than one and can be expressed as a mixed fraction.

Equivalent fractions are various fractions having the same value. Fractions with the larger terms can be reduced to ones with lower terms (10/16 reduces to 5/8). Both terms of a fraction can be multiplied or divided by the same number. The result is an equivalent fraction.

To add or subtract common fractions, the fractions must all have *common* (same) *denominators.* Several fractions to be added must be changed to equivalent fractions with common denominators. The sum is then found by adding numerators. The sum will have the same denominator as the fractions being added.

9-1 THE LEAST COMMON DENOMINATOR

For a group of fractions, the *least common denominator* (LCD) is the smallest number that can be used as a common denominator. The LCD is the smallest number that can be evenly divided by all the denominators. A *common denominator* can always be found by simply multiplying all denominators together. However, this will not be the least common denominator unless all denominators happen to be prime.

Example A Find a common denominator for the fractions: $\frac{1}{3}, \frac{5}{6}$ and $\frac{3}{4}$.

Solution: The product $3 \times 6 \times 4 = 72$ is a common denominator. It is not the LCD, however. Other numbers that all denominators evenly divide are 48, 36, 24 and 12, among others. The number 12 is the LCD.

To find the least common denominator, reduce all denominators to prime factors. Multiply the prime factors together, taking common ones only once. In some cases, the LCD can be found by inspection.

Example B Find the least common denominator for the fractions: $\frac{1}{3}, \frac{5}{6}$ and $\frac{3}{4}$.

Solution: Find the prime factors of each denominator.

3 = 3
↓
6 = 3 X 2
↓
4 = 2 X 2
↓ ↓ ↓
3 X 2 X 2 = 12.

Multiply, taking each common prime once. Thus, 12 is the LCD.

To find equivalent fractions with the LCD, find the number of times the original denominator will divide the LCD.

Example C Convert the fractions $\frac{1}{3}, \frac{5}{6}$ and $\frac{3}{4}$ to equivalent fractions with the LCD (12) as the new denominator for each.

Solution: 3 divides 12 four times. Multiply both terms of $\frac{1}{3}$ by 4.

$$\frac{4 \times 1}{4 \times 3} = \frac{4}{12}$$

6 divides 12 two times. Multiply both terms of $\frac{5}{6}$ by 2.

$$\frac{2 \times 5}{2 \times 6} = \frac{10}{12}$$

4 divides 12 three times. Multiply both terms of $\frac{3}{4}$ by 3.

$$\frac{3 \times 3}{3 \times 4} = \frac{9}{12}$$

The equivalent fractions are $\frac{4}{12}, \frac{10}{12}$ and $\frac{9}{12}$.

Example D Find the least common denominator for the fractions: $\frac{3}{4}, \frac{7}{10}$ and $\frac{9}{15}$. Convert the fractions to equivalent fractions with the LCD.

Solution: Find the prime factors:

4 = 2 X 2
↓
10 = 2 X 5
↓
15 = 5 X 3
↓ ↓ ↓ ↓
2 X 2 X 5 X 3 = 60 (LCD)

The equivalent fractions are: $\frac{3}{4} = \frac{45}{60}, \frac{7}{10} = \frac{42}{60}$ and $\frac{9}{15} = \frac{36}{60}$.

EXERCISE 9-1

Find the least common denominator of these groups of fractions. Rewrite the fractions in each group as equivalent fractions using the least common denominator.

1. $\frac{1}{2}, \frac{2}{3}, \frac{3}{4}$
2. $\frac{1}{3}, \frac{5}{6}, \frac{3}{5}$
3. $\frac{4}{9}, \frac{1}{3}, \frac{1}{6}$
4. $\frac{3}{4}, \frac{2}{5}, \frac{7}{10}$
5. $\frac{1}{9}, \frac{9}{12}, \frac{15}{24}$
6. $\frac{17}{45}, \frac{16}{25}, \frac{5}{21}, \frac{4}{15}$

9-2 ADDING AND SUBTRACTING FRACTIONS

Two or more fractions are said to be *similar* or *like fractions* when all have the same denominator. *Unlike fractions* do not have the same denominators. Like fractions are added or subtracted the same as so many nuts or bolts:

3 bolts + 5 bolts + 1 bolt = 9 bolts
3 eighths + 5 eighths + 1 eighth = 9 eighths

$$\frac{3}{8} + \frac{5}{8} + \frac{1}{8} = \frac{9}{8}$$

ADDING AND SUBTRACTING LIKE FRACTIONS

- To add like fractions, add the numerators and place the sum over the common denominator.
- To subtract like fractions, subtract the numerators and place the difference over the common denominator.

Example A Add: $\frac{3}{16} + \frac{5}{16} + \frac{1}{16}$

Solution: $\frac{3 + 5 + 1}{16} = \frac{9}{16}$

Example B Subtract: $\frac{17}{25} - \frac{12}{25}$

Solution: $\frac{17 - 12}{25} = \frac{5}{25} = \frac{1}{5}$

When the fractions to be added or subtracted are not like fractions, the fundamental rule of arithmetic applies: *only like quantities can be added or subtracted.* When unlike fractions are to be added or subtracted, the fractions must first be written as equivalent fractions with the same denominators. The equivalent like fractions are then added or subtracted.

Example C Add: $\frac{4}{5} + \frac{2}{3} + \frac{7}{15}$

Solution: The LCD is 15

$$\frac{12}{15} + \frac{10}{15} + \frac{7}{15} = \frac{12 + 10 + 7}{15} = \frac{29}{15} = 1\frac{14}{15}$$

(By inspection)

Example D Subtract: $\frac{5}{6} - \frac{19}{24}$

Solution: The LCD is 24

$$\frac{20}{24} - \frac{19}{24} = \frac{20 - 19}{24} = \frac{1}{24}$$

Reciprocals of whole numbers are proper fractions. The reciprocals are added or subtracted in the same manner as any common fractions.

Example E Add the reciprocals of these numbers: 3, 6, 8 and 12.

Solution: The problem is to add:

$$\frac{1}{3} + \frac{1}{6} + \frac{1}{8} + \frac{1}{12}$$

The LCD is 24.

$$\frac{8}{24} + \frac{4}{24} + \frac{3}{24} + \frac{2}{24} = \frac{17}{24}$$

EXERCISE 9-2

Perform the indicated operation. Express answers as mixed fractions reduced to lowest terms. Do not use a calculator.

1. $\frac{2}{5} + \frac{3}{5} + \frac{1}{5}$
2. $\frac{7}{9} - \frac{3}{9}$
3. $\frac{3}{4} + \frac{1}{2} - \frac{1}{4}$
4. $\frac{3}{5} - \frac{2}{10}$
5. $\frac{9}{15} + \frac{2}{30} - \frac{2}{15}$
6. $\frac{5}{12} - \frac{7}{24} + \frac{2}{12} + \frac{3}{24} - \frac{1}{12}$
7. $\frac{5}{12} + \frac{7}{15} + \frac{13}{30}$
8. $\frac{7}{12} + \frac{3}{10} - \frac{1}{4}$
9. $\frac{2}{3} - \frac{7}{11}$
10. $4 - \frac{2}{3} - \frac{3}{4} + \frac{1}{6} - 3$
11. $\frac{1}{6} + \frac{1}{3} + \frac{1}{4} - \frac{1}{2}$
12. $\frac{1}{8} + \frac{1}{3} - \frac{1}{6}$
13. $\frac{1}{10} + \frac{1}{12} - \frac{1}{9}$

Solve the following problems.

14. An electrician worked the following amounts of overtime one week: $2\frac{3}{4}$ hours, $1\frac{1}{3}$ hours and $2\frac{1}{2}$ hours. Find the total overtime hours the electrician worked that week.

Hint: To add mixed fractions, add the whole numbers and fractions separately, then combine.

15. An electrician had the following change expressed in dollars in his pocket: $\frac{3}{2}$ (halves), $\frac{5}{4}$ (quarters), $\frac{2}{10}$ (dimes), $\frac{3}{20}$ (nickels) and $\frac{4}{100}$ (cents). How many dollars of change does he have expressed as a mixed fraction reduced to lowest terms?

16. The following pieces of wire are required for a job: $5\frac{1}{2}''$, $9\frac{5}{8}''$, $7\frac{13}{16}''$ and $8\frac{3}{4}''$. What is the total length of wire used expressed as a mixed fraction?
17. A full 30-gal drum of oil was used to refill four pole transformers. How much oil is left in the drum if the transformers required the following amounts in gallons: $3\frac{1}{2}$, $2\frac{3}{4}$, $8\frac{1}{8}$ and $\frac{7}{4}$?
18. In a crossarm installation, a bolt must pass through a pole $9\frac{5}{16}''$ thick, a crossarm $4\frac{3}{4}''$ thick, two washers, each $\frac{1}{8}''$ thick, a nut $\frac{5}{16}''$ thick and $\frac{7}{8}''$ of thread must be left for an eye nut. Find the required length of the bolt. How many inches too long is a 16″ bolt?
19. Three light bulbs are connected in series. The resistances, in ohms, are as follows: $1\frac{1}{3}$, $\frac{7}{8}$ and $\frac{4}{5}$. Find the total resistance.
20. A line maintenance truck carries four coils of rope with the following lengths in feet: $35\frac{1}{5}$, $48\frac{3}{4}$, $92\frac{1}{3}$ and $16\frac{1}{2}$. What total length of rope is this?

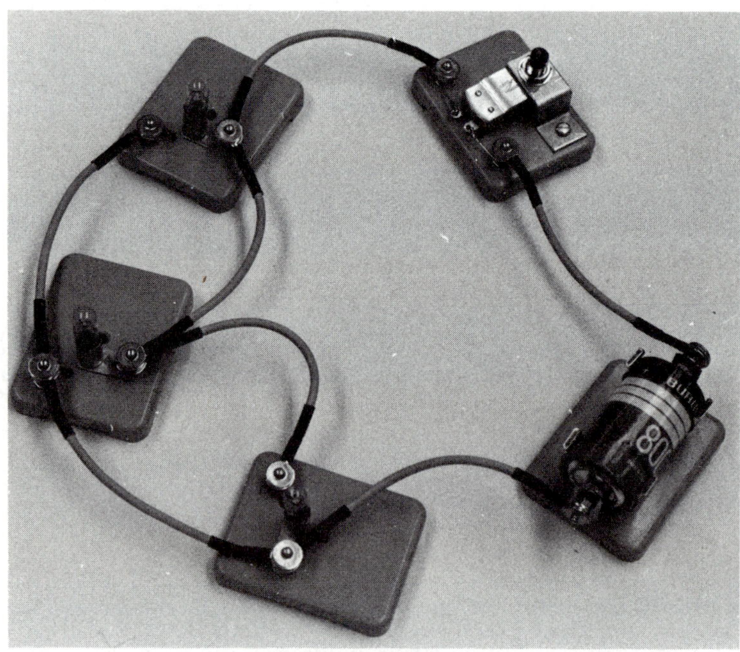

Figure 9-1 A parallel circuit. A flashlight cell and a switch with three lamps in parallel. (Courtesy of Southeast Nebraska Community College)

PARALLEL CIRCUITS

Parallel circuits were first introduced in Chapter 2 to show how voltmeters are connected in a circuit. In a parallel circuit, the current in the main circuit divides and follows separate paths through the loads, Figure 9-1. The various currents recombine as they leave the parallel branches before returning to the source.

As discussed in the beginning paragraphs of this chapter, the voltage on all loads is the same in a parallel circuit. A voltmeter across the A and B sides, Figure 9-2, of any of the loads will give the same readings. It is the same as putting the voltmeter directly across the source terminals. In fact, the circuit conductors are simply extensions of the source terminals.

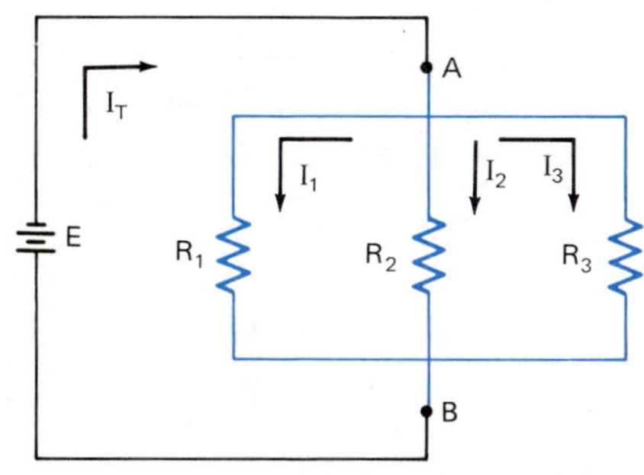

Figure 9-2 A three-load parallel circuit diagram

The current in the main conductors is greater than the current that would exist if only one of the loads was in the circuit. This means the equivalent resistance is less than the smallest of the three load resistances.

Obviously, the different branches of a parallel circuit are independent circuits. Each branch usually has its own control (switch). When one load is turned off, the others continue operating without any noticeable change. The safety device (fuse) is usually located in the main line.

9-3 RESISTANCE IN PARALLEL CIRCUITS AND EQUIVALENT CIRCUITS

As with series circuits, a simpler circuit with a single resistance can replace a circuit with a group of resistances in parallel. A circuit that is equivalent to the one in Figure 9-3(A) is shown in Figure 9-3(B).

The value of the equivalent resistance (R_T) can be derived from Kirchhoff's Current Law. The current entering point A equals the sum of the currents leaving point A, Figure 9-3:

$$I_T = I_1 + I_2 + I_3$$

Substituting Ohm's Law values gives

$$\frac{E}{R_T} = \frac{E}{R_1} + \frac{E}{R_2} + \frac{E}{R_3}$$

Divide both sides by E (division axiom).

$$\frac{1}{R_T} = \frac{1}{R_1} + \frac{1}{R_2} + \frac{1}{R_3}$$

There will be as many terms on the right as there are branches in the circuit. For a two-branch circuit, the equivalent resistance can be found by solving for R_T. The result is

$$R_T = \frac{R_1 R_2}{R_1 + R_2}$$

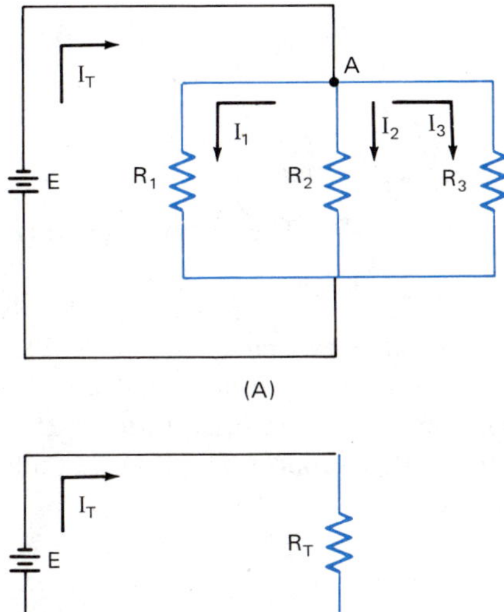

Figure 9-3 (A) Parallel circuit (B) Equivalent circuit

The equivalent resistance of any two resistances in parallel is equal to their product divided by their sum.

Example A Find the equivalent resistance of a 200-Ω resistor and a 300-Ω resistor in parallel, Figure 9-4.

Solution: Data: $R_1 = 200\ \Omega, R_2 = 300\ \Omega, R_T = ?$

Formula: $R_T = \dfrac{R_1 R_2}{R_1 + R_2}$

Substitute: $R_T = \dfrac{(200\ \Omega)(300\ \Omega)}{200\ \Omega + 300\ \Omega} = \dfrac{60\ 000}{500} = 120\ \Omega$

Check: R_T is less than either resistance.

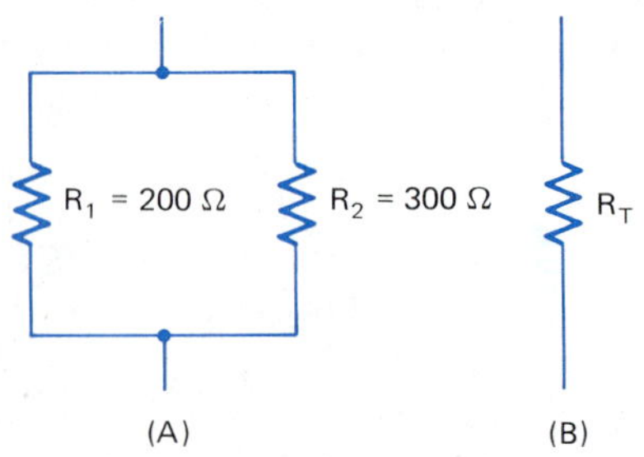

Figure 9-4

For parallel circuits having three or more branches, use the formula in which the reciprocals are added. If the numbers are simple, the methods of adding fractions manually can be used. For numbers that are not simple, use the calculator.

Example B Find the equivalent resistance of a parallel circuit, Figure 9-5, with resistances of 12 Ω, 15 Ω, 25 Ω and 32 Ω.

Solution: Data: $R_1 = 12\ \Omega, R_2 = 15\ \Omega, R_3 = 25\ \Omega, R_4 = 32\ \Omega, R_T = ?$

Formula: $\dfrac{1}{R_T} = \dfrac{1}{R_1} + \dfrac{1}{R_2} + \dfrac{1}{R_3} + \dfrac{1}{R_4}$

Substitute: $\dfrac{1}{R_T} = \dfrac{1}{12} + \dfrac{1}{15} + \dfrac{1}{25} + \dfrac{1}{32}$

12 [1/x] [+] 15 [1/x] [+] 25 [1/x] [+] 32 [1/x] [=] [1/x] , 4.5198

$R_T = 4.5\ \Omega$

Check: R_T (4.5 Ω) is less than the smallest resistance (12 Ω).

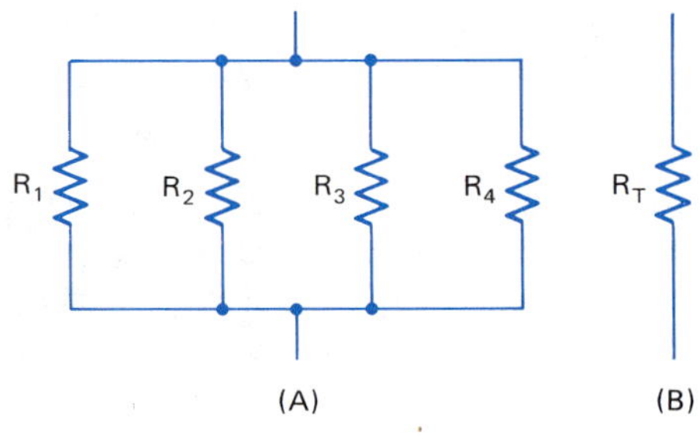

Figure 9-5

When all resistances are the same: $\dfrac{1}{R_T} = N\left(\dfrac{1}{R_1}\right) = \dfrac{N}{R_1}$ where N is the number of loads. The total resistance is

$$R_T = \frac{R_1}{N}$$

The equivalent resistance of five 35-Ω loads is 7 Ω.

Sometimes the word *conductance* is used instead of resistance. If resistance represents the opposition to the flow of current, conductance is a quantity that supports the flow. Conductance (G) is the reciprocal of resistance:

$$G = \frac{1}{R}$$

Conductance is measured in *siemens* (S). **A conductance of 1 S permits a current flow of 1 A when the voltage is 1 V.** Ohm's Law (I = E/R) is written I = EG in terms of conductance.

While the resistance of a parallel circuit is

$$\frac{1}{R_T} = \frac{1}{R_1} + \frac{1}{R_2} + \frac{1}{R_3}$$

for a three branch circuit, the formula for conductance is

$$G_T = G_1 + G_2 + G_3$$

EXERCISE 9-3

Find the equivalent resistance of these sets of resistances connected in parallel (data, diagram, formula, substitute, check).

1. 10 Ω and 10 Ω
2. 10 Ω and 20 Ω
3. 10 Ω and 40 Ω
4. 10 Ω and 100 Ω
5. 10 Ω and 1 kΩ
6. 10 Ω and 1 MΩ
7. 5600 Ω, 4700 Ω and 3800 Ω
8. Two equal resistances of 2.5 kΩ
9. Three equal resistances of 2.5 kΩ
10. Four equal resistances of 2.5 kΩ

In problems 11 and 12, find (a) the conductance of each load, (b) the total conductance and (c) the equivalent resistance.

11. 3.5 kΩ, 4.5 kΩ and 6.8 kΩ
12. 42 kΩ, 8.7 kΩ, 123 kΩ, 76 kΩ and 15.2 kΩ
13. Find the equivalent resistance of the circuit in Problem 12 if the 15.2-kΩ branch became an open circuit.
14. What resistance must be connected in parallel with a 40-Ω load to yield a total resistance of 30 Ω?
15. What resistance must be connected in parallel with a 600-kΩ load providing a total resistance of 400 kΩ?
16. One branch of a three-branch parallel circuit has a conductance of 2.5 mS. A second branch has a resistance of 600 Ω. If the total resistance is 109 Ω, find the resistance and conductance of the third branch.

9-4 VOLTAGE IN PARALLEL CIRCUITS

In parallel circuits, the voltage is common to all circuit elements. That is, for the circuit in Figure 9-6

$$E_T = E_1 = E_2 = E_3 = E_4$$

If one branch is an open circuit, each of the other loads still has the full supply voltage applied.

When equal voltage sources are connected in parallel, the voltage is unchanged. The cells shown in Figure 9-7 each have a voltage of 1.5 V. The combination also has a voltage of 1.5 V. The energy available, however, is increased. The three cells in parallel deliver energy at the same voltage. This means the combination can deliver energy at the same rate as one cell but for a period of time three times as long.

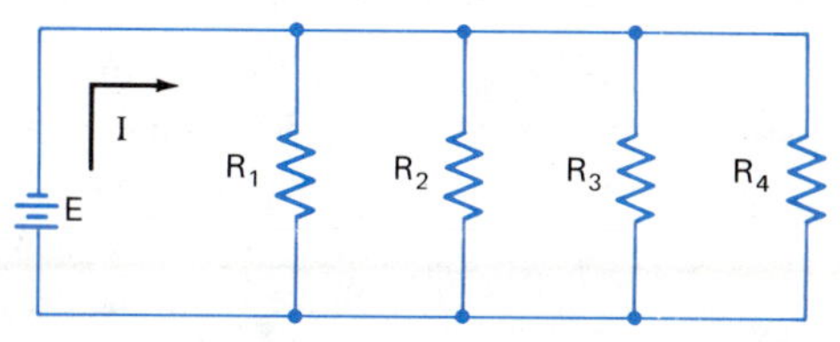

Figure 9-6 Four-branch parallel circuit

Figure 9-7 Three flashlight cells connected in parallel (Courtesy of Southeast Nebraska Community College)

9-5 DIVISION OF CURRENT IN PARALLEL CIRCUITS AND OHM'S LAW

As noted previously, current divides, flows through the parallel branches and recombines before returning to the source. For a three-branch circuit:

$$I_T = I_1 + I_2 + I_3$$

Ohm's Law applies to each branch. Since the voltage is the same:

$$\frac{E}{R_T} = \frac{E}{R_1} + \frac{E}{R_2} + \frac{E}{R_3}$$

The corresponding terms of these two equations are equal to each other: $I_1 = E/R_1$, $I_2 = E/R_2$, $I_3 = E/R_3$ and $I_T = E/R_T$. Ohm's Law applies to individual branches and to the circuit as a whole using the equivalent resistance.

In solving circuit problems, always list the knowns and unknowns (data) for each branch and the overall circuit separately. This can be done conveniently on the circuit diagram itself.

Example A A 250-Ω load and a 350-Ω load are connected in parallel to a 24-V battery. Find I_1, I_2, I_T and R_T.

Solution: Data: See diagram, Figure 9-8.

Formulas: $I = \frac{E}{R}$, $I_T = I_1 + I_2$, $R_T = \frac{R_1 R_2}{R_1 + R_2}$

Substitute: $I_1 = \frac{E}{R_1} = \frac{24\text{ V}}{250\ \Omega} = 0.096\text{ A}$

Substitute: $I_2 = \frac{E}{R_2} = \frac{24\text{ V}}{350\ \Omega} = 0.069\text{ A}$

Substitute: $I_T = 96\text{ mA} + 69\text{ mA} = 165\text{ mA}$

Substitute: $R_T = \frac{(250)(350)}{250 + 340} = 145\ \Omega$

Check: All the calculations can be checked by calculating R_T using Ohm's Law.

$$R_T = \frac{E}{I_T} = \frac{24\text{ V}}{0.165\text{ A}} = 145\ \Omega$$

Since there is agreement, all other calculations may be assumed to be correct.

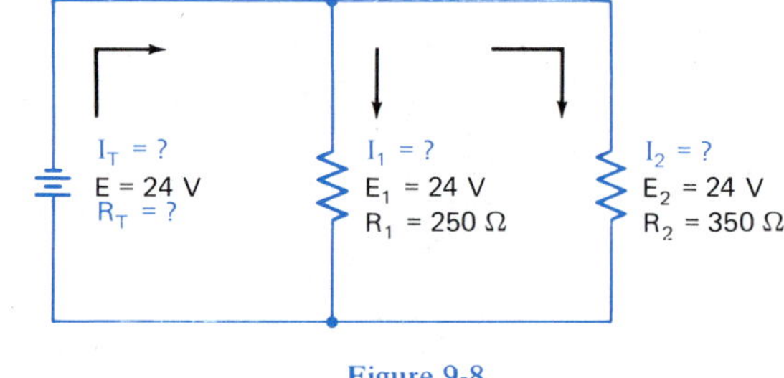

Figure 9-8

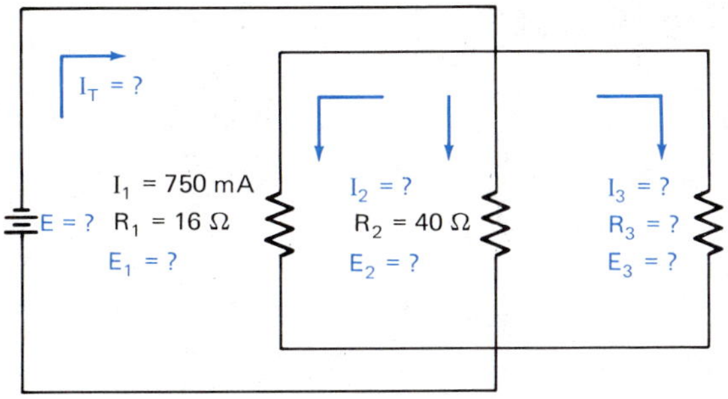

Figure 9-9

Example B In a three-branch circuit, one load has a resistance of 16 Ω and a current of 0.75 A. The second load has a resistance of 40 Ω and the equivalent resistance is 6 Ω. Find R_3, I_2, I_3, I_T and E.

Solution: Data: See diagram, Figure 9-9.

Formulas: $E = I_1 R_1, \frac{1}{R_T} = \frac{1}{R_1} + \frac{1}{R_2} + \frac{1}{R_3}, I_T = I_1 + I_2 + I_3$

Substitute: $E = I_1 R_1 = (0.75\ \text{A})(16\ \Omega) = 12\ \text{V}$

Substitute: $I_2 = \frac{E}{R_2} = \frac{12\ \text{V}}{40\ \Omega} = 3\bar{0}0\ \text{mA}$

Substitute: $I_T = \frac{E}{R_T} = \frac{12\ \text{V}}{6.0\ \Omega} = 2.0\ \text{A}$

Substitute: $I_3 = I_T - I_1 - I_2$

$I_3 = 2\ \text{A} - 0.75\ \text{A} - 0.30\ \text{A} = 950\ \text{mA}$

Substitute: $R_3 = \frac{E}{I_3} = \frac{12\ \text{V}}{0.95\ \text{A}} = 13\ \Omega$

Check: $\frac{1}{R_3} = \frac{1}{R_T} - \frac{1}{R_1} - \frac{1}{R_2} = \frac{1}{6.0} - \frac{1}{16} - \frac{1}{40}$

$\frac{1}{R_3} = 0.0792$

$R_3 = 13\ \Omega$

Note: There is usually more than one way to solve a parallel circuit. The unknowns were found in this order: E, I_2, I_T, I_3, R_3. Among others, the following order could have been used: R_3, E, I_2, I_3, I_T.

The following statements summarize simple parallel circuits:

- The voltage drop across each load is equal to the voltage source.

$$E = E_1 = E_2 = E_3 = \ldots$$

- The total current supplied by the source is the sum of the currents in the parallel branches.

$$I_T = I_1 + I_2 + I_3 + \ldots$$

- The equivalent resistance is given by the formulas:

$$R_T = \frac{R_1 R_2}{R_1 + R_2} \quad \text{(Two branch)}$$

$$\frac{1}{R_T} = \frac{1}{R_1} + \frac{1}{R_2} + \frac{1}{R_3} + \ldots \quad \text{(Polybranch)}$$

- Ohm's Law applies independently to each branch and to the equivalent circuit.

EXERCISE 9-4

Data, diagram, formula, substitute, check.

1. A parallel circuit consists of two loads of $2\bar{0}$-Ω and 35-Ω resistance. If the current through the $2\bar{0}$-Ω load is 3.0 A, find (a) I_2, (b) I_T, (c) E and (d) R_T.
2. A parallel circuit of two branches has a total resistance of $1\bar{0}$ Ω and a total current of 11 A. If one of the branches has a current of 7.5 A, (a) find the current in the other branch. Determine (b) R_1, (c) R_2 and (d) E.
3. One load of a two-branch parallel circuit has a current of 1.0 mA through it. The other has a voltage of 220 V across it. The total current is 3.0 mA. Find (a) R_1, (b) R_2, (c) R_T and (d) I_2.
4. Three resistors are in parallel and pass a total current of 1.2 A. The first resistor passes 120 mA. The second has a voltage drop of 150 V across it. The third has a resistance of 200 Ω. Find (a) the equivalent resistance, (b) R_1, (c) R_2, (d) I_2 and (e) I_3.
5. Find the total current when loads of 47 kΩ, 53 kΩ and 74 kΩ are connected in parallel to a 120-V source.
6. Four equal electric lights are connected in parallel across a 110-V source. If the total current is 4 A, find (a) the resistance of each lamp and (b) the equivalent resistance of the circuit. (c) What is the current in each lamp?
7. The circuit current must be limited to 20 A to keep a fuse from blowing. The supply voltage is 120 V. If the resistance of one load is 15 Ω, find the least resistance that can be connected in parallel.
8. A three-branch parallel circuit (16 kΩ, 23 kΩ and 33 kΩ) draws a total current of 25 mA. Find the current through (a) the 16-kΩ branch, (b) the 23-kΩ branch and (c) the 33-kΩ branch.
9. In a 12-V parallel circuit, the two headlamps of an automobile each draw 3.85 A. The tail lamps each have a resistance of 14.5 Ω. Find the (a) total resistance and (b) total current.
10. In a home kitchen circuit, these appliances are connected in parallel. A toaster with an applied voltage of 115 V and drawing 10.0 A, a coffee maker with a resistance of 11.0 Ω and a slow cooker drawing 1.35 A. Find the (a) total resistance and (b) total current in this circuit.
11. Find the voltage needed to produce a total current of 240 mA through 1-Ω, 2-Ω, and 6-Ω loads wired in parallel.
12. The resistance of an ammeter coil is 2.75 Ω. A shunt connected in parallel has a resistance of 0.025 Ω. If the total current is 10 A, (a) find the current through the coil. (b) What is the voltage drop across the ammeter?
13. Two circuit elements have conductances of 3.6 mS and 8.3 mS. They are connected in parallel with a total current of 3.0 A. Find the current through (a) the 3.6-mS element and (b) the 8.3-mS element.

14. Three loads in parallel pass 650 mA total current. The first load has a resistance of 350 Ω, the second passes a $9\bar{0}$ mA current and the third has a voltage drop of 180 V across it. Find the conductance of (a) the second load and (b) the third load.
15. A DC generator applies an emf of 24 V to three motors connected in parallel. The resistances of the motors are 2.4 Ω, 3.6 Ω and 4.8 Ω. Find the current through (a) the motor with a resistance of 2.4 Ω, (b) the motor with a resistance of 3.6 Ω and (c) the motor with a resistance of 4.8 Ω.
16. A 60-Ω load is arranged in parallel with a second load of unknown resistance across a 240-V source, Figure 9-10. If the total current is 10 A, what is the resistance of the second load?

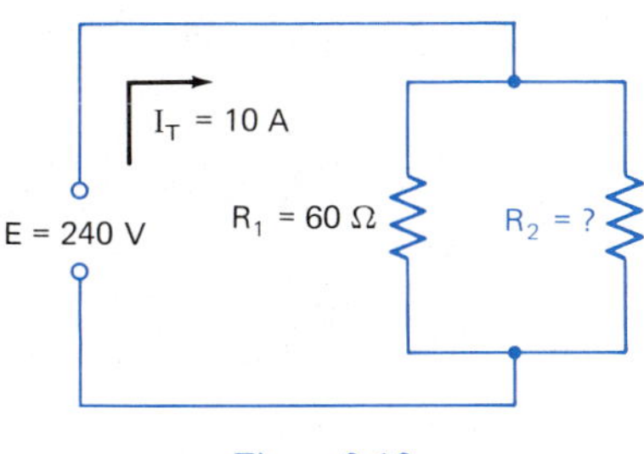

Figure 9-10

9-6 POWER DISSIPATION IN PARALLEL CIRCUITS

Regardless of how circuit loads are connected, total power dissipation is the sum of the power losses in each of the loads.

$$P_T = P_1 + P_2 + P_3 \ldots$$

Power losses in each load are calculated according to the known quantities. The formulas $P = EI$, $P = E^2/R$ and $P = I^2R$ are used.

Example A Find the total power dissipation in a circuit with three resistors in parallel. One has a resistance of 12 kΩ. The second has a current of 85 mA. The third has a resistance of 6 kΩ with a power dissipation of 15 W.

Solution: Data: See diagram, Figure 9-11.

Formulas: $P = E^2/R, P = EI, P_T = P_1 + P_2 + P_3$

$$PR = \frac{E^2 \cancel{R}}{\cancel{R}} \quad \text{(MA, × R and cancel)}$$

$$PR = E^2$$

$$\sqrt{PR} = \sqrt{E^2} \quad (\text{RA}, \sqrt{\ })$$

$$\sqrt{PR} = E$$

$$E = \sqrt{P_3 R_3}$$

Substitute: $E = \sqrt{(15\ \text{W})(6000\ \Omega)}$

$E = 300$ V

Substitute: $P_2 = EI_2 = (300\ \text{V})(0.085\ \text{A}) = 26$ W

Substitute: $P_1 = E^2/R_1 = (300\ \text{V})^2/12\,000\ \Omega = 7.5$ W

Substitute: $P_T = 15\ \text{W} + 26\ \text{W} + 7.5\ \text{W}$

$P_T = 48.5\ \text{W} \approx 49$ W

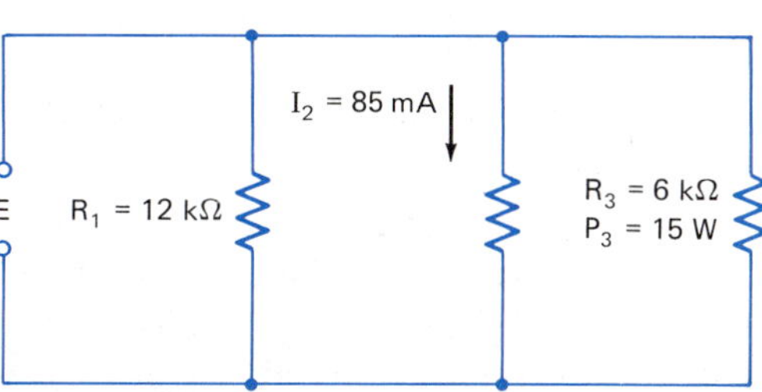

Figure 9-11

EXERCISE 9-5

Data, diagram, formula, substitute, check.

1. Two loads with resistances of 15 Ω and 20 Ω are connected in parallel to a 24-V source. Find the total power dissipation.
2. Two lamps are connected in parallel. One draws a current of 560 mA and has a resistance of 190 Ω. The second lamp draws a current of 870 mA. Find the total power dissipation in the circuit.

3. A three-resistor parallel circuit (12 kΩ, 15 kΩ and 25 kΩ) draws a total current of 15 mA. Find the total power dissipated by the circuit.
4. A 75-W and a 100-W lamp are connected in parallel in a 115-V circuit. What is the equivalent hot resistance of the circuit?
5. A heating element has a power rating of 500 W when connected to 120 V. Find the total current when three such elements are connected in parallel.
6. Three resistors are connected in parallel: R_1 = 80 kΩ, R_2 = 200 kΩ and R_3 = 5 kΩ. What is the maximum voltage these resistors can be connected to without overheating any of them? All three are rated at 0.5 W.
7. A 20-kΩ, 25-W resistor; a 50-kΩ, 100-W resistor; and a 10-kΩ, 150-W resistor are connected in parallel. (a) Find the maximum emf that can be applied without exceeding the power rating of any of them. (b) If they were connected in *series,* what total voltage could be used?
8. Three lamps are connected in parallel in a 120-V circuit. One lamp is rated at 60 W. A second lamp has a hot resistance of 192 Ω. The third lamp is passing a current of 1.25 A. Find (a) the total power, (b) the equivalent resistance and (c) the total current in this circuit.
9. A 120-V electric iron has a hot resistance of 10 Ω. The iron is connected in parallel with several lamps having a total resistance of 245 Ω. What size circuit breaker (15 A, 20 A or 30 A) should be used on this circuit?
10. An automobile's electric system operates on 12 V. Find the power consumption when all of the following devices are turned on: two headlamps each with a hot resistance of 3 Ω, two 6-W tail lamps, two brake lights drawing a current of 0.67 A each and a fan motor with a resistance of 3.2 Ω.
11. A 120-V electric house circuit is fused with a 20-A circuit breaker. The following appliances are connected to the outlets on this circuit: two 60-W lamps, one radio drawing 200 mA current, one hot plate with a hot resistance of 16.5 Ω and one 1200-W coffee pot. (a) Determine whether the circuit breaker is overloaded. (b) What is the total power?
12. The total current of three loads in parallel is 375 mA. R_1 has a conductance of 1.75 mS, I_2 is 135 mA and E_3 is 24.0 V. Find (a) the total power dissipation, (b) the current through R_1 and (c) the current through R_2.

SIMPLE COMBINATION CIRCUITS

In practice, many electric circuits have combinations of series and parallel circuits. In Figure 9-12(A), R_2 and R_3 form a pair of loads in parallel. It is also shown in different (standard) form, Figure 9-12(A′). The parallel group (R_{eq}) is in series with R_1, Figure 9-12(B).

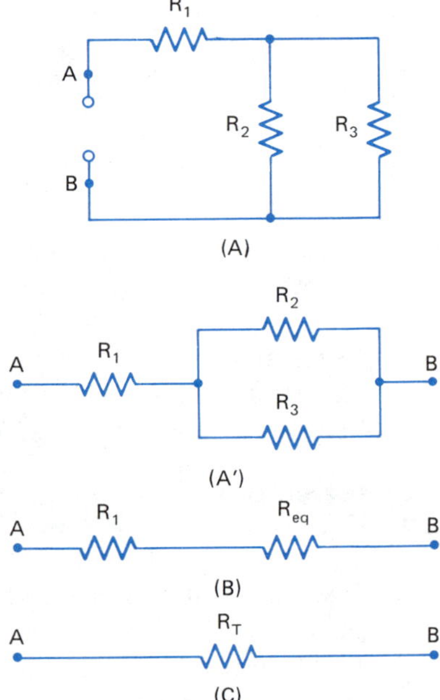

Figure 9-12 Chain of equivalent circuits (series-parallel)

A circuit of this type is sometimes called a series-parallel circuit. The parallel group is in series with another load. Another simple combination is the parallel-series circuit, Figure 9-13(A). Here, two series loads are in parallel with a third. Standard form Figure 9-13(A′) has point A on the left and point B on the right. Parallel groups are clearly identified on a circuit in standard form.

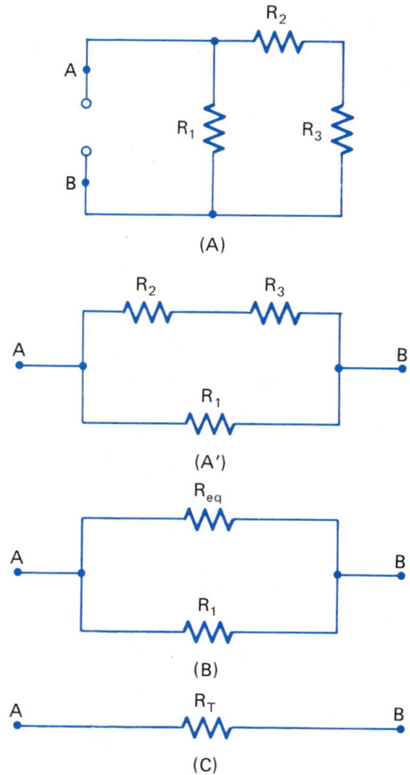

Figure 9-13 Chain of equivalent circuits (parallel-series)

9-7 EQUIVALENT-CIRCUIT METHOD

A method of solving some of the simpler circuits is suggested by Figure 9-12 and Figure 9-13. By substituting intermediate equivalent resistances for some of the groups in a chain of figures, the circuit is reduced to a single equivalent resistance.

In Figure 9-12, notice that (A) and (A′) are the same circuit. Traveling from point A in a direction toward point B, point A connects to R_1. R_1 connects to both R_2 and R_3, and both R_2 and R_3 connect to point B. Similarly, (A) and (A′) in Figure 9-13 are identical circuits.

9-8 FINDING TOTAL RESISTANCE

By constructing a chain of equivalent circuits, the total resistance is determined. Simply calculate the series or parallel group resistance at each step.

Example A Determine the total resistance for the circuit of Figure 9-13 if $R_1 = 65\ \Omega$, $R_2 = 40\ \Omega$ and $R_3 = 50\ \Omega$.

Solution: Diagram: See Figure 9-13(A).

Redraw the circuit, Figure 9-13(A′) in standard form. Redraw the circuit, Figure 9-13(B), and determine the series combination:

$$R_{eq} = R_2 + R_3$$
$$R_{eq} = 40\ \Omega + 50\ \Omega = 90\ \Omega$$

Redraw the circuit, Figure 9-13(C) as the single total resistance of the circuit and determine its value.

$$R_T = \frac{R_1 R_{eq}}{R_1 + R_{eq}}$$
$$R_T = \frac{(65\ \Omega)(90\ \Omega)}{65\ \Omega + 90\ \Omega} = 37.7\ \Omega$$

Example B Change the circuit in Figure 9-14 to a single equivalent resistance by redrawing it as a chain of equivalent circuits. Describe at each step how intermediate resistance calculations are made.

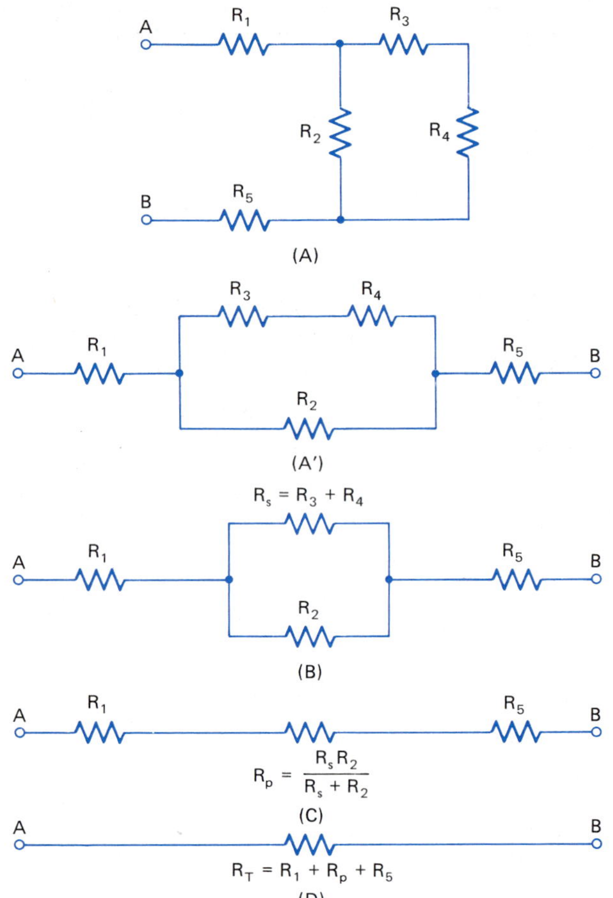

Figure 9-14 Chain of equivalent circuits

Solution: Redraw the circuit, Figure 9-14(A′) in standard form.

Determine the series combination, Figure 9-14(B):
$R_s = R_3 + R_4$

Find the parallel combination, Figure 9-14(C):

$$R_p = \frac{R_s R_2}{R_s + R_2}$$

Calculate the total resistance for the circuit, Figure 9-14(D):
$R_T = R_1 + R_p + R_5$

When individual load resistances are known, the equivalent resistance can be calculated using the procedure outlined in Example B.

Example C In the circuit of the previous example, $R_1 = 10\ \Omega$, $R_2 = 40\ \Omega$, $R_3 = R_4 = 20\ \Omega$ and $R_5 = 30\ \Omega$. Find the total resistance.

Solution: Data: See diagram, Figure 9-15(A).
As the following calculations are studied, refer to the figure indicated.

Formula: $R_s = R_3 + R_4$ — Figure 9-14(B)

Substitute: $R_s = 20\ \Omega + 20\ \Omega = 40\ \Omega$ — Figure 9-15(B)

Formula: $R_p = \dfrac{R_s R_2}{R_s + R_2}$ — Figure 9-14(C)

Substitute: $R_p = \dfrac{(40\ \Omega)(40\ \Omega)}{40\ \Omega + 40\ \Omega} = 20\ \Omega$ — Figure 9-15(C)

Formula: $R_T = R_1 + R_p + R_5$ — Figure 9-14(D)

Substitute: $R_T = 10\ \Omega + 20\ \Omega + 30\ \Omega = 60\ \Omega$ — Figure 9-15(D)

Note: The chain diagrams shown in Figure 9-15 are the same as in Figure 9-14 but have not been changed to standard form. The reader should compare corresponding drawings.

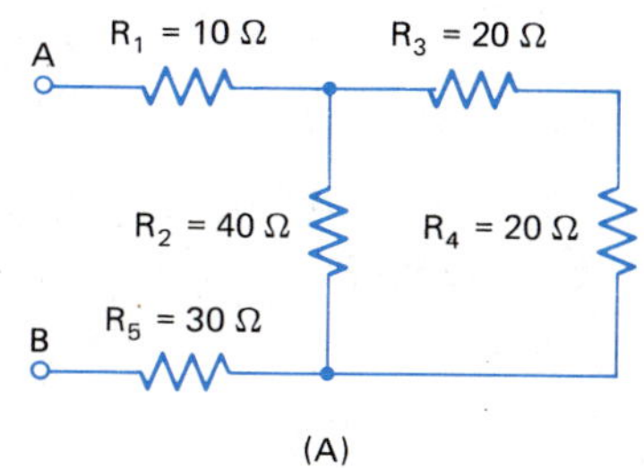

(A)

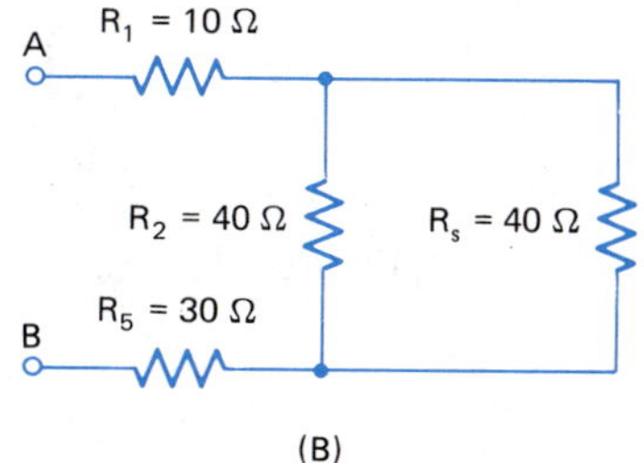

(B)

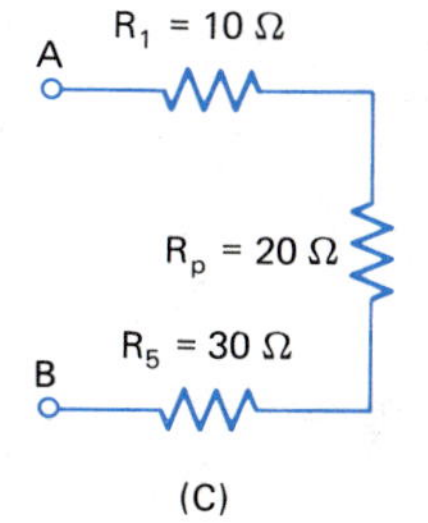

(C)

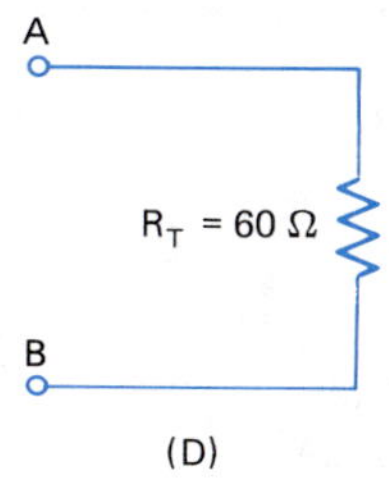

(D)

Figure 9-15 Chain of equivalent circuits

Example D Find the current, voltage drop and power dissipation across each load in the circuit shown in Figure 9-13 if the applied voltage is 240 V. Resistances are given in Example A.

Solution: Find the total current and the current in each branch:

$$I_T = \frac{E_T}{R_T} = \frac{240\ \text{V}}{37.7\ \Omega} = 6.36\ \text{A}$$

$$I_1 = \frac{E_T}{R_1} = \frac{240\ \text{V}}{65\ \Omega} = 3.69\ \text{A}$$

$$I_2 = I_3 = \frac{E_T}{R_2 + R_3} = \frac{240\ \text{V}}{90\ \Omega} = 2.67\ \text{A}$$

Find the voltage drops across each load:

$E_1 = E_T = 240\ \text{V}$

$E_2 = I_2 R_2 = (2.69\ \text{A})(40\ \Omega) = 107\ \text{V}$

$E_3 = I_3 R_3 = (2.67\ \text{A})(50\ \Omega) = 133\ \text{V}$

Find the power dissipation in each load:

$P_1 = E_1 I_1 = (240\ \text{V})(3.69\ \text{A}) = 886\ \text{W}$

$P_2 = E_2 I_2 = (107\ \text{V})(2.67\ \text{A}) = 285\ \text{W}$

$P_3 = E_3 I_3 = (133\ \text{V})(2.67\ \text{A}) = 355\ \text{W}$

Determine the total power:

$P_T = E_T I_T = (240\ \text{V})(6.36\ \text{A}) = 1526\ \text{W}$

Check: $P_T = P_1 + P_2 + P_3$

$P_T = 886\ \text{W} + 285\ \text{W} + 355\ \text{W} = 1526\ \text{W}$

EXERCISE 9-6

Data, diagram, formula, substitute, check.

Change each of the following circuits to standard form. Convert to a single equivalent load by drawing a chain of equivalent circuits. Find the value of the equivalent resistance.

1. $R_1 = 6\ \Omega$, $R_2 = 4\ \Omega$, $R_3 = 12\ \Omega$, Figure 9-16.

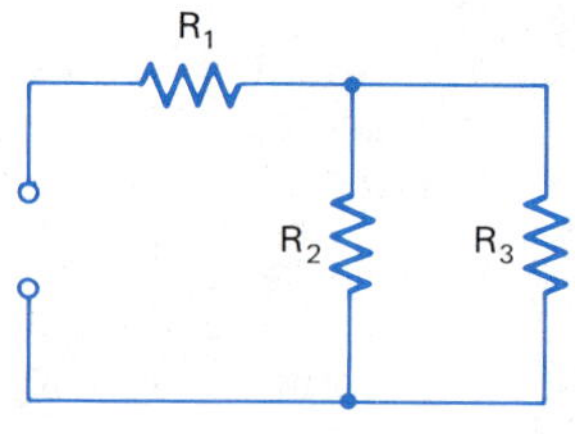

Figure 9-16

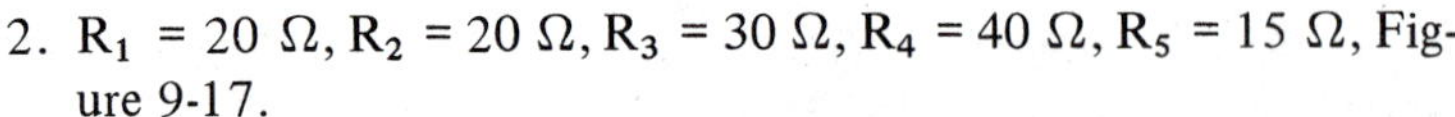

2. $R_1 = 20\ \Omega$, $R_2 = 20\ \Omega$, $R_3 = 30\ \Omega$, $R_4 = 40\ \Omega$, $R_5 = 15\ \Omega$, Figure 9-17.

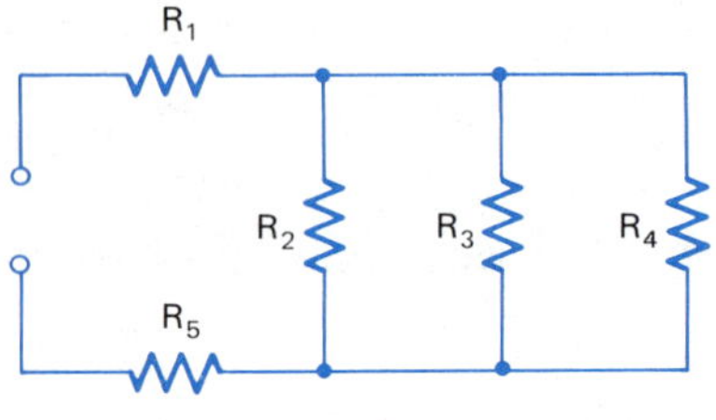

Figure 9-17

3. $R_1 = 5\ \Omega$, $R_2 = 18\ \Omega$, $R_3 = 10\ \Omega$, $R_4 = 15\ \Omega$, $R_5 = 12\ \Omega$, $R_6 = 9\ \Omega$, Figure 9-18.

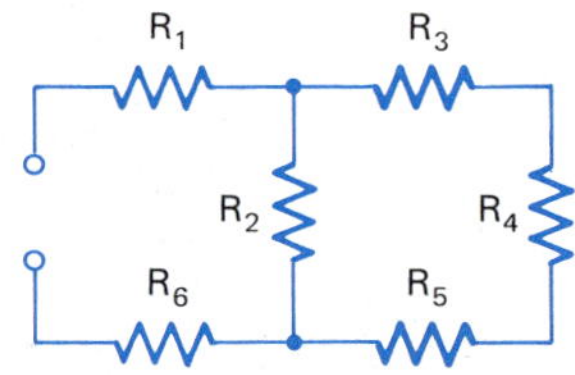

Figure 9-18

4. $R_1 = 5\ k\Omega$, $R_2 = 8\ k\Omega$, $R_3 = 3\ k\Omega$, $R_4 = 12\ k\Omega$, $R_5 = 6\ k\Omega$, Figure 9-19.

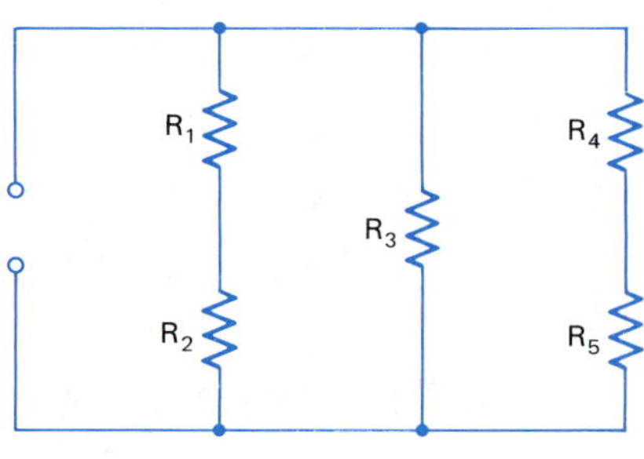

Figure 9-19

5. $R_1 = 5\ k\Omega$, $R_2 = 3\ k\Omega$, $R_3 = 4\ k\Omega$, $R_4 = 20\ k\Omega$, $R_5 = 30\ k\Omega$. Figure 9-20.

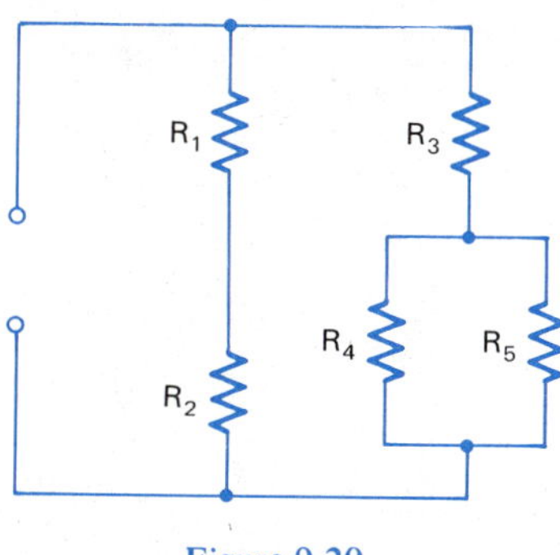

Figure 9-20

6. $R_1 = 10\ \Omega$, $R_2 = 3\ \Omega$, $R_3 = 5\ \Omega$, $R_4 = 4\ \Omega$, $R_5 = 12\ \Omega$, Figure 9-21.

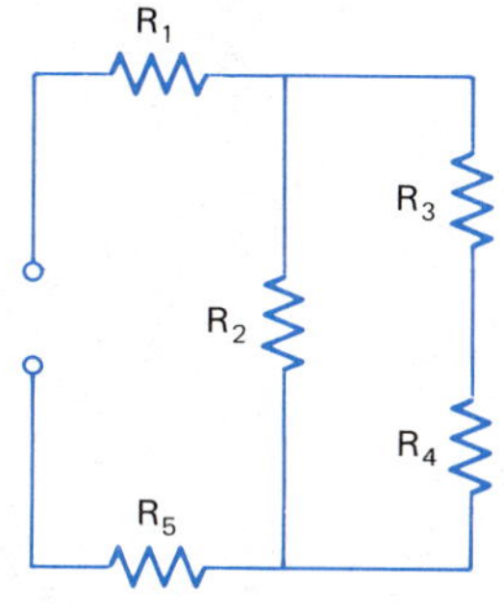

Figure 9-21

7. $R_1 = 6.8\ k\Omega$, $R_2 = 8.4\ k\Omega$, $R_3 = 9.6\ k\Omega$, $R_4 = 5.0\ k\Omega$, $R_5 = 7.5\ k\Omega$, Figure 9-22.

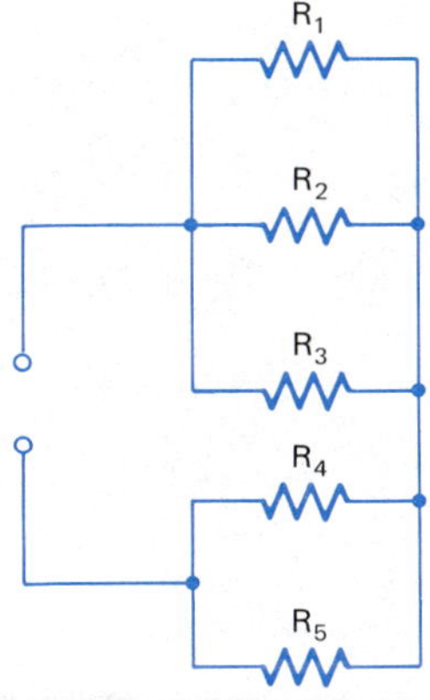

Figure 9-22

8. $R_1 = 14.3\ k\Omega$, $R_2 = 21.6\ k\Omega$, $R_3 = 19.4\ k\Omega$, $R_4 = 9.8\ k\Omega$, $R_5 = 15.2\ k\Omega$, $R_6 = 15.2\ k\Omega$, $R_7 = 15.0\ k\Omega$, Figure 9-23.

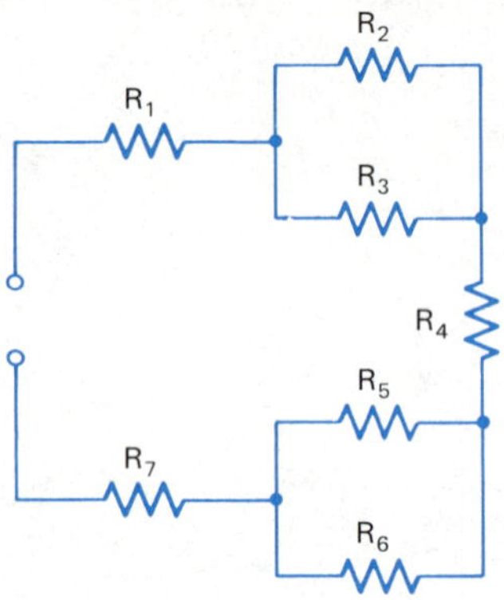

Figure 9-23

9-9 SOLVING COMBINATION CIRCUITS

Solving a circuit means finding the resistance, current, voltage drop and power dissipation of each load and the total values. The following example illustrates one method:

Example A In the circuit shown in Figure 9-15, the applied voltage is 120 V. Find the current through each resistor, the voltage drop across each resistor, the power dissipation in each resistor and the total power when $R_T = 60\ \Omega$.

Note: Knowing the total resistance, (60 Ω), the circuit current can now be found. This is the current through R_1 and R_5, Figure 9-15.

Solution: Formula: $I_T = \frac{E}{R_T}$ Figure 9-15(D)

Substitute: $I_T = I_1 = I_5 = \frac{120\ V}{60\ \Omega} = 2\ A$

It is obvious from Figure 9-15(B) that the current divides exactly in half:

$I_2 = 1\ A$ Figure 9-15(B)

$I_3 = I_4 = 1\ A$ Figure 9-15(A)

Formula: $E = IR$

Substitute: $E_1 = (2\ A)(10\ \Omega) = 20\ V$ Figure 9-15(C)

$E_5 = (2\ A)(30\ \Omega) = 60\ V$ Figure 9-15(C)

$E_2 = (1\ A)(40\ \Omega) = 40\ V$ Figure 9-15(B)

$E_3 = E_4 = (1\ A)(20\ \Omega) = 20\ V$ Figure 9-15(A)

Formula: $P = EI$

Substitute: $P_1 = (20\ V)(2\ A) = 40\ W$

$P_2 = (40\ V)(1\ A) = 40\ W$

$P_3 = (20\ V)(1\ A) = 20\ W$

$P_4 = (20\ V)(1\ A) = 20\ W$

$P_5 = (60\ V)(2\ A) = 120\ W$

Formula: $P_T = P_1 + P_2 + P_3 + P_4 + P_5$

Substitute: $P_T = 40\ W + 40\ W + 20\ W + 20\ W + 120\ W$

$P_T = 240\ W$

Check: $P_T = I^2 R_T$

$P_T = (2\ A)^2 (60\ \Omega) = 240\ W$

The table in Figure 9-24 summarizes all given and calculated quantities.

LOAD	RESISTANCE (Ω)	VOLTAGE (V)	CURRENT (A)	POWER (W)
1	10	20	2	40
2	40	40	1	40
3	20	20	1	20
4	20	20	1	20
5	30	60	2	120
T	60	120	2	240

Figure 9-24 Table of results

EXERCISE 9-7

Redraw each of these circuit diagrams in standard form. Place point A on the left and B on the right. Show the circuit as single resistors and parallel groups in series.

1.

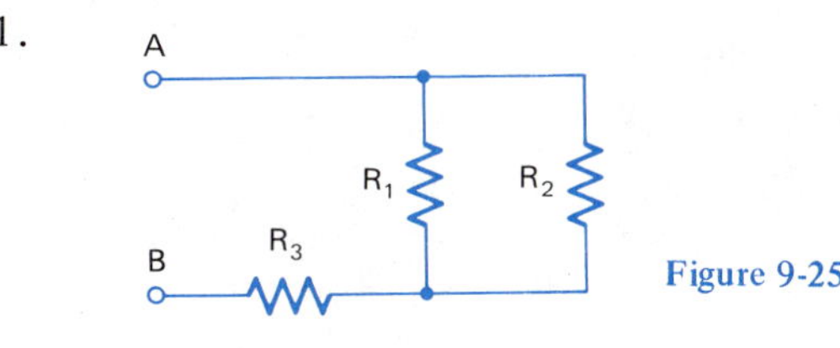

Figure 9-25

2.

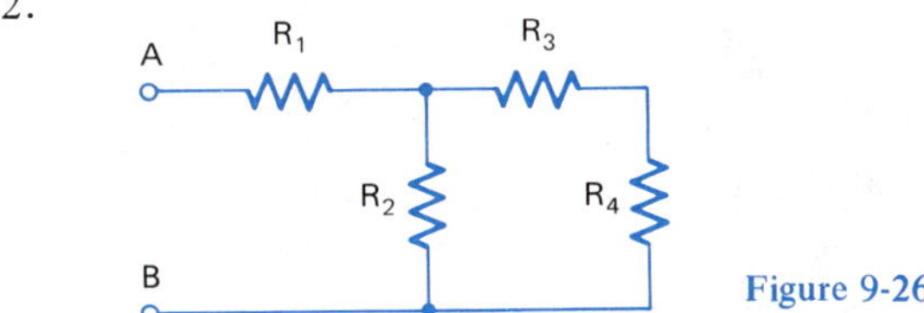

Figure 9-26

3.

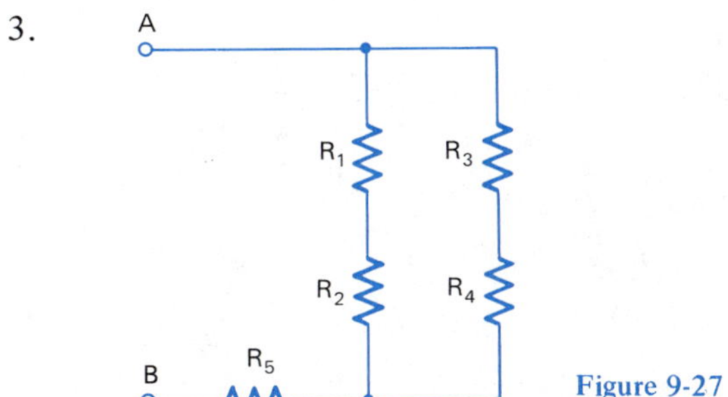

Figure 9-27

4.

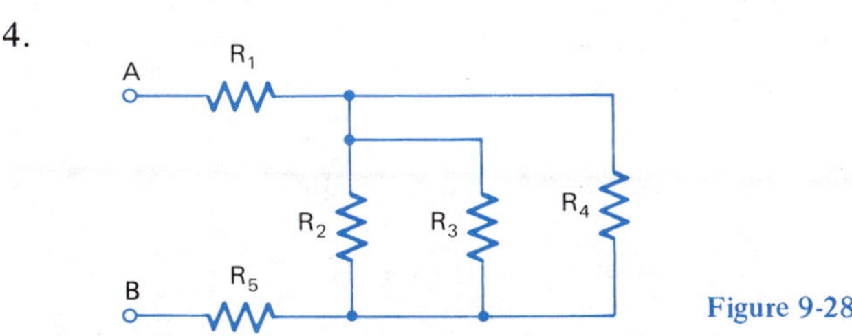

Figure 9-28

5.

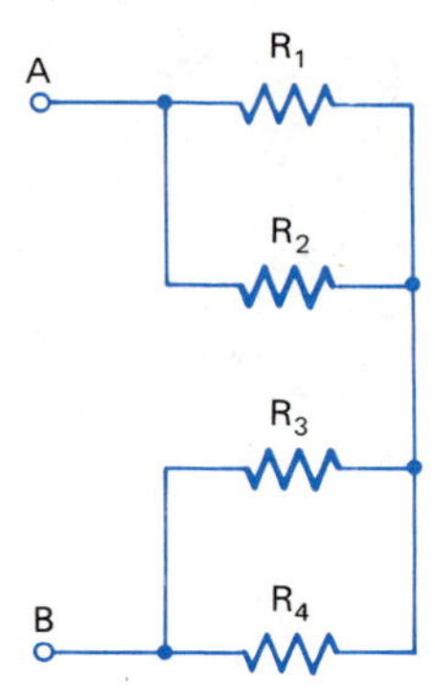

Figure 9-29

6.

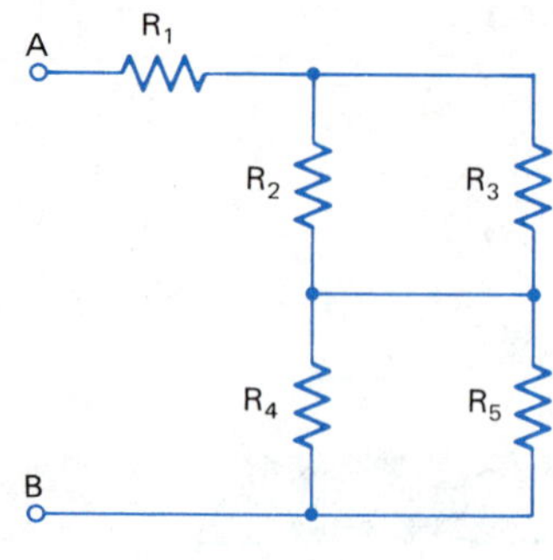

Figure 9-30

Change each circuit to a single equivalent resistance by drawing a chain of equivalent circuits. For each step, give the formula for making intermediate calculations.

7. Use Figure 9-25.
8. Use Figure 9-26.
9. Use Figure 9-27.
10. Use Figure 9-28.
11. Use Figure 9-29.
12. Use Figure 9-30.

Find the equivalent resistance, the total current, the current through each load, the voltage drop across each load, the power dissipation in each load and the total power dissipation in each of the following problems. Construct a table as shown in Figure 9-24 showing results for each problem.

13. Use Figure 9-25: $R_1 = 10\ \Omega$, $R_2 = 15\ \Omega$, $R_3 = 6\ \Omega$, $E = 12$ V.
14. Use Figure 9-26: $R_1 = 800\ \Omega$, $R_2 = 1.2\ k\Omega$, $R_3 = 500\ \Omega$, $R_4 = 900\ \Omega$, $I_T = 1$ A.
15. Use Figure 9-27: $R_1 = 50\ \Omega$, $R_2 = 70\ \Omega$, $R_3 = 40\ \Omega$, $R_4 = 120\ \Omega$, $R_5 = 30\ \Omega$, $E = 120$ V.
16. Use Figure 9-28: $R_1 = 100\ \Omega$, $R_2 = R_3 = R_4 = 300\ \Omega$, $R_5 = 200\ \Omega$, $I_T = 60$ mA.
17. Use Figure 9-29: $R_1 = 200\ \Omega$, $R_2 = 300\ \Omega$, $R_3 = 600\ \Omega$, $R_4 = 400\ \Omega$, $E = 48$ V.
18. Use Figure 9-30: $R_1 = 80\ \Omega$, $R_2 = 30\ \Omega$, $R_3 = 40\ \Omega$, $R_4 = 20\ \Omega$, $R_5 = 50\ \Omega$, $E = 16$ V.

CHAPTER 10

ADDITIONAL ELECTRICAL CONCEPTS

OBJECTIVES

After satisfactorily completing this chapter, the student should be able to:

- Describe and use direct, inverse, square and inverse square variation concepts to derive electrical formulas.
- Define percent and use the principle to describe and solve problems involving relative change, efficiency and tolerance.
- Define energy and calculate the amount of energy consumed and its cost in various electrical circuit problems.

This chapter is mainly concerned with variation, change and methods of describing change. A change in one electrical quantity is usually related to changes in another electrical quantity. Such a relationship between quantities is called *variation*. It is possible to predict what will happen if the variation relationship between the quantities can be determined.

Sometimes, a set of beginning and final values of both quantities are known. Perhaps several sets of intermediate values are also known. If such data is plotted on a graph, the relationship can be seen in the form of a curve. If the curve is a simple one, a formula can be derived. Once a formula is derived, predictions can be made about changes in the quantities involved.

The chapter also includes a discussion of percent with electrical applications. A study of electrical energy consumption in electrical circuits and the cost of energy and its relationship to efficiency of electrical devices is included.

VARIATION OF ELECTRICAL QUANTITIES

There are several words and expressions used to indicate that a variation interdependence exists between two quantities. In a single load circuit in which the resistance is constant, voltage and current are interdependent. The following statements are used to indicate that current and voltage are related.

Current depends on voltage.
Current varies as voltage.
Current varies with voltage.
Current increases with voltage.
Current is a function of voltage.
Current changes when voltage changes.

Such statements simply imply that voltage and current are related. The type of variation must be determined before a formula relating the two can be written. Of course, the relationship in this example is the well-known Ohm's Law. In the next topics, several simple types of variations are described. Two methods of determining formulas are given for the simpler types.

10-1 DIRECT VARIATION

The relationship between voltage and current in the single-load circuit in which the resistance is constant is an example of a direct variation. Suppose measurements are taken for voltage and current (12.0 V, 2.40 mA). If the voltage is doubled, the current will double (24.0 V, 4.80 mA). If voltage is tripled, current triples (36.0 V, 7.20 mA). If voltage is increased by a factor of seven, current increases by a factor of seven (84.0 V, 16.8 mA). Suppose voltage is halved, current also is halved (6.0 V, 1.20 mA). By using a 500-Ω resistor for the load, all of these readings can be easily verified.

If the data in the previous paragraph is graphed, the curve is a straight line. The curve associated with any direct variation is a straight line. Sometimes a direct variation is called a *linear relationship* because of the resulting straight line curve.

In general, if quantity A varies directly as quantity B, then the following statements characterize the *direct variation* relationship:

- If A increases (or decreases) then B increases (or decreases) by the same factor.
- If a graph of A and B is constructed, the curve will be a straight line. It will not be a direct variation if the curve is not a straight line.
- The formula relating the two variables A and B is

  $$A = kB$$

  where k is the *variation constant.* The value of the variation constant can be determined from the graph. The dependent variable equals a constant times the independent variable.

A *rheostat,* Figure 10-1, is a device in which the resistance can be varied. The rheostat (R_1) in Figure 10-2 is adjusted to yield the currents listed in

Figure 10-1 Two kinds of rheostats (Courtesy of Southeast Nebraska Community College)

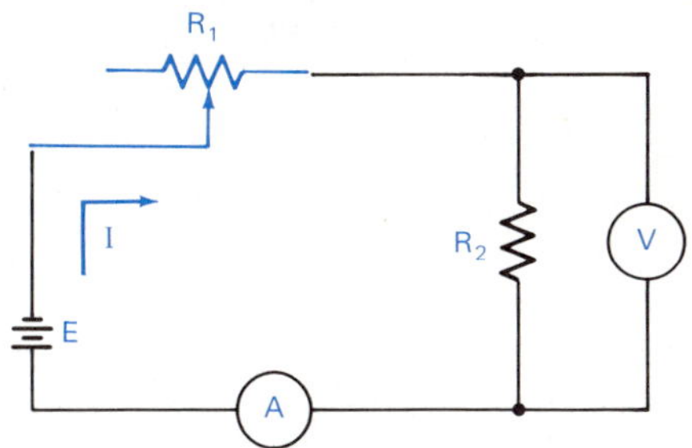

Figure 10-2 Circuit diagram. Changing the rheostat (R_1) alters the voltage drop across R_2 and the circuit current.

CURRENT (mA)	R_2 VOLTAGE DROP (V)
10	1.51
20	2.40
30	3.66
40	4.92
50	6.00
60	7.33
70	8.58
80	9.62

Figure 10-3 Recorded measurements of current through and voltage drop across the load in the circuit

Figure 10-3. A graph of the data is given in Figure 10-4. The current (I) is equal to a constant (k) times the voltage drop (E_2) across R_2:

$$I = kE_2$$

The constant (k) can be determined by locating two points on the graph, Figure 10-4, where the curve crosses exactly on an intersection of the graph paper. These two points should be as far apart as possible. Points P and Q are two such points. The constant is equal to the vertical distance between the two points divided by the horizontal distance.

Example A (1) Determine the constant from the graph of Figure 10-4. (2) Write the equation relating current and voltage drop. (3) Use the equation to find the amount of voltage drop when the current is 45 mA.

Solution: The coordinates of points P and Q, Figure 10-4 are (10 V, 0.0825 A) and (4 V, 0.0325 A), respectively.

(1) $k = \dfrac{0.0825\text{ A} - 0.0325\text{ A}}{10\text{ V} - 4\text{ V}} = 0.008\ 33\ \dfrac{\text{A}}{\text{V}}$

(2) $I = 0.008\ 33E$

(3) $E_2 = \dfrac{I}{0.008\ 33} = \dfrac{0.045\text{ A}}{0.008\ 33\ \dfrac{\text{A}}{\text{V}}} = 5.4\text{ V}$

The answer to part (3) should be checked on the graph to see that it is correct. Note that the constant 0.008 33 is the conductance of the load. The resistance $\left(R_2 = \dfrac{1}{k}\right)$ is 120 Ω.

When only one pair of measurements is given, the formula can still be written. It must be known, however, that the relationship is linear.

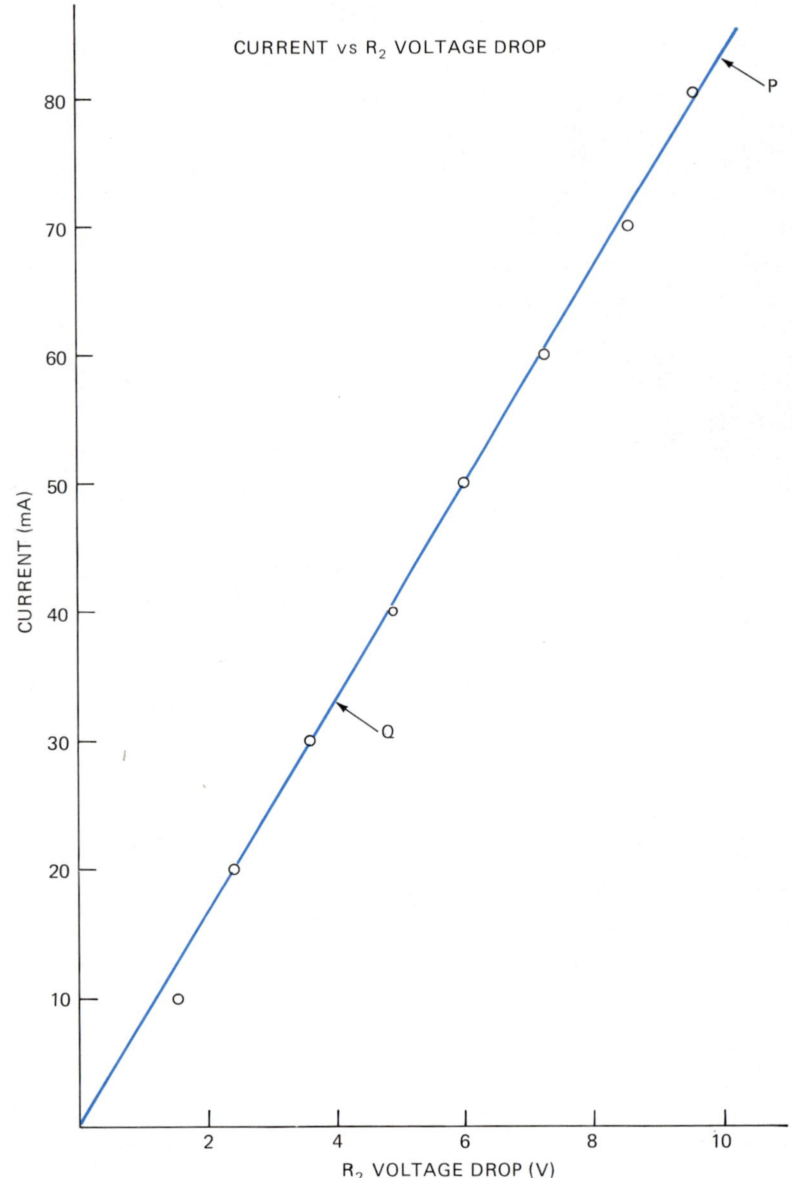

Figure 10-4

Example B The perimeter (P) of a square is in a direct variation with the length of the side (s). That is, if one doubles, the other doubles. A certain square has a perimeter of 94 cm and a side of 23.5 cm. Find the variation constant and find the side of a square warehouse building with a perimeter of 159.6 yd.

Solution: Data: $P_1 = 94$ cm, $s_1 = 23.5$ cm, $P_2 = 159.6$ yd, $k = ?$, $s_2 = ?$

Formula: $P_1 = ks_1$

$$k = \frac{P_1}{s_1}$$

Substitute: $k = \dfrac{94 \text{ cm}}{23.5 \text{ cm}} = 4$

Notice that the units cancel so this variation constant has no units.

Formula: $P = 4s$

$$s_2 = \frac{P_2}{4}$$

Substitute: $s_2 = \dfrac{159.6 \text{ yd}}{4} = 39.9 \text{ yd}$

Check: $39.9(4) = 159.6$

RADIUS (cm)	CIRCUMFERENCE (cm)
4.9	30.3
6.1	38.6
9.7	60.9
13.8	86.4
19.9	124.0
26.8	168.0

Figure 10-5 Radius and circumference of various circular objects

EXERCISE 10-1

Data, diagram, formula, substitute, check.

1. Circular objects of various sizes are measured. The radius (r) and circumference (C) for each object is given in Figure 10-5. (a) Plot a full page graph of these data. (b) If the curve is linear, write a formula relating C and r. (c) Calculate the circumference when the radius is 15 cm.
2. The data in Figure 10-6 were recorded in an experiment. (a) Plot a full page graph of the data. (b) Write a formula and determine the variation constant. (c) What is the resistance of the load? (d) Find the current when the voltage drop is 3.50 V.
3. A certain size of copper wire has resistance (R) as shown in Figure 10-7. (a) Plot a graph of the data. (b) Write a formula for resistance and find the variation constant. (c) What is the resistance of a piece of this wire that is 5.5 ft long? (d) Find the length of a wire that has a resistance of 10 Ω.

CURRENT (mA)	VOLTAGE DROP (V)
10	0.51
20	1.04
30	1.55
40	2.07
50	2.62
60	3.19
70	3.70
80	4.21

Figure 10-6

LENGTH (ft)	RESISTANCE (Ω)
1.00	1.05
2.00	2.11
3.00	3.15
4.00	4.19
5.00	5.26
6.00	6.30
7.00	7.34
8.00	8.41

Figure 10-7

4. The heat energy (W) produced in the heating element of a popcorn popper, Figure 10-8, is directly proportioned to the time (t) the appliance is used. If 216 kJ of heat is produced in 180 s (3 min), (a) write a formula for the heat produced. (b) Find the amount of heat developed in 10 min.
5. A capacitor stores 785 μC of charge (Q) when a voltage (E) of 120 V is applied to it. Charge varies directly as the applied voltage. Use the methods of variation to find the charge when a voltage of 500 V is applied.
6. A charge (Q) of 680 C passes through a switch in a time (t) of 75 s. The relationship is a direct variation. Find the variation constant and use it to determine the amount of charge that flows through the switch in 5 min.
7. A change in resistance (ΔR) of a certain piece of wire varies directly as the change in temperature (ΔT). When the temperature increases by 2.5 C°, the resistance increases by 15.0 Ω. (a) Write a formula for the change in resistance. (b) Use the formula to determine the change in temperature required to *decrease* the resistance by 210 Ω.
8. The force (F) on an electron moving across a magnetic field is directly proportional to the speed (v). The force is 3.2×10^{-15} N when the speed is 20 km/s. (a) Write a formula for the force. (b) What is the speed when the force is 8×10^{-15} N?

Figure 10-8 Popcorn popper showing heating element

10-2 INVERSE VARIATION

Sometimes one variable will increase when the other decreases. In such cases the first variable will decrease when the second increases. The first is said to vary *inversely* as the second. The relationship is an *inverse variation.* If two variables vary inversely, one varies directly with the reciprocal of the other. In symbols:

$$y = k\left(\frac{1}{x}\right)$$

$$\text{or } y = \frac{k}{x}$$

where y and x are the variables.

When several values are known, the reciprocals of the independent variable can be found using a calculator. The dependent variable is then plotted on a graph as a function of the reciprocal of the independent variable. The result will be a straight line curve. The variation constant is found the same as with direct variation.

Example A A simple circuit consists of a battery and a resistor. Power is measured with a wattmeter. Identical resistors are added in series one at a time. Power is measured each time and the data is recorded, Figure 10-9. (1) Plot a graph of the data (P and R) as given. (2) Calculate reciprocals of the independent variable and plot a graph. (3) Determine the variation constant and write a formula relating total power and total resistance of the circuit.

POWER (W)	RESISTANCE (Ω)
57.6	10
28.8	20
19.3	30
14.4	40
11.5	50
9.5	60
8.2	70
7.2	80

Figure 10-9

Solution: (1) See Figure 10-10(A)

(2) Resistance is the independent variable. Reciprocals are 0.1, 0.05, 0.033, 0.025, 0.02, 0.017, 0.014 and 0.0125. The second graph is shown in Figure 10-10(B).

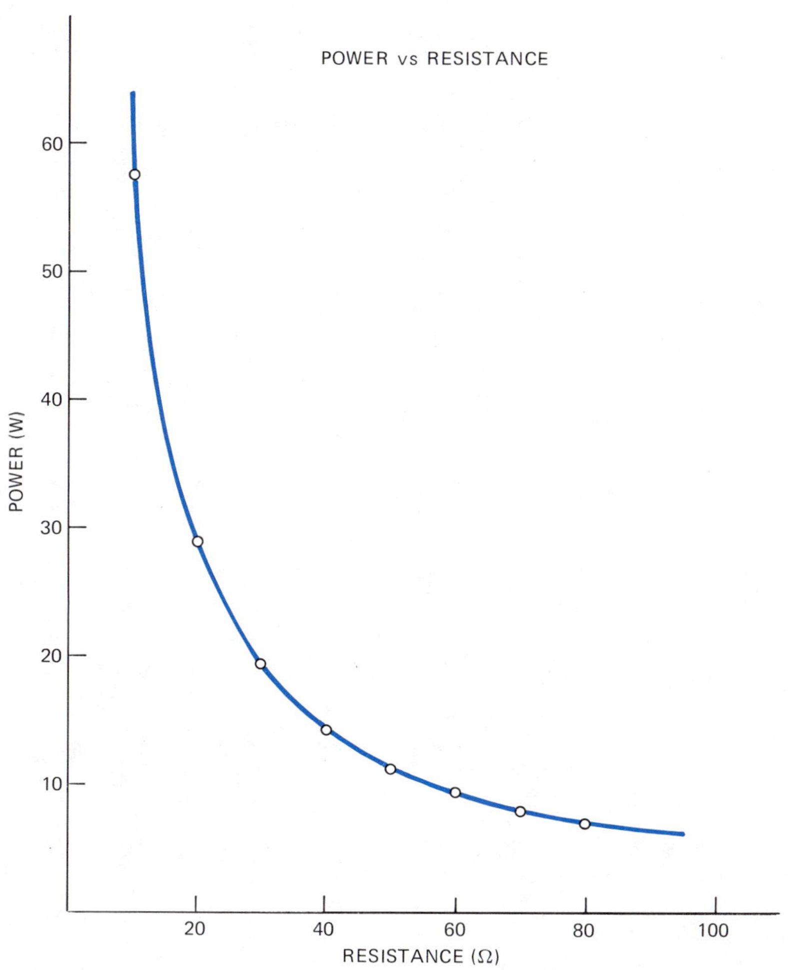

Figure 10-10(A) Graph of inverse variation

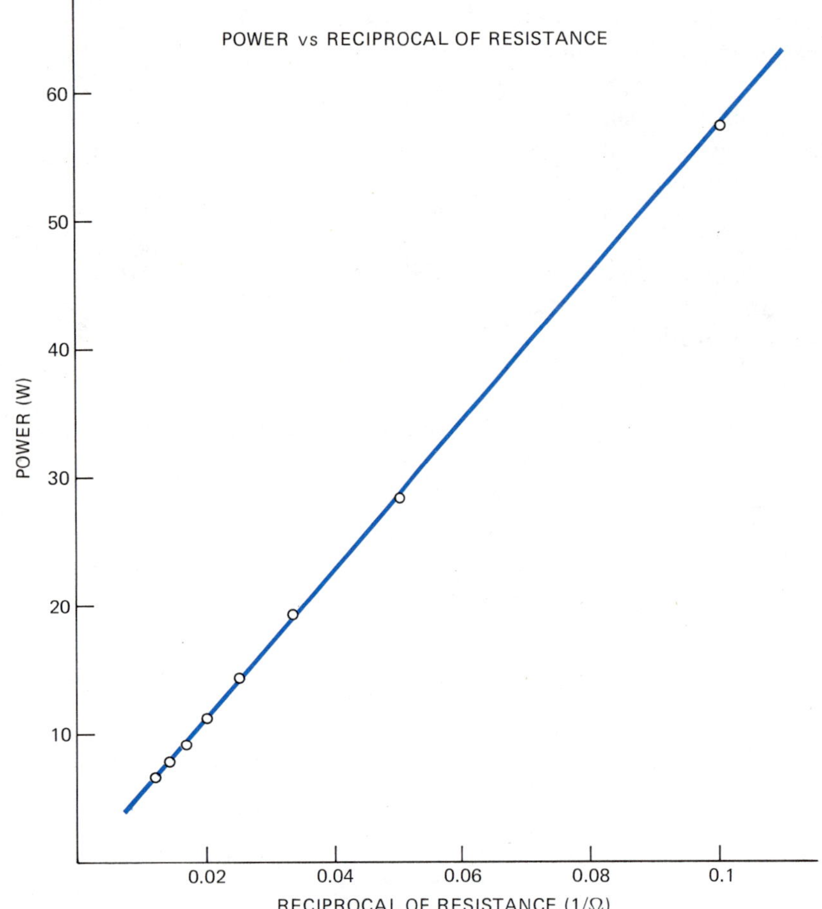

Figure 10-10(B) Graph of power and the reciprocal of the resistance

(3) Curve passes through coordinates: (0.10, 57.6) and (0.0125, 7.2).

$$k = \frac{\text{vertical}}{\text{horizontal}} = \frac{57.6 - 7.2}{0.10 - 0.0125}$$

$$k = 576$$

$$P = \frac{k}{R} = \frac{576}{R}$$

Note: Since $P = E^2/R$, it is apparent that $k = E^2$: the square of the voltage (24 V).

If it is known that the relationship between two quantities is an inverse variation, the formula can be obtained. Only one data point is needed.

Example B The altitude (h) of a triangle with an area of 150 m^2 varies inversely with the base (b), Figure 10-11. Determine the variation constant and write a formula for the altitude. The altitude is 10 m when the base is 30 m. Calculate the altitude when the base is 17.5 m.

Figure 10-11 Various triangles with equal areas

Solution: Data: $b_1 = 30\ \text{m}, h_1 = 10\ \text{m}, b_2 = 17.5\ \text{m}, h_2 = ?$

Formula: $h = \dfrac{k}{b}$

$k = b_1 h_1$

Substitute: $k = (30\ \text{m})(10\ \text{m}) = 300\ \text{m}^2$

Formula: $h = \dfrac{300\ \text{m}^2}{b}$

Notice the variation constant has units of m^2 and equals twice the area.

Substitute: $h = \dfrac{300\ \text{m}^2}{17.5\ \text{m}} = 17.1\ \text{m}$

EXERCISE 10-2

1. The circuit, Figure 10-12, contains a variable resistor. Various settings yield the following data:

I (A)	12	6.0	4.0	3.0	2.4	2.0
R (Ω)	2.0	4.0	6.0	8.0	$1\bar{0}$	12

(a) Plot a graph of the data. (b) Find the reciprocals of the resistance and plot a graph: I vs $\frac{1}{R}$. (c) Determine the variation constant from the second graph and write a formula for current. (d) Find the current when the resistance is 48 Ω.

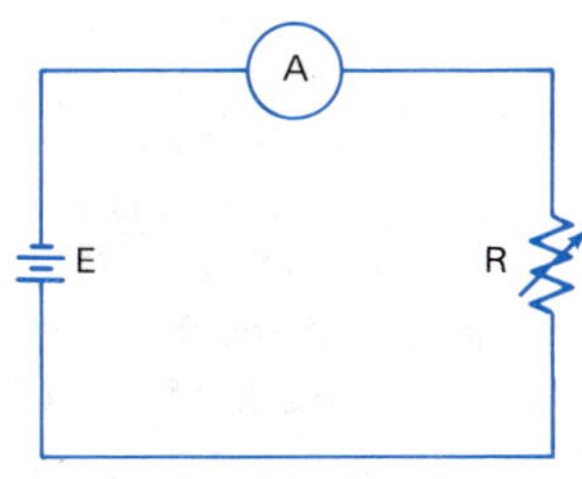

Figure 10-12

2. The amount of charge (Q) flowing through a load required to dissipate a certain amount of energy in the load depends inversely on the applied voltage (E). If 1.14 C of charge moves through the load under the influence of 220 V, find the variation constant. (a) Write the formula for the charge. (b) How much charge must pass through the load to dissipate the same amount of heat when the voltage is 48 V?
3. The time (t) required to pass a given quantity of charge through a switch varies inversely as the current (I). When the current is 2.5 mA, the time required is 2 ks. (a) What is the variation constant and the formula relating the two quantities? (b) How much time is required if the current is 175 A?
4. The time (t) required for a circuit to dissipate a certain amount of energy depends inversely on the power (P). If 15 s is required when the power is 567 W, (a) find the variation constant and the formula. (b) What is the power if 85 s is required? (c) What time is necessary when the power is 90 W?

5. Voltage and current are inversely proportional in various circuits that all dissipate heat at the same rate. The current (I) is 2.5 A when the voltage (E) is 75 V. (a) Determine the variation constant and write a formula for current in these circuits. (b) What current flows if the voltage is 220 V?

10-3 JOINT VARIATION

One variable often depends on two or more other variables. For instance, the change in resistance (ΔR) of a wire depends both on the original resistance (R) of the wire and the change in temperature (ΔT):

$$\Delta R = \alpha R \Delta T$$

where α is the variation constant and depends on the type of metal.

When one quantity depends on the product of two or more other quantities, the first quantity *varies jointly* with the others. If y varies jointly with x and z, then y varies as the product of x and z:

$$y = kxz$$

where k is the variation constant.

A common joint variation example is when one quantity varies as the square of another,

$$y = kx^2$$

Sometimes a variable depends jointly as two other variables and at the same time varies inversely with a third.

$$y = \frac{kxz}{r}$$

Example A Find the variation constant and the formula for the force acting on two small charged objects. It varies jointly with the charges (Q_1 and Q_2) on the objects and inversely with the square of the distance (r) between them. The force on an object charged with 5 nC which is 50 mm from another object with a charge of 10 nC is 180 μN.

Note: Recall that force is measured in newtons (N) in SI units.

Solution: Jointly means product and inverse means division.

Formula: $F = \dfrac{kQ_1Q_2}{r^2}$

$$k = \frac{Fr^2}{Q_1Q_2}$$

Substitute: $k = \dfrac{(180 \times 10^{-6}\ \text{N})(50 \times 10^{-3}\ \text{m})^2}{(5 \times 10^{-9}\ \text{C})(10 \times 10^{-9}\ \text{C})}$

$$k = 9 \times 10^9\ \frac{\text{Nm}^2}{\text{C}^2}$$

The formula can be written:

$$F = \frac{9 \times 10^9\ Q_1Q_2}{r^2}$$

If three of the quantities are known, the fourth one can be calculated.

Figure 10-13 is an apparatus that illustrates a square and another inverse square law. Four strings, tied at the vertex, stretch downward through three transparent surfaces. The strings form the four corners of a square on the first surface.

The second surface is twice as far from the vertex as is the first surface. Again, the strings form the corners of a square. But, the area is four times larger than the square on the first level. The third level is three times further from the vertex than the first. The square formed here is nine times the size of the square on the first surface. Obviously, the area (A) increases with the square of the distance (r) from the vertex ($A = kr^2$).

Example B If the distance (r) from the vertex to the first surface is 17.5 cm and the side of the square (s) on the first layer is 8.3 cm, find the formula for the area (A) of the square.

Solution: First, find the area of the square.

$A_1 = s_1{}^2 = (8.3\text{ cm})^2 = 68.9\text{ cm}^2$

Formula: $A_1 = kr_1{}^2$

$$k = \frac{A_1}{r_1{}^2}$$

Substitute: $k = \dfrac{68.9\text{ cm}^2}{(17.5\text{ cm})^2}$

$k = 0.225$ (Units cancel)

The formula is written:

$A = 0.225r^2$

The constant would have a larger value if the strings were spread out forming a larger area at each surface.

Example C Use the formula $A = 0.225r^2$ to find the area at the second and third surfaces. They are 35.0 cm and 52.5 cm from the vertex, respectively.

Solution: Formula: $A = 0.225r^2$

Substitute: $A_2 = 0.225(35.0\text{ cm})^2 = 276\text{ cm}^2$

Substitute: $A_3 = 0.225(52.5)^2 = 620\text{ cm}^2$

The reader should verify that A_2 and A_3 are four times and nine times the area A_1, respectively.

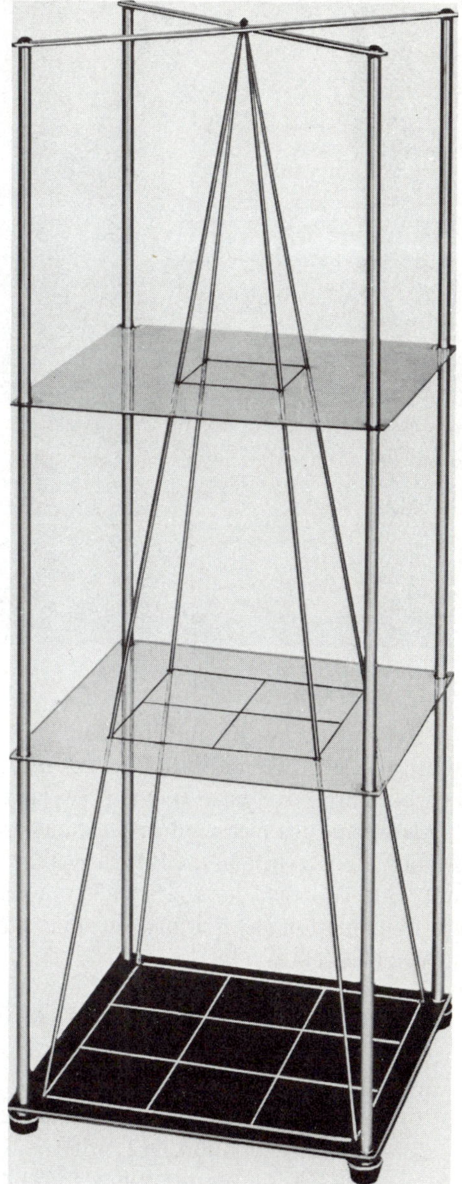

Figure 10-13 Inverse square apparatus (Courtesy of Sargent-Welch Scientific Company)

Suppose the filament of a light bulb is located at the vertex of the apparatus, Figure 10-13. If there is no reflector, light is emitted in all directions. The light emitted is described in terms of the power that is radiated. Light power is called *luminous flux* and is measured in units of *lumens*, (lm). 1 W = 668.2 lm.

Only a small part of the light power from the lamp will be radiated toward the surfaces formed by the strings. An object on the first surface will be *illuminated*. A certain amount of luminous flux will strike the first surface area within the strings. The same amount of flux spreads out over an area four times larger at the second surface. An object located here is

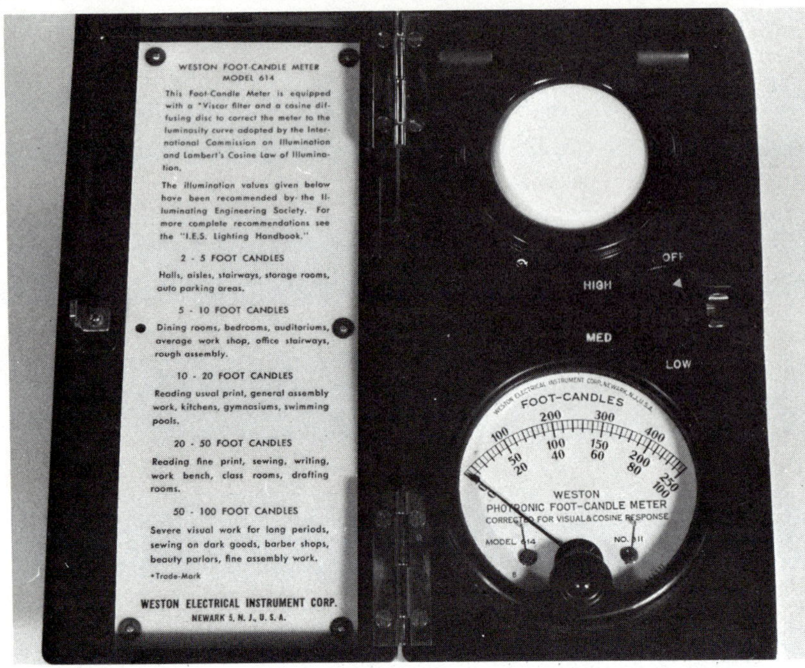

Figure 10-14 Photronic foot-candle meter (light meter) (Courtesy of Southeast Nebraska Community College)

still illuminated but by only one-fourth as much. The illumination at the third surface is one-ninth that at the first. The same amount of flux must distribute itself over an area nine times larger.

Illumination is measured in SI units of lumens per square metre which is called the lux (lx). In the British system, the unit is the lumen per square foot which is called the foot-candle. A light meter is shown in Figure 10-14. Illumination (E) is a quantity that varies inversely with the square of the distance (r) from the source.

$$E = \frac{k}{r^2}$$

Example D A certain lamp produces an illumination of 45 lm/ft^2 at a distance of 6 ft from the source. (1) Write the formula for this variation. (2) What is the illumination at a distance of 10 ft from the source? (3) At what distance will the illumination be twice the original illumination?

Solution: Formula: (1) $E = \frac{k}{r^2}$

$$k = Er^2$$

$$k = \left(45 \frac{lm}{ft^2}\right)(6\ ft)^2 = 1620\ lm$$

$$E = \frac{1620}{r^2}$$

(2) $E = \frac{1620}{(10\ ft)^2} = 16.2 \frac{lm}{ft^2}$

(3) $$E = \frac{1620}{r^2}$$

$$Er^2 = 1620 \qquad (MA, \times r^2)$$

$$r^2 = \frac{1620}{E} \qquad (DA, \div E)$$

$$\sqrt{r^2} = \sqrt{\frac{1620}{E}} \qquad (RA, \sqrt{\ })$$

$$r = \sqrt{\frac{1620}{E}}$$

$$r = \sqrt{\frac{1620 \text{ lm}}{90 \text{ lm/ft}^2}} = 4.2 \text{ ft}$$

EXERCISE 10-3

Data, diagram, formula, substitute, check.

1. The energy (W) stored in a capacitor varies jointly as the capacitance (C) and the square of the applied voltage (E). A 54-pF capacitor stores 3.89 nJ when 12 V is applied. (a) Write the formula with a numerical variation constant. (b) Find the energy stored when 48 V is applied to a 3.6-nF capacitor.
2. The resistance (R) of a wire varies directly as the length (L) and inversely with the cross-sectional area (A). The resistance is 5.37 Ω when the length of a wire is 1 km and the cross-sectional area is 5.25×10^{-6} m². (a) Write a formula for the resistance. (b) Find the resistance of a wire 150 m long and a cross-sectional area of 2.69×10^{-5} m².
3. The energy (W) stored in the magnetic field of a coil of wire varies jointly with the inductance (L) of the coil and the square of the current (I) in the wire. When the current is 12 A in a 32.5-mH coil, the energy stored is 2.34 J. (a) Write the formula for the energy stored in the magnetic field of a coil. (b) Find the current in a 600-mH coil if the energy stored is 20 J.
4. Never stand under a person working overhead. The speed (v) of dropped objects varies as the square root of the distance (d) fallen. If the speed is 64 ft/s when the object falls 64 ft, (a) write a formula for the speed. Find the speed of falling objects dropped from (b) 20 ft, (c) 30 ft, (d) 40 ft, (e) 50 ft, (f) 75 ft and (g) 100 ft.
5. The strings of a device similar to Figure 10-13 spread into a larger "solid angle." At a distance of 14 cm, a square with 16.5-cm sides is formed. (a) Write the equation for the area. (b) Find the area at 25 cm. (c) How far from the vertex will the square formed have an area of 2500 cm²?
6. The square formed by a device similar to Figure 10-13 has an area of 576 in² at a distance of 24 in. (a) Write the equation of variation. (b) How far from the vertex will the area be 2500 in²? (c) What is the area at 12 in from the vertex?

7. A lamp, 2.00 m above a workbench, produces an illumination of 30 lm/m^2 at the work surface. The illumination (E) varies inversely with the square of the distance (r) from the source. (a) Write the equation for illumination. (b) How far above the bench would this lamp double the illumination? (c) At what distance above the work surface would the lamp produce half the original illumination?
8. The illumination (E) of a surface, Figure 10-13, varies directly as the intensity (I) of the light source and inversely as the square of the distance (r) between the surface and the light source. At 4 m from a 125-cd (candela) source, a surface has an illumination of 7.81 lm/m^2 (lumen per square metre or lux). (a) Write the equation for illumination. (b) Find the illumination of a workbench 2 m below a 175-cd lamp.
9. A street lamp 10.5 m above the ground has an intensity of 800 cd. (a) What is the illumination at ground level? (b) What is the illumination 1.75 m above the ground? Assume there is no reflector and use the equation $E = I/r^2$ where E = illumination, I = intensity and r = distance between the surface and the light source.
10. What is the effect on the illumination of a workbench if a 96-cd lamp 1.5 m above it is lowered to 1 m?
11. An electrician would like to double the illumination on a workbench. The 114-cd lamp is located 2.2 m above the work surface. Where should the lamp be placed?
12. The minimum recommended illumination of a workbench for fine work is 500 lx. Find the luminous intensity of a single lamp 1.8 m above the bench.

PERCENTS

The word *percent* is a common one in everyday language. Its meaning can be understood by thinking of the metric prefix "centi" which means hundredths. Actually, the word percent comes from the Latin *per centum* meaning "by the hundred." Whenever a fraction has a denominator of 100, it is a percent.

10-4 CHANGING FRACTIONS TO PERCENTS

The following fractions (common, decimal and percent) are equivalent:

$$\frac{69}{100} = 0.69 = 69\%$$

The common fraction (69/100) is changed to percent by dropping the denominator (100) and adding the percent symbol (%). The decimal fraction (0.69) is converted to percent by moving the decimal point two places to the right and adding the percent symbol.

Example A Convert these fractions to percent:

(1) $\frac{29}{100}$, (2) 0.044, (3) $\frac{87\ 1/2}{100}$ and (4) 4.355.

Solution: (1) $\frac{29}{100} = 29\%$ (2) $0.044 = 4.4\%$ (3) $\frac{87\ 1/2}{100} = 87\ 1/2\%$

(4) $4.355 = 435.5\%$ (Improper)

Common fractions that do not have a denominator of 100 must be changed. Divide 100 by the denominator, then multiply both terms of the fraction by that number. Change to percent.

Example B Change the fraction $\frac{7}{16}$ to percent.

Solution: $100 \div 16 = 6.25$

$$\frac{7 \times 6.25}{16 \times 6.25} = \frac{43.75}{100}$$

$$\frac{7}{16} = 43.75\%$$

To convert a percent to a fraction, just reverse the procedure. Drop the percent sign. Place the number over a denominator of 100. Reduce the resulting fraction to lowest terms.

Example C Convert 62.5% to a common fraction.

Solution: $62.5\% = \frac{62.5}{100} = \frac{625}{1000} = \frac{5}{8}$

EXERCISE 10-4

Complete the following table. This table is very useful. When completed, it contains the most common proper fractions and their equivalent decimals and percents.

	COMMON FRACTION	DECIMAL FRACTION	PERCENT
1.	$\frac{1}{10}$		
2.		0.125	
3.			20%
4.	$\frac{1}{4}$		
5.		0.3	
6.			$33\frac{1}{3}\%$
7.	$\frac{3}{8}$		
8.		0.4	
9.			50%
10.	$\frac{3}{5}$		
11.			62.5%
12.		0.666...	
13.	$\frac{7}{10}$		
14.		0.75	
15.			80%
16.	$\frac{7}{8}$		
17.		0.9	
18.			100%

19. General purpose solder is $\frac{3}{5}$ tin. What percent of general purpose solder is tin?
20. Eutectic solder is 37% lead. Express this percent as a decimal.
21. In wiring a house, 0.05 of the wire is waste. What percent is waste?
22. The retail sales tax rate in a certain state is 4.5%. Express the tax rate as a common fraction reduced to lowest terms.

10-5 OPERATIONS WITH PERCENTS

When all the parts of a whole amount are together, they consist of 100% of that amount. When only some of the parts are together, they make up less than 100% of the whole. There are three quantities involved in the preceding sentences: the number of parts, the whole amount and the percent.

The number of parts is referred to as the *part* (P). The original or whole amount is referred to as the *base* (B). The third quantity is the *rate* (R) and is expressed as a percent. They are related this way:

$$R = \frac{P}{B}$$

The rate is expressed as a percent and is always changed to a decimal or common fraction before it is used in the formula.

10-6 COMMON APPLICATIONS OF PERCENTS

In various applied problems involving percents, the three basic terms (rate, part, base) may have other names. See in following formulas:

$$\text{tax rate} = \frac{\text{taxes}}{\text{base}}$$

$$\text{interest rate} = \frac{\text{interest}}{\text{principal}}$$

$$\text{discount rate} = \frac{\text{discount}}{\text{retail price}}$$

$$\text{commission rate} = \frac{\text{commission}}{\text{total sales}}$$

$$\text{profit rate} = \frac{\text{profit}}{\text{cost}}$$

Percentage problems are easy to solve after the knowns and unknowns are identified. Make a list of the data: rate (R), part (P) and base (B). The reader must decide which quantity is the base (whole), which is the number of parts and which is the rate (%). Set the calculator aside until the data is substituted into the formulas.

Example A Find the sales tax rate if an electrician collects \$52.35 sales tax on a sale of supplies costing \$1495.65.

Solution: Data: P = \$52.35, B = \$1495.65, R = ?

Formula: $R = \frac{P}{B}$

Substitute: $R = \frac{\$52.35}{\$1495.65} = 0.035$ or 3.5%

Example B How much is the discount that a wholesaler gives to an electrical shop? The discount rate is 25%. The retail price for an electric motor is \$68.85. How much will the shop be billed for this purchase?

Solution: Data: B = \$68.85, R = 25%, P = ?

Formula: $R = \frac{P}{B}$ (MA, × B)

$P = RB$

Substitute: $P = 0.25 \times \$68.85 = \$17.21$

Formula: Net cost = B − P

Substitute: Net cost = \$68.85 − \$17.21 = \$51.64

Example C The floor area of an electrical warehouse was increased by 1244 ft². Find the original floor area if the increase is 9.5%.

Solution: Data: $P = 1244 \text{ ft}^2$, R = 9.5%, B = ?

Formula: $R = \frac{P}{B}$

$B = \frac{P}{R}$ (MA, × B; DA, ÷ R)

Substitute: $B = \frac{1244 \text{ ft}^2}{0.095} = 13\,095 \text{ ft}^2$

EXERCISE 10-5

Data, diagram, formula, substitute, check.

1. \$64.50 is what percent of \$143.33?
2. What is 35.2% of 785 kW?
3. \$65.50 is 7.5% of what amount?
4. Find 84.5% of 249 m.
5. 9.5 kΩ is what percent of 6 kΩ?
6. 15 lb is 0.13% of what weight of copper wire?
7. What percent of 250 V is 185 V?
8. Find 1/4% of 3200 mL.
9. One-half is 20% of what number?
10. An electrician charges a customer \$92.50 for a lamp. What is the profit if the profit rate is 40%?
11. The selling price of electrical equipment is \$845.78. The customer is charged \$879.61. What is the sales tax rate?
12. General purpose solder is 60% tin. How much tin is in 250 kg of solder?
13. An electrical contractor bid \$1940.00 for a certain job. Another contractor bid 11.5% less. What was the second bid?

14. Out of a manufactured lot of capacitors of a certain size, 8.5% were rejected. If 425 were rejected, how many capacitors were in the lot?
15. Ceramic insulators are made of 8 parts of one kind of clay and 24 parts of a second kind. (a) What percent is the first kind? (b) What percent is the second kind?
16. An electrician bought tools costing $678.75 and pays a deposit of $175.00. The deposit is what percent of the total cost?
17. A transformer housing contains 36.5 L of oil, 19.3 L of air space and the transformer itself, Figure 10-15 occupies 15.8 L. What percent of the housing is occupied by (a) the oil, (b) the air and (c) the transformer?
18. A power company had 286 "lost time" accidents one year. The following year with a new safety program, the number of accidents decreased by 27.3%. How many accidents were there the second year?
19. From winter to summer, a span of copper power line increased in length by 0.215%. If the increase was 64.5 cm, what is the length of the span?
20. In a 24-V circuit, the current is 1.25 A. If the voltage is raised to 28 V, find the percent increase in the current and in the voltage. Assume resistance is constant.
21. An aluminum solder consists of 78% tin, 9% aluminum, 8% zinc and 5% cadmium. How many kilograms of (a) tin, (b) zinc and (c) cadmium must be added to 10 kg of aluminum to make a small batch of the solder? (d) How many kilograms of solder will there be in the batch?
22. The manufacturing specifications for a V-belt indicate it will transmit 45 hp under normal loading. If 2% is lost in slippage, what power does it actually transmit?
23. A voltmeter reads 12.25 V when it should actually read 12.5 V. Find the error in percent.

Figure 10-15 Pole transformer removed from the housing

10-7 TOLERANCE

A *tolerance* is the total range by which a measurement is allowed to vary. For example, many 450-Ω resistors are manufactured at the same time. Individual resistors in the lot may vary from a low of 405 Ω to a high of 495 Ω. The tolerance is 90 Ω for this order. The amount of tolerance is specified by the customer. The *nominal* or *specified value* is the value from which maximum and minimum limits are determined. The nominal value (450 Ω) is the midpoint between maximum and minimum limits. The *bilateral tolerance* (45 Ω) is one-half the tolerance. The resistance is expressed using the bilateral tolerance as 450 Ω ± 45 Ω.

The *relative tolerance* is the bilateral tolerance expressed as a fraction of the nominal value, usually in percent:

$$\textbf{relative tolerance} = \frac{\textbf{1/2} \times \textbf{tolerance}}{\textbf{nominal value}}$$

If the tolerance is 90 Ω and the nominal value is 450 Ω, the relative tolerance is

$$\frac{1/2 \times \text{tolerance}}{\text{nominal value}} = \frac{45\ \Omega}{450\ \Omega} = \frac{1}{10} = 10\%$$

Thus, the tolerance can be expressed as a percent (450 Ω ± 10%). Remember, it is a percent of the nominal value.

Example A A resistor has a color code that reads 26 kΩ ± 10%. What is the nominal value, the tolerance, maximum value and minimum value?

Solution: Nominal value = 26 kΩ

Bilateral tolerance = (26 kΩ)(0.1) = 2.6 kΩ

Tolerance = 2(bilateral)
= 2(2.6 kΩ) = 5.2 kΩ

Maximum = Nominal + bilateral
= 26 kΩ + 2.6 kΩ = 28.6 kΩ

Minimum = Nominal − bilateral
= 26 kΩ − 2.6 kΩ = 23.4 kΩ

Example B A batch of resistors are specified as 850 Ω. If the resistance can vary by as much as 42.5 Ω, what is the relative tolerance in percent?

Solution: Data: Tolerance = 2 × 42.5 Ω = 85 Ω, nominal value = 850 Ω

Formula: $\text{Relative tolerance} = \dfrac{1/2\ \text{tolerance}}{\text{nominal value}}$

Substitute: $= \dfrac{1/2(85\ \Omega)}{850\ \Omega}$

$= 0.05 = 5\%$

The resistors in this order would be specified as 850 ± 5% and color coded to indicate the specification. See color code, Figure 5-10.

EXERCISE 10-6

Data, diagram, formula, substitute, check.

1. A resistor is coded to read 78 kΩ ± 10%. (a) What is the nominal value? What is the (b) tolerance, (c) maximum value and (d) minimum value?
2. A resistor has a maximum value of 180 MΩ and a minimum of 120 MΩ. (a) Find the nominal value. (b) Give the specifications of the resistor using percent.
3. A capacitor has a maximum value of 94.5 nF and a minimum of 85.5 nF. Give the specifications using percent.
4. A capacitor is coded to read 640 pF ± 10%. What is the (a) tolerance, (b) maximum value and (c) minimum value?

Give the color code for each of the following described resistors (Ω) or capacitors (pF). (See Figure 5-10.)

5. Maximum value = 26.4 kΩ, minimum value = 17.6 kΩ.
6. Maximum value = 792 nF, minimum value = 648 nF.
7. Nominal value = 290 kΩ, bilateral tolerance 5.8 kΩ.
8. Nominal value = 5.0 μF, bilateral tolerance 0.5 μF.

Give (a) the nominal value, (b) the tolerance, (c) the maximum value and (d) the minimum value of the resistors and capacitors in problems 9–12. (See Figure 5-10.)

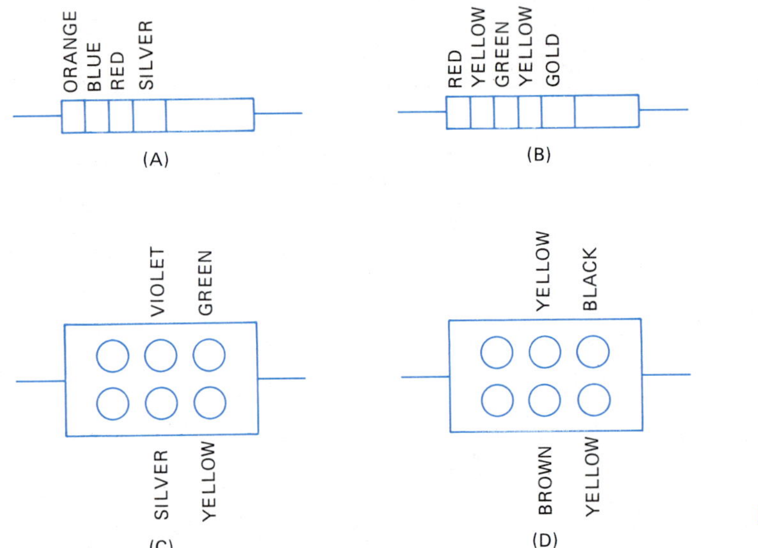

Figure 10-16

9. Use Figure 10-16(A).
10. Use Figure 10-16(B).
11. Use Figure 10-16(C).
12. Use Figure 10-16(D).

EFFICIENCY AND ENERGY CONSUMPTION

Throughout the text are photographs of various parts of the electrical generation and distribution system described in Figure 10-17. Energy in one form is converted to rotational mechanical energy by turbines or engines in a power plant, Figure 2-1. A generator, Figure 2-2, converts the rotational energy to electrical energy. Step-up transformers, Figure 14-1, increase the voltage and decrease current so it can be transported along long distance transmission lines, Figure 1-2, with fewer ($I^2 R$) losses.

At substations, Figure 7-18, the voltage is reduced by distribution transformers, Figure 6-8. Within a city, power moves along smaller conductors, Figure 16-28. Voltage is reduced further by pole transformers,

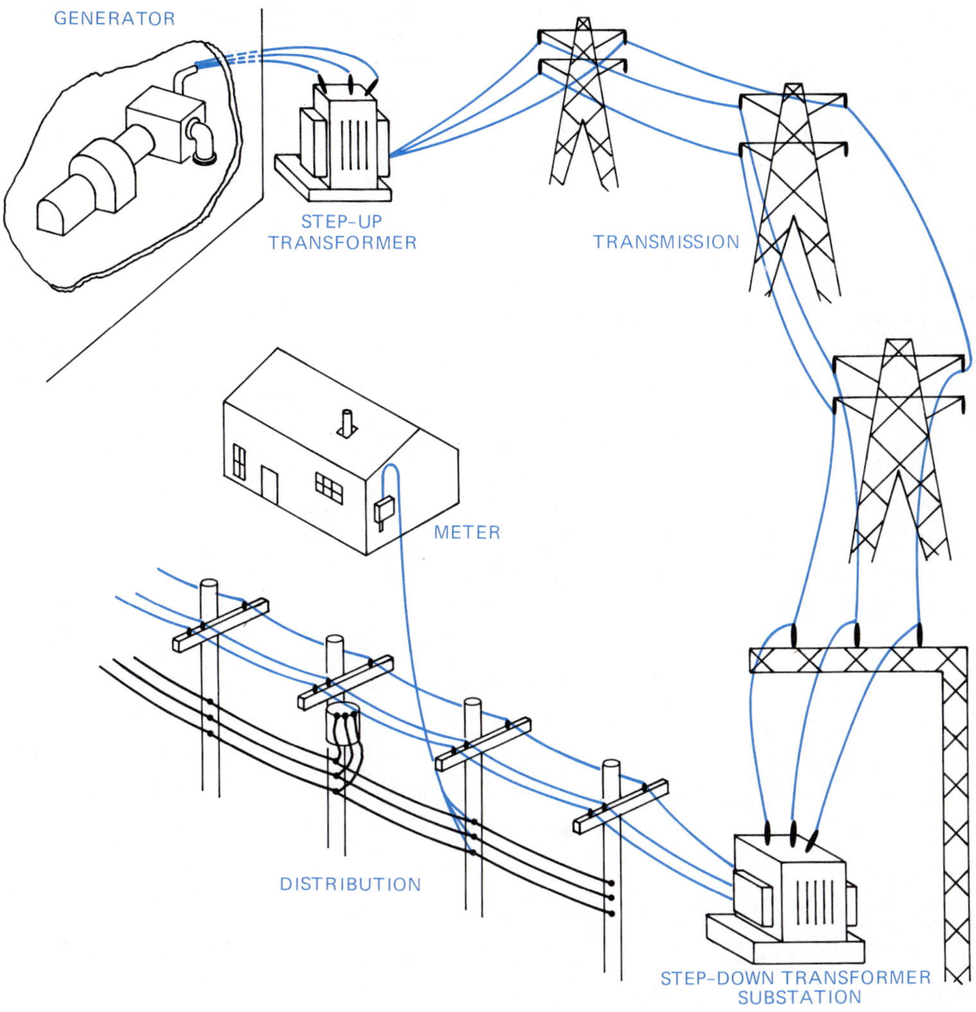

Figure 10-17 A generation, transmission and distribution system

Figure 3-13, for use in homes and factories. Control and protective devices are installed at every juncture.

The reader may ponder the billions of dollars of investment in such equipment. It is important to keep in mind the purpose of this equipment. Energy is being generated and delivered. Energy is purchased by the customer and the monthly or bimonthly electric bill reflects the amount of energy used.

Each part of the system is designed to move energy with the least amount of losses possible. Only recently, however, has the consumer exerted effort and purchased various appliances that minimize wasted energy. The rest of this chapter is devoted to the study of efficiency and energy cost.

10-8 EFFICIENCY

Suppose an object is being lifted. Its weight is 2000 N (newtons) or about 450 lb. It is to be lifted 10 m. The work done in lifting the object is its weight (w) times the height (h) it is lifted.

$$W = \text{wh}$$

If the weight is in newtons and the height in metres, the work will be in joules: (2000 N)(10 m) = 20 000 J.

Suppose a power winch (Figure 10-18) is used to do the job. An interesting question is, "Does it require more than 20 000 J of electrical energy to perform the lifting?". By measuring the current, voltage and time, the electrical energy can be calculated. Suppose these measurements are taken and found to be 12 V, 20 A and 100 s. Since P = EI, then: P = (12 V)(20 A) = 240 W. Also, P = W/t or W = Pt, then W = (240 W)(100 s) = 24 000 J.

More electrical energy is used than work done! This, of course, should not be surprising. But, where did the extra 4000 J go? It was converted to heat in the winch motor and conductors and dissipated into the atmosphere. In other words:

Energy input = work output + heat energy

Since there will be heat produced in any device that converts energy, the work output (or energy output) will always be less than the energy input.

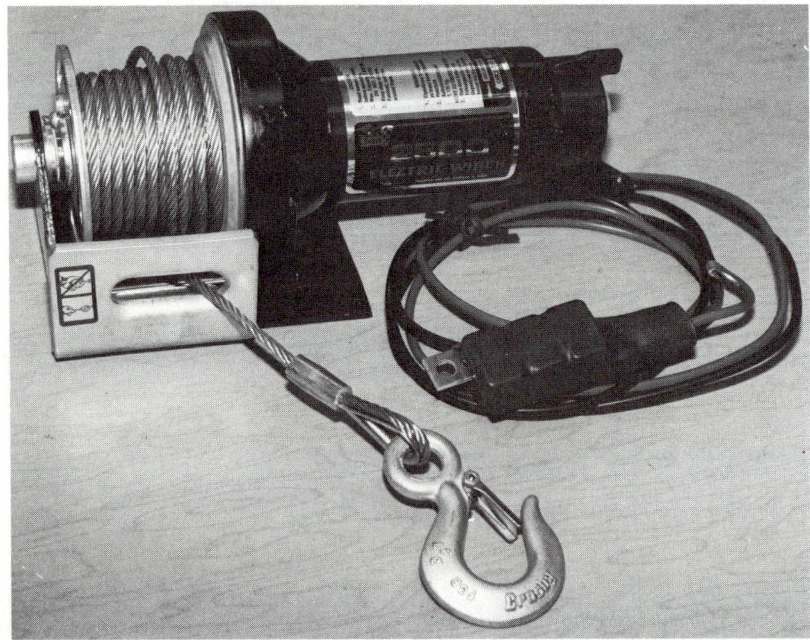

Figure 10-18 A small 12-V power winch (Courtesy of Southeast Nebraska Community College)

Figure 10-19(A) A wattmeter (Courtesy of Triplett Corporation)

Figure 10-19(B) A kilowatt-hour meter

The *efficiency* of the device, whatever it may be, is the work output (or energy output) (W_o) divided by the energy input (W_i). The fraction is usually expressed as a percent.

$$\text{Eff} = \frac{W_o}{W_i}$$

Efficiency can also be calculated by dividing the power output (P_o) by the power input (P_i):

$$\text{Eff} = \frac{P_o}{P_i}$$

The power output plus the power loss (P_L) equals the power input:

$$P_i = P_o + P_L$$

The latter formulas are usually easier to use because of the difficulty of measuring the amount of energy used. The wattmeter, Figure 10-19(A), measures power and is a relatively simple instrument. In contrast, the kilowatt-hour meter, Figure 10-19(B), which measures energy, is a complex and expensive instrument.

Example A Find the efficiency of the winch in the previous discussion. The input electrical energy is 24 000 J and the work done by the winch is 20 000 J.

Solution: Data: $W_o = 20\,000$ J, $W_i = 24\,000$ J, Eff = ?

Formula: $\text{Eff} = \dfrac{W_o}{W_i}$

Substitute: $\text{Eff} = \dfrac{20\,000 \text{ J}}{24\,000 \text{ J}} = 83.3\%$

There are many electrical, mechanical and other devices that *transform* energy from one form to another, *transfer* it from one material to another or *transport* energy from one place to another. Some examples of interest are as follows:

- *Motors.* Transform electrical energy to mechanical (rotational) energy.
- *Generators.* Transform mechanical energy to electrical energy.
- *Lamps.* Transform electrical energy to light energy.
- *Heating element.* Transform electrical energy to heat.
- *Transformer.* Transfers electrical energy from one circuit to another usually at a different voltage.
- *Conductors.* Transport electrical energy from one place to another.

In most cases, efficiency is measured in percent. This means that power output and power input must be measured in the same units. In some devices, the units must be converted. Power input to motors, for example, is given or can be calculated in watts. The output power (rated power) is expressed in *horsepower* (hp). Another example is generators. The mechanical input power is usually given in horsepower. The electrical output is in watts. Conversion factors for various power units are listed in Figure 10-20.

QUANTITY	ENERGY	POWER
Mechanical-Electrical	1 J = 1 W•s = 1 N•m 1 kW•h = 3.6 MJ	1 W = 1 J/s 1 kW = 1.341 hp 1 hp = 745.7 W
Heat*	1 kcal = 4186 J 1 Btu = 1055 J	1 kW = 0.2389 kcal/s 1 kW = 3413 Btu/h
Light		1 W = 668.4 lm (lumen)

*One kilocalorie is the heat energy required to raise the temperature of 1 kg of water 1 C°. One British thermal unit (Btu) is the heat energy required to raise the temperature of 1 lb of water 1 F°.

Figure 10-20 Power and energy conversion table

Example B A DC motor is rated at 1/3 hp. When used at the rated load, it draws 2.73 A from a 120-V line, Figure 10-21. Find (1) the power input, (2) the power losses and (3) the efficiency of the motor.

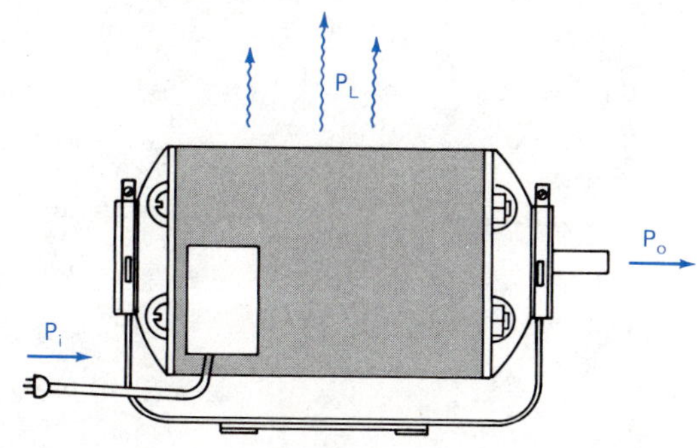

Figure 10-21 Power input equals power output plus power losses

Solution: Data: $P_o = \frac{1}{3}\text{ hp} \times 745.7\,\frac{\text{W}}{\text{hp}} = 249\text{ W, E} = 120\text{ V}$

$I = 2.73\text{ A}, P_i = ?, P_L = ?, \text{Eff.} = ?$

Formula: $P_i = EI$

Substitute: $P_i = (120\text{ V})(2.73\text{ A}) = 328\text{ W}$ (1)

Formula: $P_i = P_o + P_L$

$P_L = P_i - P_o$ $(SA, -P_o)$

Substitute: $P_L = 328\text{ W} - 249\text{ W} = 79\text{ W}$ (2)

Formula: $\text{Eff} = \frac{P_o}{P_i}$

Substitute: $\text{Eff} = \frac{249\text{ W}}{328\text{ W}} = 75.9\%$ (3)

Example C A pole transformer has a power output of 5.75 kW at a certain time. Its efficiency is 98%. If the line voltage (input) is 4160 V, find the input current. Assume the ac circuit is pure resistive.

Solution: Data: $P_o = 5750\text{ W}, \text{Eff} = 98\%, E_i = 4160\text{ V}, I_i = ?$

Formula: $\text{Eff} = \frac{P_o}{P_i}$

$P_i = \frac{P_o}{\text{Eff}}$ $(MA, \times P_i;\ DA, \div \text{Eff})$

Substitute: $P_i = \frac{5750\text{ W}}{0.98} = 5867\text{ W}$

Formula: $P = EI$

$I = \frac{P}{E}$ $(DA, \div E)$

Substitute: $I = \frac{5867\text{ W}}{4160\text{ V}} = 1.41\text{ A}$

Efficiency of lamps is an exception, Figure 10-22. Luminous efficiency is not expressed in percent. The output power of light sources is expressed in *lumens.* The luminous output power *(luminous flux)* is a small fraction of the electrical input power. Most of the energy is converted to heat. The efficiency of lamps is expressed in lumens per watt instead of percent. The input power and efficiency of several lamps are given in Figure 10-23. Notice that incandescent lamps are less efficient. The efficiency of any type increases with the size of the lamp.

Example D Find the output power of a 100-W incandescent lamp. How much power is converted to heat?

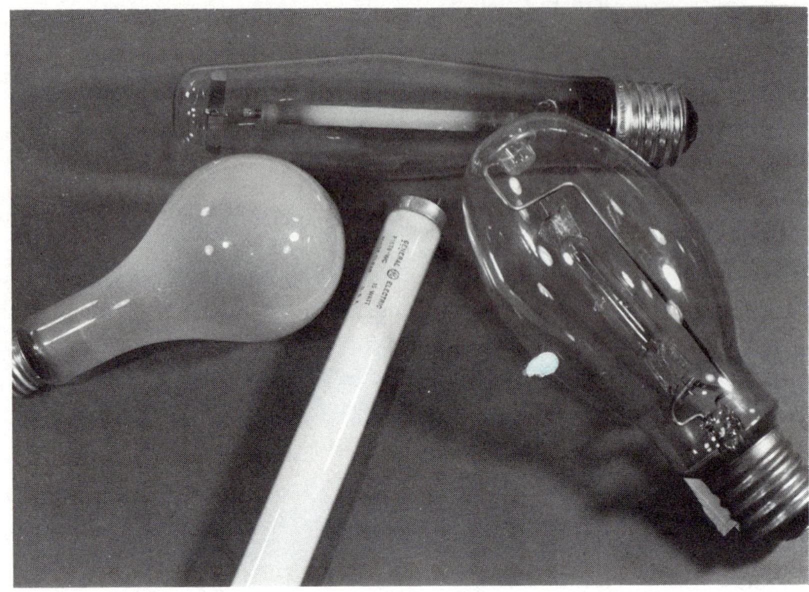

Figure 10-22 Various kinds of lamps (Courtesy of Southeast Nebraska Community College)

WATTAGE	INCANDESCENT	MERCURY	FLUORESCENT	SODIUM
10	8	–	38	–
40	11	23	59	–
75	16	30	63	75
100	17	35	68	86
1000	24	48	–	126

Figure 10-23 Power rating and efficiency in lumens per watt for various lamps

Solution: Data: Eff = 16 lm/W (Figure 10-23), P_i = 100 W, P_o = ?, P_L = ?

Formula: $\text{Eff} = \frac{P_o}{P_i}$

$P_o = \text{Eff} \times P_i$ (MA, $\times P_i$)

Substitute: $P_o = (16\ \text{lm/W})(100\ \text{W}) = 1600\ \text{lm}$

$P_o = (1600\ \text{lm})\left(\frac{1\ \text{W}}{668.4\ \text{lm}}\right) = 2.4\ \text{W}$

Formula: $P_i = P_o + P_L$

$P_L = P_i - P_o$ (SA, $- P_o$)

Substitute: $P_L = 100\ \text{W} - 2.4\ \text{W} = 97.6\ \text{W}$ (Heat)

Note: Energy is converted to heat at a rate of 97.6 J each second.

EXERCISE 10-7

Data, diagram, formula, substitute, check.

1. An electric motor is rated at 1/4 hp. What is the efficiency if the input is 220 W?
2. A pole transformer has an efficiency of 97.5%. What is the power output at a time when the input is 6.38 kW?
3. A large motor has an efficiency of 95%. What is the power output in horsepower if the input is 20 kW?

4. Find the efficiency of a 1/2-hp motor if it draws 410 W from the line.
5. Use Figure 10-23 and find the output power of a 100-W mercury lamp in (a) lumens and (b) watts. (c) Express the efficiency in percent.
6. Find the input of a mercury vapor lamp if the efficiency is 44 lm/W. The output is 22 000 lm.
7. A generator has an efficiency of 88%. Its output is 120 V and supplies 65 A. (a) What horsepower is delivered to the generator by the V-belt that drives it? (b) What is the internal power loss?
8. A 12-V automobile starter motor is 80% efficient. (a) If the motor develops 1.25 hp, what current does it draw? (b) Calculate the internal power loss.
9. The heater fan motor on an automobile has an input power of 46.6 W. If 9.3 W are converted to heat internally, (a) what is the output horsepower? (b) What is the efficiency?
10. A transmission line receives energy at 225 V and a current of 25 A. The power delivered is 5.51 kW. Find (a) the power loss and (b) the efficiency.
11. A 1200-hp diesel engine delivers power by direct-drive to a generator with an output of 2200 V and 365 A. Find the (a) power losses and (b) efficiency of the generator.
12. A football field is illuminated by six towers. Each tower has 20 lamps with luminous efficiency of 25 lm/W. Each lamp emits 50 000 lm of luminous power. (a) Find the total input power. (b) At what rate is electrical energy converted to heat?
13. A DC automotive generator is about 65% efficient. The alternator, Figure 10-24, being about 80% efficient has in recent years replaced the generator. Calculate the power input from the drive belt for the (a) generator and (b) alternator that is required for 0.5 kW output.

Figure 10-24 A modern automotive alternator (Courtesy of Southeast Nebraska Community College)

10-9 ENERGY CONSUMPTION AND COST

Energy usage can be measured with a kilowatt-hour meter, Figure 10-19(B). One can be found on the side of a building near the "electrical entrance." It is read monthly or bimonthly. The energy used is found by subtracting the previous reading from the present one. Cost is based on the amount used.

It should be remembered that using energy means converting it to a different form. In the many ways that electrical energy is used, it is not destroyed. It is merely transformed to a different kind (usually heat) and dissipated into the atmosphere. It eventually radiates to outer space and is no longer available.

Except for billing purposes and transfers between power companies, electrical energy is seldom measured. Energy meters are complex and expensive instruments. To calculate energy usage in electrical appliances for a given time, the rate (power) must be known:

$$W = Pt$$

In Chapter 7, this formula was used to calculate energy in joules (J) in problems where power (P) was in watts (W) and time in seconds (s). **One joule equals one watt-second (W•s).**

The joule is a small unit of energy. It is much too small for expressing the amount of energy used in a home or business. A 100-W lamp will consume 360 000 J each hour. A larger unit is derived from the formula W = Pt by expressing power in kilowatts and the time in hours. The kilowatt-hour (kW•h) is

$$(1000\ \text{J})(3600\ \text{s}) = 3\,600\,000\ \text{J}$$
$$1\ \text{kW·h} = 3.6\ \text{MJ}$$

In the case of the 100-W lamp, it will use 0.1 kW•h of energy each hour.

Example A Find the energy dissipated in a solid-state color television (200 W) in an average month of 183 h usage.

Solution: Data: P = 200 W = 0.2 kW, t = 183 h, W = ?

Formula: $W = Pt$

Substitute: $W = (0.2\ \text{kW})(183\ \text{h}) = 36.6\ \text{kW·h}$

Formulas $P = EI$, $P = I^2R$ or $P = E^2/R$ can be substituted into the equation $W = Pt$ to give

$$W = EIt$$
$$W = I^2Rt$$
$$W = \frac{E^2t}{R}$$

These formulas are used when power is not given.

Example B A hair dryer has a resistance of 16.1 Ω and operates on 110 V. Find the heat produced in a year's usage of 51 h. Assume 100% converts to heat.

Solution: Data: $E = 110\text{ V}, R = 16.1\ \Omega, t = 51\text{ h}, W = ?$

Formula: $W = \frac{E^2 t}{R}$

Substitute: $W = \frac{(110\text{ V})^2 \times 51\text{ h}}{16.1\ \Omega} = 38\ 300\text{ W}\cdot\text{h}$

$W = 38.3\text{ kW}\cdot\text{h}$

Figure 10-25 gives the average power rating and typical annual usage for several common appliances found in the home. The actual amount used by any appliance varies significantly depending on its size and the geographical area of use.

The cost of energy depends on the amount used. Rates vary from one power company to another. The rate is expressed in cents per kilowatt-hour. Summer rates are greater than winter rates. Residential customers are charged at a different rate than are commercial customers.

One electric company, in addition to a minimum customer charge, requires a higher price for each of the first 250 kW•h used each month, less for each of the next 500 kW•h and a smaller rate for the rest. The changes are in a ratio of about 9 to 4 to 3. A production cost charge is added on reflecting variable costs of fuel. In some states and cities, a sales tax is required. Thus, an electric bill is quite complicated. The average residential cost per kilowatt-hour may vary from 4¢ to 10¢ depending on usage. The average cost rate for a given user normally changes from month to month. Except in the air-conditioning season or in all-electric homes, normal residential usage is less than the 750 kW•h required for the lowest rate. Heavy users generally pay lower rates.

APPLIANCE	WATTAGE (W)	USAGE (h/yr)
Air Conditioner	3750	1000
Blender	385	40
Clock	2	8760
Coffee Maker	1000	70
Curling Wand	40	50
Electric Blanket	180	830
Fry Pan (cycles off and on)	1200	150 (on)
Iron	1000	140
Radio	70	1200
Razor	14	130
Refrigerator (12 ft^3)	240	3000
Television (solid state)	200	2200
Television (tube)	300	2200
Toaster	1200	35
Vacuum Cleaner	600	75
Washer	500	208
Water Heater	4475	1050

Figure 10-25 Energy requirements for various home appliances

Example C Find the cost of operating the television set in Example A for one month. Assume the usage is in a billing period when the average cost is 6.8¢/kW•h. Add 4% sales tax.

Solution: Data: W = 36.6 kW•h, rate = 6.8¢/kW•h

Formula: Cost = rate × W

Substitute: Cost = (6.8¢/kW•h)(36.6 kW•h)

Cost = \$2.49

Sales tax is 4% of the cost: (\$2.49)(0.04) = 10¢

Total bill is \$2.49 + \$0.10 = \$2.59

EXERCISE 10-8

Data, diagram, formula, substitute, check.

1. How much energy is used by a heating element rated at 900 W if it is used continuously for 24 h? Express the answer (a) in joules and (b) in kilowatt-hours.
2. An electric tank heater for a car engine operates at 120 V and draws 12.5 A of current. Find the energy used in kilowatt-hours if it is plugged in for 30 min.
3. A transmission line has a resistance of 2.5 Ω and carries a current of 80 A. Assuming the current remains constant, find the energy loss (a) in one day; (b) in one year.
4. The rating of a service entrance to a home is 50 A at 220 V. How much energy is available each day?
5. An electric furnace converts 1745 MJ of energy to heat each day if used continuously. How much heat is this (a) in kW•h? (b) In kcal?
6. A wind generator is rated at 15 kW. If it operates 19.5 days per month, find the value of its output at 6¢ per kilowatt-hour.
7. Seven 60-W lamps are used 6 h per day. Find the cost for 30 days at 6.7¢ per kilowatt-hour.
8. Find the cost of operating a clock, see Figure 10-25, for one year at 6.9¢ per kilowatt-hour.
9. A factory uses 1200 lamps (150 W) eight hours per day 25 days per month. Find the operating cost at 5.2¢ per kilowatt-hour.
10. Find the annual cost of operating a refrigerator, see Figure 10-25, at 6.2¢ per kilowatt-hour.
11. A battery charger draws 3 A from a 110-V line. Find the cost of operating 8 h/day, 6 days/wk for a four-week month. The billing rate is 6.9¢ per kilowatt-hour.
12. Find the cost of operating a motor continuously for one year (365 days). It draws 10 A from a 110-V line. The cost is based on a rate of 6.2¢ per kilowatt-hour.
13. A night-light operates on 110 V and has a hot resistance of 4.0 kΩ. Find the annual cost to operate it ten hours per day at 6.8¢ per kilowatt-hour.
14. An electric hot water heater draws a current of 19.5 A and has a resistance of 10.8 Ω. If the heater is cycled on one hour in every eight, find the annual cost at 5.9¢ per kilowatt-hour.

15. An air conditioner operates on 230 V and has a resistance of 14.1 Ω. If it is used 1000 h each year, find the cost at 6.8¢ per kilowatt-hour.
16. Meter readings of 8402 kW•h and 9424 kW•h were recorded at the beginning and at the end of a two-month billing period. Find the cost if the rate is 9.9¢/kW•h for the first 500 kw•h, 5.9¢/kW•h for the next 1000 kW•h and 4.8¢/kW•h for the remainder.
17. Meter readings at the beginning of the two-month period (A) and at the end (B) are shown in Figure 10-26. Using the rates given in Problem 16, determine the cost.

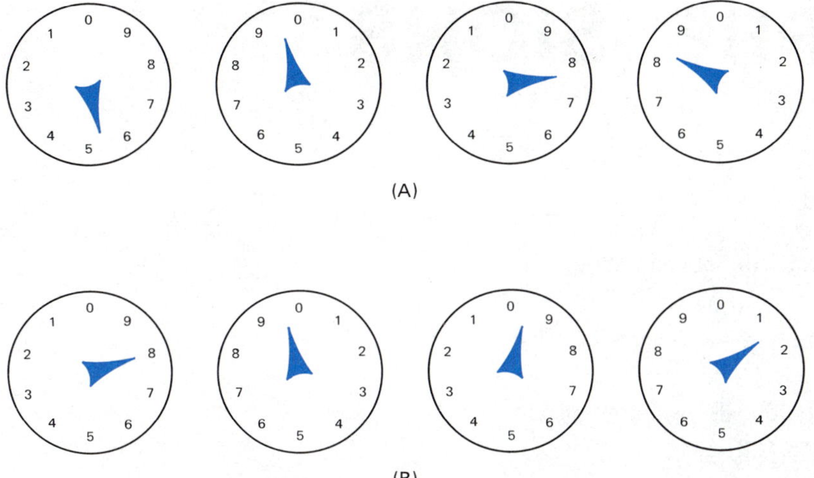

Figure 10-26

18. A 1/3-hp furnace fan motor cycles on for about 8 hours per day for a 31 day month. The motor is 91% efficient. Find the monthly bill if the billing rate is 6.4¢/kW•h.

CHAPTER 11

SIMPLIFYING ALGEBRAIC EXPRESSIONS

OBJECTIVES

After satisfactorily completing this chapter, the student should be able to:

- Combine terms in monomial and polynomial expressions by addition or subtraction.
- Multiply or divide monomials or polynomials by a monomial.
- Factor monomials from polynomial expressions.

This chapter and the one following cover some additional concepts in algebra. Both are in preparation for Chapter 13, Circuit Analysis. Many circuits have already been studied using Ohm's Law. Some circuits yield to analysis better by using various other approaches. These require a more thorough knowledge of algebra.

This chapter involves simplifying or combining algebraic expressions. It includes the concept of factoring and methods of treating certain problems manually. The ability to perform such tasks quickly and accurately can be acquired. The student is encouraged to use paper and pencil and to solve as many exercise problems as necessary to learn these skills.

POLYNOMIALS

In algebra, an expression is often changed to a different form. Of course, this must be done according to specified rules so the new form has the same value as the old. In this process, a more convenient form is obtained. Many of these processes have been studied in previous chapters. Factoring, cancellation and adding like terms are three such processes. These and other concepts will be extended to include use with expressions containing grouping symbols.

11-1 TERMS

An *algebraic expression* is a collection of various numbers and literal number symbols. The numbers are combined by one or more of the basic operations of addition, subtraction, multiplication, division and square root. Some of the numbers may be enclosed within grouping symbols.

A *term* is a part of an algebraic expression within which numbers and letters are not separated by plus or minus signs. Any quantity enclosed by parentheses is to be thought of as a single quantity. An expression such as $-4(a - 2)$ is a single term, for example. It is the negative of the product of 4 and another number.

Certain algebraic expressions are named according to the number of terms they contain:

- *Monomial.* A single term expression.
- *Binomial.* An expression with two terms.
- *Trinomial.* A three term expression.
- *Polynomial.* An expression with more than one term.

Two or more expressions with exactly identical literal factors are said to be *like* or *similar terms.* If the literal factors are not exactly the same, they are *unlike* or *dissimilar terms.* The numerical factor of a term is called the *coefficient.* Further discussion about terms can be reviewed in Topic 8-1 and Topic 8-2.

EXERCISE 11-1

How many terms are there in each of the following expressions? Is the expression monomial, binomial or polynomial?

1. R_1
2. 3E
3. $\frac{1}{2}(P + EI)$
4. $5 - 2(E - Ir)$
5. $A_1 - \frac{1}{2}b_o h_o - A_2$
6. $-[(\alpha - \beta) + 1]$
7. $8(P - 3) + 7(P - 3)$
8. $-7(a^2 b)(abc)$
9. $4\sqrt{PR} - 2I_o + \frac{E}{R} - 5$
10. $5IR + 8\frac{P}{I} - 5\sqrt{PR} + 3E_o - 8$

State the coefficient of each of these monomials.

11. W
12. 215Pt
13. $\frac{4}{3}\pi r^2$
14. $\frac{\rho L}{A}$
15. $-\alpha\beta$
16. 144
17. $2\pi^2 f^2 R^2 m$
18. $\frac{5E^2}{R}$

Indicate whether the following pairs of monomials are like or unlike terms.

19. $7a^2 b, 8ba^2$
20. $7x\sqrt{y}, 9xy$
21. $5E_1, 2E_2$
22. $X_C, 4X_L$
23. $8EI_1, 15I_1E$
24. 4R, 3r
25. −7, 128
26. $45a^2bc^3d, 2bc^3a^2d$

11-2 ADDING MONOMIALS AND POLYNOMIALS

Adding Monomials. To add monomials, add the coefficients. The sum retains the same literal factors. It must be remembered that only similar things can be added. Only like monomials can be added.

Example A Add: $9I^2R, -12I^2R$ and $+8I^2R$

Solution:

$$\begin{array}{r} 9I^2R \\ -12I^2R \\ 8I^2R \\ \hline 5I^2R \end{array}$$

Check by adding in the reverse order.

Note: Form column, combine coefficients, write answer, check.

Example B Add: $\frac{3}{4}(1 - a^2), -\frac{7}{8}(1 - a^2), +\frac{2}{3}(1 - a^2), -\frac{1}{6}(1 - a^2)$

Solution:

$$\begin{aligned} \frac{3}{4}(1 - a^2) &= \frac{18}{24}(1 - a^2) \\ -\frac{7}{8}(1 - a^2) &= -\frac{21}{24}(1 - a^2) \\ +\frac{2}{3}(1 - a^2) &= +\frac{16}{24}(1 - a^2) \\ -\frac{1}{6}(1 - a^2) &= -\frac{4}{24}(1 - a^2) \\ \hline & +\frac{9}{24}(1 - a^2) \end{aligned}$$

Check by adding in the reverse order.

Note: The least common denominator for the fractional coefficients is 24.

Adding Polynomials. To add several polynomials, arrange the terms of each in the same order. If the terms have various powers of the same literal factor, arrange them in order of decreasing powers. List the polynomials with like terms in separate columns. Add each column to find the sum of the polynomials using the rules for adding signed numbers.

Example C Add: $3a + 4b + 1$, $6a - 9b$ and $5 - 3b + 2a$

Solution:

$$\begin{array}{l} 3a + 4b + 1 \\ 6a - 9b \\ 2a - 3b + 5 \\ \hline 11a - 8b + 6 \end{array}$$

Check by adding each column in the reverse order.

Note: 1. arrange in order, 2. form columns, add and check.

Example D Add: $4\theta^3 - 4\theta + 2, \theta^2 + 2\theta^3 + \theta, 7 - \theta^3 + 5\theta - 7\theta^2$

Solution:

$$\begin{array}{l} 4\theta^3 \qquad\qquad - 4\theta + 2 \\ 2\theta^3 + \theta^2 + \theta \\ -\theta^3 - 7\theta^2 + 5\theta + 7 \\ \hline 5\theta^3 - 6\theta^2 + 2\theta + 9 \end{array}$$

Check by adding all columns in the reverse order.

EXERCISE 11-2

Add the following monomials.

1. $5a^2, -3a^2, 12a^2, -7a^2$
2. $0.5R_1, -1.7R_1, 0.7R_1, 0.9R_1$
3. $-\frac{3}{4}P, \frac{2}{3}P, -\frac{5}{8}P, \frac{5}{6}P$
4. $8(p + q), -5(p + q), -13(p + q), 7(p + q)$
5. $-\beta(\alpha + 1), 3\beta(\alpha + 1), -2\beta(\alpha + 1), 5\beta(\alpha + 1)$
6. $4\omega^2\lambda^3, -5\omega^2\lambda^3, 3\omega^2\lambda^3, 2\omega^2\lambda^3, -\omega^2\lambda^3$
7. $1.3(\theta - \phi), 2.5(\theta - \phi), -2.1(\theta - \phi), 1.8(\theta - \phi)$
8. $\frac{1}{5}(k - 0.3n), -\frac{3}{4}(k - 0.3n), \frac{5}{6}(k - 0.3n), -\frac{1}{2}(k - 0.3n)$

Add the following polynomials.

9. $8P + 7EI, 4EI - 3P, P - 2EI$
10. $5\alpha - 3\beta, 2\alpha + 7\beta, 4\beta - 4\alpha, \alpha + \beta$
11. $3\omega_o - 8\omega, 15\omega + 3\omega_o, -5\omega_o - 2\omega$
12. $9W - 4Pt - EQ, 8W + 4EQ - 4Pt$
13. $2E^2/R, 3E^2/R + 2I^2R - 2P, 7P + 2I^2R - 4E^2/R$
14. $4abc - 3a^2bc + 8ab^2c, 9ba^2c - 3cab + cab^2, 4bac + cb^2a$
15. $4(\mu + 1) - 12(\mu - 1), 8(\mu - 1) - 2(\mu + 1), 5(\mu + 1) + 6(\mu - 1)$
16. $4\mu^3 - 3\mu + 2\mu^2, 5\mu + \mu^2 - 1, 2\mu^3 - 8\mu + 4 - \mu^2$
17. $2a^3 + 3a - 4, -10a^2 + a^3 - 2, 6 + 2a + 3a^3 + 4a^2$
18. $2Q^4 - 3Q^3 + 2Q^2 + 5Q - 3, 7 - Q^3 + Q^2 - 4Q + 2Q^4, -3Q^4 + 4Q^2 + 8 - 5Q + 4Q^3$

An expression such as 5I across a circuit load can mean either the voltage drop ($R = 5\ \Omega$) or power dissipation $E = 5$ V. The expressions $7I^2$ and $\frac{E^2}{7}$ are power dissipations across a 7-Ω resistor.

19. Find the sum of all the voltage drops in the circuit of Figure 11-1.

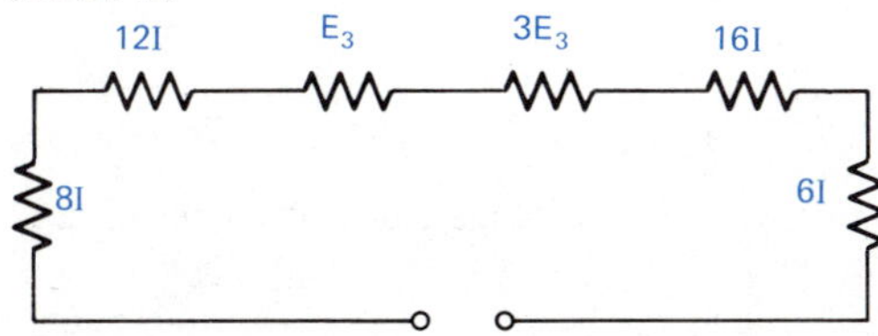

Figure 11-1

20. What is the total power dissipation in the circuit of Figure 11-2?

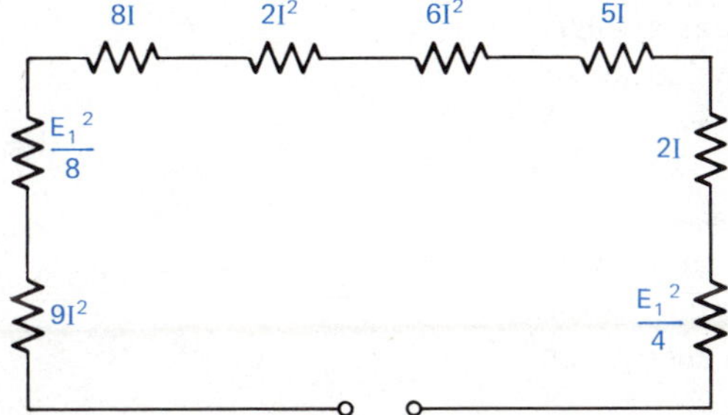

Figure 11-2

21. Find the total power dissipation in the parallel circuit of Figure 11-3.

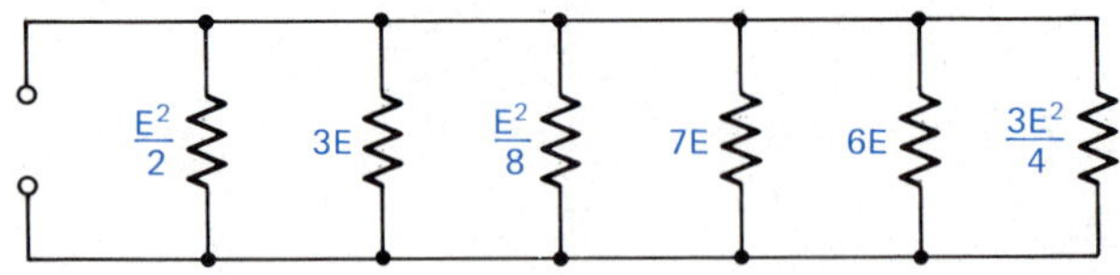

Figure 11-3

11-3 SUBTRACTING MONOMIALS AND POLYNOMIALS

As with addition, only like terms may be subtracted. The difference is found by subtracting the subtrahend from the minuend. The order is important.

> **Subtracting Polynomials.** Place the subtrahend under the minuend so that like terms are in the same column. Subtract the coefficients. The difference retains the common literal factor. When subtracting, use the rules for subtracting signed numbers.

Example A Subtract $12I^2$ from $-7I^2$

Solution: 1. Determine subtrahend, 2. arrange in columns, 3. combine and check.

The subtrahend is $12I^2$

$$\begin{array}{rr} \text{Subtract:} & -7I^2 \\ & \underline{12I^2} \\ & -19I^2 \end{array}$$

(Change sign and add manually)

Example B By how much is $5\theta^2 - 2\theta + 4$ larger than $+3\theta - \theta^2 + 5$?

Solution:

$$\begin{array}{rrrr} \text{Subtract:} & 5\theta^2 & -2\theta & +4 \\ & \underline{-\theta^2} & \underline{+3\theta} & \underline{+5} \\ & 6\theta^2 & -5\theta & -1 \end{array}$$

(Change signs of all terms, add manually)

11-4 ADDING AND SUBTRACTING HORIZONTALLY

When a polynomial is enclosed within parentheses it acts as a single quantity. By enclosing each polynomial in parentheses, an addition or subtraction problem can be written horizontally. The sum of $4ab + 2a^2b - 5ab^2$ and $3ab - 5a^2b + 2ab^2$ can be written:

$$(4ab + 2a^2b - 5ab^2) + (3ab - 5a^2b + 2ab^2)$$

To find the sum simply drop the parentheses and *collect like terms.* Collecting like terms is the process of combining like terms according to the rules of adding and subtracting signed numbers: $4ab + 2a^2b - 5ab^2 + 3ab - 5a^2b + 2ab^2 = 7ab - 3a^2b - 3ab^2$.

Subtraction is similar but the sign of the subtrahend must be changed before like terms are collected. A polynomial preceded by a minus sign can be changed to one preceded by a plus sign by changing the signs of every term. Then parentheses may be dropped and like terms collected.

Example A Subtract: $(5t^2 + 6t + 8) - (3t^2 + 5t - 3)$

Solution: $= (5t^2 + 6t + 8) + (-3t^2 - 5t + 3)$ (Change signs)

$= 5t^2 + 6t + 8 - 3t^2 - 5t + 3$

$= 2t^2 + t + 11$ (Collect terms)

Of course, most of the steps can be done mentally.

Several polynomials can be combined by addition and subtraction. Change signs of all terms in those being subtracted.

Example B Combine: $(4\theta^2\phi - 8\theta\phi^2) - (2\theta^2\phi - 4\theta\phi^2) + (7\theta^2\phi - 9\theta\phi^2)$

Solution: $= (4\theta^2\phi - 8\theta\phi^2) + (-2\theta^2\phi + 4\theta\phi^2) + (7\theta^2\phi - 9\theta\phi^2)$

$= 4\theta^2\phi - 8\theta\phi^2 - 2\theta^2\phi + 4\theta\phi^2 + 7\theta^2\phi - 9\theta\phi^2$

$= 9\theta^2\phi - 13\theta\phi^2$

EXERCISE 11-3

1. Subtract $5I^2R$ from $16I^2R$.
2. Subtract $-3a^2b^3$ from $11a^2b^3$.
3. Subtract $14x^3y^2z$ from $-23x^3y^2z$.
4. Subtract $-11\alpha\beta^2$ from $-2\alpha\beta^2$.
5. Subtract: $(184\mu N) - (37\mu N)$
6. Subtract: $[42(\rho - 1)] - [-37(\rho - 1)]$
7. Subtract: $(5R + 3Z - X) - (7Z + 12R - 14X)$
8. Subtract: $[4R^2 + 3(X_L - X_C)^2] - [17R^2 - 7(X_L - X_C)^2]$
9. Subtract: $(4\mu + 8\lambda) - (2\mu - 7\lambda + 1)$
10. Subtract: $(188\theta - 266\phi + 1822) - (64\phi - 114\theta + 1183)$

Remove grouping signs and combine similar terms. Evaluate horizontally.

11. $(\alpha + \beta + 2\gamma) + (3\beta - 4\alpha + 5\gamma) - (7\gamma + 2\beta - 8\alpha)$
12. $(4EI - 2P + 7I^2R) - (3EI - 5I^2R) - (5P + 2EI + I^2R)$
13. $(3W - EIt + 2Pt) + (8EIt - 8Pt - 11W) - (7Pt - 8W)$
14. $(7\omega + 5 + 6\mu) - (7 - 6\mu + 2\omega) - (2\omega + \mu - 2)$
15. $[5\gamma - 2(90 - \theta)] - [4\gamma + 3(90 - \theta)] + [3\gamma - 6(90 - \theta)]$
16. $(3E + 2IR - 5P/I) - (7IR - 7E - 5P/I) - (2P/I + 10E - 5IR)$
17. Find the voltage drop (E_1) across resistor R_1 in the circuit of Figure 11-4.

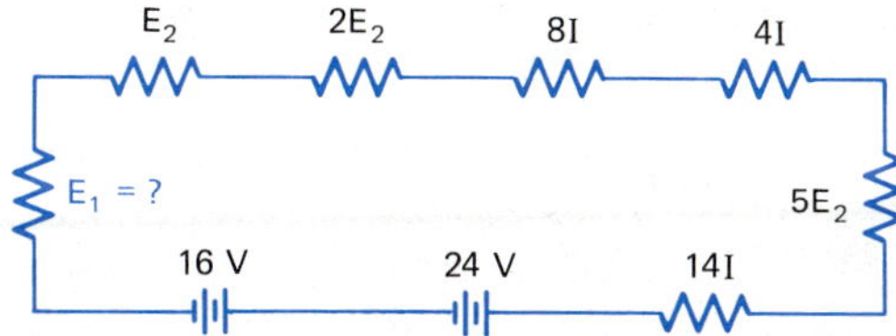

Figure 11-4

18. Determine the total power dissipation (P_1) across load R_1 in the circuit of Figure 11-5.

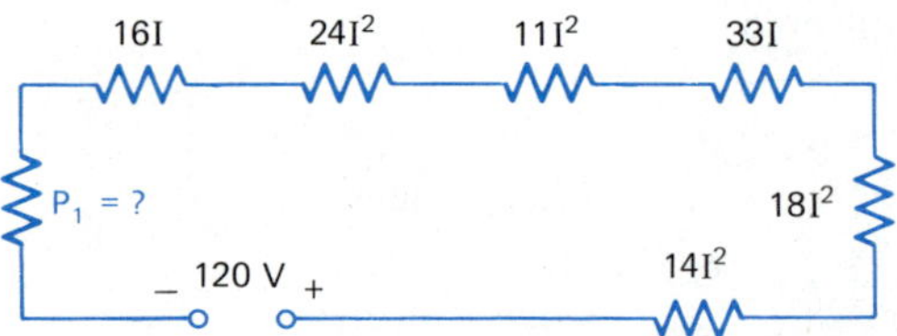

Figure 11-5

MULTIPLYING AND DIVIDING POLYNOMIALS

Any letter used to represent a number is a *literal* number. Literal numbers have the same properties as specific numbers. This fact can be easily forgotten while studying algebra. The basic concepts and operations of arithmetic apply to literal numbers as well as specific numbers. The results of adding, multiplying and performing other operations can sometimes only be indicated. Terms should be combined whenever possible. The results of combining terms are definitely easier to use than the original expressions.

The expression

$$(5t^2 + 6t + 8) - (3t^2 + 5t - 3)$$

when combined is $2t^2 + t + 11$. If the expression is to be evaluated for specific values of t, it takes less effort using the result. If $t = 2$, for example, the expression is as follows:

$$[5(2)^2 + 6(2) + 8] - [3(2)^2 + 5(2) - 3]$$
$$= (20 + 12 + 8) - (12 + 10 - 3)$$
$$= 40 - 19$$
$$= 21$$

Using the result is easier.

$$2(2)^2 + 2 + 11 = 8 + 2 + 11 = 21$$

If the expression is to be evaluated for several values of t, the time required to combine terms first is certainly worthwhile. The ability to simplify algebraic expressions quickly and accurately is a valuable skill for electricians. The rest of this chapter is devoted to simplifying algebraic expressions.

11-5 EXPONENTS IN MULTIPLICATION

In Chapter 6, the laws of exponents were introduced as they apply to powers of ten and scientific notation. Those laws are applicable to powers of any base. Recall that powers of ten are multiplied by adding their exponents:

$$(10^3)(10^2) = 10^{3+2} = 10^5$$

It is the quantities 10^3 and 10^2 that are being multiplied; not the exponents. The product of powers of any base is found by adding exponents.

Exponent Rule for Multiplying. In multiplying algebraic quantities expressed as powers of the same base, the product is found by adding the exponents:

$$(b^m)(b^n) = b^{m+n}$$

Example A Find the product: $(R^4)(R^2)$

Solution: $(R^4)(R^2) = R^{4+2} = R^6$

Note: Remember that R^4 means R•R•R•R and R^2 means R•R. The problem can be expressed as (R•R•R•R)(R•R) = R•R•R•R•R•R which is R^6.

11-6 MULTIPLYING MONOMIALS

To find the product of two monomials, simply apply the laws of multiplication and the laws of exponents.

MULTIPLYING MONOMIALS

To find the product of two or more monomials:

- Determine the sign of the product. It is positive for an even number of negative factors and negative for an odd number.
- Multiply the coefficients of all factors. This will be the coefficient of the product.
- Find the product of the literal factors.
- Write the product with the proper sign as the product of the coefficient and the literal factors.

Example A Multiply: $(-3a^2b)(7ab)$

Solution: $(-3)(7) = -21$ (Coefficients)

$(a^2b)(ab) = a^3b^2$ (Literal factors)

$(-3a^2b)(7ab) = -21a^3b^2$ (Combine)

In most cases, the steps can be done mentally.

Example B Find the product: $\theta(-3\theta^2\phi)(-8\theta\phi^2\mu)$

Solution: $(1)(-3)(-8) = +24$ (Coefficients)

$\theta(\theta^2\phi)(\theta\phi^2\mu) = \theta^4\phi^3\mu$ (Literal factors)

$\theta(-3\theta^2\phi)(-8\theta\phi^2\mu) = 24\theta^4\phi^3\mu$ (Combine)

EXERCISE 11-4

1. Express the following in exponential form.
   a. 3•3•3•3•3
   b. 12•12•12
   c. *a•a•a•a•a•a•a*
   d. a•a•b•b•b
   e. I•IR
   f. (4P)(4P)(4P)(4t)(4t)(4t)

2. Express the following as a product of individual factors. ($a^3 = a \cdot a \cdot a$)
   a. 5^3
   b. 15^5
   c. a^4
   d. $2a^2b^3$
   e. $\frac{4}{3}\pi r^3$
   f. $8^3\theta^5\phi^2\mu^3$

Find the product of the following monomials.

3. $(5a)(3ab)$
4. $(3\theta^2)(11\theta\phi)$
5. $(-2ab)(4ab)$
6. $-(4m^2n)(144mn^2)$
7. $(-3rt)(-4st)(3rs)$
8. $(4\alpha)(-4\alpha\beta)(-4\beta)$
9. $(-4a^2b)(-3ab^2)(-2a^2)$
10. $(5\omega^0 t^2)(2\omega^3 t)(3\omega^5 t^3)(\omega t)$
11. $-(-\lambda)(3\lambda^2)(-2\lambda^4)(-\lambda)$
12. $\left(\frac{1}{2}IX\right)\left(-\frac{2}{3}IX\right)\left(-\frac{1}{6}I^2Z\right)$
13. $(20x^5y^3z^4)(-5y^3x^4z^2)$
14. $(2^5R^3I)(-2^0RI^2t)(2^2R^2I^3t^5)$
15. $-P(P^2Q^3k)(k^2)(-Q^5k^0P^3)$
16. $\omega_o{}^3(2\omega_o{}^2\omega)(5\omega^3\omega_o{}^5)(5)$

11-7 MULTIPLYING A POLYNOMIAL BY A MONOMIAL

The product of two lengths can be interpreted as an area since A = Lw. The product of 3 and 5 is represented geometrically in Figure 11-6. By counting squares, one finds that 3 × 5 represents 15 units of area. A geometrical illustration of the product 3(5 + 4) = 3(5) + 3(4) is given in Figure 11-7.

$$3(5 + 4) = 3(5) + 3(4)$$
$$3(9) = 15 + 12$$
$$27 = 27$$

The illustration in Figure 11-7 gives the general rule for multiplying a polynomial (5 + 4) by a monomial (3):

$$a(b + c) = ab + ac$$

where a, b and c can be any numerical or literal terms. This equation is sometimes called the *Distributive Law* and gives rise to the following rule:

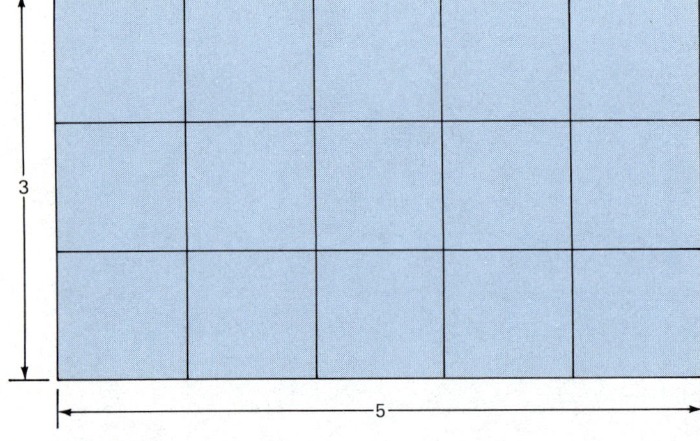

Figure 11-6 The geometric representation of a product of two numbers as an area: 3(5) = 15 squares.

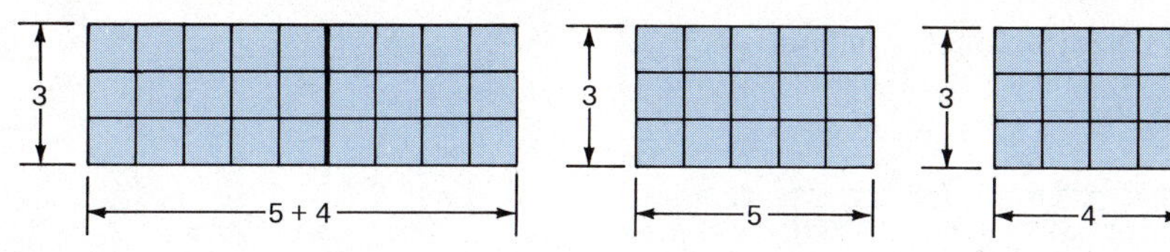

Figure 11-7 The geometric representation of the product of a binomial and monomial as an area: 3(5 + 4) = 3(5) + 3(4) = 27 squares.

Multiplying a Polynomial by a Monomial. Multiply each term of the polynomial by the monomial. Insert the proper sign for each resulting term.

Example A Multiply each of these expressions:

(1) $5(4 + b)$ (3) $R(2I + 3i)$

(2) $-2(3b - c)$ (4) $5ab(2b^2 - a^2b + 3a)$

Solution: (1) $5(4 + b) = 5(4) + 5(b) = 20 + 5b$

(2) $-2(3b - c) = -2(3b) - (-2)(c) = -6b + 2c$

(3) $R(2I + 3i) = R(2I) + R(3i) = 2RI + 3Ri$

(4) $5ab(2b^2 - a^2b + 3a) = 5ab(2b^2) - 5ab(a^2b) + 5ab(3a) = 10ab^3 - 5a^3b^2 + 15a^2b$

Note: The middle step usually does not need to be written. Perform all steps manually.

EXERCISE 11-5

Multiply each of the following by constructing freehand drawings similar to Figure 11-7.

1. $5(2 + 3)$
2. $4(6 + 3)$
3. $3(7 + 5)$
4. $6(8 + 5)$

Multiply each of the following.

5. $9(4b + 7)$
6. $2i(R_1 - 2R_2)$
7. $\theta^2(2\theta + 3)$
8. $3\alpha(\alpha^2\beta - \beta)$
9. $\theta(3\theta + 2\phi)$
10. $I^2(4R_1 + 3R_2 + 6R_3)$
11. $2\pi f(R - 5r)$
12. $(\mu^2 + 4\mu - 2)(2d^2)$
13. $5t(IR + ir + e - E)$
14. $\frac{1}{3}\lambda(9\lambda^2 - 12\mu + 18\pi^2)$
15. $4ab^2\left(\frac{1}{16}a - \frac{a^2}{12} + \frac{b}{20}\right)$
16. $(3k\theta^2\phi - 8\pi^2\phi\mu + 2\lambda\mu^2\phi)(3k\theta^2\phi)$

Simplify each of the following. (Combine like terms after multiplying.)

17. $a(2b - c) + a(b + 2c)$
18. $\omega(2\pi - 7R) - \pi(2\omega - 15R)$
19. $2r(3r^2 + 4rt - 8t^2) - 3(2r^3 - r^2t - 4rt^2)$
20. $5Z(3I^2 + i^2) + 3Z(2I^2 - 4i^2) - 2Z(5I^2 - 2i^2)$

11-8 EXPONENTS IN DIVISION

Powers of ten are divided by subtracting exponents:

$$\frac{10^5}{10^2} = 10^{5-2} = 10^3$$

The quotient of powers of any base is found by subtracting exponents in the same way.

Exponent Rule for Dividing. When one algebraic quantity expressed as a power is divided by another with the same base, the quotient is found by subtracting the denominator exponent from that of the numerator exponent:

$$\frac{b^m}{b^n} = b^{m-n}$$

If n = m, the result is b^0. Recall that $10^0 = 1$. Actually, *any* number divides itself one time:

$$\frac{b^3}{b^3} = b^{3-3} = b^0 = 1$$

Any number to the zero power equals 1: 5^0, 10^0, 185^0, b^0 R^0, θ^0 all equal 1.

If m is less than n in the equation, the resulting exponent is negative. The number b^3 divided by b^5 is b^{-2}:

$$\frac{b^3}{b^5} = b^{3-5} = b^{-2}$$

or

$$\frac{\not{b}\cdot\not{b}\cdot\not{b}}{\not{b}\cdot\not{b}\cdot\not{b}\cdot b\cdot b} = \frac{1}{b^2}$$

Since these are the same problem, then:

$$b^{-2} = \frac{1}{b^2}$$

or, in general:

$$b^{-n} = \frac{1}{b^n}$$

Recall that a number with an exponent of −1 is called the *reciprocal* of that number: $4^{-1} = 1/4$, $2^{-1} = 1/2$, $(\omega C)^{-1} = \frac{1}{\omega C}$.

Negative Exponent Rule. Any number expressed with an exponent can be moved from the numerator to the denominator or from the denominator to the numerator by changing the sign of the exponent.

Example A Evaluate: (1) $\frac{2^4}{2^3}$, (2) $\frac{2^{-3}m^3}{m^1}$, (3) $\frac{10^{-3}I^2R}{10^{-2}\frac{E^2}{R}}$

Solution: (1) $\frac{2^4}{2^3} = 2^4(2^{-3}) = 2^{4-3} = 2^1 = 2$

(2) $\frac{2^{-3}m^3}{m^1} = \frac{m^3 m^{-1}}{2^3} = \frac{m^{3-1}}{2^3} = \frac{m^2}{8}$

(3) $\frac{10^{-3} I^2 R}{10^{-2}\frac{E^2}{R}} = \frac{10^{-3} 10^2 I^2 R}{E^2 R^{-1}} = \frac{10^{-3+2} I^2 R R^1}{E^2}$

$= \frac{10^{-1} I^2 R^2}{E^2} = \frac{I^2 R^2}{10E^2}$

11-9 DIVIDING A MONOMIAL BY A MONOMIAL

To find the quotient of two monomials, apply the laws of division and the laws of exponents.

DIVIDING MONOMIALS

To find the quotient of two monomials:

- Determine the sign of the quotient. It is positive for an even number of negative factors (numerator and denominator) and negative for an odd number.
- Divide the coefficients. This will be the coefficient of the quotient.
- Find the quotient of the literal factors.
- Write the quotient with the proper sign as the *product* of the coefficient quotient and the literal factor quotient.

Example A Divide $15I^2Rt$ by $3t$

Solution: $\frac{15}{3} = 5$ (Divide coefficients)

$\frac{I^2Rt}{t} = I^2R$ (Divide literal factors)

$\frac{15I^2Rt}{3t} = 5I^2R$ (Form quotient)

Example B Divide: $18a^3b^2c^{-3}$ by $6a^5b^{-1}c$

Solution: $\frac{18}{6} = 3$ (Coefficients)

$\frac{a^3b^2c^{-3}}{a^5b^{-1}c} = a^{-2}b^3c^{-4}$ (Literal factors)

$\frac{18a^3b^2c^{-3}}{6a^5b^{-1}c} = 3a^{-2}b^3c^{-4}$ (Form quotient)

The rule for negative exponents and the process of cancellation gives an alternate method of dividing. Of course cancellation must follow the division rule for exponents. The only advantage is to change all exponents to positive numbers. The problem in Example B becomes:

$$\frac{18a^3b^2c^{-3}}{6a^5b^{-1}c} = \frac{18a^3b^2b}{6a^5cc^3} = \frac{18a^3b^3}{6a^5c^4}$$

The same result is obtained when the factors have been cancelled and combined.

EXERCISE 11-6

Express the following numbers as a product of the factors ($a^3 = a \cdot a \cdot a$). Divide by cancellation and express the final answer as a power.

1. $\frac{10^5}{10^2}$
2. $\frac{t^4}{t^3}$
3. $\frac{r^5}{r^3}$
4. $\frac{\theta^7}{\theta^8}$
5. $\frac{I^2R}{I}$
6. $\frac{\alpha^5\beta^4}{\beta^7\alpha}$

Use the division rule to solve each of the following. Divide.

7. $15R^2 \div 3R$
8. $-72P^5 \div 6P^{-2}$
9. $28a^{-5} \div 7a^{-9}$
10. $36R^3 \div -9R^{-1}$
11. $-36x^5y^{-3} \div 18x^2y^{-2}$
12. $96\theta^2\phi^5\mu^0\lambda^{-3} \div 24\theta^{-2}\phi^2\mu^{-2}\lambda^2$
13. $-33E^2I^{-3}R^5t^{-9} \div -22t^{-15}R^3E^2I^{-5}$
14. $23Z^8R^2X_C{}^{-5}X_L{}^2 \div 46X_C{}^3X_L{}^{-5}Z^4R$
15. $27\omega^3f^{-1}\pi^{12}R^{-4} \div 9\omega^2f^3\pi^8R^{-2}$
16. $3^2\alpha\beta^2\gamma^3 \div -36\alpha^{-2}\beta^4\gamma^2$

Convert all exponents to positive numbers before dividing. Cancel where possible. Place all factors of the quotient in the numerator. Divide.

17. $64r^5s^{-3}t^{-1} \div 8r^{-3}s^2t^{-1}$
18. $-9a^2b^{-1}c^4d^2 \div 54a^5b^{-6}c^4d^{-3}$
19. $-16\pi^4km^2 \div -64\pi^3k^3k^2m^3$
20. $34(-q^3)(p)(-r^5) \div -17(p^2)(-q)(-r^{-4})$

11-10 DIVIDING A POLYNOMIAL BY A MONOMIAL

When a polynomial is multiplied by a monomial, each term of the polynomial must be multiplied by the monomial.

$$p(q + r) = Pq + pr$$

Similarly, when a polynomial is divided by a monomial, each term must be divided. Division is the inverse of multiplication.

> **Dividing a Polynomial by a Monomial.** Divide each term of the polynomial by the monomial.

Example A Divide $12ab + 9b^2$ by $3b$.

Solution: $\frac{12ab + 9b^2}{3b} = \frac{12ab}{3b} + \frac{9b^2}{3b} = 4a + 3b$

Note: $3b$ is the common denominator for the two numerator terms.

Example B Divide $12a^2x^3 + 6a^3x^2 - 3a^3x$ by $3ax$.

Solution: $\dfrac{12a^2x^3 + 6a^3x^2 - 3a^3x}{3ax}$

$$\frac{12a^2x^3}{3ax} + \frac{6a^3x^2}{3ax} - \frac{3a^3x}{3ax} = 4ax^2 + 2a^2x - a^2$$

Note: $3ax$ is the common denominator or divisor for each of the three terms in the numerator.

Usually the problem is solved manually with the second step omitted.

EXERCISE 11-7

Divide.

1. $(8R + 12) \div 4$
2. $(15C - 10) \div 5$
3. $(40\theta^2 + 25\theta) \div 5\theta$
4. $(20E^2 - 25E) \div 5E$
5. $(-24\alpha^3 + 18\alpha^2\beta - 12\alpha\beta^2) \div -6\alpha$
6. $(3X_L{}^2 + 4X_L{}^2) \div X_L{}^2$
7. $(1.5I^2R^3 - 20I^3R^2 + 2.5I^4R) \div 0.5I^2R$
8. $(\theta^5 + \theta^3 + \theta^2 - \theta - 1) \div \theta^2$
9. $(\pi\lambda^3 - \pi^2\lambda^2 + \pi^3\lambda) \div \pi^2\lambda^2$
10. $(8\pi\omega^2 f + 32\pi^2\omega^3 f^2 - 56\pi^{-2}\omega^{-4}f^4) \div 16\pi\omega^2 f^2$
11. $9.3Z^{-3}X_C{}^{-2}X_L{}^2 - 18.6ZX_C{}^{-3}X_L{}^3 + 27.9Z^2X_C{}^4X_L) \div -9.3Z^2X_CX_L$
12. $(46\alpha^8\beta^6 + 69\alpha^{12}\beta^{-2} - 23\alpha^{-5}\beta^{-9}) \div 23\alpha^{-5}\beta^5$
13. Write an expression for the total power dissipation in the circuit of Figure 11-8. After collecting terms, divide by I to get an expression for the total voltage drop.

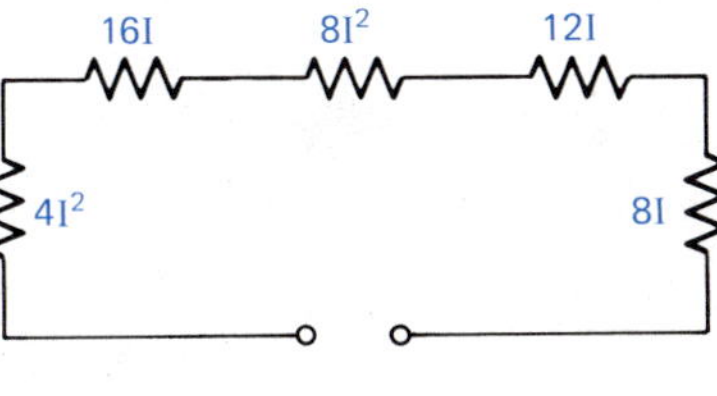

Figure 11-8

11-11 POWERS AND ROOTS OF MONOMIALS

There are two laws of exponents that are important in finding powers and roots. One law is for finding the *power of a product:*

$$(ab)^n = a^n b^n$$

The other one is the *power of a power* law:

$$(b^m)^n = b^{mn}$$

These equations are used in combination to raise a monomial to a higher power. Only squares and cubes are important in the electrical industry.

Finding Powers of a Monomial. To square a monomial, square the coefficient and double the power of each literal factor. To cube a monomial, cube the coefficient and triple the power of each literal factor.

Example A Square the monomial: $5P^2Q$

Solution: $(5P^2Q)^2 = 5^2(P^2)^2Q^2 = 25P^{2\times 2}Q^2 = 25P^4Q^2$

Check: Since $a^2 = a \cdot a$, the problem can be written $(5P^2Q)(5P^2Q)$ and the answer is evaluated using the rule for multiplying monomials.

$(5P^2Q)(5P^2Q) = 25P^4Q^2$

Example B Cube: $-3a^2b^3c$

Solution: $(-3a^2b^3c)^3$

$= (-3)^3(a^2)^3(b^3)^3c^3 = -27a^{2\times 3}b^{3\times 3}c^3 = -27a^6b^9c^3$

Note: There are three factors of −3 so the answer is negative. Most powers of monomials can be found without writing out the second step. Do the problems manually.

Example C Find the square of $2I^2R^3t$.

Solution: $(2I^2R^3t)^2$

↓↓ ↓ ↓

$= 4I^4R^6t^2$

The rule for finding powers of monomials also holds when the power is negative (reciprocal).

Example D Find $(4\theta^3\phi^{-2}\omega)^{-2}$.

Solution: $4^{-2}(\theta^3)^{-2}(\phi^{-2})^{-2}\omega^{-2} = \frac{1}{16}\theta^{-3\times 2}\phi^{2\times 2}\omega^{-2}$

$= \frac{1}{16}\theta^{-6}\phi^4\omega^{-2}$

Check: Rewrite the problem with positive powers:

$\frac{1}{(4\theta^3\phi^{-2}\omega)^2} = \frac{1}{16}\theta^{-6}\phi^4\omega^{-2}$

Recall that square roots as well as higher roots can be expressed as exponents, Topic 3-16. For example, a square root sign is replaced with an exponent of 1/2:

$$\sqrt{a} = a^{1/2}$$

$$\sqrt[3]{a} = a^{1/3}$$

Thus, the rule for finding powers of monomials can be used for finding roots of monomials.

> **Finding a Root of a Monomial.** Find the square root of the coefficient and divide the exponents of literal factors by 2. The cube root is found by finding the cube root of the coefficient and dividing the exponents of literal factors by 3.

Example E Find the square root of $16I^2R^4t^6$

Solution: $\sqrt{16I^2R^4t^6}$

$= (16I^2R^4t^6)^{1/2} = 16^{1/2}I^{2\times 1/2}R^{4\times 1/2}t^{6\times 1/2}$

$= 4IR^2t^3$

Example F Evaluate the cube root of $27a^3b^9$

Solution: $(27a^3b^9)^{1/3} = 27^{1/3}a^{3 \times 1/3}b^{9 \times 1/3}$
$= 3ab^3$

EXERCISE 11-8

Find the following powers and roots.

1. $(ab)^2$
2. $(-\alpha\beta)^2$
3. $-(\alpha^2\beta\gamma)^2$
4. $(3\theta^{-1}\phi^2)^3$
5. $(2\pi fX_C)^2$
6. $(2\pi fX_C)^{-2}$
7. $\sqrt{25I^2P^4}$
8. $\sqrt[3]{125N^6B^3}$
9. $(3iR)^{-3}$
10. $-5(-a^2b^{-3}c)^2$
11. $\sqrt{169\pi^2f^4c^8}$
12. $\sqrt{4 \times 10^8E^2R^4}$
13. $(-3 \times 10^4I^2R)^3$
14. $(3 \times 10^8ts^2r^{-1})^3$
15. $-(64a^2b^{-4}c^8)^{1/2}$
16. $(-216R^3P^{-6})^{-1/3}$

11-12 FACTORING POLYNOMIALS

The *Distributive Law of multiplication over addition* or simply the *Distributive Law* can be expressed as follows:

$$a(b + c) = ab + ac$$

The process of changing the expression on the left to the one on the right is multiplying a polynomial by a monomial.

When the expression on the right is changed to the one on the left, the process is called factoring.

$$\rightarrow\text{multiplying}\rightarrow$$
$$a(b + c) = ab + ac$$
$$\leftarrow\text{factoring}\leftarrow$$

Factoring is an important process. It can easily be learned by just reversing the multiplication process. Only monomial factors are considered.

FACTORING

- By inspection, determine the largest factor that is common to all terms of the polynomial.
- Divide every term of the polynomial by the largest common factor.
- The factored expression is the product of the common factor and the quotient.

Example A Factor 5R + 15.

Solution: The common factor is 5.

$$\frac{5R}{5} + \frac{15}{5} = R + 3$$

$$5R + 15 = 5(R + 3)$$

Usually, factoring is a manual process.

Example B Factor the expression $3a^2bc - 9ab^2 + 12ab$.

Solution: $3ab$ is the largest common factor (by inspection).

$$\frac{3a^2bc}{3ab} - \frac{9ab^2}{3ab} + \frac{12ab}{3ab} = ac - 3b + 4$$

$$3a^2bc - 9ab^2 + 12ab = 3ab(ac - 3b + 4)$$

Again, the process can be done manually.

Example C Factor $\frac{1}{5}rt^2 - \frac{rt}{15}$.

Solution: Notice the second term can be written $\frac{1}{15}rt$. Also, the factor $\frac{1}{5}$ in the first term can be written $\frac{3}{15}$. An equivalent expression is $\frac{3}{15}rt^2 - \frac{1}{15}rt$.

The largest common factor is $\frac{1}{15}rt$. The quotient is $3t - 1$.

$$\frac{1}{5}rt^2 - \frac{rt}{15} = \frac{1}{15}rt(3t - 1)$$

When a polynomial is divided by a monomial, an alternate method with factoring can be used. Instead of dividing each term of the polynomial by the monomial, the polynomial is factored. The monomial factor is then divided by the monomial divisor using cancellation.

Example D Divide: $(7a^2b + 21ab^2) \div 7ab$.

Solution: Factor the polynomial:

$$7a^2b + 21ab^2 = 7ab(a + 3b)$$

Now divide and cancel:

$$\frac{\cancel{7ab}(a + 3b)}{\cancel{7ab}} = a + 3b$$

EXERCISE 11-9

Factor the following polynomial expressions. Check by multiplying.

1. $7R + 21$
2. $8i - 32$
3. $-27\mu^2 + 9$
4. $2L + 2w$
5. $\pi d + 2\pi r$
6. $P^2 - 3P$
7. $\theta^3\phi - \theta\phi^3$
8. $35E^2/R - 21EI$
9. $\frac{1}{2}B_1h + \frac{1}{2}B_2h$
10. $54R^2S^3T - 36RS^2T^3$
11. $35m^3n^5 + 70m^4n^4 - 28m^5n^3$
12. $12uvw - 16v^2w + 28uv^2$
13. $4\alpha^2\beta + 8\alpha\beta\gamma - 20\alpha\beta^2$
14. $24\mu^2\pi^3\lambda - 18\pi\lambda^2\mu^3 + 30\lambda\pi^2\mu^3 - 54\lambda^2\pi^2\mu^2$

15. $42a^3 b + 28ab^3 - 14a^2 b^2 - 35a^5 b$
16. $\frac{1}{3}r^2 s^3 t^5 - \frac{r^5 s^2 t^3}{9} + \frac{5r^3 s^5 t^2}{12}$

Divide the following by using the factoring method.

17. $(25R^2 - 75R) \div 25R$
18. $(24\alpha^3 + 18\alpha^2 \beta - 12\alpha\beta^2) \div 6\alpha$
19. $(2 \times 10^5 Q^2 t^3 - 5 \times 10^5 Qt^4 + 3 \times 10^5 Q^3 t^2) \div 10^5 Q^2 t^2$
20. $(5\theta^2 \phi + 20\theta\phi^2 - 15\theta^3) \div 5\theta^2 \phi$
21. $(\pi\omega^2 t^2 + 3\pi\omega t^3 - 2\pi\omega^3 t) \div 2\pi\omega t$
22. $[2\pi fL - (2\pi fC)^{-1}] \div 2\pi f$

CHAPTER 12

PREPARING FOR CIRCUIT ANALYSIS

OBJECTIVES

After satisfactorily completing this chapter, the student should be able to:

- Rearrange the terms and factors of an equation or formula using algebra to solve for any one of the literal factors including equations with grouping symbols and fractions.
- Plot graphs of linear equations and solve two simultaneous linear equations for two unknowns by graphical means.
- Solve a system of linear equations by the addition-subtraction method, the substitution method and the comparison method.

The terms and factors of an equation or formula are often rearranged. The equation may be transformed into a mathematically equivalent equation with the unknown isolated on one side of the equal sign. Such a process is known as *solving an equation.*

Solving simple equations was introduced in Chapters 7 and 8. By using the procedures for simplifying algebraic expressions in the previous chapter, the study can be extended. Equations with grouping symbols and fractional terms can now be worked.

This chapter also includes the study of solving simultaneous equations. A system of such equations includes more than one unknown. Several procedures for finding the unknowns are described.

SOLVING EQUATIONS

The study of solving equations was first presented in Chapter 7. Only simple equations without plus and minus signs and without grouping symbols were considered. The multiplication and division axioms (MA and DA) were the only algebraic processes required. In Chapter 8, the addition and subtraction axiom (AA and SA) were introduced. Simple equations with plus and minus signs could then be solved. Also in Chapter 8, the power and root (PA and RA) axioms were used to solve for unknowns carrying an exponent or situated within a radical.

In the present chapter, equations with grouping symbols and fractional components will be studied. It begins with a review of simpler equations and formulas.

12-1 REVIEW OF SIMPLE EQUATIONS

Some of the simplest formulas studied previously are those without plus or minus signs. The formulas consist of four types:

• $C = \pi d$	Solve for (d)	(DA)
• $I = \frac{E}{R}$	Solve for (E)	(MA)
• $P = \frac{W}{t}$	Solve for (t)	(MA and DA)
• $R = \frac{\rho L}{\pi r^2}$	Solve for (L)	(MA and DA)

Solved examples can be found in Topic 7-5.

Formulas with plus and minus signs and combined operations were studied in Topic 8-6. The following types were discussed:

• $E_T = E_1 + E_2$	Solve for (E_2)	(SA)
• $R_2 = E_T - E_1$	Solve for (E_T)	(AA)
• $E = IR_1 + IR_2$	Solve for (R_1)	(SA and DA)

Formulas in which the unknown is raised to a power or located under a radical sign can be reviewed in Topic 8-12. Formulas such as the following are typical:

• $P = I^2 R$	Solve for (I)	(DA and RA)
• $Z = \sqrt{R^2 + X^2}$	Solve for (R)	(PA, SA and RA)

Before equations with grouping symbols and fractions are introduced, the student may wish to review the topics mentioned in the previous paragraphs. The following exercise is intended for a review of that material.

EXERCISE 12-1

Solve the following equations or rearrange the formulas for the indicated unknown enclosed in parentheses. Solved examples can be found in the reference topic mentioned.

Simple equations, Topic 7-5.

1. $C = \pi d$ (d)
2. $d = \frac{m}{V}$ (m)
3. $P = EI$ (E)
4. $I = \frac{Q}{t}$ (t)
5. $P = \frac{W}{t}$ (W)

6. $c = \lambda f$ (λ)
7. $\omega = 2\pi f$ (f)
8. $E = \dfrac{W}{Q}$ (Q)
9. $R = \dfrac{\rho L}{A}$ (L)
10. $P = I^2 Rt$ (t)
11. $X_L = 2\pi fL$ (f)
12. $X_C = \dfrac{1}{2\pi fC}$ (C)
13. $\mu = \dfrac{B^2 AL}{16\pi f}$ (f)

Equations with plus and minus signs, Topic 8-6.

14. $R - 5 = 4$
15. $P + 3 = 7$
16. $3.5 = \alpha - 6.5$
17. $10 + Q = 32$
18. $17 - \theta = -90$
19. $\theta + \phi = 90$ (ϕ)
20. $P - 2\omega = 15$ (ω)
21. $\Delta f + 1624 = 1681$
22. $5X - 14 = 11$
23. $e = E - Ir$ (r)
24. $\Delta R = R_2 - R_1$ (R_1)
25. $v^2 = v_0{}^2 + 2as$ (s)
26. $\alpha + \beta + \gamma = 180$ (β)
27. $P = E_1 I + E_2 I + E_3 I$ (E_2)

Equations with exponents and radicals, Topic 8-12.

28. $\sqrt{x} = 25$
29. $R^2 = 121$
30. $E^{1/2} = 4 \times 10^3$
31. $\gamma^2 = 169$
32. $\sqrt{I + 7} = 2$
33. $(D - 5)^2 = 36$
34. $r = \sqrt{\dfrac{A}{\pi}}$ (A)
35. $W = \dfrac{1}{2} LI^2$ (I)
36. $P = I^2 R$ (I)
37. $Z = \sqrt{X^2 + R^2}$ (R)
38. $W = \dfrac{1}{2} CE^2$ (E)
39. $E = \sqrt{PR}$ (P)
40. $I = \left(\dfrac{P}{R}\right)^{1/2}$ (P)
41. $I = \sqrt{\dfrac{W}{Rt}}$ (t)
42. $f = \dfrac{1}{2\pi\sqrt{LC}}$ (L)

12-2 EQUATIONS WITH UNKNOWN IN SEVERAL TERMS

On occasion, one uses equations in which the unknown is found in more than one term. An example is

$$3x - 4 = 2x + 8 - 5x$$

The best procedure is to transpose all terms containing the unknown to one side. All other terms are taken to the other side. The addition and subtraction axioms are used. When terms are combined, the equation is reduced to a simple one.

Example A Solve for x: $3x - 4 = 2x + 8 - 5x$

Solution:

$$3x - 4 = 2x + 8 - 5x$$

$$3x - 4 = 8 - 3x \qquad \text{(Collect terms)}$$

$$3x - 4 + 4 = 8 - 3x + 4 \qquad \text{(AA, + 4)}$$

$$3x = 12 - 3x$$

$$3x + 3x = 12 - 3x + 3x \qquad \text{(AA, + 3x)}$$

$$6x = 12$$

$$\frac{6x}{6} = \frac{12}{6} \qquad \text{(DA, ÷ 6)}$$

$$x = 2$$

Check: $3(2) - 4 = 2(2) + 8 - 5(2)$

$$2 = 2$$

The problem in the previous example could be solved mentally. Combine $2x$ and $-5x$ on the right $(-3x)$, transpose (change signs) and combine with $3x$. The result is $6x$ on the left. Transpose -4 and combine with 8 which is 12 on the right, $(6x = 12)$. This process is called *collecting terms* and results in a simple equation.

SOLVING EQUATIONS WITH SEVERAL TERMS

- Transpose all terms containing the unknown to one side and all others to the opposite side of the equal sign.
- Collect terms.
- Remove the coefficient of the unknown using the division axiom.

In formulas containing several literal factors, the process is the same. Of course, the sums, differences, products, etc. are indicated instead of being performed.

Example B Solve: $E = IR_1 + IR_2 + IR_3$ for I.

Solution: The three like terms are collected into one term by the process of factoring.

$E = IR_1 + IR_2 + IR_3$

$E = I(R_1 + R_2 + R_3)$ (Factor I)

Divide by the group to solve:

$$\frac{E}{R_1 + R_2 + R_3} = I \qquad [DA, \div (R_1 + R_2 + R_3)]$$

$$I = \frac{E}{R_1 + R_2 + R_3}$$

EXERCISE 12-2

Solve for the unknown in each of the following equations.

1. $4R - 9 = R$
2. $10 - \mu = 4\mu$
3. $P - 8 = 4 + 5P$
4. $9\theta - 2 = 2\theta + 5$
5. $7W - 5 = 3W + 11$
6. $6 - E = 8E - 12$
7. $5\rho - 1 = 14 + 2\rho$
8. $8\alpha + 5 + 3\alpha = 6\alpha$
9. $3C + 4 - 2.75C = 8 - 0.25C$
10. $0.4I + 6 - 0.1I - 2 = 0.7I - 11 + 0.5I - 3$

Solve the following formulas for the indicated unknowns.

11. $A = \frac{1}{2}B_1h + \frac{1}{2}B_2h$ (h)
12. $E = IR + Ir$ (I)
13. $IE_1 = P - IE_2$ (I)
14. $W = E_1It + I^2R_2t + \frac{E_3{}^2t}{R_3}$ (t)
15. $R_sI_T - R_sI_m = I_mR_m$ $(I_m), (R_s)$
16. $E = IR_s + IR_m$ (I)

17. $C_TC_1 + C_TC_2 = C_1C_2$ (C_1)
18. $P = I^2R_1 + I^2R_2 + I^2R_3$ (I) Hint: Factor I^2, use (RA)
19. $R_TR_1 + R_TR_2 = R_1R_2$ (R_2)
20. $pr = 1 + r$ (r)

12-3 EQUATIONS WITH UNKNOWN IN PARENTHESES

Many equations and formulas have the unknown quantity contained within parentheses. To solve this type, the first step is to remove the parentheses. Formulas that are important to electricians probably will not have more than one or two sets of grouping symbols. Those with more than one are discussed in the next topic.

Parentheses can be removed in two ways:

1. Multiplication method (Distributive Law)
   a. Group preceded by + sign. Simply drop the parentheses.
   b. Group preceded by – sign. Change signs of all terms and drop the parentheses.
   c. Group preceded by a number. Multiply all terms by the number and drop the parentheses.
2. Axiom method
   If the group is isolated on one side of the equation as a monomial, use the division or multiplication axiom to move the coefficient to the other side of the equal sign. Drop the parentheses.

SOLVING EQUATIONS CONTAINING PARENTHESES

- Remove parentheses by either the multiplication method or the axiom method.
- Transpose and collect all terms containing the unknown on one side. Transpose and evaluate all others on the opposite side.
- Remove the coefficient using the division axiom.

Example A Solve: $8(E + 4) = 56$

Solution 1:

$8(E + 4) = 56$ (Multiplication method)

$8E + 32 = 56$ (Multiply)

$8E = 24$ (SA, – 32)

$E = 3$ (DA, ÷ 8)

Solution 2:

$8(E + 4) = 56$ (Axiom method)

$\frac{8(E + 4)}{8} = \frac{56}{8}$ (DA, ÷ 8)

$E + 4 = 7$

$E = 3$ (SA, – 4)

When there are several groups in parentheses, use the multiplication method on each group.

Example B Solve: $5 - 2(Q + 7) - 5Q = 2Q + (3Q - 3) - 18$

Solution: $5 - 2Q - 14 - 5Q = 2Q + 3Q - 3 - 18$ (Remove parentheses)

$-12Q = -12$ (Collect terms)

$Q = 1$ (DA, ÷ −12)

Example C Solve the formula $P = 2(L + w)$ for L

Solution 1:

$P = 2(L + w)$ (Multiplication method)

$P = 2L + 2w$ (Multiply)

$P - 2w = 2L$ (SA, − 2w)

$\frac{P - 2w}{2} = L$ (DA, ÷ 2)

$L = \frac{P - 2w}{2}$

Solution 2:

$P = 2(L + w)$ (Axiom method)

$\frac{P}{2} = L + w$ (DA, ÷ 2)

$\frac{P}{2} - w = L$ (SA, − w)

$L = \frac{P}{2} - w$

The two answers to the previous example do not have the same form. If one can be changed algebraically to the other, then the answers are equivalent.

If the numerator of the answer in Solution 1 is divided by the denominator, the result is

$$\frac{P - 2w}{2} = \frac{P}{2} - \frac{2w}{2} = \frac{P}{2} - w$$

which is the same as the answer in Solution 2.

The answer in Solution 2 can be changed by multiplying the second term by 1 in the form $\frac{2}{2}$. This does not change the value of the term. Now the two terms have a common denominator and can be combined into the form of the answer in Solution 1:

$$\frac{P}{2} - w = \frac{P}{2} - \frac{2w}{2} = \frac{P - 2w}{2}$$

The use of the bar as a grouping symbol is discussed in the next topic.

EXERCISE 12-3

Use the multiplication method to aid in solving these equations.

1. $2(C + 4) = 13\,C - 3$
2. $8(2\phi - 3) = 4\phi + 12$
3. $5(2\beta - 2) = 6\beta + 2$
4. $5(1 - 2P) = 2P + 17$
5. $-8E - 7 = 3(7 - 5E)$
6. $10(I - 6) = -5 - 3(I + 1)$

Use the division axiom to help solve the following equations.

7. $-8 = 7 - 3(2\omega + 5)$
8. $2(7 - 3W) = 26$
9. $1 - 3(L + 4) = 4$
10. $8 - 5(4 - Z) = -7$

Solve the following formulas for the indicated unknown enclosed in parentheses.

11. $E = I(R + r)$ (R)
12. $A = \frac{1}{2}h(b_1 + b_2)$ (b_2)
13. $C = \frac{5}{9}(F - 32)$ (F)
14. $P = I(E_1 + E_2 + E_3)$ (E_1)
15. $E = I(R_s + R_m)$ (R_m)
16. $\Delta R = aR(T_2 - T_1)$ (T_2)
17. $A = P(rt + 1)$ (r)
18. $s = (v + v_o)\frac{t}{2}$ (v)
19. $\Delta L = \alpha L(T_2 - T_1)$ (T_1)
20. $f_o(V - v_s) = f_s(V + v_o)$ (V)

In the following problems show that one expression is equivalent to the other.

21. $\frac{\beta + 1}{\beta}, 1 + \frac{1}{\beta}$
22. $\frac{5}{9}C + 32, \frac{5C + 288}{9}$
23. $\frac{1 - r}{r}, \frac{1}{r} - 1$
24. $\frac{\Delta R}{aR} + T_1, \frac{\Delta R + aRT_1}{aR}$

12-4 EQUATIONS WITH FRACTIONS

A *fractional equation* is one that contains fractions with the unknown in the denominator. Simple fractional equations have already been used in previous chapters. Most had only one term with a fraction. Some fractional formulas seen before are

$$I = \frac{E}{R}$$

$$R = \frac{\rho L}{\pi r^2}$$

These equations are easy to solve using the multiplication and division axioms.

Some more complex fractional formulas seen before are

$$R_T = \frac{R_1 R_2}{R_1 + R_2}$$

$$\frac{1}{R_T} = \frac{1}{R_1} + \frac{1}{R_2} + \frac{1}{R_3}$$

Although values for some of the quantities were substituted and the unknown was evaluated, no attempt was made at that time to solve the equation for the unknown.

To solve the more complex fractional equations, one additional step is required. The first step is called *clearing the equation of fractions.*

Clearing an equation of fractions requires two previously learned concepts.

- Find the least common denominator (LCD) of all fractional terms in the equation.
- Multiply all terms on both sides by the LCD. In other words: (MA, × LCD).

When the equation is cleared of fractions, it is solved by methods used previously.

SOLVING EQUATIONS CONTAINING FRACTIONS

- Clear fractions by multiplying by the LCD.
- If the resulting equation contains parentheses, remove them.
- Transpose and collect all terms containing the unknown on one side. Collect and evaluate all others on the opposite side.
- Remove the coefficient using the division axiom.

Example A Solve: $2 + \frac{2}{3}M = \frac{1}{4}M - \frac{1}{2}$

Solution: The LCD is 12 (by inspection).

$$12(2) + 12\left(\frac{2}{3}M\right) = 12\left(\frac{1}{4}M\right) - 12\left(\frac{1}{2}\right) \qquad \text{(MA, × 12)}$$

$$24 + 8M = 3M - 6 \qquad \text{(Simplify)}$$

$$5M = -30 \qquad \text{(Collect terms)}$$

$$M = -6 \qquad \text{(DA, ÷ 5)}$$

Check: $2 + \frac{2}{3}(-6) = \frac{1}{4}(-6) - \frac{1}{2}$

$$-2 = -2$$

A fractional equation has the unknown in the denominator of at least one term. Solve them by using the same method as shown in Example A. Clear fractions by multiplying every term by the LCD.

Example B Solve: $1 - \frac{3}{Z} = \frac{3}{4} - \frac{1}{Z}$

Solution: The LCD is 4Z. Multiply by the LCD.

$$4Z - 4Z\left(\frac{3}{Z}\right) = 4Z\left(\frac{3}{4}\right) - 4Z\left(\frac{1}{Z}\right)$$

$$4Z - 12 = 3Z - 4 \qquad \text{(Simplify)}$$

$$Z = 8 \qquad \text{(Collect terms)}$$

Check: $1 - \frac{3}{8} = \frac{3}{4} - \frac{1}{8}$

$$\frac{5}{8} = \frac{5}{8}$$

Example C Solve: $\frac{2P}{P + 4} = 2 + \frac{2}{P}$

Solution: The LCD is the product of all the denominators P(P + 4). Clear fractions by multiplying by the LCD.

$$\frac{P\cancel{(P + 4)}(2P)}{\cancel{(P + 4)}} = P(P + 4)(2) + \frac{\cancel{P}(P + 4)(2)}{\cancel{P}}$$

$$2P^2 = 2P^2 + 8P + 2P + 8 \quad \text{(Cancel and simplify)}$$

$$2P^2 - 2P^2 - 10P = 8$$

$$-10P = 8 \quad \text{(Collect terms)}$$

$$P = -\frac{8}{10} = -0.8 \quad (\text{DA}, \div -10)$$

Check: $\frac{2(-0.8)}{-0.8 + 4} = 2 + \frac{2}{-0.8}$

$$\frac{-1.6}{3.2} = 2 - 2.5$$

$$-0.5 = -0.5$$

Sometimes, the unknown is found in the numerator and the denominator of formulas. Again, the solution is found in the same way.

Example D Solve: $R_T = \frac{R_1 R_2}{R_1 + R_2}$ for R_1

Solution: The LCD is $R_1 + R_2$ since that is the only term having a denominator. Multiply both sides by $R_1 + R_2$.

$$R_T(R_1 + R_2) = \frac{R_1 R_2 \cancel{(R_1 + R_2)}}{\cancel{(R_1 + R_2)}}$$

$$R_T(R_1 + R_2) = R_1 R_2 \quad \text{(Cancel)}$$

$$R_T R_1 + R_T R_2 = R_1 R_2 \quad \text{(Remove parentheses)}$$

$$R_T R_2 = R_1 R_2 - R_T R_1 \quad \text{(Collect } R_1 \text{ terms)}$$

$$R_T R_2 = R_1(R_2 - R_T) \quad \text{(Factor)}$$

$$\frac{R_T R_2}{R_2 - R_T} = \frac{R_1 \cancel{(R_2 - R_T)}}{\cancel{R_2 - R_T}} \quad [\text{DA}, \div (R_2 - R_T)]$$

$$R_1 = \frac{R_T R_2}{R_2 - R_T}$$

Notice that the process of clearing fractions is a method of removing the bar or bars as a grouping symbol. As in the previous topic, this step is performed first.

Example E Solve for C_1 in the formula: $\frac{1}{C_T} = \frac{1}{C_1} + \frac{1}{C_2}$

Solution: The LCD is the product of all the denominators $C_T C_1 C_2$. Multiply all terms by the LCD.

$$\frac{C_T C_1 C_2}{C_T} = \frac{C_T C_1 C_2}{C_1} + \frac{C_T C_1 C_2}{C_2}$$

$$C_1 C_2 = C_T C_2 + C_T C_1 \quad \text{(Cancel)}$$

$$C_1 C_2 - C_T C_1 = C_T C_2 \quad \text{(Collect } C_1 \text{ terms)}$$

$$C_1(C_2 - C_T) = C_T C_2 \quad \text{(Factor)}$$

$$C_1 = \frac{C_T C_2}{C_2 - C_T} \quad [\text{DA}, \div (C_2 - C_T)]$$

EXERCISE 12-4

Solve for the unknown in each of the following equations. Clear fractions as the first step in each problem.

1. $\frac{1}{4}I = -2$
2. $\frac{P}{6} = \frac{1}{2}$
3. $\frac{3}{8}R = \frac{27}{32}$
4. $\frac{\alpha}{4} - 2 = 5 - \frac{\alpha}{3}$
5. $\frac{2\theta + 5}{7} + \frac{\theta}{4} = 5$
6. $\frac{2}{P} - \frac{3}{2P} = -1$
7. $\frac{9}{D - 1} = 3$
8. $\frac{1}{2(E - 1)} = \frac{2}{E - 1} + 1$

Solve the following formulas for the indicated unknown enclosed in parentheses. Clear fractions as the beginning step.

9. $P = \frac{W}{t}$ (t)
10. $I = \frac{Q}{t}$ (t)
11. $\alpha = \frac{1}{\beta} + \frac{1}{\gamma}$ (β)
12. $I = \frac{E}{R + r}$ (r)
13. $p = \frac{1 + r}{r}$ (r)
14. $\alpha = \frac{\beta}{\beta + 1}$ (β)
15. $\frac{1}{R_T} = \frac{1}{R_1} + \frac{1}{R_2}$ (R_1)
16. $\frac{1}{Z_2} = \frac{1}{Z_T} - \frac{1}{Z_1}$ (Z_1)
17. $Z_1 = \frac{Z_T Z_2}{Z_2 - Z_T}$ (Z_2)
18. $\mu = \frac{1}{\gamma} + \frac{1}{t}$ (t)
19. $I_1 = \frac{I_2 R}{R + Z}$ (R)
20. $C_T = \frac{C_1 C_2}{C_1 + C_2}$ (C_2)

SIMULTANEOUS EQUATIONS

The single-load circuit was introduced previously. In such a circuit, suppose the resistance (R) of the load is known. Also assume the power dissipation (P) in the load is known. Two equations that apply are Ohm's Law ($E = IR$) and the formula for power, ($P = EI$). Together, the two formulas contain all the unknown and the known quantities.

If no other formulas are allowed, could one find the voltage and current? Yes, but a new method of solving the equations is required. When two or more equations apply to a situation at the same time, they are called *simultaneous equations.*

There are special ways of solving simultaneous equations. Four methods will be described. A system of simultaneous equations may consist of many equations. When two equations apply, two unknowns can be found. If three equations apply simultaneously, three unknowns can be evaluated. Generally speaking, one must have as many equations as there are unknowns to find the values of them all.

12-5 GRAPHING EQUATIONS

An equation can be graphed on a rectangular coordinate system. If an equation gives a straight line curve when graphed, it is called a *linear equation.* If x and y are the variables, all linear equations have the form:

$$ax + by + c = 0$$

where a, b and c are *constants*. Constants, of course, can be any positive or negative number.

To graph an equation, one must determine several values of x and y that satisfy the equation. These combinations are the data points used to construct the graph. Usually, the equation is solved for y. Various values of x are chosen and the corresponding values of y are evaluated. If such pairs of numbers are listed in a table, they can then be plotted and the curve can be constructed.

GRAPHING AN EQUATION

- Solve the equation for y.
- Choose several values of x (independent variable) and determine the corresponding values of y (dependent variable). List the pairs of numbers in a table. A minimum of three pairs is required for linear equations.
- Plot the points on a coordinate system (freehand). Draw the curve through the points.

Example A Plot a graph of the equation: $3x - 6y + 18 = 0$

Solution: $3x - 6y + 18 = 0$

$$y = \frac{1}{2}x + 3 \quad \text{(Solve)}$$

Choose several values of x such as −10, −5, 0, +5 and +10. Substitute these values of x into the solved equation and find the corresponding values of y.

$$y = \frac{1}{2}(-10) + 3 = -2 \qquad y = \frac{1}{2}(5) + 3 = 5.5$$

$$y = \frac{1}{2}(-5) + 3 = 0.5 \qquad y = \frac{1}{2}(10) + 3 = 8$$

$$y = \frac{1}{2}(0) + 3 = 3$$

Pairs are listed and plotted in Figure 12-1.

At least three points should be evaluated and plotted for linear equations. If all lie on a straight line then one can assume the graph is correct. Sometimes it is easier to solve the equation for x, choose values of y and evaluate the values of x. If points are determined this way, y must still be the dependent variable and be plotted vertically.

Example B Plot the graph of $7y - x = 4$

Solution: $x = 7y - 4$ (Solve for x)

Find points: $x = 7(2) - 4 = 10$

$x = 7(1) - 4 = 3$

$x = 7(0) - 4 = -4$

Points are listed and plotted in Figure 12-2.

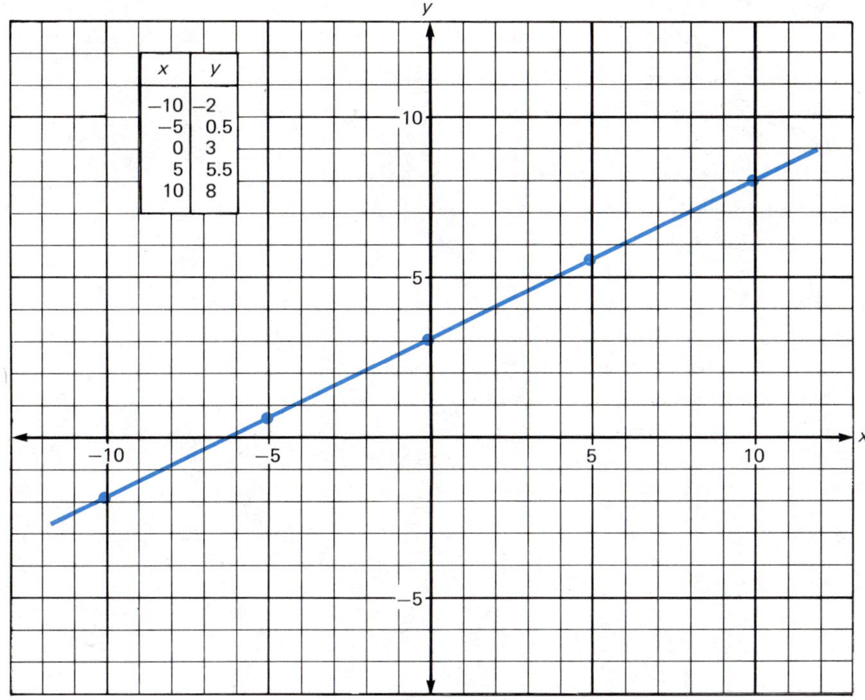

x	y
−10	−2
−5	0.5
0	3
5	5.5
10	8

Figure 12-1 Data points and graph for the equation $3x - 6y + 18 = 0$.

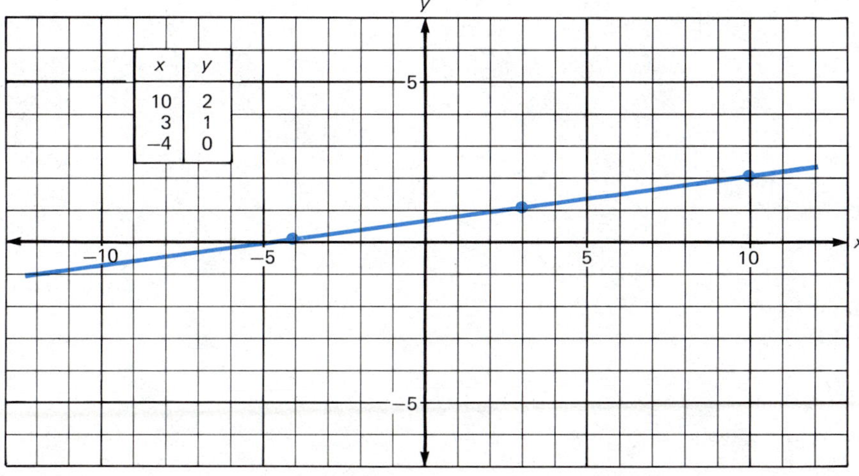

x	y
10	2
3	1
−4	0

Figure 12-2 Table and graph of equation $7y - x = 4$.

12-6 GRAPHICAL METHOD

A linear equation such as $y = 2x + 3$ has many possible solutions. The values $x = 1$ and $y = 5$, that is (1,5) is one solution. Some other solutions are: (2,7), (0,3) and (−1,1). One can easily find hundreds of solutions. In fact, there is an infinite number of solutions. Any linear equation has an infinite number of solutions. This is not to say that any pair of numbers will work. The pair (7,−2), for example, is not a solution. The coordinates of any point not on the line will not satisfy the equation.

When two linear equations represent conditions in a practical problem situation, both can be plotted on the same graph. Each equation has an infinite number of solutions, but there is only one solution that satisfies both equations simultaneously. The solution is the coordinates of the point of intersection.

FINDING THE GRAPHICAL SOLUTION

- Prepare a table of values for each equation.
- Plot the points for each equation on graph paper. Plot a minimum of three points for each line.
- Draw the two lines with a straightedge.
- Determine the coordinates of the point where the two lines intersect.
- Check the solution by substituting both numbers into both original equations. Both must be satisfied.

Example A Solve graphically: $y = -x - 1$

$$2x - y = 4$$

Solution: Prepare tables; Figure 12-3. The two equations have been plotted in Figure 12-3. The coordinates of the intersection is (1,−2). Thus, $x = 1$, $y = -2$ is the solution.

Check: $y = -x - 1$

$$-2 = -1 - 1$$

$$2x - y = 4$$

$$2(1) - (-2) = 4$$

Although there are many solutions to each equation separately, there is only one that satisfies both. The point of intersection is the only point that lies on both curves. It is the only solution to both equations.

Of course, the two equations cannot be *equivalent*. Such a pair of equations representing the same line would be satisfied by any point on the line. An example of two equivalent equations is

$$y + 3x = 7$$

$$3y + 9x = 21$$

Notice the second equation is just the first multiplied by 3.

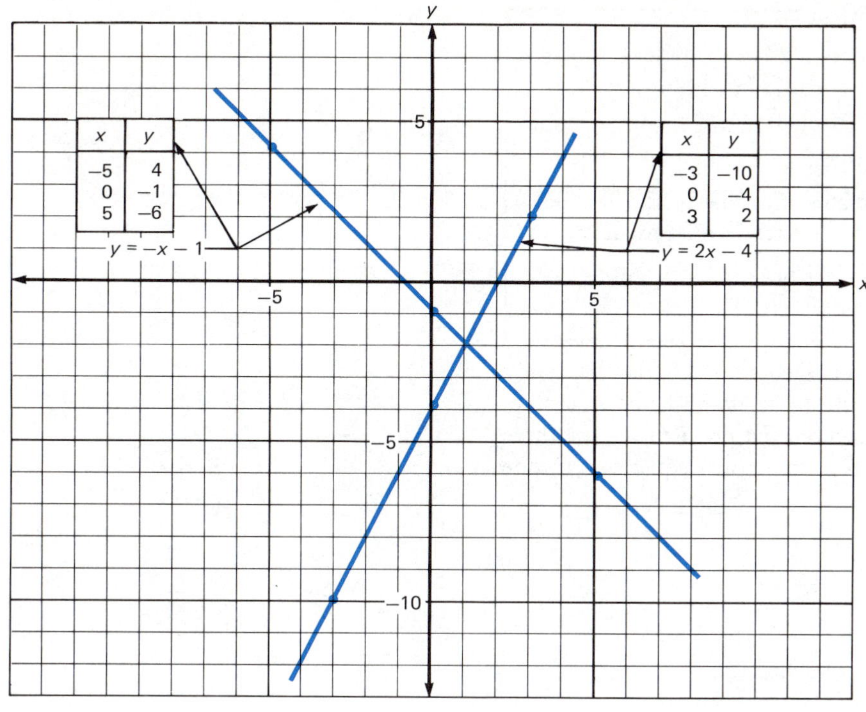

Figure 12-3 Tables and graphs of equations $y = -x - 1$ and $y = 2x - 4$. The solution, $x = 1$ and $y = -2$, is the coordinates of the point of intersection.

Another pair of equations may yield no solution at all. This occurs if the equations, when graphed, represent parallel lines. The two would never intersect. Two such equations are said to be *inconsistent.*

EXERCISE 12-5

Solve the following simultaneous equations by the graphical method.

1. $y = 3x - 1$
   $y = -\frac{1}{2}x + 6$
2. $y = -x - 1$
   $x = -5y + 7$
3. $R = 3P - 1$
   $2R = 3P + 13$
4. $5\theta - 3\phi = 7$
   $4\theta + \phi = 9$
5. $s - t = 3$
   $2s + t = 12$
6. $2E + 3I = 3$
   $E - I = 4$
7. $2\alpha - 3\beta = 1$
   $5\alpha + 2\beta = 12$
8. $y = 2x + 4$
   $3y - 6 = 6x$

12-7 ALGEBRAIC METHODS

A pair of simultaneous equations contains two unknowns. The values of the unknowns can be determined by several algebraic methods. Basically, the process is this:

SOLVING SIMULTANEOUS EQUATIONS

- Combine the two equations by eliminating one of the unknowns. The result of this step is one equation with one unknown. Methods of doing this are described in subsequent topics.
- Use the resulting equation to evaluate the first unknown.
- Substitute the value of the first variable into one of the original equations (either one). Now evaluate the second unknown.
- Check the answer by substituting the values of both unknowns into the equation not used in the previous step.

The last three steps are the same for all algebraic methods. The only difference in methods used is the procedure to accomplish the first step. The names of the various methods are based on that procedure. The three methods used to reduce the system of two (or more) equations to one equation with one unknown are as follows:

- Addition-subtraction method.
- Substitution method.
- Comparison method.

Each of these methods is described in the following topics.

12-8 ADDITION-SUBTRACTION METHOD

Suppose an object on the left side of a balance scale is balanced by standard weights on the right pan, Figure 12-4. With another balance scale, sand on the left pan is used to balance a rock on the right, Figure 12-5. Now, suppose the sand and rock are transferred to the pans of the first scale, Figure 12-6. The scale is still balanced.

Equal weights have been added to equal weights. Four different things were involved: object, standard weights, rock and sand. Everything, however, balanced at the end.

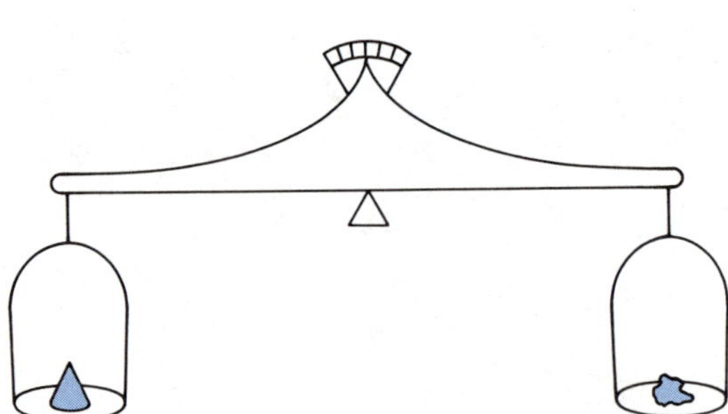

Figure 12-4 Object balanced by a standard weight

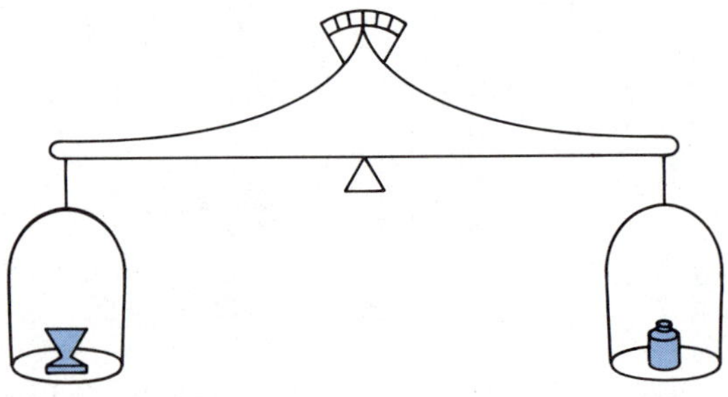

Figure 12-5 Rock on the right pan is balanced by sand on the left pan.

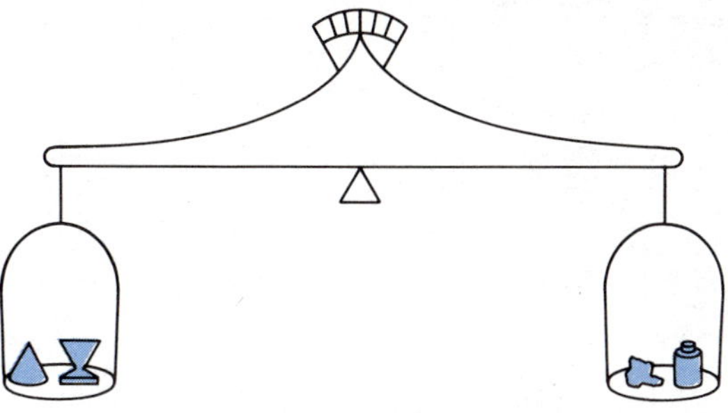

Figure 12-6 Object and sand are balanced by rock and standard weight.

The addition and subtraction axioms have often been used to add the same number or literal number to both sides of an equation. The new equation is still balanced. The question now arises, "can an algebraic expression be added to one side of an equation and a second expression be added to the other side?". The answer is yes if the two expressions are equal. In other words, if $A = B$ and $C = D$, then $A + C = B + D$. If equals are added to (or subtracted from) equals, the results are equal.

One equation of a pair of simultaneous linear equations can be added to or subtracted from the other. If the coefficients of one of the variables in both equations have the same absolute value, that variable will be eliminated from the resulting equation. This will occur if the matched coefficients have like or unlike signs. With unlike signs, the equations are added. With like signs, they are subtracted.

Example A Solve for x and y in the equations: $x + y = 8$

$$x - 2y = 2$$

Solution: The coefficient of x is the same in both equations and have like signs. By subtracting, x will be eliminated.

$$\begin{array}{l} x + y = 8 \\ \underline{x - 2y = 2} \\ 0 + 3y = 6 \\ \quad\; y = 2 \end{array}$$

(Eliminate one variable)

Substitute the value of y into one of the original equations and solve for x:

$$x + 2 = 8$$
$$x = 6$$

The solution is: $x = 6, y = 2$

Check: Substitute both values into the other original equation: $6 - 2(2) = 2$

When the coefficients of one of the variables do not have the same absolute value, multiply one or both equations by some nonzero number. Every term on both sides is multiplied using the multiplication axiom. The number must be chosen so the new coefficients of one variable will have the same absolute value.

ELIMINATING A VARIABLE BY ADDITION-SUBTRACTION

- Multiply one or both equations (if necessary) by a number or numbers that make the absolute value of the coefficients of one variable equal.
- Eliminate that variable by adding or subtracting.

Example B Solve for x and y in the equations: $2x - 3y = 5$
$4x + 7y = -3$

Solution: By inspection, notice the first equation can be multiplied by 2. Then, both x-coefficients will be the same. Eliminate x by subtraction:

$4x - 6y = 10$ (MA, X 2)
$4x + 7y = -3$
$0 - 13y = 13$ (Eliminate x)
$y = -1$
$2x - 3(-1) = 5$ (Substitute)
$x = 1$
Check: $4(1) + 7(-1) = -3$

$x = 1, y = -1$ is the solution.

Example C Solve for x and y in the equations: $3x - 5y = 18$
$7x + 2y = 1$

Solution: Multiply the first equation by 2, the second by 5 to eliminate y by addition.

$6x - 10y = 36$ (MA, X 2)
$35x + 10y = 5$ (MA, X 5)
$41x + 0 = 41$ (Eliminate y)
$x = 1$

$3(1) - 5y = 18$ (Substitute)
$y = -3$
Check: $7(1) + 2(-3) = 1$

$x = 1, y = -3$ is the solution.

EXERCISE 12-6

Solve for the two variables in the following problems using the addition-subtraction method. If there is no solution, indicate whether the equations are equivalent or inconsistent.

1. $x + y = 12$
   $x - y = 2$
2. $I + i = 9$
   $2I - i = 6$
3. $E_1 + 5E_2 = 7$
   $E_1 + E_2 = -1$
4. $5Z = 5R + 2$
   $2R = 2Z - 5$
5. $3\theta + 5\phi = 1$
   $\theta + 2\phi = -2$
6. $2P + 3Q = 10$
   $4P = Q + 6$

7. $2\lambda = 3\mu + 12$
   $\mu = 13 - 5\lambda$
8. $3X_L + 2X_C = 1$
   $5X_C + 4X_L = -8$
9. $5\alpha + 15\beta = -45$
   $2\alpha + 5\beta = -14$
10. $2 - 4I + 3i = 0$
    $5i - 7I + 5 = 0$

12-9 SUBSTITUTION METHOD

The first step in solving simultaneous equations is to eliminate one of the variables. This results in one equation with one unknown. The second method for doing this is the substitution method.

ELIMINATING A VARIABLE BY SUBSTITUTION

- By inspection, determine which variable of which equation has the "simplest" coefficient.
- Solve that equation for that variable.
- Substitute the expression obtained in the previous step into the other equation for that variable. This yields one equation with one unknown.

The substitution method is illustrated in the following examples. Two equations are reduced to one equation with one unknown. The procedure for evaluating the variables is the same as before and will not be shown in complete detail. Checking the answers will not be shown to conserve space. When working exercise problems, however, the student should always check the solution.

Example A Solve for x and y using the substitution method:

$x + 3y = 4$
$2x + 5y = 7$

Notice the variable x in the first equation has the simplest coefficient. Solve for that variable.

Solution: $x = 4 - 3y$

Substitute $4 - 3y$ for x in the other equation.

$2(4 - 3y) + 5y = 7$ (Substitute)
$8 - 6y + 5y = 7$ (Remove parentheses)
$-y = -1$ (Collect terms)
$y = 1$ (Evaluate)
$x = 1$

Example B Solve for x and y:

$5x - 7y = 46$
$3x + 4y = 3$ (3 is the smallest coefficient)

Solution: $x = \dfrac{3 - 4y}{3}$

$5\left(\dfrac{3 - 4y}{3}\right) - 7y = 46$ (Substitute)

$\dfrac{3(5)(3 - 4y)}{3} - 3(7)y = 3(46)$ (MA, X 3)

$5(3 - 4y) - 21y = 138$ (Clear fractions)

$15 - 20y - 21y = 138$ (Remove parentheses)

$-41y = 123$ (Collect terms)

$y = -3$ (Evaluate)

$x = 5$

EXERCISE 12-7

Solve for the two variables in each of the following problems using the substitution method. If there is no solution, indicate whether the equations are equivalent or inconsistent.

1. $x - y = 6$
   $x + y = 4$
2. $2R - X = 0$
   $2X + R = 10$
3. $3\alpha + 2\beta = -4$
   $\alpha - 3\beta = -5$
4. $7\theta + 2\phi = 11$
   $5\theta - \phi = 3$
5. $I - 3i = -1$
   $2I + i = 12$
6. $2E_1 - 3E_2 = 1$
   $5E_1 + 2E_2 = 12$
7. $3Z_1 - 6 = 6Z_2$
   $Z_1 = 2Z_2 + 4$
8. $5P + 2Q = 9$
   $15P + 5Q = 23$
9. $5\mu - 8 = 3\omega$
   $7\mu - 8 = 5\omega$
10. $21 + 3R_1 = 5R_2$
    $8 - 5R_1 = 6R_2$

12-10 COMPARISON METHOD

ELIMINATING A VARIABLE BY COMPARISON

- Solve both equations for the same variable. Choose the unknown with the simpler pair of coefficients.
- Set the two expressions equal to each other. The result is one equation with one unknown.

Now proceed in the usual way.

Example A Solve for x and y using the comparison method:

$5x + y = 13$

$7x - y = 11$

Solution: Note that y has the simpler coefficients.

$y = 13 - 5x$

$y = 7x - 11$

$13 - 5x = 7x - 11$ (Equate expressions)

$12x = 24$ (Collect terms)

$x = 2$ (Evaluate)

$y = 3$

Example B Solve for x and y: $5x + 3y = -1$
$2x - 4y = 10$

Solution: Solve for x

$$x = \frac{-3y - 1}{5}$$

$$x = \frac{10 + 4y}{2}$$

$$\frac{-3y - 1}{5} = \frac{10 + 4y}{2} \quad \text{(Equate)}$$

$$\frac{10(-3y - 1)}{5} = \frac{10(10 + 4y)}{2} \quad \text{(MA, X 10)}$$

$$2(-3y - 1) = 5(10 + 4y) \quad \text{(Clear fractions)}$$

$$-6y - 2 = 50 + 20y \quad \text{(Remove parentheses)}$$

$$-26y = 52 \quad \text{(Collect terms)}$$

$$y = -2 \quad \text{(Evaluate)}$$

$$x = 1$$

Sometimes, one of the variables has the same coefficient in both equations. In this case, solve both equations for that combination of coefficient and variable.

Example C Solve for x and y: $3x - 5y = 4$
$7x + 5y = 3$

Solution: Solve for $5y$ in both equations

$$5y = 3x - 4$$

$$5y = 3 - 7x$$

$$3x - 4 = 3 - 7x \quad \text{(Equate)}$$

$$10x = 7 \quad \text{(Collect terms)}$$

$$x = \frac{7}{10} = 0.7 \quad \text{(Evaluate)}$$

$$y = -\frac{19}{50} = -0.38$$

EXERCISE 12-8

Solve for both variables in the following problems using the comparison method.

1. $x + y = 2$
   $x - y = 4$
2. $2T + 3t = 5$
   $t + T = 2$
3. $2I + 3i = 0$
   $4I - 5i = 22$
4. $4R_1 + 5 = 3R_2$
   $5R_2 = 16 - R_1$
5. $2\theta + \phi = 5$
   $3\phi - 2\theta = 7$
6. $5X_L + 2X_C = 11$
   $4X_L - X_C = 1$
7. $2\alpha = 3 - \beta$
   $6 + 4\alpha = 2\beta$
8. $4p + 3\omega = 10$
   $8p - 6\omega = 20$
9. $7r - 8d = -10$
   $8 - 7r = -2d$
10. $E_1 - 2.5E_2 = 10$
    $2.5E_2 + 2E_1 = 5$

12-11 WORD PROBLEMS

Lists of words meaning multiplication or division were given previously, Topic 7-4. Lists of other words meaning addition and subtraction were presented in Topic 8-4. If the student has difficulty translating from verbal to algebraic statements, a review of those topics is suggested.

Problems in this topic contain two unknowns and two equations. They must be translated and written in equation form. Use the following procedure:

TRANSLATING AND SOLVING WORD PROBLEMS

- Read the problem carefully. Exactly what are the unknowns?
- Choose symbols to represent the two unknowns. Write down exactly what each symbol represents.
- Write two equations according to the situation described in the problem. Write in symbols exactly what the problem says in words. Construct a diagram if this will help clarify the situation.
- Choose one of the methods for solving simultaneous equations and solve for the two unknowns. Be sure to check the answers.

After reading the problem, take the time and make the effort to write down the names of the unknowns and the letters decided upon. It would be very easy to become confused and lose track of the letters chosen if they were not recorded. Writing down the unknowns not only saves time but it also represents a starting point. Do not neglect this very important step.

Example A The cost of a light fixture, wire and other parts is three-fifths the labor cost for installing it. Find the cost of the parts and the cost of labor if the total bill is $124.

Solution: Let L = the cost of labor
P = the cost of parts

Write equations:

$P + L = \$124$ (Sum = total cost)

$P = \frac{3}{5}L$ (First sentence)

$P = 124 - L$ (First equation)

$\frac{3}{5}L = 124 - L$ (Equate)

$L = \$77.50$

$P = \$46.50$

Check: $\$46.50 = \frac{3}{5}(\$77.50)$

$\$46.50 + \$77.50 = \$124$

Of course, a different method could be used.

Example B Eighteen resistors connected in series have a total resistance of 125 Ω. Some resistors have a resistance of 5 Ω and others, 10 Ω. How many of each kind are there?

Solution: Let x = number of 5-Ω resistors
y = number of 10-Ω resistors

$x + y = 18$	(Sum = total number)
$5x + 10y = 125$	(Total resistance)
$y = 18 - x$	(First equation)
$5x + 10(18 - x) = 125$	(Substitute)
$5x + 180 - 10x = 125$	[Remove ( )]
$-5x = -55$	(Collect terms)
$x = 11$	
$y = 7$	
Check: $11 + 7 = 18$	(Number)
$55 + 70 = 125$	(Resistance)

EXERCISE 12-9

Solve each of the following problems. Practice using each of the three algebraic methods.

1. The total cost of an electrical shop and the lot was \$84 800. The building costs \$830 more than eight times the cost of the lot. Find the cost of the building and of the lot.
2. A conductor 65 ft long is cut into two pieces. One piece is 5 ft longer than one-half the other. How long is each piece?
3. The current through one branch of an electric circuit, Figure 12-7, is 0.36 A less than twice the current through the other parallel branch. The total current when the branches join is 1.74 A. Find the current in each branch.
4. Two resistors have a total resistance of 5.8 MΩ when connected in series. The first has a resistance of 0.6 MΩ more than three times the other. Find the two resistances.
5. An electric current of 22.5 mA branches off so that one branch carries a current of 0.15 mA more than twice the other, Figure 12-7. Find the current in each branch.
6. A variable resistor is set at 3 kΩ more than four times another. If both are decreased by 5 kΩ, their sum is 28 kΩ. What was the original resistance of each resistor?
7. The sum of the three angles in any triangle is always 180°, regardless of the size and shape. In a right triangle (one 90° angle), one acute angle is twice the other, Figure 12-8. What are the two acute angles?

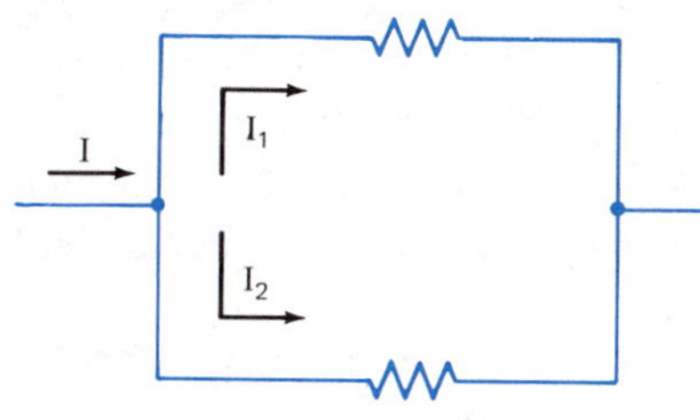

Figure 12-7

A
B

Figure 12-8

8. In an isosceles triangle, two angles are equal. In one isosceles triangle, Figure 12-9, angles A and B are equal. Angle C is 15.8° more than angle A. Find the three angles.

Note: The sum of the three angles of a triangle is 180°.

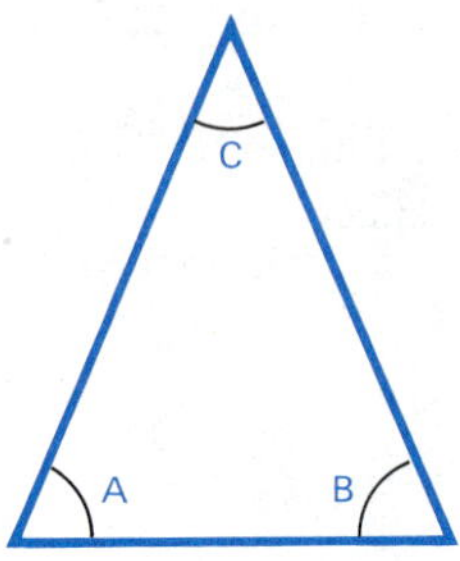

Figure 12-9 The sum of the three angles of a triangle is 180°. This triangle is an isosceles triangle.

9. A storage room is 2.5 times longer than it is wide, Figure 12-10. The perimeter is 28 m. Find the length and width.
10. There are 10 resistors in series. Some are 5-Ω resistors and some are 10-Ω resistors. The total resistance is 65 Ω. Find the number of each.
11. A group of 10-Ω and 25-Ω resistors are connected in series. There are 69 resistors making a total of 1245 Ω. Find the number of each kind.
12. An electrician owns life insurance policies with two different companies. The policy with Company A has a face value of $3125 more than 1.25 times the other (Company B). Together, they have a face value of $42 500. Find the face value of each policy.
13. Two kinds of batteries are connected in two series circuits as shown in Figure 12-11. Find the emf of each kind. Notice some batteries are connected so they oppose the others in one circuit.
14. Two sizes of resistors are connected in two series circuits as shown in Figure 12-12. What is the resistance of each size?

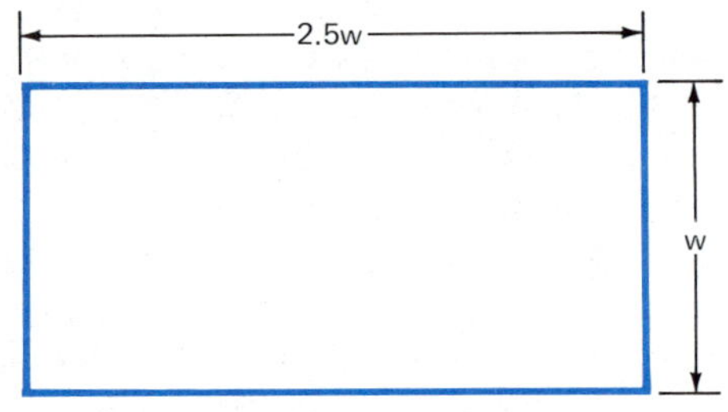

Figure 12-10

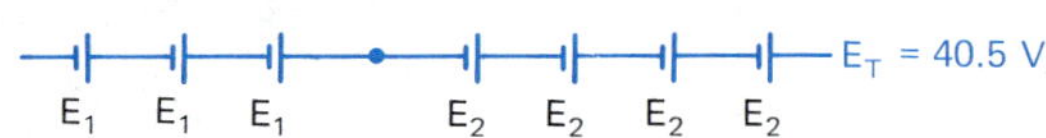

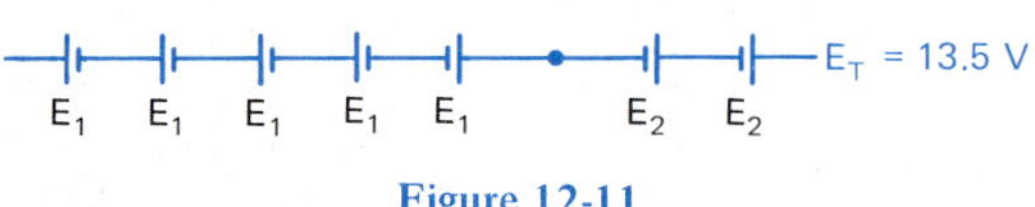

Figure 12-11

Figure 12-12

15. Two sizes of lamps are connected as shown in Figure 12-13. Find the power rating of each size.

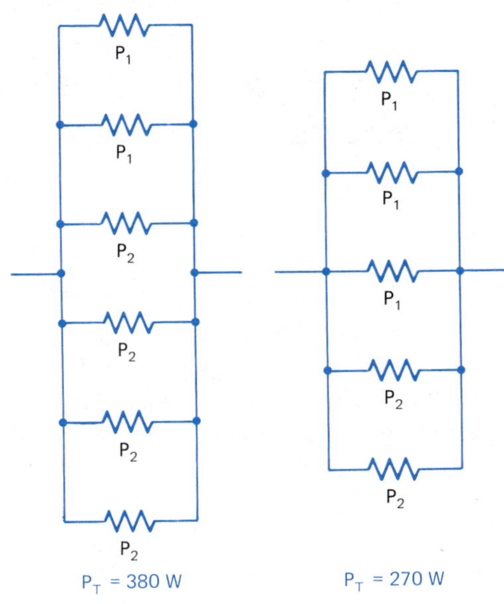

Figure 12-13

12-12 DERIVING FORMULAS

The methods of solving simultaneous equations also applies to simultaneous formulas. The methods of substitution and comparison are used most often. When a variable is eliminated, a new equation (or formula) appears. It contains only the one unknown and can be solved.

Suppose the power dissipation and the resistance of a circuit load are known. Can the current and voltage be determined? The formulas P = EI and E = IR both contain both unknowns. Notice the power formula does not give a linear relationship between current and voltage. Either the substitution or the comparison method can be used to eliminate one of the variables in nonlinear equations of this type.

Example A Derive a formula for current (I) from the equations P = EI and E = IR. Eliminate voltage (E) using the comparison method.

Solution: $E = \frac{P}{I}$ and $E = IR$ (Solve for E)

$\frac{P}{I} = IR$ (Equate)

$\frac{P}{R} = I^2$ (MA, × I; DA, ÷ R)

$\sqrt{\frac{P}{R}} = \sqrt{I^2}$ (RA, $\sqrt{\ }$)

$I = \sqrt{\frac{P}{R}}$ (Solve for I)

Example B Derive a formula for time (t) from the formulas $E = \frac{W}{Q}$ and $I = \frac{Q}{t}$ by eliminating charge (Q). Use the substitution method.

Solution: One formula is solved for Q. The expression is substituted into the other equation, then the resulting equation is solved for t.

$Q = It$ (Second equation)

$E = \frac{W}{It}$ (Substitute)

$t = \frac{W}{EI}$ (MA, × t; DA, ÷ E)

EXERCISE 12-10

Use the method indicated to derive the following formulas.

1. Use E = IR and P = EI to find a formula for E. Eliminate I by comparison.

2. Combine $P = EI$ and $W = Pt$ to find a formula for W. Eliminate P by substitution.

3. Derive a formula for W from the formulas $C = \frac{Q}{E}$ and $W = \frac{1}{2}QE$. Eliminate Q by substitution.

4. If $L = \frac{X_L}{2\pi f}$ and $T = \frac{1}{f}$, write a formula for X_L. Eliminate f by comparison.

Use the substitution or the comparison method to solve the formulas indicated in the following problems.

5. Find a formula for the area (A) of a circle in terms of the diameter. Use the formulas $A = \pi r^2$ and $d = 2r$. Eliminate r.

6. Find a formula for the area (A) of a circle in terms of the circumference (C). Eliminate r from the formulas. $A = \pi r^2$ and $C = 2\pi r$.

7. If $c = \frac{\lambda}{T}$ and $f = \frac{1}{T}$ find a formula for λ in terms of c and f. Eliminate T.

8. Eliminate E from the formulas $Q = \frac{W}{E}$ and $I = \frac{E}{R}$. Solve for W.

9. Use the formula derived in Problem 8 and the formula $I = Q/t$ to find a formula for W after eliminating Q.

10. Eliminate ω from the equations $\omega L = \frac{1}{\omega C}$ and $f = \frac{\omega}{2\pi}$. Solve for f.

CHAPTER 13

CIRCUIT ANALYSIS

OBJECTIVES

After satisfactorily completing this chapter, the student should be able to:

- Solve systems of equations consisting of three equations with three unknowns.
- Analyze various circuits by applying Kirchhoff's Current and Voltage Laws.
- Apply the principle of superposition as a method of analyzing various circuits.
- Apply Thévenin's Theorem as a means of simplifying complex circuits.

The analysis of compound circuits was discussed in Chapter 9. Only circuits with a single voltage source were considered at that time. The method used was to reduce the compound circuit resistance in a series or chain of increasingly simpler combinations. An equivalent circuit was found that contained only one load.

When the equivalent resistance was determined, other circuit quantities could be calculated. When the resistance, voltage drop and power dissipation of each load and the current in each branch was found, the analysis was complete.

In this chapter, circuits with more than one voltage source will be considered. Several methods can be used. In some, systems of simultaneous equations are employed. In others, the circuit is reduced to simpler circuits. Three methods are described in this chapter.

KIRCHHOFF'S LAWS

Kirchhoff's Current Law, Topic 8-8 and Voltage Law, Topic 8-10, can be used to write several simultaneous circuit equations. Complex circuits result in many equations with many unknowns. Such systems of equations are very tedious to solve. The laws will be used in this text to solve only simple circuits.

Simple two-branch parallel circuits yield three equations with three unknowns. Kirchhoff's Laws apply to such circuits even when they contain more than one voltage source. These circuits will be studied after explaining the mathematics of three simultaneous equations with three unknowns.

13-1 THREE EQUATIONS WITH THREE UNKNOWNS

A third order system of equations is a group of three equations with three unknowns. An example of *third order simultaneous equations* is

$$\begin{aligned} x + y - z &= 7 \quad \text{(a)} \\ 2x - y - z &= 3 \quad \text{(b)} \\ x - 2y + 2z &= -8 \quad \text{(c)} \end{aligned}$$

To solve the system of equations means to find the values of x, y and z that satisfy all three equations. One of the unknowns is eliminated from two of the equations. For example, z can be eliminated from the first two equations by subtracting. The same unknown is then eliminated from another pair of equations which must include the third equation. This results in two equations with two unknowns. These can then be solved by methods developed in the previous chapter.

SOLVING THIRD ORDER SIMULTANEOUS EQUATIONS

- Choose any two of the three equations. Eliminate any one of the three unknowns by the addition-subtraction, substitution or comparison methods. The result is one equation with two unknowns.
- Choose two of the three equations including the one not used in the first step. Eliminate the same unknown eliminated in the first step.
- Solve the resulting system of two equations with two unknowns.
- Substitute the values found in the previous step into any one of the three original equations. Solve for the third unknown.
- Check the solution by substituting into both of the other equations.

Example A Solve the system of equations listed in the previous discussion.

Solution: Eliminate y from equations (a) and (b).

$$\begin{aligned} x + y - z &= 7 \quad \text{(a)} \\ 2x - y - z &= 3 \quad \text{(b)} \\ \hline 3x \quad - 2z &= 10 \quad \text{(d)} \end{aligned} \qquad \text{(Add)}$$

Eliminate y from equations (b) and (c).

$$\begin{array}{rlr} 4x - 2y - 2z = 6 & \text{(b)} & \text{(MA, X 2)} \\ x - 2y + 2z = -8 & \text{(c)} & \\ \hline 3x \quad\quad - 4z = 14 & \text{(e)} & \text{(Subtract)} \end{array}$$

Solve for x and z in equations (d) and (e).

$$\begin{array}{rr} 3x - 2z = 10 & \\ 3x - 4z = 14 & \\ \hline 2z = -4 & \text{(Subtract)} \\ z = -2 & \end{array}$$

$$3x - 2(-2) = 10 \quad \text{(Substitute in d)}$$
$$x = 2$$

Substitute $x = 2$, $z = -2$ into any one of the three original equations and solve for y.

$$(2) + y - (-2) = 7 \quad \text{(a)}$$
$$y = 3$$

Check: $2(2) - (3) - (-2) = 3$ (b) $(3 = 3)$

$(2) - 2(3) + 2(-2) = -8$ (c) $(-8 = -8)$

Thus, $x = 2, y = 3$ and $z = -2$ is the solution.

A typical system of equations resulting from Kirchhoff's Laws is

$$\begin{array}{rr} I_1 - I_2 - I_3 = 0 & \text{(a)} \\ 7I_1 - 5I_2 \quad\quad = 13 & \text{(b)} \\ 5I_2 + 4I_3 = 19 & \text{(c)} \end{array}$$

The coefficients of equation (a) are either 1 or −1. Only two of the three currents appear in equations (b) and (c). For these reasons, such equations are easier to solve than the set in the preceding example.

Example B Solve the system of equations given in the previous paragraph.

Solution: Eliminate I_1 from equations (a) and (b).

$$\begin{array}{rlr} 7I_1 - 7I_2 - 7I_3 = 0 & \text{(a)} & \text{(MA, X 7)} \\ 7I_1 - 5I_2 \quad\quad = 13 & \text{(b)} & \\ \hline -2I_2 - 7I_3 = -13 & \text{(d)} & \text{(Subtract)} \end{array}$$

Equations (d) and (c) together form a second order system:

$$\begin{array}{rr} -2I_2 - 7I_3 = -13 & \text{(d)} \\ 5I_2 + 4I_3 = 19 & \text{(c)} \end{array}$$

$$\frac{13 - 7I_3}{2} = \frac{19 - 4I_3}{5} \quad \text{(Comparison)}$$

$$65 - 35I_3 = 38 - 8I_3 \quad \text{(MA, X LCD)}$$
$$27I_3 = 27 \quad \text{(Collect terms)}$$
$$I_3 = 1$$

$$-2I_2 - 7(1) = -13 \quad \text{(Substitute in d)}$$
$$I_2 = 3$$

The third unknown can now be found.
Substitute $I_3 = 1, I_2 = 3$ into (a)
$I_1 - 3 - 1 = 0$
$I_1 = 4$
Check: $7(4) - 5(3) = 13$ (b)
$5(3) + 4(1) = 19$ (c)
Therefore, the solution is $I_1 = 4, I_2 = 3$ and $I_3 = 1$

The three equations in a system must each be *unique*. This means they cannot be equivalent or inconsistent. One must not be *derived* from another. The equation $x + 2y + z = 2$ when multiplied by 3 is $3x + 6y + 3z = 6$. The resulting equation is derived from and is *equivalent* to the first.

One of the three equations cannot be derived from the other two. When $x - 2y - z = 2$ is subtracted from $x + y + z = 1$, the result is $3y + 2z = -1$. The result is derived and cannot be combined with the other two to form a system of three equations. The three are *inconsistent* and cannot be solved.

EXERCISE 13-1

Solve the following systems of equations.

1. $x + y + z = 3$
   $2x + 3y = 4$
   $2y + 3z = 10$
2. $I_1 - I_2 - I_3 = 7$
   $-3I_2 + I_3 = -10$
   $2I_1 - I_2 = 8$
3. $I_1 - I_2 - I_3 = -8$
   $2I_1 + 4I_3 = 12$
   $2I_1 - 7I_2 = -18$
4. $I_1 - I_2 - I_3 = -5$
   $2I_1 + 4I_2 = 20$
   $4I_2 + 3I_3 = 25$
5. $\alpha + \beta - \gamma = -6$
   $2\alpha + 3\beta + 5\gamma = 7$
   $3\alpha + \gamma = 0$
6. $2a + 3b + c = 1$
   $a - 2b + 3c = 10$
   $3a + 3b - c = 2$

13-2 KIRCHHOFF'S LOOP METHOD

Kirchhoff's Laws were introduced in previous chapters. They are repeated here for convenience:

KIRCHHOFF'S LAWS

- *Current Law.* At any branch point in a circuit, the sum of all currents flowing toward the point is equal to the sum of all currents flowing away from the point.
- *Voltage Law.* Around any complete circuit loop, the sum of all voltage rises is equal to the sum of all voltage drops.

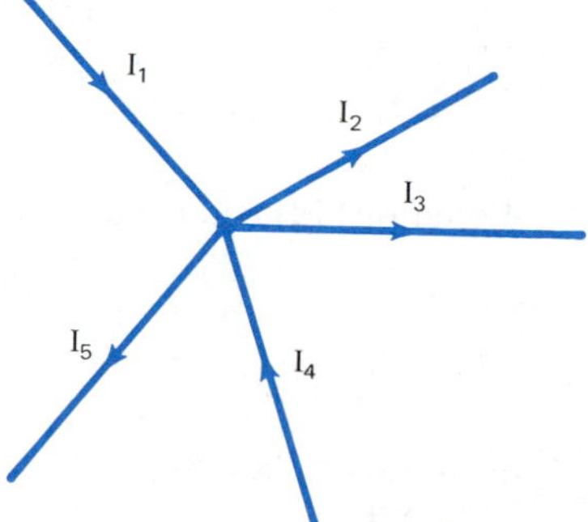

Figure 13-1 Kirchhoff's Current Law. The sum of all currents flowing into a point equals the sum of all currents flowing out of the point. $I_1 + I_4 - I_2 - I_3 - I_5 = 0$

In writing equations using the current law, one can write "the sum of incoming currents minus all outgoing current equals zero." Sometimes it is desirable to have all currents on one side of the equal sign, Figure 13-1: $I_1 + I_4 - I_2 - I_3 - I_5 = 0$.

When writing equations using the voltage law, the signs of the voltage changes is of great importance. Remember, voltage changes occur in both sources and loads. One observes the changes during an imaginary "walk" completely around the circuit loop. In some parts of the circuit, one walks with the current, in others, against it.

For example, pretend to walk clockwise around the left loop, Figure 13-2. Starting at point A, you walk facing into the current (I_2) to point B. In the lower, left and upper parts of the loop, you move with the current (I_1). If the loop is transversed counterclockwise, the situation is reversed.

In applying Kirchhoff's Laws to a circuit, carry out the following procedure step by step.

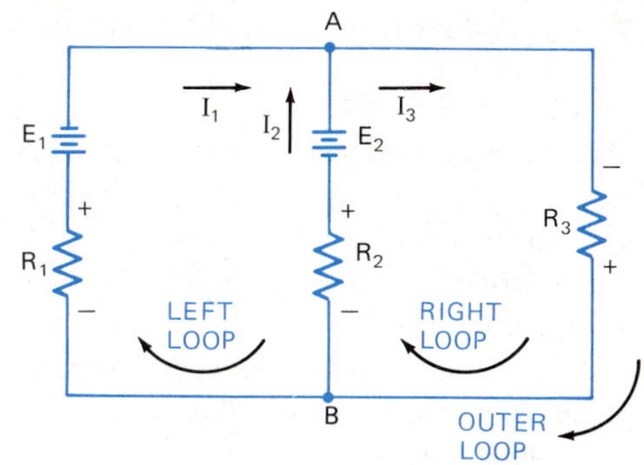

Figure 13-2 Circuit analysis using Kirchhoff's loop method

APPLYING KIRCHHOFF'S LAWS

1. Assign a symbol to the current in each branch of the circuit. Indicate the direction with an arrow. If the direction chosen is wrong, it will be indicated as a negative current when the current values have been determined.
2. Place plus and minus signs at the terminals of each load. Remember that electrons flow from the negative terminal to the positive terminal in a load, Figure 13-3. The current directions already chosen determine the terminal signs *(polarity)* of all loads, Figure 13-4. The polarity of a battery is indicated by its circuit symbol. The polarity of power supplies and generators must be indicated.
3. Use the current law at enough points so each current appears in at least one equation.
4. Use the voltage law in as many closed loops as necessary to include every current in at least one equation.
5. Solve the resulting simultaneous equations.

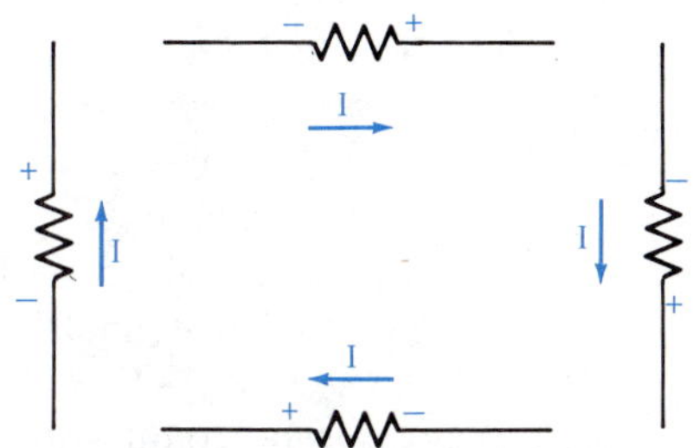

Figure 13-3 Assigning polarity to resistors according to the direction of current flow

In Step 1, assign the symbols I_1, I_2 and I_3 for the currents in various parts of the circuit, Figure 13-2. Indicate the direction with arrows. Sometimes, it is easy to tell which direction the current will flow. In some branches of the current, it may be difficult to decide. However, this should not cause concern. When the currents are solved in later steps, they will have positive values if the correct direction was chosen. If the wrong choice was made, the answers are negative but the values of current are determined correctly.

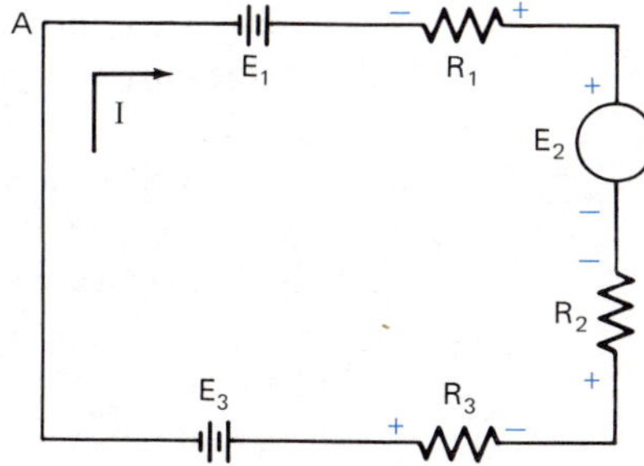

Figure 13-4 The polarity of batteries is indicated by the circuit symbol. It must be indicated on other voltage sources and loads. The two cell battery symbol used in circuits with more than one source represents a battery of any voltage. This is similar to the use of the resistor symbol to represent a load of any resistance.

In Step 2, current always flows through a load from the negative side to the positive side. Assign plus and minus polarities to each load as shown in Figure 13-3. Use the directions of the current assigned in Step 1. In Figure 13-2, current I_1 is assumed to flow upward through load R_1. Place a minus sign at the bottom and a plus sign at the top as shown.

The polarity of R_2 is also negative on the bottom and positive on the top since current I_2 flows upward through it. Load R_3 is different because the assumed direction of current I_3 flows through it in a downward direction. It is negative on top and positive on the bottom as shown.

The two-cell battery symbol represents batteries of any voltage. It consists of two long lines and two short ones. These represent the plates of a battery. The end with the longer plate is always the positive end and the one with the shorter plate is negative as shown (E_1 in Figure 13-2). Since the symbol itself indicates the polarity, positive and negative signs are unnecessary.

When the first two steps are completed, Kirchhoff's Voltage and Current Laws are applied, Steps 3 and 4. This results in a system of simultaneous equations. When solved, the magnitude (and direction) of the unknown currents are determined.

In writing the voltage equation, take the imaginary walk. When you follow a path passing a circuit element from negative to positive, a potential or voltage rise is observed. If you move from positive to negative, there is a voltage drop. The sum of the voltages rises minus all voltage drops equals zero.

Example A Start at point A, travel (1) clockwise, and (2) counterclockwise around the loop, Figure 13-4. Record the voltage rises and drops according to Kirchhoff's Voltage Law.

Solution: The polarity of the loads is indicated using the assumed current direction.

(1) The first circuit element encountered is the battery E_1. A voltage rise is observed when moving from negative to positive. The same observation is made when passing R_1, R_2 and R_3. A voltage drop is observed in E_2 and E_3. The walk past these elements is from positive to negative. Kirchhoff's Voltage Law gives $E_1 + IR_1 - E_2 + IR_2 + IR_3 - E_3 = 0$

(2) Traveling counterclockwise, the voltage law gives $E_3 - IR_3 - IR_2 + E_2 - IR_1 - E_1 = 0$

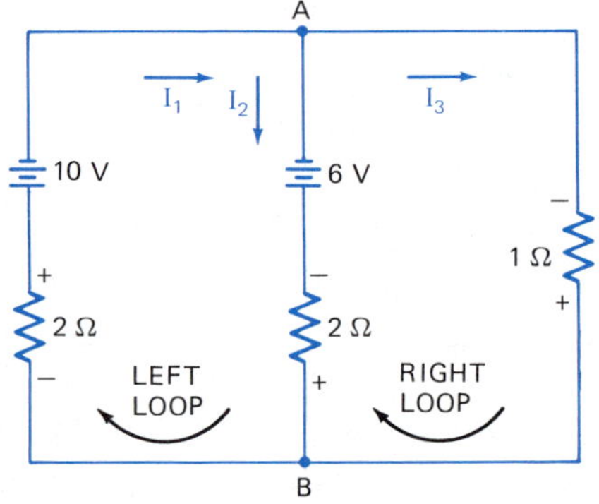

Figure 13-5

Example B Find the current in each branch of the circuit, Figure 13-5.

Solution:

1. Current directions are chosen as indicated.
2. Plus and minus signs are indicated at the terminals of each load according to current directions.
3. Apply the current law at point A:

$$I_1 - I_2 - I_3 = 0 \quad \text{(a)}$$

4. Apply the voltage law to the left loop starting at point A:

$$6 + 2I_2 + 2I_1 - 10 = 0 \quad \text{(b)}$$

Apply the voltage law to the right loop starting at point A:

$$I_3 - 2I_2 - 6 = 0 \quad \text{(c)}$$

5. Find the value of I_3 from equation (a) and substitute into equation (c):

$$I_1 - 3I_2 - 6 = 0 \quad \text{(d)}$$

The resulting equations (d) and (b) form a second order system. Multiply (d) by 2:

$$2I_1 - 6I_2 = 12 \quad \text{(d)}$$
$$2I_1 + 2I_2 = 4 \quad \text{(b)}$$
$$-8I_2 = 8 \quad \text{(Subtract)}$$
$$I_2 = -1 A$$

Substitute $I_2 = -1$ into equation (d):
$2I_1 - 6(-1) = 12$
$I_1 = 3$ A
When $I_1 = 3$ A and $I_2 = -1$ A are substituted into equation (a), the result is $I_3 = 4$ A. Check to see that the answers satisfy equations (b) and (c).

The negative sign in the value for I_2 indicates the current actually flows in the opposite direction through the center branch. The magnitude of the current in that branch is 1 A. If the problem is reworked with the correct direction for I_2, the same answers are obtained with I_2 positive. It is not necessary to worry about which direction to assign to the current in any branch. Any wrong choice will be indicated by a minus sign.

A third equation could have been obtained by applying the voltage law to the outer loop. The equation for a clockwise walk around the outer portion of the circuit starting at point A is:

$$I_3 + 2I_1 - 10 = 0$$

If equations (b) and (c) from the two inner loops are combined, eliminating I_2, the same equation is obtained. Therefore, the third equation obtained by the voltage law is a derived one. Only two of the three loop voltage equations can be used.

The required third equation must be written using the current law. The current law could have been applied at point B, Figure 13-5:

$$I_2 + I_3 - I_1 = 0$$

This equation can be derived from equation (a) by multiplying by −1. Therefore this equation and equation (a) are not unique.

To obtain a set of unique equations, both of Kirchhoff's Laws must be used. Each must be applied only as many times as necessary to include all currents.

EXERCISE 13-2

1. Write an equation relating the currents in both A and B, Figure 13-6.

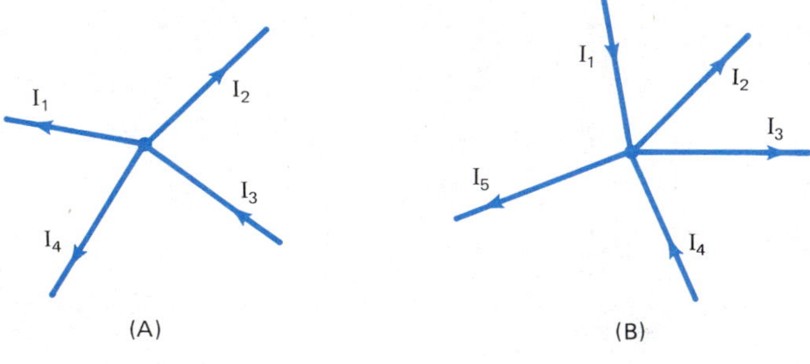

Figure 13-6

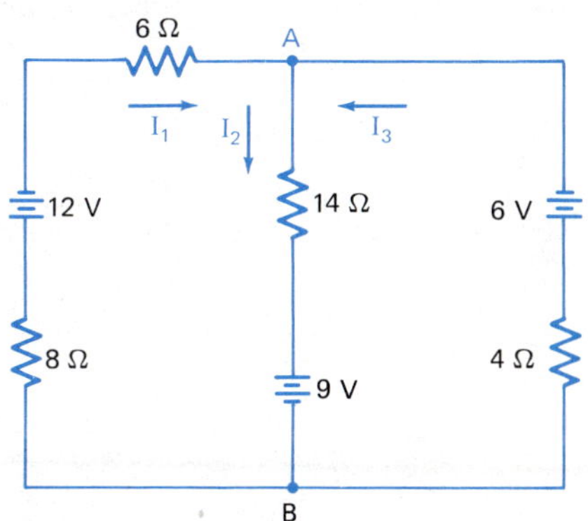

Figure 13-7

2. Use Figure 13-7. (a) Write an equation relating the currents at point A. (b) Write an equation relating the same currents at point B. (c) Are these two equations unique, equivalent or inconsistent?

3. Redraw Figure 13-7 on notebook paper and assign plus and minus signs to each load according to the directions of the currents shown. Write the left loop voltage equation.
4. Assign plus and minus signs to the loads in Figure 13-8 according to the currents indicated. Start at A, travel clockwise, write the left loop voltage equation.

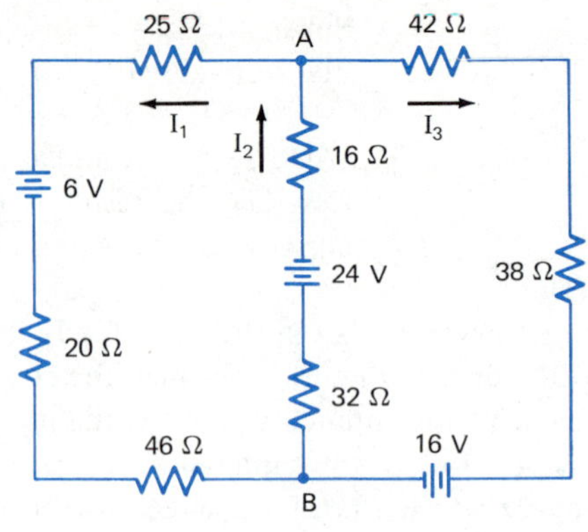

Figure 13-8

5. Find the currents in each branch of the circuit, Figure 13-9.

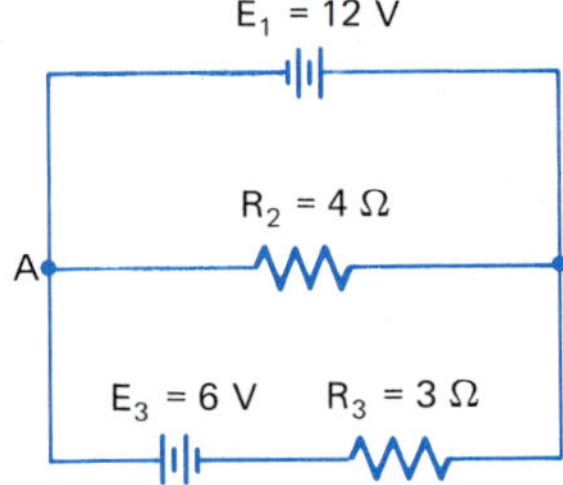

Figure 13-9

6. Find the currents in each branch of the circuit, Figure 13-10.

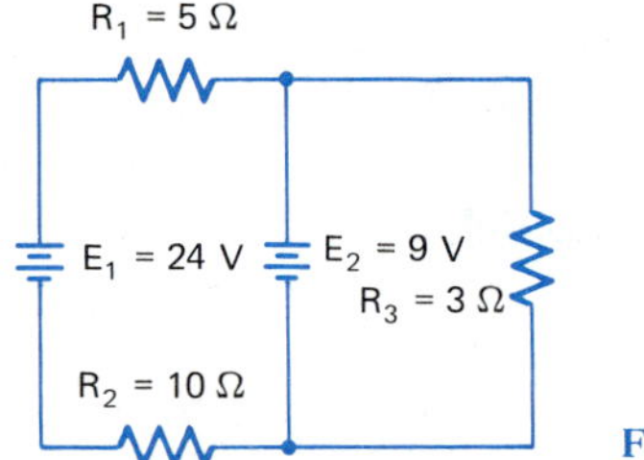

Figure 13-10

7. Find the current in each branch of the circuit, Figure 13-11.

8. Find the current in each branch of the circuit, Figure 13-12.

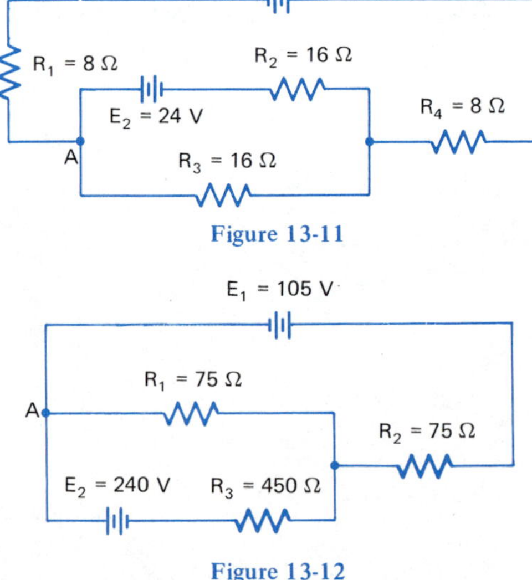

Figure 13-11

Figure 13-12

9. Find the currents in the circuit, Figure 13-13.
10. Find the unknowns in the circuit, Figure 13-14: I_2, I_3 and E_1.

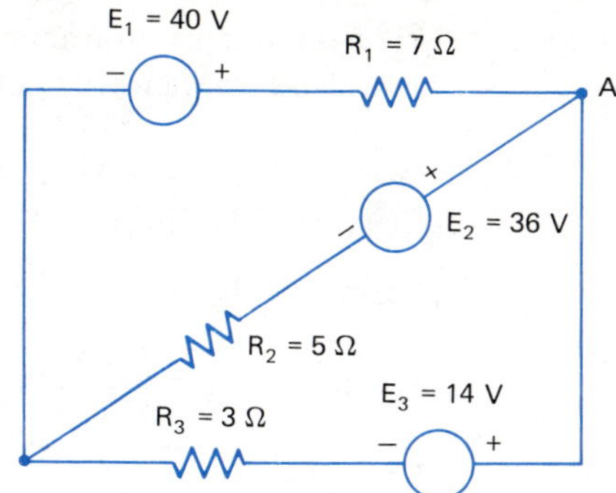

Figure 13-13

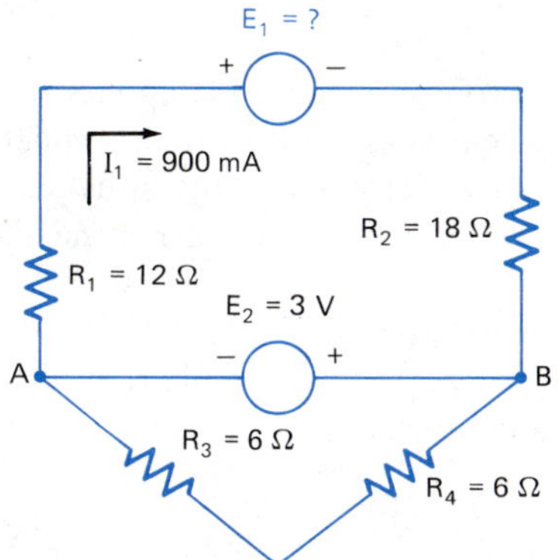

Figure 13-14

OTHER METHODS

There are several methods of analyzing electrical circuits. Two additional ones using the principle of superposition and Thévenin's Theorem are described in the following topics. The *principle of superposition* method applies a simplifying procedure before it is used. In contrast, *Thévenin's method* results in a much simplified circuit which is easily analyzed.

The former method often results in circuits that can be reduced to simple parallel circuits. For this reason, a quick way to find currents in each branch is desirable. Thévenin's method results in simple series circuits. Here, it is helpful to use a quick method for finding the voltage drop across each element.

The voltage divider rule for series circuits and the current divider rule for parallel circuits provide the welcome shortcut methods. These will be discussed first.

13-3 VOLTAGE AND CURRENT DIVIDER RULES

The current is the same through all series components, Figure 13-15. In other words:

$$I_T = I_1 = I_2$$

If E_1 and E_2 are the voltage drops across R_1 and R_2 respectively, then

$$\frac{E_T}{R_T} = \frac{E_1}{R_1} = \frac{E_2}{R_2}$$

Since $R_T = R_1 + R_2$ in a series circuit then

$$\frac{E_1}{R_1} = \frac{E_T}{R_1 + R_2}$$

and

$$\frac{E_2}{R_2} = \frac{E_T}{R_1 + R_2}$$

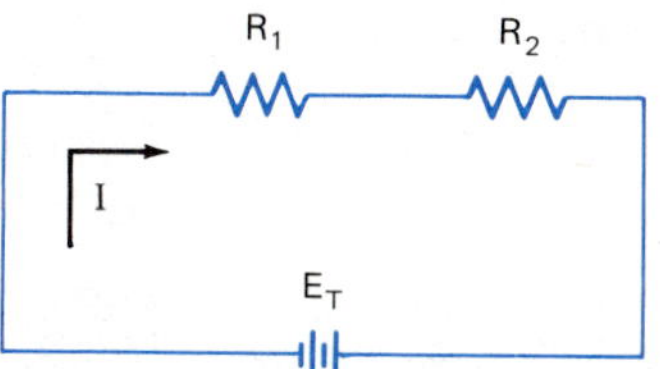

Figure 13-15 Simple series circuit illustrating the voltage divider rule

When these formulas are solved for E_1 and E_2, the result is the voltage divider rule:

$$E_1 = E_T\left(\frac{R_1}{R_1 + R_2}\right)$$

$$E_2 = E_T\left(\frac{R_2}{R_1 + R_2}\right)$$

Voltage Divider Rule. To find the voltage drop across any load in a series circuit, multiply the applied voltage by the resistance of the load and divide by the total circuit resistance. This rule applies to any number of loads in series.

Example A Find the voltage drop across each of two loads in series. The loads have resistances of 500 Ω and 300 Ω. The applied voltage is 24 V.

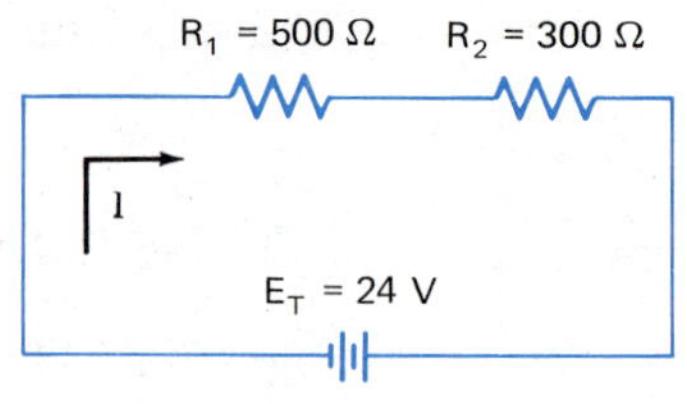

Figure 13-16

Solution: Data: $E_T = 24\ V, R_1 = 500\ \Omega, R_2 = 300\ \Omega, E_1 = ?, E_2 = ?$

Diagram: See Figure 13-16

Formula: $E_1 = E_T\left(\frac{R_1}{R_1 + R_2}\right)$

Substitute: $E_1 = \frac{(24\ V)(500\ \Omega)}{500\ \Omega + 300\ \Omega} = 15\ V$

Formula: $E_2 = E_T\left(\frac{R_2}{R_1 + R_2}\right)$

Substitute: $E_2 = \frac{(24\ V)(300\ \Omega)}{500\ \Omega + 300\ \Omega} = 9\ V$

Check: $E_T = E_1 + E_2 = 15\ V + 9\ V = 24\ V$

The voltage divider rule applies to series circuits. The advantage is that the current does not have to be evaluated.

For parallel circuits, the situation is different. It is the current that divides. The voltage drop across each load is the same, Figure 13-17.

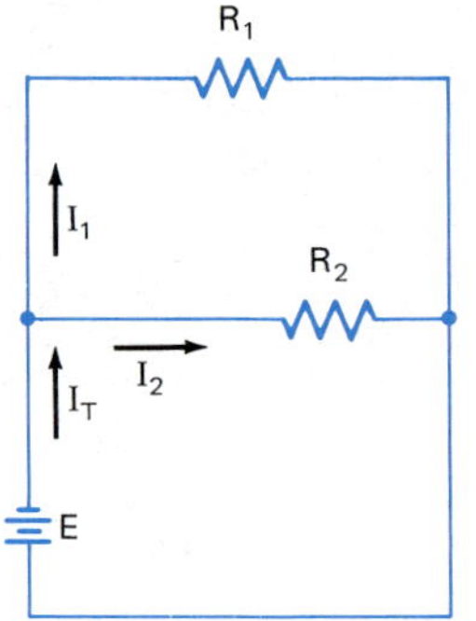

Figure 13-17 Simple parallel circuit illustrating the current divider rule

$$E_T = E_1 = E_2$$

Substituting (Ohm's Law):

$$I_T R_T = I_1 R_1 = I_2 R_2$$

The total resistance is

$$R_T = \frac{R_1 R_2}{R_1 + R_2}$$

Therefore:

$$I_1 R_1 = I_T \frac{R_1 R_2}{R_1 + R_2}$$

R_1 cancels: (DA, $\div R_1$)

$$\mathbf{I_1 = I_T\left(\frac{R_2}{R_1 + R_2}\right)}$$

Similarly:

$$\mathbf{I_2 = I_T\left(\frac{R_1}{R_1 + R_2}\right)}$$

The last two formulas allow one to find the current through each branch without evaluating the voltage drop. The following rule describes the method:

> **Current Divider Rule.** The current in one branch of a parallel circuit is found by multiplying the total current by the resistance of the *other* branch and dividing by the sum of the two resistances. This rule applies to two-branch circuits only.

Example B Find the current in each branch of the circuit, Figure 13-17. R_1 is 500 Ω and R_2 is 1000 Ω. The total current is 45 mA.

Solution: Data: $R_1 = 500\ \Omega$, $R_2 = 1000\ \Omega$, $I_T = 45$ mA, $I_1 = ?$, $I_2 = ?$

Diagram: See Figure 13-17

Formula: $I_1 = I_T\left(\dfrac{R_2}{R_1 + R_2}\right)$

Substitute: $I_1 = \dfrac{(45\text{ mA})(1000\ \Omega)}{500\ \Omega + 1000\ \Omega} = 30\text{ mA}$

Formula: $I_2 = I_T\left(\dfrac{R_1}{R_1 + R_2}\right)$

Substitute: $I_2 = \dfrac{(45\text{ mA})(500\ \Omega)}{500\ \Omega + 1000\ \Omega} = 15\text{ mA}$

Check: $I_T = I_1 + I_2 = 30\text{ mA} + 15\text{ mA} = 45\text{ mA}$

Note that current is larger through the smaller resistance.

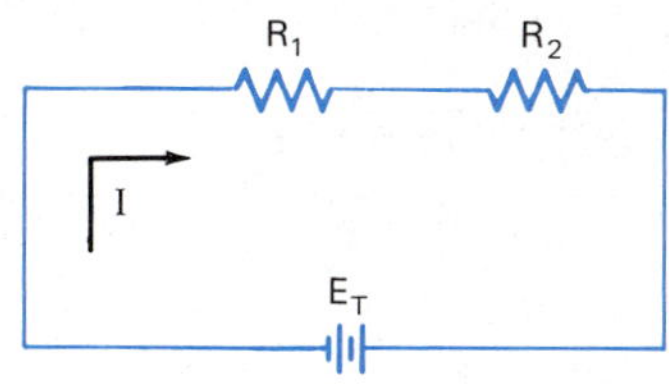

Figure 13-18

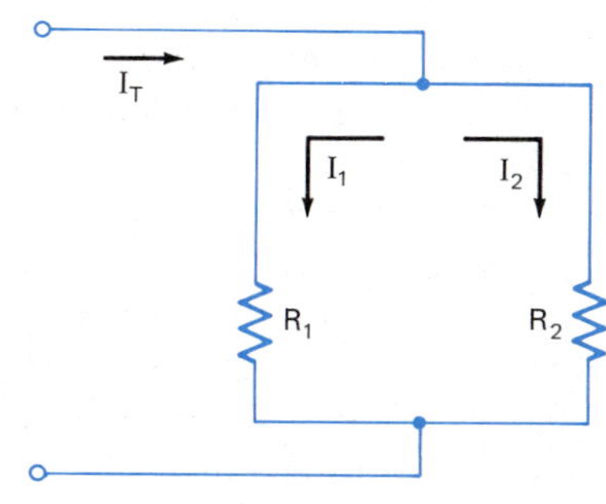

Figure 13-19

EXERCISE 13-3

Find the voltage drop across each load in the following series circuits using the voltage divider rule. Refer to Figure 13-18.

1. $R_1 = 50\ \Omega$, $R_2 = 150\ \Omega$, $E_T = 12$ V
2. $R_1 = 900\ \Omega$, $R_2 = 500\ \Omega$, $E_T = 105$ V
3. $R_1 = 35\ k\Omega$, $R_2 = 65\ k\Omega$, $E_T = 48$ V
4. $R_1 = 4.4\ k\Omega$, $R_2 = 6.6\ \Omega$, $E_T = 220$ V

Find the current through each load in the following parallel circuits using the current divider rule. Refer to Figure 13-19.

5. $R_1 = 70\ \Omega$, $R_2 = 210\ \Omega$, $I_T = 140$ mA
6. $R_1 = 800\ \Omega$, $R_2 = 1400\ \Omega$, $I_T = 44$ mA
7. $R_1 = 95\ k\Omega$, $R_2 = 135\ k\Omega$, $I_T = 2.5$ A
8. $R_1 = 25\ \Omega$, $R_2 = 40\ \Omega$, $I_T = 800$ mA
9. Find the voltage drop across each load in a three-load series circuit using the voltage divider rule. $R_1 = 50\ \Omega$, $R_2 = 75\ \Omega$, $R_3 = 110\ \Omega$, $E_T = 64$ V.
10. Find the voltage drop across each element of a series-parallel circuit, Figure 13-20. Find the current through each branch of the parallel group. Solve this problem using the two divider rules. $R_1 = 60\ \Omega$, $R_2 = 100\ \Omega$, $R_3 = 150\ \Omega$, $I_T = 2$ A, $E_T = 240$ V.

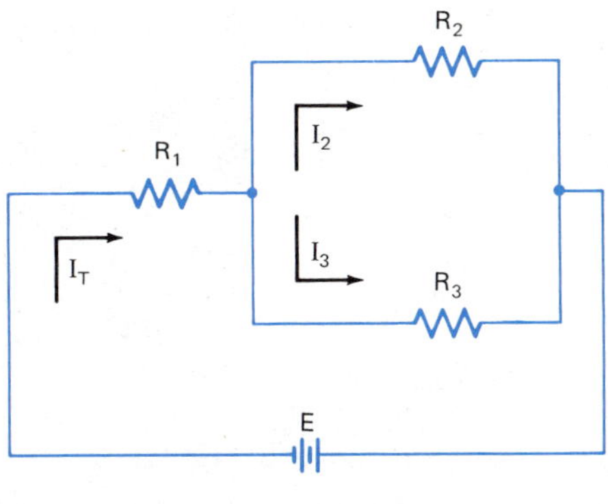

Figure 13-20

13-4 THE PRINCIPLE OF SUPERPOSITION

The principle of superposition is useful in circuits having more than one source of emf:

The Principle of Superposition. In a circuit with more than one emf source, the current through a given branch is the arithmetic sum of the currents produced independently by each source. That is, the currents due to each source are superimposed upon each other to yield the total current through any branch of the circuit.

The principle is applied in a series of steps given in the following rule:

APPLYING THE PRINCIPLE OF SUPERPOSITION

- Choose one source of emf and replace all others with their internal resistances. If the internal resistances are neglected, replace them with a short circuit.
- Determine the current including its direction in the branch or branches as required.
- Repeat the first two steps for each source of emf.
- Add the currents in each branch to find the total current. Pay particular attention to current directions. Opposing currents are subtracted.

In the calculations of the second step, any method can be used: Ohm's Law, formulas for equivalent resistance and/or Kirchhoff's Current and Voltage Laws. All are valid. Often, one needs to find the current through only one load or branch of the circuit. The superposition principle is convenient to use for this purpose.

Example A Calculate the current through load R_2, Figure 13-21(A), using the principle of superposition.

Solution: EFFECT OF E_1 (12 V).

Figure 13-21(B) shows the circuit when source E_2 is short circuited, with the loads shown in standard form in Figure 13-21(C).

$$R_T = R_2 + \frac{R_1 R_3}{R_1 + R_3}$$

$$R_T = 10\ \Omega + \frac{(5\ \Omega)(20\ \Omega)}{5\ \Omega + 20\ \Omega} = 14\ \Omega$$

$$I_T = I_2 = \frac{E_T}{R_T} = \frac{12\ \text{V}}{14\ \Omega} = 857\ \text{mA}$$

EFFECT OF E_2 (24 V)

Figure 13-21(D) shows the circuit with E_1 shorted and the loads are shown in standard form in Figure 13-21(E).

$$R_T = R_3 + \frac{R_1 R_2}{R_1 + R_2}$$

$$R_T = 20\ \Omega + \frac{(5\ \Omega)(10\ \Omega)}{5\ \Omega + 10\ \Omega} = 23.33\ \Omega$$

$$I_T = I_3 = \frac{E_T}{R_T} = \frac{24\ \text{V}}{23.33\ \Omega} = 1029\ \text{mA}$$

Use the current divider rule to find I_2:

$$I_2 = I_T\left(\frac{R_1}{R_1 + R_2}\right)$$

$$I_2 = (1029\ \text{mA})\frac{5\ \Omega}{5\ \Omega + 10\ \Omega} = 343\ \text{mA}$$

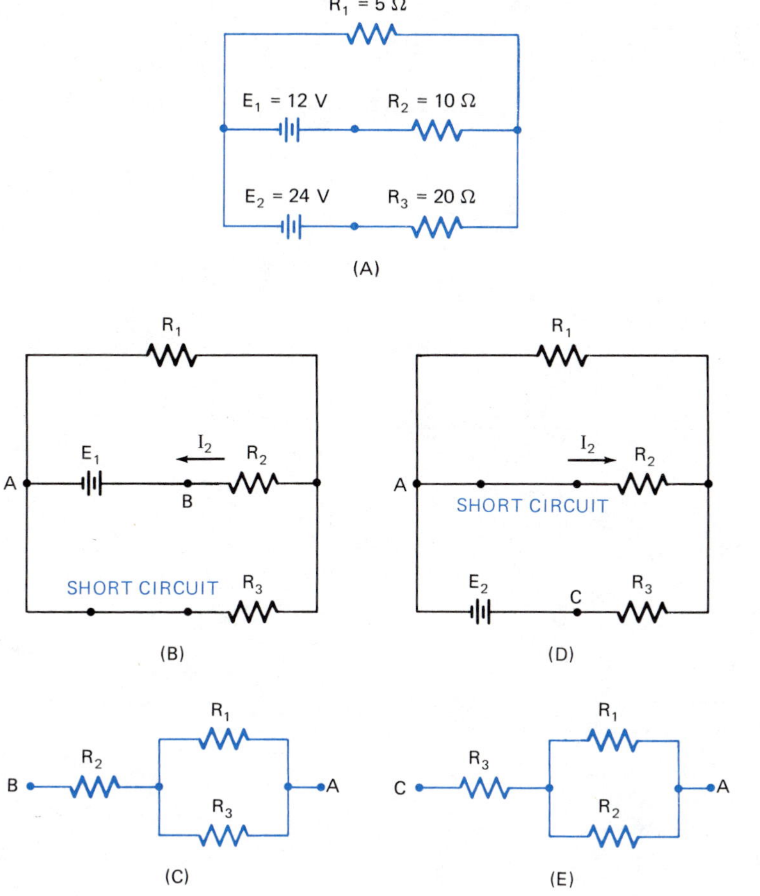

Figure 13-21

EFFECT OF BOTH E_1 AND E_2

Assume current flowing rightward is positive and leftward through the load is negative. The current due to E_1 is −857 mA and that due to E_2 is +343 mA. Adding the currents gives a result of −514 mA. The minus sign means the resultant current is 514 mA and flows leftward through the load R_2, Figure 13-21(A).

Naturally, the principle of superposition can be used for finding all the currents in a circuit. To do so requires about the same amount of effort as does Kirchhoff's method.

Example B Calculate the current through each branch of the circuit, Figure 13-22(A), using the principle of superposition.

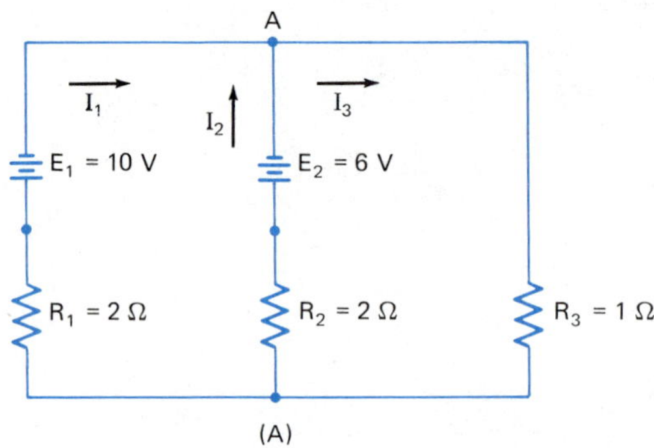

(A)

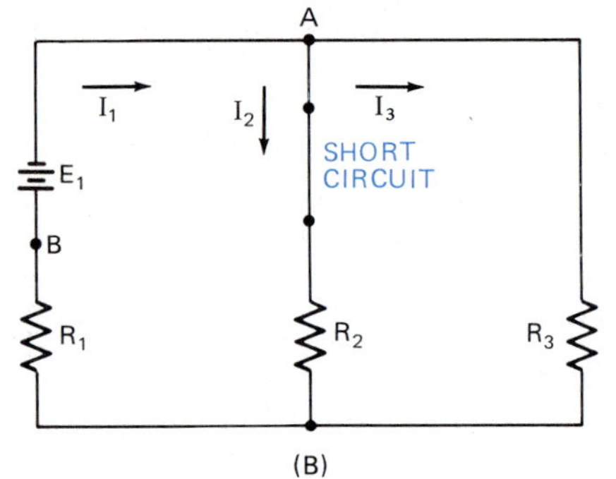

(B)

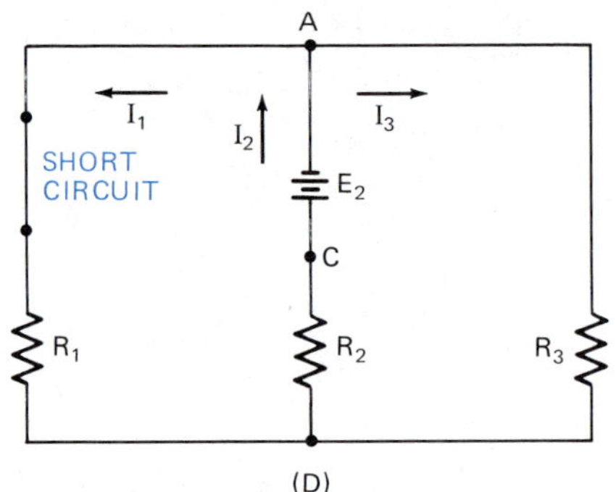

(D)

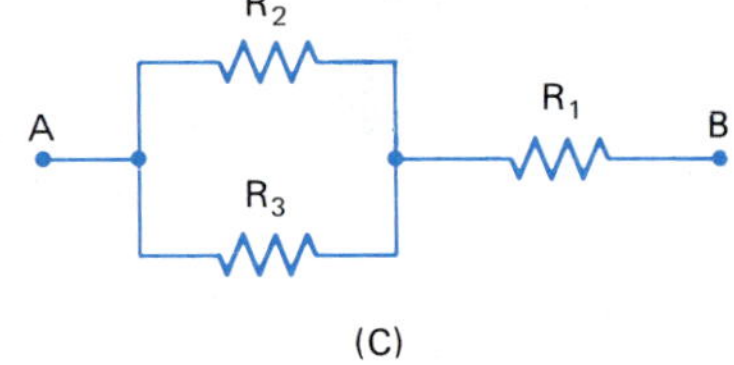

(C)

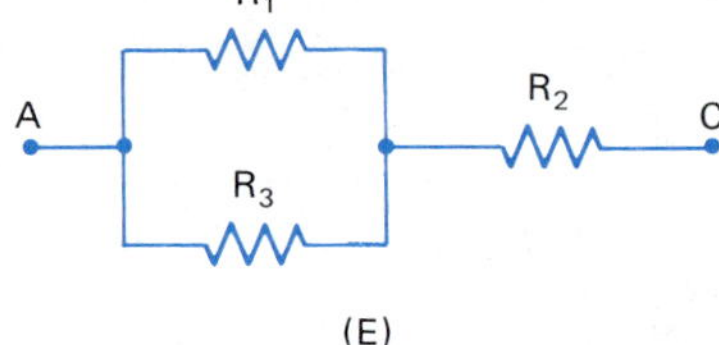

(E)

Figure 13-22

Solution: EFFECTS OF E_1 (10 V)

Figure 13-22(B) with E_2 short circuited is shown in standard form in Figure 13-22(C). The total resistance is

$$R_T = R_1 + \frac{R_2 R_3}{R_2 + R_3}$$

$$R_T = 2\ \Omega + \frac{(2\ \Omega)(1\ \Omega)}{2\ \Omega + 1\ \Omega} = 2.67\ \Omega$$

$$I_T = I_1 = \frac{E_T}{R_T} = \frac{10\ \text{V}}{2.67\ \Omega} = 3.75\ \text{A}$$

Use the current divider rule to find I_2 and I_3.

$$I_2 = I_T\left(\frac{R_3}{R_2 + R_3}\right) = \frac{(3.75\ \text{A})(1\ \Omega)}{1\ \Omega + 2\ \Omega} = 1.25\ \text{A}$$

$$I_3 = I_T\left(\frac{R_2}{R_2 + R_3}\right) = \frac{(3.75\ \text{A})(2\ \Omega)}{1\ \Omega + 2\ \Omega} = 2.5\ \text{A}$$

These results are posted in the first line of Figure 13-23. Current directions upward and rightward are positive and downward and leftward are negative as indicated in Figure 13-22(B) in the vicinity of point A.

EFFECTS OF E_2 (6 V)
Figure 13-22(D) is shown in standard form in Figure 13-22(E). The total resistance is

$$R_T = R_2 + \frac{R_1 R_3}{R_1 + R_3}$$

$$R_T = 2\ \Omega + \frac{(2\ \Omega)(1\ \Omega)}{2\ \Omega + 1\ \Omega} = 2.67\ \Omega$$

$$I_T = I_2 = \frac{E_T}{R_T} = \frac{6\ V}{2.67\ \Omega} = 2.25\ A$$

Again, use the current divider rule to find I_1 and I_3.

$$I_1 = I_T\left(\frac{R_3}{R_1 + R_3}\right) = \frac{(2.25\ A)(1\ \Omega)}{2\ \Omega + 1\ \Omega} = 0.75\ A$$

$$I_3 = I_T\left(\frac{R_1}{R_1 + R_3}\right) = \frac{(2.25\ A)(2\ \Omega)}{2\ \Omega + 1\ \Omega} = 1.5\ A$$

Currents are entered in Line 2 of Figure 13-23. Again, upward and rightward currents are positive. Downward and leftward currents are negative as indicated in Figure 13-22(D) in the vicinity of point A.

EFFECTS OF BOTH E_1 AND E_2
To find the current in each branch due to both sources, add the currents due to each source. Add each column in Figure 13-23. All resultant currents are positive. The directions are shown correctly in Figure 13-22(A).

Check: Apply Kirchhoff's Current Law at point A in Figure 13-22(A) using the resultant currents.
$I_1 + I_2 = I_3$
$3\ A + 1\ A = 4\ A$

In a circuit such as that shown in Figure 13-24, part of the problem is simple. In problems such as this, the internal resistance of the sources are neglected, when E_2 is shorted across, the entire circuit from A to B is shorted. In this case, $I_1 = I_3$ and $I_2 = 0$. These are the currents due to E_1

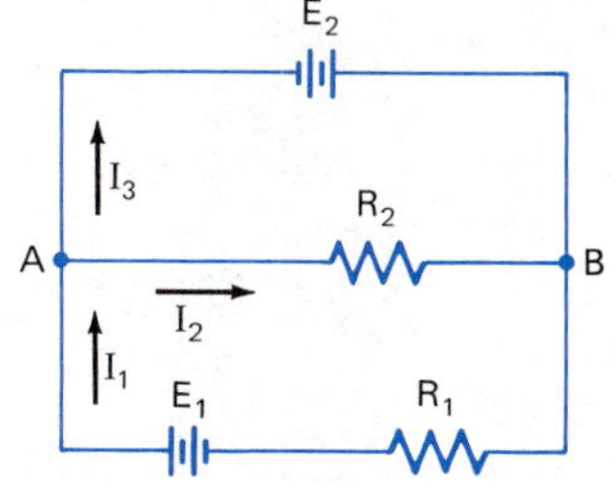

Figure 13-24 If E_2 is short-circuited, the entire circuit from A to B is shorted. In this case, $I_1 = I_3 = \frac{E_1}{R_1}$ and $I_2 = 0$.

EMF SOURCE	I_1	I_2	I_3
E_1	+3.75 A	−1.25 A	+2.5 A
E_2	−0.75 A	+2.25 A	+1.5 A
E_1 & E_2	+3 A	+1 A	+4 A

Figure 13-23 Summary of currents due to each source and composite

and are easily found: $I_2 = 0, I_1 = I_3 = \frac{E_1}{R_1}$. These, of course, must be added to those due to source E_2. See Problem 1 in the following exercise.

If the situation is right, the total current in one branch can be zero. If the current due to one source is equal and opposite of that due to the other source, their sum is zero. See Problem 2 in the following exercise.

EXERCISE 13-4

Solve each of the following problems using the principle of superposition.

1. Find the current through load R_2 only, Figure 13-25.

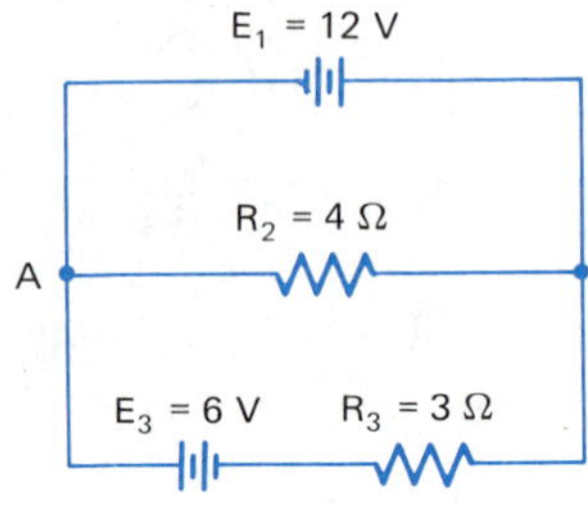

Figure 13-25

2. Find the current through load R_3 only, Figure 13-26.

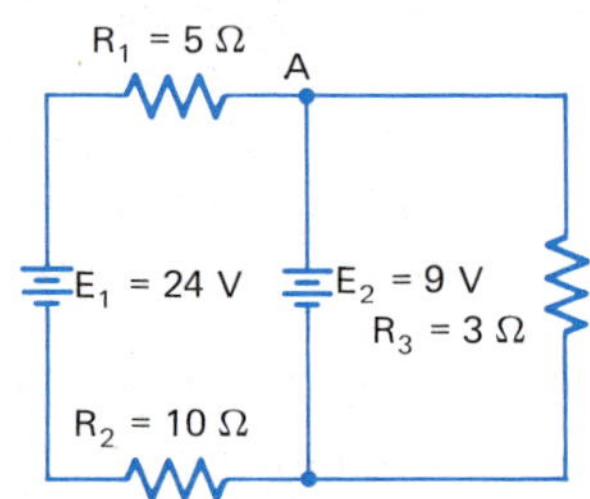

Figure 13-26

3. Find the current through the load R_1, Figure 13-27.

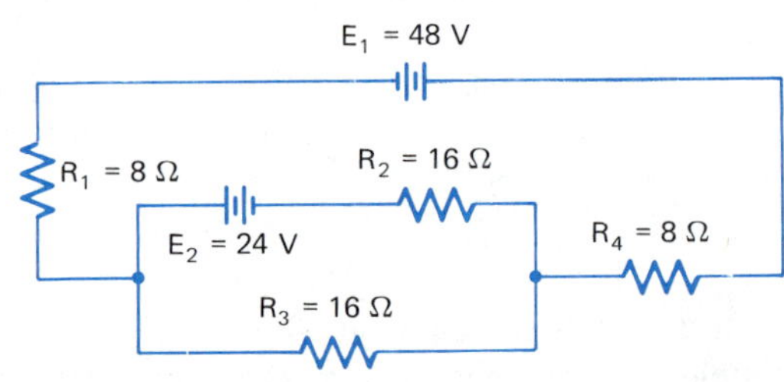

Figure 13-27

4. Find the current through the load R_3, Figure 13-28.

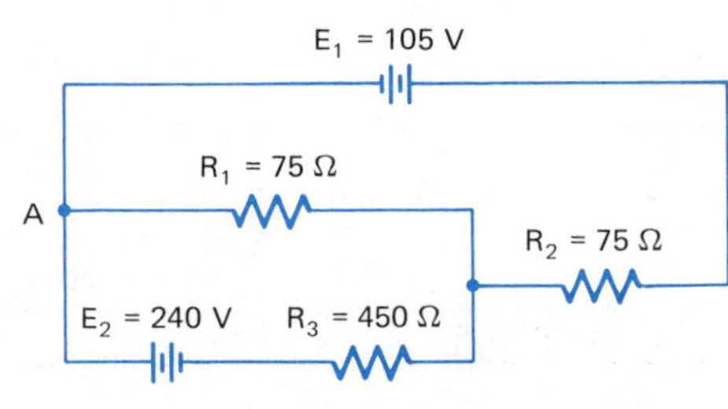

Figure 13-28

5. Find the current through each load, Figure 13-29.

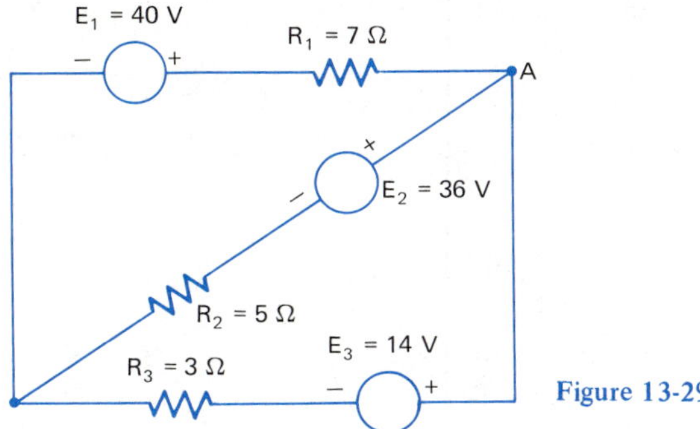

Figure 13-29

6. Find the unknowns I_1, I_2 and I_3, Figure 13-30.

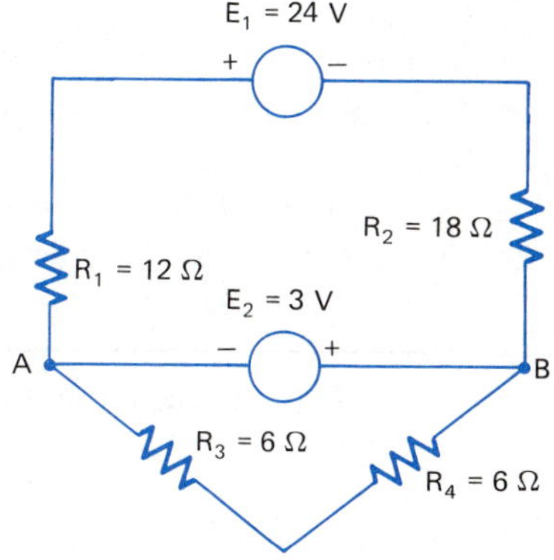

Figure 13-30

13-5 THÉVENIN'S THEOREM

Thévenin's Theorem is used to find the current in one load of a circuit. This is accomplished by reducing a complex circuit, Figure 13-31(A), to a single source of emf (E_{TH}) and a single resistance (R_{TH}) in series with the load (R_L) in question, Figure 13-31(B). E_{TH} and R_{TH} are the equivalent Thévenin's voltage and resistance, respectively.

> **Thévenin's Theorem:** Any complex circuit can be divided into two parts. One part (the load) is retained. The other part (the rest of the circuit) is replaced by a single battery and a single resistance connected in series. The battery has a voltage (E_{TH}) equal to the emf of the open terminals of the part being replaced. The resistance (R_{TH}) is the resistance between the open terminals of the part being replaced when all voltage sources have been short circuited.

Thévenin's Theorem can be applied to any circuit solved so far by treating the resistance of one branch as a load. The rest of the circuit is treated as a two-terminal circuit containing one or more voltage sources. The current in the load is found by applying the following rule in sequence.

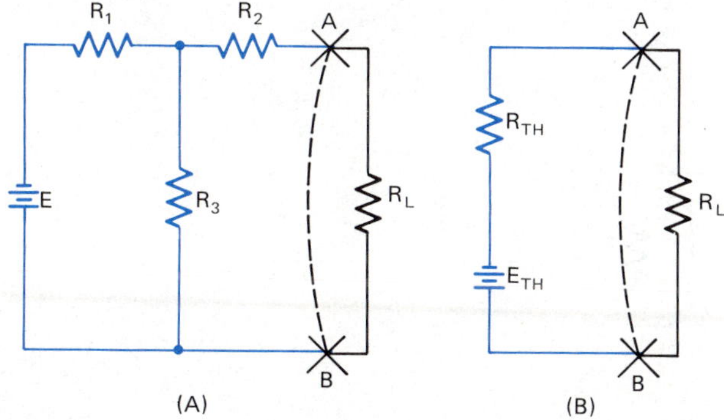

Figure 13-31 Any circuit to the left of breaks A and B can be replaced with the Thévenin's equivalent source E_{TH} and resistance R_{TH} in series with the load (B).

APPLYING THÉVENIN'S THEOREM

- Remove the load resistance from the circuit and calculate the open circuit terminal voltage (E_{TH}).
- Redraw the circuit replacing each voltage source with a short circuit or with the internal resistance if it cannot be neglected.
- Find the equivalent resistance (R_{TH}) of the circuit.
- Draw the equivalent Thévenin circuit consisting of the source E_{TH} and resistance R_{TH} in series.
- Reconnect the load across the output terminals of the equivalent Thévenin circuit. The current in the load can now be determined. Find the total resistance ($R_{TH} + R_L$) and use Ohm's Law to find the current. If the load resistance is changed, the new load current is easily calculated.

Example A Find the current in the load (R_L) using Thévenin's Theorem, Figure 13-32(A).

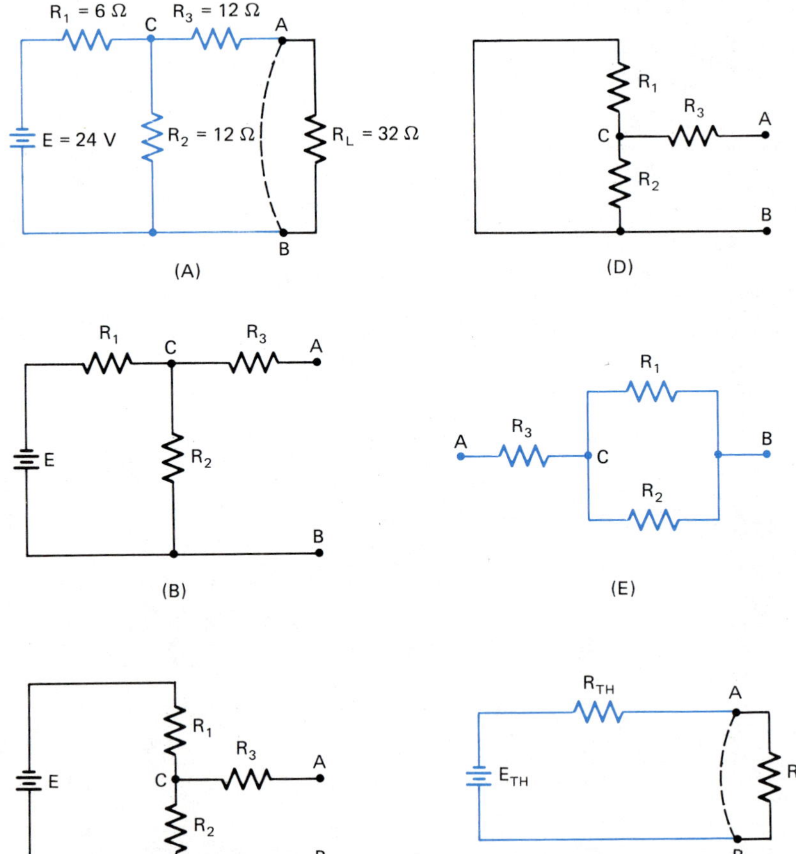

Figure 13-32

Solution: Remove the load resistance (R_L), Figure 13-32(B). If a high resistance voltmeter is placed across points A and B, the reading is the open circuit voltage E_{TH}. Since there is no current flowing through R_3, the voltmeter would read the same if it is connected across points C and B (across R_2). Thus, the voltage drop across R_2 is E_{TH}. Notice that Figures 13-32(B) and 13-32(C) are the same circuit. Find E_{TH} using the voltage divider rule.

$$E_2 = E_{TH} = E_T\left(\frac{R_2}{R_1 + R_2}\right)$$

$$E_{TH} = \frac{(24\ V)(12\ \Omega)}{6\ \Omega + 12\ \Omega} = 16\ V$$

Find the equivalent resistance of the open circuit when the battery is short-circuited, Figure 13-32(D). Figure 13-32(E) shows the circuit in standard form.

$$R_{TH} = R_3 + \frac{R_1 R_2}{R_1 + R_2}$$

$$R_{TH} = 12\ \Omega + \frac{(6\ \Omega)(12\ \Omega)}{6\ \Omega + 12\ \Omega} = 16\ \Omega$$

The equivalent Thévenin circuit is shown in Figure 13-32(F). The total resistance is

$$R_T = R_{TH} + R_L = 16\ \Omega + 48\ \Omega = 64\ \Omega$$

and the current through the load is

$$I = \frac{E_{TH}}{R_T} = \frac{16\ V}{64\ \Omega} = 250\ mA$$

Current flows from A to B through R_2.

Example B Find the current through the load in the previous problem if the load resistance is reduced to 16 Ω.

Solution: The change in the load resistance has no effect on the values of E_{TH} and R_{TH}. Figure 13-32(F) is unchanged except for the value of R_L.

$$R_T = R_{TH} + R_L = 16\ \Omega + 16\ \Omega = 32\ \Omega$$

$$I = \frac{E_{TH}}{R_T} = \frac{16\ V}{32\ \Omega} = 500\ mA$$

Thévenin's Theorem can also be used to find current in the load in circuits with more than one voltage source.

Example C Find the equivalent Thévenin's circuit for the circuit in Figure 13-33(A).

Solution: Assume the current flows clockwise in the circuit loop. The value of the current can be found using simple circuit laws or Kirchhoff's Voltage Law. Since there is only one loop:

$$+6\ V + 2I + 2I - 10\ V = 0$$

$$I = \frac{4\ V}{4\ \Omega} = 1\ A$$

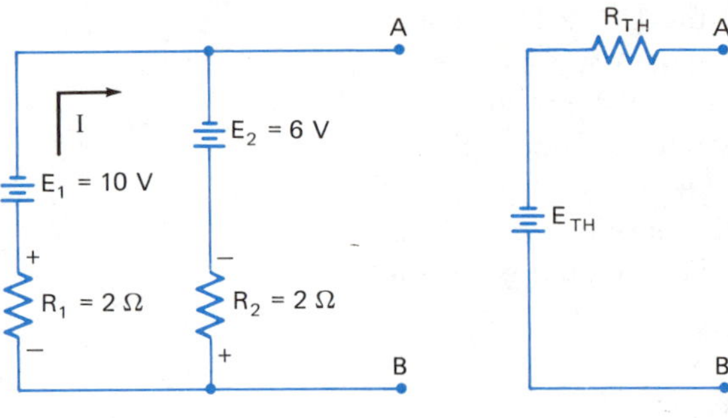

Figure 13-33

To find the Thévenin's voltage E_{TH}, add all voltage rises and drops along either path from A to B using Kirchhoff's voltage method. This is not a complete circuit loop so the sum will be the net voltage E_{TH} across A and B.

$$E_{TH} = 6\text{ V} + (1\text{ A})(2\ \Omega) = 8\text{ V}$$

Short circuit the voltage sources to find the equivalent resistance R_{TH}.

$$R_{TH} = \frac{R_1 R_2}{R_1 + R_2}$$

$$R_{TH} = \frac{(2\ \Omega)(2\ \Omega)}{2\ \Omega + 2\ \Omega} = 1\ \Omega$$

Thévenin's equivalent circuit, Figure 13-33(B) has a voltage E_{TH} of 8 V and a resistance R_{TH} of 1 Ω. If the load resistance is 1 Ω, the current is 4 A.

A circuit such as the one shown in Figure 13-34 is a special case. When one branch of the circuit contains only a voltage source and is applied directly to the load, then E_{TH} = E and R_{TH} = 0. See Problems 3 and 4 in the following exercise. If the load resistance is small, then the battery's internal resistance should not be neglected.

A circuit such as Figure 13-35 is also easy. When the battery is short-circuited, the only resistance is R_2. In this case $R_{TH} = R_2$. However, if R_1 is small, the internal resistance of the source cannot be neglected. See Problem 1 in the following exercise.

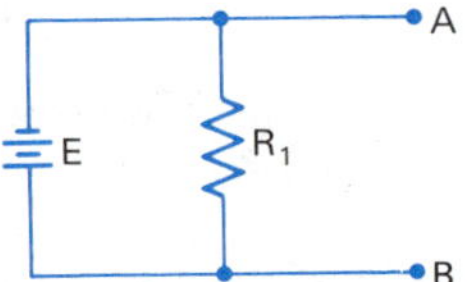

Figure 13-34 When one branch consists of a voltage source applied directly to the load, then E_{TH} = E and R_{TH} = 0.

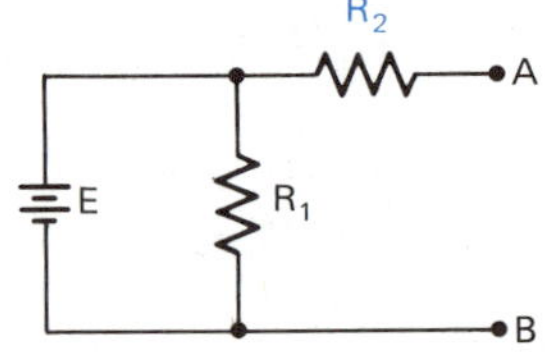

Figure 13-35 When the battery is short-circuited, the only resistance between points A and B is R_2. That is, $R_{TH} = R_2$.

EXERCISE 13-5

1. In Figure 13-35, E_1 = 250 V, R_1 = 2.0 kΩ and R_2 = 1.25 kΩ. Find the Thévenin equivalent (a) R_{TH} and (b) E_{TH} of the circuit. The load is placed across points A and B. Find the current through the load, if the resistance R_L is (c) 750 Ω, (d) 2.25 kΩ. Find the power dissipated in each load if the resistance R_L is (e) 750 Ω, (f) 2.25 kΩ.

2. In Figure 13-36, E = 48 V, R_1 = 8 Ω, R_2 = 4 Ω and R_3 = 24 Ω. Find the Thévenin equivalent (a) R_{TH} and (B) E_{TH} of the circuit. The current through the load is 2 A. (c) Find the resistance of the load. (d) Find the energy dissipated in the load in one hour.

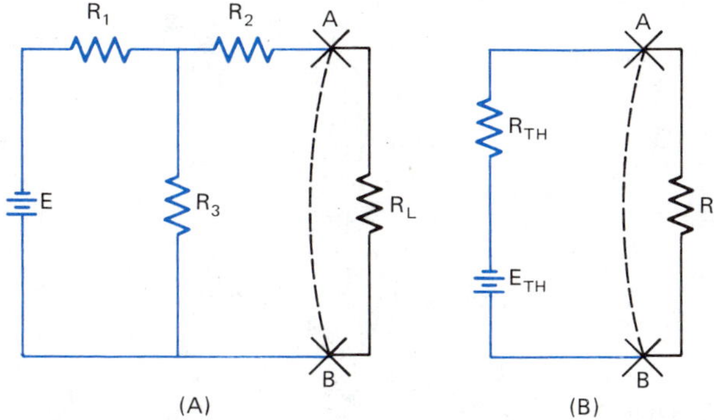

Figure 13-36

3. Using Thévenin's Theorem and the circuit shown in Figure 13-37, find the (a) current and (b) power dissipation in load R_2.

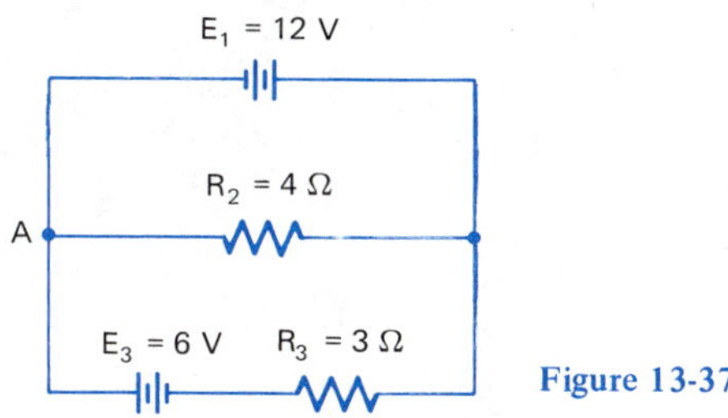

Figure 13-37

4. Using Thévenin's Theorem and the circuit shown in Figure 13-38, find the (a) current and (b) power dissipation in load R_3.
5. Using Thévenin's Theorem and the circuit shown in Figure 13-39, (a) find the current through the load R_L. (b) What is the power dissipation in the load? Notice this circuit can be redrawn in the form of Figure 13-36.

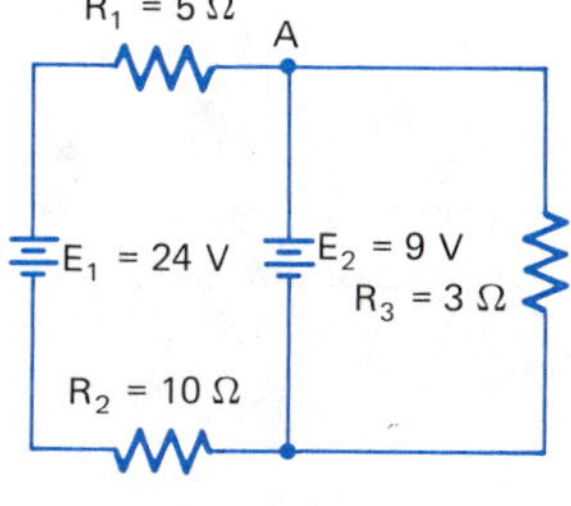

Figure 13-38

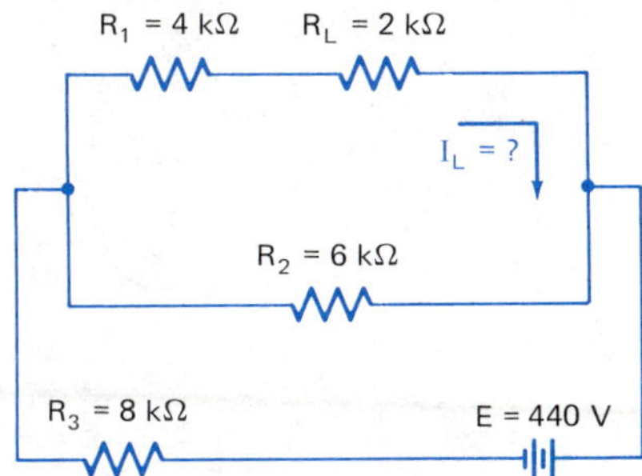

Figure 13-39

6. Using Figure 13-40 and Thévenin's Theorem, (a) find the current through the load R_L. (b) How much charge flows through the circuit in one hour? (c) How much energy is dissipated in one hour?

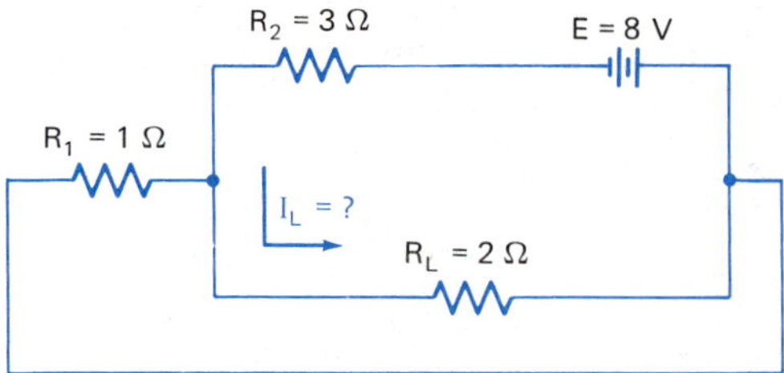

Figure 13-40

7. Using Figure 13-41, (a) find the current using Thévenin's Theorem in the load R_3. (b) What is the voltage drop across R_3?

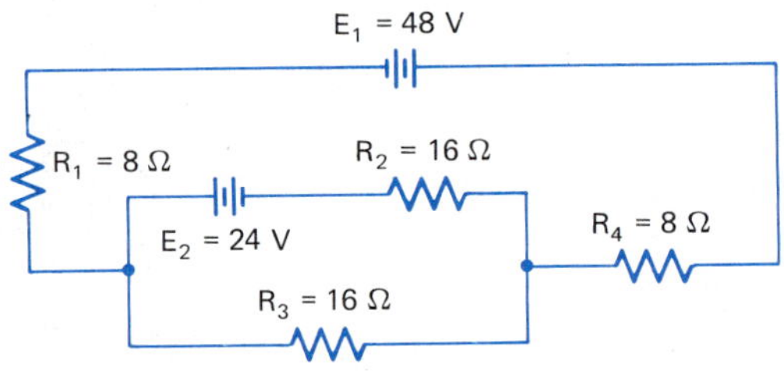

Figure 13-41

8. Using Figure 13-42 and Thévenin's Theorem, (a) determine the current in load R_1. (b) How much energy is dissipated by this load in one hour?

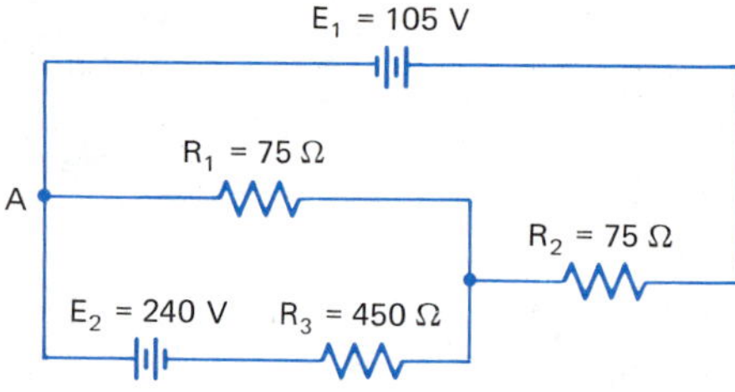

Figure 13-42

SECTION 2 REVIEW

Chapter 13 concludes Section 2: DC Circuits. This review highlights the many topics in Section 2.

- *Algebra.* Solving simple equations and formulas using the addition, subtraction, square and square root axioms.
- *Arithmetic.* Adding and subtracting fractions. Lowest common denominator (LCD).
- *Circuits.* Series, parallel and combination circuits. Total current, total resistance, total voltage, total power, internal source resistance.
- *Variation.* Direct, joint, inverse and inverse square.
- *Problem solving.* Percent, efficiency, relative change and tolerance. Energy consumption and cost.
- *Complex algebra.* Combining monomials and polynomials, factoring. Solving equations with grouping signs and fractions. Simultaneous equations, graphing equations.
- *Circuit analysis.* Ohm's Law, voltage divider rule, current divider rule, Kirchhoff's Laws, principle of superposition, Thévenin's Theorem.

SECTION 3 INTRODUCTION TO ac CIRCUITS

CHAPTER 14

RATIO AND PROPORTION

OBJECTIVES

After satisfactorily completing this chapter, the student should be able to:

- Define various words and terms introduced in this chapter.
- Compare numbers as ratios and reduce them to lowest form as fractions, decimals or percents.
- Change verbal expressions into ratios or proportions (mathematical expressions) and solve direct and inverse proportion problems with electrical applications.

Among electricians, one often hears the word efficiency. Efficiency is used in reference to electric motors, generators, transformers, Figure 14-1, and other electrical devices. Efficiency compares the output power of the device to the input power. In this chapter, several ratios including efficiency will be reviewed and other new ones are introduced.

Pairs of numbers that change so that their quotient always stays the same will be studied. Two such pairs form a proportion. The ability to set up proportion problems and solve them is a practical way of working certain types of electrical problems.

RATIO

The *ratio* of two numbers is the quotient obtained by dividing the first number by the second. For example, the ratio of 9 to 3 is 3 to 1. In contrast, the ratio of 3 to 9 is 1 to 3. Ratios can be expressed in three forms: the word form (a to b), the colon form (a:b) and the fractional form: $\left(\frac{a}{b}\right)$. The first number (represented by a) in all three forms is called the *first*

Figure 14-1 A very large (1 300 000 kVA, 345 000 V) generator step-up transformer (Courtesy of Westinghouse Electric Corporation)

term. The second number (represented by b) is the *second term.* The fractional form is preferred by electricians. However, familiarity with all three forms is essential.

14-1 SIMPLIFYING RATIOS

Ratios should be reduced to lowest terms. This is accomplished by dividing both terms by the largest common factor, Topic 3-2. The terms of the ratio $\frac{16}{24}$ have a largest common factor of 8. Division of both terms by 8 yields a reduced ratio of $\frac{2}{3}$. This process reduces the ratio (fraction) to lowest terms.

A ratio is simply a number; therefore both terms must be the same kind of measurable quantity. It is meaningful to compare the output voltage with the input voltage of a transformer. It is not meaningful, however, to compare the output voltage with the weight of the transformer. Both terms of a ratio must also be expressed in the same units of measure. In a ratio such as 15 000 W:45 kW, convert 45 kW to 45 000 W. The ratio is now 15 000 W:45 000 W which reduces to 1:3.

Ratios of two fractions can be reduced by dividing the first term by the second. Remember, when dividing fractions, the numerator is multiplied by the reciprocal of the denominator.

Example A Express the ratio $\frac{3}{4}:\frac{5}{3}$ in fractional form and reduce to lowest terms.

Solution: $\frac{3/4}{5/3} = \frac{3}{4} \times \frac{3}{5} = \frac{9}{20}$

EXERCISE 14-1

Express these ratios in fractional form and simplify (reduce to lowest terms) without using a calculator.

1. 64:24
2. 35:75
3. 16:12
4. 20 h:3 days
5. 600 μA:0.04 mA
6. 650 kΩ:32.5 MΩ

Express these ratios in colon form and simplify without using a calculator. Conversion factors are given in the Appendix.

7. $\frac{75}{45}$
8. $\frac{450 \text{ V}}{0.675 \text{ kV}}$
9. $\frac{18 \text{ W}}{24\,000 \text{ mW}}$
10. $\frac{1 \text{ hp}}{1.492 \text{ kW}}$
11. $\frac{12 \text{ in}}{15.24 \text{ cm}}$
12. $\frac{8 \text{ h}}{7200 \text{ s}}$

Express these ratios in fractional form and simplify without using a calculator.

13. $\frac{3}{8}$ to $\frac{4}{9}$
14. $\frac{1}{4}$ inch to $\frac{1}{3}$ foot
15. 200¢ to $0.25
16. $\frac{7}{4}$ oz to $\frac{1}{32}$ lb

14-2 INVERSE, DECIMAL AND PERCENT RATIOS

An *inverse ratio* is the reciprocal of the given ratio. The inverse ratio of 2:1 is 1:2. The ratio $\frac{5}{3}$ has an inverse ratio of $\frac{3}{5}$.

Sometimes ratios are expressed as decimals. This is accomplished by dividing the first term by the second. The ratio $\frac{3}{4}$ becomes $\frac{0.75}{1}$ or simply 0.75. The ratio 25:8 in decimal form is 3.125:1 or simply 3.125.

A ratio can be expressed as a percent by first changing to decimal form. The decimal form is then multiplied by 100 to express it as a percent.

Example A Express the ratio $\frac{3}{5}$ as a percent.

Solution: $\frac{3}{5} = 0.6 = 60\%$

Example B Express these ratios as a simplified ratio in fractional form: (1) 0.44:1, (2) 62.5%

Solution: (1) $0.44:1 \rightarrow \frac{44}{100} = \frac{11}{25}$

(2) $62.5\% \rightarrow \frac{62.5}{100} = \frac{625}{1000} = \frac{5}{8}$

EXERCISE 14-2

(a) Express each of the following ratios in decimal form correct to three decimal places. (b) Give the inverse ratio of each in reduced fractional form.

1. $\frac{6}{16}$
2. 16:20
3. 8:2
4. $\frac{9}{48}$
5. 50:4
6. 5 to 25

Express each of the following ratios in percent form correct to three significant figures.

7. $\frac{7}{8}$
8. $\frac{8}{40}$
9. 1 to 16
10. $\frac{5}{32}$
11. $2\frac{3}{4}$ lb:11 oz
12. 845 mW:0.003 92 kW

Express each of the following decimals or percents in reduced fractional form.

13. 0.75
14. 0.375
15. 0.1875
16. 87.5%
17. 45%
18. 6.25%

Express each of the following in percent as indicated.

19. A 150-m span of copper power line increases to 150.18 m from winter to summer due to thermal expansion. What is the percent increase?
20. An electrician in a small shop purchases small motors for $35.00 wholesale and sells them for $49.00 retail. What is the percent markup?
21. The resistance of a motor field winding is 3.0524 Ω at room temperature (20°C). At operating temperature (70°C), the resistance is 3.6324 Ω. What is the percent increase in the resistance?
22. The voltage across a certain resistor is 110 V. At a later time, the voltage has changed to 120 V. What is the percent change?
23. General purpose solder for electrical use has three parts tin and two parts lead. What percent of each metal is in a sample of solder?
24. Ceramic insulators are made of 18 parts of one kind of clay and 55 parts of a second kind. What percent is each kind?
25. A pole transformer housing, Figure 14-2, contains 36.5 L of oil for heat transfer purposes. There is 19.3 L of air above the oil for thermal expansion. The transformer is submerged in the oil and occupies 15.8 L of space. What percent of the volume of the housing is occupied (a) by oil? (b) by air? (c) by the transformer?

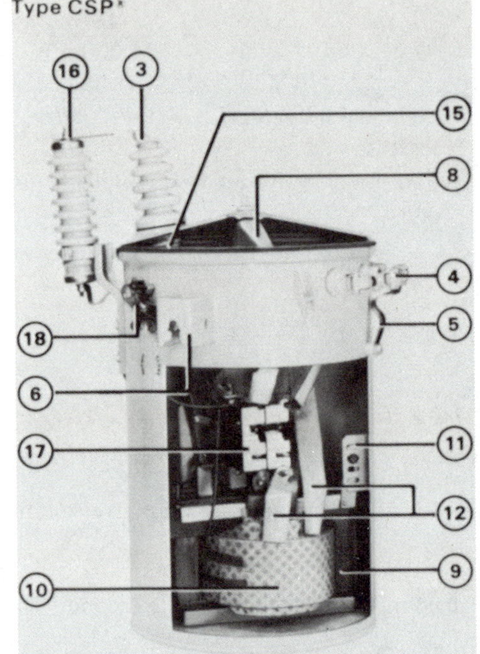

Figure 14-2 Cutaway of a pole transformer housing: (3) porcelain bushing, (4) low voltage porcelain bushing, (5) grounding strap, (6) hanger bracket, (8) self-venting cover, (9) core, (10) coil, (11) core/coil support bracket, (12) low-voltage leads, (15) protective link, (16) surge arrester, (17) secondary circuit breaker, (18) breaker reset handle and warning light. (Courtesy of Westinghouse Electric Corporation)

14-3 ELECTRICAL APPLICATIONS OF RATIOS

Ratios are used in a number of applications in the electrical industry. Many have special names. Some have been discussed previously and are reviewed briefly. Efficiency, mechanical advantage, speed ratios and percent change are just a few. Ratios of two sides of a right triangle are some of the most important ratios in electrical work. Right triangle applications are discussed in Chapter 16.

14-4 EFFICIENCY

Electric motors, generators, transformers and many other devices are rated in terms of efficiency, Topic 10-8. Less energy is used in a device if its efficiency is high. *Efficiency* is defined as the ratio of work output to work input. A more useful formula for electricians is

$$\mathbf{Eff} = \frac{\mathbf{P_o}}{\mathbf{P_i}}$$

where P_o is the power output and P_i is the power input.

Usually, efficiency is expressed in percent but not always. Unfortunately, the power output of electric motors is expressed in horsepower while the input is in watts. A conversion of units is, therefore, necessary (1 hp = 746 W). The efficiency of any device cannot exceed 100%.

Example A An electric motor is rated at $\frac{1}{4}$ hp. What is the efficiency to the nearest percent if the input is 220 W?

Solution: First, convert $\frac{1}{4}$ hp to watts

$$\frac{1}{4}\,\cancel{\text{hp}}\left(746\,\frac{\text{W}}{\cancel{\text{hp}}}\right) = 187\text{ W}$$

$$\text{Eff} = \frac{P_o}{P_i}$$

$$\text{Eff} = \frac{187\,\cancel{\text{W}}}{220\,\cancel{\text{W}}} = 0.85 = 85\%$$

Example B A pole transformer is rated at 98% efficiency. The input at a certain time is 8.52 kW. What is the power output of the transformer?

Solution: $P_o = \text{Eff} \times P_i$

$P_o = 0.98(8.52\text{ kW}) = 8.35\text{ kW}$

EXERCISE 14-3

Data, diagram, formula, substitute, check.

1. A DC car generator has an output power of 50 W. This load requires an input power of 77 W transmitted by the V belt. What is the efficiency of the generator?
2. A car alternator requires an input power of 63 W to produce an output of 50 W. (a) What is the efficiency of the alternator? (b) If the efficiency of a DC generator is 65%, what is the ratio of the generator's efficiency to that of the alternator. Express the answer as a percent.

3. A large three-phase electric motor has an efficiency of 95%. What is the power output in horsepower if the input is 19.6 kW?
4. The output power of lightbulbs is measured in lumens (1 lumen = 0.001 496 W). Efficiency of lamps, Topic 10-8, is always expressed in lumens per watt rather than percent. A 40-W tungsten lamp has an output of 480 lumens. What is the efficiency (a) in lumens per watt? (b) in percent?
5. What is the luminous efficiency of a 40-W fluorescent lamp if its output is 3000 lumens?
6. A pole transformer has an efficiency of 97.8%. If the output power at a certain time is 5000 W, what is the input power?

14-5 MECHANICAL ADVANTAGE AND SPEED RATIO

Many hydraulic and mechanical machines are used in the electrical and electromechanical industries. Some machines are described in terms of mechanical advantage. Rotating machines are described in terms of speed ratio.

Mechanical advantage (MA) of a machine is defined as the ratio of the output force (F_o) to the input force (F_i):

$$MA = \frac{F_o}{F_i}$$

Do not confuse mechanical advantage with efficiency. Mechanical advantage is a ratio of forces while efficiency is a ratio of powers or energies.

Example A The input force on the handle (a lever) of a hydraulic conduit bender, Figure 14-3, is 12 lb. The output is 182 lb. What is the mechanical advantage of the handle?

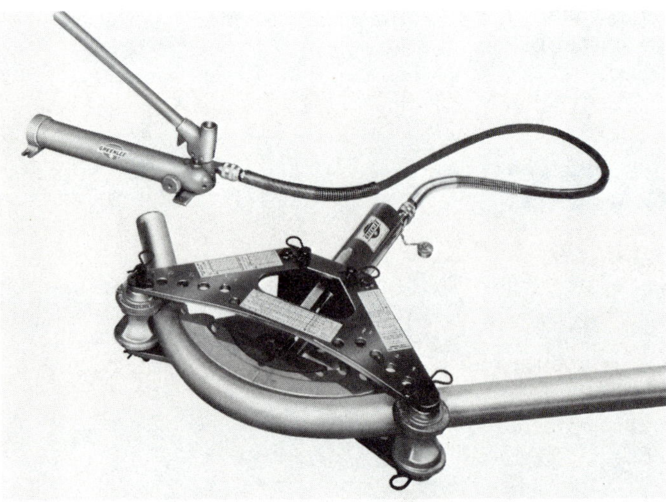

Figure 14-3 A typical lightweight hydraulic conduit bender (Courtesy of Greenlee Tool Division, Ex-Cell-O Corporation)

Solution: $MA = \frac{F_o}{F_i}$

$MA = \frac{182 \cancel{lb}}{12 \cancel{lb}} = 15.2$

Speed ratio (SR) is a shortened expression for *speed reduction ratio.* If the speed ratio is greater than one, the *driven* gear (output) rotates slower than the *driver* (input). It is defined as the ratio of the *angular speed* (ω_i) of the driver gear or pulley (input) to the angular speed (ω_o) of the driven gear or pulley (output). Angular speed is usually measured in revolutions per minute (r/min). The letter ω is the lowercase Greek letter *omega.*

$$\mathbf{SR} = \frac{\omega_i}{\omega_o}$$

Example B Find the speed reduction ratio if the driven pulley rotates at 500 r/min when the driver pulley is rotating at 1750 r/min.

Solution: $SR = \frac{\omega_i}{\omega_o}$

$SR = \frac{1750 \cancel{r/min}}{500 \cancel{r/min}} = 3.5$

Speed ratios of gears can also be determined by finding the inverse of the number-of-teeth ratio:

$$\mathbf{SR} = \frac{N_o}{N_i}$$

where N_o and N_i are the number of teeth on the driven and driver gears, respectively. The ratio of angular speeds is driver to driven but the ratio of the number of teeth is driven to driver. The teeth ratio must be the inverse since the smaller gear rotates faster. For pulleys, the ratio of diameters is used:

$$\mathbf{SR} = \frac{d_o}{d_i}$$

Figure 14-4 A two-gear system (Courtesy of Southeast Nebraska Community College)

Example C Gear A, Figure 14-4, has 45 teeth and gear B has 25. If gear A is the driver, what is the speed ratio?

Solution: $SR = \frac{N_o}{N_i}$

$SR = \frac{25}{45} = 0.56$

Example D The speed ratio in the gear system in Example C (0.56) is less than 1. This means the driven gear rotates faster than the driver. If gear A rotates at 1200 r/min, what is the angular speed of gear B?

Solution: $SR = \frac{\omega_i}{\omega_o}$

$$\omega_o = \frac{\omega_i}{SR}$$

$$\omega_o = \frac{1200 \text{ r/min}}{0.56} = 2140 \text{ r/min}$$

EXERCISE 14-4

Data, diagram, formula, substitute, check.

1. A hydraulic conduit bender consists of a handle and a hydraulic cylinder, Figure 14-3. An input force of 200 N (newtons) on the handle creates an output force of 24 300 N executed by the cylinders. What is the overall mechanical advantage?
2. A large pipe cutter for cutting electrical conduit pipe consists of three cutting wheels, Figure 14-5. Find the force on the cutting wheel if a 10-lb force is applied on the handle. The overall mechanical advantage is 240.

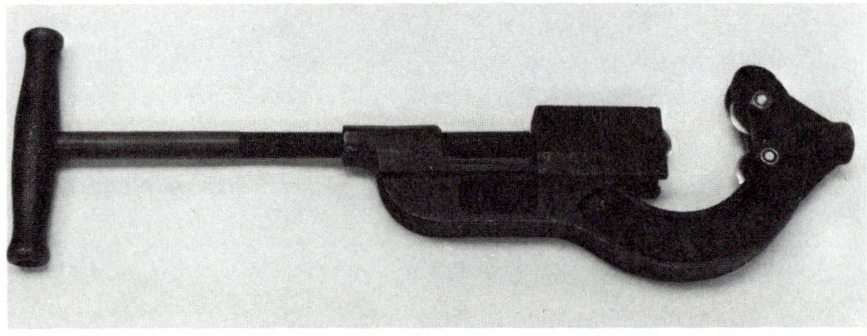

Figure 14-5 A typical conduit pipe cutter (Courtesy of Southeast Nebraska Community College)

Figure 14-6 A vacuum pump. The safety cover has been removed to show the pulley system. (Courtesy of Southeast Nebraska Community College)

3. A block and tackle is sometimes used by power companies as a hoist. What input force is required to lift a load of 5000 N if the mechanical advantage is 6?
4. A vacuum pump, Figure 14-6, is driven by a motor rotating at 1720 r/min. The pump rotates at 478 r/min. What is the speed ratio?

Use Figure 14-7 and the following information for Problems 5 and 6. Gear A is the driver and has 30 teeth. Gear B has 45 teeth and Gear C has 25 teeth.

5. What is the speed ratio (a) between gears A and B? (b) between B and C?
6. Gear A rotates at 800 r/min. What is the angular speed (a) of gear B? (b) of gear C? (c) What is the overall speed ratio?

Figure 14-7 A three-gear system (Courtesy of Southeast Nebraska Community College)

Use Figure 14-8 for Problems 7 and 8.

7. (a) Find the speed ratio of the alternator-crankshaft pair. (b) If the crankshaft is rotating at 2800 r/min, find the angular speed of the alternator.
8. (a) Determine the speed ratio of the fan-crankshaft pair. (b) What engine (crankshaft) speed is required to rotate the fan at 3430 r/min?
9. A gear box is driven by an electric motor rotating at 1720 r/min. If the speed reduction ratio is 4.75:1, find the output shaft speed.
10. A pulley is driven by a 4.5-inch drive pulley rotating at 425 r/min. (a) What is the speed ratio if the driven pulley rotates at 1275 r/min? (b) What is the diameter of the driven pulley?

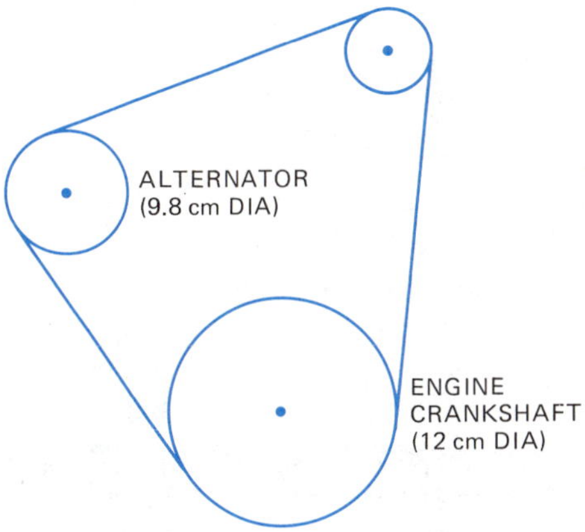

FAN (7.6 cm DIA)
ALTERNATOR (9.8 cm DIA)
ENGINE CRANKSHAFT (12 cm DIA)

Figure 14-8 A pulley on the engine crankshaft drives the fan and the alternator.

PROPORTION

When two ratios are equal to each other, the resulting equation is called a *proportion*. A proportion such as $\frac{3}{4} = \frac{21}{28}$ in fractional form can be written in colon form: 3:4::21:28. In either case, the proportion is read, "3 is to 4 as 21 is to 28." Usually, one of the four terms is an unknown quantity and the problem is to solve for the unknown.

14-6 CROSS-MULTIPLICATION

If the proportion is expressed in colon form, rewrite it in fractional form. It may be helpful, at this point, to cross-multiply. *Cross-multiplying* means to set the product of the numerator on the left and the denominator on the right equal to the product of the numerator on the right and the denominator on the left. Divide both sides by the coefficient of the unknown to solve.

CROSS-MULTIPLYING

- Multiply the two known numbers of one cross pair.
- Divide by the known number of the other pair.

This rule will work every time, regardless of where the unknown is located within the proportion.

In colon form, the *law of proportionality* can be stated as: "The product of the means equals the product of the extremes." The means are the two internal numbers. The means of the proportion 5:8::y:96 are 8 and y. The extremes are the two outer numbers: 5 and 96. Thus

$$8y = 5 \times 96$$

This result is the same as that found by cross-multiplying in the fractional form of the proportion.

Example A Solve the proportion: 5:8::y:96.

Solution: $\frac{5}{8} = \frac{y}{96}$ (Rewrite)

$8y = 5 \times 96$ (Cross-multiply)

$y = \frac{5 \times 96}{8}$ (Solve)

$y = 60$

Example B Solve the proportion: $\frac{624}{N} = \frac{894}{2862}$

Solution: $624 \times 2862 = 894N$ (Cross-multiply)

$N = \frac{624 \times 2862}{894}$ (Solve)

$N = 20\bar{0}0$ (Answer rounded)

EXERCISE 14-5

Solve for the unknown in each of the following proportions using cross-multiplication.

1. $\frac{E}{12} = \frac{26}{156}$
2. 15:I::225:105
3. $\frac{5}{7} = \frac{60}{R}$
4. 440:66::Z:110
5. 9 is to 25 as 35 is to X
6. 0.5:X::0.44:3.8
7. $\frac{P}{17} = \frac{14}{119}$
8. I:40::4.80:$\frac{1}{4}$
9. 120 is to E as 5 is to 40
10. $\frac{54}{25} = \frac{C}{225}$

14-7 DIRECT AND INVERSE PROPORTION

Proportions can be direct or inverse. In a direct proportion, a change in one quantity produces a similar change in another quantity. An increase in one causes an increase in the other. A decrease in the first causes a decrease in the second. Direct proportions have the following form:

$$\frac{A_1}{A_2} = \frac{B_1}{B_2}$$

where A_1 and B_1 are values of quantities related to one object or situation or are beginning values. A_2 and B_2 relate to the second object or situation or are the final values.

Example A Two wires of the same diameter and made of the same material have resistances proportional to their lengths. One wire is 100 ft long and has a resistance of 0.2525 Ω. The second wire is 750 ft long. What is the resistance of the second wire?

Solution:

$$\frac{R_1}{R_2} = \frac{L_1}{L_2}$$

$$\frac{0.2525\ \Omega}{R_2} = \frac{100\ \cancel{ft}}{750\ \cancel{ft}}$$

$$100R_2 = 750 \times 0.2525\ \Omega$$

$$R_2 = \frac{750 \times 0.2525\ \Omega}{100}$$

$$R_2 = 1.8938\ \Omega$$

Sometimes, one quantity increases as the other decreases. When the first quantity decreases, the other increases. These variations are called *inverse* or *indirect proportions.* Inverse proportions have the following form:

$$\frac{A_1}{A_2} = \frac{B_2}{B_1}$$

Notice the subscripts are inverted in the ratio on the right.

Example B In Figure 14-9, gear A, the driver gear, has 25 teeth and rotates at 1200 r/min. Gear B is the driven gear and has 40 teeth. What is its angular speed?

Solution: Refer to the speed ratio formula in Topic 14-5. The two ratios are both equal to the speed ratio so they are equal to each other. Notice the subscripts. This is an example of an inverse proportion.

$$\frac{\omega_i}{\omega_o} = \frac{N_o}{N_i}$$

$$\frac{1200\ r/min}{\omega_o} = \frac{40\ \cancel{teeth}}{25\ \cancel{teeth}}$$

$$\omega_o = \frac{25 \times 1200\ r/min}{40}$$

$$\omega_o = 750\ r/min$$

Figure 14-9 A two-gear system (Courtesy of Southeast Nebraska Community College)

EXERCISE 14-6

Solve each of the following by setting up a proportion.

1. A motor repair shop uses 450 m of copper wire every 5 1/2 days. How many kilometres of wire are used in 137.5 days?
2. Double 0 gage copper wire has a mass of 0.4747 kg for each metre of length. What is the length of 1000 kg of this wire?

3. A gear has a diameter of 6 cm and has 22 teeth. How many teeth does a gear with a 15-cm diameter have? The two gears are in mesh.
4. The driver gear is 6 cm and is rotating at 600 r/min. Find the angular speed of a 15-cm gear if the two gears are in mesh.
5. In a factory, 250 lamps use 6250 kW•h of energy each month. An efficiency study suggests the factory eliminate 75 lamps. What is the new monthly energy usage?
6. A capacitor has 0.000 88 C of charge when a voltage of 120 V is impressed on it. What is the charge in coulombs when 550 V are applied to the same capacitor? Voltage (E) and charge (Q) are directly proportional.
7. A charge of 150 C passes through a resistor in 3 min. How much charge in coulombs will pass through the same resistor in 1 h? Assume the current is constant.
8. A vacuum pump is driven by a motor rotating at 1750 r/min. If the motor pulley has a 5-cm diameter and the pump pulley an 18-cm diameter, what is the pump's angular speed?

14-8 ELECTRICAL APPLICATIONS OF PROPORTION

The proportion has many applications in electrical fields. The Wheatstone bridge, transformers and similar triangles involve proportions. Many electrical quantities such as voltage, current and resistance also form proportion problems. These will be discussed one at a time.

14-9 TRANSFORMERS

The transformer is an ac (alternating current) device. It is studied briefly in this chapter as an application of proportion. A transformer consists of two coils of wire wrapped around a common iron yoke, Figure 14-10. The first coil is called the *primary* and has N_p turns of wire. The second coil is the *secondary* and has N_s turns.

A magnetic field is always produced around a current-carrying wire. If the wire is wrapped into a coil, the magnetic field passes through its center. One end of the coil is a north pole, the other a south pole. A laminated

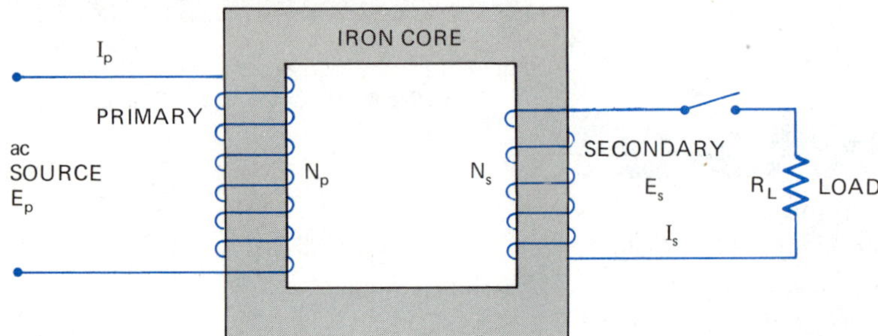

Figure 14-10 Transformer showing primary and secondary coils

iron core placed in the coil makes the field very strong, Figure 14-11. On the other hand, whenever a magnetic field passing through a coil is increasing or decreasing in strength, it will induce a current in the wire. A constant-strength magnetic field will not induce such electric currents.

If the primary coil of a transformer carries an alternating current, the induced magnetic field in the iron core will also alternate. It is continuously increasing and decreasing and reversing direction just as the current does. Since the changing magnetic field passes through the secondary coil, an alternating current will be produced in a closed secondary circuit.

The secondary coil acts as a new source of emf and is completely insulated from the primary, Figure 14-12. The primary-secondary voltage ratio is in direct proportion to the ratio of the number of turns:

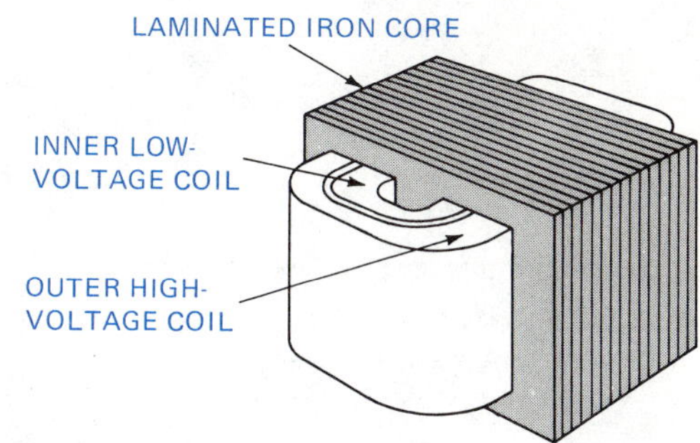

Figure 14-11 Transformer showing laminated iron core

$$\frac{E_P}{E_S} = \frac{N_P}{N_S}$$

where E_P is the primary voltage and E_S is that of the secondary.

Transformers provide a means of changing voltage in a simple and efficient way. The voltage can be increased or decreased. If the voltage across the secondary is greater than that of the primary, the transformer is called a *step-up transformer.* Voltage is decreased in a *step-down transformer.* A one-to-one transformer has the same voltage on the secondary as the primary. These are called *isolation transformers.* They isolate the secondary circuit from the primary source.

Step-up transformers are used to raise generated voltage (2400 V) for cross-country transmission (120 000 V). When the voltage is increased, the current is decreased by a like factor. Power losses ($P = I^2 R$) are reduced in high voltage transmission. Step-down transformers, Figure 14-13, reduce cross-country voltage for power lines within cities (2400 V). A further step-down (pole mounted) transformer is used before the power is delivered to

Figure 14-12 Various small transformers (Courtesy of Southeast Nebraska Community College)

Figure 14-13 A substation transformer

the home (120 V). Even further step-downs are used in the home for thermostat (24 V) and doorbell circuits (12 V). Of course, other voltages are possible and are often used at all stages of distribution.

Small transformers have an efficiency well over 90%. Large ones may exceed 99%. The power output is approximately equal to the input ($E_sI_s = \text{Eff} \cdot E_pI_p$). The current ratio is approximately inversely proportional to the turns ratio.

$$\frac{I_p}{I_s} = \frac{N_s}{N_p} \text{ (approximately)}$$

Example A The secondary coil of a transformer has 100 turns and the primary has 2000 turns. The effective output current is 5.0 A at 120 V. Find the effective input voltage and current. Assume an efficiency of 100%.

Solution: Data: $N_s = 100$, $N_p = 2000$, $I_s = 5.0$ A, $E_s = 120$ V, $I_p = ?$, $E_p = ?$

Formula: $$\frac{E_p}{E_s} = \frac{N_p}{N_s}$$

$$E_p = \frac{E_sN_p}{N_s}$$

Substitute: $$E_p = \frac{(120 \text{ V})(2000)}{100} = 2400 \text{ V}$$

Formula: $\frac{I_p}{I_s} = \frac{N_s}{N_p}$

$I_p = \frac{I_s N_s}{N_p}$

Substitute: $I_p = \frac{(5.0 \text{ A})(100)}{2000} = 250 \text{ mA}$

EXERCISE 14-7

Data, diagram, formula, substitute, check.

1. A pole transformer, Figure 14-14, is designed to reduce city power line voltage from 2400 V to 120 V. How many turns does the primary have if the secondary has 85 turns?

Figure 14-14 Various pole transformers

2. A step-up transformer changes 20 kV from the generator to 345 kV for cross-country transmission. If the secondary has 1120 turns, find the number of turns in the primary.
3. A transformer has 1000 turns on the primary winding and 50 turns on the secondary. (a) What is the output voltage if the input is 4400 V? (b) What is the primary current if the secondary current is 9.09 A. Assume 100% efficiency.
4. Household voltage (120 V) is stepped down to 24 V to operate the furnace-thermostat circuit. (a) If there are 500 turns on the primary coil, how many are on the secondary coil? (b) If the secondary circuit has a current of 0.25 A, what is the primary current? Assume 100% efficiency.
5. A doorbell transformer has a primary coil of 1125 turns, and a secondary of 150 turns. What voltage operates the doorbell if the primary connects to a 120-V line?

6. A transformer used for an electric toy delivers 12 V when connected to a 120-V source. Find the number of turns in the secondary if there are 400 in the primary.
7. Find the current in the primary of a pole transformer which is designed to reduce city power line voltage from 2400 V to 120 V. The secondary current is 50 A.

14-10 SERIES AND PARALLEL CIRCUITS

Electrical problems are usually solved by using Ohm's Law. However, many simple ones can be solved by using proportions. In the series circuit shown in Figure 14-15, the current through each resistor is the same. The ratio of the voltage drops must equal the ratio of the resistances:

$$\frac{E_1}{E_2} = \frac{R_1}{R_2}$$

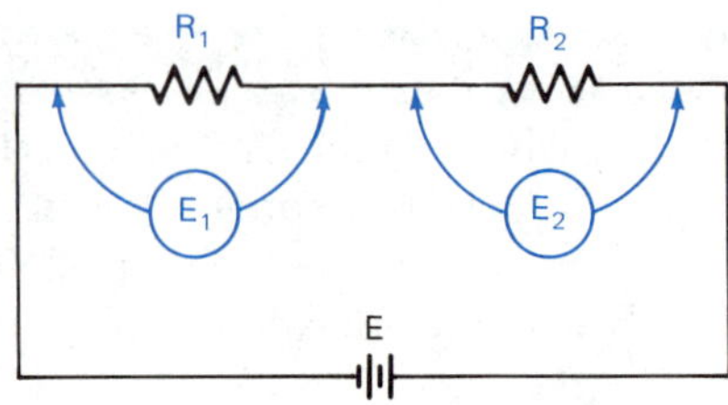

Figure 14-15 In a series circuit, the voltage drops across the loads are directly proportional to the resistance.

where E_2 and R_2 can be either the voltage drop and resistance of second loads or the overall voltage drop and the overall resistance.

In series circuits, the voltage drops are proportional to the resistances. A circuit such as that shown in Figure 14-15 is an example of a simple *voltage divider*. An example of inverse proportion occurs in resistors connected in parallel. Since the voltage is the same across each resistor, it is the current that is divided, Figure 14-16. The current is inversely proportional to the resistance:

$$\frac{I_1}{I_2} = \frac{R_2}{R_1}$$

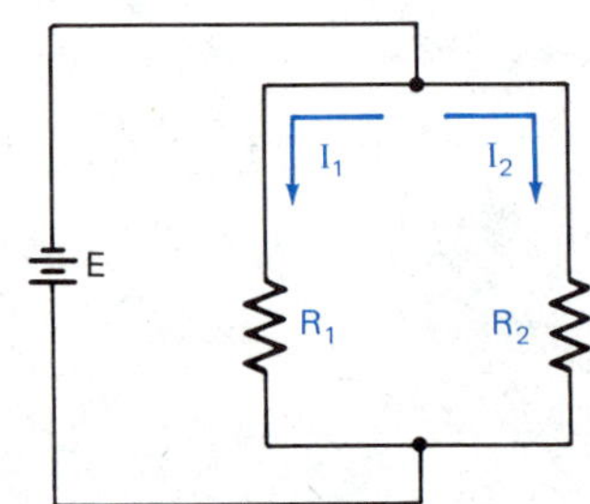

Figure 14-16 Current through loads connected in parallel is inversely proportional to the resistance.

I_2 and R_2 can be either the current and resistance of the second branch or the overall current and overall resistance, respectively.

Example A In the series circuit of Figure 14-15, the voltage drop across resistor R_1 is 42 V. Across R_2 it is 78 V. If R_1 is 3400 Ω, find R_2.

Solution: $\frac{42\,\text{V}}{78\,\text{V}} = \frac{3400\ \Omega}{R_2}$

$$R_2 = \frac{78 \times 3400\ \Omega}{42}$$

$$R_2 = 6300\ \Omega$$

(Answer rounded)

Example B In the parallel circuit of Figure 14-16, R_1 is 750 Ω. R_2 is 420 Ω. If the current through R_1 is 32 mA, (1) find I_2. (2) Find the total resistance.

Solution: (1) $\frac{32\ \text{mA}}{I_2} = \frac{420\ \Omega}{750\ \Omega}$ (Inverse)

$$I_2 = \frac{32\ \text{mA} \times 750}{420}$$

$$I_2 = 57\ \text{mA}$$

(2) $I_1 + I_2 = 89$ mA

$$\frac{32\ \text{mA}}{89\ \text{mA}} = \frac{R_T}{750\ \Omega} \quad \text{(Inverse)}$$

$$R_T = \frac{32 \times 750\ \Omega}{89}$$

$$R_T = 269\ \Omega$$

Note: The answer to part (2) can be checked using the formula:

$$R_T = \frac{R_1 R_2}{(R_1 + R_2)}$$

EXERCISE 14-8

Set up proportions to solve these problems. Do not use Ohm's Law. Data, diagram, formula, substitute, check.

1. Find the resistance R_2 in a series circuit, Figure 14-17. R_1 is 420 Ω. The voltage drop across R_2 is 7 V. The source voltage is 24 V.

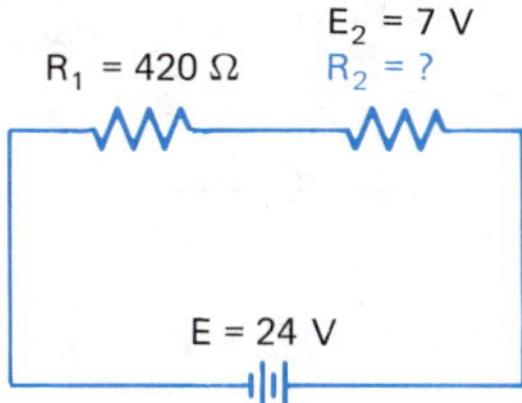

Figure 14-17

2. Find E_1 and E_2 in Figure 14-17 if R_1 is changed to 310 Ω.
3. Find E_1, Figure 14-15, if E_2 is 14 V, R_1 = 650 Ω and R_2 = 870 Ω.
4. Find the current I_2 in the parallel circuit, Figure 14-18. R_1 is 480 Ω, R_2 is 1140 Ω and I_1 is 229 mA.

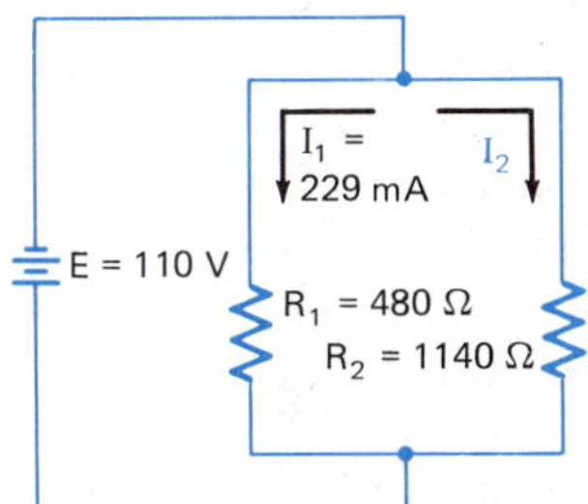

Figure 14-18

5. Find the total resistance of the circuit in Figure 14-18. Solve by using proportions. Check by using regular parallel circuit formulas.
6. Find R_2, Figure 14-16, if R_1 = 2400 Ω, I_1 = 45 mA and I_2 = 75 mA.

14-11 AMMETERS AND VOLTMETERS (DC)

The basic movement used in (DC) ammeters, voltmeters and other measuring instruments is the stationary permanent-magnet moving-coil meter.

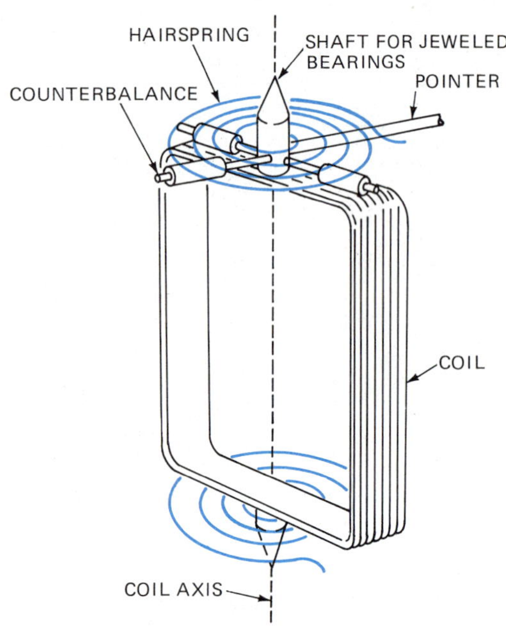

Figure 14-19 Meter coil

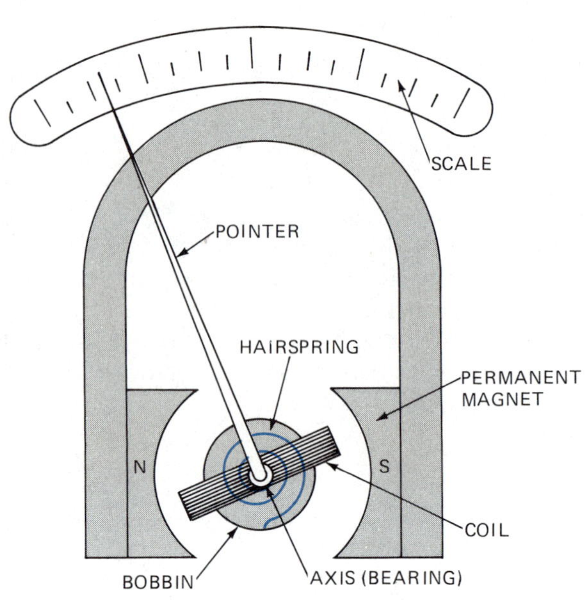

Figure 14-20 Microammeter or galvanometer with a D'Arsonval movement

The coil, Figure 14-19, is mounted on a shaft fitted between jeweled bearings. A lightweight pointer is attached to and turns with the coil. This type of movement is called the *D'Arsonval movement* since it was first used by D'Arsonval in France.

Current in the coil produces a magnetic field. Thus, the coil acts as an electromagnet having a north and south pole. The poles reverse when the current reverses. The magnet thus formed reacts with a permanent magnet, Figure 14-20. The coil rotates against the restraining force of the hairsprings. The pointer indicates the current flow on a scale. An aluminum bobbin mounted within the coil dampens the movement and prevents oscillation.

The D'Arsonval movement along with the pointer and scale is referred to as a *microammeter* or a *galvanometer.* The small size of the coil wire limits the amount of current that can pass through it. The maximum amount for full-scale deflection is determined by the stiffness of the hairspring, the strength of the permanent magnets and the number of turns in the coil. Only very small currents (a few microamperes) may be used. Any larger current will drive the pointer off the scale and damage the movement.

Ammeters. To measure larger currents, a shunt is used with the meter. A *shunt* is a heavy low-resistance conductor placed in parallel with the coil. The shunt carries practically all of the load current. The required resistance of the shunt for a desired current range can be calculated. The resistance of the coil and the current for full-scale deflection must be known.

Example A Find the resistance (R_s) of a shunt, Figure 14-21, that will change a 75-Ω (R_m) microammeter to an ammeter reading 1.5 A full-scale. The coil deflects full-scale when it carries a current (I_m) of 50 μA.

Solution: Since the voltage is the same across the two parallel branches (coil and shunt), the current and resistance form an inverse proportion (current divider):

$$\frac{I_m}{I_s} = \frac{R_s}{R_m}$$

The 50 μA of meter current (I_m) is very small compared to the total current of 1.5 A. It is assumed the shunt current (I_s) is the same as the total.

$$R_s = \frac{R_m I_m}{I_s}$$

$$R_s = \frac{(50 \times 10^{-6}\ \text{A})(75\ \Omega)}{1.5\ \text{A}} = 2.5\ \text{m}\Omega$$

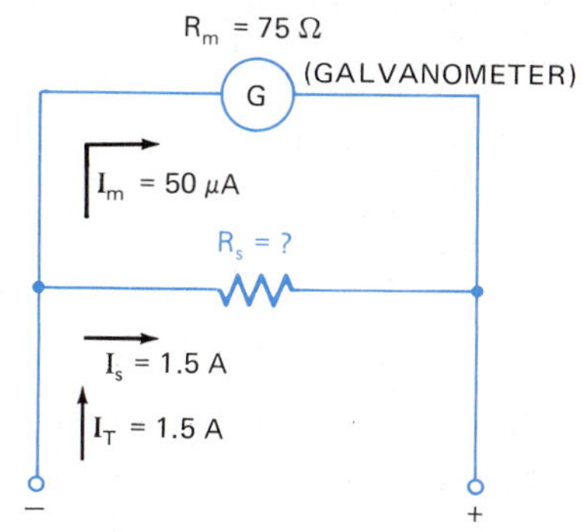

Figure 14-21 An ammeter consists of a shunt in parallel with the galvanometer. The shunt carries almost the entire circuit current. Assume that $I_T = I_s$.

Example B Find the full-scale reading of the microammeter in the previous example if the shunt is replaced with one of 375 μΩ.

Solution:

$$\frac{I_m}{I_s} = \frac{R_s}{R_m}$$

$$I_s = \frac{I_m R_s}{R_m}$$

$$I_s = \frac{(50 \times 10^{-6}\ \text{A})(75\ \Omega)}{0.000\,375\ \Omega} = 10\ \text{A}$$

Always observe the correct polarity before connecting an ammeter. If reversed, the coil deflects backwards and may be damaged. In addition, ammeters must always be connected in series with the circuit load. Never connect the ammeter in parallel with the load. To do so will form a short-circuit through the ammeter shunt and the resulting higher voltage across the coil will damage the movement.

Voltmeters. The microammeter can also be converted to a voltmeter. Voltmeters are connected in parallel with the circuit load. Therefore, the full voltage of the load is applied to the meter. To limit the current through the meter coil, a high resistance is placed in series with it.

The current through the series resistor and the coil is directly proportional to the applied voltage. For this reason, the meter scale can be calibrated directly in volts for a fixed series resistance. The resistance required for the desired range can be calculated. A simple direct proportion (voltage divider) is set up with the known resistance (R_m) of the coil and the voltage drop (E_m) required to deflect the needle full-scale.

Example C Find the series resistance (R_s) required to convert the microammeter of Example A to a voltmeter that reads 150 mV (E_T) full-scale. The coil resistance is 75 Ω and 50 μA deflects the needle full-scale.

Solution: Find the coil voltage required for full-scale deflection using Ohm's Law:

$$E_m = I_m R_m$$
$$E_m = (50 \times 10^{-6}\ \text{A})(75\ \Omega) = 3.75\ \text{mV}$$

The total resistance of the meter can now be determined:

$$\frac{E_T}{E_m} = \frac{R_T}{R_m}$$
$$R_T = \frac{(0.15\ \cancel{\text{V}})(75\ \Omega)}{0.003\ 75\ \cancel{\text{V}}} = 3000\ \Omega$$

The series resistance is

$$R_s = R_T - R_m$$
$$R_s = 3000\ \Omega - 75\ \Omega \approx 2900\ \Omega \qquad \text{(Rounded)}$$

If the desired range is larger, the total resistance and the series resistance can be assumed to be the same. For the previous example, if the range is to be 150 V, the required series resistance is 3 MΩ. In comparison, the 75-Ω coil resistance is negligibly small. Multiple-range ammeters and voltmeters, Figure 14-22, are discussed in Problems 5–10 in the following exercise.

Figure 14-22 A multimeter (Courtesy of Triplett Corporation)

EXERCISE 14-9

Data, diagram, formula, substitute, check.

1. A galvanometer has a resistance of 80 Ω and requires 150 μA to produce full-scale deflection. What series resistance will convert the galvanometer into a voltmeter reading 10 V full-scale?
2. What shunt resistance will convert the galvanometer in Problem 1 into an ammeter reading 5 A full-scale?
3. A microammeter has a coil resistance of 110 Ω. A current of 75 μA will deflect the needle full-scale. What shunt resistance must be inserted to convert the microammeter into an ammeter that measures 150 A full-scale?
4. If the microammeter in Problem 3 is converted to a voltmeter reading 300 V full-scale, what series resistance is required?
5. Figure 14-23 shows a triple-scale voltmeter. Switch S converts from one scale to another. If the switch is in position A, the full-scale reading is 1.5 V. In position B, it is 15 V. Find R_1 and R_2.
6. If the voltmeter in Problem 5 is set in position C, it connects R_3 in series with the coil. If R_3 = 3.75 MΩ, what is the full-scale reading?
7. A double-scale ammeter is shown in Figure 14-24(A). Switch S converts from one scale to the other. In position A, the full-scale reading is 100 mA. Find the total shunt resistance ($R_1 + R_2$).

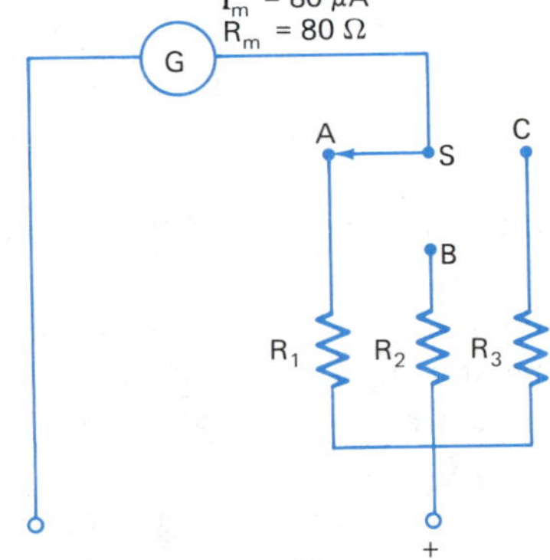

Figure 14-23

8. Suppose the switch on the ammeter shown in Figure 14-24(A) is moved to position B. Resistance R_2 becomes a part of the coil resistance. If the full-scale current is 100 mA or more, R_2 is negligible. Resistor R_1 becomes the shunt. (a) Find the value of R_1 if the full-scale current is 1 A. (b) Find the value of R_2 if the total shunt resistance is 0.064 Ω.
9. An alternate shunt connection for multiple-range ammeters is shown in Figure 14-24(B). (a) What would happen to the current during switching when contact is broken with point A and before contact is made with point B? Remember, the ammeter is connected in series with the load. (b) Is this a good arrangement for switching to a different scale in live circuits?
10. (a) Answer Problem 9 in regard to the multiple-range ammeter in Figure 14-24(A). (b) Answer Problem 9 in regard to the voltmeter in Figure 14-23.

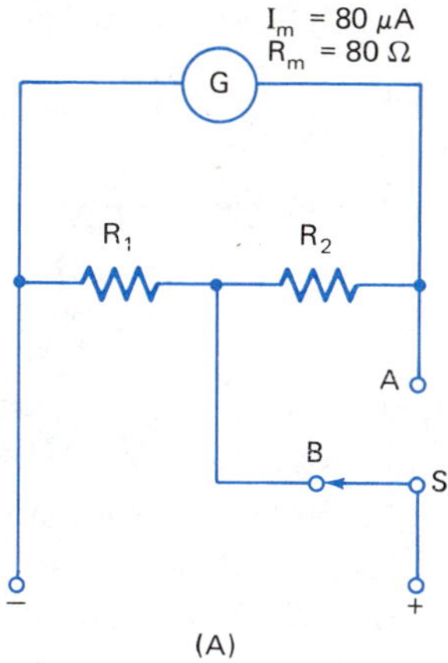

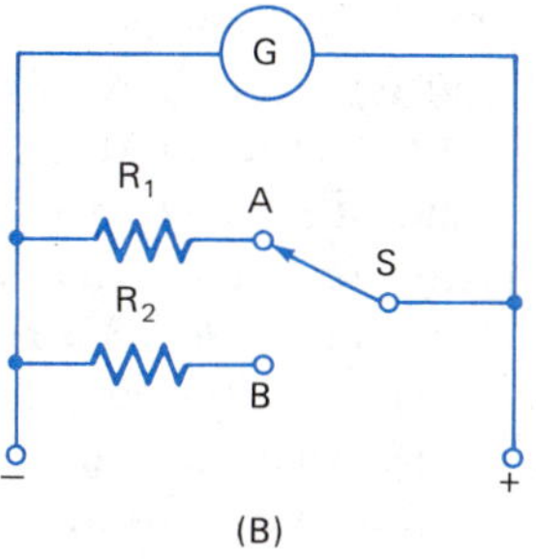

Figure 14-24

14-12 THE WHEATSTONE BRIDGE

The Wheatstone bridge is a circuit as shown in Figure 14-25. It is used for measuring resistance when great precision is required. Three of the four resistors have known values. Resistors R_2 and R_3 are adjustable in ratios of 1:1, 1:10, 1:100, etc. Standard resistors are correct to four significant figures. A third resistor R_1 is adjustable in small steps. The fourth resistor R_X is the unknown. The unknown is measured by adjusting R_1 until the galvanometer (G) reads zero. This balances the bridge.

In a balanced bridge, the voltage drop across R_1 and R_X must be exactly the same. Otherwise, current would flow through the galvanometer. This is also true across R_2 and R_3. These statements form the proportion:

$$\frac{E_1}{E_2} = \frac{E_X}{E_3}$$

Voltage drops E_1 and E_2 are on branch A of the circuit. E_3 and E_X are on branch B. Therefore:

$$\frac{I_A R_1}{I_A R_2} = \frac{I_B R_X}{I_B R_3}$$

Since currents cancel on both sides:

$$\mathbf{\frac{R_1}{R_2} = \frac{R_X}{R_3}}$$

The four resistances form a simple proportion. This happens only when the bridge is balanced.

One should not memorize this equation since different notation may be used. Learn to set up the proportion as follows: set the ratio of resistance on top, Figure 14-25, left to right equal to those on the bottom, left to right. An alternate way is to set the ratio of resistance on the left, top to bottom equal to those on the right, top to bottom.

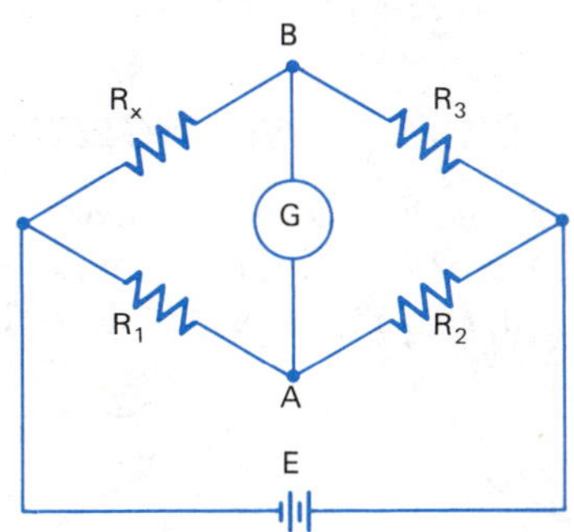

Figure 14-25 Circuit diagram of a Wheatstone bridge

Example A In the circuit of Figure 14-25, R_1 is 10 Ω, R_2 is 1000 Ω and R_3 is 18.1 Ω. What is the unknown resistance?

Solution:

$$\frac{R_1}{R_2} = \frac{R_X}{R_3}$$

$$\frac{10\ \Omega}{1000\ \Omega} = \frac{R_X}{18.1\ \Omega}$$

$$R_X = \frac{10 \times 18.1\ \Omega}{1000}$$

$$R_X = 0.181\ \Omega$$

EXERCISE 14-10

1. In a balanced Wheatstone bridge (Figure 14-25), R_1 is 0.01 Ω, R_2 is 10 Ω and R_3 is 79.3 Ω. What is the unknown resistance?
2. In a Wheatstone bridge, R_1 is 0.1 Ω, R_3 is 46.6 Ω and the resistance R_X was found to be 4.66 Ω. What was the setting of R_2 when the bridge was balanced?
3. The field winding of an electric motor at room temperature (20°C) is placed in a Wheatstone bridge, branch R_X. R_1 is set at 10 Ω and R_2 at 10 000 Ω. When the bridge balances, R_3 has a value of 3052 Ω. What is the resistance of the motor winding?
4. After operating under load for several hours, the motor resistance in Problem 3 is remeasured. R_1 is set at 0.01 Ω and R_2 at 10 Ω. Find the new resistance if R_3 is set at 3632 Ω.
5. Some Wheatstone bridges use a resistance wire of uniform cross section and uniform material. The wire (AC) is shown in Figure 14-26. The galvanometer is connected at point B by a slider mechanism. At the point where the balance is achieved:

$$\frac{L}{100 - L} = \frac{R_3}{R_X}$$

This proportion is correct since the resistance is proportional to the length (L). If R_3 is 44.9 Ω and the bridge balances when L is 28.9 cm, find R_X.

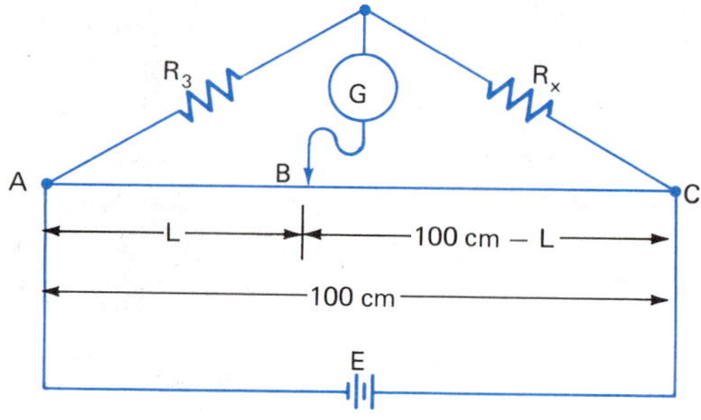

Figure 14-26 Circuit diagram of a slide-wire Wheatstone bridge

14-13 SIMILAR TRIANGLES

A triangle has six parts: three sides and three angles. Two triangles are said to be *similar triangles* if any two corresponding angles of both triangles are equal. To illustrate this definition, refer to Figure 14-27. The two

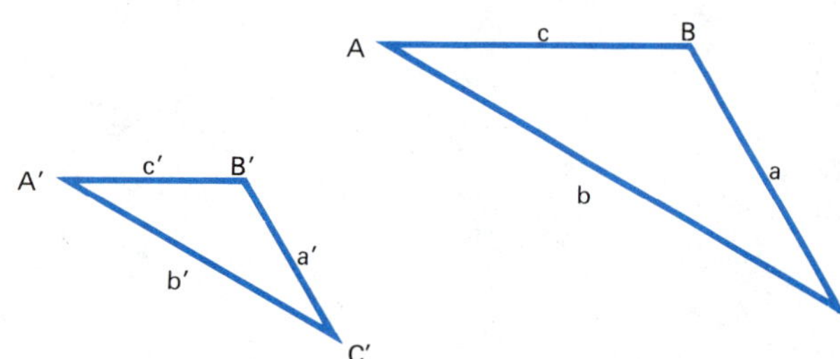

Figure 14-27 Similar triangles

triangles are ΔABC and $\Delta A'B'C'$. Angles A and A', B and B' and C and C' are corresponding angles. In these triangles:

$$\text{Angle A} = \text{Angle A}'$$
$$\text{Angle B} = \text{Angle B}'$$
$$\text{Angle C} = \text{Angle C}'$$

The two triangles are similar even though they are not the same size. The test for similarity is that corresponding angles are equal.

One triangle can be very small (postage stamp size) and the other can have sides that are many miles long. The triangles are similar if any two of their corresponding angles are equal. It should be noted that if two corresponding angles are equal, the third pair will also be equal.

In similar triangles: **a ratio of any two sides of a triangle forms a proportion with the corresponding ratio of any similar triangle.** In Figure 14-27:

$$\frac{a}{b} = \frac{a'}{b'}$$
$$\frac{a}{c} = \frac{a'}{c'}$$
$$\frac{b}{c} = \frac{b'}{c'}$$

Example A Triangle ABC in Figure 14-27 has sides with the following lengths: a = 11 cm, b = 24 cm and c = 15 cm. If side c' of triangle A'B'C' is 39 cm long, how long are sides a' and b'?

Solution: Formula: $\frac{a}{c} = \frac{a'}{c'}$

Substitute: $\frac{11 \text{ cm}}{15 \text{ cm}} = \frac{a'}{39 \text{ cm}}$

$$a' = \frac{11 \times 39 \text{ cm}}{15} = 28.6 \text{ cm}$$

Formula: $\frac{b}{c} = \frac{b'}{c'}$

Substitute: $\frac{24 \text{ cm}}{15 \text{ cm}} = \frac{b'}{39 \text{ cm}}$

$$b' = \frac{24 \times 39 \text{ cm}}{15} = 62.4 \text{ cm}$$

Check: Side b' can also be calculated using the value found for a':

Formula: $\frac{a}{b} = \frac{a'}{b'}$

Substitute: $\frac{11 \text{ cm}}{24 \text{ cm}} = \frac{28.6 \text{ cm}}{b'}$

$$b' = \frac{24 \times 28.6 \text{ cm}}{11} = 62.4 \text{ cm}$$

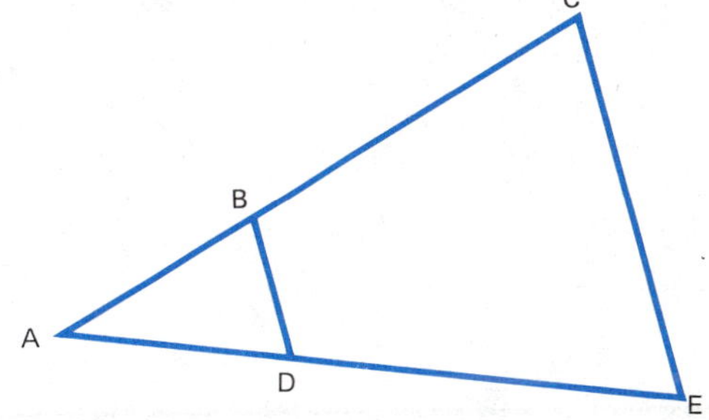

Figure 14-28 Triangles ABD and ACE are similar if lines BD and CE are parallel.

Two triangles such as those formed as shown in Figure 14-28 are similar if line BD and CE are parallel. Triangle ABD is similar to triangle ACE.

EXERCISE 14-11

1. Triangles ABC and XYZ, Figure 14-29, are similar. Set up proportions to find the lengths XZ and YZ.

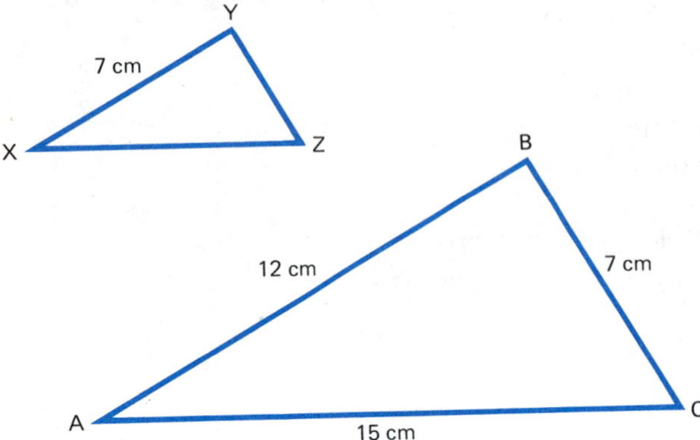

Figure 14-29

2. Triangles ABC and XYZ, Figure 14-30, are similar. Find the lengths XY and XZ by proportion.

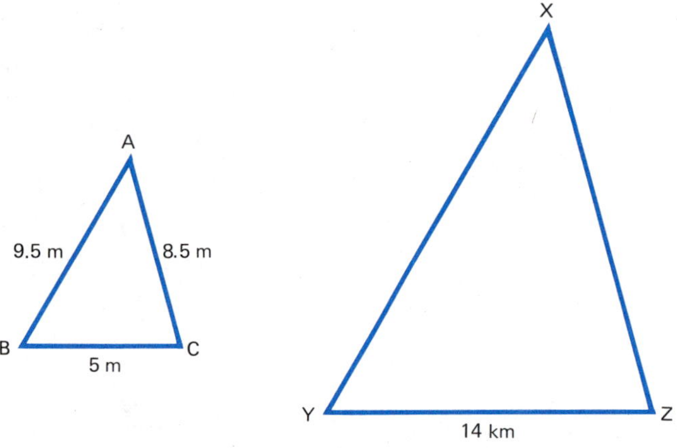

Figure 14-30

3. A pole of unknown height (CB in Figure 14-31) casts a 22-m shadow (AB). A metre stick (EF) casts a 125-cm shadow (DE). What is the height of the pole?

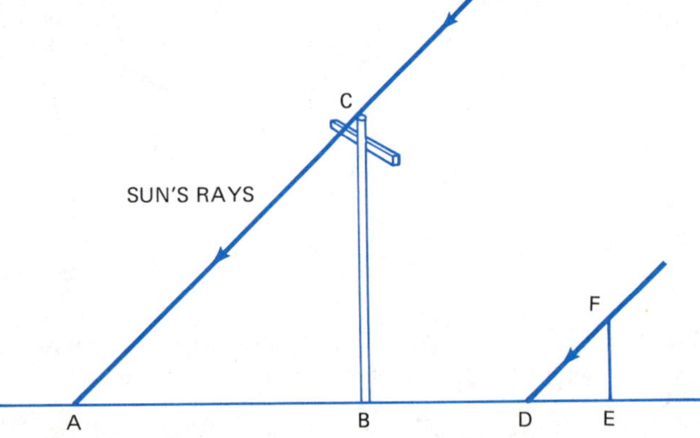

Figure 14-31

4. Poles are placed at points B and C on each side of a river. Refer to Figure 14-32. The span (BC) must be determined accurately so the correct wire size and sag can be installed. Surveying equipment is used to locate points A, D and E so that lines BD and CE are parallel. What is the span if AB = 240 m, AD = 190 m and DE = 175 m?

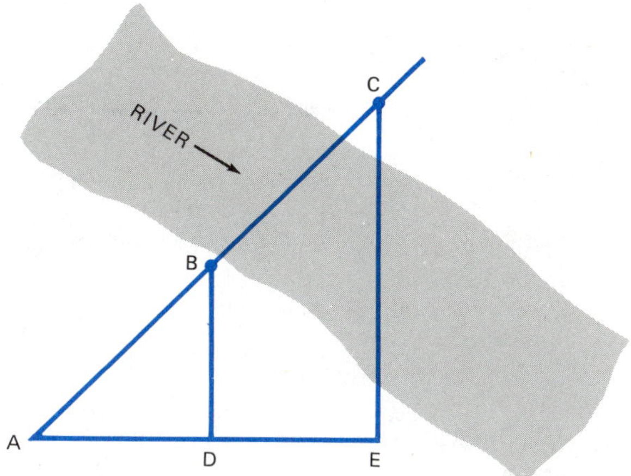

Figure 14-32

5. A pole is set and guyed as shown in Figure 14-33. A plumb bob is tied at point D so the line BD is vertical and, therefore, parallel to the pole. The distance AB, BC and AD are measured and found to be 160 cm, 8.3 m and 210 cm, respectively. What is the length of the guy wire (AE)?

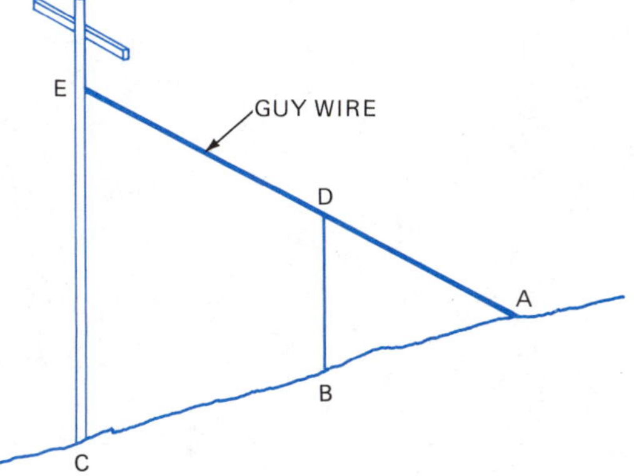

Figure 14-33

Use Figure 14-34 and the following information for Problems 6–8. If a right triangle has one acute angle of 30°, the ratios of the sides are constant values as follows:

$$\frac{a}{b} = 0.577$$

$$\frac{a}{c} = 0.500$$

$$\frac{b}{c} = 0.866$$

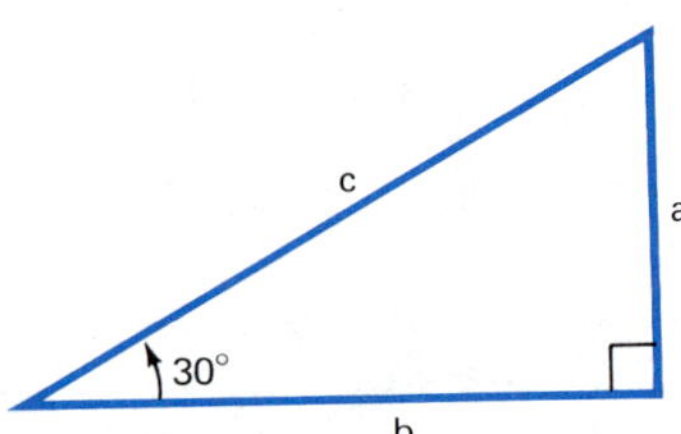

Figure 14-34 The right triangle. Side a is the altitude, b is the base and c is the hypotenuse.

Assume the three right triangles in the following problems have the required 30° angle. Use two of the formulas to find the unknowns. Use the third formula to check the answers.

6. If side c is 18 in, what are the lengths of a and b?
7. If side b is 450 cm, find sides a and c.
8. If side a is 1000 m, find sides b and c.

CHAPTER

15

ANGLES AND RIGHT TRIANGLES

OBJECTIVES

After satisfactorily completing this chapter, the student should be able to:

- Express angles in various units of measurement.
- Generate angles of any size by rotating a radius vector either clockwise or counterclockwise.
- Relate the angles of a right triangle to each other and use the Pythagorean Theorem to find the third side of a right triangle given the other two.

The study of alternating current and ac circuits requires a thorough knowledge of the right triangle. The right triangle is widely used in ac circuits to represent phase relationships. The more one learns about the right triangle, the easier it will be to understand alternating currents.

This chapter begins with the study of angular measurement and introduces the right triangle. Relationships involving the angles and sides of right triangles are presented. Electrical applications are described in many of the exercise problems.

ANGLES

Angles are formed by two intersecting lines, Figure 15-1. Lines AO and BO intersect at the point O. The intersecting point O is called the *vertex* of the angle.

In the study of electricity, angles are usually symbolized with lowercase Greek letters. The letters θ (theta), ϕ (phi), α (alpha), β (beta) and γ (gamma) are the most common. Another way of representing angles is with the

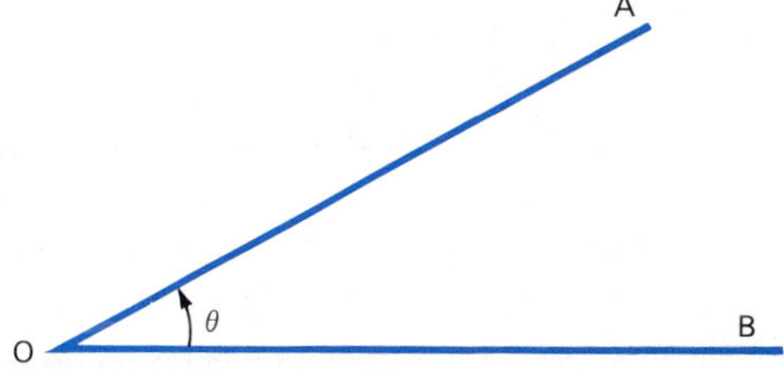

Figure 15-1 An angle (θ) is formed by two intersecting lines (AO and BO).

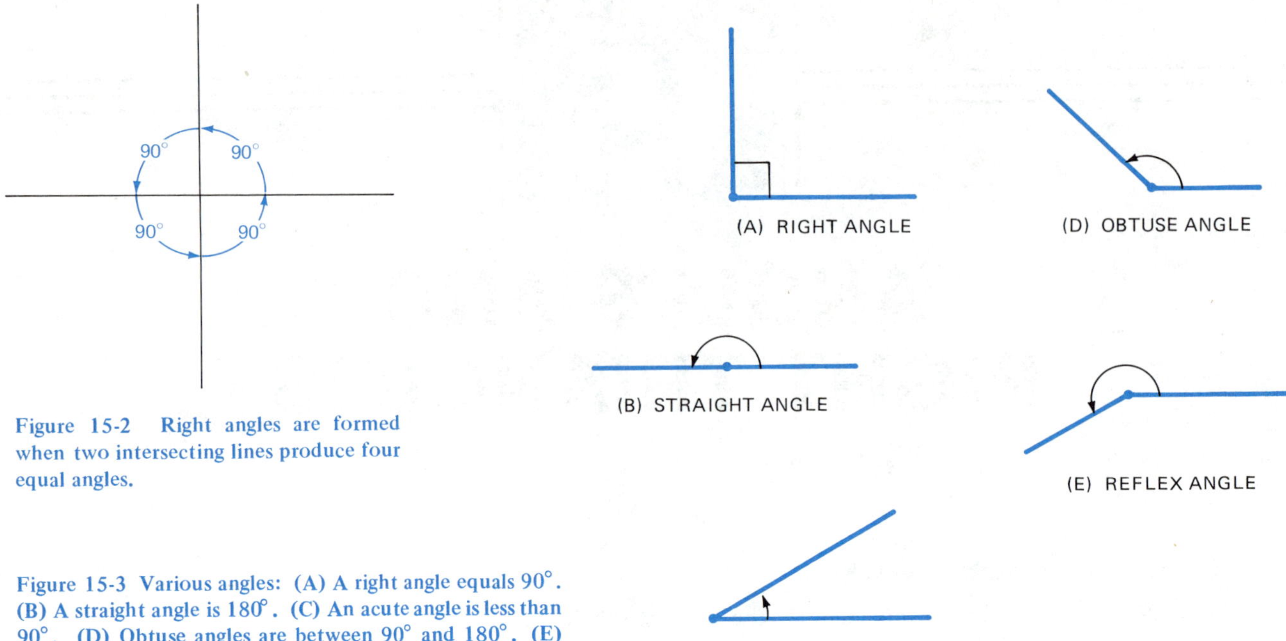

Figure 15-2 Right angles are formed when two intersecting lines produce four equal angles.

Figure 15-3 Various angles: (A) A right angle equals 90°. (B) A straight angle is 180°. (C) An acute angle is less than 90°. (D) Obtuse angles are between 90° and 180°. (E) Reflex angles are greater than 180°.

letters that define the lines forming the angle. In Figure 15-1, lines AO and BO form ∠AOB, ∠BOA or ∠O where O is the vertex of the angle.

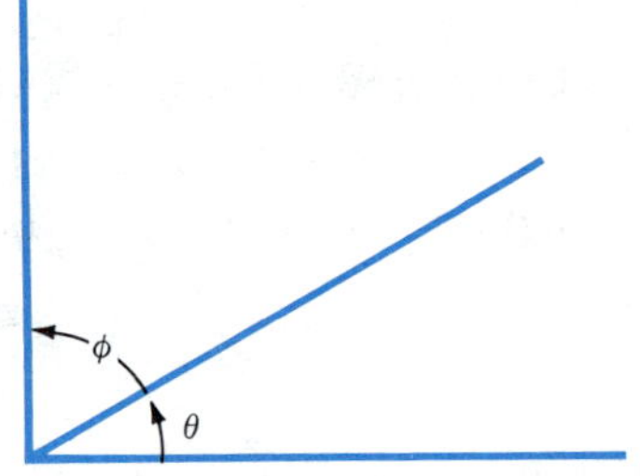

Figure 15-4 Complementary angles ($\theta + \phi = 90°$)

15-1 THE DEGREE AND SPECIAL ANGLES

Figure 15-2 shows two intersecting lines that form four equal angles. Such angles are called *right angles,* Figure 15-3(A). The two lines forming right angles are said to be *perpendicular.* A *straight angle* is formed by a straight line, Figure 15-3(B). A straight angle equals two right angles.

Acute angles are angles that are less than a right angle, Figure 15-3(C). An angle between a right angle and a straight angle in size is called an *obtuse angle,* Figure 15-3(D). An angle greater than a straight angle is a *reflex angle,* Figure 15-3(E).

For most electrical purposes, angles are measured in units called *degrees* (°). A right angle contains 90°. The sizes of other special angles are given in the caption of Figure 15-3. An angle of less than one degree is expressed as a decimal fraction. In this system of measurement, a straight angle contains 180°. A full circle contains 360°.

When a pair of angles have a sum of 90°, they are called *complementary angles,* Figure 15-4. One is the complement of the other. Another important pair of angles are *supplementary angles.* If two angles have a sum of 180°, they are supplementary. One is the supplement of the other, Figure 15-5.

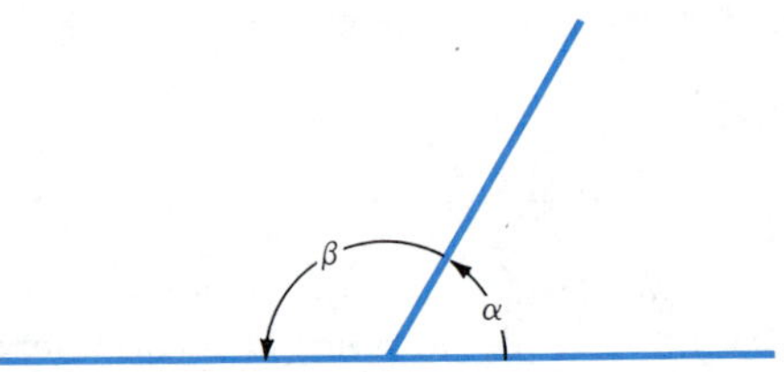

Figure 15-5 Supplementary angles ($\alpha + \beta = 180°$)

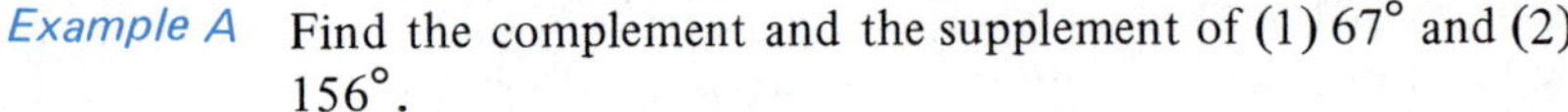

Example A Find the complement and the supplement of (1) 67° and (2) 156°.

Solution: (1) $\phi = 90° - \theta$
$\phi = 90° - 67° = 23°$ (Complement)
$\beta = 180° - \alpha$
$\beta = 180° - 67° = 113°$ (Supplement)
(2) $\phi = 90° - 156° = -66°$ (Complement)
$\beta = 180° - 156° = 24°$ (Supplement)

EXERCISE 15-1

Match the following angles with the name that correctly identifies the angle.

1. 37.5°	a. Obtuse
2. 346°	b. Straight
3. 90°	c. Reflex
4. 137.5°	d. Acute
5. 180°	e. Right

Find the complement of the following angles.

6. 30°	8. 15°	10. 37.5°	12. 125°
7. 45°	9. 65°	11. 72.68°	13. −58°

Find the supplement of the following angles.

14. 90°	16. 135°	18. −45°	20. −180°
15. 100°	17. 4°	19. 265°	21. 315°

How many right angles are contained in each of these angles?

22. 135°	23. 270°	24. 315°	25. 22.5°

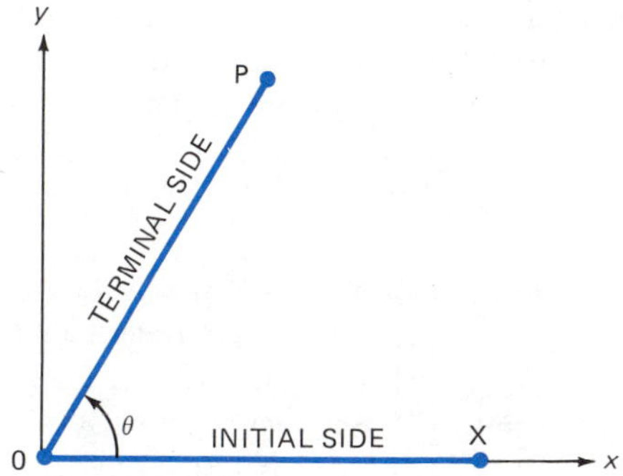

Figure 15-6 A positive generated angle in standard position

15-2 GENERATING ANGLES

The hands of a clock rotate through various angles. The size of the angle depends on which hand is being observed and for what length of time. In thirty minutes, for example, the minute hand rotates 180°, the hour hand rotates only 15° and the second hand rotates thirty revolutions or 10 800°.

A line such as OP, Figure 15-6, can *generate* an angle if it rotates about a point (O) called the vertex. The point (O) is usually the origin of a rectangular coordinate system. Assume the line begins its rotation from the *initial position* OX. It has rotated to its *terminal position* OP. The lines OX and OP are called the *initial side* and the *terminal side*, respectively. The initial side and the terminal side form the generated angle θ which is indicated by a curved arrow, Figure 15-6.

The terminal side is often represented as an arrow and is called a *radius vector.* Its length is the radius of the circle generated by the movement of the arrow tip, Figure 15-7. If the terminal side lies in the second quadrant, Figure 15-8(A), the generated angle is called a second quadrant angle. Third and fourth quadrant angles have the radius vector in the third and fourth quadrants, respectively, Figures 15-8(B) and 15-8(C).

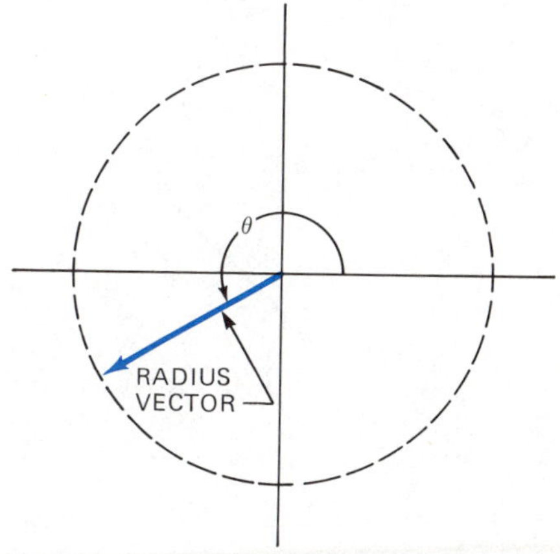

Figure 15-7 The terminal side of a generated angle is called a radius vector (arrow).

An angle is in *standard position* if the vertex is at the origin of a rectangular coordinate system. The initial side is along the *x*-axis. An angle is

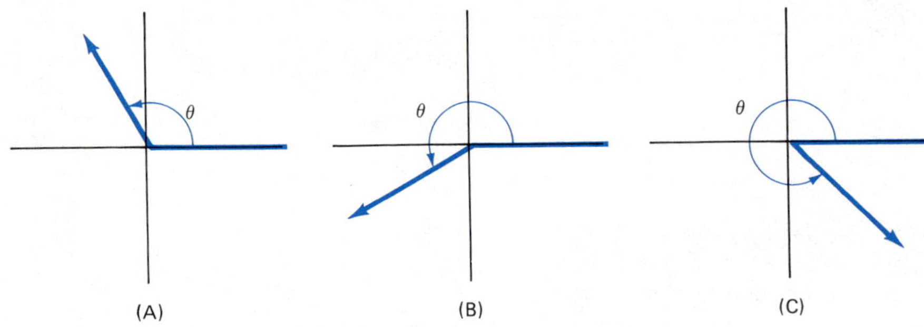

Figure 15-8 Positive generated angles: (A) 2nd quadrant angle (B) 3rd quadrant angle (C) 4th quadrant angle.

said to be *positive* if the radius vector rotates counterclockwise (ccw), Figure 15-8. An angle is *negative* if the radius vector rotates clockwise (cw), Figure 15-9. The hands of a running clock sweep out negative angles.

Naturally, a radius vector can rotate through any angle including those greater than 360°, Figure 15-10(A). Two angles in standard position are said to be *equivalent* if their terminal sides lie in the same position. An angle of 360° is equivalent to 0°, for example.

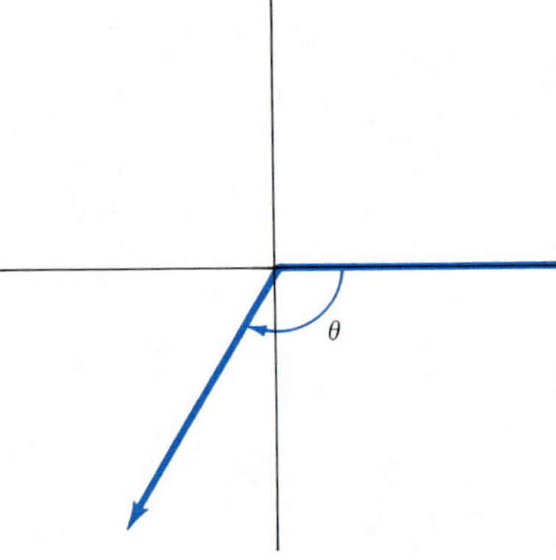

Figure 15-9 A negative angle is generated when the radius vector rotates clockwise.

Example A (1) Make a drawing of a positive angle of 765°.
(2) Find an equivalent angle between 0° and 360°.
(3) Find an equivalent negative angle between 0° and –360°.

Solution: (1) See Figure 15-10(A).
(2) 765° is a first quadrant angle formed after two full revolutions (720°) of the radius vector.

$$\theta = 765° - 720° = 45°$$

(3) Refer to Figure 15-10(B).

$$\phi = 45° - 360° = -315°$$

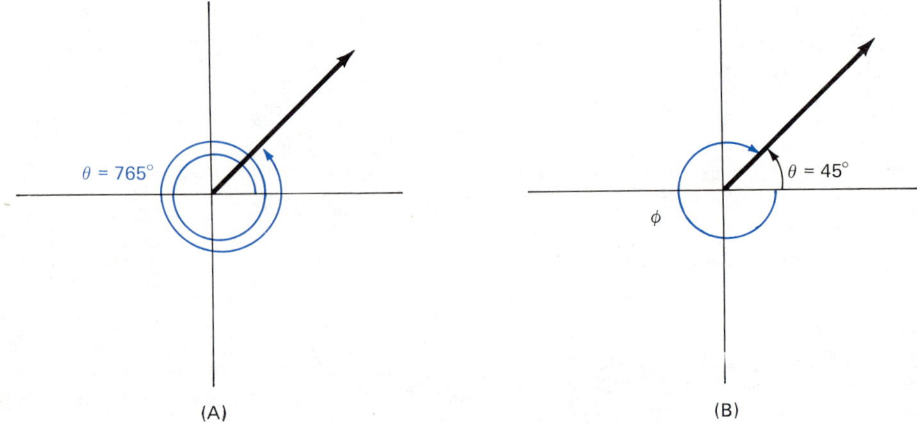

Figure 15-10

EXERCISE 15-2

Make freehand drawings similar to Figure 15-10(A) showing the generation of the following angles.

1. 30°
2. 45°
3. 60°
4. 135°
5. 300°
6. 495°
7. −660°
8. 930°
9. −780°
10. 1320°

In which quadrant does the terminal side of these angles fall? The initial side is in standard position.

11. 298°
12. −265°
13. −317°
14. 485°
15. −680°
16. 1340°

Give the equivalent positive angle less than 360° for the following angles. In which quadrant does the radius vector lie?

17. −30°
18. −115°
19. 395°
20. 510°
21. −510°
22. 798°
23. −1115°
24. 1325°

Find the angle in revolutions, degrees and right angles generated by the minute hand during these times.

25. 15 min
26. 20 min
27. 42.5 min
28. 1 h 10 min
29. 3 h 35 min
30. 7 h 50 min
31. 12 h 25 min
32. 18 h 04 min

33. From 8:15 AM to 11:45 AM (same day)
34. From 9:30 AM to 4:45 PM (same day)
35. From 8:45 PM to 7:15 AM (the following day)

15-3 ANGULAR MEASUREMENT (DEGREE AND RADIAN)

Generated angles and angles of triangles are commonly measured in *degrees.* Fractions of a degree are usually expressed as decimals. Other units are also used for electrical purposes. Besides the degree unit, revolutions, radians (natural system) and π-radians are also used.

It is easy to convert from degrees to revolutions (r) since 1 r = 360°.

Example A Convert 5000° to revolutions.

Solution: $5000°\left(\frac{1\ \text{r}}{360°}\right) = 13.9\ \text{r}$

To convert from revolutions to degrees, multiply by 360°. Usually, one would choose to use degrees if the angle is less than one revolution.

The radian unit is sometimes more difficult to work with. For some purposes, however, it is the easiest unit to use. The *radian* (rad) unit of angular measurement is defined by the equation:

$$\theta = \frac{a}{R}$$

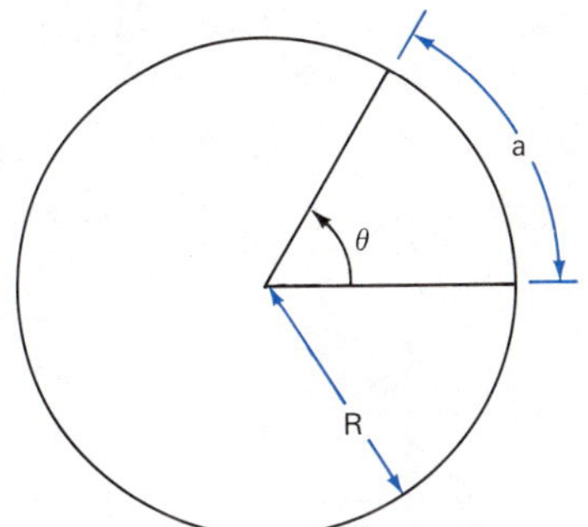

Figure 15-11 The radian is an angular unit defined as the ratio of a to R.

where a is the length of the arc subtended (cut) by the angle and R is the radius of the circle, Figure 15-11. Arc length and radius must be measured in the same units. It does not matter whether the units are English or metric but they must be the same for both measurements.

An angle of 1 rad is the angle subtended by an arc with a length equal to the radius, Figure 5-12(A). In one complete revolution, the arc is the circumference of the circle. Since the circumference is C = 2πR, the angle is as follows:

$$\theta = \frac{2\pi \cancel{R}}{\cancel{R}} = 2\pi \text{ rad}$$

for one revolution. This means 2π rad (6.28 rad) is equal to 360°, Figure 15-12(B). From this, 1 rad = 57.3°, approximately.

Another important unit for measuring angles is the *π-radian* (π•rad). A π-radian is simply π (3.14) times larger than the radian. It is equal to a half-revolution (1 π•rad = π rad = $\frac{1}{2}$ r). An angle of 45 revolutions can be expressed as either 90π rad or 90 π•rad. Notice the difference! 90 π•rad is the same as 90 half-revolutions while 90π rad is approximately 282.7 rad.

Usually, degrees and radians are used for angles less than one full rotation. Revolutions and π-radians are usually reserved for larger angles. Figure 15-13 shows some important relationships between these units.

Many calculators are equipped for converting between radians and degrees. These are one-number functions. Rounding should indicate the accuracy of the original measurement. Since 1° ≈ 0.02 rad, an angle measured to the nearest degree should be rounded to two decimal places.

Example B Convert 125° to radians.
Solution: 125 [D→R] , 2.18 rad

Example C Convert 4.35 rad to degrees.
Solution: 4.35 [INV] [D→R] , 249°

Note: Function keys on some calculator brands are [→DEG] and [→RAD] among others.

Example D Convert 197° to π-radians.
Solution: Convert to radians first, then divide by π.
197 [D→R] , 3.44

$$(3.44 \cancel{\text{rad}})\left(\frac{1\ \pi\cdot\text{rad}}{\pi\ \cancel{\text{rad}}}\right) = 1.09\ \pi\cdot\text{rad}$$

If the calculator doesn't have this capability, it is best to convert to revolutions first. It is usually a simple matter to then convert to the desired units.

DEGREES	RADIANS	π-RADIANS	REVOLUTIONS
57.3°	1	—	—
180°	π	1	$\frac{1}{2}$
360°	2π	2	1

Figure 15-13 Conversion factors between various angular units

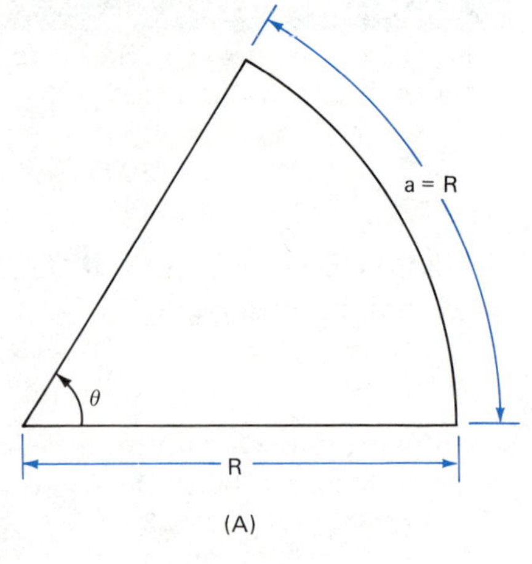

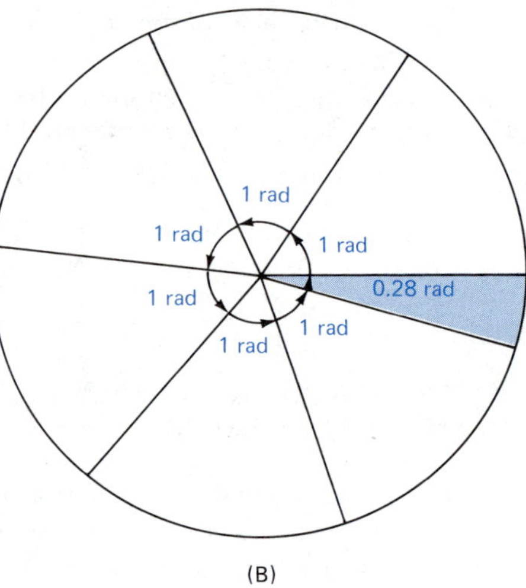

Figure 15-12 (A) When a = R, the angle is 1 radian (1 rad = 57.3°). (B) Radian divisions of a circle

Example E Convert 227° to radians.

Solution: $(227\ \cancel{\text{deg}})\left(\frac{1\cancel{r}}{360\ \cancel{\text{deg}}}\right)\left(\frac{2\pi\ \text{rad}}{1\cancel{r}}\right) = 3.96\ \text{rad}$

Note: The first conversion factor converts degrees to revolutions. The second converts revolutions to radians.

Example F Convert 3.74 π•rad to degrees.

Solution: $(3.74\cancel{\pi\cdot\text{rad}})\left(\frac{1\cancel{r}}{2\cancel{\pi\cdot\text{rad}}}\right)\left(\frac{360\ \text{deg}}{1\cancel{r}}\right) = 673°$

Example G Find the angle generated by the second hand of a watch in 2 min 40 s. Express the angle in (1) revolutions, (2) degrees, (3) radians and (4) π•radians.

Solution: First determine the number of revolutions. The second hand makes one revolution each minute. In 2 min 40 s, it rotates $2\frac{40}{60}$ r.

(1) $2\frac{40}{60} = 2\frac{2}{3}$ r

The bottom line of Figure 15-13 gives the conversion factors to convert to other units.

(2) $(2\frac{2}{3}\cancel{r})\left(\frac{360\ \text{deg}}{1\cancel{r}}\right) = 960°$

(3) $(2\frac{2}{3}\cancel{r})\left(\frac{2\pi\ \text{rad}}{1\cancel{r}}\right) = 16.76\ \text{rad}$

(4) $(2\frac{2}{3}\cancel{r})\left(\frac{2\ \pi\cdot\text{rad}}{1\cancel{r}}\right) = 5.33\ \pi\cdot\text{rad}$

15-4 OTHER ANGULAR UNITS

Sometimes the degree is subdivided into *minutes of arc* (sexagesimal system) instead of decimals. One degree equals 60 min. A minute of arc is $\frac{1}{60}$ of a degree. The minute of arc is further subdivided into *seconds of arc* (s). There are 60 s in 1 min or 3600 s in one degree:

$$1° = 60\ \text{min} = 3600\ \text{s}$$
$$1\ \text{min} = 60\ \text{s}$$

The symbols (′) and (″) are often used for minutes and seconds, respectively.

Electricians prefer to express fractions of a degree as a decimal. The conversion is the same as converting time units (hours, minutes, seconds). The calculator conversion key is [D→DMS] (degree, minute, second). Other calculators use the symbol [→H.MS] (hour, minute, second).

Example A Convert 23.38° to the sexagesimal system.

Solution: 23.38 [D→DMS], 23.22 48

Note: The display 23.22 48 means 23°22′48″

Another system used in some European countries uses the *grad* unit (grd). In this system, the right angle is divided into 100 grd. This system will not be used in this text.

EXERCISE 15-3

Convert each of the following angles to radians and to π-radians.

1. 35°
2. −60°
3. 142°
4. −246°
5. 194°
6. 325°
7. 57.3°
8. −800°

Express each of the angles that follow in degrees and π-radians.

9. π rad
10. 0.524 rad
11. 20π rad
12. −1.36 rad
13. −1 rad
14. 0.0175 rad
15. −1.57 rad
16. 25.65 rad

Change the following angles to degrees and radians.

17. 1 π•rad
18. 0.25 π•rad
19. −0.167 π•rad
20. 1.5 π•rad
21. 20 π•rad
22. −0.0833 π•rad
23. 0.005 56 π•rad
24. −0.3183 π•rad

Write these angles in degrees using decimal form.

25. 44 min 35 s
26. 29′48″
27. 3°7′10″
28. 25° 15 min
29. 88° 25 min 37 s
30. 156°53′02″

Complete the following table.

	DEGREES	RADIANS	π-RADIANS	REVOLUTIONS
31.	0°			
32.		0.5236		
33.			0.25	
34.				0.1667
35.	90°			
36.		3.1416		
37.			2	
38.				2

THE RIGHT TRIANGLE

A triangle is a plane figure with three angles. Such a figure also has three sides. The various classifications were discussed in Topic 7-8. These are repeated here for convenience.

- A *scalene triangle* has three unequal angles and unequal sides.
- An *obtuse triangle* has one obtuse angle.
- An *isosceles triangle* has two equal angles and two equal sides.
- An *equilateral triangle* has three equal angles and three equal sides.
- A *right triangle* has one right angle.

The right triangle has a special application for electricians. It is used in ac circuit theory to represent phase relationships. It is used extensively in most of the chapters that follow.

15-5 ANGLES OF A RIGHT TRIANGLE

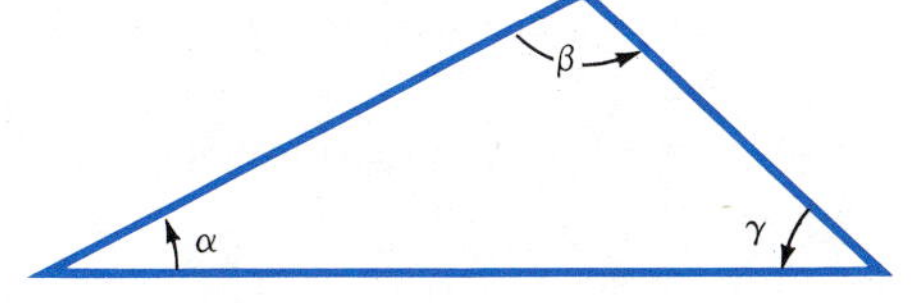

Figure 15-14 Angles of a triangle ($\alpha + \beta + \gamma = 180°$)

The sum of the three angles of a triangle always equals a straight angle (180°). This is true for all triangles, large or small. It is true regardless of the triangle's shape or classification. In equation form, Figure 15-14, the previous statement is

$$\alpha + \beta + \gamma = 180°$$

Of course, if angles are expressed in radians, the sum is π rad.

If any two angles are known, their sum is the supplement of the third angle. Thus, the third angle can easily be found.

Example A A triangle has angles of 32.3° and 94.6°. Find the third angle.

Solution:

$$\alpha + \beta + \gamma = 180°$$
$$32.3° + 94.6° + \gamma = 180°$$
$$\gamma = 180° - 126.9°$$
$$\gamma = 53.1°$$

Example B A certain triangle has angles of 1.23 rad and 1.04 rad. Find the third angle.

Solution:

$$\alpha + \beta + \gamma = \pi \text{ rad}$$
$$1.23 \text{ rad} + 1.04 \text{ rad} + \gamma = \pi \text{ rad}$$
$$\gamma = \pi \text{ rad} - 2.27 \text{ rad}$$
$$\gamma = 0.87 \text{ rad}$$

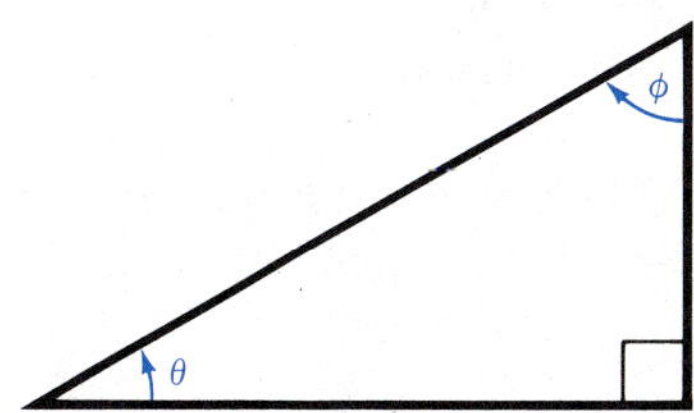

Figure 15-15 The acute angles of a right triangle are complementary ($\theta + \phi = 90°$).

One of the angles of a right triangle is, of course, a right angle (90°). The two acute angles, Figure 15-15, and the right angle have a sum of 180°.

$$\theta + \phi + 90° = 180°$$

or

$$\theta + \phi = 90°$$

The acute angles are complementary.

Example C Find the second acute angle of a right triangle if the first is 37.3°.

Solution:

$$\theta + \phi = 90°$$
$$\theta + 37.3° = 90°$$
$$\theta = 90° - 37.3°$$
$$\theta = 52.7°$$

EXERCISE 15-4

Find the missing angle, Figure 15-16, in each of the following problems and name the type of each triangle.

1. $\alpha = 28°, \beta = 107°$
2. $\beta = 78°, \gamma = 78°$
3. $\alpha = 37°, \beta = 53°$
4. $\gamma = 60°, \beta = 60°$
5. $\alpha = 73°, \beta = 62°$
6. $\gamma = 1$ rad, $\alpha = 1$ rad

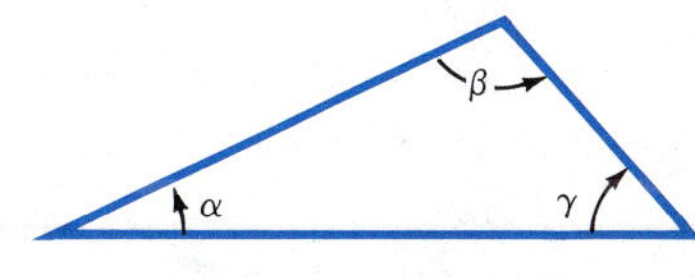

Figure 15-16

7. $\alpha = 1.02$ rad, $\beta = 0.55$ rad
8. $\alpha = 0.333\ \pi \cdot$rad, $\beta = 0.333\ \pi \cdot$rad
9. $\beta = 42^\circ, \gamma = 96^\circ$
10. $\gamma = 1.75$ rad, $\alpha = 0.97$ rad
11. $\beta = \frac{1}{3}\pi \cdot$rad, $\gamma = \frac{1}{6}\pi \cdot$rad
12. $\alpha = \frac{1}{3}\pi \cdot$rad, $\beta = \frac{1}{3}\pi \cdot$rad

Use Figure 15-17 and find the missing acute angle in each of the following right triangles.

13. $\theta = 15.3^\circ$
14. $\phi = \frac{1}{3}\pi \cdot$rad
15. $\phi = 1$ rad
16. $\theta = 68.2^\circ$
17. $\theta = \frac{1}{12}\pi \cdot$rad
18. $\phi = 45^\circ$
19. $\theta = 0.785$ rad
20. $\phi = 57.3^\circ$

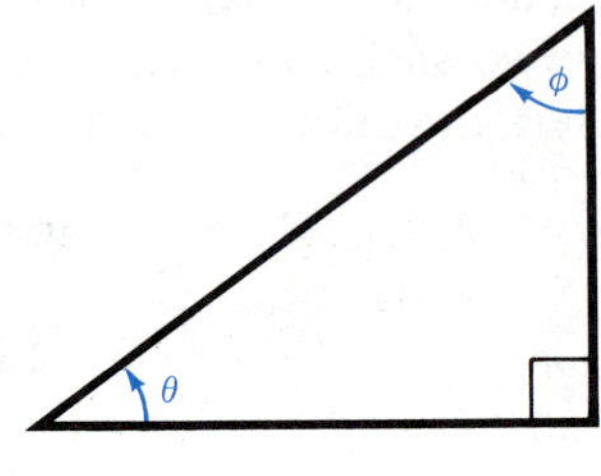

Figure 15-17

Construct right triangle diagrams for each situation in the following problems.

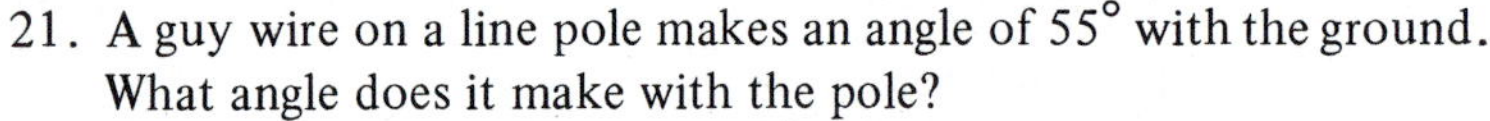

21. A guy wire on a line pole makes an angle of 55° with the ground. What angle does it make with the pole?
22. A ladder is set at an angle of 32° with the vertical wall of a building. What angle does it make with the ground?
23. A cross-country transmission line makes an angle of 22° with a north-south road. What angle does it make with an east-west road?
24. A diagonal brace on a line tower makes an angle of 45° with the horizontal cross member. What angle does it make with the vertical side?

15-6 SIDES OF A RIGHT TRIANGLE

Additional properties of triangles relate to the sides. Suppose the lengths of two sides of a triangle such as a and b, figure 15-18, are known. The larger side (b) is drawn. The smaller side can then be drawn connecting points C and B.

Since the shape of the triangle is not known, side a can rotate about point C. All possibilities are covered. Point B must lie somewhere on the semicircle (dashed). Think of side a as a radius vector rotating about point C. The third side must connect points A and B.

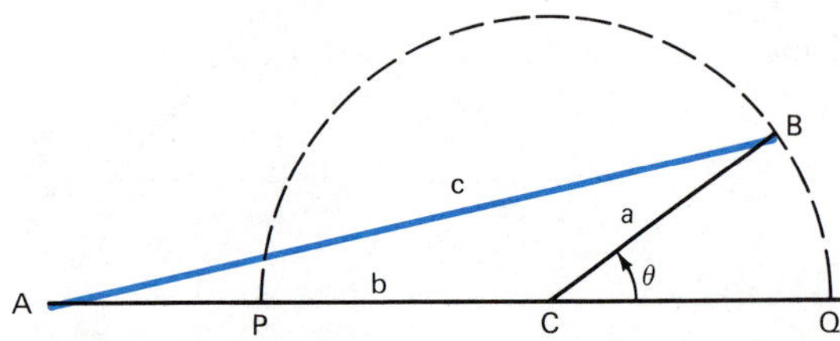

Figure 15-18 Side c is less than b + a but greater than b − a.

The third side (c) cannot be greater than the sum of the two known sides (b + a). Further, the third side cannot be less than the difference of the two knowns (b – a). These statements are properties of all triangles.

Example A If two sides of a triangle are b = 5 cm and a = 2 cm, respectively, find the range of values for the third side (c).

Solution: Side c is less than b + a:

$b + a = 5 \text{ cm} + 2 \text{ cm} = 7 \text{ cm}$

Side c is greater than b – a:

$b - a = 5 \text{ cm} - 2 \text{ cm} = 3 \text{ cm}$

Side c is greater than 3 cm but less than 7 cm.

Right triangles have one right angle. The side opposite the right angle is the longest of the three sides and is called the *hypotenuse.* In Figure 15-19, side c is the hypotenuse. The triangle is in *standard position* with the sides vertical (a) and horizontal (b). In this orientation, a is the *altitude* and b is the *base.*

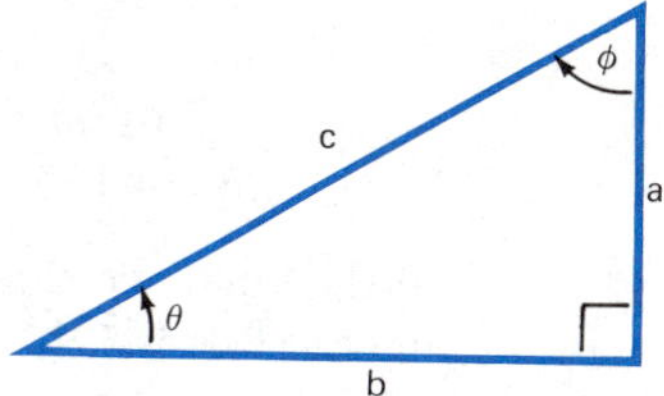

Figure 15-19 Side c is the hypotenuse, a is the altitude and b is the base.

The three sides of a right triangle, Figure 15-19, are related by the Pythagorean Theorem. It was discovered by the early Greek philosopher, Pythagoras.

The Pythagorean Theorem. The square of the hypotenuse of a right triangle equals the sum of the squares of the other two sides.

$$c^2 = a^2 + b^2$$

Once before, the product of two numbers was interpreted as an area. Thus, a^2 is taken to mean the area of a square with sides of length a. In this way, the Pythagorean Theorem can be visualized, Figure 15-20.

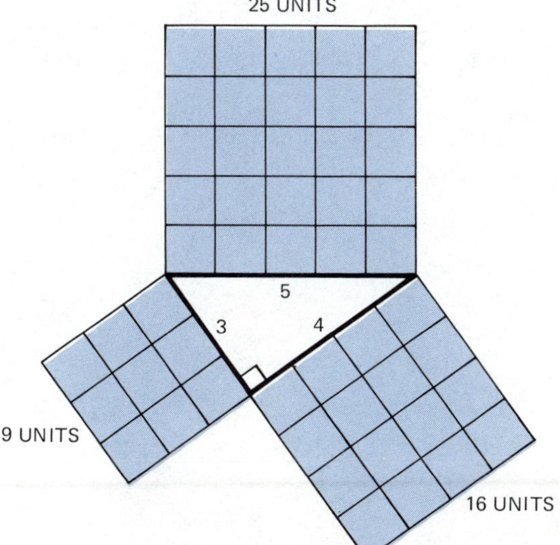

Figure 15-20 The square of side c is an area (25 units) equal to the sum of the areas (9 units and 16 units) formed by the squares of the other two sides.

The square of side c is an area equal to the areas of the squares formed from the other two sides of a right triangle.

Example B Find the length of the hypotenuse in a right triangle having sides of 8 m and 6 m.

Solution: Solve the Pythagorean Theorem ($c^2 = a^2 + b^2$) for c by taking the square root.

$$c = \sqrt{a^2 + b^2}$$

Substitute: $c = \sqrt{8^2 + 6^2}$

Note: 8 $\boxed{x^2}$ $\boxed{+}$ 6 $\boxed{x^2}$ $\boxed{=}$ $\boxed{\sqrt{x}}$, 10

$$c = \sqrt{64 + 36} = \sqrt{100} = 10 \text{ m}$$

Check: The hypotenuse must be greater than the larger of the other two sides (8) but less than the sum of the two sides. This kind of a check is quick but obviously does not give an exact check. For an exact check, substitute the answer and the given data into the Pythagorean Theorem: $10^2 = 8^2 + 6^2$ or $100 = 64 + 36$.

Example C Find the unknown side of a triangle if the hypotenuse is 13 ft and one side is 5 ft.

Solution: $c^2 = a^2 + b^2$

$b^2 = c^2 - a^2$ (SA, $-a^2$)

$b = \sqrt{c^2 - a^2}$ (RA, $\sqrt{\ }$)

Substitute: $b = \sqrt{13^2 - 5^2}$

$$b = \sqrt{169 - 25} = \sqrt{144} = 12 \text{ ft}$$

Check: 12 is greater than $13 - 5 = 8$ but less than $13 + 5 = 18$. Of course, it must be less than the hypotenuse (13) also.

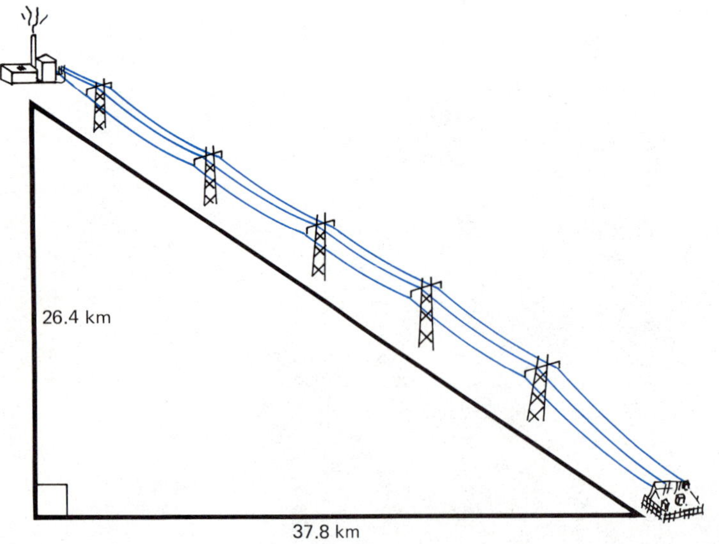

Figure 15-21

Example D A transmission line goes from a power plant to a substation following a straight path. To travel from the power plant to the substation by car, one must travel 26.4 km straight south and 37.8 km straight east. (1) How long is the transmission line? (2) If there are three conductors, how much cable is saved by not following the roadway?

Solution: (1) Data and diagram: See Figure 15-21.

Formula: $c^2 = a^2 + b^2$

Substitute: $c = \sqrt{26.4^2 + 37.8^2}$

$c = 46.1$ km

(2) The total distance along the roadway is: 26.4 km + 37.8 km = 64.2 km. The distance saved is 64.2 km − 46.1 km = 18.1 km. The length of cable saved is 3(18.1 km) = 54.3 km.

EXERCISE 15-5

Find the maximum and minimum lengths of the third (unknown) side in each of the following triangles, Figure 15-22

1. a = 7 cm, b = 12 cm
2. a = 27 m, b = 17 m
3. a = 58 in, b = 79 in
4. b = 15.4 km, a = 28.9 km
5. a = 69.4 cm, b = 69.4 cm
6. b = 7.34 ft, a = 3.67 ft

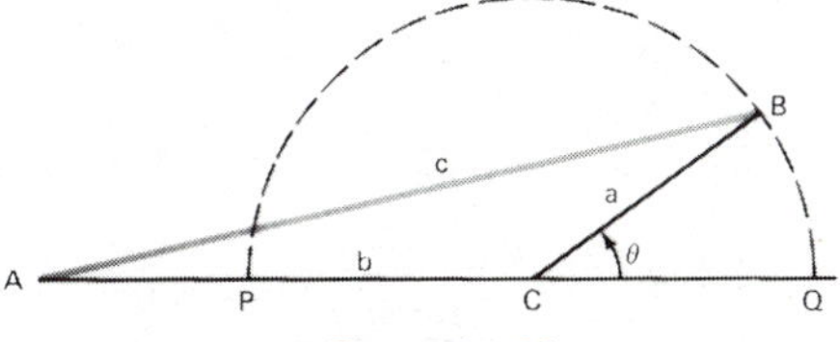

Figure 15-22

Find the length of the unknown side of the following right triangles, Figure 15-23. Construct a diagram for each triangle with the known sides indicated. Check all answers.

7. a = 3 in, b = 4 in
8. a = 10 m, b = 24 m
9. c = 27 cm, a = 21 cm
10. b = 46.3 ft, c = 91.7 ft
11. b = 27 m, a = 35 m
12. a = 44 cm, c = 64 cm
13. a = 9.92 mi, b = 7.22 mi
14. b = 44.3 km, c = 71.3 km

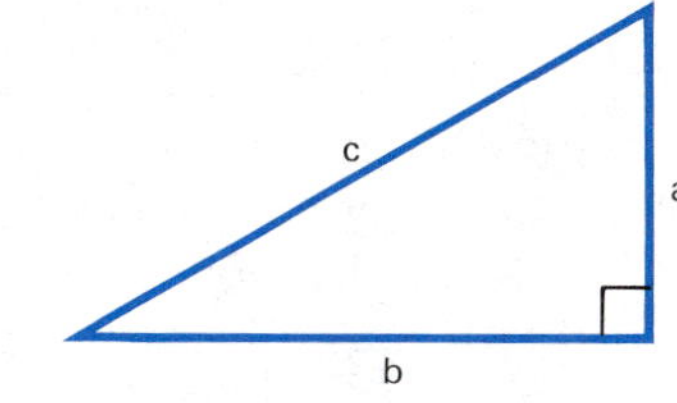

Figure 15-23

Construct a right triangle indicating the situation in each of the following problems. Find the unknown side in each. Data, diagram, formula, substitute, check.

15. A guy wire is attached to a pole at a point 15 m above the ground. It is anchored at a point 17.2 m from the base. How long is the guy wire?
16. An electrician places a 22-ft ladder against a building. The ladder touches the building 17.3 ft above ground. How far is the base of the ladder from the building?
17. An electrical conduit pipe is bent, Figure 15-24. (a) Find the length of conduit between bends. (b) Find the total length of conduit.

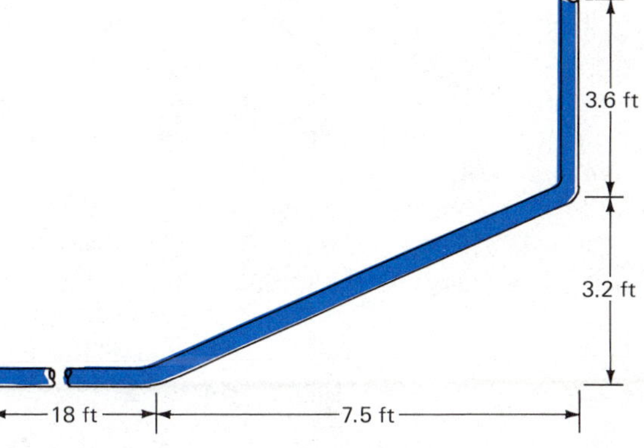

Figure 15-24

18. Find the length of conduit between bends, Figure 15-25.

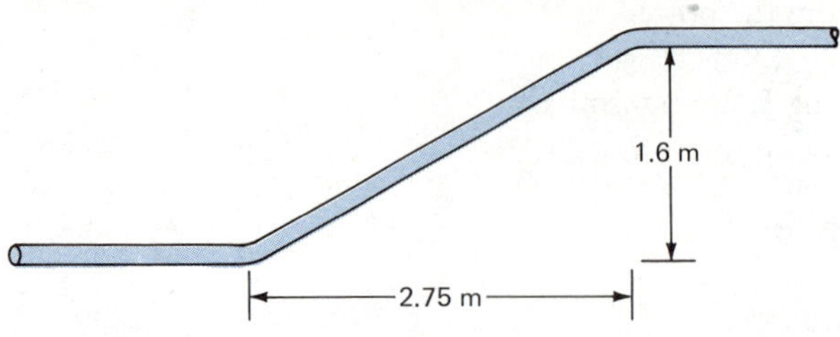

Figure 15-25

19. A helicopter is used to raise towers in the construction of a power line. It leaves the marshaling yard and travels 24.5 mi in a straight line to the job site. The work crew travels 18.4 mi straight north, then goes straight west to the job site. How much further did the work crew have to travel?
20. A *grade* or *slope* is the ratio of rise (vertical) to the horizontal distance (run). It is usually expressed in percent. A 30% grade rises 30 m for each horizontal distance of 100 m. What length of transmission line cable (three wires) is needed for a 30% grade with a horizontal run of 3.7 km?
21. The diagonal distance from one corner of a shop room to the opposite corner is 13.6 m. The length is 11.2 m. Determine the length of conduit for outlet boxes (surface mounted) if it is to be installed along one side and the back of the room. Assume corner to corner installation.
22. A *service drop* is the service conductor from an outdoor pole to the service entrance of a building. A service drop connects at the pole 38 ft above ground. It connects to the service entrance 12 ft above ground and 75 ft from the pole. (a) Find the length of the service drop by neglecting sag. (b) What length is required if 5% is added for sag and splicing?

15-7 DRAWING RIGHT TRIANGLES

The longest side of a right triangle is called the hypotenuse. It is always the side opposite the right angle. It will prove convenient to name the other two sides also. The words altitude (rise) and base (run) have served their purpose. Such terms have little meaning if the right triangle is not in standard position. New words must be found that describe the sides in reference to one of the acute angles.

A right triangle is formed by three sides and includes a right angle and two acute angles. Usually, one of the acute angles is of greater importance. The other is the complement of the first. It can always be evaluated if needed. The known angle is called the *reference angle.*

Imagine a right triangle being drawn in the following way:

- Two intersecting lines are drawn to include the correct reference angle (θ), Figure 15-26(A). These lines can be extended as far as desired.

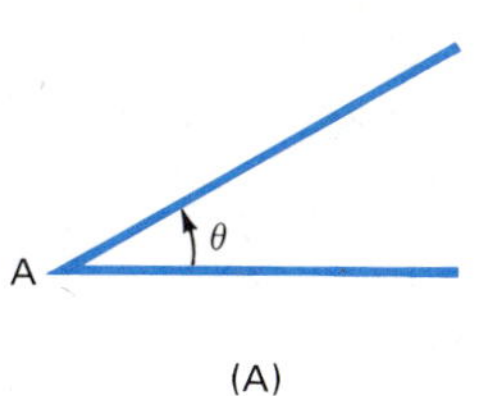

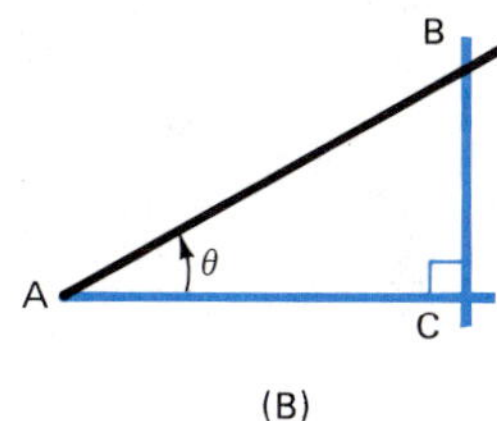

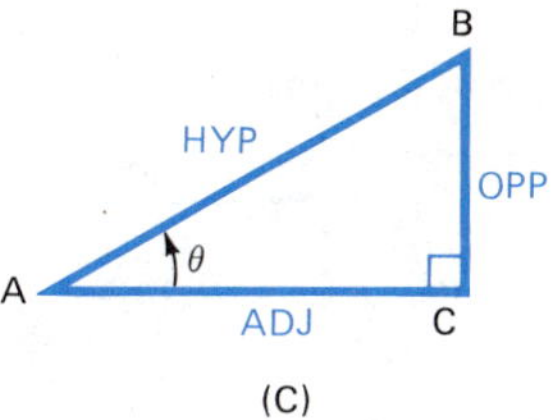

Figure 15-26 Drawing a right triangle

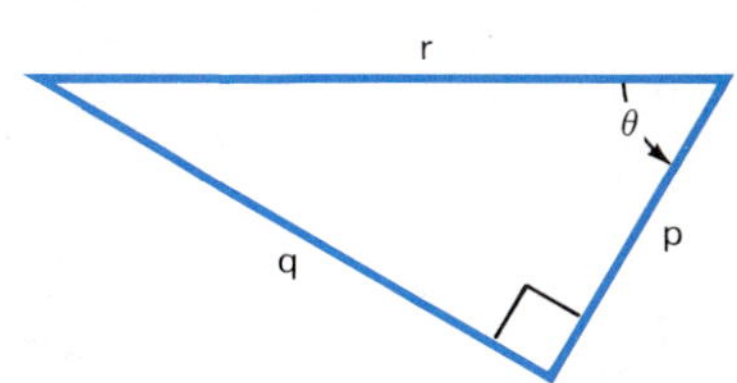

Figure 15-27 A right triangle not in standard position

- At the correct distance AC, a third line is drawn to intersect both of the first two lines, Figure 15-26(B). The third line intersects line AC at a right angle.

In this triangle, side AC is adjacent to the reference angle. It is not only adjacent to the angle, it helps form the angle. It is called the *adjacent side* (ADJ), Figure 15-26(C). The adjacent side and the hypotenuse (HYP) are the two sides that always create the reference angle.

Side BC is opposite the reference angle. It intersects the hypotenuse and the adjacent side but is not a part of the reference angle. Side BC is called the *opposite side* (OPP).

Using these definitions, the opposite and adjacent sides can be determined regardless of how a right triangle is oriented. In Figure 15-27, for example, θ is the reference angle. The opposite side is q, the adjacent side is p and r is the hypotenuse.

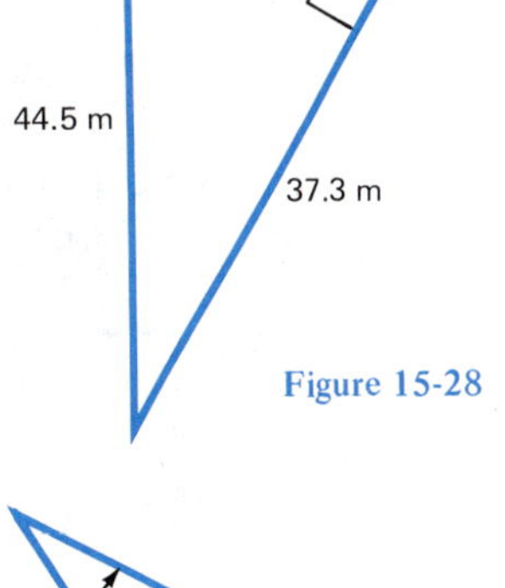

Figure 15-28

Example A State the lengths of (1) the hypotenuse, (2) the adjacent side and (3) the opposite side, Figure 15-28, in reference to the angle θ.

Solution: (1) HYP = 44.5 m (The longest side)
(2) ADJ = 24.2 m (Helps form angle θ)
(3) OPP = 37.3 m (Opposite angle θ)

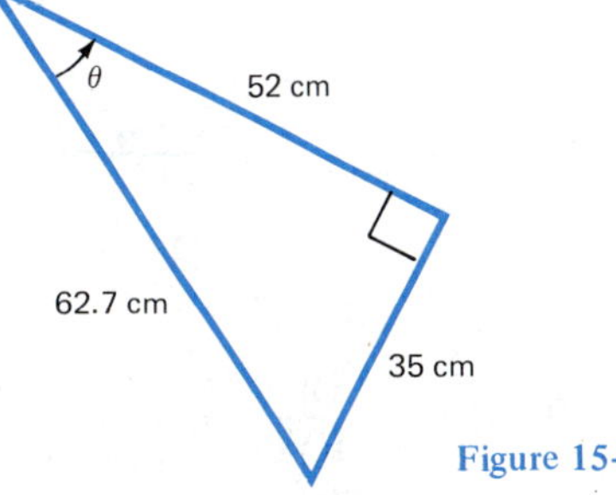

Figure 15-29

Example B Find the following ratios, Figure 15-29, in reference to angle θ: (1) $\frac{\text{OPP}}{\text{HYP}}$, (2) $\frac{\text{ADJ}}{\text{HYP}}$, (3) $\frac{\text{OPP}}{\text{ADJ}}$.

Solution: (1) $\frac{\text{OPP}}{\text{HYP}} = \frac{35}{62.7} = 0.558$

(2) $\frac{\text{ADJ}}{\text{HYP}} = \frac{52}{62.7} = 0.829$

(3) $\frac{\text{OPP}}{\text{ADJ}} = \frac{35}{52} = 0.673$

EXERCISE 15-6

1. In reference to Figure 15-30, what letter represents (a) the hypotenuse, (b) the side adjacent to angle θ, and (c) the side opposite to angle θ?
2. In Figure 15-30, with reference to angle ϕ, what letter represents (a) the hypotenuse, (b) the adjacent side, and (c) the opposite side?

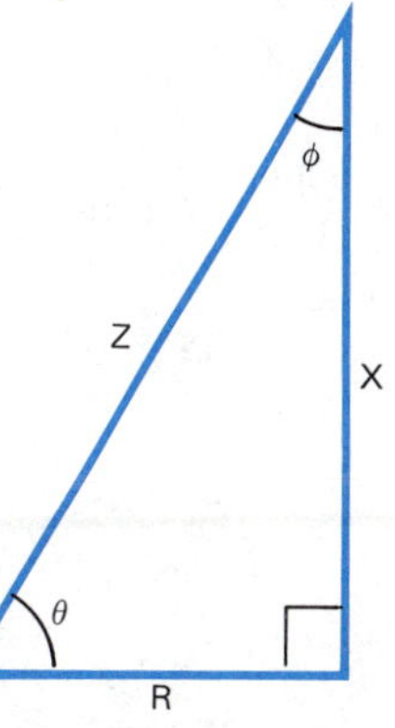

Figure 15-30

Find the ratios (a) $\frac{\text{OPP}}{\text{HYP}}$, (b) $\frac{\text{ADJ}}{\text{HYP}}$, and (c) $\frac{\text{OPP}}{\text{ADJ}}$ for each of the following right triangles in reference to the angle indicated.

3.

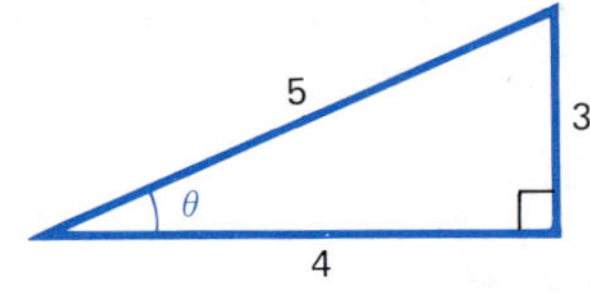

4.

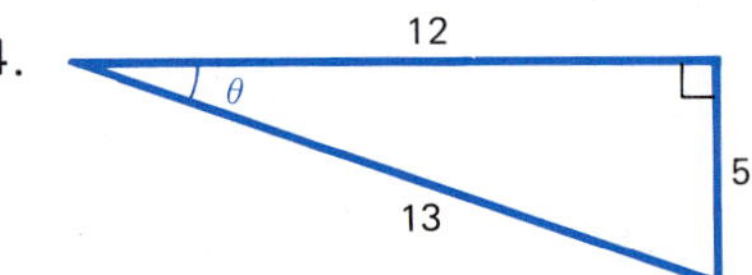

5.

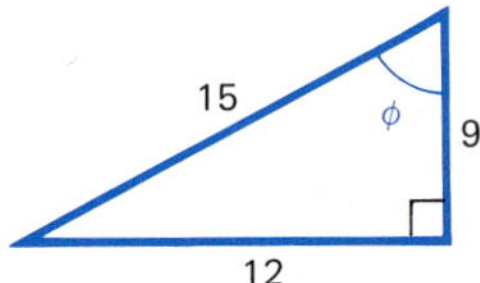

6.

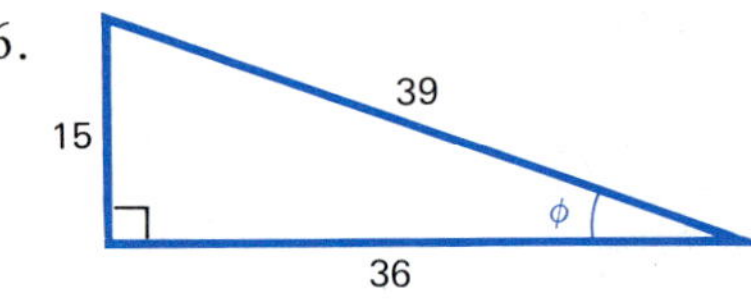

7.

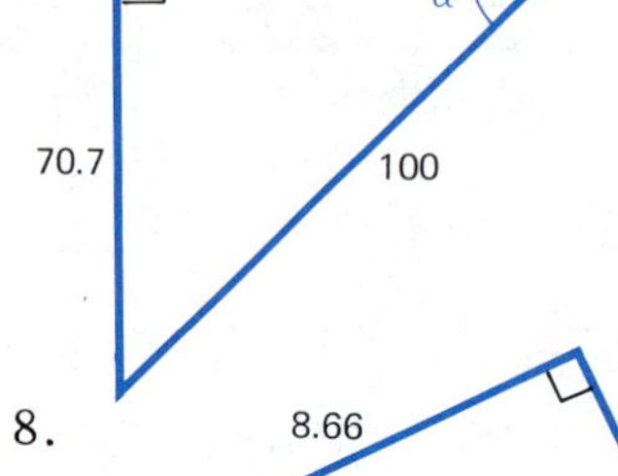

8.

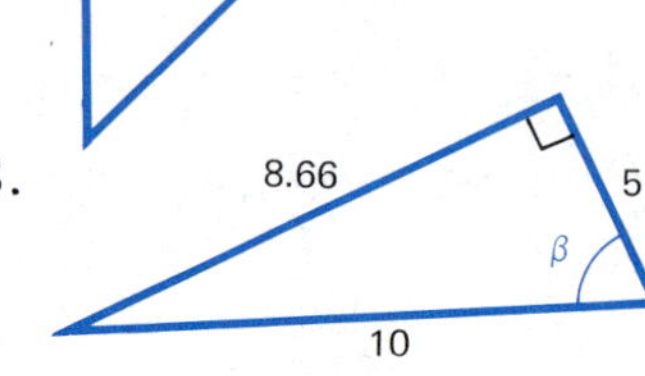

9.

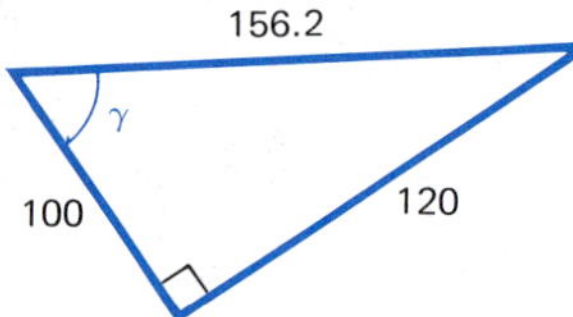

10.

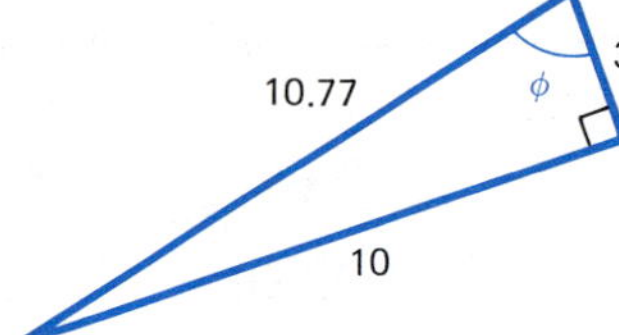

11. If the ratio $\frac{\text{OPP}}{\text{HYP}} = 0.5$ and $\frac{\text{ADJ}}{\text{HYP}} = 0.866$, Figure 15-31, find the lengths of sides x and y.

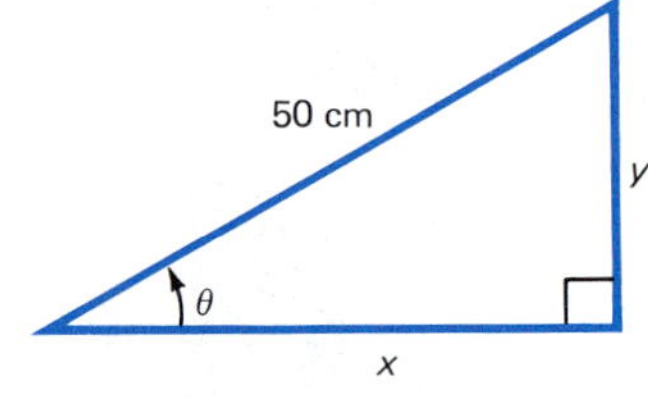

Figure 15-31

12. If the ratio $\frac{\text{OPP}}{\text{HYP}} = 0.8$ and $\frac{\text{OPP}}{\text{ADJ}} = 1.333$, Figure 15-32, find the lengths of R and Y.

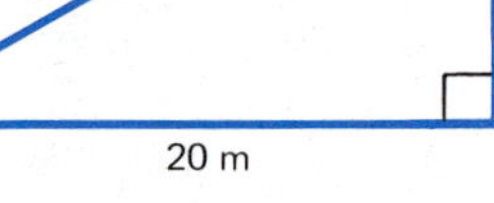

Figure 15-32

CHAPTER 16

TRIGONOMETRY

OBJECTIVES

After satisfactorily completing this chapter, the student should be able to:

- Define the trigonometric and inverse trigonometric functions.
- Evaluate trigonometric and inverse trigonometric functions using a calculator.
- Solve right triangle problems given two sides or one side and an angle including practical problems.

It is enlightening to learn the meaning of the word trigonometry. In the center is the base "gon" which comes from the Greek word for angle. The words pentagon, hexagon, octagon and polygon are common words using this base. The prefix "tri" means three. "Trigon," the first half of the word trigonometry, therefore, means three angled figures (triangles). The suffix is "metry." It means the "study or science of measuring." Trigonometry then, means "the study of measuring triangles." In this text, only one special kind of triangle will be studied. The right triangle was introduced in the previous chapter and is extended in this chapter.

TRIGONOMETRIC FUNCTIONS

Some students may be surprised to learn that they have already studied most of the basic concepts of trigonometry. Those concepts for right triangles are as follows:

- The two acute angles are complementary, Topic 15-5.
- The three sides are related by the Pythagorean Theorem, Topic 15-6.
- The ratios of two corresponding sides in similar triangles are proportional, Topic 14-13.
- Angles of any size can be generated by a rotating radius vector in either positive or negative directions, Topic 15-2.

Solving right triangles means to determine the length of the three sides and the size of the two acute angles. Previously, this could be done only if two sides and one of the angles were known. To do so, one used the first two concepts of the previous list.

What remains to be studied is the interconnection between the sides and angles. This will allow the student to solve a right triangle when only two sides or one side and one angle are known. When two sides are given, it is said to be a side-side (SS) problem. If one side and one angle are given, it is a side-angle (SA) problem. Both types will be discussed and applied to ordinary right triangles.

16-1 SINE, COSINE AND TANGENT

An angle (θ) is generated by lines representing the initial position and terminal position of a radius vector. In other terminology, the angle is formed by the adjacent side (ADJ) and the hypotenuse (HYP). The opposite side (OPP) is perpendicular to the adjacent side, Figure 16-1.

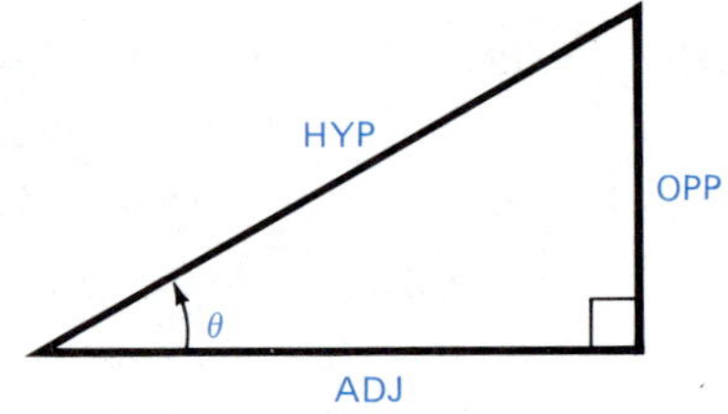

Figure 16-1 Right triangle nomenclature

There are three ratios of interest involving the three sides. These are called the *trigonometric ratios.* Since the values of the three ratios depend on the angle, they are usually called *trigonometric functions.* Each has a special name relating it to the angle as follows:

- The ratio $\frac{\text{OPP}}{\text{HYP}}$ is the *sine* of angle θ.
- The ratio $\frac{\text{ADJ}}{\text{HYP}}$ is the *cosine* of angle θ.
- The ratio $\frac{\text{OPP}}{\text{ADJ}}$ is the *tangent* of angle θ.

The names of the three ratios are abbreviated by using the first three letters of each. They should be pronounced as though they were not shortened, however. The following equations give the definitions of the trigonometric functions in abbreviated form. It is essential that these definitions be memorized.

$$\sin \theta = \frac{\text{OPP}}{\text{HYP}}$$

$$\cos \theta = \frac{\text{ADJ}}{\text{HYP}}$$

$$\tan \theta = \frac{\text{OPP}}{\text{ADJ}}$$

It should be noted that there are three additional named trigonometric ratios. They are the reciprocals of the three just mentioned and are called the *cosecant, secant* and *cotangent,* respectively. These will not be used in this textbook.

The values of trigonometric functions have no dimensions since they are pure ratios. The three sides must all be measured in the same units. In problems, a sketch is always desirable.

Example A Find the value of the sine, cosine and tangent for an angle if the opposite side is 20 cm and the adjacent side is 15 cm.

Solution: Data-diagram: See Figure 16-2. Find the hypotenuse (Pythagorean Theorem):

Formula: $c^2 = a^2 + b^2$

Substitute: $c = \sqrt{20^2 + 15^2}$

$c = \sqrt{400 + 225} = \sqrt{625} = 25 \text{ cm}$

Evaluate ratios: $\sin\theta = \dfrac{\text{OPP}}{\text{HYP}} = \dfrac{20 \text{ cm}}{25 \text{ cm}} = 0.800$

$\cos\theta = \dfrac{\text{ADJ}}{\text{HYP}} = \dfrac{15 \text{ cm}}{25 \text{ cm}} = 0.600$

$\tan\theta = \dfrac{\text{OPP}}{\text{ADJ}} = \dfrac{20 \text{ cm}}{15 \text{ cm}} = 1.33$

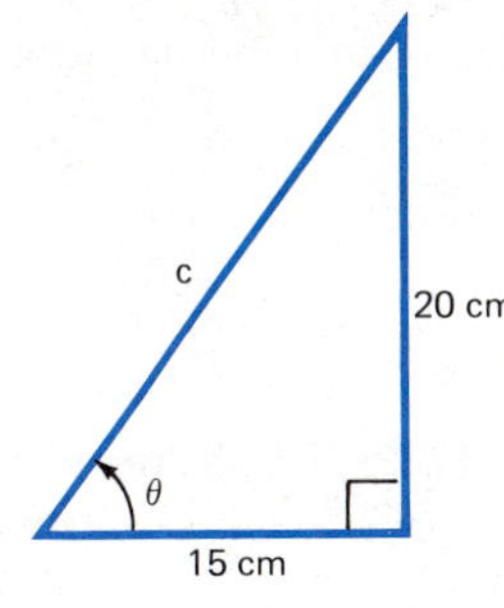

Figure 16-2

Example B Find the value of the sine, cosine and tangent for an angle if the hypotenuse is 26 m and the adjacent side is 10 m.

Solution: Data-diagram: See Figure 16-3.

Formula: $c^2 = a^2 + b^2$

$a^2 = c^2 - b^2$

Substitute: $a = \sqrt{26^2 - 10^2}$

$a = \sqrt{676 - 100} = \sqrt{576} = 24 \text{ m}$

Evaluate ratios: $\sin\theta = \dfrac{\text{OPP}}{\text{HYP}} = \dfrac{24}{26} = 0.923$

$\cos\theta = \dfrac{\text{ADJ}}{\text{HYP}} = \dfrac{10}{26} = 0.385$

$\tan\theta = \dfrac{\text{OPP}}{\text{ADJ}} = \dfrac{24}{10} = 2.40$

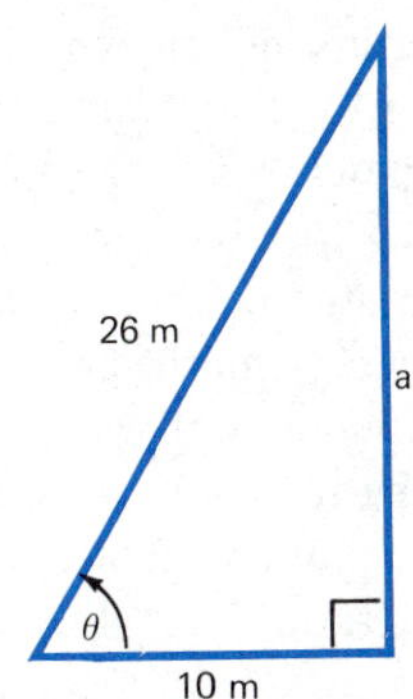

Figure 16-3

EXERCISE 16-1

Data, diagram, formula, substitute, check. Make a sketch of each of the following triangles indicating the angle with the letter θ. Determine the third side and evaluate the sine, cosine and tangent for each.

1. ADJ = 4, OPP = 3
2. OPP = 8, ADJ = 6
3. HYP = 15, ADJ = 9
4. ADJ = 12, HYP = 20
5. OPP = 56, HYP = 65
6. HYP = 53, OPP = 28
7. OPP = 8, ADJ = 15
8. ADJ = 7, OPP = 24
9. ADJ = 16, HYP = 34
10. HYP = 41, ADJ = 9
11. HYP = 29, OPP = 20
12. OPP = 12, HYP = 37

16-2 FUNCTIONS OF AN ANGLE

The sine, cosine and tangent ratios are functions of the angle. Their values do not depend on the length of the sides. Consider the similar triangles formed by several vertical lines, Figure 16-4. In the 3-4-5 right triangle, ADJ = 3, OPP = 4 and HYP = 5. The three ratios are

$$\sin\theta = \frac{4}{5} = 0.800$$

$$\cos\theta = \frac{3}{5} = 0.600$$

$$\tan\theta = \frac{4}{3} = 1.33$$

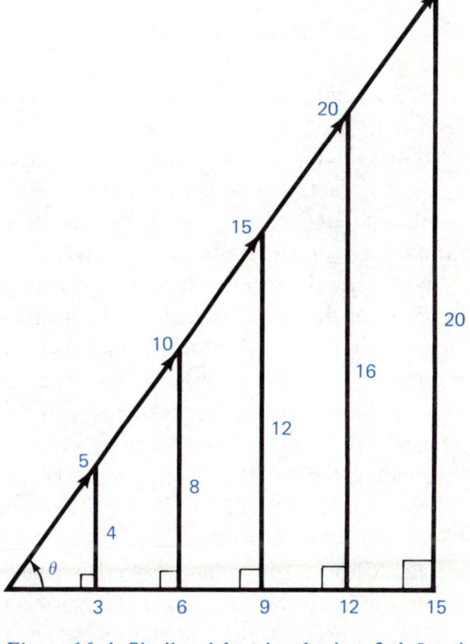

Figure 16-4 Similar right triangles in a 3:4:5 ratio

In the 6-8-10 triangle, all sides are twice as long as those in the 3-4-5 triangle. The trigonometric ratios are the same as before.

$$\sin\theta = \frac{8}{10} = 0.800$$

$$\cos\theta = \frac{6}{10} = 0.600$$

$$\tan\theta = \frac{8}{6} = 1.33$$

Triangles 9-12-15, 12-16-20 and 15-20-25 give the same results.

$$\sin\theta = 0.800$$

$$\cos\theta = 0.600$$

$$\tan\theta = 1.33$$

As one moves from one triangle to another, the values of the three trigonometric ratios remain constant as does the angle itself. That would still be true if the lines were extended to form even larger similar triangles.

If the angle changes, one should expect the sine, cosine and tangent to change. Using the 3-4-5 triangle, imagine the hypotenuse as a radius vector having rotated counterclockwise by a small amount, Figure 16-5. The result is an increase in the angle.

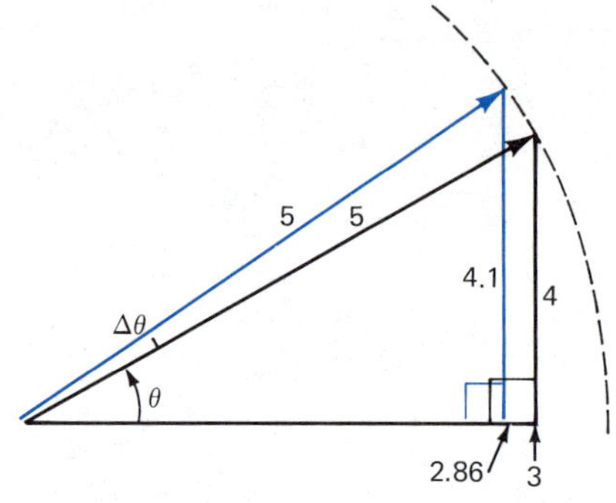

Figure 16-5 Effect of a small change of the angle

Suppose the amount was just enough to change the opposite side from 4 units of length to 4.1. At the same time, the adjacent side decreases in length from 3 units to 2.86. The hypotenuse is the same as before. The trigonometric ratios have all changed.

$$\sin\theta_2 = \frac{4.1}{5} = 0.820$$

$$\cos\theta_2 = \frac{2.86}{5} = 0.572$$

$$\tan\theta_2 = \frac{4.1}{2.86} = 1.43$$

The sine and tangent increased along with the angle while the cosine decreased.

Every angle has its own associated sine, cosine and tangent. Until a few years ago, the values were listed in most mathematics and science textbooks. These were usually given in a one-page table to four decimal places for each degree between 0 and 90°. Some mathematical and other handbooks had 45 pages of "trigonometric tables." These tables gave the ratios to five decimal places for each minute of the arc.

With the electronic calculator, tables are not necessary. Some calculators have the capability of a nine-decimal-place readout. These can distinguish between the sine of an angle and that of one that is one ten-millionth of a degree larger. To store such information in tables would require about 20 000 books the size of this text. With the calculator, it is all available at one's fingertips.

Finding the value of a trigonometric ratio for an angle is a one-number function of the calculator. It must first be set in the correct angular mode. If the given angle is expressed in degrees, set the calculator in the degree mode. Choose the radian mode for angles given in radians. Some calculators use a selector switch for this, others use the DRG or other key. Some use the display to indicate the mode setting; others do not. If a selector switch is not used, the calculator will automatically be in the degree mode when it is turned on. Refer to the owners manual for mode selection.

Once the angular mode is set, enter the angle. Press the correct trigonometric function key, [Sin], [Cos] or [Tan]. The value of the function will now be displayed. Round all function values to four significant figures. For many electrical applications, three significant figures may be adequate.

Example A Determine the sine, cosine and tangent for an angle of 25° using a calculator.

Solution: 25 [Sin], 0.4226
25 [Cos], 0.9063
25 [Tan], 0.4663
Remember, the displayed answer always follows the comma.

Example B Find the sine, cosine and tangent for an angle of 0.5 rad.

Solution: Set the calculator in the radian mode.
0.5 [Sin], 0.4794
0.5 [Cos], 0.8776
0.5 [Tan], 0.5463

Example C Find the three trigonometric functions for an angle of $\frac{1}{6}\pi \cdot \text{rad}$.

Solution: In these problems, the angle must be converted to radians or degrees. In radians, this can be done by multiplying π by $\frac{1}{6}$ or dividing by 6. Set the calculator in the radian mode.
6 [1/x] [×] [π] [=] [Sin], 0.5000 (Using ×)
[π] [÷] 6 [=] [Cos], 0.8660 (Using ÷)
[π] [÷] 6 [=] [Tan], 0.5774

Note: Since $1\ \pi \cdot \text{rad} = 180°$, $\frac{1}{6}\pi \cdot \text{rad}$ is 180 [÷] 6 [=], 30°. To find the sine (calculator in degree mode), for example: 180 [÷] 6 [=] [Sin], 0.5000

Every angle has its own associated sine, cosine and tangent. The values of each function are listed in Appendix B for angles between 0° and 90°. It is expected that most students will use the calculator to determine the functions of a given angle. The table, however, gives an overall perspective which the calculator is unable to display. Instructions for using the table are given in Appendix B.

Refer to Appendix B and note the following: As the angle changes from 0° to 90°, the sine changes from 0 to 1. It is always a fraction which increases as the angle increases. The tangent also increases as the angle increases. It is a fraction only up to 45°, however.

For right triangles of larger angles, the opposite side is greater than the adjacent side and the tangent is greater than 1. The tangent becomes very large as the angle approaches 90°. In contrast to the sine and tangent, the cosine *decreases* as the angle increases. It changes from 1 to 0 as the angle changes from 0° to 90°.

EXERCISE 16-2

Find the sine, cosine and tangent for each of the following angles. Do not convert units. Set the calculator in the unit mode as given.

1. 8°
2. 40°
3. 68°
4. 1 rad
5. 0.25 rad
6. 77.4°
7. 0.333 rad
8. 0.75 rad
9. 31.2°
10. 1.25 rad
11. 56.8°
12. 1.5 rad

Use the radian mode to find the sine, cosine and tangent of the following angles.

13. $\frac{1}{3}\pi\cdot$rad
14. $0.25\,\pi\cdot$rad
15. $0.125\,\pi\cdot$rad
16. $\frac{1}{6}$ r

Use the degree mode to find the trigonometric functions of these angles.

17. $0.375\,\pi\cdot$rad
18. $\frac{1}{12}\pi\cdot$rad
19. $\frac{1}{8}$ r
20. $0.0625\,\pi\cdot$rad

16-3 ESTIMATING SIDES AND ANGLES

In side-angle (SA) problems, one side and one angle are known. The main problem is to find the other two sides. The other angle is simply the complement of the known angle. An accurate freehand sketch should always be constructed first. It not only gives a visual check but good estimates of the unknown sides can be found. This provides a means of checking calculated answers.

An accurate triangle can be drawn in the following way:

PROCEDURE FOR DRAWING ORDINARY SA TRIANGLES

- Draw two coordinate axes emphasizing the first quadrant.
- The x-axis will represent the adjacent side of the triangle.
- Draw the hypotenuse so the angle between it and the x-axis is the reference (known) angle.
- Draw the opposite side as a vertical line intersecting the hypotenuse at least 1 1/2 inches from the origin. It will be perpendicular to the x-axis.

The above procedure can be used for any triangle as long as one angle is known. It does not make any difference which side is the known side since they are proportional. The third step is the most important. The angle must be estimated as closely as possible.

The purpose of drawing the coordinate axis is to give a reference (90°) from which the angle can be estimated. Remember, a bisected quadrant produces two 45° angles, Figure 16-6(A). Bisecting again gives 22.5° angles, Figure 16-16(B). Trisecting a quadrant produces three 30° angles, Figure 16-7. If one of these is bisected, the result is 15° angles. These references will allow good estimates. Be sure to draw the two axes perpendicular. If the angle can be drawn within 5° to 10° of the given angle, estimates of the lengths of sides will be adequate.

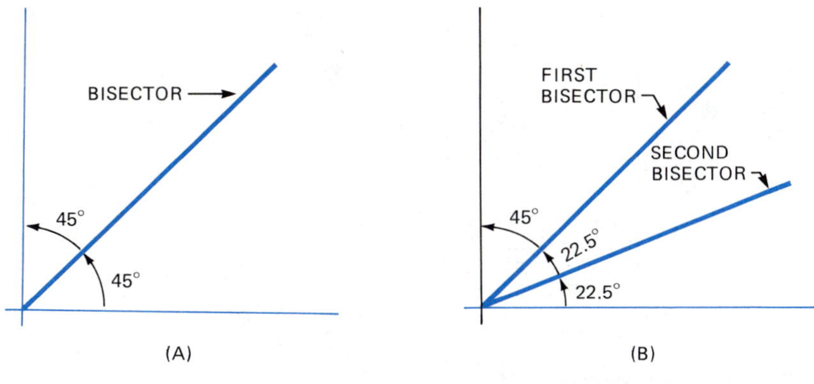

Figure 16-6 Bisecting 90° angles and 45° angles

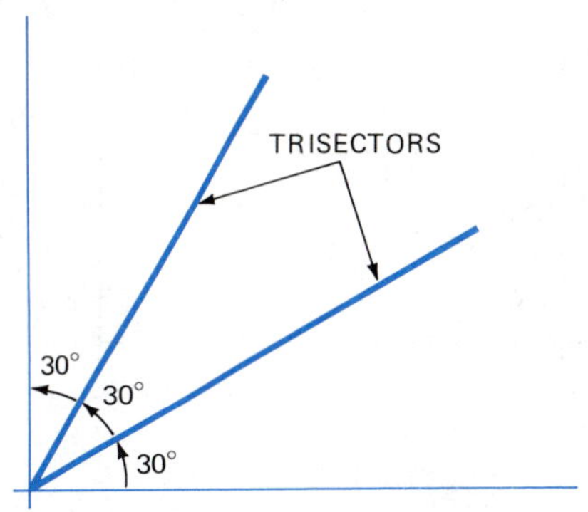

Figure 16-7 Trisecting 90° angles

Rotate small sides onto larger sides, Figure 16-8(A). The hypotenuse can now be estimated. After OQ is rotated into position OR, estimate RP: 24.7 cm + RP = estimate of the hypotenuse. RP is about $\frac{1}{5}$ of OR, so estimate the hypotenuse as ≈ 25 cm + 5 cm or ≈ 30 cm. The symbol (≈) means *approximately equal to.*

In Figure 16-8(B), side PQ is rotated onto side OQ. It is more than half as long as OQ. A good estimate then is ≈ 14 cm.

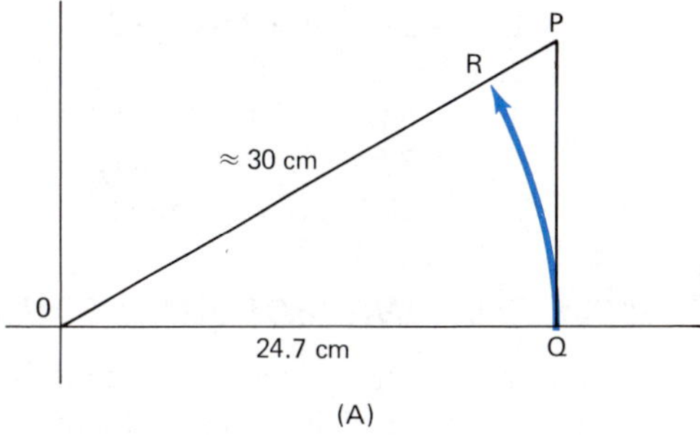

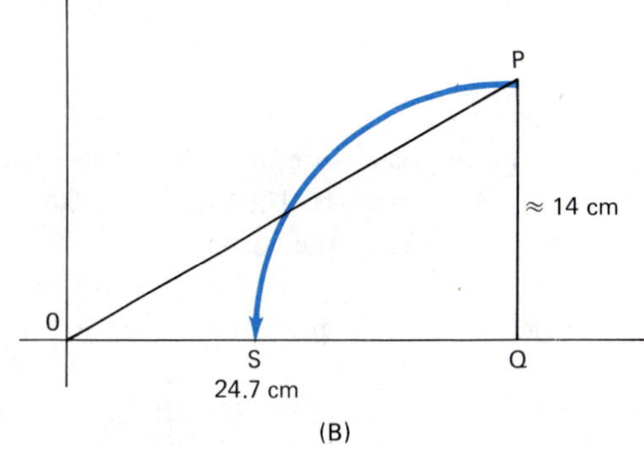

Figure 16-8 Estimating sides of a triangle. (A) Side OQ is rotated into position OR. (B) Side PQ is rotated into position SQ.

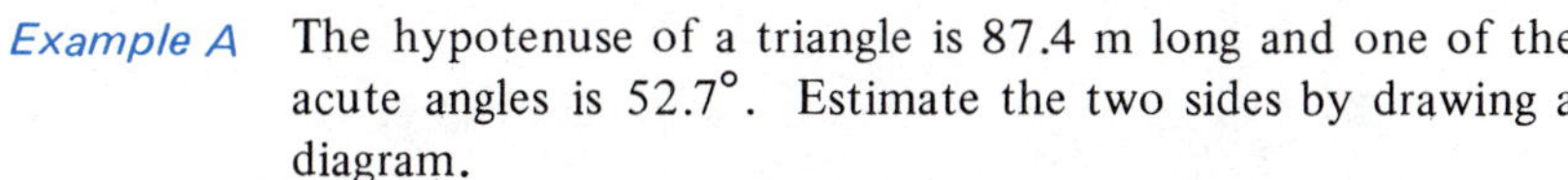

Example A The hypotenuse of a triangle is 87.4 m long and one of the acute angles is 52.7°. Estimate the two sides by drawing a diagram.

Solution: The angle is a few degrees larger than a 45° angle. Draw the axes, hypotenuse and vertical side as shown, Figure 16-9(A). Label the known quantities.

Opposite side: Rotate the opposite side onto the hypotenuse using P as the pivot point, Figure 16-9(B). OPP ≈ $\frac{3}{4}$(90 m) or ≈ 67 m.

Adjacent side: Rotate the adjacent side onto the hypotenuse, Figure 16-9(C), using point O as the pivot. ADJ is more than half but less than two-thirds the hypotenuse. Estimate at ≈ 55 m.

Note: The actual values are OPP = 69.5 m and ADJ = 53.0 m.

With practice, good estimates can be made on freehand drawings. The triangle can be drawn in other quadrants, if desired. Estimating is similar.

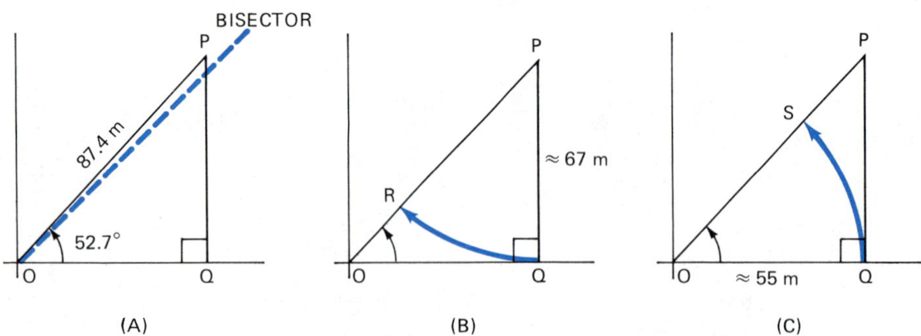

Figure 16-9

EXERCISE 16-3

With the following right triangle data, draw a freehand diagram estimating the angle accurately. Obtain reasonable values from the diagram for the two unknown sides. Do not use drawing equipment. The names of the sides are in reference to the angle given.

1. $\theta = 21.3°$, HYP = 29.3 cm
2. $\phi = 78.1°$, OPP = 55.2 km
3. $\theta = 64.8°$, ADJ = 7.12 in
4. $\phi = 37.4°$, HYP = 44.6 ft
5. $\theta = 58.9°$, OPP = 356 m
6. $\phi = 17.5°$, ADJ = 17.3 km
7. $\theta = 84.6°$, HYP = 52.8 ft
8. $\phi = 33.3°$, OPP = 4.2 mm
9. $\theta = 68.3°$, ADJ = 2.5 cm
10. $\phi = 41.7°$, HYP = 430 m

16-4 SOLVING SIDE-ANGLE RIGHT TRIANGLES

The three trigonometric functions (sine, cosine and tangent) are functions of angles of right triangles, Figure 16-10:

$$\sin\theta = \frac{a}{c}$$

$$\cos\theta = \frac{b}{c}$$

$$\tan\theta = \frac{a}{b}$$

a is the opposite side, b is the adjacent side and c is the hypotenuse. As drawn in Figure 16-10, side a is often called the *altitude* and side b is called the *base* of the triangle.

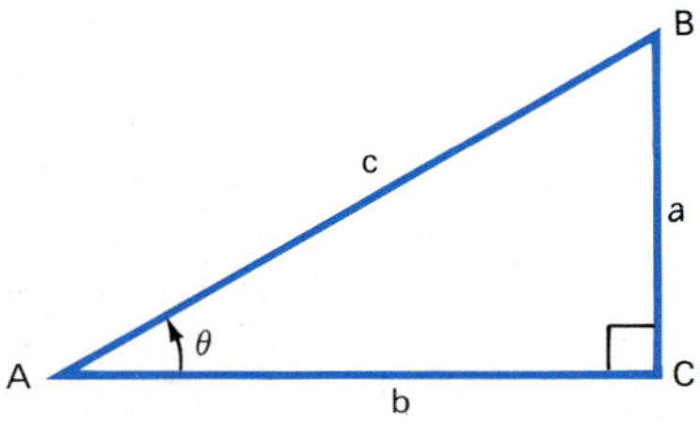

Figure 16-10 The ABC right triangle

These functions can be used to find the unknown parts of a right triangle in side-angle problems. Use the following procedure:

PROCEDURE FOR SOLVING SA RIGHT TRIANGLES

- Draw an accurate freehand sketch of the right triangle.
- Label the known parts with the given values and indicate the unknown parts with a proper symbol.
- Choose the trigonometric function containing the two knowns and one of the unknowns. Remember the two acute angles are complementary.
- Solve for the unknown value.

- Repeat for the other unknowns.
- Compare calculated values with the sketch for a quick check. Use the Pythagorean Theorem if an accurate check is required.

Example A Find sides a and b in Figure 16-11. Side c is 12.5 cm and angle θ is 28.7°.

Solution: Decide which function contains the two known values and one of the unknowns.

Formula: $\sin \theta = \frac{a}{c}$ (Unknown: a)

$a = c(\sin \theta)$

Substitute: $a = 12.5(\sin 28.7°)$

$a = 12.5(0.4802) = 6.00$ cm

or: 12.5 [×] 28.7 [Sin] [=], 6.00 cm

(Calculator in degree mode.)

Formula: $\cos \theta = \frac{b}{c}$ (Unknown: b)

$b = c(\cos \theta)$

Substitute: $b = 12.5(\cos 28.7°)$

$b = 12.5(0.8771) = 11.0$ cm

or: 12.5 [×] 28.7 [Cos] [=], 11.0 cm

Check: $c^2 = a^2 + b^2$

$12.5^2 = 6^2 + 11^2$

$12.5 = \sqrt{36 + 121} = \sqrt{157} = 12.5$

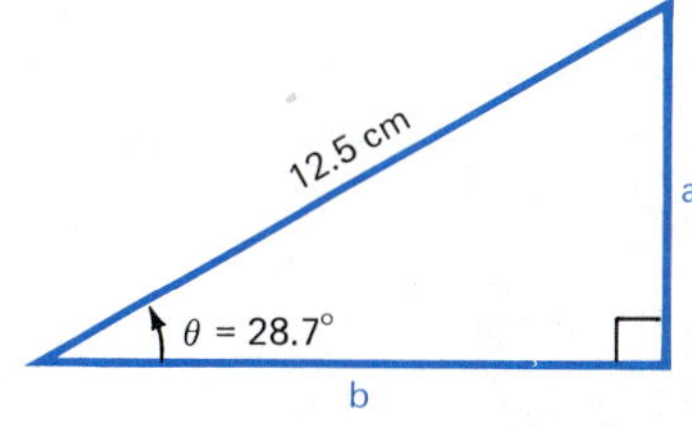

Figure 16-11

Example B Find the hypotenuse, Figure 16-12, side b and angle ϕ if a = 46.7 ft and θ = 59.4°

Solution: Angle ϕ can be found at any time.

Formula: $\theta + \phi = 90°$

Substitute: $59.4° + \phi = 90°$

$\phi = 30.6°$

Find side c (Formula must contain a, c and θ)

Formula: $\sin \theta = \frac{a}{c}$

$c = \frac{a}{\sin \theta}$

Substitute: $c = \frac{46.7}{\sin 59.4} = \frac{46.7}{0.8607} = 54.3$ ft

or: 46.7 [÷] 59.4 [Sin] [=], 54.3 ft

Find side b (Formula contains a, b and θ)

Formula: $\tan \theta = \frac{a}{b}$

$b = \frac{a}{\tan \theta}$

Substitute: $b = \frac{46.7}{\tan 59.4°} = \frac{46.7}{1.691} = 27.6$ ft

or: 46.7 [÷] 59.4 [Tan] [=], 27.6 ft

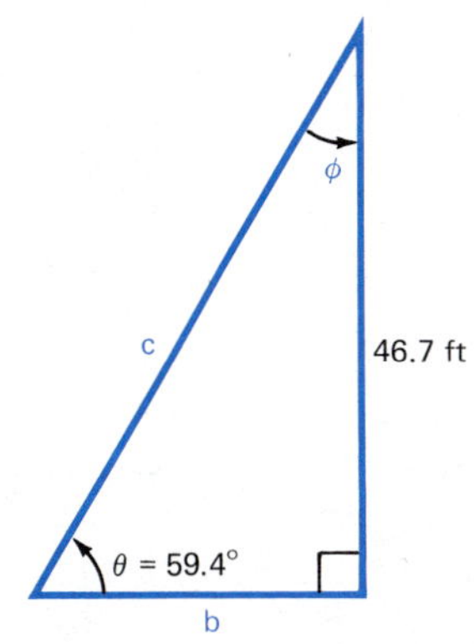

Figure 16-12

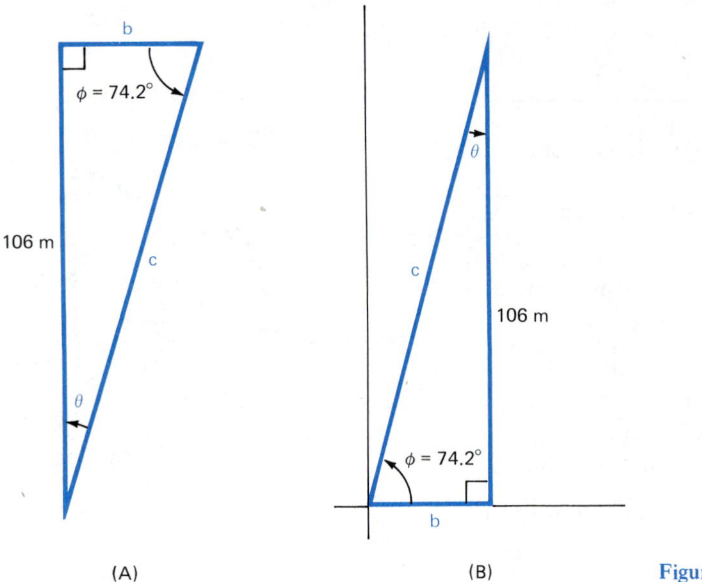

Figure 16-13

Example C Find side c, b and θ, Figure 16-13(A). Side a is 106 m and $\phi = 74.2°$. An alternative sketch is shown in Figure 16-13(B) for estimating purposes.

Solution: Solve for c

Formula: $\sin \phi = \frac{a}{c}$

$$c = \frac{a}{\sin \phi}$$

Substitute: $c = \frac{106 \text{ m}}{\sin 74.2°} = \frac{106 \text{ m}}{0.962} = 110 \text{ m}$

or: 106 [÷] 74.2 [Sin] [=], 110 m

Solve for b:

Formula: $\tan \phi = \frac{a}{b}$

$$b = \frac{a}{\tan \phi}$$

Substitute: $b = \frac{106 \text{ m}}{\tan 74.2°} = \frac{106 \text{ m}}{3.534} = 30.0 \text{ m}$

or: 106 [÷] 74.2 [Tan] [=], 30.0 m

Solve for θ:

Formula:

$$\theta + \phi = 90°$$
$$\theta + 74.2° = 90°$$
$$\theta = 15.8°$$

In many word problems, it is necessary to find only one of the unknowns. Making a sketch of the right triangle is especially important in stated problems. Make sure it indicates the situation as described. Otherwise, the procedure is the same.

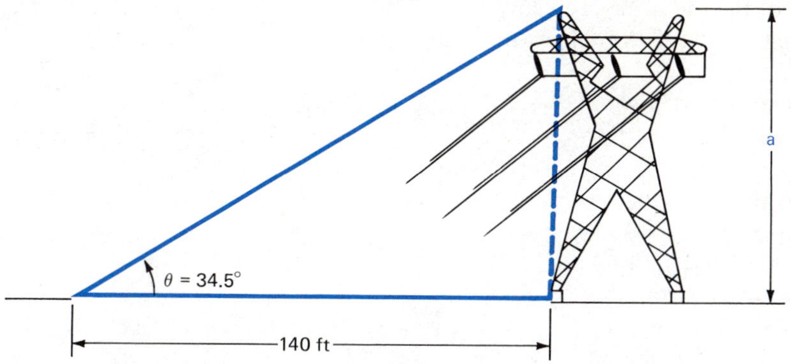

Figure 16-14

Example D A transmission-line tower has an unknown height. A surveying party finds the top of the tower makes an angle of 34.5° above the horizon at a distance of 140 ft from the base. How tall is the tower?

Solution: Data: See Figure 16-14.

Formula: $\tan\theta = \frac{a}{b}$

$$a = b(\tan\theta)$$

Substitute: $a = 140(0.6873) = 96$ ft

An observer often looks at an object above or below the horizon. The angle between the *line of sight* and the horizontal is called the *angle of elevation* for objects above the horizon. For objects below the horizon, it is called the *angle of depression.*

The angle between an inclined surface or plane and the horizontal is the *angle of incline.* The *grade* or *slope* of an incline is the rise to run ratio or simply the tangent of the angle of incline. Grade is often expressed in percent. A 25% grade rises 25 ft for each 100 ft of horizontal run.

EXERCISE 16-4

Construct a freehand diagram and estimate the unknowns. Solve each triangle for all unknown parts. Check by comparing estimated and calculated answers.

1. $a = 44$ m, $\theta = 23.7°$
2. $a = 645$ cm, $\theta = 62.1°$
3. $b = 8.8$ ft, $\theta = 41.2°$
4. $b = 5280$ ft, $\theta = 45°$
5. $c = 64.3$ km, $\theta = 18.4°$
6. $c = 29\bar{0}0$ m, $\theta = 71.6°$

Construct a freehand diagram and solve for all unknown parts using the trigonometric functions. Check by using the Pythagorean Theorem.

7. $a = 32.6$ m, $\phi = 47.6°$
8. $b = 1.35$ cm, $\phi = 36.8°$
9. $c = 94.0$ mi, $\phi = 29.2°$
10. $a = 7.39$ km, $\phi = 68.7°$
11. $b = 5.84 \times 10^4$ m, $\theta = 1.00$ rad
12. $c = 82.6 \times 10^6$ in, $\theta = 0.500$ rad

Construct a diagram of the situation in each of the following problems. Estimate the value of the unknown. Calculate the value of the unknown: data, diagram, formula, substitute, check.

13. How long is a guy wire if it makes an angle of 55° with the ground and is anchored 8.5 m from the base of the pole?
14. An electrical conduit pipe is bent as shown in Figure 16-15. It is 5.3 ft between bends horizontally. How much is the rise if the angle with the horizontal is 40°?
15. A helicopter travels 32.8 km in a straight line from the marshaling yard to the job site. It travels at an angle of 52.3° with a north-south road. How far south and how far east is the job site?
16. A guy wire is attached to a pole at a point 54.3 ft above the ground. It makes a 35° angle with the pole. How far is the anchor from the base of the pole?
17. A 30% grade makes an angle of 16.7° with the horizontal. What length of transmission line cable (three wires) is required for a 30% grade with a horizontal run of 5.3 km?
18. A diagonal brace on a line tower makes an angle of 45° with the horizontal cross member. How long is the cross member if the brace is 2.75 m long?
19. A cross-country transmission line, Figure 16-16, makes an angle of 67° with an east-west road. (a) What north-south distance is covered if the east-west distance is 43 mi? (b) How long is the transmission line?
20. A tall pole is guyed with two cables parallel to each other, Figure 16-17(A). How far apart are the two anchors if the longer guy is 25 m long and the shorter is tied 14.5 m above the ground? Each cable makes an angle of 55° with the ground.

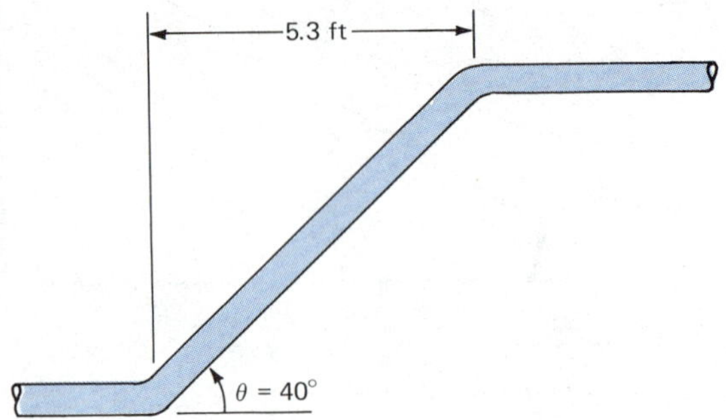

Figure 16-15

Figure 16-16 A 115 000-V high-voltage transmission line

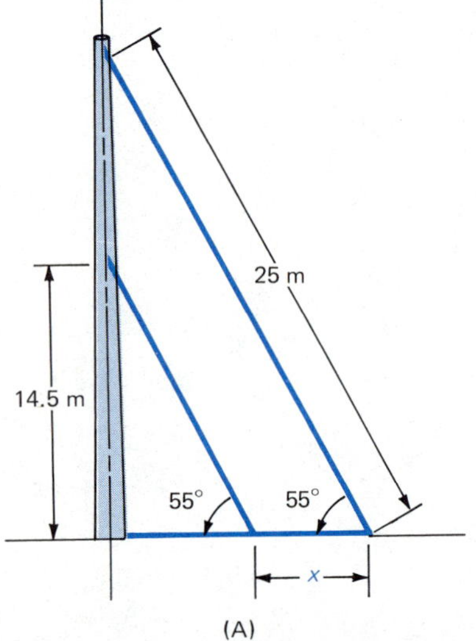

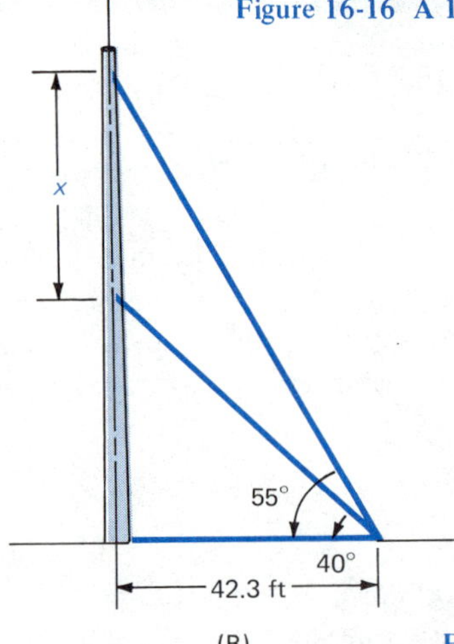

Figure 16-17

21. A tower is guyed with two cables both anchored 42.3 ft from the base, Figure 16-17(B). The longer cable makes an angle of 55° with the ground and the shorter a 40° angle. How far below the top is the shorter one attached to the tower?
22. An electrician places a ladder against a building. It touches 5.85 m above the ground and makes a 27.5° angle with the building. (a) How far is the base of the ladder from the building? (b) How long is the ladder?
23. A transmission line crosses a river, Figure 16-18. A surveying crew locates point C at a distance of 57.0 m from the near tower. They measure the angle at C to be 62.84°. Find the distance, AB, between towers. Assume the angle at A is a right angle.

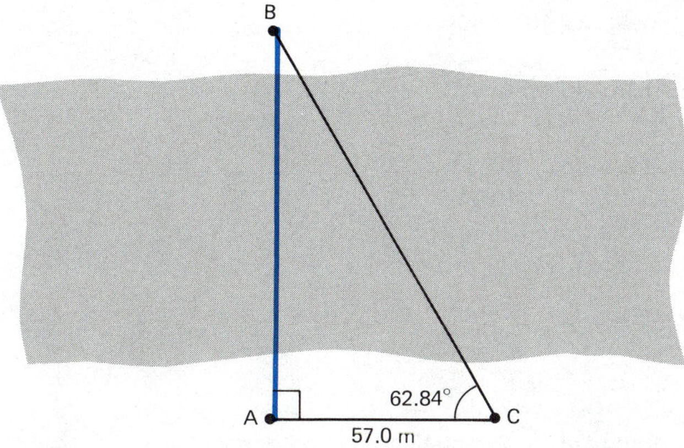

Figure 16-18

24. A substation is located on a right-triangular piece of land. One of the acute angles is 52.5° and the hypotenuse is 108 m. Find the area of the piece of land ($A = \frac{1}{2}bh$).
25. The shadow of a smokestack at a power plant is 38.5 m long. The angle of elevation of the sun is 65.2°. How tall is the smokestack?

INVERSE TRIGONOMETRIC FUNCTIONS

In side-angle problems, a side and an angle are the known quantities. The other acute angle is easily found.

$$\theta + \phi = 90°$$

The other two sides have definite lengths which can be determined using the trigonometric functions.

$$\sin \theta = \frac{\text{OPP}}{\text{HYP}} = \frac{a}{c}$$

$$\cos \theta = \frac{\text{ADJ}}{\text{HYP}} = \frac{b}{c}$$

$$\tan \theta = \frac{\text{OPP}}{\text{ADJ}} = \frac{a}{b}$$

The third side of a right triangle can be found if two sides are known (side-side problems) using the Pythagorean Theorem.

$$c^2 = a^2 + b^2$$

When two sides are known, the triangle can be drawn. The acute angles will have definite values. However, the Pythagorean Theorem and the trigonometric functions do not provide a means by which the angles can be calculated. The inverse trigonometric functions are introduced next for just that purpose.

16-5 ARC SINE, ARC COSINE AND ARC TANGENT

Each of the trigonometric functions contains two sides and an angle. If two sides are known, (side-side problems) the angle is one of the unknowns. The value of one of the trigonometric functions can be calculated. From this, the angle can be determined. The process of doing it is an inverse procedure.

Suppose the sine of an angle is 0.866 calculated from the known hypotenuse and opposite side. The problem is to find:

"the angle (θ) whose sine is 0.866"

There are many problems of this nature, so more convenient methods of writing such statements have been devised. The angle indicated is called the *arc sine* and is abbreviated in one of the following three ways:

$$\theta = \arcsin 0.866$$
$$\theta = \sin^{-1} 0.866$$
$$\theta = \text{invsin } 0.866$$

Each case means "θ is the angle whose sine is 0.866." All three symbols are in common use. The *arc tangent* and *arc cosine* are indicated in similar ways.

The second symbol uses an exponent notation and is the simplest to write. Students should be cautioned that the −1 symbol does not mean reciprocal. The reciprocal of the sine of an angle would be written $(\sin x)^{-1} = \frac{1}{\sin x}$. To be meaningful, the number x must be an angle. With the −1 acting on the function symbol, $\sin^{-1} y$, it means inverse sine (invsin y), the same as the third symbol. In this case, the number y is simply a number and represents a ratio of two sides.

The angle is easily determined for any proper value of the sine, cosine or tangent by using a calculator. Most calculators use two keystrokes such as [INV] [Sin]. Others are marked [Sin⁻¹] but require a second function keystroke first. Read the owners guide.

The angle will be displayed in the units that have been preselected. If the degree is the desired unit, set the calculator in the degree mode. For radians, set it in the radian mode. Since the sine and cosine ratios have the hypotenuse (the longest side) in the denominator, their values are always less than 1. The display will indicate an error for numbers greater than 1. The tangent ratio has no such limits. The student should evaluate the following examples with the calculator as the problem is read.

Example A Find the angle with a sine of 0.866. Express the angle in degrees.

Solution: Set the calculator in the degree mode. Enter the number 0.866, press the inverse key, then the sine key. The angle is now displayed:
0.866 [INV] [Sin] , 60.0°

Example B Find the arc tangent of 0.5774 in radians.
Solution: Set the calculator in the radian mode.
0.5774 [INV] [Tan] , 0.524 rad

Example C Evaluate: $\cos^{-1} 0.707$ in degrees.
Solution: Move again to the degree mode.
0.707 [INV] [Cos] , 45°

Example D Evaluate: arctan 2.704 in radians.
Solution: Change to the radian mode.
2.704 [INV] [Tan] , 1.217 rad

EXERCISE 16-5

Use the calculator to evaluate each of the following angles in degrees.

1. The angle with a sine of 0.2354
2. The angle with a cosine of 0.4395
3. The angle with a tangent of 0.3852

4. arcsin 0.3195
5. arccos 0.5219
6. arctan 1.752
7. $\sin^{-1} 0.6785$
8. $\cos^{-1} 0.8391$
9. $\tan^{-1} 8.483$
10. invsin 0.9115
11. invcos 0.9831
12. invtan 5.731
13. $\tan^{-1} 100$

Find the following angles in radians.

14. The angle with a sine of 0.0225
15. The angle with a cosine of 0.7500
16. The angle with a tangent of 0.5774

17. $\sin^{-1} 0.5055$
18. arccos 0.4852
19. invtan 0.6797
20. arcsin 0.8431
21. $\cos^{-1} 0.1852$
22. arctan 2.851
23. invsin 0.7852
24. invcos 0.2975
25. $\tan^{-1} 3.484$
26. $\tan^{-1} 1000$

Find the following angles in the units indicated.

27. $\sin^{-1} \frac{4}{5}$ (deg)
28. $\cos^{-1} \frac{3}{8}$ (rad)
29. $\tan^{-1} \frac{13}{15}$ (rad)
30. $\arcsin \frac{25}{32}$ (deg)
31. $\arccos \frac{7}{16}$ (deg)
32. $\arctan \frac{25}{12}$ (rad)

Evaluate each of the following using a calculator.

33. sin (arcsin 0.4521)
34. cos (invcos 0.9955)
35. tan ($\tan^{-1}$ 4.535)
36. $\sin^{-1}$ (sin 30°)
37. arccos (cos 45°)
38. invtan (tan 60°)

16-6 SOLVING SIDE-SIDE RIGHT TRIANGLES

A trigonometric function contains three quantities: two sides and an angle. The corresponding inverse trigonometric function contains the same two sides and the same angle. In terms of a standard ABC right triangle, Figure 16-19:

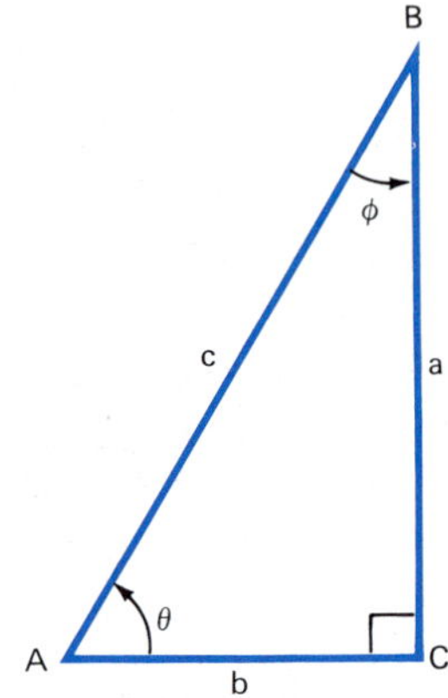

Figure 16-19 The ABC right triangle

Trigonometric Function	**Inverse Trigonometric Function**
$\sin\theta = \frac{a}{c}$	$\theta = \arcsin\frac{a}{c}$
$\cos\theta = \frac{b}{c}$	$\theta = \arccos\frac{b}{c}$
$\tan\theta = \frac{a}{b}$	$\theta = \arctan\frac{a}{b}$

In side-side right triangles, two of the three sides are known. One of the acute angles can be found using the inverse trigonometric functions. The third side can then be evaluated with one of the trigonometric functions. Use the following procedure:

PROCEDURE FOR SOLVING SS RIGHT TRIANGLES

- Draw a freehand sketch of the triangle.
- Label all known parts with the given values and indicate the others using symbols.
- Choose the inverse trigonometric function containing the two known sides.
- Solve for the acute angle (θ).
- Use either of the trigonometric functions containing the unknown side.
- Solve for the unknown side.
- Compare calculated values with the sketch for a quick check. For a more accurate check, use the Pythagorean Theorem.

If the opposite and adjacent sides are known, an accurate freehand drawing can be drawn, Figure 16-20. Make the adjacent side (OB) at least an inch long on the *x*-axis. Draw a vertical line (BC). Estimate its length accurately in proportion to the adjacent side. Connect points O and C. Estimate the length of the hypotenuse and size of the angle.

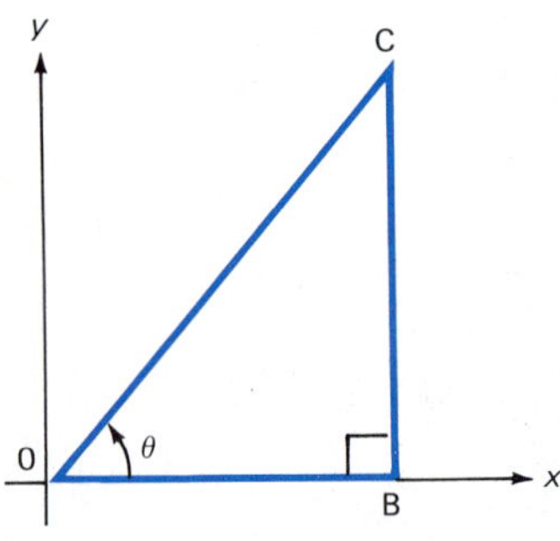

Figure 16-20 Drawing a right triangle

If the hypotenuse is one of the given sides, an accurate construction is more difficult. A simple diagram with the knowns and unknowns labeled is adequate. The following examples show how to calculate the unknown angles and the unknown side in SS problems.

Example A Find the base length (b) and the two angles in the right triangle of Figure 16-21. The hypotenuse (c) has a length of 84.8 m and side a is 45.2 m.

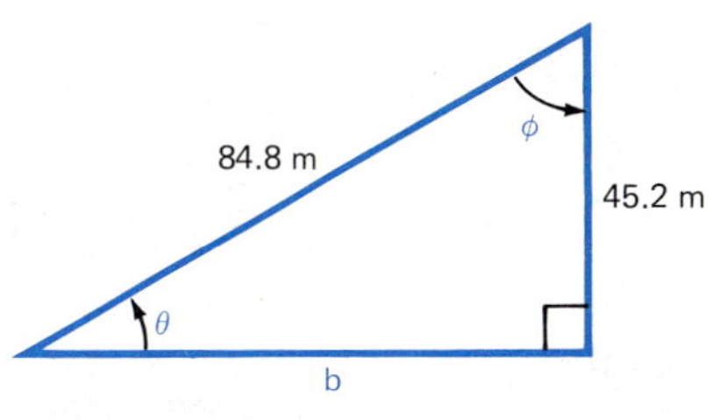

Figure 16-21

Solution: Find angle θ by using the arc sine function:

Formula: $\theta = \arcsin \frac{a}{c}$

Substitute: $\theta = \arcsin \frac{45.2}{84.8}$

45.2 [÷] 84.8 [=] [INV] [Sin] , 32.2°

Formula: $\theta + \phi = 90°$

Substitute: $\phi = 90° - 32.2° = 57.8°$

Formula: $\cos \theta = \frac{b}{c}$ (b is the unknown)

$b = c(\cos \theta)$

Substitute: $b = 84.8(\cos 32.2°)$

84.8 [×] 32.2 [Cos] [=], 71.7 m

Check: $c^2 = a^2 + b^2$

$84.8 = \sqrt{45.2^2 + 71.7^2} = \sqrt{7184} = 84.8$

Note: Side b could have been found using the tangent function.

It would be better to delay the calculation of the other acute angle. The reason is the angle (32.2^0) is displayed in the calculator. The second calculation, $b = 84.8(\cos 32.2^0)$ is 32.2 [Cos] [×] 84.8 [=], 71.7 m. This is much easier since the angle (32.2^0) does not need to be cleared and re-entered.

Example B Find a, θ and ϕ in the right triangle of Figure 16-22. The hypotenuse is 144 ft long and side b is 95.0 ft long. Express the angle in radians.

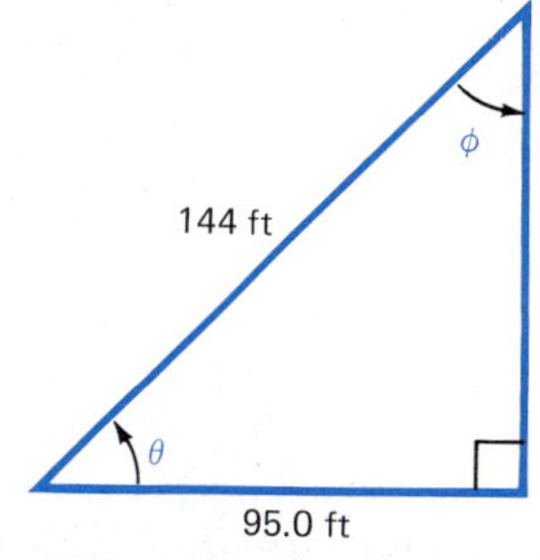

Figure 16-22

Solution: Set the calculator in the radian mode. Choose the arc function containing θ, b and c.

Formula: $\theta = \cos^{-1} \frac{b}{c}$

Substitute: $\theta = \cos^{-1} \frac{95.0 \cancel{ft}}{144 \cancel{ft}}$

95.0 [÷] 144 [=] [INV] [Cos] , 0.850 rad

Do not clear the calculator!

Formula: $\tan \theta = \frac{a}{b}$

$a = b(\tan \theta)$

$a = (95.0 \text{ ft})(\tan 0.850)$

$a = (\tan 0.850)(95.0 \text{ ft})$

0.850 [Tan] [×] 95.0 [=], 108 ft

The angle (0.850 rad) did not need to be re-entered. The sine function could have been used to find a.

Formula: $\theta + \phi = \frac{1}{2}\pi \text{ rad}$

$\phi = \frac{1}{2}\pi - 0.850 = 0.721 \text{ rad}$

Note: Here, the angle must be re-entered. When working with degrees, this step can often be done manually.

Check: $c^2 = a^2 + b^2$

$$144 = \sqrt{108^2 + 95^2} = \sqrt{20\,689} = 144$$

Example C Find θ and ϕ in degrees and the length of side c if side a is 246 km and side b is 188 km.

Solution: In this problem an accurate drawing can be made, Figure 16-23. Reasonable estimates are: $c \approx 300$ km and $\theta \approx 50°$.

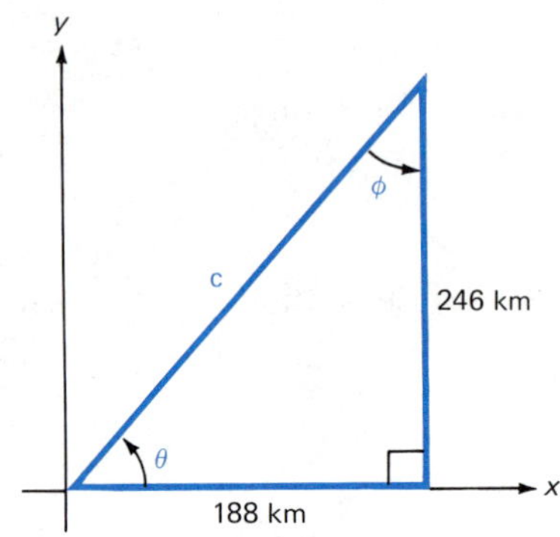

Figure 16-23

Formula: $\theta = \text{invtan}\ \frac{a}{b}$ (Degree mode)

Substitute: $\theta = \text{invtan}\ \frac{246\ \text{km}}{188\ \text{km}}$

246 [÷] 188 [=] [INV] [Tan] , 52.6°

Do not clear!

Formula: $\cos\theta = \frac{b}{c}$ (c is the unknown)

$$c = \frac{b}{\cos\theta}$$

Substitute: $c = \frac{188\ \text{km}}{\cos 52.6°}$

Since the angle (52.6°) is still in the calculator, it is best to evaluate the cosine, take the reciprocal and multiply by the numerator. The student may wish to review Topic 3-14.

52.6 [Cos] [1/x] [×] 188 [=] , $31\bar{0}$ km

Check: Compare estimated and calculated values.

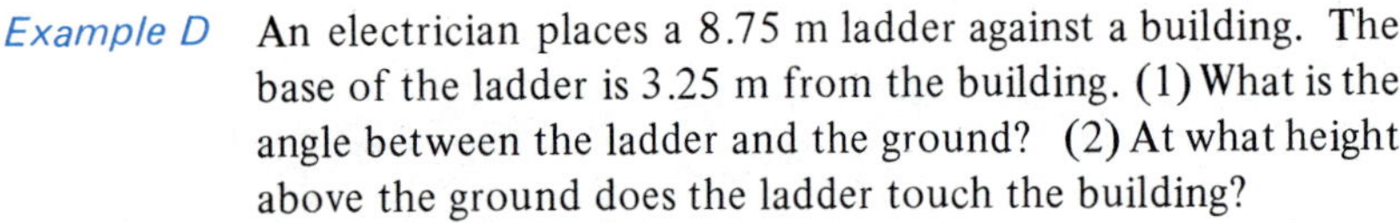

Example D An electrician places a 8.75 m ladder against a building. The base of the ladder is 3.25 m from the building. (1) What is the angle between the ladder and the ground? (2) At what height above the ground does the ladder touch the building?

Solution: Data-diagram: See Figure 16-24.

Formula (1): $\theta = \cos^{-1}\frac{b}{c}$

Substitute: $\theta = \cos^{-1}\frac{3.25\ \text{m}}{8.75\ \text{m}}$

3.25 [÷] 8.75 [=] [INV] [Cos] , 68.2°

Do not clear!

Formula (2): $\sin\theta = \frac{a}{c}$

$$a = c(\sin\theta)$$

Substitute: $a = 8.75(\sin 68.2°)$

68.2° [Sin] [×] 8.75 [=] , 8.12 m

Check: Use Pythagorean Theorem.

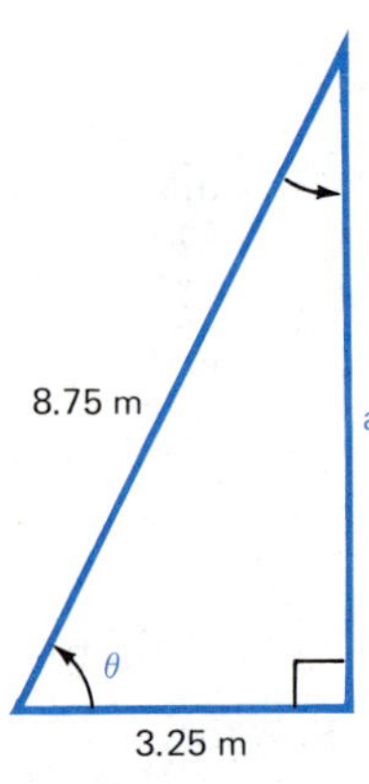

Figure 16-24

EXERCISE 16-6

Determine one of the acute angles in the units indicated for each of the following triangles using one of the arc functions. Find the third side using one of the trigonometric functions. Make a sketch for each problem and check. Make an accurate drawing for those marked with an asterisk and estimate answers.

1. b = 8, a = 15 (deg)*
2. c = 37, b = 12 (deg)
3. b = 20, c = 29 (rad)
4. a = 9, c = 41 (rad)
5. c = 34, a = 16 (deg)
6. a = 7, b = 24 (deg)*
7. c = 53, b = 28 (rad)
8. b = 56, c = 65 (rad)
9. a = 12, b = 16 (deg)*
10. a = 24, b = 7 (deg)*

Data, diagram, formula, substitute, check.

11. *A guy wire, Figure 16-25, is attached to a pole at a point 16.5 m above the ground. It is anchored at a point 12.2 m from the base. (a) What angle does it make with the ground? (b) How long is the wire?
12. A helicopter leaves the marshaling yards to go to a job site. It travels 62.4 km in a straight line. The work crew travels by van straight east, then 41 km south. (a) What angle with the north-south road did the helicopter travel? (b) How much farther did the crew have to travel?
13. An electrical conduit pipe is bent, Figure 16-26. Find (a) the angle θ and (b) the vertical rise.

Figure 16-25 Wires to the right begin on this pole. The guy wire balances the tension pulling to the right.

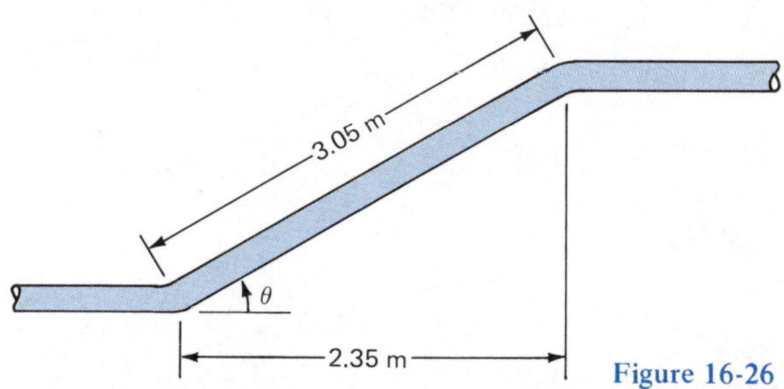

Figure 16-26

14. *A 40% grade rises 40 m for each horizontal distance of 100 m. (a) What is the angle of incline? (b) What length of transmission line (3 wires) is required on a 40% grade with a horizontal run of 3.1 km?
15. *A service drop connects at a pole 12.5 m above the ground. It connects to the service entrance of a house 3.85 m above ground and 30.2 m from the pole. (a) Find the angle the service drop makes with the horizontal (neglect sag). (b) Neglecting sag, how long is the service drop? (c) What length is required if 5% is added for sag and splicing?
16. *The dimensions of a room are 60 ft by 42 ft. A three-wire cable runs from one corner to the opposite corner in the attic. How long is the cable?

17. A transmission line from a power plant to a substation follows a straight path 42.5 mi long. The substation is 31.8 mi east of the plant. (a) What angle does the line make with a north-south road? (b) How far north is the substation?
18. *A pole is guyed with two wires, Figure 16-27, both anchored 14.2 m from the base. The shorter one is attached 11.9 m above the ground and the longer 22.3 m. (a) Find the angle between the guy wires. (b) What is the total length of guy wire used?

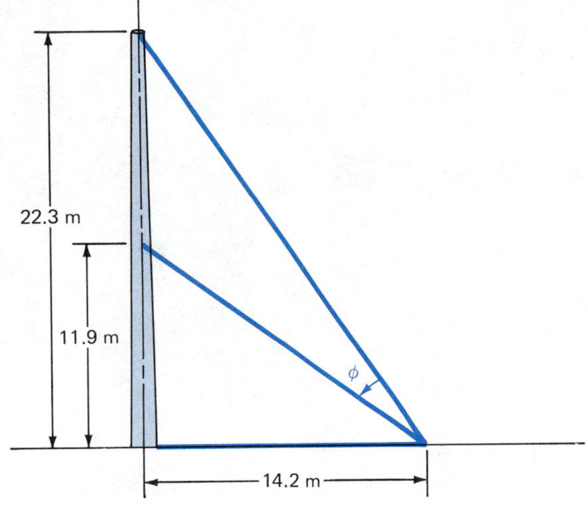

Figure 16-27

19. A 32-ft ladder touches the side of a building 28 ft above the ground. (a) What angle does it make with the building? (b) How far is the base from the building?
20. A diagonal brace on a crossbar of a pole, Figure 16-28, is 1.00 m long. It is attached to the crossber 65 cm from the center of the pole. (a) What angle does it make with the pole? (b) it is attached to the pole how far below the center of the crossbar?

Figure 16-28 Crossbars are braced with steel bars. The four lines attached to insulators carry 4160 V from the substation. The three wires just below carry 120/240 V. These are the secondary lines of the pole transformer on the next pole. The two cables angling to the right are service drops. Near the bottom are cable TV and telephone lines (large).

21. A storage yard is in the shape of a right triangle. One side is 65 m and the hypotenuse is 95 m long. What is the area ($A = \frac{1}{2}bh$) of the yard?

22. A transmission line tower is 120 m from the base of a cliff 50 m high, Figure 16-29. A second tower is set back 20 m from the edge of the cliff. Neglecting sag, how long is the line that connects them? Assume the two towers are of equal height.

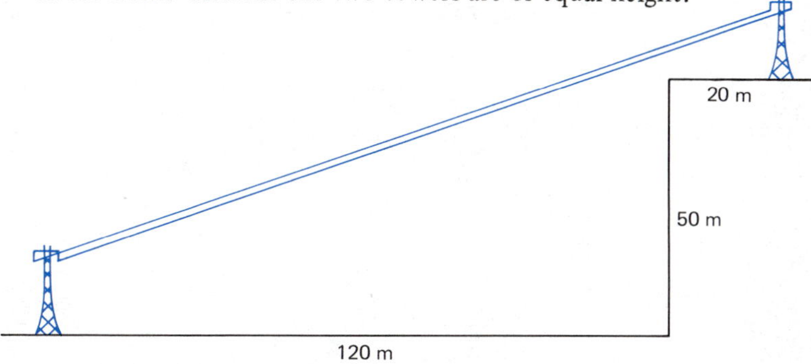

Figure 16-29

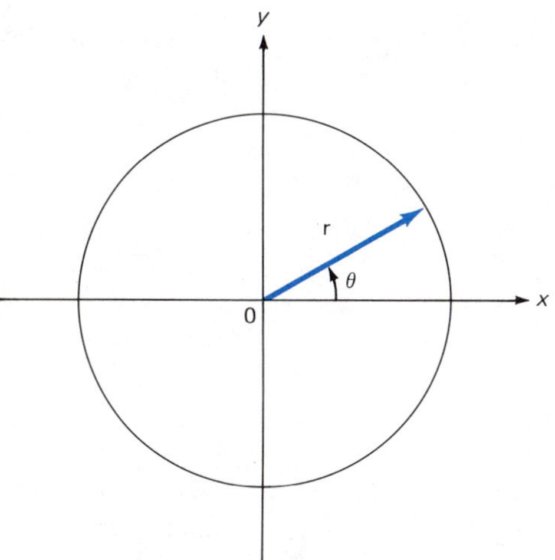

Figure 16-30 Radius vector

FUNCTIONS OF GENERATED RIGHT TRIANGLES

A radius vector, Topic 15-2, sweeps out angles as it rotates. Angles of any size, positive or negative, are possible. With one additional step, the radius vector can become the hypotenuse of a right triangle. Such triangles are functions of the angle generated. The sine, cosine and tangent, therefore, become functions of generated angles of any size, also. Their values repeat after each complete rotation.

This is an important concept in the study of electricity. Alternating currents are generated by rotating equipment. The expansion of the trigonometric functions to include angles greater than 90° allows alternating currents to be described mathematically.

16-7 GENERATING RIGHT TRIANGLES

A radius vector generates a positive angle as it rotates counterclockwise. It is represented as an arrow, Figure 16-30. It has a length (r), the radius of the circle the arrow tip generates. The radius vector represents the terminal side of the angle generated. The angle is in standard position when the initial side is along the positive x-axis.

These ideas are extended by adding a vertical line through the arrow tip downward to the x-axis. The vertical line is perpendicular to the x-axis and forms a right triangle, Figure 16-31. The right triangle has sides of length, x and y, which are the coordinates (x,y) of the point at the arrow tip. Side x is called the *x-component* and y is the *y-component.*

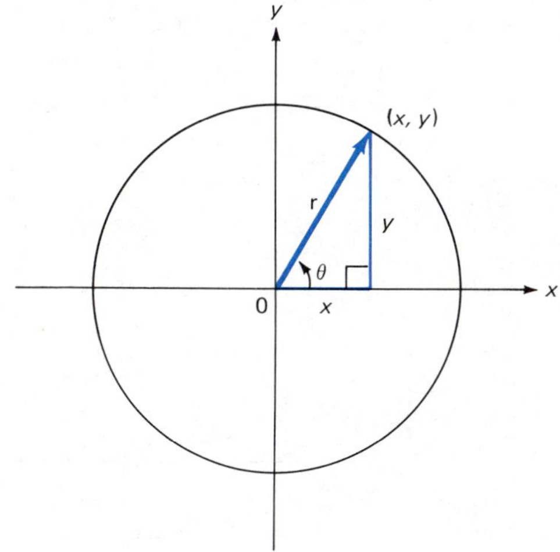

Figure 16-31 Generated right triangle. The triangle is formed by drawing a vertical line from the arrow point to the x-axis.

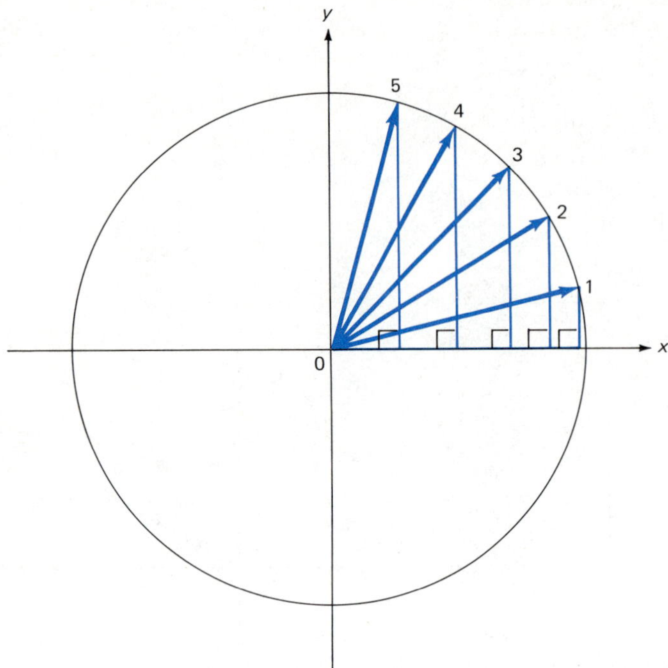

Figure 16-32 Radius vector shown in several positions as it rotates counterclockwise. Side x decreases and y increases in the triangles formed in the sequence.

Imagine the radius vector rotating from standard position to position 1, Figure 16-32. Side x is much greater than side y. As the radius vector rotates past position 2 and on toward position 3, y gets larger and x becomes smaller. In position 3, the angle is 45° and $x = y$. Beyond position 3, y is greater than x. Beyond position 5, it is much greater than x as the angle approaches 90°.

What happens when the radius vector rotates beyond the 90° mark? Little is changed. It continues to generate right triangles, Figure 16-33(A). The triangle so formed is called a *second quadrant triangle*. The x-component has changed to a negative value since it extends to the left of the origin. The angle θ is now greater than 90°.

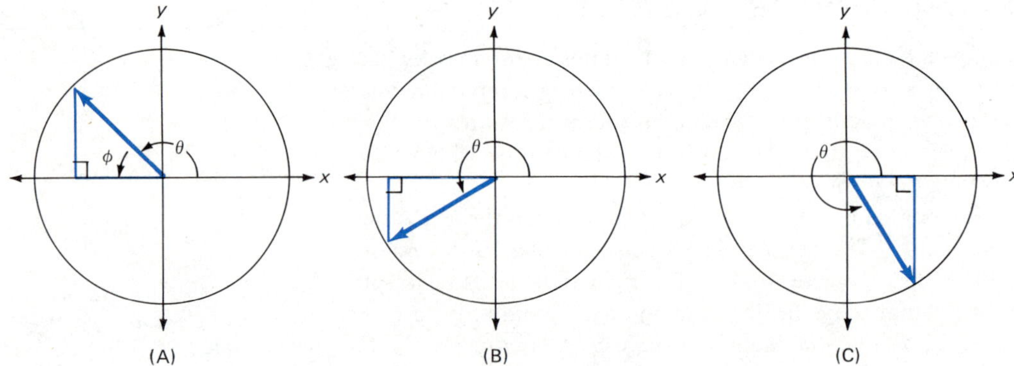

Figure 16-33 (A) Second quadrant triangle (B) Third quadrant triangle (C) Fourth quadrant triangle

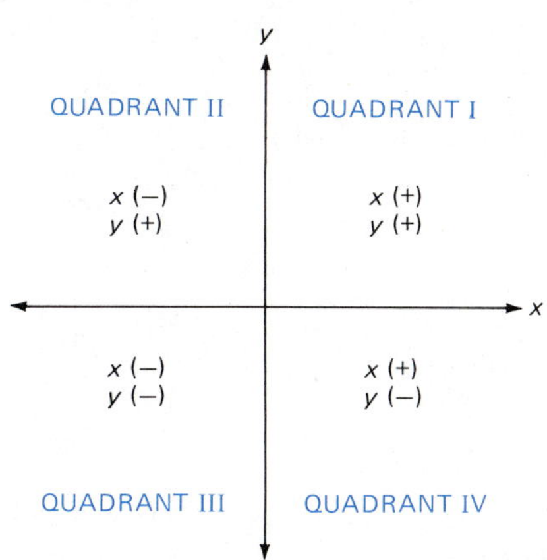

Figure 16-34 Signs of *x* and *y* in various quadrants

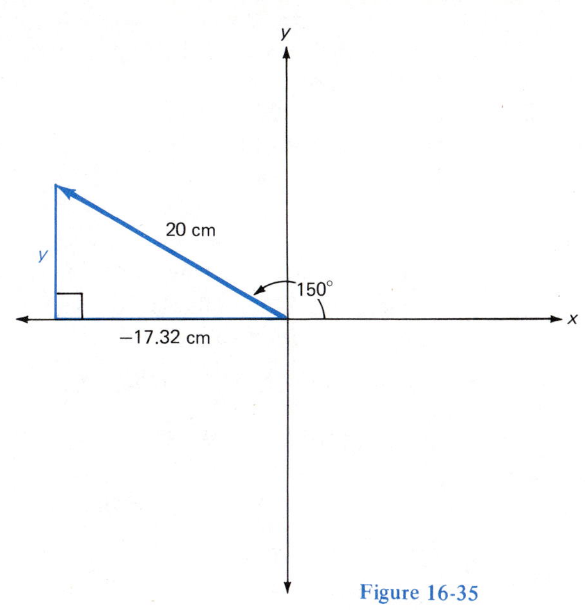

Figure 16-35

When the radius vector is in the third quadrant, Figure 16-33(B), the right triangle is still formed by a vertical line through the arrow tip. It forms a right angle with the x-axis. In the third quadrant, x and y both have negative values. Fourth quadrant triangles, Figure 16-33(C), have positive x-components and negative y-components. A summary is shown in Figure 16-34.

Example A Draw a right triangle if the radius vector is rotated 150° from standard position and has a length of 20 cm. In this position x has a value of −17.32 cm. Find y.

Solution: See Figure 16-35.

Formula: $r^2 = x^2 + y^2$ (Pythagorean Theorem)

$$y = \sqrt{r^2 - x^2}$$

Substitute: $y = \sqrt{(20)^2 - (-17.32)^2}$

$$y = \sqrt{400 - 300} = \sqrt{100} = \pm 10 \text{ cm}$$

Note: In this case the positive value is chosen because side y extends upward from the x-axis: $y = 10$ cm.

16-8 SIGNS OF THE FUNCTIONS

The length of the radius vector (r) always has a positive value. In terms of previous notation (OPP, ADJ and HYP), the following statements are always true:

- Radius vector (r) = hypotenuse (HYP)
- x-coordinate (x) = adjacent side (ADJ)
- y-coordinate (y) = opposite side (OPP)

Example A Give the values of (1) the hypotenuse, (2) adjacent side and (3) the opposite side for the triangle in Figure 16-35.

Solution: (1) HYP = r = 20 cm
(2) ADJ = x = −17.32 cm (Note the negative sign)
(3) OPP = y = 10 cm

Example B Calculate the sine, cosine and tangent from the previous example using (1) the values of the sides of the triangle, and (2) using the function keys on the calculator for the given angle:

Solution: (1) $\sin 150° = \frac{y}{r} = \frac{10}{20} = 0.500$

$\cos 150° = \frac{x}{r} = \frac{-17.32}{20} = -0.866$ (Note the negative signs on the cosine and tangent.)

$\tan 150° = \frac{y}{x} = \frac{10}{-17.32} = -0.577$

(2) 150 [Sin] , 0.500
150 [Cos] , −0.866
150 [Tan] , −0.577
Either method gives the correct sign.

When the radius vector terminates in the second or third quadrant, the x-component of the triangle formed is negative. The y-component is negative for third and fourth quadrant triangles, Figure 16-33. The trigonometric functions, therefore, have positive or negative values accordingly. These are summarized in Figure 16-36. When the radius vector rotates clockwise through a negative angle, triangles are generated in a similar way.

Example C Through what negative angle would the radius vector have to rotate to produce the same right triangle as that in Figure 16-35?

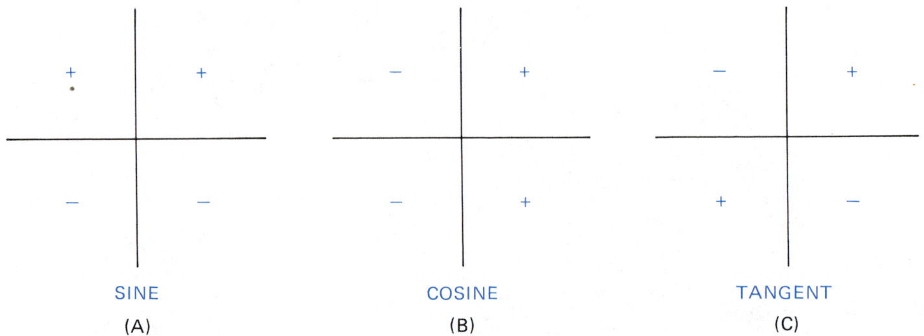

Figure 16-36 Signs of the trigonometric functions in various quadrants

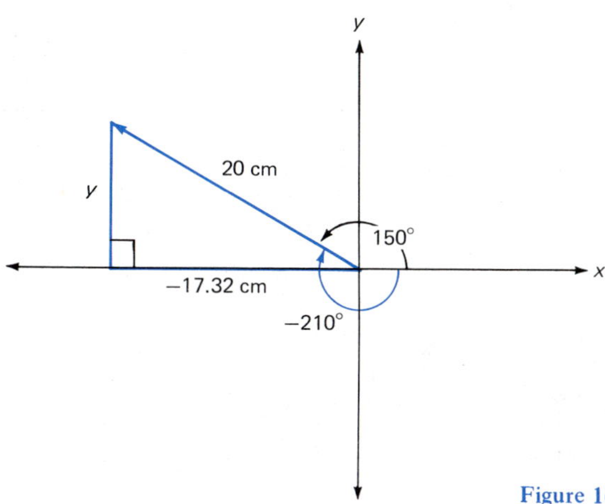

Figure 16-37

Solution: To produce an equivalent right triangle, the radius vector must terminate in exactly the same position, Figure 16-37. It must rotate clockwise through a negative angle of

$$150° - 360° = -210°$$

Such a right triangle is identical in every respect. The sine, cosine and tangent will have the same values and signs as in the previous example.

EXERCISE 16-7

Data is given for several generated right triangles in the following problems. Make a freehand drawing of each, estimating lengths and angles as accurately as possible. For each problem, (a) determine the third side of the triangle with the correct sign, (b) calculate the sine, cosine and tangent using ratios of sides and check with the calculator using the given angle.

1. $r = 25$ m, $x = 15$ m, $\theta = 53.1°$
2. $r = 44$ cm, $x = -22$ cm, $\theta = 120°$
3. $x = -23$ in, $y = 23$ in, $\theta = 135°$
4. $x = -52$ mi, $r = 75$ mi, $\theta = 226°$
5. $y = -25$ km, $r = 55$ km, $\theta = -27°$
6. $x = -143$ cm, $y = 143$ cm, $\theta = -225°$
7. Find the equivalent angle for each of the following if the radius vector rotates in the opposite direction.
   a. 75°
   b. $\frac{1}{2}\pi\cdot\text{rad}$
   c. −150°
   d. −1.57 rad
   e. 240°
   f. −45°
   g. $\frac{3}{4}\pi\cdot\text{rad}$
   h. 320°
8. In what quadrant is:
   a. x positive and y negative?
   b. y negative and x negative?
   c. x negative and y positive?
   d. x positive and y positive?

9. In what quadrant is:
   a. the sine positive and the tangent negative?
   b. the cosine positive and the tangent negative?
   c. the sine and cosine both negative?
   d. the cosine and tangent both positive?
10. Find the sine, cosine and tangent for each of the following angles using the calculator and state the quadrant in which each generated triangle lies.

a. 239°	g. $\frac{1}{2}\pi\cdot$rad	*l.* −740°
b. 1.57 rad	h. −52.5°	m. $\frac{3}{4}\pi\cdot$rad
c. 316°	i. 193.5°	n. 1039°
d. −143°	j. $\frac{1}{8}\pi\cdot$rad	o. −3785°
e. −3.00 rad	k. 5.00 rad	p. −7.00 rad
f. 255°		

16-9 INVERSE FUNCTIONS OF GENERATED TRIANGLES

A generated triangle of 150°, Figure 16-38(B), is similar to one of 30°, Figure 16-38(A). The only difference is that the x-component is negative. The sine, cosine and tangent have the same value for both angles except that the cosine and tangent are negative. Comparable statements can be made about a third-quadrant triangle, Figure 16-38(C), formed by an angle of 210° and a fourth-quadrant one of −30°, Figure 16-38(D). Notice in the four cases, the angle between the radius vector and the x-axis inside the triangle is 30°.

The sines of the four angles, Figure 16-38, all have the same magnitude (0.500). It is positive in the first and second quadrant and negative in the third and fourth, the same as the y-component, Figure 16-36. This creates a difficulty because the calculator cannot distinguish between the sine of 30° and that of 150° (0.500). Nor can it distinguish between the sine of 210° and that of −30° (−0.500). The calculator simply chooses 30° and −30° for the arc sine of 0.500 and −0.500, respectively. The student should try it on the calculator. These angles are called the *principal values* of the arc sine.

As the radius vector rotates from 0° to 90°, the sine changes from 0 to 1. From 90° to 180°, it changes back from 1 to 0. In the 180° to 270° range, the sine starts at 0 and changes to −1. It changes from −1 to 0 as

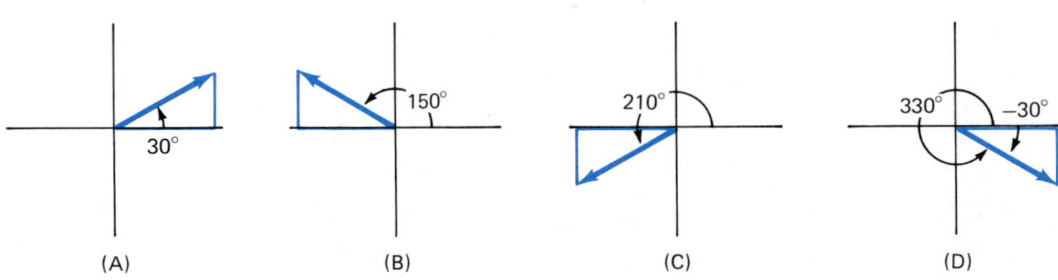

Figure 16-38

QUADRANT	ARC FUNCTIONS (°) (ANGLES)	TRIGONOMETRIC FUNCTIONS SINE	COSINE	TANGENT
IV	−90° to 0°	[−1 to 0]	0 to 1	[−∞ to 0]
I	0° to 90°	[0 to 1]	[1 to 0]	[0 to ∞]
II	90° to 180°	1 to 0	[0 to −1]	∞ to 0
III	180° to 270°	0 to −1	−1 to 0	0 to −∞

Figure 16-39 Range of values of trigonometric functions and arc functions. Boxed-in ranges yield the corresponding principal values of the arc functions (angles).

the radius vector completes the last quarter revolution from 270° to 360°, Figure 16-39.

In a similar way, the cosine has a value of 1 at 0°, 0 at 90°, −1 at 180°, 0 at 270° and returns to 1 at 360°. The value of the tangent changes from 0 to ∞ (infinity), to 0, to −∞ and returns to 0 in one revolution. A summary is shown in Figure 16-39.

The calculator can only give the principal values of the arc functions. For the arc sine and arc tangent, these are angles from −90° to +90° (fourth and first quadrants). The principal values of the arc cosine are angles from 0° to 180° (first and second quadrants).

If the value of the sine, cosine or tangent is positive, the principal values (angles) are all in the first quadrant, Figure 16-40. The principal values of the arc sine and arc tangent for negative sines and tangents are in the fourth quadrant. The principal values of the arc cosine for negative cosines are in the second quadrant. Sometimes the "a" in arc and "i" in inverse are capitalized to indicate the principal value.

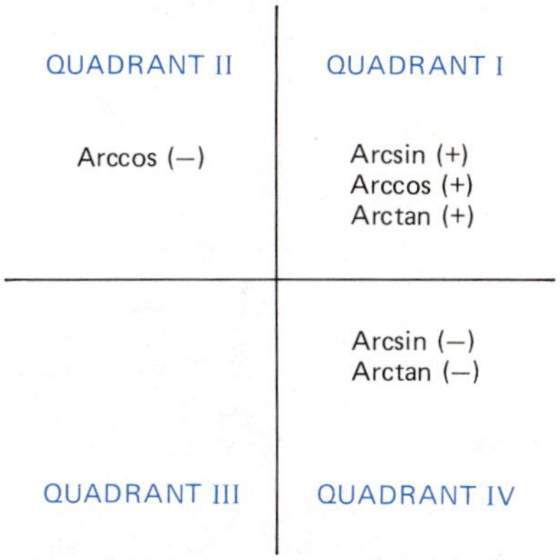

Figure 16-40 Location of the principal values of the arc functions

EXERCISE 16-8

Use the calculator to find the principal values of the arc function indicated. Give all answers in degrees.

1. Sin^{-1} 0.866
2. Arcsin 0.382
3. Cos^{-1} 0.500
4. Arccos 0.853
5. Tan^{-1} 0.750
6. Arctan 2.539
7. Arcsin (−0.566)
8. Invsin (−0.739)
9. Arccos (−0.222)
10. Invcos (−0.463)
11. Arctan (−7.399)
12. Invtan (−14.365)
13. Invsin (−0.0931)
14. Sin^{-1} (−0.931)
15. Invcos (−0.905)
16. Cos^{-1} (−0.101)
17. Invtan (−175)
18. Tan^{-1} (−1.000)

CHAPTER 17

ALTERNATING CURRENT

OBJECTIVES

After satisfactorily completing this chapter, the student should be able to:

- Construct graphs of the trigonometric functions.
- Define various concepts concerning cyclic motion and express alternating voltage and current as sine functions of time both in equation and graphic form.
- Define inductance and capacitance and calculate inductive and capacitive reactance.
- Define effective voltage and effective current and use Ohm's Law in purely resistive, purely inductive and purely capacitive circuits.

Over 99% of all power stations in the United States generate electricity in the form of alternating current. The main reasons are because it is easy to generate and economical to transmit over long distances. Transformers permit the distribution and use at safe voltages. In contrast, DC voltages are not readily changed.

This chapter begins with a discussion of the graphs of trigonometric functions generated by a rotating radius vector at larger angles. These concepts are then applied immediately to the description of alternating current. The chapter concludes by describing two important circuit quantities. In an ac circuit, *inductance* and *capacitance,* in addition to resistance, tend to oppose the flow of current. Their combination with *resistance* is the basis of most of the chapters that follow.

VOLTAGE GENERATION

A machine used for producing an emf by electromagnetic induction is called a *generator.* It has two parts: a *rotor* (rotating part) and a *stator* (stationary part). An engine or turbine causes the rotor to turn. The energy source might be a fossil fuel such as coal or diesel fuel. Other sources are the heat energy from concentrated nuclear piles, from the earth or the sun and the kinetic energy of moving water or air, Figure 17-1.

Figure 17-1 A small 2-kW wind generator

The discussion about transformers, Topic 14-9, involved a changing magnetic field in the secondary coil. When the strength of a magnetic field through the coil changes, a current is induced. In the case of transformers, the changing magnetic field was produced by an alternating current in the primary. In other words, the increasing, decreasing and reversing current of the primary caused an increasing, decreasing and reversing magnetic field through the secondary coil. The induced current in the secondary increases, decreases and reverses in the same time sequence as that of the primary.

Following a brief discussion about trigonometric graphs, other methods of changing the magnetic field strength through a coil are described. This and the resulting currents induced in the coil are the fundamental actions that make a generator work.

17-1 GRAPHS OF THE TRIGONOMETRIC FUNCTIONS

The range of values for the trigonometric functions was described in Topic 16-9 and summarized in Figure 16-39. The values of the sine and cosine vary between +1 and −1. The tangent, however, can have values from negative infinity ($-\infty$) to positive infinity (∞).

The relationship between any of the three functions and the angle is not linear. The sine function, for example, can be plotted on a graph in the following way. Choose a radius vector (hypotenuse) with a length of one unit, Figure 17-2. Such a vector is called a *unit vector.* Since r = 1 and $\sin \theta = y/\text{r}$, then:

$$y = \sin \theta$$

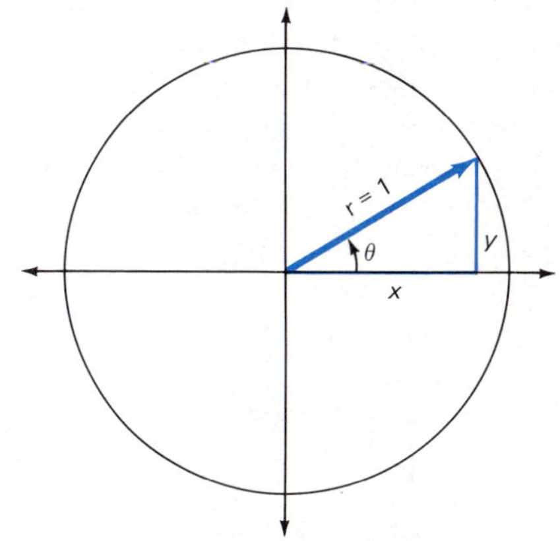

Figure 17-2 The unit vector. When a radius vector has a length of one unit, $y = \sin \theta$ and $x = \cos \theta$.

A graph of y vs θ is plotted in Figure 17-3. It shows how the sine function varies as the angle increases from 0° to 360°. If the radius vector continued to rotate, the graph would repeat itself over and over. This is an example of a *periodic function.*

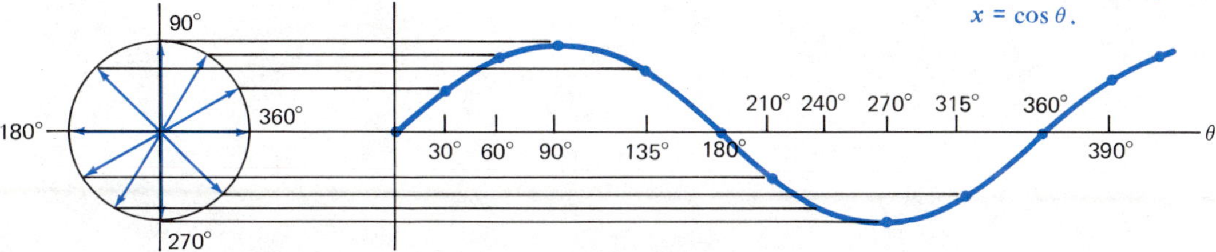

Figure 17-3 The sine wave, a periodic function

θ (°)	$\cos\theta$
0	1.00
15	0.97
30	0.87
45	0.71
60	0.50
75	0.25
90	0.00
105	−0.25
120	−0.50
135	−0.71
150	−0.87
165	−0.97
180	−1.00

θ (°)	$\cos\theta$
180	−1.00
195	−0.97
210	−0.87
225	−0.71
240	−0.50
255	−0.25
270	0.00
285	0.25
300	0.50
315	0.71
330	0.87
345	0.97
360	1.00

Figure 17-4 Table of cosine values

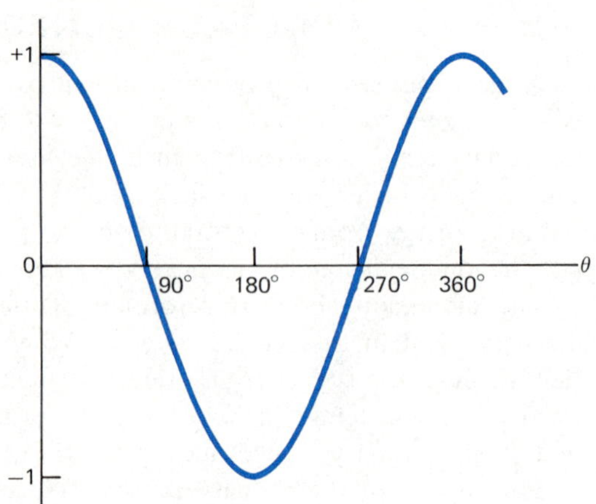

Figure 17-5 The cosine curve

Example A Plot a graph of the cosine function ($x = \cos\theta$). It represents the variation of side x, Figure 17-2, as the unit vector rotates.

Solution: Construct a table of values of the cosine for all angles that are multiples of 15°, Figure 17-4. Plot the graph of the data points, Figure 17-5. Draw a smooth curve through the points as described in Topic 4-2.

Notice that the values of the sine and cosine do not exceed +1 and are never less than −1. The cosine is another example of a periodic function since it too will repeat itself every 360°. The shape of the tangent function is shown in Figure 17-6. The value of the tangent is unlimited. Notice that it repeats itself every 180°. Even though it is *discontinuous* at 90° and at 270°, it is still considered a periodic function because of its repeating nature.

The graphs of the three trigonometric functions can be shifted to the right or left by adding or subtracting a constant to the angle. The function

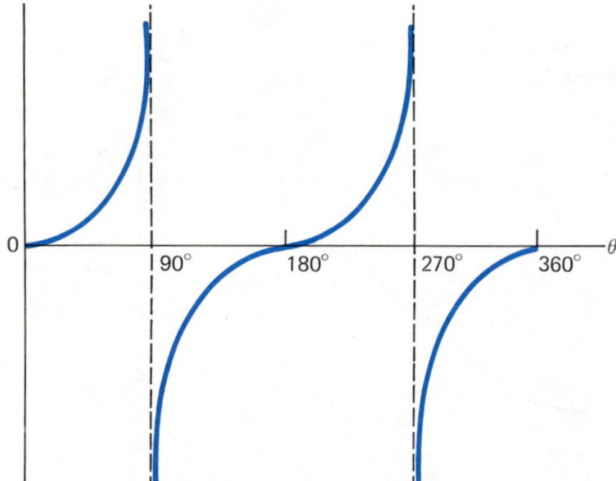

Figure 17-6 The tangent function

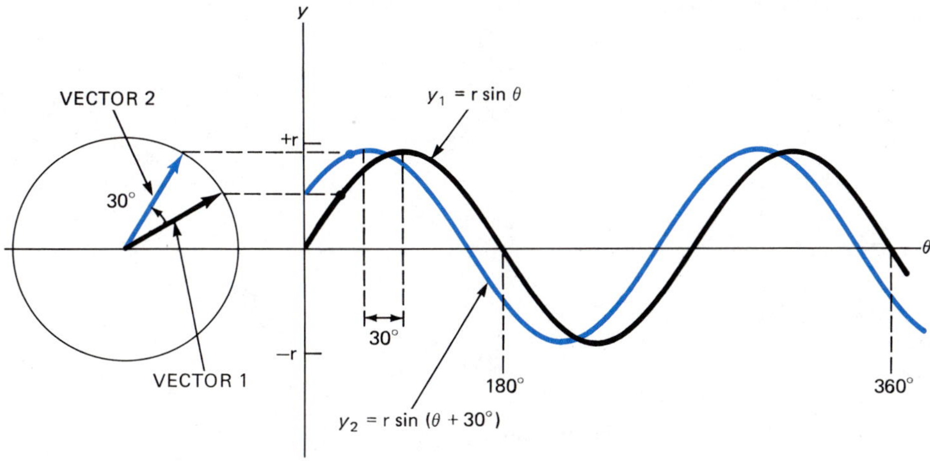

Figure 17-7 Sine waves showing phase angle. Wave y_2 leads wave y_1 by 30°. The two vectors rotate counterclockwise at the same speed.

$y_2 = \sin(\theta + 30°)$ shifts the entire curve 30° to the left, for example, Figure 17-7. The radius vectors shown in the circle indicates the reason for the shift. Such a constant is referred to as a *phase angle*. Function y_2 is said to *lead* y_1 by 30°. Function y_1 *lags* y_2 by 30°. If the plus sign in the equation for y_2 is replaced with a minus sign, the curve will shift to the right.

If the function is multiplied by a constant, the height of the curve can be changed, Figure 17-8. In this example, all values of the sine are multiplied by 2.5. Refer to the accompanying circles for an explanation. The value of the function $y_2 = 2.5 \sin \theta$ varies from +2.5 to −2.5.

EXERCISE 17-1

Make a table of values for each of the following functions. Plot a graph of each.

1. $y = \sin \theta$ (Every 15° between 0° and 360°)
2. $y = \sin \theta$ (Every 30° between 0° and 720°)
3. $y = \cos \theta$ (Every 30° between 0° and 720°)

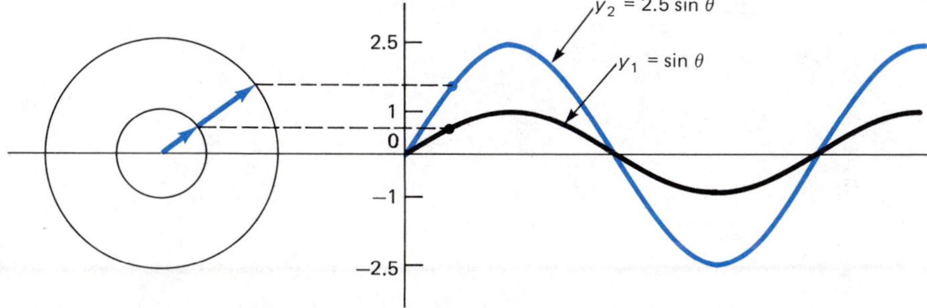

Figure 17-8 Sine waves with different heights

Plot the following pairs of functions on the same graph.

4. $y_1 = \sin\theta$ and $y_2 = 3.5 \sin\theta$
   (Every 30° between 0° and 360°)
5. $y_1 = \sin\theta$ and $y_2 = 2 \sin(\theta - 45°)$
   (Every 30° between 0° and 450°)
6. $y_1 = \cos\theta$ and $y_2 = \sin(\theta + 90°)$
   (Every 30° between 0° and 360°)
7. $y = 5 \sin\theta$
   (Every 90° between 0° and 1440°)

17-2 MAGNETIC FLUX, MOTION AND INDUCED CURRENT

Chemical, thermal and other forms of energy are converted to rotational energy by engines or turbines. It is this form of mechanical energy that drives the rotor of a generator. A very simple generator is shown in Figure 17-9. Even though it is too simple to be of practical use, it is easy to understand.

The rotor, the rectangular loop ABCD, rotates about its axis driven by an external source of power. The loop, which could be a coil of many turns, rotates within a strong magnetic field. The magnetic flux indicated by "lines of force" pass through the coil in the position shown. As the coil rotates, the amount of magnetic flux linking the coil changes and an emf is induced. Current will flow if a load is connected to the coil.

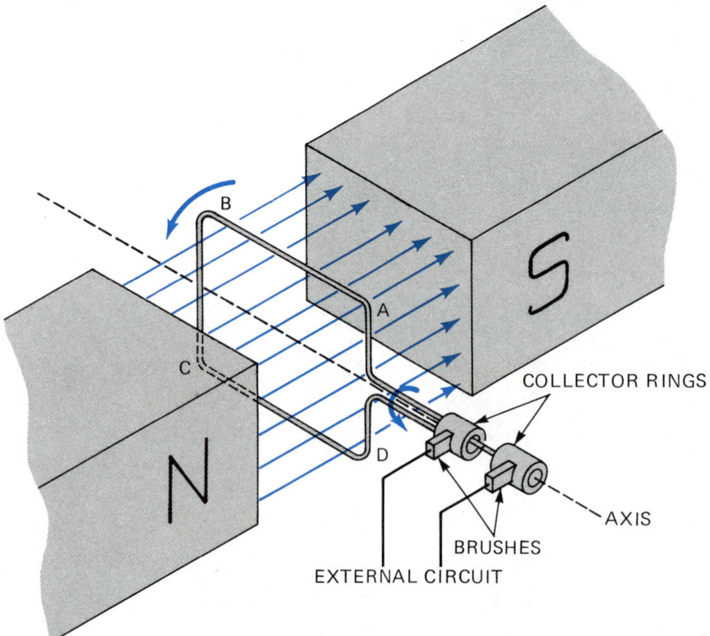

Figure 17-9

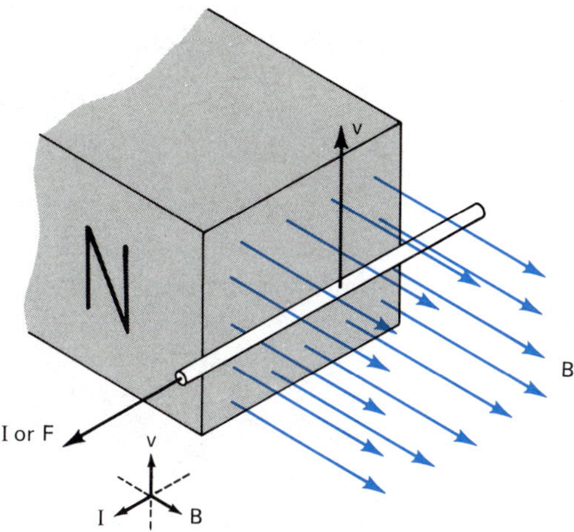

Figure 17-10 Current induced in a wire moving through a magnetic field. Current will flow only if the ends of the wire are connected.

One can see some similarities between the changing magnetic linkage of the generator coil and that of a transformer secondary. The changing flux in a transformer is produced by the alternating current in the primary without any moving parts. In the generator, it is produced by rotation of the coil in a magnetic field of constant strength. In both cases, the induced voltage is a function of the rate at which the magnetic linkage changes.

The rate of change of magnetic linkage is a difficult concept to work with. However, one can regard the induced emf as being the result of a wire cutting the magnetic flux. Sections AB and CD cut the magnetic flux lines as the coil rotates. Both concepts are equally correct.

Figure 17-10 shows the magnetic flux lines near a north pole of a magnet. The direction of the magnetic flux lines is the direction of the force on a north pole in the field. The north pole of a small compass will point away from the larger north pole shown. A horizontal wire is moved upward in the direction of v, the velocity vector. Electrons in the wire are forced to move along the wire setting up an induced emf between the wire ends. The wire acts as a source of emf. If the ends are connected to an external circuit, current will flow in the direction shown. The direction of the induced current is given by the left-hand rule.

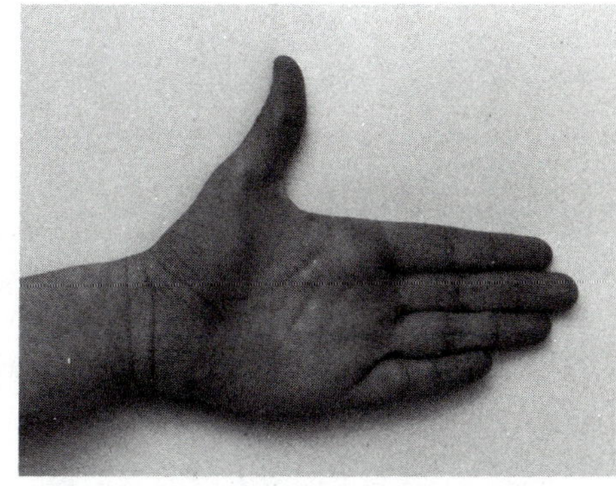

Figure 17-11 The open palm left-hand rule. Point four fingers in the direction of B, the thumb in the direction of v and the force on an electron is in the direction the palm normally exerts a force, F.

LEFT-HAND RULE

- Hold the left hand flat as shown in Figure 17-11.
- Point the four fingers in the direction of the magnetic flux (B) lines.
- Point the thumb in the direction of electron or wire velocity (v) (imposed movement).
- The direction of the force (F) on the electrons gives the direction of the induced current (I) in the wire. It is indicated by the direction the palm normally exerts a pushing force. Induced current is due to the force on the electrons in the wire.

The six possible directions in space are indicated by arrows that represent the three vectors: velocity (v), magnetic field (B) and force (F) or current (I). On paper, they are left (←), right (→), up (↑), down (↓), into (X) and outward (•). The symbol (X) means a direction "into the paper." It represents the feathered tail of an arrow just disappearing into the paper. The dot symbol (•) represents the tip of the arrow just emerging from the paper. On a map, the equivalent directions are west, east, north, south, downward (into the ground) and skyward, respectively.

Example A Find the direction of the induced current or other unknown in each of the five cases of Figure 17-12.

Solution: Use the left hand in all cases.

Figure 17-12(A)
Point the four fingers (B) into the paper. Rotate the hand until the thumb (v) points toward the bottom of the page. The palm (F) points leftward so the induced current is to the left in the wire (←).

Figure 17-12(B)
Point the four fingers (B) in a direction out of the page. Rotate the hand so the thumb (v) points to the left. The induced current is in the direction the palm (F) faces which is toward the bottom of the page (↓).

Figure 17-12(C)
Point the four fingers (B) to the right. The X in the center of the wire means the wire is moving into the paper. Point the thumb (v) in that direction. The palm (F) faces upward toward the top of the page. That is the direction of the induced current (↑).

Figure 17-12(D)
Current in the wire is into the paper as shown by the symbol (X). What direction is the wire moving? Point four fingers (B) toward the bottom of the page with palm (F) pointed into the page. The thumb points in the direction of motion (left) (←).

Figure 17-12(E)
No current is induced in this case because the wire does not cut across the magnetic flux lines.

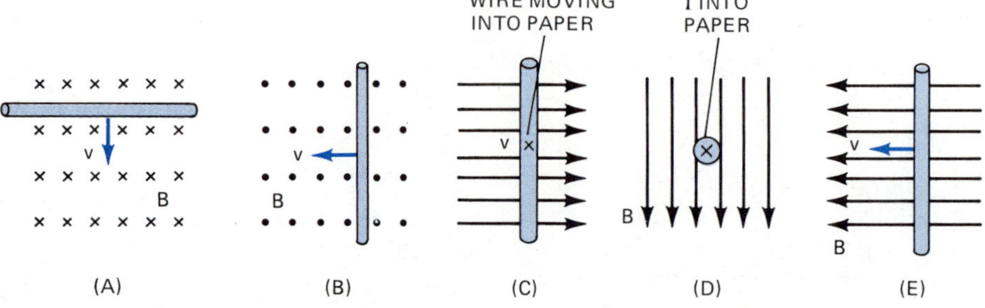

Figure 17-12

There are several other ways to apply the left-hand rule. Figure 17-13 illustrates another method with the thumb, forefinger and middle finger representing v, B and F, respectively. They can also represent F, v and B, respectively. A third scheme is to let them represent B, F and v, respectively. Other similar finger combinations do not work. In addition, some memory aids that do not involve the left hand have been devised.

The open-palm rule, however, is especially easy to remember. The many magnetic flux lines are represented by four (many) fingers. The force that induces and gives the direction of current flow is represented by the palm which is often used to exert a pushing force. The remaining quantity is the imposed electron or wire motion represented by the thumb. "Going somewhere? Use your thumb!"

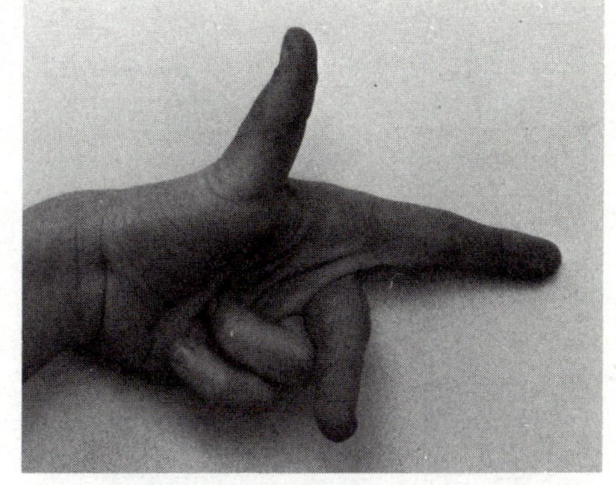

Figure 17-13 Three-finger left-hand rule. Point forefinger and thumb in the direction of B and v, respectively. The middle finger points in the direction of the force on an electron, F.

EXERCISE 17-2

The directions of two of the three quantities [movement velocity (v), magnetic flux (B) and force or current (I)] are given. Find the unknown direction for the indicated quantity in each of the following.

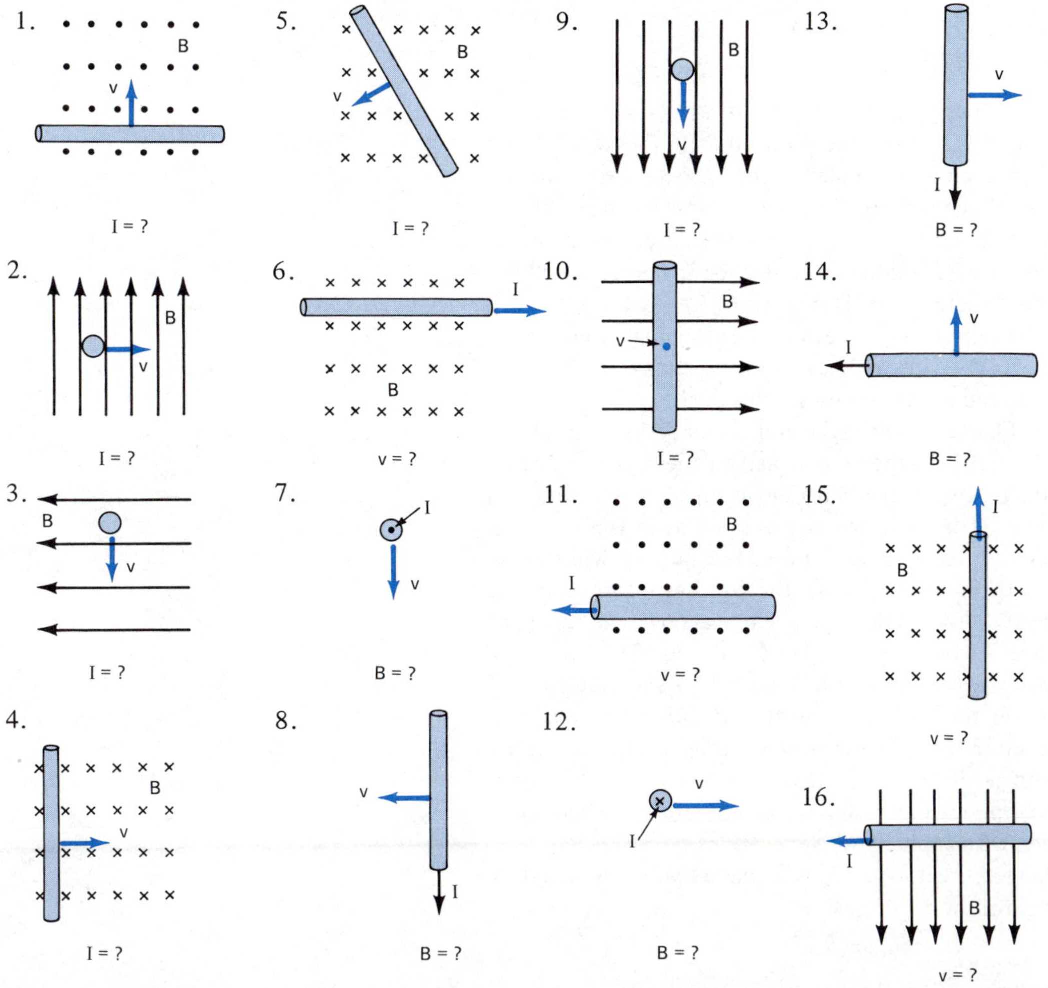

1. I = ?
2. I = ?
3. I = ?
4. I = ?
5. I = ?
6. v = ?
7. B = ?
8. B = ?
9. I = ?
10. I = ?
11. v = ?
12. B = ?
13. B = ?
14. B = ?
15. v = ?
16. v = ?

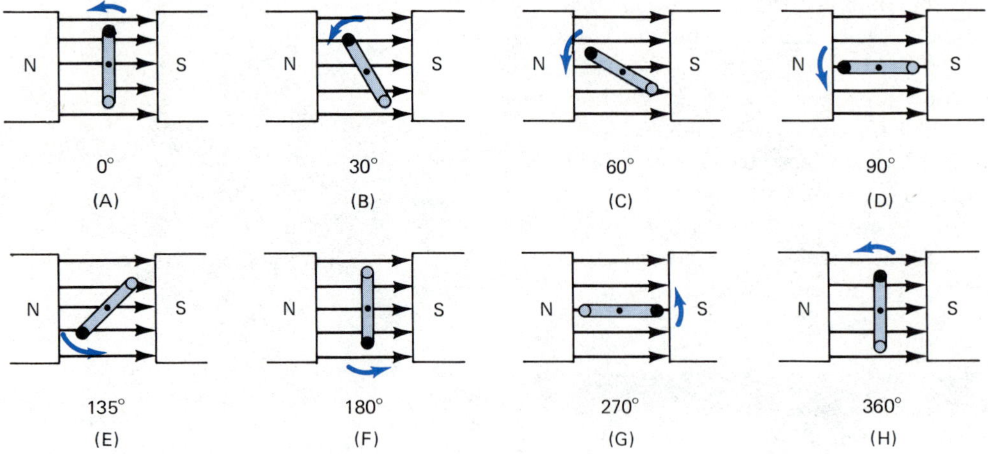

Figure 17-14 End views of the loop generator as it rotates counterclockwise at a constant speed.

17-3 MECHANICAL PRODUCTION OF ALTERNATING VOLTAGE

Figure 17-14 shows a sequence of end-view pictures as the coil rotates. In Figure 17-14(A), the coil is in the same position as in Figure 17-9. At that instant, both wires are moving parallel to the flux lines so none are being cut. The induced voltage is zero. In Figure 17-14(B), after 30° rotation, the wires are cutting across the flux lines and an emf is induced. The emf is induced in both wires (aiding). Under load, a current will circulate in the coil in the direction ABCD, Figure 17-9. At the 60° position, Figure 17-14(C), lines are being cut even faster so a higher emf is developed. At 90°, Figure 17-14(D), the lines are cut at a maximum rate so the maximum emf is produced as the coil passes this position.

In the 135° position, Figure 17-14(E) the emf is decreasing and falls to zero at 180°, Figure 17-14(F). In the second half of the rotation, Figure 17-14(F) through Figure 17-14(H), the wires move through the flux field in the opposite direction. The sequence is the same as in the first half revolution except the polarity is reversed. Under load, current will flow in the opposite direction. In Figure 17-14(H), the coil has returned to the starting position, Figure 17-14(A). The sequence repeats over and over.

A graph of the voltage (e) versus the angle of rotation (θ) is shown in Figure 17-15. Each position of the coil in Figure 17-14 is indicated by a letter on the graph. On the vertical scale, the *variable voltage* is indicated by the lowercase letter (e). The *maximum voltage* is represented by the symbol (E_m) in both positive and negative directions. This choice is in keeping with the usual practice of using lowercase letters for variables. Constants are symbolized with capital letters.

The graph of the voltage, Figure 17-15, is a sine curve. The equation for the voltage curve is as follows:

$$e = E_m \sin \theta$$

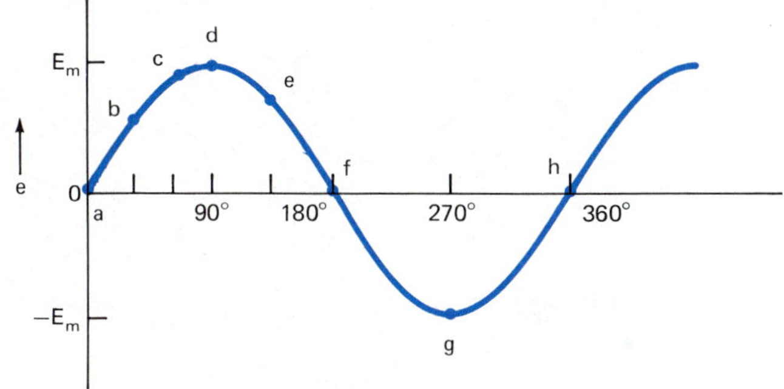

Figure 17-15 Graph of voltage (e) versus the angle rotated (θ).

The voltage (e) is the voltage at a particular position of the coil. It is usually called the *instantaneous voltage.* The maximum voltage is sometimes called the *amplitude.* There is no such thing as "maximum amplitude" (a redundant expression). Amplitude is simply the largest value above or below the zero point. The maximum voltage depends on the *angular velocity* (rate of rotation) of the coil and the strength of the magnetic field.

Example A (1) Write a formula for the instantaneous voltage if the amplitude is 160 V. (2) Find the instantaneous voltage when the coil has rotated through 148°. (3) Is the voltage increasing or decreasing?

Solution: (1) $e = 160 \sin \theta$

(2) $e = 160 \sin 148° = 160(0.530)$

$e = 84.8$ V

(3) The coil is in the second quarter of rotation so the voltage is decreasing. Its value at 150°, for example, is 80 V.

The voltage wave can also be thought of as a rotating voltage vector, Figure 17-16. The vector has a length equal to the amplitude (E_m). The vertical side (e) of the triangle formed represents the instantaneous voltage. Its length alternates in sine wave fashion according to the equation for the voltage curve. The voltage vector turns at a constant rotational speed.

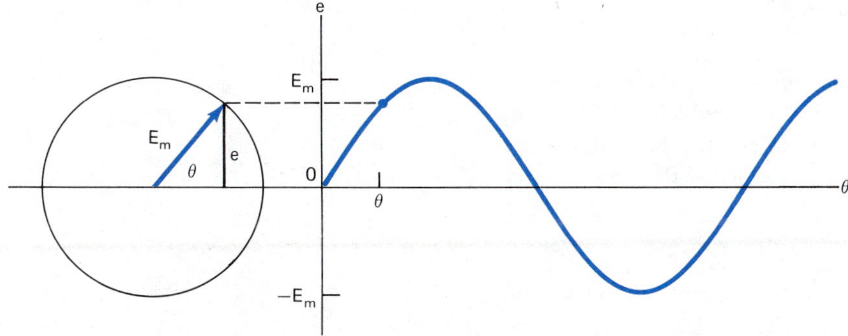

Figure 17-16 Voltage wave as a rotating vector

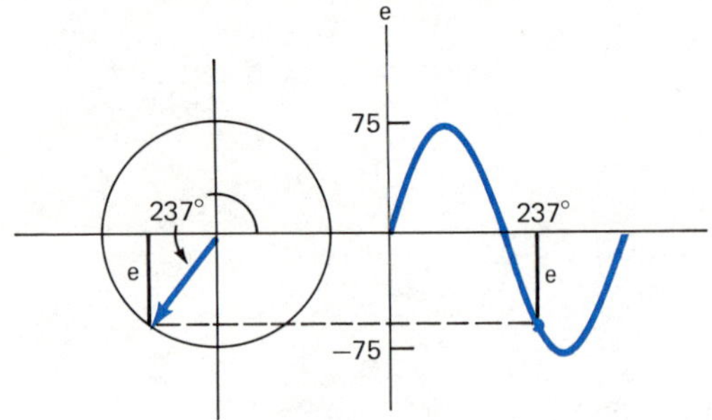

Figure 17-17

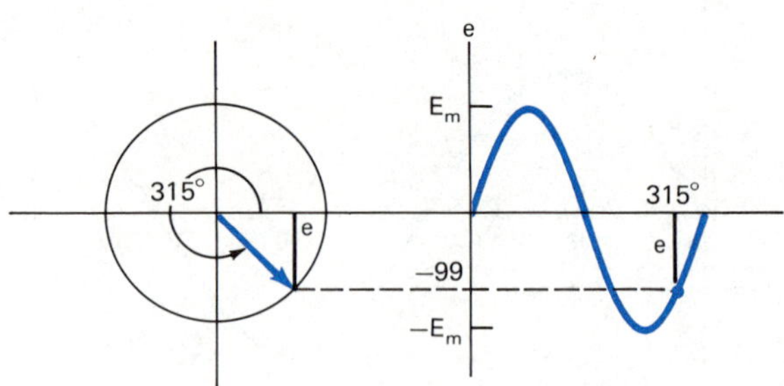

Figure 17-18

Example B What is the instantaneous voltage at 237° if the maximum voltage is 75 V. Is it increasing or decreasing?

Solution: Data-diagram: Figure 17-17

Formula: $e = E_m \sin \theta$

Substitute: $e = (75\text{ V})(\sin 237°)$

$e = (75\text{ V})(-0.8387)$

$e = -62.9\text{ V}$

The voltage is increasing in the negative direction. At 270° it will be at the negative maximum.

Example C At 315°, the instantaneous voltage is −99 V. What is the amplitude of this voltage wave?

Solution: Data-diagram: Figure 17-18

Formula: $e = E_m \sin \theta$

$$E_m = \frac{e}{\sin \theta}$$

Substitute: $$E_m = \frac{-99\text{ V}}{\sin 315°} = \frac{-99\text{ V}}{-0.707} = 140\text{ V}$$

or: 99 [+/-] [÷] 315 [Sin] [=] , 140 V (Degree mode)

The current produced in a pure resistive circuit will follow the same variation pattern as the voltage:

$$i = I_m \sin \theta$$

where i is the instantaneous current and I_m is the maximum current or current amplitude. An alternating current is one in which the electrons flow first in one direction, then in the opposite direction. For this reason, ac generators are often called *alternators.* The electrons flow only a very short distance before reversing directions.

Example D If the maximum current is 2.5 A, find the instantaneous current at 1125°.

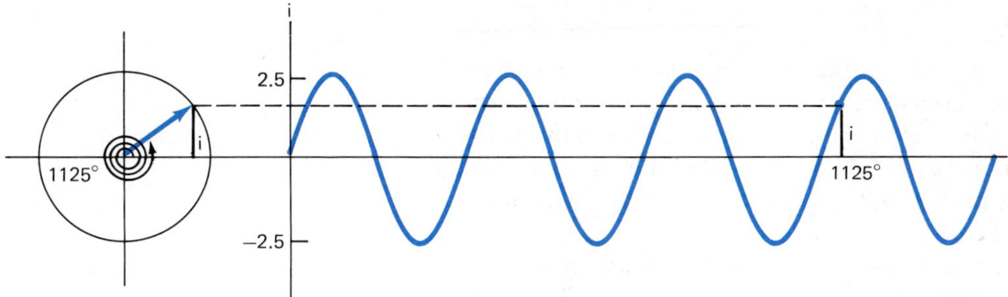

Figure 17-19

Solution: Data-diagram: Figure 17-19
Formula: $i = I_m \sin \theta$
Substitute: $i = (2.5\ \text{A})(\sin 1125^\circ) = (2.5\ \text{A})(0.707) = 1.77\ \text{A}$

Note: The generator coil is in the 45° position (first quadrant) after having completed three full revolutions. The current is increasing in this position.

EXERCISE 17-3

Data, diagram, formula, substitute, check.

1. An alternating emf has a maximum voltage of 170 V. Find the instantaneous voltage at each of the following positions: (a) 35°, (b) 100°, (c) 165°, (d) 225°, (e) 325°.
2. An alternating current has an amplitude of 500 mA. Find the instantaneous current at each of these angles: (a) 65°, (b) 145°, (c) 245°, (d) 270°, (e) 335°.

Draw a vector diagram and a freehand sine wave similar to Figure 17-16 for each of the following problems. Indicate the voltage or current on both the vector and wave diagrams.

3. Find the instantaneous voltage at 150° if the amplitude is 170 V. Is the voltage increasing or decreasing?
4. Find the voltage amplitude if the instantaneous voltage is -65 V at 240°.
5. Find the current amplitude if the instantaneous current is -250 mA at 310°.
6. What is the instantaneous current at 110° if the current amplitude is 4.00 A? Is the current increasing or decreasing?
7. Find the current at 1000° if the amplitude is 5.00 A. Is the current increasing or decreasing?
8. An alternating emf has an amplitude of 165 V. In the first cycle, what angles would produce an instantaneous voltage of 100 V?
9. (a) Write the voltage equation for an alternating current if the amplitude is 2.00 V. (b) Find the voltage at 800°. (c) Is the voltage increasing or decreasing?
10. An alternating current has an amplitude of 900 mA. In the first cycle, what angles would produce a current of -300 mA?

— FUNDAMENTALS OF THE PERIODIC CYCLE —

Sometimes it is necessary to know the time required for a complete cycle of current alternation. Even more important is the number of cycles completed in each second. Real alternators usually have several magnetic poles. More than one current cycle will be completed during each revolution of the rotor. Thus, the number of current and voltage cycles does not correspond to the number of revolutions of the rotor.

To completely describe the voltage wave, words such as period, frequency, angular velocity and others must be fully understood. These are discussed in detail in the following paragraphs.

17-4 PERIOD, FREQUENCY AND ANGULAR VELOCITY

Cycle. A *cycle* is a complete set of events that recur regularly and in the same sequence and usually lead back to the beginning. The voltage cycle, for example, begins at zero, increases, reaches a maximum, decreases, returns to zero, increases in the negative direction, reaches a negative maximum, decreases and returns to zero. The cycle repeats over and over.

Period. The *period* (T) of a cycle of events is the amount of time required for one complete sequence to occur. For electrical events, the period is usually expressed in seconds, milliseconds, etc. The symbol used for the period is the capital letter (T) since the period is a constant. The lowercase (t) represents the variable *time* as measured on a clock. Be sure to distinguish between the two symbols. Capital T is the period; lowercase t is the variable time. If N is the number of cycles completed in time t, then:

$$T = \frac{t}{N}$$

Example A Find the period for each of the following occurrences:

(1) 360 voltage cycles in 6 s
(2) The rotation of a second hand of a clock
(3) The rotation of a motor rotating at 1720 r/min
(4) Spin rotation of the earth

Solution: (1) $T = \frac{t}{N} = \frac{6\text{ s}}{360\text{ cycles}} = 0.0167\text{ s} = 16.7\text{ ms}$

(2) $\frac{60\text{ s}}{1\text{ cycle}} = 60\text{ s}$

(3) $\frac{60\text{ s}}{1720\text{ cycles}} = 0.0349\text{ s} = 34.9\text{ ms}$

(4) 1 day = 24 h = 1440 min = 86 400 s

$\frac{86\,400\text{ s}}{1\text{ cycle}} = 86\,400\text{ s} = 86.4\text{ ks}$

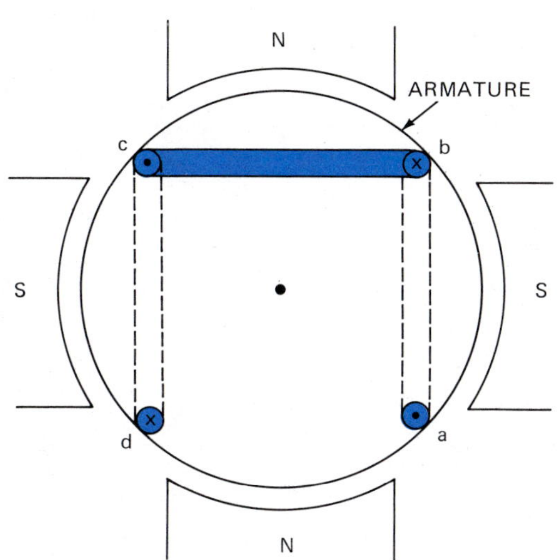

Figure 17-20A Four-pole alternator. The four wires (a, b, c and d) are placed in slots 90° apart.

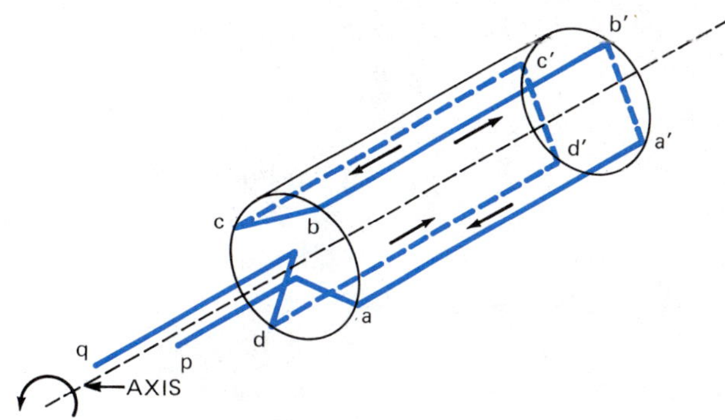

Figure 17-20B Armature winding for a 4-pole alternator. Current follows the path: p-a-a′-b′-b-c-c′-d′-d-q when wires aa′ and cc′ are cutting across a south pole. High voltage alternators have the armature wound on the stator with electromagnetic poles on the rotor. Only very low power alternators use permanent magnets for pole pieces.

Frequency. The *frequency* (f) of a cyclic sequence such as alternating current or voltage is the number of complete cycles occurring in each second of time. The unit of frequency is the *hertz* (Hz). One hertz is one cycle per second.

$$f = \frac{N}{t}$$

A comparison of the formulas for periods and frequency shows that the period and frequency are reciprocals of each other.

$$f = \frac{1}{T}$$

When a single coil rotates between two magnetic poles, a voltage cycle is completed each revolution. Most alternators have more than two poles. Figure 17-20 shows one with four. A single coil of the type shown in Figure 17-9 will not work in such a field. In a four-pole alternator, two coils connected in series are mounted on the rotor as shown. Most real alternators have the armature coils mounted on the stator. It is the electromagnetic field coils that are mounted on the rotor.

In multiple-pole alternators, the frequency is the number of *pairs* of poles (P) times the rotational speed in revolutions per second. If the rotational speed (S) is given in r/min, the formula for frequency is

$$f = \frac{PS}{60}$$

Example B Find the frequency and period of an 8-pole alternator rotating at 900 r/min.

Solution: Formula: $f = \frac{PS}{60}$

Substitute: $f = \frac{4(900 \text{ r/min})}{60}$

$f = 60 \text{ Hz}$

Formula: $f = \frac{1}{T}$

$T = \frac{1}{f}$

Substitute: $T = \frac{1}{60 \text{ Hz}} = 0.0167 \text{ s} = 16.7 \text{ ms}$

Example C Find the frequency of each of the following:

(1) 1800 voltage cycles in 4.5 s
(2) The rotation of the second hand on a clock
(3) The vibration of a pendulum with a period of 2.5 s

Solution: (1) $f = \frac{N}{t} = \frac{1800 \text{ cycles}}{4.5 \text{ s}} = 400 \text{ Hz}$

(2) $f = \frac{1 \text{ cycle}}{60 \text{ s}} = 0.0167 \text{ Hz}$

(3) $f = \frac{1}{T} = \frac{1}{2.5 \text{ s}} = 0.4 \text{ Hz}$

Angular Velocity. When a rotating device turns through one revolution, it also turns through 2π radians. The rotational speed is sometimes called *angular velocity*. The lowercase Greek letter omega (ω) is used to represent angular velocity. The quantity is measured in radians per second (rad/s) and is defined by the formula:

$$\omega = \frac{\theta}{t}$$

where θ is the angle in radians and t is the time required.

Example D Find the angular velocity for each of the following:

(1) A rotation of 350 rad in 15 s
(2) A motor that turns 4375 r in 2.5 s
(3) The second hand on a clock

Solution: (1) $\omega = \frac{\theta}{t} = \frac{350 \text{ rad}}{15 \text{ s}} = 23.3 \frac{\text{rad}}{\text{s}}$

(2) $\omega = \frac{2\pi(4375 \text{ r})}{2.5 \text{ s}} = 11\ 000 \frac{\text{rad}}{\text{s}}$

(3) $\omega = \frac{2\pi(1 \text{ r})}{60 \text{ s}} = 0.105 \frac{\text{rad}}{\text{s}}$

There are 2π radians in one revolution and 60 s in 1 min. The rotational speed (r/min) can be converted to angular velocity $\left(\frac{\text{rad}}{\text{s}}\right)$ by using a unit conversion:

$$1\ \text{r/min} = \frac{2\pi}{60}\left(\frac{\text{rad}}{\text{s}}\right)$$

The fraction $\frac{2\pi}{60}$ is 0.105. To check a conversion from r/min to $\frac{\text{rad}}{\text{s}}$, just remember the number of $\frac{\text{rad}}{\text{s}}$ is slightly more than 0.1 the number of r/min.

Example E Convert a motor speed of 1720 r/min to angular velocity units.

Solution: $\left(1720\ \frac{\text{r}}{\text{min}}\right)\left(\frac{2\pi}{60}\right) = 180\ \frac{\text{rad}}{\text{s}}$

Check: 1720(0.1) = 172
180 is slightly greater than 172

EXERCISE 17-4

1. Find the period for each of the following:
   (a) 1024 voltage cycles in 51.2 s
   (b) The minute hand on a clock
   (c) A motor that turns 24 750 revolutions in 7.5 min
   (d) A generator armature rotating at 1800 r/min
   (e) An alternating current with a frequency of 400 Hz
2. Find the frequency of each of the following:
   (a) 2440 current cycles in 12.2 s
   (b) The hour hand of a clock
   (c) A motor that turns 8610 revolutions in 5.25 s under load
   (d) An alternating voltage with a period of 5 ms
3. An alternator has four poles and has an aramture speed of 1800 r/min. Find (a) the frequency and (b) the period.
4. An alternator has 8 poles and has a rated frequency of 60 Hz. At what rotational speed does it operate?
5. A generator armature operates at 2400 r/min and produces an alternating voltage with a period of 2.5 ms. How many poles does it have?
6. Find the angular velocity in radians per second of a generator operating at (a) 1800 r/min, (b) 2400 r/min, (c) 4000 r/min.
7. Find the rotational speed in r/min if the angular velocity is (a) $500\ \frac{\text{rad}}{\text{s}}$, (b) $100\pi\ \frac{\text{rad}}{\text{s}}$, (c) $1400\ \frac{\pi\cdot\text{rad}}{\text{s}}$, (d) $628\ \frac{\text{rad}}{\text{s}}$.
8. Through what angle does a motor turn in 6.25 ms if its angular velocity is 1700 r/min?
9. Through what angle does a generator armature turn in 750 μs if its angular velocity is 1800 r/min?
10. How much time is required for an alternator rotor to turn 1.57 rad if the angular velocity is 2400 r/min?
11. If the frequency is 400 Hz, how many cycles are completed (a) in 125 ms? (b) in 25 ms? (c) in 1 min?
12. How many cycles will a 60-Hz voltage complete (a) in 25 ms? (b) in 1 min? (c) in 1 h?

17-5 ANGULAR FREQUENCY AND THE VOLTAGE EQUATION

One seldom sees the alternators that produce the ac power used in homes and industry. There is no angular velocity involved in lighting a lamp. Angular velocity has little or no meaning in such a situation. In a like manner, the voltage equation, $e = E_m \sin \theta$, has little meaning either. There are no angles (θ) associated with light bulbs, heating elements, etc.

If the formula $\omega = \theta/t$ is solved for θ, it becomes

$$\theta = \omega t$$

The clock that measures time (t) begins when $\theta = 0$. The angle turned is proportional to time. If the right side is substituted for θ in the voltage equation, it becomes

$$\mathbf{e = E_m \sin \omega t}$$

which indicates that voltage is a function of time. This seems more reasonable when discussing light bulbs. The instantaneous current formula is

$$\mathbf{i = I_m \sin \omega t}$$

Even the constant ω (angular velocity) will not concern electricians if they call it *angular frequency* instead. It is always measured in $\frac{\text{rad}}{\text{s}}$ or $\frac{\pi \cdot \text{rad}}{\text{s}}$, the same as before. It is easily calculated with the formula

$$\omega = 2\pi f$$

where f is the frequency. Now the voltage and current can be described without knowing anything about the source or sources.

Example A What is the angular frequency of an alternating voltage if the frequency is 400 Hz?

Solution: Formula: $\omega = 2\pi f$

Substitute: $\omega = 2\pi(400\ \text{Hz})$

$$\omega = 800\pi\ \frac{\text{rad}}{\text{s}} = 2513\ \frac{\text{rad}}{\text{s}}$$

Example B Find the frequency and period of an alternating current if the angular frequency is $1200\pi\ \frac{\text{rad}}{\text{s}}$.

Solution: Formula: $\omega = 2\pi f$

$$f = \frac{\omega}{2\pi}$$

Substitute: $f = \dfrac{1200\pi\ \frac{\text{rad}}{\text{s}}}{2\pi\ \text{rad}} = 600\ \text{Hz}$

Formula: $T = \dfrac{1}{f}$

Substitute: $T = \dfrac{1}{600\ \text{Hz}} = 0.001\ 67\ \text{s} = 1.67\ \text{ms}$

The equation $e = E_m \sin \omega t$ is for the instantaneous voltage. It has two purposes. It can be used as a formula to determine the instantaneous voltage at a particular time. It also is a convenient shorthand method of describing all the necessary information about a particular voltage waveform.

It is the second purpose that is most important. The instantaneous voltage seldom needs to be known. The amplitude and angular frequency are displayed in the equation. The frequency and period can be calculated with little effort. The voltage waveform can be constructed by locating zeros and peaks.

Example C A certain generated alternating voltage has the equation $e = 400 \sin 800\pi t$. What is the (1) amplitude, (2) angular frequency, (3) frequency and (4) period? (5) Construct a sine curve that represents the voltage.

Solution: Compare the given equation with the general form:

$e = E_m \sin \omega t$

$e = 400 \sin 800\pi t$

(1) $E_m = 400 \text{ V}$

(2) $\omega = 800\pi \frac{\text{rad}}{\text{s}}$

Formula: $\omega = 2\pi f$

(3) $f = \frac{\omega}{2\pi} = \frac{800\pi}{2\pi} = 400 \text{ Hz}$

Formula: (4) $T = \frac{1}{f} = \frac{1}{400 \text{ Hz}} = 2.5 \text{ ms}$

(5) A graph of the voltage can be plotted by locating the times when the voltage is zero and at peak (positive and negative) values. Zeros occur every half cycle: t = 0, 1.25 ms, 2.5 ms, 3.75 ms, 5 ms, etc. The peaks are halfway between the zeros: t = 0.625 ms, 1.875 ms, 3.125 ms, 4.375 ms, etc. A template can be used to draw the curve if desired. Scales can be added after the curve is formed, Figure 17-21.

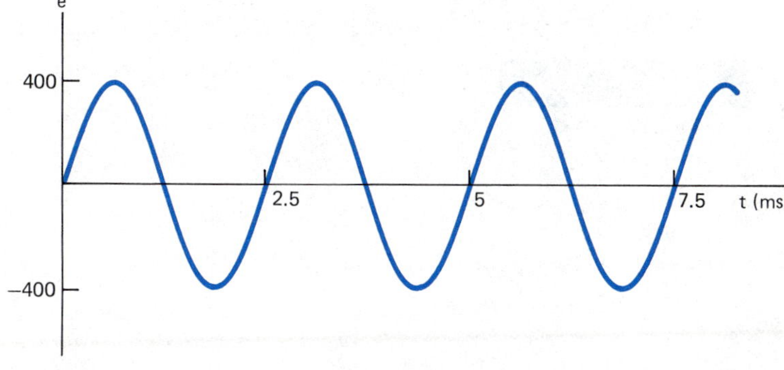

Figure 17-21

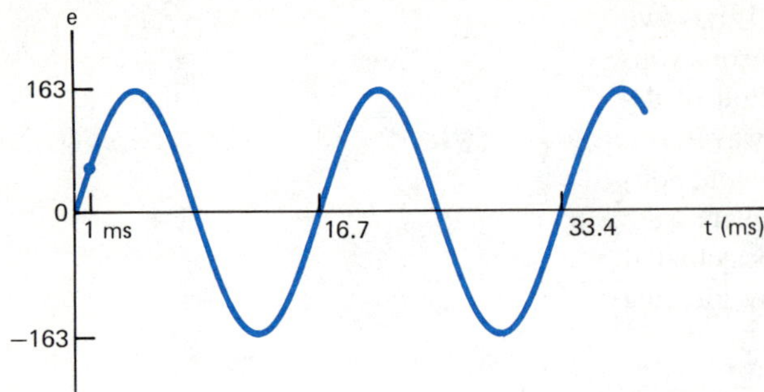

Figure 17-22

Example D A certain alternating voltage has an amplitude of 163 V and a frequency of 60 Hz. (1) Write an equation that represents these facts in the form of the voltage equation, (2) plot a graph of the voltage function, and (3) find the instantaneous voltage at the end of 1 ms.

Solution: (1) $\omega = 2\pi f = 2\pi(60 \text{ Hz}) = 120\pi \frac{\text{rad}}{\text{s}}$

The values of the amplitude and angular frequency are simply substituted into the voltage equation to obtain the desired equation:

$e = E_m \sin \omega t$

$e = 163 \sin 120\pi t$

(2) The period is $\frac{1}{60 \text{ Hz}} = 16.7$ ms. Zeros occur at t = 16.7 ms, 25.05 ms, 33.4 ms, etc. Peaks are at t = 4.18 ms, 12.53 ms, 20.88 ms, etc. These values of time are plotted and a sine curve is sketched in, Figure 17-22. Since there are 60 cycles each second, only a small part of one second is graphed.

(3) $e = 163 \sin (120\pi \frac{\text{rad}}{\text{s}})(0.001 \text{ s})$

$e = 163 \sin (0.377 \text{ rad})$

$e = 163(0.3681) = 60 \text{ V}$

This instant is indicated on Figure 17-22.

EXERCISE 17-5

1. Find the angular frequency for each of the following.
   (a) a 600-Hz voltage
   (b) an 800-Hz current
   (c) a pendulum vibrating at 4 Hz
2. Find the frequency for each of these angular frequencies.
   (a) $2600 \frac{\pi \cdot \text{rad}}{\text{s}}$
   (b) $1000 \frac{\text{rad}}{\text{s}}$
   (c) $750\pi \frac{\text{rad}}{\text{s}}$
   (d) $2500 \frac{\text{rad}}{\text{s}}$

Find the amplitude, angular frequency, frequency and period for each of these voltages and currents.

3. v = 167 sin 377t
4. i = 2.5 sin 1257t
5. v = 673 sin 1250πt
6. i = 0.075 sin 800πt

Write the voltage or current equation for each of the following.

7. A peak voltage of 310 V and a frequency of 60 Hz.
8. A peak current of 3.5 A and a frequency of 1.5 kHz.
9. A voltage amplitude of 157 V and a period of 0.01 s.
10. A current amplitude of 750 mA and a period of 100 ms.

Plot a graph of each of these waveforms (two cycles).

11. v = 45 sin 160πt
12. i = 0.45 sin 80πt
13. v = 167.3 sin 377t
14. Figure 17-23 is a graph of a certain voltage. What is the (a) period, (b) frequency, (c) angular frequency, (d) amplitude? (e) Write an equation for this voltage.

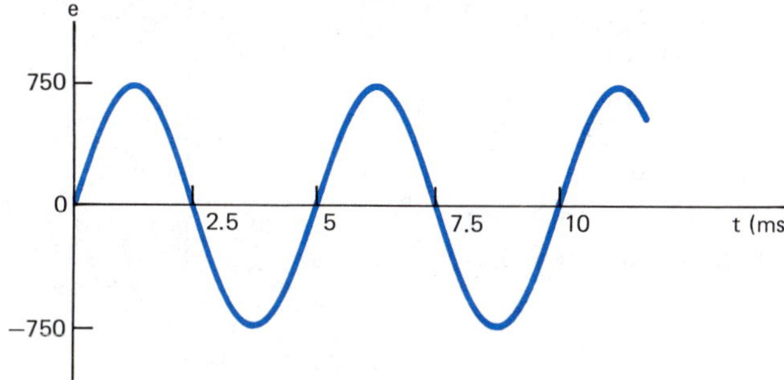

Figure 17-23

15. Find the instantaneous voltage if the equation is v = 673 sin 1257t (a) at 600 μs, (b) at 2 ms.
16. Find the instantaneous current for i = 2.5 sin 120πt at (a) 1.39 ms, (b) 5.56 ms, (c) 27.8 ms. (d) Is the voltage in the first, second or third cycle at 27.8 ms?

17-6 EFFECTIVE ALTERNATING VOLTAGE AND CURRENT IN RESISTIVE CIRCUITS

Electricians have sensibly defined certain values of alternating voltage and current in terms of the heat they produce in a resistive circuit. Such a definition is sensible because simple DC circuit formulas can be used with ac resistive circuits:

- Ohm's Law **E = IR**
- Power **P = EI**
- Energy ***W* = Pt**

In a given DC circuit, voltage, current, resistance and power are constants. The energy consumed is a function of time.

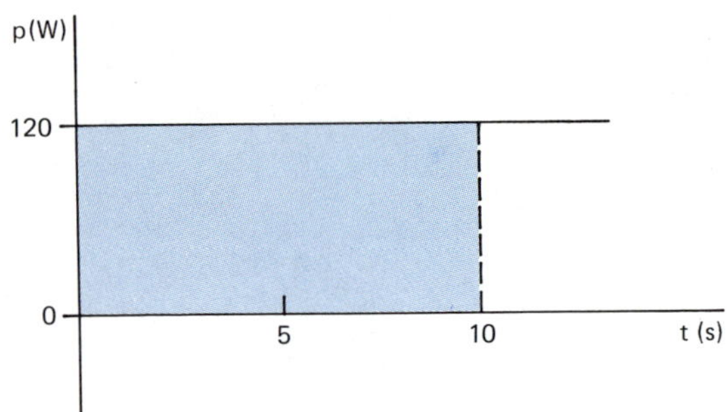

Figure 17-24 Energy as the area (shaded) under a curve. The rectangle is 10 s long and 120 W high. The area is (120 W)(10 s) = 1200 J.

Energy in a DC circuit can be represented graphically, Figure 17-24, as the area under the curve (shaded). In this example, power is 120 W, the vertical side of the rectangle. The horizontal side is the time. Area is the length times the width. In 10 s, a total of (120 W)(10 s) = 1200 J (joules) of energy is converted to heat. In 20 s, it would be twice as much.

In ac circuits, the situation is not so simple. Voltage and current both increase, decrease and reverse directions with the same frequency. In resistive circuits, the changes in both occur in phase with each other, Figure 17-25(A). Two *in phase* waves attain peak values and zeros simultaneously.

Power obviously changes instantaneously also. It is zero whenever e and i are zero and maximum when e and i are maximum. When e and i switch directions, power remains positive. The product of two negative quantities is positive, Figure 17-25(B). The instantaneous power (p) is the product of $I_m \sin \omega t$ and $E_m \sin \omega t$

$$P = E_m I_m \sin^2 \omega t$$

Since the *maximum power* (P_m) is the product $E_m I_m$, this can be written:

$$P = P_m \sin^2 \omega t$$

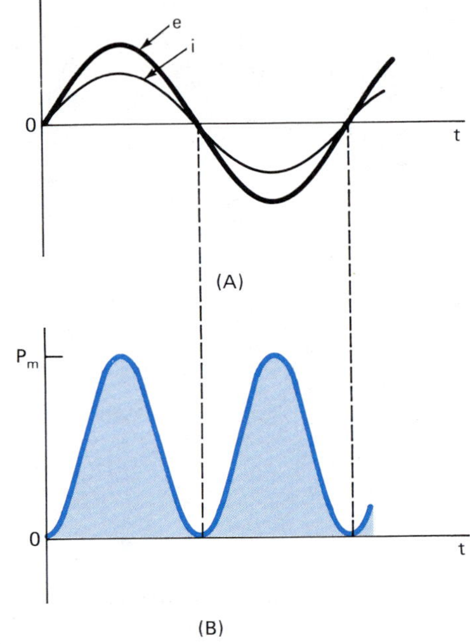

Figure 17-25 Voltage, current and power in a resistive circuit

Maximum power (P_m) is sometimes called the *peak power* or *power amplitude*. The symbol $\sin^2 \omega t$ is the square of $\sin \omega t$, the same as $(\sin \omega t)^2$.

The shaded area under the power curve, Figure 17-26, represents the energy converted to heat in one cycle. Note that the same amount of

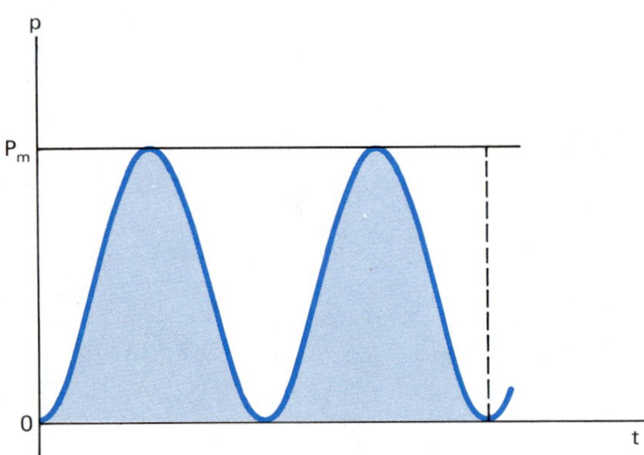

Figure 17-26 Power curve for a resistive circuit

heat is produced in the second half when current flows in the negative direction. Moving electrons produce heat in a resistance regardless of the direction of flow.

In order to use DC circuit formulas, an equivalent DC power must be found that produces the same heating effect as the ac circuit. This would be a constant power, similar to that of Figure 17-24. It would have the same area as the shaded area of Figure 17-26 in the same amount of time.

If the area of the first quarter-cycle can be determined, the total area for one period will be four times that amount. The area can be approximated by dividing the quarter-cycle into rectangular sections of equal widths, Figure 17-27. Figure 17-27(A) divides it into three sections and Figure 17-27(B) shows it divided into nine. Obviously, adding the section areas of either figure will yield a one-cycle area that is larger than that under the curve.

The three-section area sum indicates the equivalent DC power to be $0.667P_m$ and the nine sections, $0.556P_m$. If eighteen sections are used, the sum is $0.528P_m$. If the error is eliminated altogether (using calculus) it is exactly:

$$P = \frac{1}{2}P_m$$

This means the unshaded area to the left of the dashed line, Figure 17-26, equals the shaded area. One can almost tell by observation that the shaded and unshaded areas are equal.

The equivalent DC power (P) is the average power of the ac circuit. Now the *effective voltage* (E) and the *effective current* (I) that produce the average power can be determined. Since $P = I^2R$,

$$I^2R = \frac{1}{2}I_m{}^2R$$

To solve for I, cancel R and apply the root axiom:

$$I = \frac{I_m}{\sqrt{2}} = 0.707I_m$$

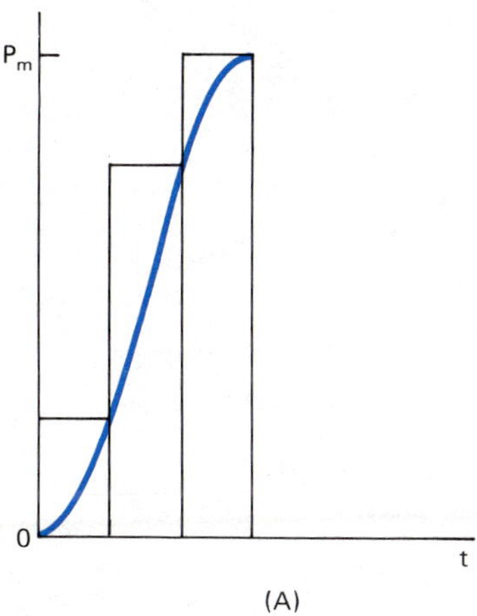

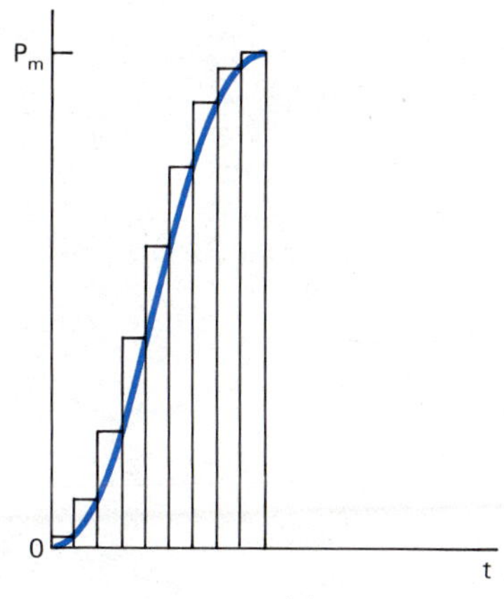

Figure 17-27 Approximating area under a curve

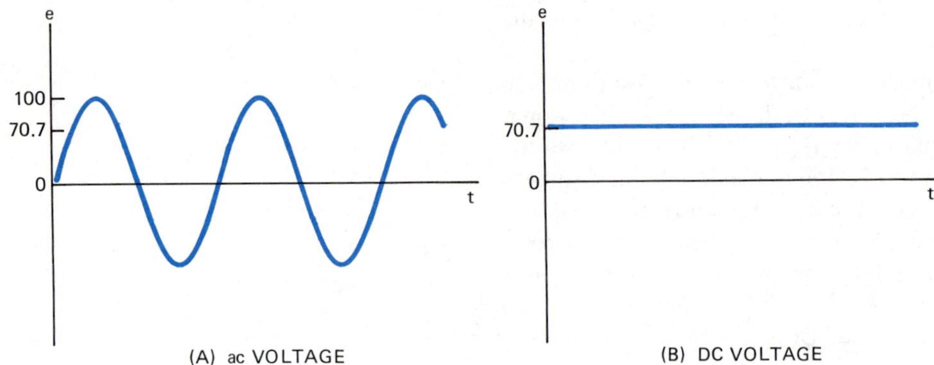

Figure 17-28 Equivalent ac and DC voltages produce equal heating effects when applied to equal resistances. The effective voltage is 70.7 V.

Similarly, since $P = \frac{E^2}{R}$:

$$E = \frac{E_m}{\sqrt{2}} = 0.707E_m$$

These facts are indicated graphically in Figure 17-28 for a 100-V (ac) wave.

The curve of Figure 17-26(B) can also represent an i^2 (or e^2) curve. The height of each area section represents the values of the square of i (or e). If these are averaged and the square root is taken, the two previous equations are obtained again. For this reason, the effective current and effective voltage are often called *root-mean-square* or *rms* values. Effective or rms values are not simple averages but the product EI is the average power. Average power is often (incorrectly) called effective power.

The symbols E and I are effective values and do not carry a subscript. Unless specified, alternating voltage and current are always effective values. The 120-V household voltage is the effective voltage. Voltmeters and ammeters designed for ac measurements are calibrated to read effective values. The cost of energy consumption is also based on effective values.

Example A An alternating voltage is given by the equation $e = 167 \sin 120\pi t$. What is the effective voltage?

Solution: $E = 0.707E_m$

$E = 0.707(167\ V) = 118\ V$

Example B Find the effective current if the current amplitude is 2.75 A.

Solution: $I = 0.707I_m$

$I = 0.707(2.75\ A) = 1.94\ A$

Example C What is the maximum voltage of an alternating voltage wave if the effective voltage is 230 V?

Solution: $E = 0.707E_m$

$$E_m = \frac{E}{0.707} = 1.414E$$

$E_m = 1.414(230\ V) = 325\ V$

Example D A voltage wave, $e = 283 \sin 120\pi t$ is applied to a heating element with a hot resistance of 12 Ω, Figure 17-29. Find (1) the effective voltage, (2) the effective current, (3) the current equation, (4) the effective power.

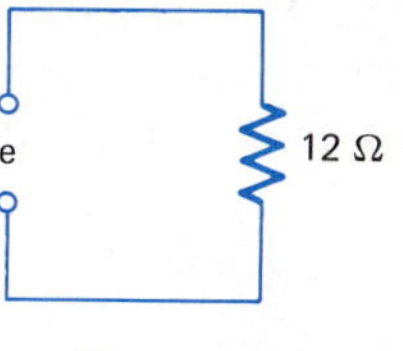

Figure 17-29

Solution: (1) $E = 0.707E_m$

$E = 0.707(283\ \text{V}) = 200\ \text{V}$

(2) $I = \frac{E}{R}$

$I = \frac{200\ \text{V}}{12\ \Omega} = 16.7\ \text{A}$

(3) $I = 0.707I_m$

$I_m = \frac{I}{0.707} = \frac{16.7\ \text{A}}{0.707} = 23.6\ \text{A}$

$i = 23.6 \sin 120\pi t$

(4) $P = EI$

$P = (200\ \text{V})(16.7\ \text{A}) = 3330\ \text{W}$

EXERCISE 17-6

Find the effective value of each of the following maximums.

1. 622 V
2. 1200 V
3. 6.85 mA
4. 375 mA
5. $e = 1245 \sin 120\pi t$
6. $i = 7.85 \sin 120\pi t$
7. $e = 5.30 \times 10^5 \sin 120\pi t$
8. $i = 0.200 \sin 800\pi t$

Find the maximum value for each of the following effective values.

9. 460 V
10. 880 V
11. 50.0 A
12. 2.55 μA

Data, diagram, formula, substitute, check.

13. A very large power transformer has a secondary circuit voltage of 1100 kV. Find the voltage amplitude.
14. A certain arc welder has a current range from 38 A to 275 A. Express this range in terms of maximum currents.
15. A furnace control circuit uses an alternating voltage with an amplitude of 34 V. What is the effective voltage?
16. A doorbell circuit uses an rms voltage of 16 V. What is the peak voltage? Write the voltage equation for the 60-Hz wave.
17. An alternating voltage, $e = 170 \sin 120\pi t$, is applied to a 65-W heating pad. Find (a) the effective voltage, (b) the effective current, (c) the resistance, (d) the current equation.
18. A 240-V clothes dryer has a power rating of 4800 W. Find (a) the effective current, (b) the voltage amplitude, (c) the peak current, (d) the current and voltage equations for the 60-Hz wave.
19. A 60-Hz wind generator produces 120 V (effective) and effective currents up to 16.7 A. What is (a) the highest effective power, (b) the instantaneous voltage equation, (c) the instantaneous current equation and (d) the load resistance that will produce the 16.7 A current?

20. A power station generates an alternating voltage of $e = 21\,210 \sin 377t$. It is stepped up to an effective voltage of 345 kV. At a substation it is combined with other sources and stepped down to a voltage with an amplitude of 170 kV. At the outskirts of a city, it is reduced to 2400 V rms. For safety in the home, the root-mean-square voltage is dropped to 118 V. It is used to operate a doorbell circuit at 17 V (peak). Construct a table and list both the effective voltage and the maximum voltage for each of the six steps.

REACTANCE

Many ac circuit problems involve lamps, heating elements and wire resistance. In these, load resistance is the only circuit quantity that tends to oppose the current. If effective values are used, the problems are the same as DC circuit problems. Series, parallel and combination circuits would be solved just as though they were DC circuits.

There are two additional circuit quantities, however, that cause an opposition to alternating current flow. They are capacitance and inductance. Actually, they do not oppose the flow of electrons as resistance does. Instead, they act as new sources of voltage and tend to always reduce the effective voltage that creates the flow. This, of course, reduces the current just as though they were retarding elements.

Of the two points of view, it is more convenient to think of capacitance and inductance as opposing or current-limiting factors. When an alternating voltage is applied to an inductor or a capacitor, the devices "react" by setting up an opposition to the flow of current. The opposition is called the *reactance*. Inductive reactance and capacitive reactance affect the current in different ways. Each kind is discussed separately.

17-7 INDUCTANCE AND INDUCTIVE REACTANCE

When a magnetic field passing through a coil of wire changes, an emf is produced in the coil. If the magnetic field alternates in direction through the coil, the emf will alternate. The coil becomes a source of emf and will produce an alternating current when connected to a load. The secondary winding of a transformer and the armature of a generator are two examples of that effect.

In the case of the transformer, the alternating magnetic field is produced by the alternating current in the primary. An iron core extends through both coils and greatly increases the strength of the magnetic field through both, Figure 17-30. The magnetic field builds up and collapses first in one direction; then the other. The alternating emf induced in the secondary changes with the same frequency as the alternating current in the primary. The effect is called *mutual induction*.

The changing magnetic field originates with the alternating current in the primary coil. The field that induces emf in the secondary also passes through the primary coil. Is an emf induced in it, too? Indeed there is. In fact, the presence of the secondary coil is not needed to produce the

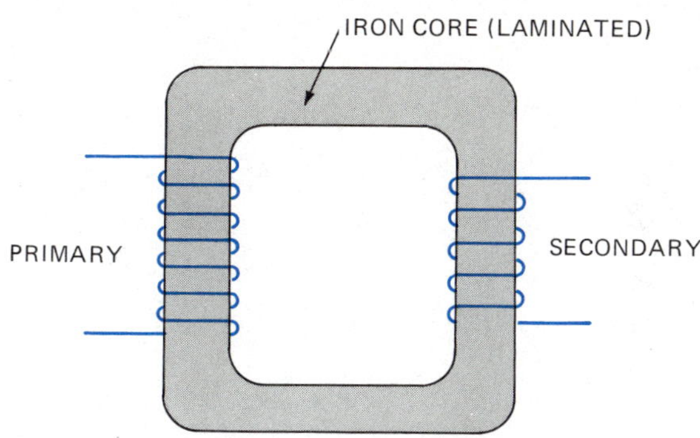

Figure 17-30 A transformer with both primary and secondary windings wound on an iron core

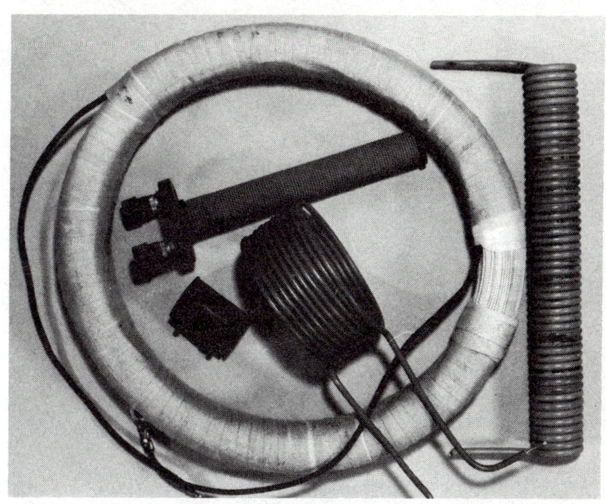

Figure 17-31 A coil has a diameter that is greater than the length. A solenoid's length is greater than its diameter. The wrapping is partially removed from the large coil to show the windings. (Courtesy of Southeast Nebraska Community College)

effect. Any coil (inductor), Figure 17-31, carrying a changing current induces an emf in itself. This effect is called *self-induction.*

Self-induction is very important. The emf produced by self-induction always opposes the change in the current. An increasing current induces a counter-emf (or back-emf) that tends to drive the current in the opposite direction. Thus, the externally produced current is limited. A decaying current induces a forward emf tending to prevent or delay the decrease in current.

Remember that an equivalent constant direct current through the same coil produces no self-induction. If the coil wire has little resistance, the direct current will be dangerously high. The coil will become very hot and destroy itself (or a fuse will be blown). Effectively, the coil short-circuits the DC source.

If an alternating voltage is applied to the coil, it will set up an alternating current. The potential difference at any instant is the sum of the applied voltage and the back-emf of self-induction. These two voltages are never in phase with each other. They always oppose (have opposite signs) each other during at least a part of the cycle.

The maximum current never comes at the same time as the maximum of the externally applied voltage. The time sequence of the entire current wave is displaced along the time axis. It is said to be *out-of-phase* with the voltage wave. If the instantaneous voltage is:

$$e = E_m \sin \omega t$$

The instantaneous current is:

$$\mathbf{i = I_m \sin(\omega t - \theta)}$$

where θ is the *phase angle,* Figure 17-32. It is usually expressed in degrees. In a circuit with inductance, the current *lags* the voltage. The word *lead*

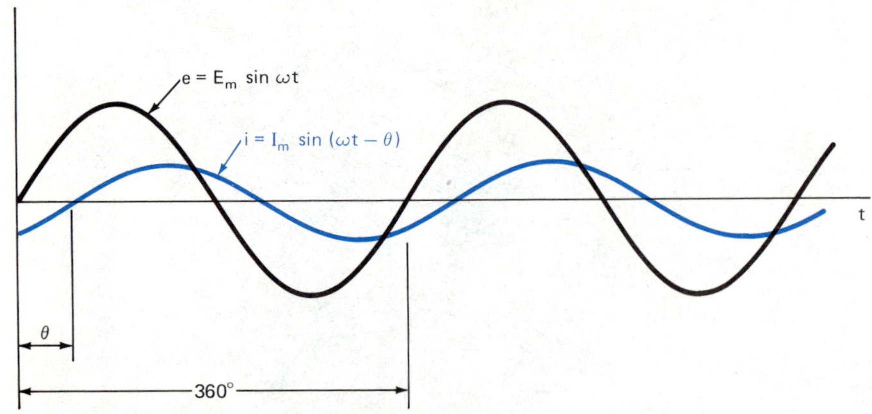

Figure 17-32 Voltage-current graphs showing phase angle. For a pure inductive circuit the phase angle (θ) is 90° with the current lagging the voltage.

can also be used: the voltage leads the current. In inductive circuits with negligible resistance, the phase angle is 90°. In circuits with inductance and resistance, the angle is less than 90° but current still lags the voltage.

Many elements in an electrical circuit contain coils of wire: transformers, motors, generators, solenoids, etc. The magnitude of the effect of self-induction is known as *self-inductance* or simply *inductance*. It is measured in units of *henrys* (H). **The inductance of a coil is one henry if an induced emf of one volt is produced when the current is changing at a rate of one ampere per second.** Basically, inductance (L) depends on the number of turns in the coil, the diameter of the coil and the core material.

17-8 THE INDUCTIVE CIRCUIT

Another quantity called *inductive reactance* is used with circuit formulas. Its symbol is (X_L). It is given by the formula:

$$X_L = 2\pi fL$$

or

$$X_L = \omega L$$

Inductive reactance is used with the applied voltage to determine the current in purely inductive circuits. If resistance is negligible, Ohm's Law takes the following form:

$$I = \frac{E}{X_L}$$

Inductive reactance is a current-opposing factor similar to resistance and is measured in ohms (Ω). Inductance is a property of the coil. Inductive reactance depends on the applied frequency and the inductance.

Example A A 60-Hz alternating voltage of 120 V is applied to a coil with an inductance of 0.45 H, Figure 17-33. Find (1) the inductive reactance and (2) the current. Assume the coil has negligible resistance.

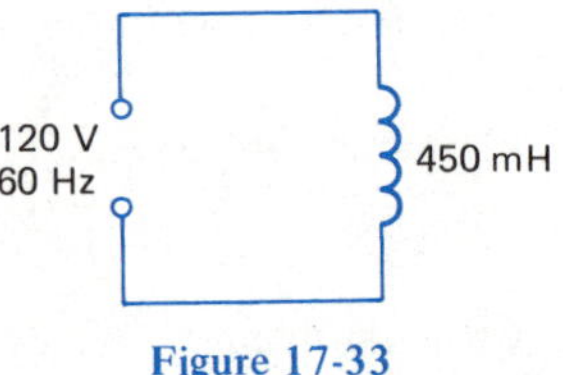

Figure 17-33

Solution: (1) $X_L = 2\pi fL$

$X_L = 2\pi(60\ \text{Hz})(0.45\ \text{H})$

$X_L = 170\ \Omega$

(2) $I = \dfrac{E}{X_L}$

$I = \dfrac{120\ \text{V}}{170\ \Omega} = 0.70\ \text{A}$ (Effective)

Example B Find the effective current, inductive reactance and the effective voltage across a 0.05-H inductor Figure 17-34. The current is given by the equation $i = 3.5 \sin(800\pi t - 90°)$.

Solution: Data: $I_m = 3.5\ \text{A},\ \omega = 800\pi\dfrac{\text{rad}}{\text{s}},\ L = 0.05\ \text{H},\ I = ?$

$X_L = ?,\ E = ?$

Formula: $I = 0.707 I_m$

Substitute: $I = 0.707(3.5\ \text{A}) = 2.5\ \text{A}$

Formula: $X_L = \omega L$

Substitute: $X_L = \left(800\pi\ \dfrac{\text{rad}}{\text{s}}\right)(0.05\ \text{H})$

$X_L = 126\ \Omega$

Formula: $E = IX_L$

Substitute: $E = (2.5\ \text{A})(126\ \Omega)$

$E = 315\ \text{V}$

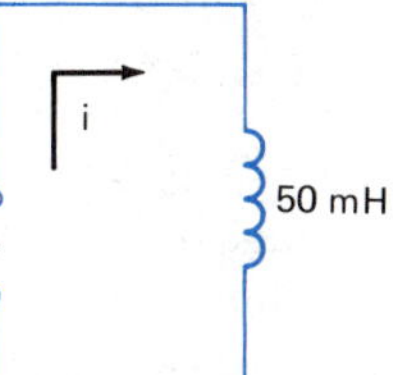

Figure 17-34

Example C A coil, Figure 17-35, produces 70 Ω of reactance when an alternating voltage of 60 Hz is applied. What is the reactance if a 400-Hz alternating voltage is used?

Solution: First find the inductance (L).

Formula: $X_{L1} = 2\pi f_1 L$

$L = \dfrac{X_{L1}}{2\pi f_1}$

Substitute: $L = \dfrac{70\ \Omega}{2\pi(60\ \text{Hz})} = 186\ \text{mH}$

Formula: $X_{L2} = 2\pi f_2 L$

Substitute: $X_{L2} = 2\pi(400\ \text{Hz})(0.186\ \text{H}) = 467\ \Omega$

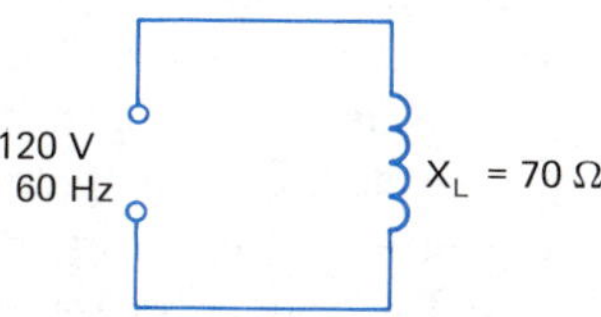

Figure 17-35

If a circuit has more than one coil, the inductances are combined in the same way as resistances. In a series circuit, the total inductance is a simple sum:

$$L_T = L_1 + L_2 + \ldots$$

They are combined by adding reciprocals in parallel circuits:

$$\frac{1}{L_T} = \frac{1}{L_1} + \frac{1}{L_2} + \ldots$$

Example D Find the total inductance of a 150-mH inductor and a 75-mH inductor connected (1) in series and (2) in parallel.

Solution: (1) $L_T = L_1 + L_2$ [Figure 17-36(A)]

$$L_T = 150 \text{ mH} + 75 \text{ mH} = 225 \text{ mH}$$

(2) $\dfrac{1}{L_T} = \dfrac{1}{L_1} + \dfrac{1}{L_2}$ [Figure 16-36(B)]

$$\frac{1}{L_T} = \frac{1}{150} + \frac{1}{75}$$

$$\frac{1}{L_T} = \frac{1}{150} + \frac{2}{150} = \frac{3}{150}$$

$$L_T = \frac{150}{3} = 50 \text{ mH}$$

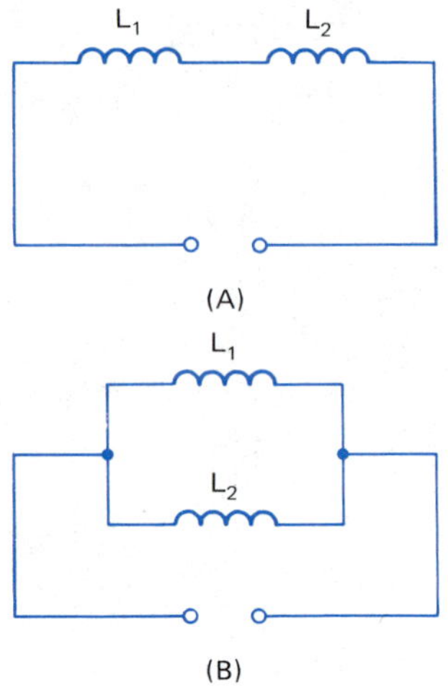

(A)

(B)

Figure 17-36

Example E Find the resistance required to replace the coil in Figure 17-35 in order to limit the current to the same amount (1.71 A).

Solution: $R = \dfrac{E}{I}$

$$R = \frac{120 \text{ V}}{1.71 \text{ A}} = 70 \ \Omega$$

The purpose of the last example was to introduce the subject of energy in inductive circuits. If a resistor is used to limit the current in the previous example, the power requirement (I^2R) is 206 W. This is the rate at which energy is changed to heat in the resistor.

Inductive reactance is used in Ohm's Law as though it was a factor that opposes current. Actually, inductance simply induces a back emf that reduces the effective voltage. Current is limited because there is less voltage. Heat is produced only in resistive circuit elements. The coil in Figure 17-35 probably has a maximum of 1 Ω of resistance. The rate at which heat is produced in the coil is no more than 2.9 W, hardly enough to mention.

A variable inductance consists of a long coil. The inductance is changed by moving an iron core into or out of the coil. Used as a light dimmer, it uses very little energy, Figure 17-37. Much more energy is consumed if the variable inductor is replaced with a rheostat.

When the secondary of an ordinary transformer is open, no current flows in it. The current in the primary is limited by self-inductance. Since there is very little resistance in the primary, the applied voltage and the back emf are 180° out of phase. They all but cancel each other, Figure 17-38. The effective current is very small and lags the applied voltage by 90°. The I^2R heating loss is minimal. It draws significant energy only when the secondary is under load.

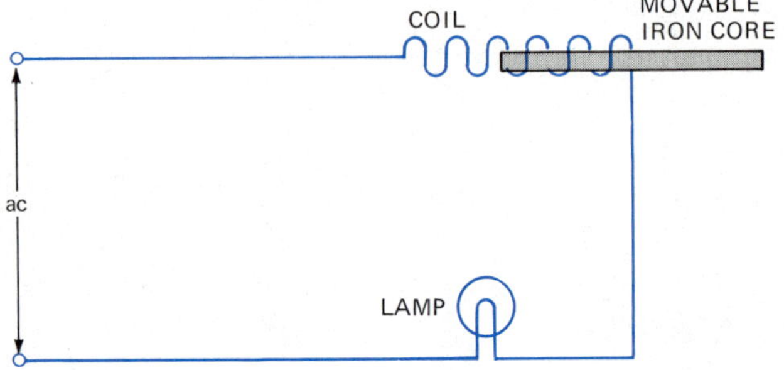

Figure 17-37 Variable inductive light dimmer once commonly used in theaters

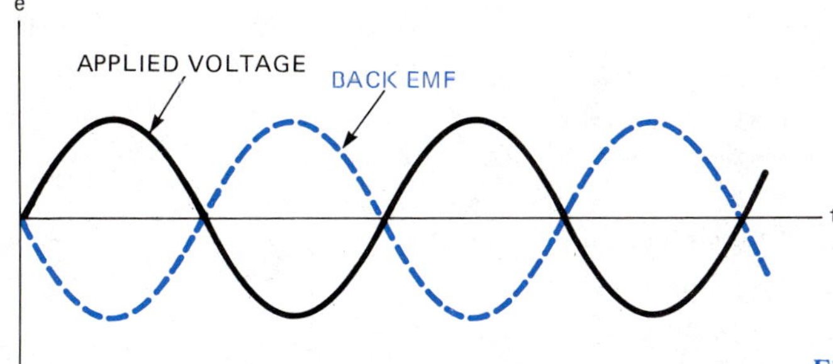

Figure 17-38 Applied voltage and the back emf. Combined by Kirchhoff's Voltage Law, they effectively cancel.

EXERCISE 17-7

Data, diagram, formula, substitute, check.

1. Find the inductive reactance for each of the following inductors when a frequency of 60 Hz is applied. (a) 2.5 mH, (b) 35 mH, (c) 500 mH, (d) 2.5 H
2. Find the inductive reactance of each of the following inductors when the frequency is 400 Hz. (a) 2.5 mH, (b) 35 mH, (c) 500 mH, (d) 2.5 H
3. Find the frequency required for each of the following inductors to produce a reactance of 100 Ω. (a) 2.5 mH, (b) 35 mH, (c) 500 mH, (d) 2.5 H
4. Find the inductance required if 100 Ω of reactance is produced (a) at 60 Hz, (b) at 400 Hz.
5. Find the inductive reactance for a 75-mH coil if the angular frequency is $400\pi \frac{\text{rad}}{\text{s}}$.
6. What is the inductance of a coil producing 500 Ω reactance when the angular frequency is $2512 \frac{\text{rad}}{\text{s}}$?
7. An 800-Hz voltage of 220 V is applied to a coil with an inductance of 35 mH. Find (a) the inductive reactance and (b) the effective current. Neglect resistance.
8. A 60-Hz voltage of 120 V creates a current of 850 mA in a coil. Find (a) the inductive reactance and (b) the inductance of the coil.
9. Two coils of 250 mH and 320 mH are connected in series and a 60-Hz voltage of 115 V is applied. Find (a) the total inductance, (b) the total inductive reactance and (c) the effective current.
10. Find (a) the total inductance, (b) the total reactance and (c) the effective circuit current for the two coils in Problem 9 if they are connected in parallel.
11. Find (a) the effective current, (b) the inductive reactance and (c) the effective voltage across a 450-mH coil. The current is $i = 0.92 \sin(120\pi t - 90°)$.
12. Find (a) the effective voltage, (b) the inductance and (c) the effective current across a 60-Ω (reactance) coil if the voltage is $e = 170 \sin 120\pi t$.

13. Write the voltage equation for the circuit of Problem 11.
14. What is the current equation for the circuit of Problem 12?
15. Two coils have reactances of 80 Ω and 120 Ω and are connected in series across a 115-V line. Find (a) the total current and (b) the voltage drop across each coil.
16. Two coils having reactances of 80 Ω and 120 Ω are connected in parallel across a 115-V line. Find (a) the total current and (b) the current through each branch.

17-9 CAPACITANCE AND CAPACITIVE REACTANCE

A brief introduction to capacitors was given in Topic 5-13. A capacitor consists of two metal plates, Figure 17-39, separated by a thin layer of insulation *(dielectric)*. A DC voltage applied to the capacitor by moving the switch (S) to position 1 drives electrons onto one plate. Electrons drain off the other plate as the capacitor is charging. There is no current through the dielectric. A counter or back emf develops and the current decreases.

The capacitor is fully charged when the electrons cease transferring from one side to the other (I = 0). The voltage across the capacitor is now the same as that of the voltage source, but opposite in direction. The switch can be moved to position 2 and the capacitor remains charged, Figure 17-40(A). In this position, the capacitor is a source of voltage, similar to a charged battery.

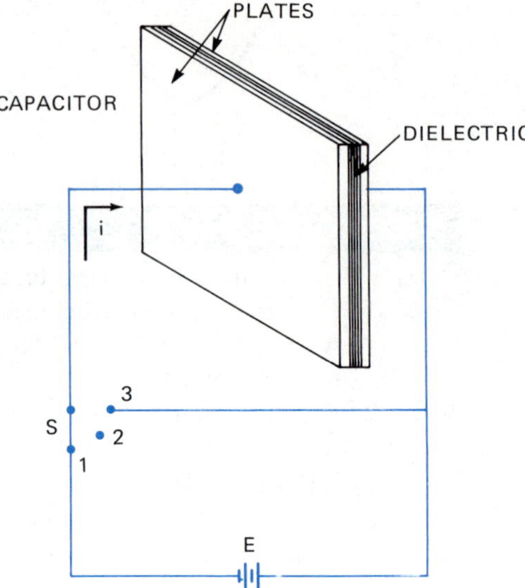

Figure 17-39 A capacitor in a DC circuit

In the charged state, an electric field (force) exists between the plates within the dielectric. There is a strong force of attraction between the plates which have opposite charges. The dielectric must be strong enough to withstand the squeeze exerted on it. In position 3, Figure 17-40(B), the capacitor discharges: voltage drops to zero, charges rebalance (equalize) on the plates and the electric field collapses (force relaxes).

When the capacitor is a part of an ac circuit, electrons first flow onto one plate and out of the other. One-half cycle later, the flow is in the opposite direction. There is alternating current in all parts of the circuit except through the dielectric (a nonconductor). Under the influence of alternating current, the capacitor continuously charges, discharges and recharges in the opposite direction.

Inductance is a property of an electric circuit that opposes *the change in the current.* Inductance delays the change in current so current lags the voltage. *Capacitance,* on the other hand, is a property of a circuit that resists a *change in the voltage.* Capacitance delays the change in the voltage so it lags the current.

The capacitance (C) of a capacitor is defined as the ratio of the charge (Q) stored on the plates to the voltage (E) applied across the plates.

$$C = \frac{Q}{E}$$

If the charge is measured in coulombs (C) and the voltage in volts, the capacitance will have the SI units of farads (F). The farad is a very large unit.

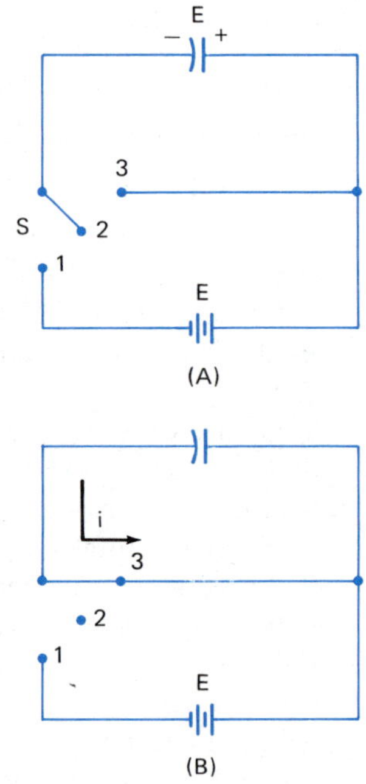

Figure 17-40 Capacitive circuits: (A) Charged (B) Discharging

Figure 17-41 Large electrical capacitors (Courtesy of Southeast Nebraska Community College)

Even larger capacitors, Figure 17-41, are usually measured in microfarads (μF) and smaller ones in picofarads (pF).

A capacitor is charged to its maximum when the voltage reaches its maximum (Q = CE). Current is zero in this situation. When the voltage begins to drop, the capacitor begins to discharge. Current increases as the voltage continues to drop. The maximum current in the negative direction is reached at the moment voltage passes through zero as it changes to negative. For a pure capacitance (no resistance), the current leads the voltage by 90°.

The instantaneous voltage is the reference:

$$e = E_m \sin \omega t$$

and the instantaneous current is:

$$i = I_m \sin(\omega t + 90^\circ)$$

The plus sign indicates that current leads the voltage. For a purely capacitive circuit, the phase relationship of current and voltage are shown in Figure 17-42. There is no heating effect associated with a pure capacitance.

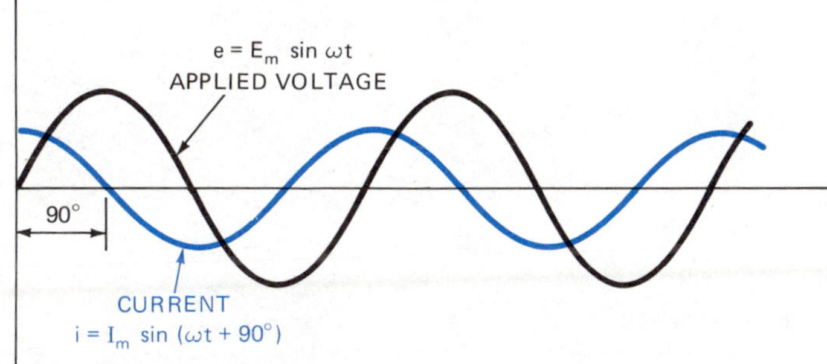

Figure 17-42 Phase relationship of current and voltage in a capacitive circuit. In a pure capacitive circuit, current leads voltage by 90°.

17-10 THE CAPACITIVE CIRCUIT

The *capacitive reactance* (X_C) is defined by the formula:

$$X_C = \frac{1}{2\pi fC}$$

or

$$X_C = \frac{1}{\omega C}$$

If C is in farads and f in hertz, the capacitive reactance is in ohms (Ω). The effective current and voltage in a purely capacitive circuit are related by Ohm's Law:

$$I = \frac{E}{X_C}$$

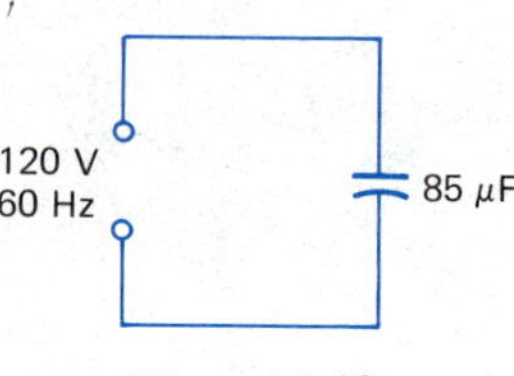

Figure 17-43

Example A A 60-Hz voltage of 120 V is applied to an 85-μF capacitor, Figure 17-43. Find (1) the inductive reactance and (2) the effective current. Neglect resistance.

Solution: (1) $X_C = \frac{1}{2\pi fC}$

$$X_C = \frac{1}{2\pi(60\ \text{Hz})(85 \times 10^{-6}\ \text{F})} = 31.2\ \Omega$$

(2) $I = \frac{E}{X_C}$

$$I = \frac{120\ \text{V}}{31.2\ \Omega} = 3.85\ \text{A}$$

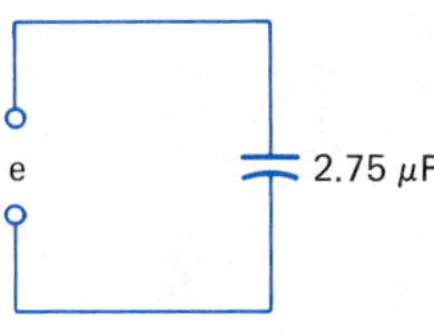

Figure 17-44

Example B Find the effective current, the capacitive reactance and the effective voltage across a 2.75-μF capacitor, Figure 17-44. The current is given by the equation $i = 0.85 \sin(800\pi t + 90°)$.

Solution: Data: $I_m = 85$ mA, $\omega = 800\pi \frac{\text{rad}}{\text{s}}$, $C = 2.75\ \mu\text{F}$, $I = ?$,

$X_C = ?, E = ?$

Formula: $I = 0.707 I_m$

Substitute: $I = 0.707(850\ \text{mA}) = 600\ \text{mA}$

Formula: $X_C = \frac{1}{\omega C}$

Substitute: $X_C = \frac{1}{\left(800\pi \frac{\text{rad}}{\text{s}}\right)(2.75 \times 10^{-6}\ \text{F})} = 145\ \Omega$

Formula: $E = IX_C$

Substitute: $E = (0.60\ \text{A})(145\ \Omega) = 87\ \text{V}$

Example C A capacitor produces 70-Ω reactance when an alternating voltage of 60 Hz is applied. What is the reactance when a 400-Hz alternating voltage is used?

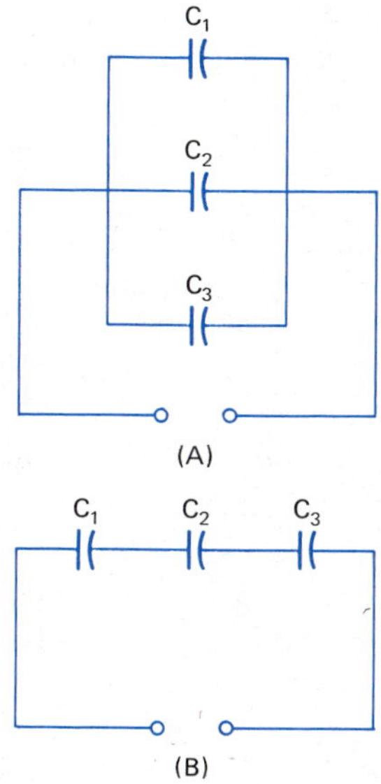

Figure 17-45 Capacitors in series and parallel

Solution: First find the capacitance (C)

Formula: $X_{C1} = \dfrac{1}{2\pi f_1 C}$

$C = \dfrac{1}{2\pi f_1 X_{C1}}$

Substitute: $C = \dfrac{1}{2\pi(60\ \text{Hz})(70\ \Omega)} = 38\ \mu\text{F}$

Formula: $X_{C2} = \dfrac{1}{2\pi f_2 C}$

Substitute: $X_{C2} = \dfrac{1}{2\pi(400\ \text{Hz})(38 \times 10^{-6}\ \text{F})} = 10.5\ \Omega$

If a higher frequency is used with an inductor, a higher inductive reactance results. The opposite effect is observed when higher frequencies are applied to capacitors. The more often a capacitor is charged, the greater the total amount of charge that must be moved in the same length of time.

Capacitances are not combined in the way resistances and inductances are in series and parallel circuits. In Figure 17-45(A), the voltage is the same across all capacitors. The total charge is the sum of the individual charges, $Q_T = Q_1 + Q_2 + Q_3$. If each term of the equation is divided by the voltage, then:

$$C_T = C_1 + C_2 + \ldots$$

In parallel, the total capacitance is the sum of all the individual capacitances. Large capacitors are usually combinations of several smaller ones in parallel.

In a series circuit, Figure 17-45(B), the sum of the voltages must equal the applied voltage. Since current is the same in all parts of the circuit at any time, the capacitors must charge to the same value. If each term of the equation $E_T = E_1 + E_2 + E_3$ is divided by the corresponding charge, then each term equals the reciprocal of the capacitance:

$$\frac{1}{C_T} = \frac{1}{C_1} + \frac{1}{C_2} + \ldots$$

Example D Find the total capacitance of a 7.5-μF capactior and a 12-μF capacitor connected in (a) parallel and (2) in series.

Solution: (1) $C_T = C_1 + C_2$ [Figure 17-46(A)]

$C_T = 7.5\ \mu\text{F} + 12\ \mu\text{F} = 19.5\ \mu\text{F}$

(2) $\dfrac{1}{C_T} = \dfrac{1}{C_1} + \dfrac{1}{C_2}$ [Figure 17-46(B)]

$\dfrac{1}{C_T} = \dfrac{1}{7.5} + \dfrac{1}{12} = 0.217$

$C_T = \dfrac{1}{0.217} = 4.6\ \mu\text{F}$

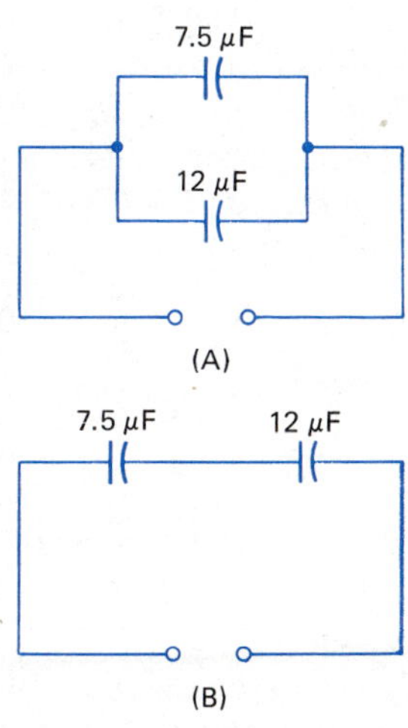

Figure 17-46

17-11 IMPEDANCE

An alternating current can be impeded by resistance (R), inductive reactance (X_L), and/or capacitive reactance (X_C). The word *impedance* is used as the collective name for any combination of the three quantities. The symbol Z is used for impedance and it is also measured in ohms.

In previous topics, only idealized circuits have been described. In purely resistive circuits, voltage and current are in phase. Ohm's Law takes the form of $E = IR$ using effective values. In purely inductive circuits, current lags voltage by 90°. Ohm's Law is $E = IX_L$. In purely capacitive circuits, current leads voltage by 90° and Ohm's law has the form $E = IX_C$. Power drawn from an ac line is called *true power* or *active power.* It is given by the formula $P = EI$ for the resistive circuit. For either kind of purely reactive circuit, true power is zero.

Most ac circuits contain impedance that is a combination of resistance and reactance. The phase angle is neither −90° or +90°, but has some value in between. The impedance of such circuits is represented by the hypotenuse of a right triangle. The sides represent the resistance and reactance. The detailed discussion about ac generation, cyclic terminology and idealized circuits in this chapter has been necessary. It is the basis of most of the chapters that follow and they include the study of realistic impedance circuits.

EXERCISE 17-8

Data, diagram, formula, substitute, check. Neglect resistance in all problems.

1. Find the capacitive reactance for each of the following capacitors when a frequency of 60 Hz is applied: (a) 850 pF, (b) 7.8 μF, (c) 119 μF, (d) 2.5 mF.
2. Find the capacitive reactance for each of the following capacitors if the frequency is 400 Hz. (a) 850 pF, (b) 7.8 μF, (c) 119 μF, (d) 2.5 mF.
3. What frequency is required for each of the following capacitors to produce a reactance of 10 Ω? (A) 850 pF, (b) 7.8 μF, (c) 119 μF, (d) 2.5 mF.
4. What capacitance is required to produce 100 Ω of reactance (a) at 60 Hz? (b) at 400 Hz?
5. Determine the capacitive reactance for a 119-μF capacitor if the angular frequency is $200\pi \frac{\text{rad}}{\text{s}}$.
6. Find the capacitance that produces 25.0-Ω reactance when the angular frequency is $5024 \frac{\text{rad}}{\text{s}}$.
7. A 50-Hz (Europe) voltage of 220 V is applied to a capacitance of 970 pF. Find (a) the capacitive reactance and (b) the effective current.
8. A 60-Hz voltage of 120 V creates a current of 250 mA in a capacitive circuit. Find (a) the capacitive reactance and (b) the capacitance.

9. Two capacitors of 25 μF and 40 μF are connected in parallel and a 60-Hz voltage of 120 V is applied. Calculate (a) the total capacitance, (b) the total capacitive reactance and (c) the effective current.
10. Find (a) the total capacitance, (b) the total capacitive reactance and (c) the effective current for the two capacitors in Problem 9 if they are connected in series.
11. Find (a) the effective current, (b) the capacitive reactance and (c) the effective voltage across a capacitor of 425 μF. The instantaneous current is $i = 0.58 \sin(120\pi t + 90°)$.
12. What is (a) the effective voltage, (b) the capacitance and (c) the effective current across a 60-Ω (reactance) capacitor? The instantaneous voltage is $e = 170 \sin 120\pi t$.
13. Write the voltage equation for the circuit of Problem 11.
14. What is the current equation for the circuit of Problem 12?
15. Two capacitors have reactances of 45 Ω and 75 Ω and are connected in series across a 115-V line. Find (a) the effective current and (b) voltage drop across each.
16. Two capacitors have reactances of 45 Ω and 75 Ω and are connected in parallel across a 115-V line. Find (a) the total effective current and (b) the effective current through each branch.

CHAPTER 18

THE j-OPERATOR

OBJECTIVES

After satisfactorily completing this chapter, the student should be able to:

- Multiply one binomial by another and find the product of special binomials.
- Solve quadratic equations using the quadratic formula.
- Express complex numbers graphically and perform addition, subtraction, multiplication and division operations on complex numbers.

This chapter introduces some additional algebra. It begins with the multiplication of one binomial by another with emphasis on special kinds of products. The quadratic equation (containing an x^2 term) is introduced. A unique formula is used to find the two solutions of any quadratic equation.

When unusual answers are obtained with the quadratic formula, a whole new set of numbers is introduced. The real signed numbers can represent arrows (vectors) along a line, Chapter 5. The new numbers, called complex numbers, are less restricted. They can represent arrows in any direction whatsoever.

The complex numbers provide a means of representing vectors analytically. They are used to represent ac voltages, currents and impedances. Ohm's Law and other circuit formulas require the addition, subtraction, multiplication and division of those quantities. The notation and the operations on complex numbers simplify the arithmetic involved.

This is the last chapter of Section 3. It introduces most of the mathematics essential for mastering ac circuits.

BINOMIAL PRODUCTS

The simplification of algebraic expressions was introduced in Chapter 11. The following facts were discussed:

- Algebraic expressions consist of both numerical and literal symbols. Such expressions represent numbers and must obey all the rules of arithmetic.
- Terms are separated by plus or minus signs and may be like or unlike. Only like terms may be added or subtracted.
- Single term expressions are monomials. Binomials have two and polynomials have many terms.
- Monomials are multiplied by (1) finding the product of numerical factors, (2) adding exponents of the same literal factors, and (3) finding the correct sign of the product.
- A polynomial is multiplied by a monomial by multiplying each term of the polynomial by the monomial.

These concepts are extended in the next topic to include the product of two binomials.

18-1 MULTIPLYING TWO BINOMIALS

The rule for multiplying a binomial by a monomial is called the Distributive Law:

$$a(b + c) = ab + ac$$

Both terms of the binomial must be multiplied by the monomial. If the number (represented by a) happened to be a binomial such as a = p + q, the product can be found by substituting into the equation for the Distributive Law:

$$(p + q)(b + c) = (p + q)b + (p + q)c$$

This gives the rule for multiplying two binomials.

MULTIPLYING TWO BINOMIALS

- Multiply each term of the multiplicand by one term of the multiplier.
- Multiply each term of the multiplicand by the other term of the multiplier.
- Combine like terms of the results and simplify when possible.

Multiplication can be performed by the long method or the horizontal method.

Example A Multiply $x - 5$ and $3x + 2$ by the long method.

Solution: The long method is the usual long method of multiplying numbers. Set the multiplier under the multiplicand.

$x - 5$	(Multiplicand)
$3x + 2$	(Multiplier)
$3x^2 - 15x$	$3x(x - 5)$
$2x - 10$	$2(x - 5)$
$3x^2 - 13x - 10$	(Add like terms)

Note: Like terms are arranged in columns to facilitate the addition.

Example B Multiply $2x + y$ and $x - 2y$ by the long method.

Solution:

$2x + y$	
$x - 2y$	
$2x^2 + xy$	$x(2x + y)$
$- 4xy - 2y^2$	$-2y(2x + y)$
$2x^2 - 3xy - 2y^2$	(Combine terms)

Example C Multiply $5E + 6$ and $4E - 2$ by the horizontal method.

Solution: Multiply both terms of the multiplicand by each term of the multiplier.

$(5E + 6)(4E - 2) = 20E^2 + 24E - 10E - 12$

$= 20E^2 + 14E - 12$ (Combine like terms)

Note: Cover the -2 and multiply $(5E + 6)$ by $(4E)$ to obtain the first two terms of the answer. Now cover the 4E and multiply $(5E + 6)$ by (-2) to get the last two terms.

The result of multiplying one binomial by another is an expression with four terms. When possible, like terms are then combined to get the final product. The four terms can also be found by the FOIL method. The four letters of the word FOIL indicate the order in which the terms of the two binomials are multiplied.

The F represents the product of the *first* terms. The letter O indicates the product of the *outside* terms. I is the product of the *inside* terms and L is the product of the *last* terms. These are highlighted in boldface print in the following:

$(\mathbf{3P} - 5)(\mathbf{3P} + 4)$	(**F**irsts)
$(\mathbf{3P} - 5)(3P + \mathbf{4})$	(**O**utsides)
$(3P - \mathbf{5})(\mathbf{3P} + 4)$	(**I**nsides)
$(3P - \mathbf{5})(3P + \mathbf{4})$	(**L**asts)

Example D Multiply $3P - 5$ and $3P + 4$ using the FOIL method.

F ↓ O ↓ I ↓ L ↓

Solution: $(3P - 5)(3P + 4) = 9P^2 + 12P - 15P - 20$

$= 9P^2 - 3P - 20$

EXERCISE 18-1

Multiply the following binomials using the long method.

1. $(x + 2)(x + 1)$
2. $(x - 5)(x + 7)$
3. $(P + 8)(P - 3)$
4. $(-I - 2)(2I + 3)$
5. $(3E + 4)(E - 6)$
6. $(4\theta + 2)(4\theta + 2)$
7. $(-2\phi + 7)(-4\phi + 2)$
8. $(-5R^2 - 3)(6R^2 - 4)$
9. $(5E + 2I)(7E + 3I)$
10. $(2\alpha - \beta)(2\beta - \alpha)$

Use the horizontal method or the FOIL method to multiply the following binomials.

11. $(2 - E)(5 - 2E)$
12. $(\frac{1}{2}B + 5)(\frac{1}{2}B + 5)$
13. $(Q + 3 \times 10^4)(2Q + 4 \times 10^4)$
14. $(7 + 2d)(8 + 3d)$
15. $(6C^2 - 2)(4 - 3C^2)$
16. $(5\theta - 2\phi)(5\theta - 2\phi)$
17. $(2E + 5I)(5I + 2E)$
18. $(2P + Q)(2P - Q)$
19. $(a + 2b)(c - d)$
20. $(2 - 3x)(4y - z)$

18-2 SPECIAL PRODUCTS

A *special product* is formed when the product of two expressions can be determined without the aid of long multiplication. An example is given by the Distributive Law.

$$a(b + c) = ab + ac$$

The answer is simply written down.

Example A Multiply: $3x^2(x^2 - 2y^2)$

Solution: $3x^2(x^2) - 3x^2(2y^2) = 3x^4 - 6x^2y^2$

There are three other special products that are easy to remember because they form a pattern. One is the square of the sum of two numbers $(a + b)^2$. The second is the square of a difference $(a - b)^2$. The third is the product of a sum and a difference $(a + b)(a - b)$.

The square of a number is the product of the number and itself.

$$\begin{array}{r} a + b \\ \underline{a + b} \\ a^2 + ab \qquad \\ \underline{+ ab + b^2} \\ a^2 + 2ab + b^2 \end{array}$$

The square of a sum is the square of the first term, twice the product of the two terms and the square of the second term:

$$(a + b)^2 = a^2 + 2ab + b^2$$

The square of a difference is similar except the two ab terms in the answer are negative:

$$(a - b)^2 = a^2 - 2ab + b^2$$

The product of the sum and difference also is similar except one ab term is positive. The other is negative so they cancel. The square of the b^2 term is negative:

$$(a + b)(a - b) = a^2 - b^2$$

Example B Square the binomial $3x + 4y$.

Solution: Use the equation $(a + b)^2 = a^2 + 2ab + b^2$

$a = 3x$, $b = 4y$

Square $3x$: $(3x)^2 = 9x^2$

Multiply: $2(3x)(4y) = 24xy$

Square $4y$: $(4y)^2 = 16y^2$

$(3x + 4y)^2 = 9x^2 + 24xy + 16y^2$

Usually the answer can be written without any intermediate steps.

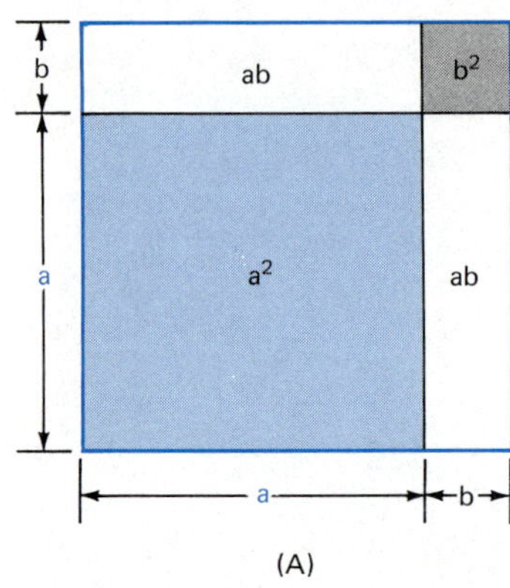

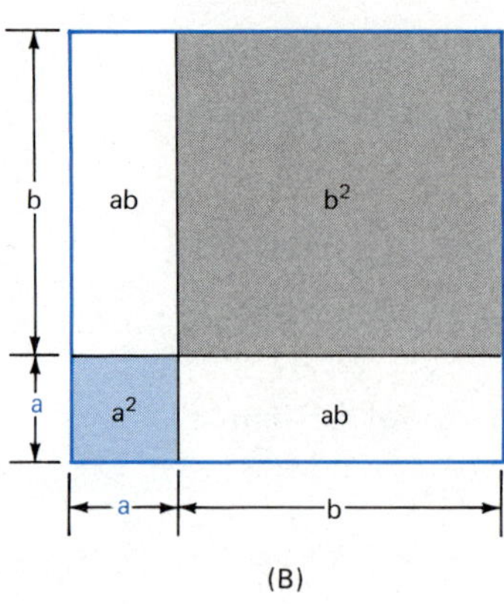

Figure 18-1 The area and subareas of a square having sides equal to a + b. $(a + b)^2 = a^2 + 2ab + b^2$. If a is greater than b, the subareas are shown in (A) and in (B) when b is greater. All four subareas are equal when a = b.

Example C Square: $6p - 5q$

Solution: Use the equation $(a - b)^2 = a^2 - 2ab + b^2$

$a = 6p, b = 5q$

Square 6p: $(6p)^2 = 36p^2$

Multiply: $(-2)(6p)(5q) = -60pq$

Square 5q: $(5q)^2 = 25q^2$

$(6p - 5q)^2 = 36p^2 - 60pq + 25q^2$

Example D Multiply: $(5R + 2D)(5R - 2D)$

Solution: Use the equation $(a + b)(a - b) = a^2 - b^2$

$a = 5R, b = 2D$

Square 5R: $(5R)^2 = 25R^2$

Square 2D: $(2D)^2 = 4D^2$

Form the difference of the squares:

$(5R + 2D)(5R - 2D) = 25R^2 - 4D^2$

An interpretation of the binomial square as an area is shown in Figure 18-1. This is similar to the area interpretation of a(b + c) in Figure 11-7.

EXERCISE 18-2

Determine the answers to each of the following.

1. $(x + 5)^2$
2. $(2R + 4)^2$
3. $(5E - 2)^2$
4. $(3I - 2)^2$
5. $(-P + 7)^2$
6. $(4D - 3)^2$
7. $(2B + 1)(2B - 1)$
8. $(\frac{1}{2}T + \omega t)^2$
9. $(2\theta - \pi)^2$
10. $(5C - 1)(5C + 1)$
11. $(5\theta + \frac{1}{4})^2$
12. $(\alpha - \beta)(\alpha + \beta)$
13. $(-3L + 2W)^2$
14. $(\omega t - 15)(\omega t + 15)$
15. $(Pt - W)(Pt + W)$
16. $(5L - 10^5)^2$
17. $(10^3 + 3E)^2$
18. $(\frac{1}{2}k - \frac{1}{4}m)(\frac{1}{2}k + \frac{1}{4}m)$

19. The voltage drops of the two loads are shown in Figure 18-2. Note: $R_T = 6\ \Omega + 2\ \Omega = 8\ \Omega$. The total voltage drop is $\frac{1}{4}E + 6I$. The total power dissipation $\left(P = \frac{E^2}{R}\right)$ is $P_T = \frac{(\frac{1}{4}E + 6I)^2}{8}$. Find the total power dissipation.

20. The total current in the parallel circuit of Figure 18-3 is $\frac{1}{4}I_T + \frac{E}{6}$. Power $(P = I^2R)$ dissipation is $P_T = \left(\frac{1}{4}I_T + \frac{E}{6}\right)^2(1.5)$. Evaluate this expression and simplify.

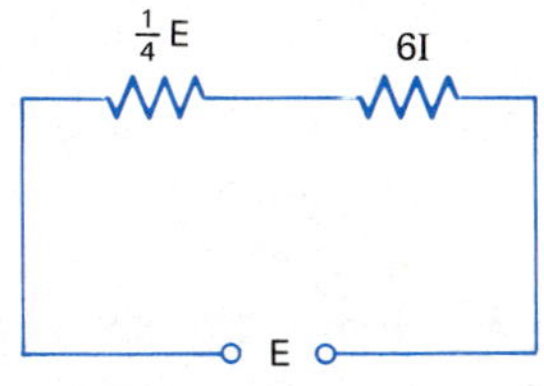

Figure 18-2

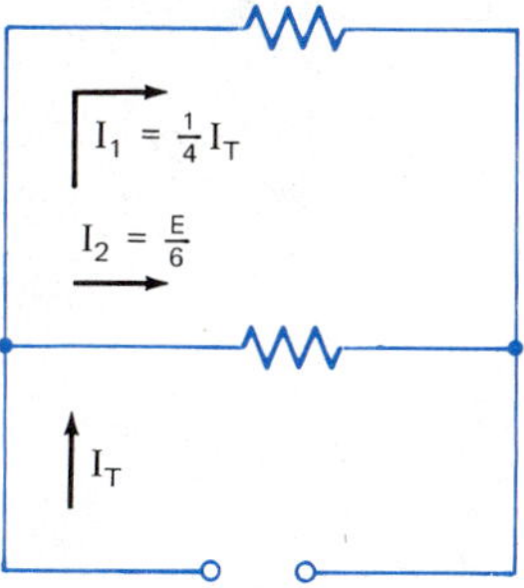

Figure 18-3

18-3 FACTORING SPECIAL PRODUCTS

Factoring is the reverse of multiplying. When an expression such as a(b + c) is changed to ab + ac, the process is multiplication. If the second expression is changed to the first, it is factoring:

→Multiplying→

$$a(b + c) = ab + ac$$

←Factoring←

Expressions such as those on the right of the following equations can also be factored.

$$(a + b)^2 = a^2 + 2ab + b^2$$
$$(a - b)^2 = a^2 - 2ab + b^2$$
$$(a + b)(a - b) = a^2 - b^2$$

Example A If possible, factor $9x^2 - 24xy + 16y^2$.

Solution: The expression has the form of the equation $(a - b)^2 = a^2 - 2ab + b^2$ but is must be tested to see if it is exactly a perfect square. The first and third terms are perfect squares: $a = 3x$, $b = 4y$. The middle term must be exactly 2ab: $2(3x)(4y) = 24xy$. The expression is a perfect square and can be factored. Usually perfect squares can be recognized by inspection: $9x^2 - 24xy + 16y^2 = (3x - 4y)^2$. It has two equal factors of $3x - 4y$.

Example B If possible, factor $4p^2 - 25q^2$

Solution: This one has the form of the equation $(a + b)(a - b) = a^2 - b^2$ in which $a = 2p$ and $b = 5q$. As long as the two terms are both perfect squares, no further test is necessary. $4p^2 - 25q^2 = (2p + 5q)(2p - 5q)$. It does not make any difference which factor is written first.

If all terms have a common monomial factor, it should be removed first.

Example C If possible, factor $8rs^2 + 48rst + 18rt^2$.

Solution: By inspection, it can be seen that 2r is a common monomial factor. When factored the expression is: $2r(4s^2 + 24st + 9t^2)$. Now test the middle term of the three-term factor to see if it is a perfect square: $a = 2s$, $b = 3t$. The middle term must be exactly $2ab = 2(2s)(3t) = 12st$ which it is not! The expression cannot be factored further. Therefore:
$8rs^2 + 48rst + 18rt^2 = 2r(4s^2 + 24st + 9t^2)$.

EXERCISE 18-3

If possible, factor each of the following expressions.

1. $x^2 - 8x + 16$
3. $P^2 + 10P + 25$
3. $4R^2 - 36$
4. $4\theta^2 - 12\theta\phi + 9\phi^2$
5. $2R^2 + 12Rr + 18r^2$
6. $25C^2 - 16$
7. $\frac{1}{4}P^2 + 4P + 64$
8. $4\alpha^2 - 16\alpha\beta + 16\beta^2$
9. $4\theta^2 - \pi^2$
10. $\omega^2L^2 - \frac{2L}{C} + \frac{1}{\omega^2C^2}$
11. $81t^2 - 144T^2$
12. $I^2 + EI + \frac{1}{4}E^2$
13. $8L^2 + 24LW + 18W^2$
14. $50kD^2 - 70kdD + 98kd^2$
15. $49E^2 - 121I^2R^2$
16. $25\theta^2 + 5\theta\pi + \frac{\pi^2}{4}$

QUADRATIC EQUATIONS

Quadratic equations are equations that contain the square of the unknown as the highest power. Some examples of quadratic equations are as follows:

- $3x^2 - 27 = 0$
- $2R^2 = 16R$
- $I^2 - 2I - 6 = 0$

A *root* of an equation is any number that will satisfy the equation. A quadratic equation has two roots.

18-4 PURE QUADRATIC EQUATIONS

The standard form of a quadratic equation is

$$ax^2 + bx + c = 0$$

The letters a and b are the coefficients of x^2 and x, respectively. The letter c is a constant. In the example $3x^2 - 4x + 18 = 0$, $a = 3$, $b = -4$ and $c = 18$.

The letter a cannot equal zero. In such a case, the equation would not be quadratic. The letters b and c or both can equal zero in which case the term or terms would drop out. If $b = 0$, the equation is called a *pure quadratic equation.* These are solved by methods developed in earlier chapters.

SOLVING PURE QUADRATIC EQUATIONS

- Transpose the constant isolating the squared term on the left.
- Remove the coefficient of the squared term using the division axiom.
- Take the square root of both sides.
- One root takes a plus sign and the second root takes a minus sign.
- Check both roots by substituting into the original equation.

Example A Solve the quadratic equation $3x^2 - 27 = 0$

Solution:

$3x^2 - 27 = 0$

$3x^2 = 27$ (AA, + 27)

$x^2 = 27/3$ (DA, ÷ 3)

$x^2 = 9$

$x = \pm\sqrt{9}$ (RA, square root)

$x = \pm 3$ (Two solutions)

$x = +3$ or $x = -3$

Check: $3(+3)^2 - 27 = 0$

$0 = 0$

$3(-3)^2 - 27 = 0$

$0 = 0$

EXERCISE 18-4

Solve the following pure quadratic equations without the aid of a calculator.

1. $x^2 = 49$
2. $P^2 = 144$
3. $R^2 - 81 = 0$
4. $\theta^2 - 121 = 0$
5. $36 - \alpha^2 = 0$
6. $5I^2 = 0$
7. $7D^2 = 252$
8. $4\phi^2 - 16 = 0$
9. $27Q^2 - 27 = 0$
10. $5E^2 - 320 = 0$

18-5 THE QUADRATIC FORMULA

A complete quadratic equation is as follows:

$$ax^2 + bx + c = 0$$

where a, b and c are all nonzero numbers. The roots depend on the values of these numbers. All that is needed is a formula in which x equals a combination of a, b and c. The formula is derived by starting with the standard form.

$$ax^2 + bx + c = 0$$

$$x^2 + \frac{b}{a}x + \frac{c}{a} = 0 \qquad (\text{DA}, \div \text{a})$$

$$x^2 + \frac{b}{a}x = -\frac{c}{a} \qquad \left(\text{SA}, -\frac{c}{a}\right)$$

$$x^2 + \frac{b}{a}x + \frac{b^2}{4a^2} = -\frac{c}{a} + \frac{b^2}{4a^2} \qquad \left(\text{AA}, +\frac{b^2}{4a^2}\right)$$

Since $(x + y)^2 = x^2 + 2xy + y^2$, the second term on the left $\left(\frac{b}{a}x\right)$ must equal $2xy$. Therefore $y = \frac{b}{2a}$. To make the left side a perfect square, the square of $\frac{b}{2a}$ must be added to both sides $\left(\frac{b^2}{4a^2}\right)$.

$$\left(x + \frac{b}{2a}\right)^2 = \frac{b^2 - 4ac}{4a^2} \qquad \text{(Factor left side)}$$

$$x + \frac{b}{2a} = \frac{\pm\sqrt{b^2 - 4ac}}{2a} \qquad \text{(RA, square root)}$$

$$x = \frac{-b}{2a} \pm \frac{\sqrt{b^2 - 4ac}}{2a} \qquad \left(\text{SA}, -\frac{b}{2a}\right)$$

$$x = \frac{-b \pm \sqrt{b^2 - 4ac}}{2a} \qquad \text{(Combine terms)}$$

This result is the *quadratic formula.* The three constants (a, b and c) must be identified correctly with the right signs. Write the equation in standard form for identification. To learn the formula, write it down every time it is used. Several examples are illustrated to show how the formula is applied.

Example A Find the roots of $x^2 + 5x + 4 = 0$

Solution: a = +1, b = +5, c = +4

Formula: $x = \frac{-b \pm \sqrt{b^2 - 4ac}}{2a}$

Substitute: $x = \frac{-5 \pm \sqrt{5^2 - 4(1)(4)}}{2(1)}$

$$x = \frac{-5 \pm \sqrt{25 - 16}}{2} = \frac{-5 \pm \sqrt{9}}{2}$$

$$x = \frac{-5 \pm 3}{2}$$

$$x = \frac{-5 + 3}{2} = -1 \text{ or } x = \frac{-5 - 3}{2} = -4$$

−1 and −4 are the two roots

Check: $(-1)^2 + 5(-1) + 4 = 0$

$0 = 0$

$(-4)^2 + 5(-4) + 4 = 0$

$0 = 0$

If the equation in the previous example is set equal to y:

$$y = x^2 + 5x + 4$$

instead of zero, a graph can be constructed. Substitute various values of x to find the corresponding values of y. The curve, known as a *parabola,* is shown in Figure 18-4. The roots are the two points on the x-axis intercepted by the curve (when $y = 0$). The minimum point on the curve is located at a value of x midway between the roots (at $x = -2.5$).

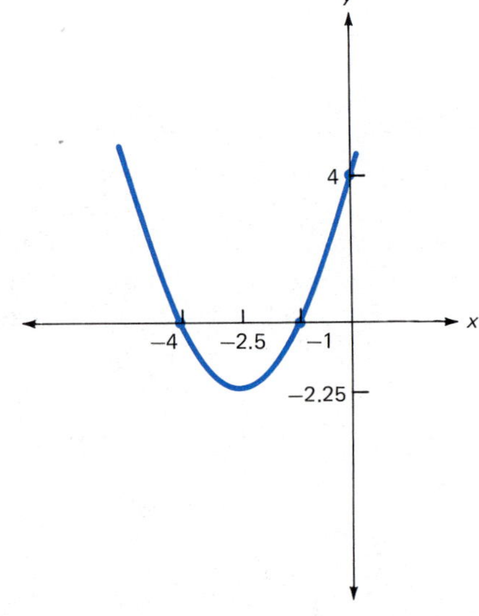

Figure 18-4 Curve (parabola) of $y = x^2 + 5x + 4$

Example B Solve the equation $3x^2 - 27 = 0$

Solution: a = 3, b = 0, c = −27

Formula: $x = \dfrac{-b \pm \sqrt{b^2 - 4ac}}{2a}$

Substitute: $x = \dfrac{-0 \pm \sqrt{0 - 4(3)(-27)}}{2(3)}$

$x = \dfrac{0 \pm \sqrt{324}}{6} = \dfrac{\pm 18}{6}$

$x = +3$

$x = -3$

Check: $3(+3)^2 - 27 = 0$

$3(-3)^2 - 27 = 0$

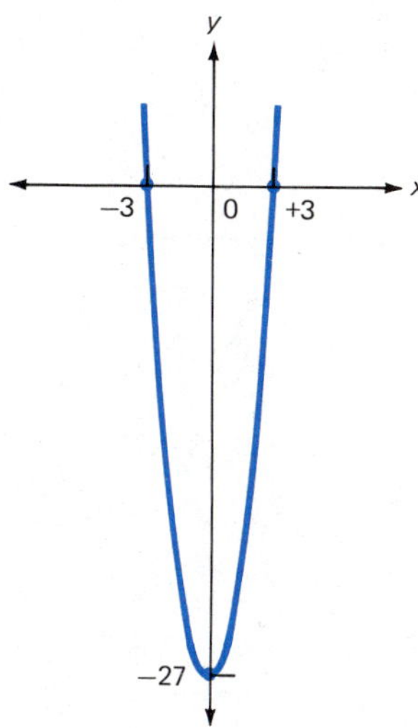

Figure 18-5 Parabolic curve for $y = 3x^2 - 27$

Whenever b = 0, the parabolic curve is symmetrical about the y-axis, as shown in Figure 18-5. The minimum point is on the y-axis ($x = 0$). The absolute values of the roots are equal with one positive and one negative.

Example C Solve $3x^2 - 6x = 0$

Solution: a = 3, b = −6, c = 0

Formula: $x = \dfrac{-b \pm \sqrt{b^2 - 4ac}}{2a}$

Substitute: $x = \dfrac{-(-6) \pm \sqrt{(-6)^2 - 4(3)(0)}}{2(3)}$

$x = \dfrac{6 \pm \sqrt{36 - 0}}{6} = \dfrac{6 \pm 6}{6}$

$x = \dfrac{0}{6} = 0 \text{ or } x = \dfrac{12}{6} = 2$

Check: $3(0)^2 - 6(0) = 0$

$0 = 0$

$3(2)^2 - 6(2) = 0$

$0 = 0$

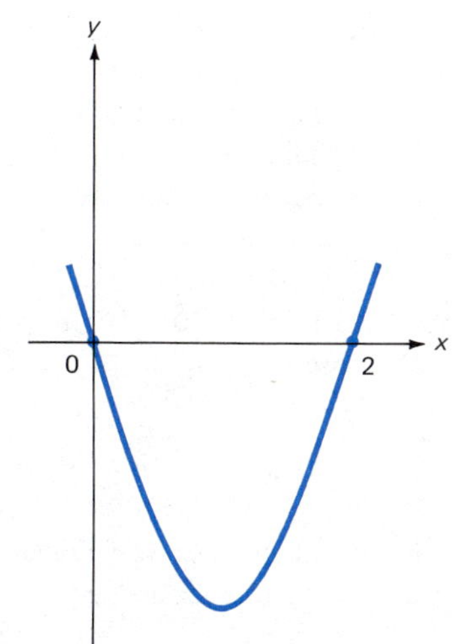

Figure 18-6 Parabolic curve of $y = 3x^2 - 6x$

If c = 0, one of the roots will always be zero, Figure 18-6. The other may be positive or negative and has a value of $x = -\dfrac{b}{a}$.

Example D Solve: $6x - 3x^2 = 3$

Solution: Rearrange to place terms in standard position.

$-3x^2 + 6x - 3 = 0$ (SA, − 3)

a = −3, b = 6, c = −3

Formula: $x = \dfrac{-b \pm \sqrt{b^2 - 4ac}}{2a}$

Substitute: $x = \dfrac{-6 \pm \sqrt{(6)^2 - 4(-3)(-3)}}{2(-3)}$

$x = \dfrac{-6 \pm \sqrt{36 - 36}}{-6} = \dfrac{-6 \pm 0}{-6}$

$x = 1$

Note that both roots are equal.

Check: $-3(1)^2 + 6(1) - 3 = 0$

$0 = 0$

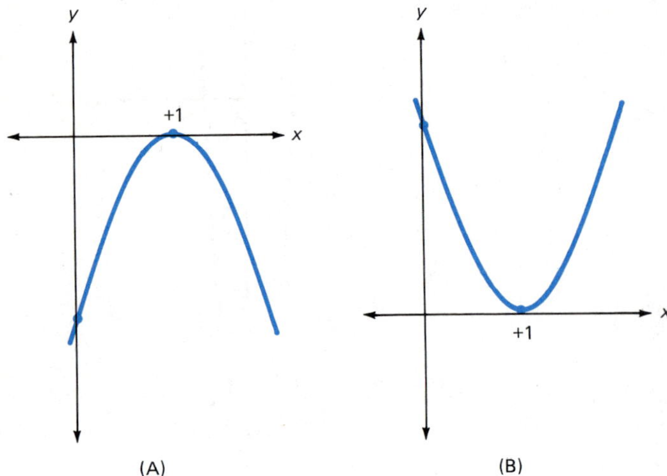

Figure 18-7 Curves showing the effect of a negative sign on the x^2 term: (A) $y = -3x^2 + 6x - 3$ and (B) $y = +3x^2 - 6x + 3$.

When the coefficient of the squared term is negative, the graph is flipped upside down. It has one maximum point instead of a minimum. If the expression under the radical ($b^2 - 4ac$) is equal to zero, both roots will be equal. The curve, Figure 18-7(A), touches the x-axis in only one place. If the original equation $-3x^2 + 6x - 3 = 0$ is multiplied by -1, all terms on both sides change signs: $3x^2 - 6x + 3 = 0$. A graph of this curve is shown in Figure 18-7(B). The roots are the same.

In some applied problems, one of the two roots may not be a meaningful answer. In these, only common sense can be used to decide which answer to choose.

Example E In a series circuit, Figure 18-8, the total power dissipated in the two loads is 28 W. The voltage drop across R_1 is 6 V and load R_2 has a resistance of 4 Ω. Find the current.

Solution: The total power equals the power dissipated in both loads. $P_T = E_1 I + I^2 R_2$. After substituting and setting the equation in standard form, the equation is $4I^2 + 6I - 28 = 0$.

Formula: $I = \dfrac{-b \pm \sqrt{b^2 - 4ac}}{2a}$

Substitute: $I = \dfrac{-6 \pm \sqrt{(6)^2 - 4(4)(-28)}}{2(4)}$

$$I = \frac{-6 \pm \sqrt{36 + 448}}{8} = \frac{-6 \pm 22}{8}$$

$$I = \frac{16}{8} = 2\text{ A or } I = \frac{-28}{8} = -3.5\text{ A}$$

Only the first answer is meaningful. The second answer would require that energy be produced in load R_1 instead of being consumed. In other words, R_1 would have to be a negative resistance.

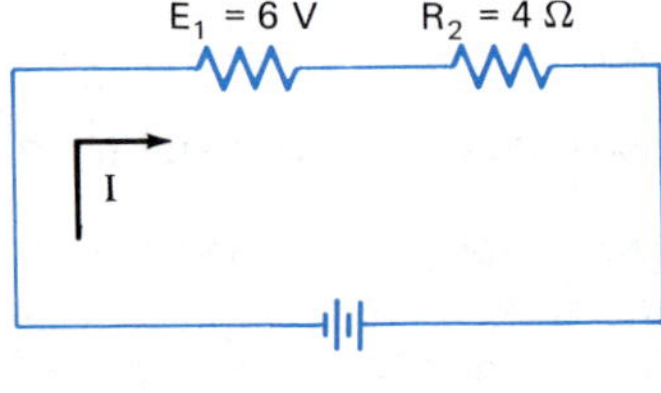

Figure 18-8

EXERCISE 18-5

Find the roots of each of the following quadratic equations using the quadratic formula.

1. $x^2 = 36$
2. $x^2 - 5x = 0$
3. $I^2 + 5I + 4 = 0$
4. $P^2 + 6P + 5 = 0$
5. $2R^2 - 12R = 0$
6. $2\theta^2 - 50 = 0$
7. $2t^2 = 8t - 8$
8. $\phi^2 + 8 = 9\phi$
9. $2E^2 = 4 - 7E$
10. $3D^2 = 48D$
11. $\alpha^2 - 4\alpha = 5$
12. $9 + 6r + r^2 = 0$
13. $2Q^2 - 4 = -7Q$
14. $\beta^2 - 8\beta = 9$
15. $P^2 - 6P + 5 = 0$
16. $2T^2 + 6T = 8$

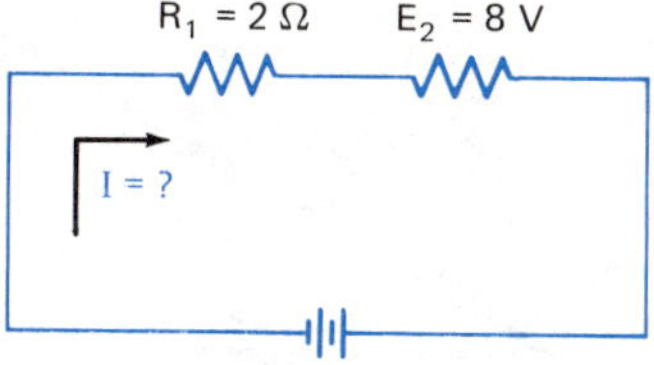

Figure 18-9

17. Find the current in the circuit of Figure 18-9. The total power dissipation in the loads is 18 W. Load R_1 has a resistance of 2 Ω and R_2 has a voltage drop of 8 V.

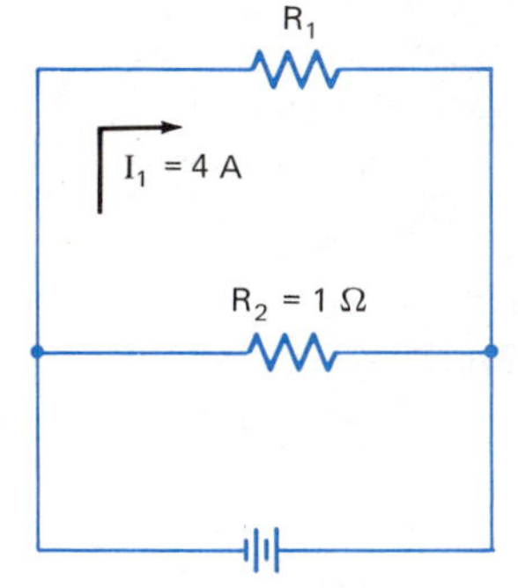

Figure 18-10

18. Find the voltage applied to the circuit of Figure 18-10. The current through R_1 is 4 A and the resistance of load R_2 is 1 Ω. The total power dissipated is 12 W.

19. An object dropped from a height (h) falls according to the formula $h = 16t^2$ where h is in feet and t is in seconds. Find the time required for a wrench dropped from a pole to reach the ground if the height of the pole is (a) 25 ft, (b) 40 ft and (c) 60 ft. Choose positive answers only.

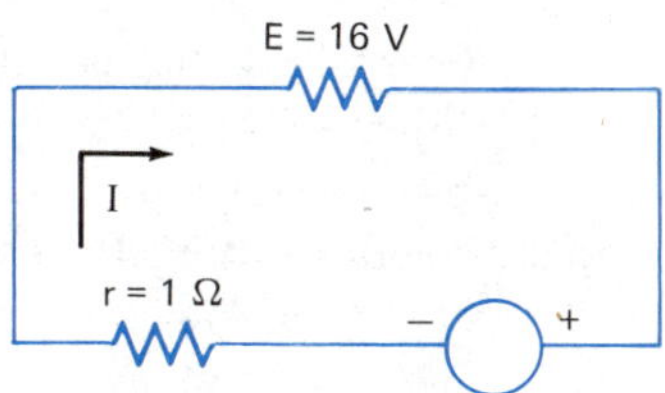

Figure 18-11

20. If an object is bumped or thrown downward with an initial speed (v_0), it falls according to the formula $h = v_0t + 4.9t^2$ where h is in metres and v_0 is in metres per second. How much time is required for a wrench to fall 25 m if it was thrown with an initial downward speed of 5 m/s?

21. A variable electric current changes according to the formula $i = t^2 - 9t + 8$. At what two times is current equal to zero? Time is measured in seconds.

22. The output power of a generator, Figure 18-11, is given by the formula $P = EI - rI^2$. E is the generator voltage and r is the internal resistance. (a) Find the current if P = 15 W, E = 16 V and r = 1 Ω. (b) Are both answers possible? (c) What is the resistance of the load for each current?

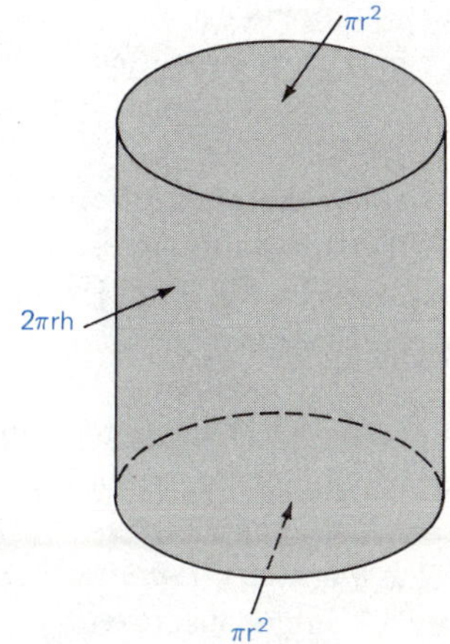

Figure 18-12

23. The surface area of a pole transformer, Figure 18-12, including top and bottom, must be 10 500 cm² for proper heat dissipation. The area is given by the formula $A = 2\pi r^2 + 2\pi rh$. If the height (h) is 75 cm, what is the radius and diameter to the nearest centimetre?

24. The base of a right triangle is 3 cm longer than its altitude, Figure 18-13. The hypotenuse is 15 cm. Find the altitude and the base. The formula is $c^2 = (a + 3)^2 + a^2$.

Note: Square (a + 3), combine terms, substitute and evaluate.

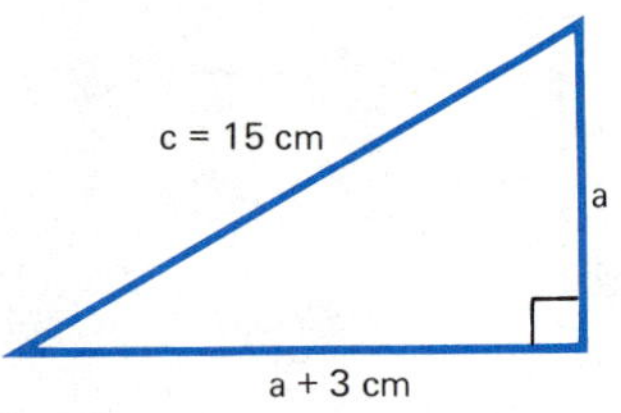

Figure 18-13

18-6 j-NUMBERS

One kind of quadratic equation results in unusual answers. It is the kind formed when the combination $b^2 - 4ac$ is less than zero. The curves of such equations never cross or touch the x-axis, Figure 18-14. The minimum point of the parabola is a positive number if the value of a is positive.

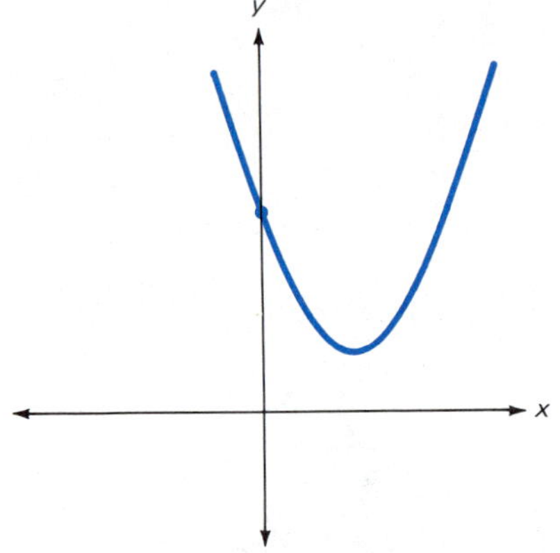

Figure 18-14 Curve of $y = x^2 - 6x + 25$. When $b^2 - 4ac$ is less than zero, the roots are complex numbers.

Example A Solve for the roots of $x^2 - 6x + 25 = 0$

Solution: $a = 1, b = -6, c = 25$

Formula: $x = \dfrac{-b \pm \sqrt{b^2 - 4ac}}{2a}$

Substitute: $x = \dfrac{+6 \pm \sqrt{6^2 - 4((1)(25)}}{2(1)}$

$$x = \frac{6 \pm \sqrt{36 - 100}}{2} = \frac{6 \pm \sqrt{-64}}{2}$$

It is the number $\sqrt{-64}$ that is unusual. The square root of −64 is not −8 or +8. If either −8 or +8 is squared, the answer is +64. The actual meaning of $\sqrt{-64}$ and other such numbers need not be a concern. It can be simplified using ordinary mathematics. The number −64 can be thought of as the product (−1)(+64):

$$\sqrt{-64} = \sqrt{(-1)(+64)}$$
$$= \sqrt{-1} \cdot \sqrt{64}$$
$$= \sqrt{-1} \cdot 8$$

$(\sqrt{ab} = \sqrt{a} \cdot \sqrt{b})$

All such numbers can be reduced to this form. The final answer to Example A can now be simplified:

$$\frac{6 \pm \sqrt{-64}}{2} = \frac{6 \pm \sqrt{-1} \cdot 8}{2} = 3 \pm \sqrt{-1} \cdot 4$$

The two roots, therefore, are: $3 + \sqrt{-1} \cdot 4$ and $3 - \sqrt{-1} \cdot 4$. The symbol $\sqrt{-1}$ and the dot multiplier consists of four distinct symbols. They are usually replaced by the letter j, a much simpler symbol. In final form, therefore, the roots of Example A are as follows:

$$x = 3 + j4 \text{ and } x = 3 - j4$$

Such numbers are called *complex numbers.*

Complex numbers consist of a real number (first term) and a *j-number* (second term). Mathematicians use the letter i instead of j and refer to numbers such as 8i as *imaginary numbers.* Electricians use the letter j, partly because i represents instantaneous current. The word imaginary is not used in this text because it conveys the mistaken impression that such numbers are unreal.

Notice that the letter j is placed in front of the number when j-numbers are formed. This is similar to placing a minus sign in front of a number to form a negative number. The properties of j-numbers and negative numbers have some remarkable similarities. Their properties will be discussed in the next topic.

Example B Express the following square roots using the letter j to represent $\sqrt{-1}$: (1) $\sqrt{-16}$, (2) $-\sqrt{-49}$, (3) $-\sqrt{-50}$.

Solution: (1) $\sqrt{-16} = \sqrt{(-1)(16)} = \sqrt{-1} \cdot \sqrt{16} = \sqrt{-1} \cdot 4 = j4$

(2) $-\sqrt{-49} = -\sqrt{(-1)(49)} = -\sqrt{-1} \cdot \sqrt{49} = -\sqrt{-1} \cdot 7 = -j7$

(3) $-\sqrt{-50} = -\sqrt{-1} \cdot \sqrt{50} = -j\sqrt{50} = -j7.07$ (Rounded)

Note: Intermediate steps are usually performed mentally.

EXERCISE 18-6

Express the following square roots in the form of j-numbers. Use calculators only when necessary.

1. $\sqrt{-81}$
2. $\sqrt{-25}$
3. $\sqrt{-144}$
4. $\sqrt{-1}$
5. $-\sqrt{-100}$
6. $-\sqrt{-3600}$
7. $\sqrt{-4}$
8. $\sqrt{-10}$
9. $-\sqrt{-75}$
10. $-\sqrt{-2}$

Find the roots of the following quadratic equations. Express all answers as complex numbers using the letter j.

11. $P^2 = -121$
12. $3R^2 = -3$
13. $5\theta^2 + 80 = 0$
14. $2E^2 + 128 = 0$
15. $2E^2 + 25 = 9$
16. $\omega^2 - 6\omega = -9$
17. $t^2 = 2t - 2$
18. $X_L{}^2 + 2X_L = -5$

COMPLEX NUMBERS

Before the properties of j-numbers and complex numbers are studied, the overall numbering system will be scanned. Complex numbers are a natural extension of the real number system. The real numbers can be thought of as points along a scale. Another interpretation is to think of them as vectors with the tail at the origin. The tip of the arrowhead ends on the point represented by the number.

The *natural numbers* are the counting numbers (1, 2, 3 . . .). On a number line, they represent spaced points, Figure 18-15(A). *Negative whole numbers* extend the scale to the left, Figure 18-15(B). The *zero* separates the two sets of numbers. Together, they are called *integers,* Figure 18-16.

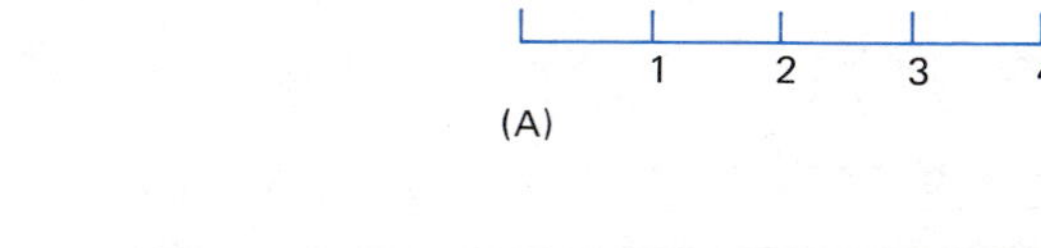

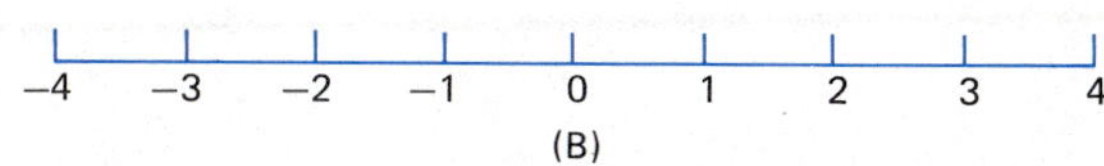

Figure 18-15 Points on a number scale: (A) Natural numbers and (B) Integers

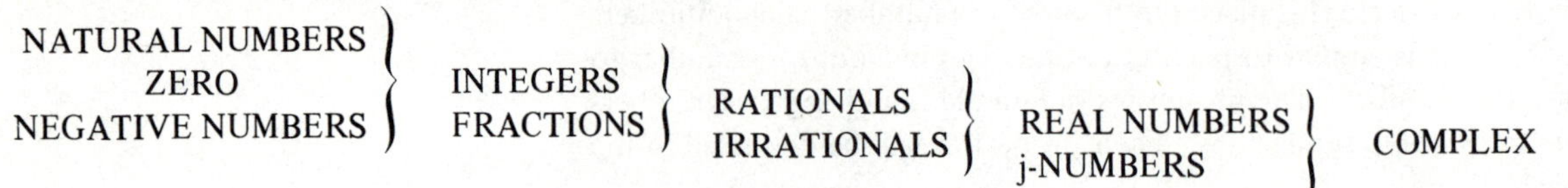

Figure 18-16 The complex number system

Fractions, including both proper and improper, fill in most of the spaces between integers. Together, fractions and integers form the set of *rational numbers.* Rationals can be written in the form $\frac{a}{b}$ where $b \neq 0$. If $b = 1$, the numbers are integers. In decimal form, the rational numbers come out even $\left(\frac{1}{8} = 0.125\right)$ or they repeat $\left(\frac{4}{3} = 1.333\ldots, \frac{14}{27} = 0.518\ 518\ 518\ldots.\right)$

Other numbers never repeat and never come out even. The number π (3.14159+) is an example. The square root of all prime numbers are other examples. These cannot be written in the form of $\frac{a}{b}$ so they are not rational. They are referred to as *irrational numbers.* Rational and irrational numbers form the set of *real numbers.* Real numbers fill in all points in a number line.

There are just as many *j-numbers* as there are real numbers. The letter j can be placed in front of any real number. The real numbers, the j-numbers and combinations of the two (a + jb) form the set of *complex numbers.* The entire scheme of the number system is shown in Figure 18-16.

18-7 THE j-OPERATOR

In Topic 5-4, the concept of directed numbers (vectors) was introduced. The reader may wish to review that topic before continuing. A number such as +5 represents an arrow five units long and points toward the right. Its tail is located at the origin (zero) of the scale, Figure 18-17(A).

The minus sign of a negative number is interpreted as an operator. The number −4 is thought of as −(+4). The +4 represents an arrow four units long pointed to the right. The operator (−) indicates the arrow should be rotated 180° counterclockwise. Thus, the number −4 is an arrow pointing left, Figure 18-17(B).

The letter j has a value of $\sqrt{-1}$. It, too, can represent an operator. It specifies a rotation of 90°. A number such as j5 can be written j(+5). On a rectangular coordinate system, Figure 18-18(A), the vector +5 is rotated onto the *y*-axis and labeled j5.

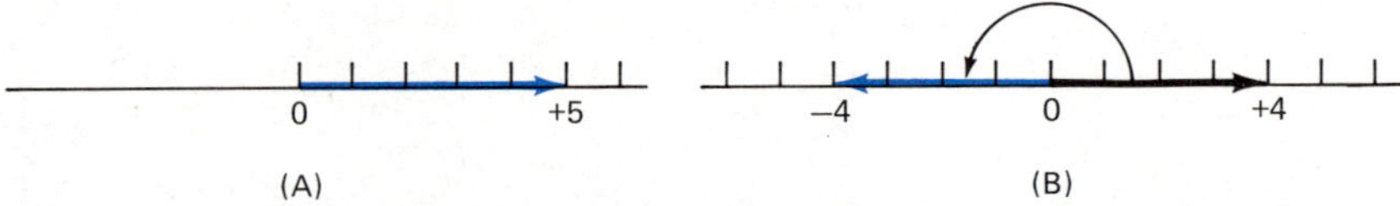

Figure 18-17 Signed numbers as vectors

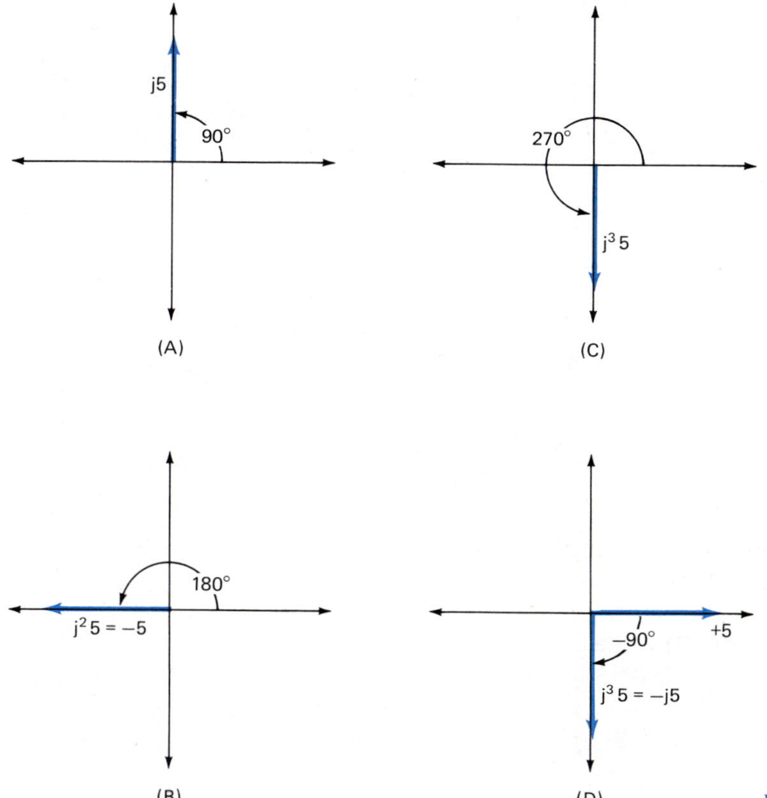

Figure 18-18 j as a rotation operator

If the j5 vector undergoes another j operation, it rotates another 90°. The vector j(j5) is written $j^2 5$. Since $j = \sqrt{-1}$, then $j^2 = -1$, mathematically. Both j^2 and –1 represent a 180° rotation, Figure 18-18(B).

A third j operation simply rotates the vector another 90°, Figure 18-18 (C). The operator $j^3 = j^2 j = -j$, so it represents 180° + 90° = 270° rotation. Of course, –j can also mean a -90° rotation, Figure 18-18(D).

In order to see the pattern more clearly, the powers of j are listed below:

$$j = \sqrt{-1} = j$$
$$j^2 = \sqrt{-1} \cdot \sqrt{-1} = -1$$
$$j^3 = j^2 \cdot j = -j$$
$$j^4 = j^2 \cdot j^2 = +1$$
$$j^5 = j^4 \cdot j = j$$

All higher powers reduce to simple j notation. They repeat above the fourth power.

Complex numbers have the form a + jb. When b = 0, the expression is a real number. If a = 0, it is a j-number. Real numbers and j-numbers represent vectors directed along the *x*-axis and *y*-axis, respectively. The set of complex numbers includes many other cases when neither a and b are zero.

A complex number such as 5 + j3 represents the sum of the vectors +5 and j3. Its direction is easily found. The tail is placed at the origin

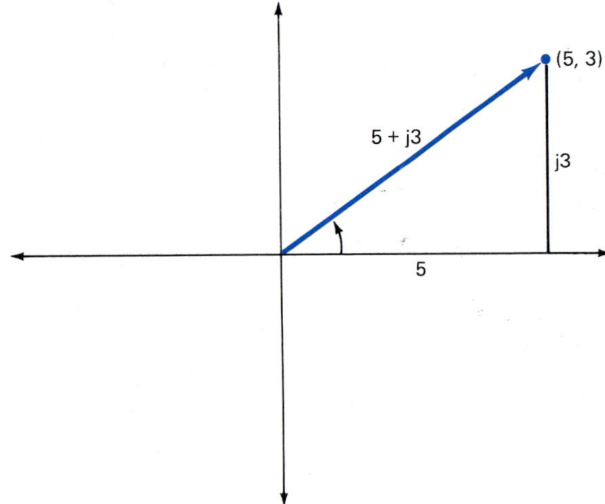

Figure 18-19 Plotting 5 + j3 as a vector

as usual. Its tip is located at the point (5,3), Figure 18-19. Its length can be found using the Pythagorean Theorem. Be sure to distinguish between the two notations. 5 + j3 is a complex number and represents a vector indicated as an arrow. The two disconnected numbers (5,3) is the point in the plane on which the vector terminates.

Example A On a coordinate system draw the vectors (a) −8 + j4, (2) 2 − j6 and (3) −3 − j.

Solution: (1) Locate the point (−8,4). Draw the vector from the origin to that point, Figure 18-20.
(2) Draw this vector from the origin to the point (2,−6), Figure 18-20.
(3) This vector terminates at point (−3,−1), Figure 18-20.

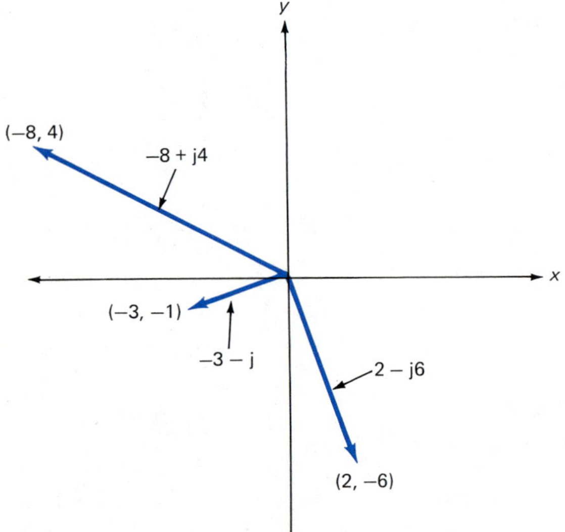

Figure 18-20 Drawing vectors

EXERCISE 18-7

Perform the indicated operation. Reduce powers of j to the simplest form. Example: $(j2)^3 = j^3 2^3 = j^3 8 = j^2 j8 = -j8$.

1. $(j3)^2$
2. $(j5)^2$
3. $(-j4)^2$
4. $(j3)^3$
5. $(-j3)^3$
6. $-(-j2)^4$
7. $(-j2)^4$
8. $-(j2)^3$
9. $-(j2)^4$
10. $-(-j2)^5$

Construct a coordinate system. Plot the following vectors.

11. $5 + j6$
12. $8 - j2$
13. $-5 + j6$
14. $-7 - j$
15. $2 + j$
16. $3 - j5$
17. $-7 + j3$
18. $-4 - j8$

The vectors $a + jb$ and $a - jb$ are known as *complex conjugates*. One is the conjugate of the other. Plot the following pairs of complex conjugates on one coordinate system.

19. $5 + j3$ and $5 - j3$
20. $2 - j6$ and $2 + j6$
21. $4 + j4$ and $4 - j4$
22. $-5 + j7$ and $-5 - j7$

Plot the following pairs on one coordinate system.

23. $5 - j6$ and $-5 + j6$
24. $2 + j8$ and $-2 - j8$
25. $-3 + j$ and $3 - j$
26. $5 + j2$ and $-5 - j2$

18-8 ADDING AND SUBTRACTING VECTORS

Vectors can be added both graphically and analytically. The rule for adding vectors graphically was given in Topic 5-4. It is repeated here for convenience:

ADDING VECTORS GRAPHICALLY

- In an x-y coordinate system, plot the first vector with its tail at the origin.
- Place the tail of the second vector at the tip of the first. Point it in its normal direction.
- Lay out the third vector similarly with its tail at the tip of the second vector. Place the fourth vector's tail at the tip of the third, etc.
- The vector sum called the *resultant* is another vector. Its tail is at the origin. Its tip is at the tip of the last vector plotted. Determine its value in the form of a complex number.

When plotting the second vector, pretend a new coordinate system exists at the tip of the first vector. Plot the second vector in the new coordinate system. Similarly pretend another coordinate system exists at the tip of the second vector when plotting the third, etc.

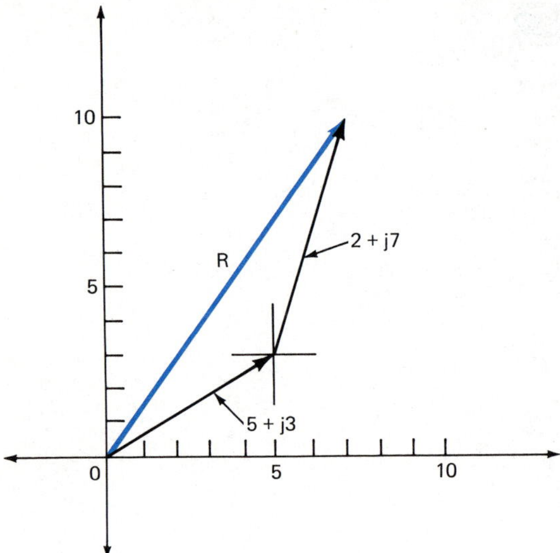

Figure 18-21

Example A Add the vectors 5 + j3 and 2 + j7 graphically.

Solution: Plot 5 + j3 with its tail at the origin, Figure 18-21. A new coordinate system is lightly drawn at its tip. Plot 2 + j7 with reference to the new axis. The resultant (labeled R) is an arrow with the tail at the origin and its tip at the tip of 2 + j7. On a carefully drawn figure, the resultant can be estimated accurately. It has a value of 7 + j10.

Example B Combine the vectors -5 + j2 and 4 + j3 by adding.

Solution: Plot -5 + j2 starting at the origin. Plot 4 + j3 starting at the tip of the first vector. The resultant has a value of -1 + j5, Figure 18-22.

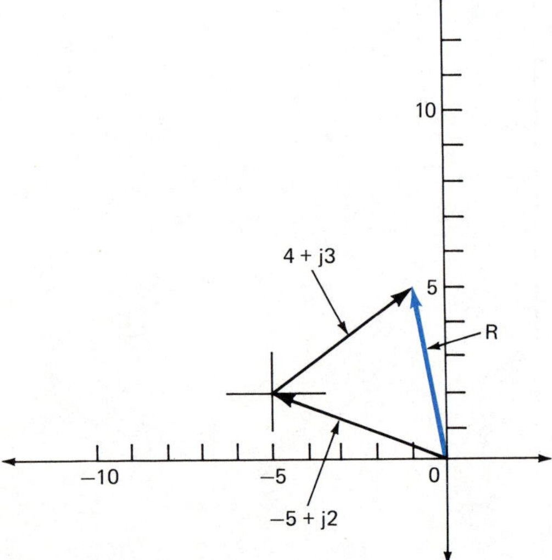

Figure 18-22

Example C Find the resultant of A = −2 −j4, B = 5 − j2, C = 4 + j5 and D = 0 + j3.

Solution: These four vectors are laid out tail to tip until all four have been plotted. The resultant is a vector from the origin to the last one plotted. Its value is 7 + j2, Figure 18-23(A). The order in which the vectors are plotted is not important. They may be laid out in any order. The reverse order is shown in Figure 18-23(B). The resultant is the same.

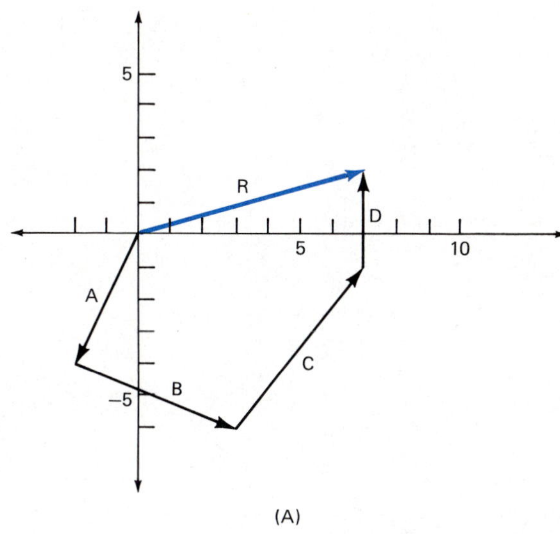

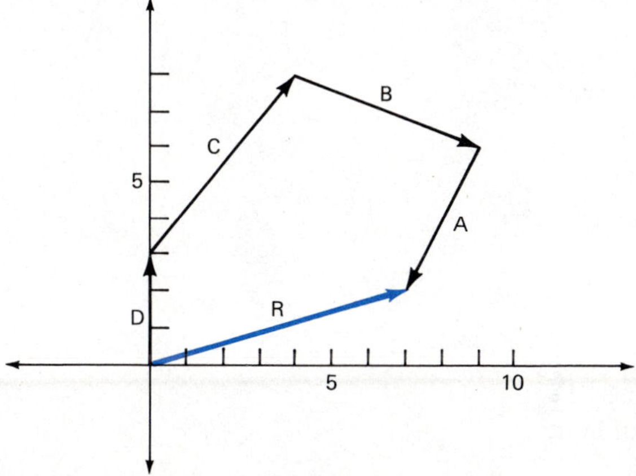

Figure 18-23 Adding vectors: (A) In the order given and (B) In reverse order

To subtract one vector from another, use the following rule, repeated from Topic 5-6:

SUBTRACTING VECTORS GRAPHICALLY

- Rotate the vector being subtracted (subtrahend) 180°. This is accomplished by changing the sign of both the real number and the j-number before it is plotted.
- Add the vectors according to the rule for graphical addition.

Example D Subtract B = 5 – j3 from A = 3 + j4.

Solution: Note that B is the subtrahend and is to be subtracted from A: A – B = ?

Plot A first. Rotate B 180°, then plot B from the tip of A, Figure 18-24. The vector difference is –2 + j7. The sum is also shown on the same figure.

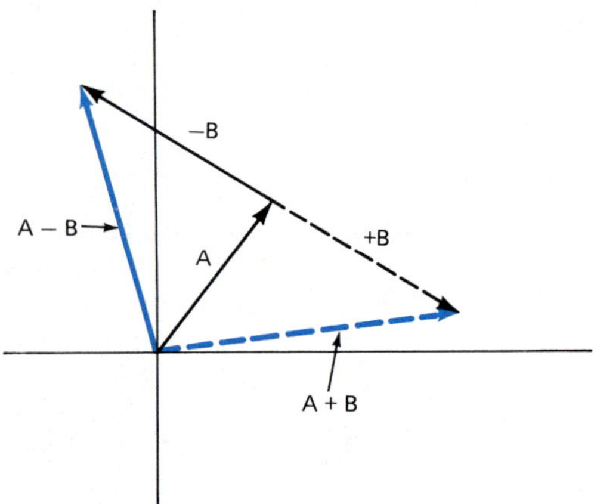

Figure 18-24

Analytical addition is easily performed by arranging the complex numbers in columns. Real numbers are added separately from the j-numbers. They are unlike terms. Each is added according to rules for adding signed numbers.

Example E Add 5 + j3, –2 + j6, j3, 2 – j8 and +3.

Solution: Note that there are five vectors being added. Arrange them in columns:

```
 5 + j3
-2 + j6
     j3
 2 - j8
 3
-------
 8 + j4        (Sum)
```

Subtracting is performed in columns also. Just remember to change the sign of the subtrahend. This is accomplished by changing the signs of both terms. The two columns are then added.

Example F Subtract $4 - j6$ from $1 + j3$.

Solution: $4 - j6$ is the subtrahend. Its signs must be changed, then the two complex numbers are added.

$$\begin{array}{r} 1 + j3 \\ \underline{-4 + j6} \\ -3 + j9 \end{array} \qquad \text{(Difference)}$$

Addition and subtraction can also be done horizontally by collecting like terms.

EXERCISE 18-8

Add the vectors in the following problems graphically. Make carefully drawn freehand diagrams and estimate the answers.

1. $(4 + j9) + (5 - j5)$
2. $(-2 + j3) + (4) + (3 - j7)$
3. $(4 + j8) + (-4 - j8)$
4. $(4 + j8) + (4 + j8)$
5. $(-2 + j6) + (5 + j2) + (-7 - j3)$
6. $(-5 - j7) + (1 - j) + (6 + j5)$
7. $(6 - j3) + (6 + j3) + (2 - j8)$

Subtract the second vector from the first graphically. Estimate the answers.

8. $(-4 + j2) - (7 - j5)$
9. $(3 - j6) - (4 + j3)$
10. $(-5 - j7) - (2 - j6)$
11. $(7 + j3) - (-5 + j2)$

Add the vectors in the following problems analytically.

12. $(4 + j9) + (5 - j5)$
13. $(-2 + j3) + (4) + (3 - j7)$
14. $(4 + j8) + (-4 - j8)$
15. $(4 + j8) + (4 + j8)$
16. $(-2 + j6) + (5 + j2) + (-7 - j3)$
17. $(-5 - j7) + (1 - j) + (6 + j5)$
18. $(6 - j3) + (6 + j3) + (2 - j8)$
19. $(15.9 + j16.3) + (-24.2 + j11.9) + (-18.4 - j21.3)$
20. $(56 - j82) + (35 - j61) + (-41 + j93) + (-87 - j34)$

The following pairs are complex conjugates. Find their sum both graphically and analytically.

21. $(44 + j53), (44 - j53)$
22. $(-16 - j14), (-16 + j14)$

Subtract the second vector from the first both graphically and analytically.

23. $(44 + j53), (44 - j53)$
24. $(-16 - j14), (-16 + j14)$

Subtract the second vector from the first analytically.

25. $(-4 + j2) - (-7 - j5)$
26. $(3 - j6) - (4 + j3)$
27. $(-5 - j7) - (2 - j6)$
28. $(7 + j3) - (-5 + j2)$

18-9 MULTIPLYING AND DIVIDING VECTORS

The product and quotient of complex numbers (vectors) will not be interpreted graphically. The following statements imply that graphical descriptions can be made, however:

- A vector multiplied by −1 rotates the vector 180° ccw.
- A vector multiplied by j rotates the vector 90° ccw.
- A vector multiplied by a real number changes its length.
- A vector multiplied by another vector changes the length and causes rotation.

Vectors in the form of complex numbers (a + jb) act as simple binomials. They can be multiplied by the long method or horizontally (FOIL).

Example A Multiply 5 + j7 by 4 − j3.

Solution: Use the long method.

$$\begin{array}{l} 5 + j7 \\ \underline{4 - j3} \\ 20 + j28 \\ \underline{\quad - j15 - j^2 21} \\ 20 + j13 - j^2 21 \end{array}$$

Notice the last term $(-j^2 21)$. Remember that $j^2 = (\sqrt{-1})^2 = -1$. The answer becomes $20 + j13 - (-21)$.

$41 + j13$ (Collect terms)

Note: Change $-j^2 21$ to $+21$. Place 21 under the 20 before adding columns.

Example B Multiply −3 + j6 and 5 + j3.

Solution: Multiply horizontally by the FOIL method.

$(-3 + j6)(5 + j3) = -15 - j9 + j30 + j^2 18$

Simplify j^2, then collect like terms.

Never leave higher powers of j in any expression.

$-15 - j9 + j30 - 18$

$-33 + j21$ (Collect terms)

Division of complex numbers is done by multiplying both the numerator and denominator by the *complex conjugate* of the denominator. The conjugate of a complex number is the same as the complex number except the sign of the j-number is changed. The numbers a + jb and a − jb are conjugates of each other.

The properties of a complex conjugate pair are as follows:

- $(a + jb) + (a - jb) = 2a$ (A real number)
- $(a + jb) - (a - jb) = j2b$ (A j-number)
- $(a + jb) \times (a - jb) = a^2 - j^2 b^2 = a^2 + b^2$ (A real number)

It is the latter property that is used in division of complex numbers. All j-numbers are removed from the denominator. Notice that the product of a complex number and its conjugate is similar to the algebraic special product $(a + b)(a - b) = a^2 - b^2$.

Example C Divide $3 - j2$ by $4 + j2$.

Solution: Express the division in the form of a fraction:

$$\frac{3 - j2}{4 + j2}$$

Multiply both the numerator and denominator by the conjugate of the denominator. Multiply horizontally (FOIL).

$$\left(\frac{3 - j2}{4 + j2}\right)\left(\frac{4 - j2}{4 - j2}\right) = \frac{12 - j6 - j8 + j^2 4}{16 - j^2 4}$$

$$= \frac{12 - j6 - j8 - 4}{16 + 4} \qquad \text{(Remove } j^2\text{)}$$

$$= \frac{8 - j14}{20} \qquad \text{(Collect terms)}$$

$$= \frac{8}{20} - j\frac{14}{20}$$

$$= 0.4 - j0.7$$

Example D Divide $4 - j8$ by $-3 - j4$

Solution:

$$\frac{4 - j8}{-3 - j4} \qquad \text{(Fractional form)}$$

$$\left(\frac{4 - j8}{-3 - j4}\right)\left(\frac{-3 + j4}{-3 + j4}\right) = \frac{-12 + j16 + j24 - j^2 32}{9 - j^2 16} \qquad \text{(Multiply)}$$

$$= \frac{-12 + j16 + j24 + 32}{9 + 16} \qquad \text{(Remove } j^2\text{)}$$

$$= \frac{20 + j40}{25} \qquad \text{(Combine terms)}$$

$$= \frac{20}{25} + j\frac{40}{25}$$

$$= 0.8 + j1.6$$

EXERCISE 18-9

Multiply each of the following using the long method. Remove all j^2 factors.

1. $(3 + j4)(5)$
2. $(5 - j3)(j6)$
3. $(-3 + j4)(2 - j7)$
4. $(2 - j6)(-1 - j3)$
5. $(5 + j4)(3 + j2)$
6. $(-2.6 - j3.1)(5.7 - j4.6)$

Multiply each of the following horizontally. Remove all j^2 factors. Use the FOIL method for binomial products.

7. $(2 + j3)(-3)$
8. $(3 - j)(-j4)$
9. $(7 + j4)(4 + j5)$
10. $(-2 + j8)(7 - j4)$
11. $(-5 - j2)(-6 - j4)$
12. $(5.3 - j1.8)(3.7 + j8.3)$

Divide each of the following. Remove all j^2 factors.

13. $(2 - j3) \div (4 - j2)$
14. $(5 + j6) \div (1 - j3)$
15. $(-3 + j7) \div (4 - j3)$
16. $(-8 - j) \div (6 + j2)$
17. $(4 + j6) \div (-1 - j)$
18. $(5 - j4) \div (3 + j)$
19. $(1) \div (3 + j4)$
20. $(j) \div (2 - j6)$

SECTION 3 REVIEW

Chapter 18 concludes Section 3: Introduction to ac Circuits. This review highlights the many topics in Section 3.

- *Arithmetic.* Adding, subtracting, multiplying and dividing complex numbers.
- *Algebra.* Ratio, direct and inverse proportion, cross-multiplying, special products, factoring quadratic equations.
- *Trigonometry.* Angles, similar triangles, the right triangle, generated triangles, sine, cosine, tangent, radians, arc functions, trigonometric graphs, vectors.
- *Electrical concepts.* Efficiency, tolerance, period, frequency, angular velocity, inductance, capacitance, reactance, impedance, instantaneous current and voltage.
- *Electrical devices.* Transformers, ammeters, voltmeters, generators, inductors, capacitors.
- *Electric circuits (ac).* The pure resistive circuit, effective current, effective voltage, the pure inductive circuit, the pure capacitive circuit.

SECTION 4
ANALYZING ac CIRCUITS

CHAPTER

19

PHASORS

OBJECTIVES

After satisfactorily completing this chapter, the student should be able to:

- Define the word phasor and express voltage-current phase relationships in radius vector form, sine wave form, wave equation form, rectangular (j-operator) form and polar form and relate each form to ac circuits.
- Transform a phasor back and forth from rectangular coordinates to polar coordinates, both graphically and analytically.
- Express the impedance of simple ac circuits in both rectangular and polar form, state whether the current leads or lags the voltage and determine the power factor of such circuits.

Alternating current fundamentals were introduced in Chapter 17 and the notation of complex numbers in Chapter 18. A new notation for complex numbers, called polar form, is presented in this chapter. Polar form is especially useful for multiplying and dividing vectors.

Most ac circuits contain reactance and resistance. The current lags or leads the voltage in such circuits by an acute angle. A particular alternating current or voltage can be represented by a sine wave graphically or by the wave equation. Both forms, however, are tedious to work with. It is easier to portray them as rotating radius vectors and use complex number notation to represent them.

Only single impedance loads are discussed in this chapter. Series and parallel circuits will be dealt with in succeeding chapters.

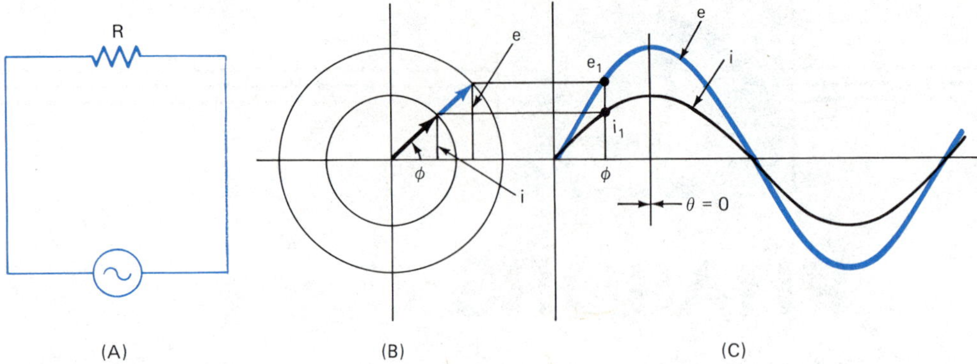

Figure 19-1 Circuit, phasor and sine wave diagrams for a pure resistive circuit. Voltage and current are in phase.

PHASE RELATIONSHIPS IN ac CIRCUITS

Voltage and current are in phase in a pure resistive circuit, Figure 19-1(A). An "in phase" relationship means the voltage and current, represented by radius vectors, always point in the same direction at any instant, Figure 19-1(B). At a later time, the rotating vectors will point in some other direction.

The radius vectors generate triangles as they rotate. The instantaneous current or voltage is indicated by the length of the vertical sides. These can be calculated at any instant using the equations:

$$e = E_m \sin \omega t$$
$$i = I_m \sin(\omega t + 0^\circ)$$

Both quantities are functions of time. The letter ω is the angular frequency. The symbol E_m is the maximum voltage and I_m is the maximum current.

The sine wave curves are shown in Figure 19-1(C). The "in phase" relationship is pictured in the two curves. Both voltage and current are zero at the same times. They attain peak values at the same time. They increase, decrease and change from positive to negative together.

In a pure inductive circuit, Figure 19-2(A), the current lags the voltage by 90°. The voltage and current vectors are shown in Figure 19-2(B). The angle (θ) between them is 90°. The instantaneous values are given by the equations:

$$e = E_m \sin \omega t$$
$$i = I_m \sin(\omega t - 90^\circ)$$

The voltage and current waves are shown in Figure 19-2(C). The current reaches its peak value a quarter cycle (90°) after the voltage does. The phase angle is shown on the diagram.

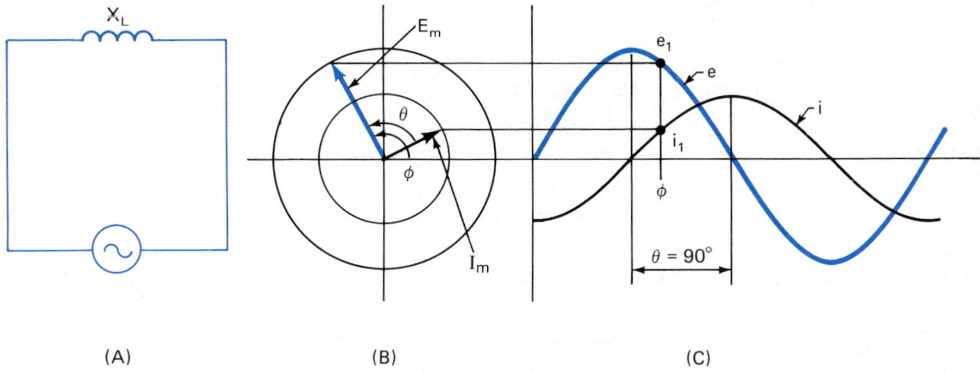

Figure 19-2 Circuit, phasor and sine wave diagrams for a pure inductive circuit. Current lags the voltage by 90°.

The pure capacitive circuit is shown in Figure 19-3(A). The corresponding radius vectors and sine waves are pictured in Figures 19-3(B) and 19-3(C), respectively. In this circuit, current leads the voltage by 90°. The current reaches its peak a quarter cycle (90°) before the voltage does. The instantaneous voltage and current equations are

$$e = E_m \sin \omega t$$
$$\mathbf{i = I_m \sin(\omega t + 90^\circ)}$$

The reader should compare the three instantaneous current equations. The three voltage equations are the same because voltage is the phase reference in this case.

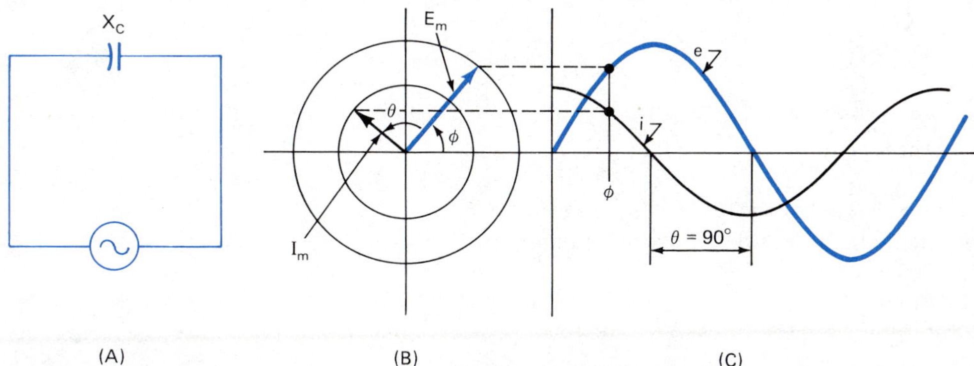

Figure 19-3 Circuit, phasor and sine wave diagrams for a pure capacitive circuit. Current leads voltage by 90°.

19-1 *CIRCUITS CONTAINING RESISTANCE AND REACTANCE*

Every real circuit contains some resistance. It may be no more than that of the wires that connect the various parts. All electric circuits also contain reactance. The circuit itself is a complete loop. Its self-inductance is very small but does exist. The wire connecting source and load will be at a different potential on one side than it is on the other. Thus, the wires act as very small capacitor plates. The wires may be separated by large distances and the dielectric is often air. In any case, a circuit will have small amounts of capacitance, also.

Usually, such resistance and reactance can be ignored. Most ac circuits, however, do contain both resistance and reactance of significant amounts. In ac circuits of this type, the current will lag or lead the voltage according to the rule:

FINDING THE PHASE ANGLE

- The current will *lag* the voltage in any circuit in which the reactance is inductive.
- The current *leads* the voltage when the reactance is capacitive.
- The phase angle is given by:

$$\theta = \tan^{-1}\frac{X}{R}$$

where X is the reactance.

In the current equation, $i = I_m \sin(\omega t \pm \theta)$, the phase angle takes a negative sign in inductive circuits where the current lags the voltage. It is positive in capacitive circuits.

Example A Find the phase angle in a circuit containing 25.2 Ω of inductive reactance and 41.7 Ω resistance, Figure 19-4.

Solution: $X_L = 25.2\ \Omega$, $R = 41.7\ \Omega$

Formula: $\theta = \tan^{-1}\frac{X}{R}$

Substitute: $\theta = \tan^{-1}\frac{25.2\ \Omega}{41.7\ \Omega}$

25.2 [÷] 41.7 [=] [INV] [Tan] , 31.1°

The current lags the voltage by 31.1°.

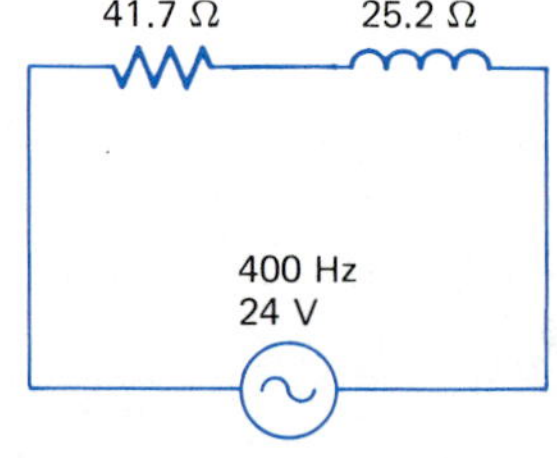

Figure 19-4

Suppose a 400-Hz voltage with a maximum voltage of 33.9 V (24 V effective) is applied to the circuit in Example A. It will produce a maximum current of 697 mA. The instantaneous equations are

$$e = 33.9 \sin 800\pi t$$

$$i = 0.697 \sin(800\pi t - 31.1^\circ)$$

The voltage vector is shown, Figure 19-5(A), in the 0° position. Since rotation is counterclockwise, the current vector is 31.1° behind. The voltage wave is placed in the reference position, Figure 19-5(B). The current

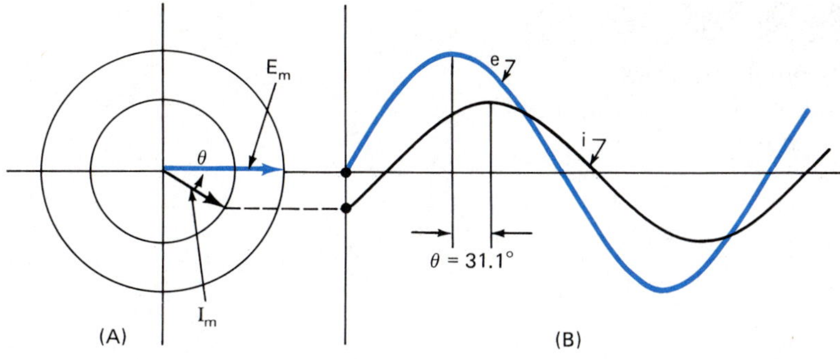

Figure 19-5 Phasor and sine wave diagrams for an inductive circuit. Current lags voltage by an angle θ of 31.1°.

lags the voltage by 31.1° at any instant. The current peaks and zeros occur 31.1° after the corresponding voltage peaks and zeros. Note that 31.1° is approximately one-third of a quarter cycle, for drawing purposes.

Example B What is the phase angle if a 120-V, 60-Hz (effective) voltage is applied to a circuit containing a 1.47-μF capacitor and a 1250-Ω resistor?

Solution: f = 60 Hz, C = 1.47 μF, R = 1250 Ω, Figure 19-6. Find the capacitive reactance first.

Formula: $X_C = \dfrac{1}{2\pi fC}$

Substitute: $X_C = \dfrac{1}{2\pi(60\ \text{Hz})(1.47 \times 10^{-6}\ \text{F})} = 1800\ \Omega$

Formula: $\theta = \tan^{-1}\dfrac{X}{R}$

Substitute: $\theta = \tan^{-1}\dfrac{1800}{1250}$

1800 [÷] 1250 [=] [INV] [Tan] , 55.2°

Current *leads* voltage by 55.2°

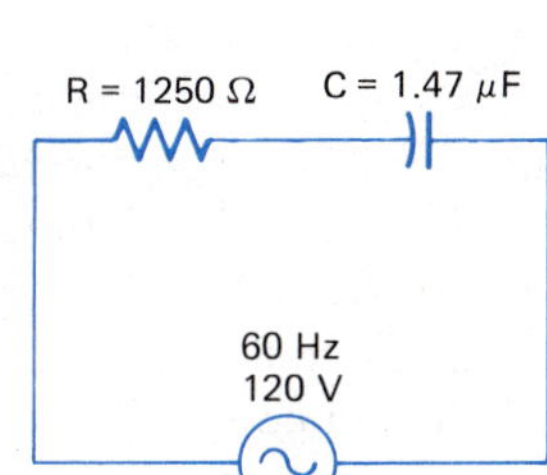

Figure 19-6

The maximum voltage produced by the source is 170 V and the maximum current is 77.5 mA. The equations for instantaneous values are

$$e = 170 \sin 120\pi t$$
$$i = 0.0775 \sin(120\pi t + 55.2°)$$

The radius vector diagram and sine wave patterns are shown in Figure 19-7. Notice how current lead is indicated in both diagrams.

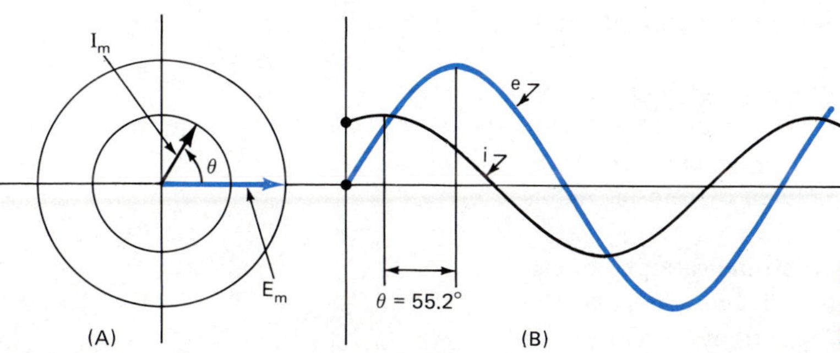

Figure 19-7 Phasor and sine wave diagrams of a capacitive circuit. Current leads voltage by an angle θ of 55.2°.

EXERCISE 19-1

Find the phase angle for each of the following problems. Does the current lead or lag the voltage? Construct a radius vector diagram such as Figure 19-7(A) for each with the voltage vector in the 0° position.

1. $R = 7.3\ \Omega, X_L = 5.3\ \Omega$
2. $R = 64.9\ \Omega, X_C = 24.2\ \Omega$
3. $R = 892\ \Omega, X_C = 1340\ \Omega$
4. $R = 628\ \Omega, X_L = 1280\ \Omega$
5. $X_C = 4380\ \Omega, R = 3120\ \Omega$
6. $X_L = 20.2\ \Omega, R = 38.1\ \Omega$
7. $R = 425\ \Omega, C = 3.85\ \mu F, f = 60\ Hz$
8. $R = 205\ \Omega, L = 390\ mH, f = 60\ Hz$
9. $R = 23.2\ \Omega, L = 5.9\ mH, f = 400\ Hz$
10. $R = 500\ k\Omega, C = 0.0052\ \mu F, f = 400\ Hz$

19-2 PHASORS

There are two kinds of measured quantities: scalars and *vectors*. The value of a scalar quantity is described by a number and a unit. This combination is called the *magnitude*. Four hours is a period of time, for example. The amount of time is completely described. The questions, "how many?" and "of what?"are both answered. Nothing more is needed. Some other examples of scalars are mass (140 kg), money ($50.35), energy (125 kW·h) and power (2.5 hp).

Vectors require more than a number and a unit. In addition to the magnitude, a direction must be specified. A force of 500 lb acting on a car, for example, is not enough to describe what will happen. Will the car be pushed forward or backward? Will it be skidded sideways? Will it be lifted upward? The force is not completely described by the magnitude.

Displacement (distance) is another vector. A meeting place 25 km from school would be difficult to find. It could be anywhere on a 157-km circumference. Here again, a direction is required. Velocity and acceleration are other examples of vector quantities.

Voltage and current are restricted in terms of direction. A current can only flow along a wire in one direction or the other. A voltage can be a forward emf or a back emf. In this sense, voltage and current act as scalars.

Vector notation is used to represent alternating voltage and current, even though they are not true vector quantities. This is done because it is a convenient way to represent the phase relationship between them. This idea is not new. Voltage and current were pictured by rotating radius vectors in the previous topic.

Voltage and current are represented as vectors to indicate phase relationships in ac circuits. For this reason, they are called *phasor quantities* or simply *phasors*. Alternating voltage, alternating current (effective) and impedance are examples of phasors. They are represented by arrows. Analytically, they will be represented as complex numbers $(x + jy)$, Figure 19-8. The complex number $x + jy$ is called the *rectangular form* of a phasor. The values x and y are the coordinates of the point at the arrowhead tip as defined by a rectangular coordinate system.

The phasor can be described just as accurately in another way. This is done by specifying the length of the arrow (r) and its direction (θ). These two numbers are called the *polar coordinates* of the phasor or vector. They are written in *polar form* as $r\underline{/\theta}$. The two analytical forms are as follows:

$$x + jy \qquad \text{(Rectangular form)}$$
$$r\underline{/\theta} \qquad \text{(Polar form)}$$

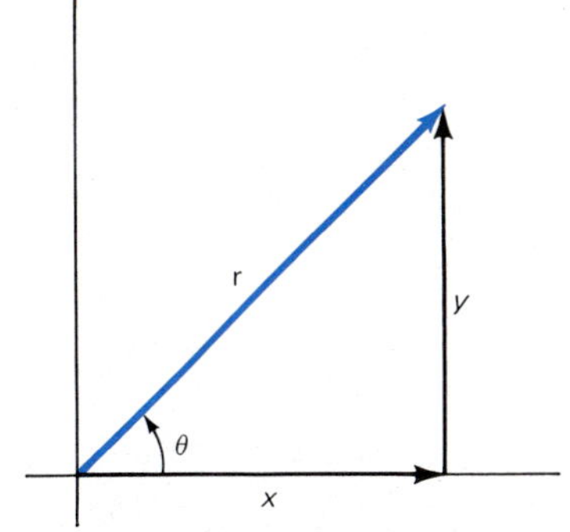

Figure 19-8 Vector diagram $x + jy = r\underline{/\theta}$

The rectangular form specifies the rectangular coordinates, Figure 19-9. When the rectangular coordinates are known, the arrow can be plotted graphically on ordinary (rectangular) graph paper. When the polar coordinates are known, the arrow can be plotted on polar graph paper. A *polar coordinate system* is shown in Figure 19-10.

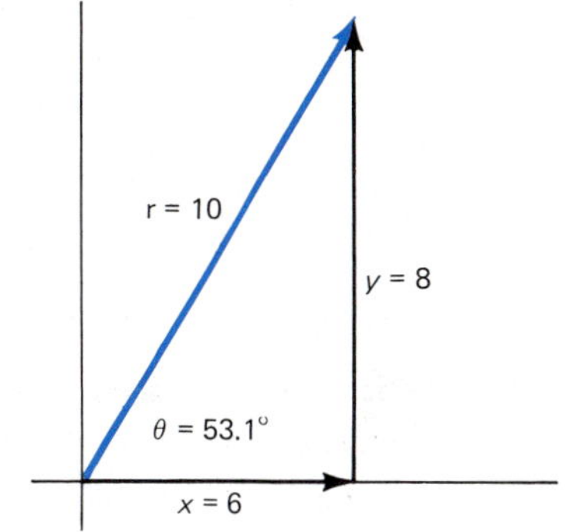

Figure 19-9 A vector diagram: $6 + j8 = 10\underline{/53.1^\circ}$

Example A Plot the phasors A = $5\underline{/53.1^\circ}$, B = $6.7\underline{/126.6^\circ}$, and C = $6.5\underline{/-63.4^\circ}$ on a polar coordinate system.

Solution: Phasor A: Move outward from the *pole* (origin) to the right along the reference axis to the fifth concentric circle, Figure 19-11. Any point on this circle is five units from the pole. Move counterclockwise to the 53.1° angle. This point is labeled A. The phasor $5\underline{/53.1^\circ}$ is drawn from the pole to position A, Figure 19-11.

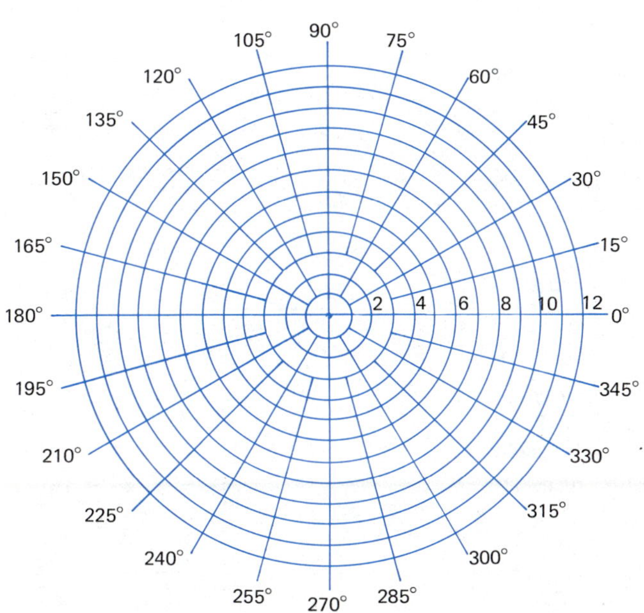

Figure 19-10 A polar coordinate system

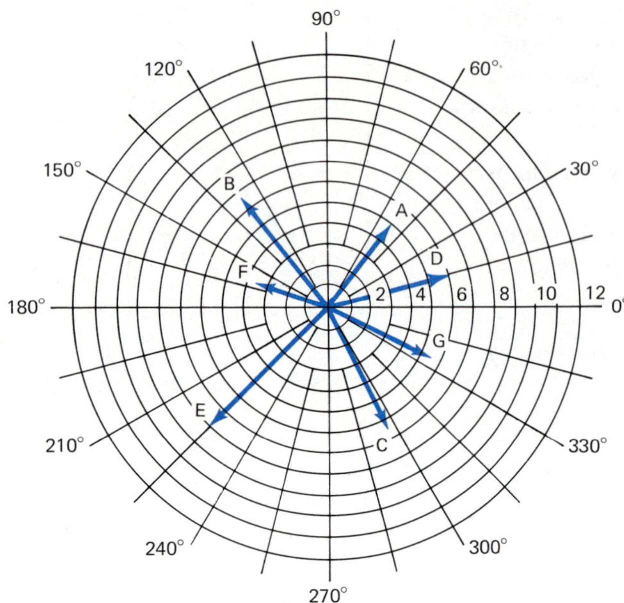

Figure 19-11 Phasors plotted on a polar coordinate system

Phasor B: Locate 6.7 units on the reference axis. Move in a direction 126.6° ccw from the reference direction. Phasor B is shown in Figure 19-11.

Phasor C: After locating the 6.5 position on the reference axis, follow cw to the -63.4° direction. Phasor C is also shown in Figure 19-11.

Example B Give the polar form ($r\angle\theta$) of phasors D, E, F and G in Figure 19-11.

Solution: Phasor D: The magnitude of phasor D is six units. Its direction is at 15°. The polar form is $6\angle 15^\circ$.

Phasor E: It is a third quadrant phasor at an angle of 225° (or −135°) and a length of eight units. The phasor can be expressed as $8\angle 225^\circ$ or $8\angle -135^\circ$. Both answers are equivalent.

Phasor F: The length of phasor F is 3.5 units and the angle is approximately 160°. The phasor is $3.5\angle 160^\circ$.

Phasor G: Fourth quadrant phasors almost always have the direction specified as a negative angle: phasor G is $5.5\angle -25^\circ$.

EXERCISE 19-2

Construct a polar coordinate system similar to the one in Figure 19-10. Plot each of the following phasors (arrows) on the drawing.

1. $6\angle 30^\circ$
2. $4\angle 165^\circ$
3. $3\angle -75^\circ$
4. $1\angle 240^\circ$
5. $5\angle 120^\circ$
6. $4\angle -30^\circ$
7. $2\angle 75^\circ$
8. $5\angle -135^\circ$

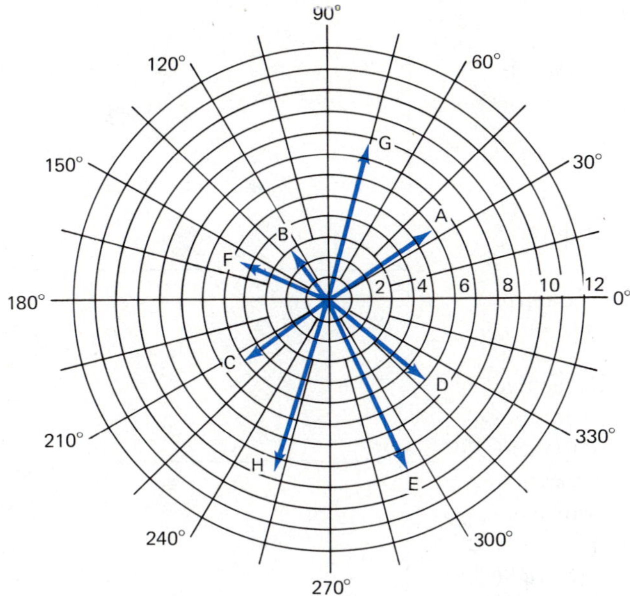

Figure 19-12

Express the following named phasors of Figure 19-12 in polar form.

9. A	11. C	13. E	15. G
10. B	12. D	14. F	16. H

TRANSFORMATION OF COORDINATES

When voltage, current and impedance phasors are related by Ohm's Law, phasor arithmetic is involved. The addition, subtraction, multiplication and division of phasors (complex numbers) in rectangular form $(x + jy)$ was described in Chapter 18. Addition and subtraction were relatively easy. Multiplication and division were more difficult processes. If, however, multiplication and division are carried out in polar form $r\underline{/\theta}$, the process is much simpler.

The most efficient way to work with phasors is to add and subtract using the notation of rectangular form. To multiply and divide, use polar form notation. The latter processes are described in the next chapter. It is necessary to learn how to transform (convert) from one form to the other. This process is called a *transformation of coordinates.*

19-3 IMPEDANCE AND ESTIMATING TRANSFORMATIONS

Impedance is the total opposition to the flow of electrons in an ac circuit. It is symbolized using the letter Z and measured in ohms. Impedance is a word that can describe the opposition to alternating current in any ac circuit: resistance only, inductive reactance only, capacitive reactance only or any combination of resistance and reactance.

Impedance is a phasor quantity. In polar form it is expressed as:

$$Z\angle\theta$$

where θ is the phase angle between the alternating voltage and current. In rectangular form, the components of impedance are resistance (R) and reactance (X):

$$R \pm jX$$

Inductive reactance (X_L) takes the positive sign. The negative sign is assigned to capacitive reactance (X_C). The rectangular form, therefore, is either:

$$R + jX_L \qquad \text{Inductive}$$

or

$$R - jX_C \qquad \text{Capacitive}$$

To construct an impedance phasor diagram, plot the resistance component along the positive x-axis. At its tip, plot the reactance component with X upward (+j) if it is inductive. R and X are added by the tail-to-tip method. The impedance is the resultant. It (Z) is a phasor measured from the tail of the resistance component (origin) to the tip of the reactance component, Figure 19-13(A). It is the hypotenuse of the right triangle. Resistance is horizontal; reactance is vertical. If the reactance is capacitive, it is plotted downward (−j), Figure 19-13(B).

Resistance and impedance are never negative. Reactance is negative only in capacitive circuits and the phase angle is negative only in capacitive circuits. When either the polar form or the rectangular form of impedance contains a negative sign, it should automatically be associated with capacitance. A positive sign in either form means inductance. The word impedance is often used to denote either the impedance magnitude (Z) or the impedance phasor which is indicated by both magnitude and direction ($Z\angle\theta$ or $R + jX$).

Estimating procedures have been explained earlier, Topic 16-3. The reader should review that topic before studying the following examples.

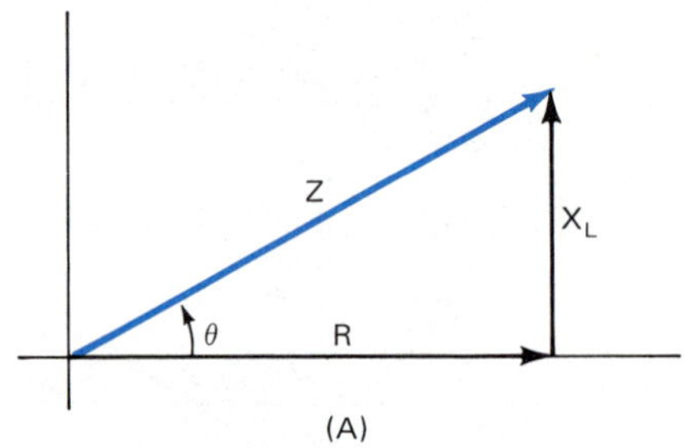

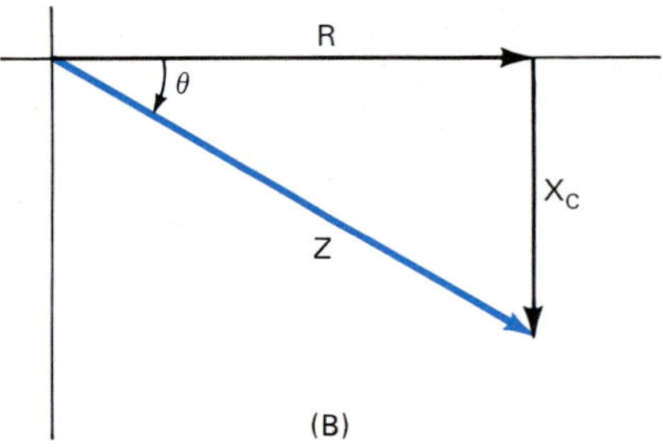

Figure 19-13 Impedance phasor diagrams: (A) Inductive (B) Capacitive

Example A An ac circuit has a resistance of 64.2 Ω and a capacitive reactance of 48.3 Ω. (1) Draw a circuit diagram, (2) construct a phasor diagram, (3) write the phasor in rectangular form, (4) estimate the values of Z and θ and write the polar form.

Solution: (1) The circuit consists of a resistor and a capacitor, Figure 19-14(A).

(2) Plot the resistance along the x-axis and the reactance downward since it is capacitive, Figure 19-14(B). Construct the lengths as accurately as possible to the same scale. The impedance connects the origin and the tip of the reactance phasor.

(3) $64.2 - j48.3$ [Use (−) with capacitive reactance.]

(4) Use the methods of Topic 16-3 to estimate the length of Z and the angle θ: $Z \approx 80\ \Omega$, $\theta \approx -35°$. Using estimated values, the polar form is $Z\angle\theta \approx 80\angle-35°$. (The calculated value is $80.3\angle-37.0°$).

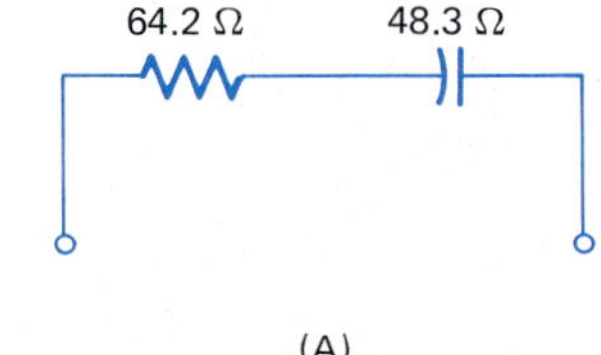

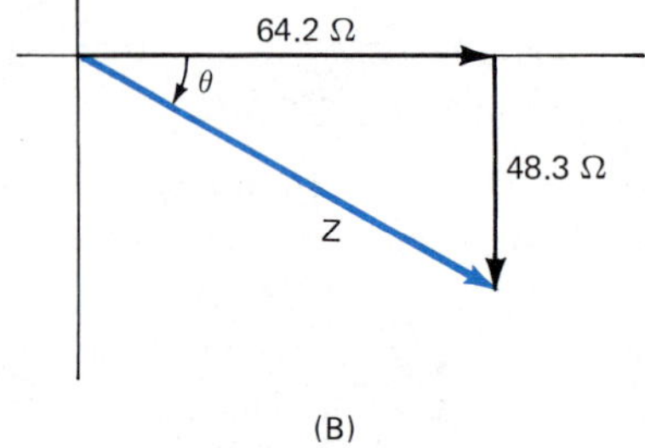

Figure 19-14

Example B An inductive circuit has a reactance of 118 Ω and a resistance of 94.3 Ω. Give (1) the circuit diagram, (2) the phasor diagram, (3) the rectangular form and (4) the polar form using estimated values.

Solution: (1) The circuit consists of a resistor and an inductor, Figure 19-15(A)

(2) Construct lengths accurately so estimates of Z and θ can be made, Figure 19-15(B).

(3) $94.3 + j118$ [Use (+) with inductive reactance.]

(4) Estimates: $Z \approx 150, \theta \approx 50°$
Polar form is: $Z\underline{/\theta} \approx 150\underline{/50°}$
(Calculated value is $151\underline{/51.4°}$)

Example C An ac circuit has an impedance of $13.6\underline{/28.3°}$. Give (1) the circuit diagram, (2) the phasor diagram and (3) the rectangular form using estimated values of R and X.

Solution: (1) A positive angle means an inductive circuit, Figure 19-16(A).

(2) Draw the impedance phasor at $28.3° \approx 30°$ (the quadrant is approximately trisected), Figure 19-16(B).

(3) Estimates: $R \approx 12\ \Omega, X \approx 6\ \Omega$
The rectangular form is $R + jX \approx 12 + j6$
(Calculated answer is $12.0 + j6.45$)

Example D The phasor $187\underline{/-66.2°}$ is the impedance of an ac circuit. Give (1) the circuit diagram, (2) the phasor diagram and (3) the rectangular form using estimated values of R and X.

Solution: (1) A negative angle means a capacitive circuit, Figure 19-17(A).

(2) See Figure 19-17(B). Note negative angle.

(3) $R \approx 80\ \Omega, X \approx -170\ \Omega$
Rectangular form: $R - jX \approx 80 - j170$
(Calculated answer is $75.5 - j171$)

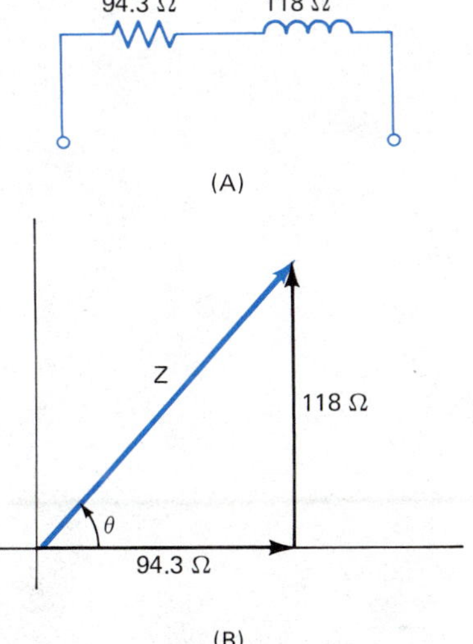

Figure 19-15

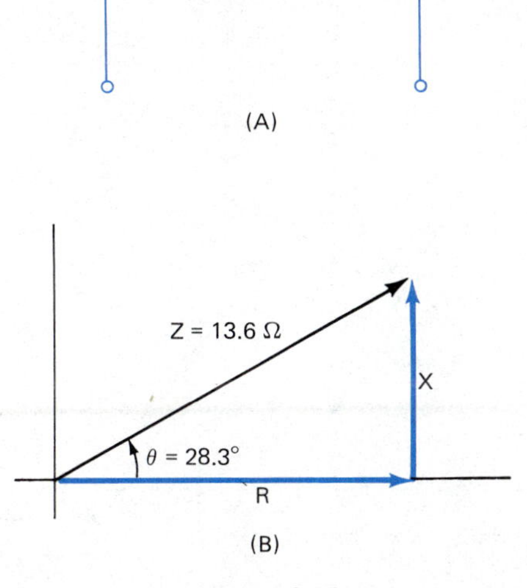

Figure 19-16

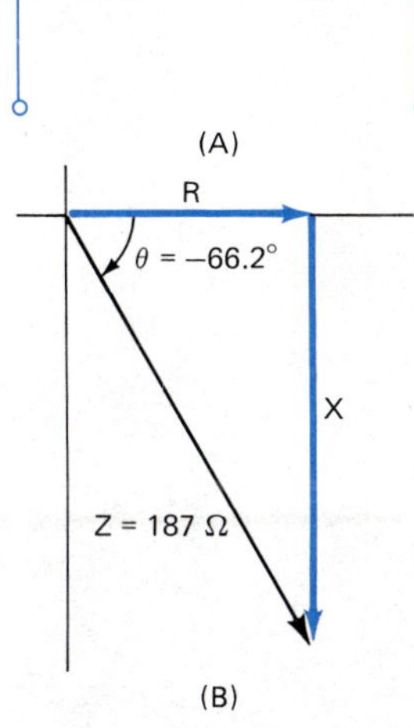

Figure 19-17

Sometimes the product and quotient of electrical phasors are phasors that extend into the second and third quadrants. The circuit phasors (alternating voltage, alternating current and impedance), however, are always first and fourth quadrant phasors. Resistance cannot be negative. Current cannot lead or lag by more than 90°.

EXERCISE 19-3

Reconstruct the following matrix of problems on notebook paper. For each problem, (a) draw the electrical circuit that the given phasor represents, (b) construct a carefully drawn freehand phasor diagram according to the methods of Topic 16-3, (c) write the unknown polar or rectangular form using estimated values. Fill in all blank spaces. Assume all units are ohms.

	CIRCUIT DIAGRAM	PHASOR DIAGRAM	RECTANGULAR ($R \pm jX$)	POLAR ($Z\angle\theta$)
1.			8 + j6	
2.			9 + j12	
3.				$15\angle 45^\circ$
4.				$25\angle 60^\circ$
5.			32 − j27	
6.			48 − j58	
7.				$12\angle -28.2^\circ$
8.				$49\angle -44.5^\circ$
9.			98 + j113	
10.			1080 + j2300	
11.				$63\angle 51.4^\circ$
12.				$348\angle -63.8^\circ$
13.	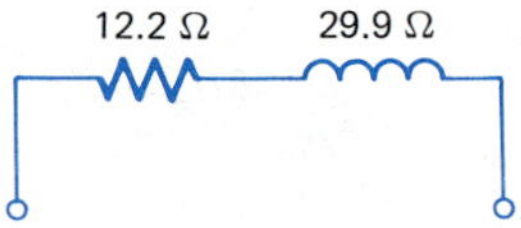 12.2 Ω 29.9 Ω			
14.		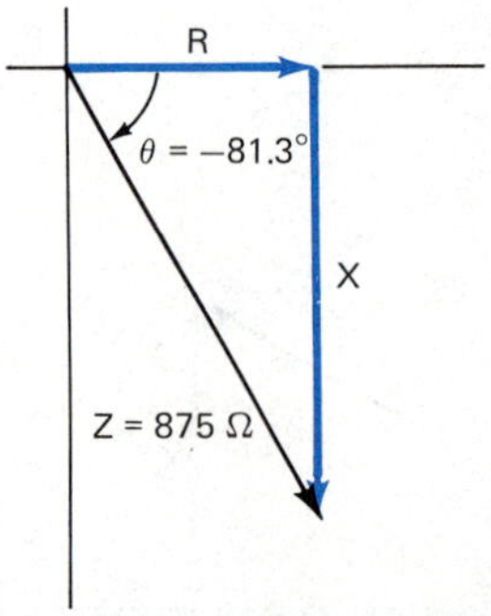 R $\theta = -81.3^\circ$ X Z = 875 Ω		
15.	XXX		−40.2 + j62.9	

19-4 RECTANGULAR TO POLAR TRANSFORMATION

The coordinates of a phasor are usually given in one form (rectangular or polar) or the other. The process of finding the coordinates of the other form is called *transforming coordinates* or a transformation of coordinates. In this topic, the rectangular form $(x + jy)$ is transformed to polar form $(r\underline{/\theta})$, Figure 19-18. The same methods are used to transform any phasor or vector quantity.

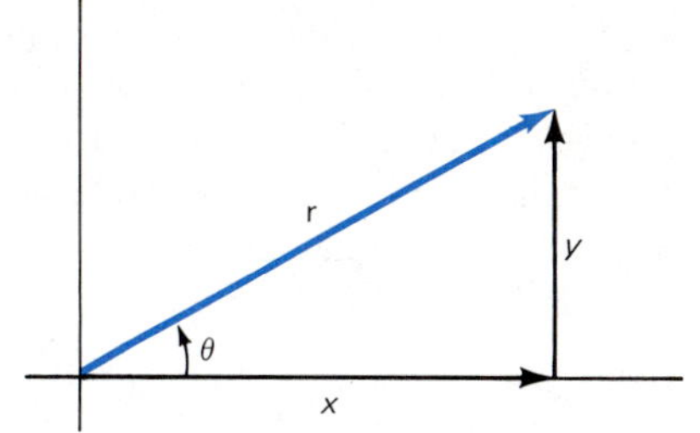

Figure 19-18 Vector diagram: $x + jy = r\underline{/\theta}$

The easiest way is to use the rectangular to polar transformation key on the calculator. This key is indicated in different ways on various calculators. The more common ones are [→P], [→Rθ] and [INV] [P→R]. Two numbers must be entered. The reader will have to consult the Owner's Manual for the exact method.

Determine the answers to these questions:

Which number must be entered first (x or y)?

How is the first number entered?

What is the exact procedure for the transformation?

Which transformed number is displayed first (r or θ)?

Which key switches the display to the other transformed number?

One calculator uses the following procedure:

x [x⇄y] y [INV] [P→R] , θ
[x⇄y] , r

The polar-rectangular transformation function on a calculator makes the procedure quick and easy.

If the calculator does not contain the R → P function, use the following rule:

RECTANGULAR TO POLAR TRANSFORMATION ($x \pm jy \rightarrow r\underline{/\theta}$):

- Calculate the angle:

$$\theta = \tan^{-1}\frac{y}{x}$$

If x is negative, add 180° to the principal value.

- Find the magnitude (hypotenuse):

$$r = \frac{x}{\cos\theta}$$

- Write the phasor in polar form $r\underline{/\theta}$ rounded to three significant digits. Angles are usually carried out to the nearest tenth of a degree.

Of course, the magnitude can be found using the Pythagorean Theorem: $r = \sqrt{x^2 + y^2}$ or the sine function: $r = \frac{y}{\sin\theta}$. The choice of the cosine function is preferable because x is never negative for circuit phasors. Secondly, the angle is already displayed and does not need to be re-entered. The following examples use circuit impedance to illustrate the process.

Always draw an accurate phasor diagram. Estimates do not need to be made but calculated answers should be compared with the diagram. It is not difficult to tell if the answers are reasonable.

When the R → P transformation rule is applied to circuit impedance, the letters in the formulas are different. The general forms $x \pm jy$ and $r\underline{/\theta}$ become $R \pm jX$ and $Z\underline{/\theta}$, respectively. Thus:

$$\theta = \tan^{-1}\frac{X}{R}$$

$$Z = \frac{R}{\cos\theta}$$

are the impedance forms. Distinguish carefully between capital and lowercase letters: x, y and r in the general formulas and R, X and Z used for circuit impedance.

Example A An ac circuit has an impedance of R + jX = 115 + j145. (1) Draw a circuit diagram, (2) construct an accurate phasor diagram, (3) transform the phasor to polar form using the transformation [P→R] key and (4) using the R → P rule.

Solution: (1) The circuit is inductive since the rectangular form carries a plus sign, Figure 19-19(A).

(2) The phasor diagram is shown in Figure 19-19(B).

(3) 115 [x⇄y] 145 [INV] [P→R], 51.6°
[x⇄y], 185 Ω

$Z\underline{/\theta} = 185\underline{/51.6^\circ}$

(4) Formula: $\theta = \tan^{-1}\frac{X}{R}$

Substitute: $\theta = \tan^{-1}\frac{145}{115}$

145 [÷] 115 [=] [INV] [Tan], 51.6°

Do not clear!

Formula: $Z = \frac{R}{\cos\theta}$

Substitute: $Z = \frac{115}{\cos 51.6^\circ}$ Note: the angle 51.6° is already displayed.

51.6° [Cos] [1/x] [×] 115 [=], 185 Ω

$Z\underline{/\theta} = 185\underline{/51.6^\circ}$

Check: Compare with Figure 19-19(B). Is the answer reasonable?

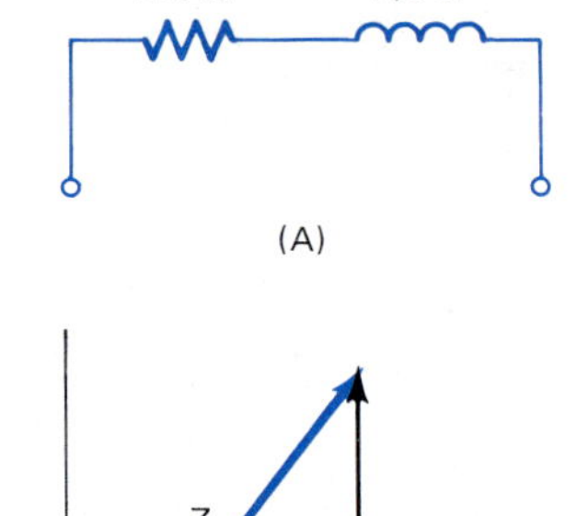

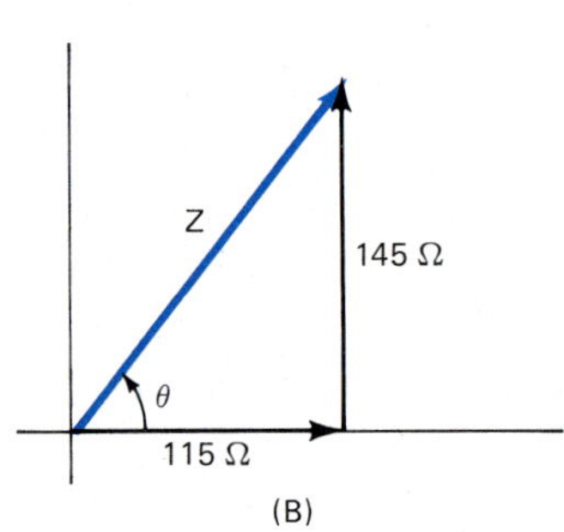

Figure 19-19

Example B An ac circuit is shown in Figure 19-20(A). (1) Find the rectangular form, (2) draw a phasor diagram, (3) find the polar form by using transformation keys and (4) find the polar form using the R → P rule for the ac circuit.

Solution: (1) R ± jX = 26.3 − j17.6 Note: Use negative sign for capacitive reactance.

(2) See Figure 19-20(B). Note the negative angle.

(3) 26.3 [x⇄y] 17.6 [+/−] [INV] [P→R], −33.8°
[x⇄y], 31.6 Ω

$Z\underline{/\theta} = 31.6\underline{/-33.8^\circ}$

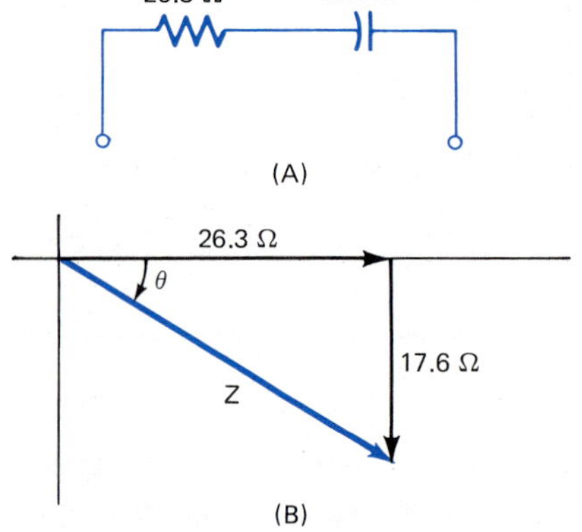

Figure 19-20

(4) Formula: $\theta = \tan^{-1}\dfrac{X}{R}$

Substitute: $\theta = \tan^{-1}\dfrac{-17.6}{26.3}$

17.6 [+/-] [÷] 26.3 [=] [INV] [Tan] , −33.8°

Do not clear!

Note: The angle will be displayed as a negative number if the reactance is entered as a negative number (−17.6).

Formula: $Z = \dfrac{R}{\cos\theta}$

Substitute: $Z = \dfrac{26.3}{\cos(-33.8^\circ)}$

−33.8° [Cos] [1/x] [×] 26.3 [=] , 31.6 Ω

$Z\underline{/\theta} = 31.6\underline{/-33.8^\circ}$

Check: See Figure 19-20(B).

A complex number such as −4 + j3 cannot be interpreted as a circuit phasor (R + jX). This would mean the resistance is negative. Many other vector quantities, however, can be second or third quadrant vectors.

Example C The position of a substation is 24.3 km west and 37.9 km south of a power station. This position expressed as a complex number is −24.3 − j37.9. (1) Draw a phasor (vector) diagram, (2) transform the vector to polar form by the [P→R] method and (3) transform the vector to polar form by the transformation rule.

Solution: (1) See Figure 19-21.

(2) 24.3 [+/-] [x⇄y] 37.9 [+/-] [INV] [P→R] , −123°

[x⇄y] , 45.0 km

$r\underline{/\theta} = 45.0\underline{/-123^\circ}$ km

Note: Some calculators display the angle as positive (237°) for the third quadrant angles. $45.0\underline{/-123^\circ}$ and $45.0\underline{/237^\circ}$ are equivalent phasors.

(3) Formula: $\theta = \tan^{-1}\dfrac{y}{x}$

Substitute: $\theta = \tan^{-1}\dfrac{-37.9}{-24.3}$

$= 57.3^\circ$ (Principal value)

Since x is negative, add 180° to the principal value: $\theta = 237^\circ$

Do not clear!

Formula: $r = \dfrac{x}{\cos 237}$

Substitute: $r = \dfrac{-24.3}{\cos 237} = 45.0$ km

$r\underline{/\theta} = 45.0\underline{/237^\circ}$ km

Check: Compare with Figure 19-21.

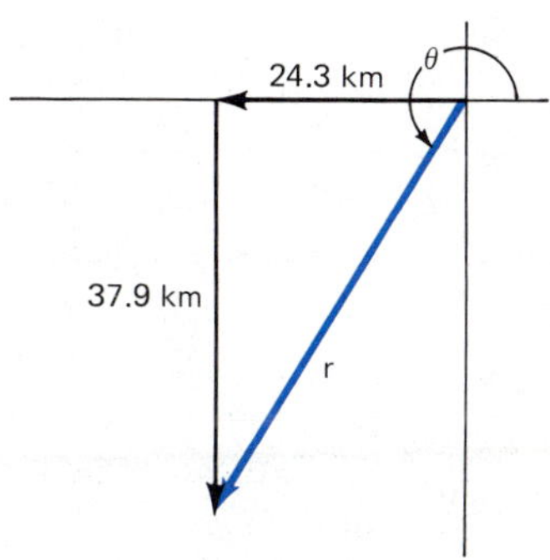

Figure 19-21

EXERCISE 19-4

Assume each of the following problems represent the impedance of an ac circuit except those with negative x values. Reproduce the problem matrix on notebook paper. Draw a circuit diagram and an accurate freehand phasor diagram. Transform the phasor to polar form using [P→R] or the transformation rule. Fill in all blank spaces in the matrix.

	CIRCUIT DIAGRAM	PHASOR DIAGRAM	RECTANGULAR (R ± jX)	POLAR ($Z\angle\theta$)
1.			12 + j9	
2.			36 − j27	
3.			1 + j	
4.			1 − j	
5.			68 − j50	
6.			47 + j39.5	
7.			78 − j138	
8.			53 + j87	
9.			j	
10.			1	
11.			0.388 − j0.462	
12.			20.8 + j2.77	
13.		12.2 Ω; θ; 29.9 Ω; Z		
14.	675 Ω; 548 Ω			
15.	XXX		−5.84 + j2.36	
16.	XXX		−112 − j69.1	

Data, diagram, formula, substitute, check.

17. A helicopter travels 14.6 km north and 10.3 km west to a job site. Express its new position from the starting point in polar form.
18. The force on an electron has a downward component of 3.19×10^{-26} N (newtons) and a leftward component of 7.16×10^{-26} N. Express the force in polar form.

19. A guy wire exerts a side force (pull) on the pole of 2250 lb to the east and a downward force of 3440 lb. Express the force in polar form.
20. A construction crew traveled −10.7 − j13.2 mi to a job site. Express the displacement in polar form. Can the crew pick up a radio transmission from the starting point if the maximum range for the transmission is 20 mi?

19-5 POLAR TO RECTANGULAR TRANSFORMATION

In this topic, the polar form ($r\underline{/\theta}$) is transformed to rectangular form ($x + jy$). As with rectangular to polar transformation, the easiest way is to use the transformation key on the calculator. The most common key symbols are [→R], [→xy] and [P→R]. Again, every calculator uses a different procedure. Consult the Owner's Manual for answers to these questions:

- Which number must be entered first, (r or θ)?
- How is the first number entered?
- What is the exact procedure?
- Which transformed number is displayed first, (x or y)?

One calculator uses the following procedure:

r [x⇄y] θ [P→R] , y
[x⇄y] , x

If the calculator does not contain the [P→R] function, use the following rule:

POLAR TO RECTANGULAR TRANSFORMATION ($r\underline{/\theta} \rightarrow x \pm jy$):

- Calculate side x:
  $x = r\cos\theta$
- Calculate side y:
  $y = r\sin\theta$
- Write the phasor in rectangular form ($x \pm jy$) (general) or ($R \pm jX$) (impedance). Round to three significant digits.

In both calculations, enter the angle exactly as given. Do not neglect negative signs. They must be entered into the calculator. The following examples use circuit impedance examples. As before, always construct a carefully drawn phasor diagram and compare it with calculated answers.

Example A An ac circuit has an impedance of $Z\underline{/\theta} = 435\underline{/65.2^\circ}$. (1) Draw a circuit diagram, (2) construct a phasor diagram, transform the phasor to rectangular form using (3) the [P→R] key and (4) by using the P → R rule.

Solution: (1) The circuit is inductive since the phase angle is positive, Figure 19-22(A).

(2) Construct phasor, Figure 19-22(B).

(3) 435 [x⇄y] 65.2 [P→R] , 395 Ω (X)

[x⇄y] , 182 Ω (R)

$R + jX = 182 + j395$

(4) Formula: $R = Z \cos\theta$

Substitute: $R = 435 \cos 65.2°$

65.2 [Cos] [×] 435 [=] , 182 Ω

Formula: $X = Z \sin\theta$

Substitute: $X = 435 \sin 65.2$

65.2 [Sin] [×] 435 [=] , 395 Ω

$R + jX = 182 + j395$

Check: Compare with Figure 19-22(B).

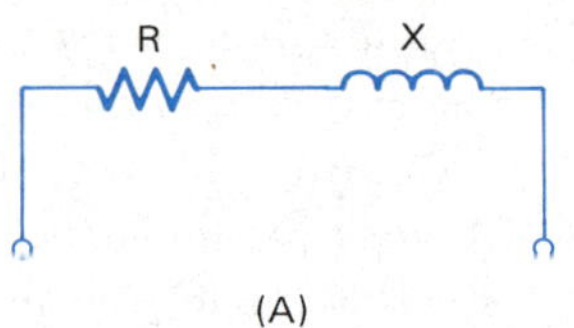

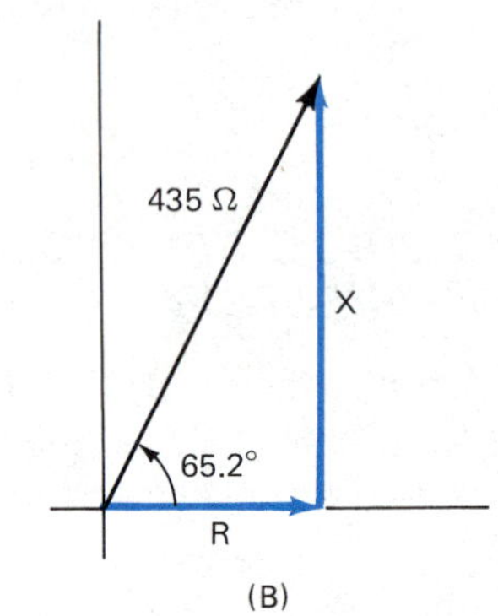

Figure 19-22

Example B Transform $93.5\angle -22.8°$ to rectangular form by constructing (1) circuit and (2) phasor diagrams using (3) the [P→R] key and (4) the P → R rule.

Solution: (1) A negative angle means capacitance, Figure 19-23(A).

(2) Construct phasor diagram, Figure 19-23(B).

(3) 93.5 [x⇄y] 22.8 [+/-] [P→R] , −36.2 Ω (X)

[x⇄y] , 86.2 Ω (R)

$R - jX = 86.2 - j36.2$

(4) Formula: $R = Z \cos\theta$

Substitute: $R = 93.5 \cos(-22.8°) = 86.2\ \Omega$

Formula: $X = Z \sin\theta$

Substitute: $X = 93.5 \sin(-22.8°) = -36.2\ \Omega$

$R - jX = 86.2 - j36.2$

Check: Compare answer with phasor diagram.

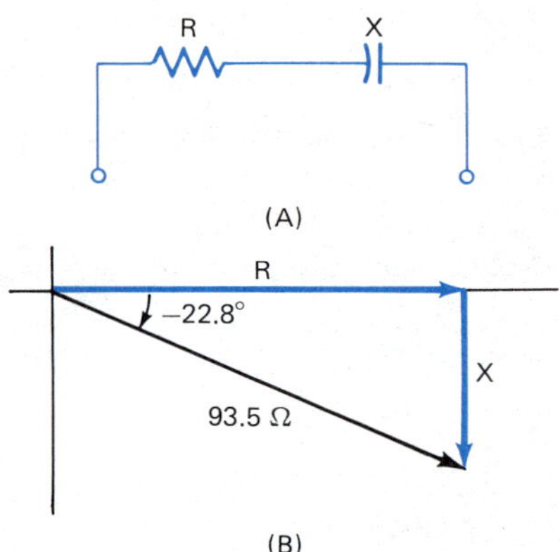

Figure 19-23

A complex number expressed in polar form such as $5\angle 150°$ cannot be interpreted as an impedance phasor ($Z\angle\theta$). Impedance phasors always lie in the first or fourth quadrant. Other vectors such as displacement, velocity or force can be second or third quadrant vectors, however.

Example C The velocity vector ($v\angle\theta$) of a helicopter is $145\angle 155°$ (mi/h). Express the vector in rectangular form: $v_x \pm jv_y$.

Solution: Construct a vector diagram, Figure 19-24.

Use the [P→R] key:

145 [x⇄y] 155° [P→R] , 61.3 mi/h

[x⇄y] , −131 mi/h

$v_x + jv_y = -131 + j61.3$ mi/h

Use P → R rule.

Formula: $v_x = v \cos\theta$

Substitute: $v_x = 145 \cos 155° = -131$ mi/h

Formula: $v_y = v \sin\theta$

Substitute: $v_y = 145 \sin 155° = 61.3$ mi/h

$v_x + jv_y = -131 + j61.3$ mi/h

Check: Compare with diagram.

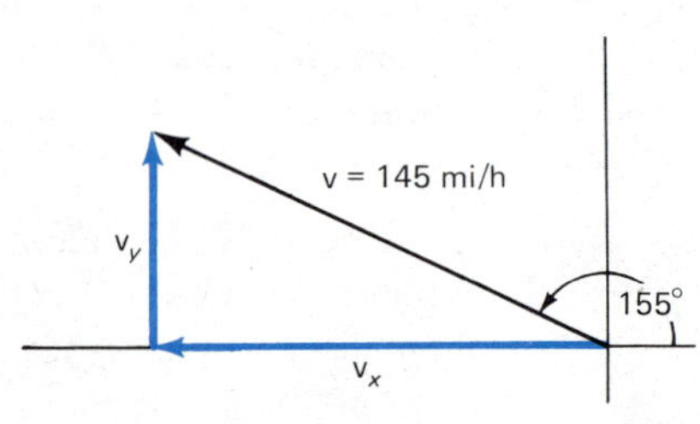

Figure 19-24

When a vector in polar form is transformed to rectangular form, the process is sometimes called a *resolution of vectors.* The vector is *resolved* into its components. In rectangular form $(x + jy)$, the components of the vector are x and y.

EXERCISE 19-5

Assume all of the following first and fourth quadrant phasors are ac circuit impedances. Reproduce the problem matrix on notebook paper. Draw a circuit and phasor diagram. Transform the phasor to rectangular form using the P→R key or the P → R rule. Fill in all blank spaces in the matrix.

	CIRCUIT DIAGRAM	PHASOR DIAGRAM	RECTANGULAR ($R \pm jX$)	POLAR ($Z\angle\theta$)
1.				$5\angle -36.9^\circ$
2.				$10\angle 53.1^\circ$
3.				$12\angle 28.2^\circ$
4.				$49\angle -45.4^\circ$
5.				$63\angle 51.0^\circ$
6.				$348\angle -68.5^\circ$
7.				$875\angle -81.0^\circ$
8.				$32.3\angle 67.8^\circ$
9.				$1.41\angle -45^\circ$
10.				$1\angle 90^\circ$
11.				$1\angle 0^\circ$
12.				$1\angle -45^\circ$
13.		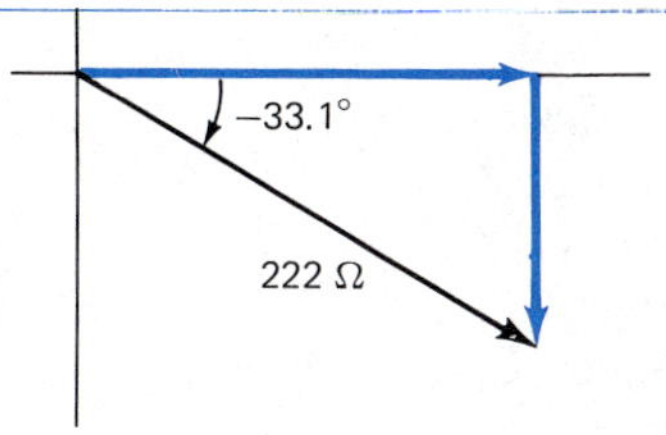 −33.1° 222 Ω		
14.	XXX	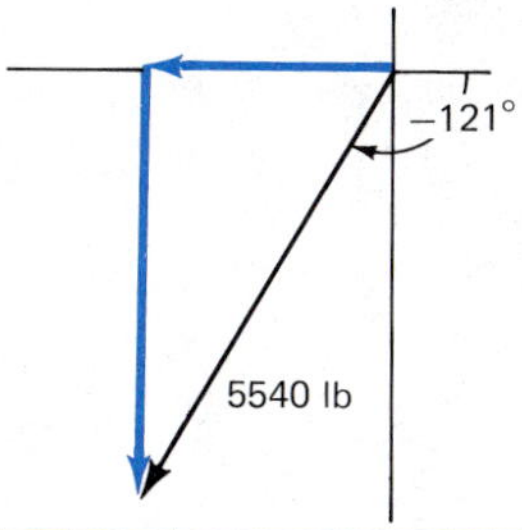 −121° 5540 lb		
15.	XXX	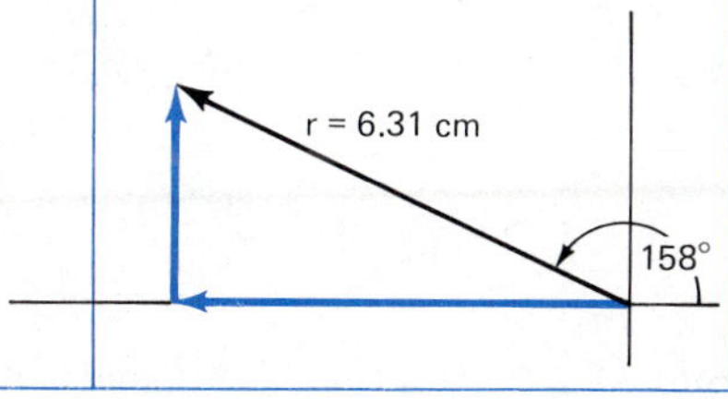 r = 6.31 cm 158°		

Data, diagram, formula, substitute, check.

16. The side pull on the top of a corner pole is $1560\angle{-118.5^\circ}$ (lb) exerted by the lines. Express the side pull in rectangular form.
17. A high voltage transmission line connecting two substations crosses a highway. Using the intersection of the line and the highway as the origin, the position vectors of the two substations are $24.8\angle{66.3^\circ}$ (km) and $37.6\angle{246.3^\circ}$ (km). Express both position vectors in rectangular form.
18. The force on an electron is $8.39 \times 10^{-28}\angle{238^\circ}$ N (newtons). Resolve the force into its components and express the force vector in rectangular form.

IMPEDANCE IN ac CIRCUITS

The transformation of coordinates of circuit phasors is an important process. The reader should learn to do them quickly and accurately. It is unwise to try to learn the process (both P → R and R → P) in one or two sittings. Instead, practice with a few problems of both types each day for a period of several days. Learning is guaranteed to be more permanent that way.

The remaining exercises in this chapter provide additional transformation practice. At the same time, a review of several aspects of ac circuit impedance is included.

19-6 PHASE AND POWER FACTOR

The phase of an ac circuit refers to the time by which the current leads or lags the voltage. As discussed previously, inductance causes the current to lag. In inductive circuits, the impedance phasor carries a plus (+) sign in both rectangular and polar form. The j-number (inductive reactance) and the phase angle are positive. In inductive circuits, Figure 19-25:

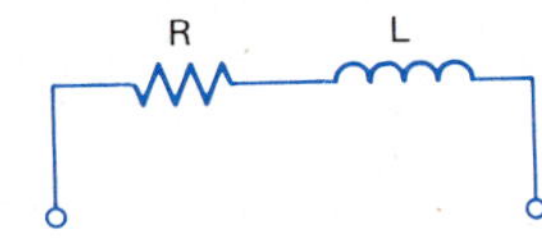

Figure 19-25 Circuit containing R and L elements

- the phasor is in the *first quadrant.*
- the rectangular form has a *positive j-number* (R + jX).
- the polar form contains a *positive phase angle* ($Z\angle\theta$).
- the current *lags* the voltage.

Capacitive circuits are different in all respects. The impedance phasor carries a minus (–) sign in both rectangular and polar form. The current leads. In capacitive circuits, Figure 19-26:

- the phasor is in the *fourth quadrant.*
- the rectangular form carries a *negative j-number* (R – jX).
- the polar form contains a *negative phase angle* $Z\angle{-\theta}$.
- the current *leads* the voltage.

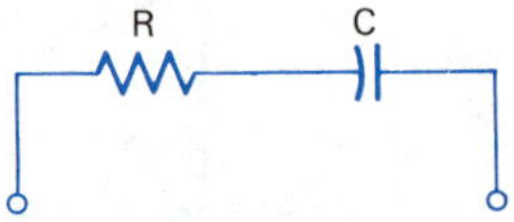

Figure 19-26 Circuit containing R and C elements

Power dissipation occurs only in the resistive elements of the circuit. In a pure resistive circuit, power is the product of the effective values of voltage and current. In this case the voltage and current phasors always point in the same direction. In reactive circuits, current leads or lags the voltage.

The true power in an ac circuit is the product of the effective voltage (E) and the component of the effective current (I) in the direction of the voltage, Figure 19-27:

$$P = EI \cos \theta$$

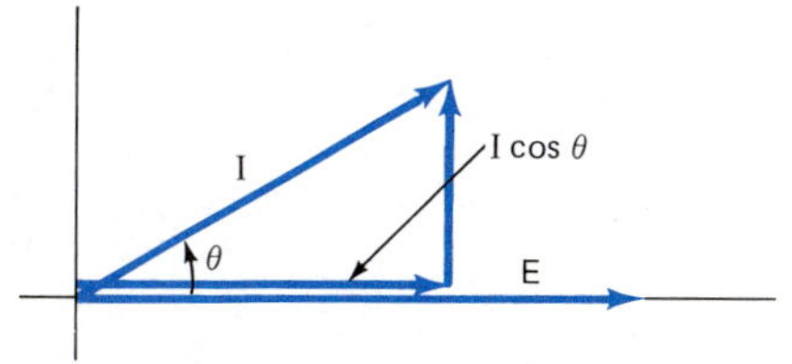

Figure 19-27 Voltage and current phasors

The cos θ factor of the expression is called the *power factor* (Pf). Do not confuse the letters Pf with pF (picofarad). The power factor is the cosine of the phase angle:

$$Pf = \cos \theta$$

Power in ac circuits is discussed in more detail in the next two chapters.

Example A (1) Transform $946\underline{/35.7^\circ}$ to rectangular form. In addition to constructing the usual drawings for this type of problem, (2) state whether the current leads or lags the voltage and (3) calculate the power factor.

Solution: See drawings, Figure 19-28.

(1) 946 [x⇄y] 35.7° [P→R] , 552 Ω (X)
[x⇄y] , 768 Ω (R)

$R + jX = 768 + j552$

Using the P → R rule:

$R = Z \cos \theta = 946 \cos 35.7^\circ = 768\ \Omega$

$X = Z \sin \theta = 946 \sin 35.7^\circ = 552\ \Omega$

$R + jX = 768 + j552$

(2) Current lags voltage (Inductance)

(3) Formula: $Pf = \cos \theta$

Substitute: $Pf = \cos 35.7^\circ = 0.812$

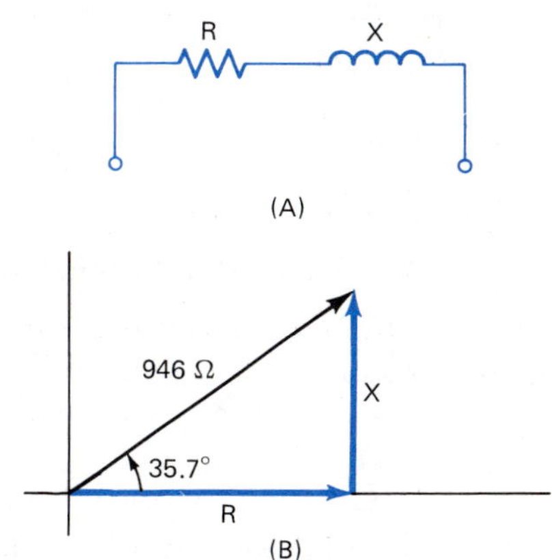

Figure 19-28

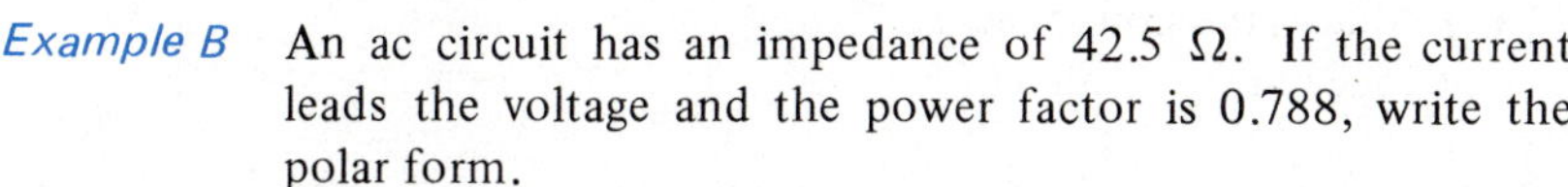

Example B An ac circuit has an impedance of 42.5 Ω. If the current leads the voltage and the power factor is 0.788, write the polar form.

Solution: Since the current leads, the circuit is capacitive; this means the phase angle is negative.

Formula: $Pf = \cos \theta$

$\theta = \cos^{-1} Pf$

Substitute: $\theta = \cos^{-1} 0.788$

0.788 [INV] [Cos] , 38.0°

The angle calculated is the principal value.

$Z\underline{/\theta} = 42.5\underline{/-38.0^\circ}$

If desired, the circuit and phasor diagrams can now be drawn and the rectangular form can be determined.

EXERCISE 19-6

Construct a phasor diagram, state whether the current leads or lags, calculate the power factor and transform the coordinates of the phasor.

	LAG OR LEAD? Pf = ?	PHASOR DIAGRAM	RECTANGULAR (R ± jX)	POLAR ($Z\angle\theta$)
1.				$3.78\angle -63.9°$
2.				$75.6\angle 29.3°$
3.			0.0311 + j0.0843	
4.			1.41 − j1.41	
5.				$5.34 \times 10^3 \angle 19.3°$
6.			766 − j603	
7.	I leads E Pf = 0.570 Z = 29.3 Ω			
8.	I lags E Pf = 0.876 Z = 74.9 Ω			
9.	I lags E Pf = 0.902 Z = 5430 Ω			
10.	I leads E Pf = 0.711 Z = 130 Ω			

19-7 INDUCTANCE AND CAPACITANCE

Both inductive and capacitive reactance depends on the frequency (f) of the voltage source:

$$X_L = 2\pi f L$$

$$X_C = \frac{1}{2\pi f C}$$

The inductance (L) or capacitance (C) can be calculated when frequency is given and the rectangular form of the impedance is known.

Example A (1) Transform 427 + j284 to polar form. Construct diagrams. (2) Does the current lead or lag? (3) What is the power factor? (4) Determine the inductance or capacitance of the circuit if the frequency is 400 Hz.

Solution: Construct diagrams, Figure 19-29.

(1) 427 [x⇄y] 284 [INV] [P→R] , 33.6°
[x⇄y] , 513 Ω

$Z\angle\theta = 513\angle 33.6°$

Using R → P rule:

$$\theta = \tan^{-1}\frac{X}{R} = \tan^{-1}\frac{284}{427} = 33.6°$$

$$Z = \frac{R}{\cos\theta} = \frac{427}{\cos 33.6°} = 513\ \Omega$$

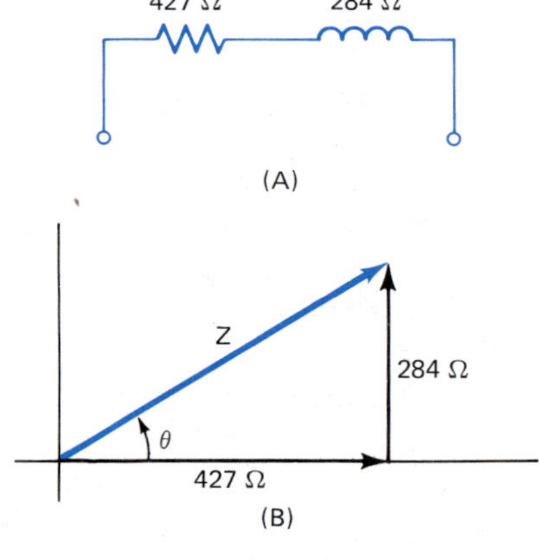

Figure 19-29

(2) The (+) in the given rectangular form means the current lags and the circuit is inductive.

(3) $\text{Pf} = \cos\theta = \cos 33.6° = 0.833$

(4) Formula: $X_L = 2\pi fL$

$$L = \frac{X_L}{2\pi f}$$

Substitute: $L = \dfrac{284\ \Omega}{2\pi(400\ \text{Hz})} = 113\ \text{mH}$

Example B A 60-Hz source is applied to a circuit containing 81.7 Ω of resistance and 22.5 μF of capacitance. (1) Construct the usual diagrams. (2) Does the current lag or lead? (3) Find the capacitive reactance and (4) write the rectangular form of the impedance.

Solution: (1) Construct diagrams, Figure 19-30. The phasor cannot be drawn accurately until the reactance is calculated.

(2) Current leads voltage.

(3) Formula: $X_C = \dfrac{1}{2\pi fC}$

Substitute: $X_C = \dfrac{1}{2\pi(60\ \text{Hz})(22.5 \times 10^{-6}\ \text{F})} = 118\ \Omega$

(4) $R - jX = 81.7 - j118$

If required, the polar form can now be determined and the power factor can be calculated.

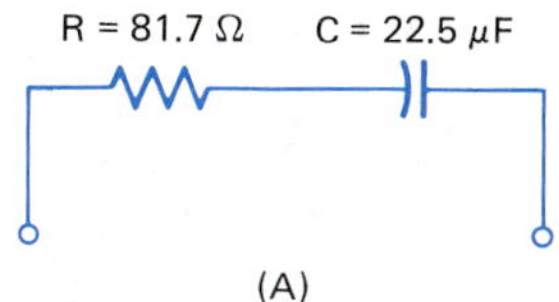

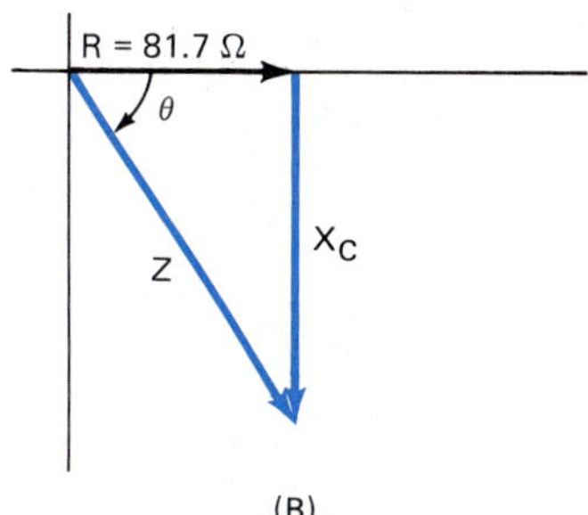

Figure 19-30

EXERCISE 19-7

Construct the phasor diagram. State whether the current leads or lags. Transform the coordinates of the phasor. Determine the power factor and find the capacitance or inductance.

	LAG OR LEAD? Pf = ? C or L = ?	PHASOR DIAGRAM	RECTANGULAR FORM, f = ?	POLAR FORM
1.			17.9 + j23.6 f = 60 Hz	
2.			f = 400 Hz	$64.8\angle -32.4°$
3.			f = 60 Hz	$358\angle 29.6°$
4.			37 100 + j25 600 f = 400 Hz	
5.		Z 38.2 Ω θ 27.3 Ω	f = 400 Hz	

	LAG OR LEAD? Pf = ? C or L = ?	PHASOR DIAGRAM	RECTANGULAR FORM, f = ?	POLAR FORM
6.		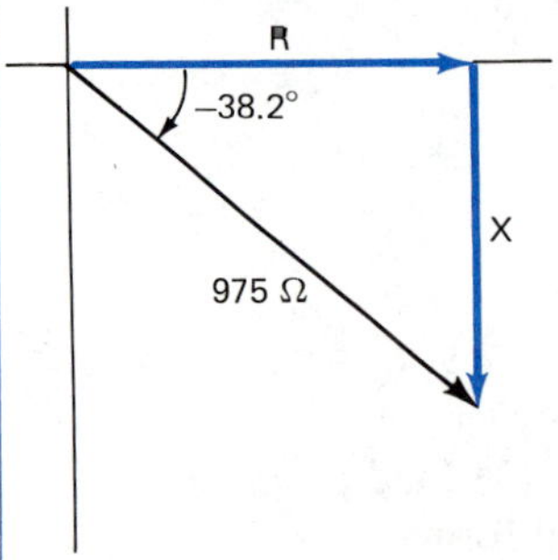 R; −38.2°; 975 Ω; X	f = 60 Hz	
7.	I leads E Pf = 0.370		f = 400 Hz	Z = 29.3 Ω
8.	L = 0.50 H		R = 4500 Ω f = 1000 Hz	
9.	C = 12.5 μF		R = 185 Ω f = 60 Hz	
10.	L = 8 H		1850 + j_____ f = 60 Hz	
11.	I leads E Pf = 0.899 C = 240 μF			Z = 25.3 Ω
12.	Pf = 0.515		74.9 − j_____ f = 400 Hz	

Hint: Draw phasor diagram. Use SA trigonometry.

CHAPTER 20

ac SERIES CIRCUITS

OBJECTIVES

After satisfactorily completing this chapter, the student should be able to:

- Add and subtract impedances both graphically and analytically in series circuits using rectangular form.
- Multiply, divide and find reciprocals using polar form.
- Use Ohm's Law and Kirchhoff's Voltage Law as they apply to single-impedance loads and to series circuits using effective voltages and currents.
- Evaluate power in ac series circuits.

In DC circuits, voltage, current and resistance were treated as scalar quantities. The use of Ohm's Law, Kirchhoff's Voltage and Current Laws, the current and voltage divider rules and formulas for finding total circuit resistance were relatively easy. Numbers and units were simply substituted in formulas. The evaluation of the unknown required mostly simple arithmetic. Addition, subtraction, multiplication, division and reciprocals were made even easier by using a calculator.

The laws, rules and formulas just mentioned also apply to ac circuits. The effective voltage, effective current and impedance phasors replace the DC scalar quantities. Otherwise, the relationships are identical. In this chapter, applications to series circuits are described. Phasor arithmetic is reviewed first and a study of power in ac series circuits concludes the chapter.

PHASOR OPERATIONS

When DC quantities are substituted in formulas, they are evaluated using ordinary arithmetic. To evaluate an unknown ac quantity, they are substituted in formulas that are similar to the DC formulas. The ac quantities, however, are phasors (complex numbers). The evaluation is performed by using complex number arithmetic, Topics 18-8 and 18-9.

Since the ac and DC formulas are similar, a means of distinguishing the two kinds of quantities is necessary. The formula E = IR (Ohm's Law), for example, is used in both ac and DC circuits. The quantities involved can be treated as scalars in DC circuits but ac circuit quantities are phasors. For DC circuits, ordinary arithmetic is all that is necessary. Complex number (polar or rectangular form) arithmetic is required, however, for ac circuit problems.

In this and the next few chapters, symbols in formulas that represent phasor quantities are printed in *boldface type* (**E, I, Z,** etc.). For scalar quantities, regular face type will continue to be used. An equation such as $E_T = E_1 + E_2 + E_3$ will represent arithmetic addition (DC circuits). The equation is written $\mathbf{E_T} = \mathbf{E_1} + \mathbf{E_2} + \mathbf{E_3}$ for ac circuits and when phasor (complex number) arithmetic is required.

Regular face type is also used to represent the magnitude or components of a phasor:

$$\mathbf{Z} = Z\underline{/\theta}$$
$$\mathbf{Z} = R + jX$$

The phasor **Z** has polar coordinates Z and θ and rectangular coordinates R and X.

20-1 ADDING PHASORS

When two or more sine waves having the same frequency are added, the resultant wave is a sine wave of the same frequency. This is true even if the waves being added are out of phase. The waves are represented as phasors so complex number (vector) addition must be used. Ordinary scalar addition cannot be applied.

Each of the three concepts of vectors, phasors and complex numbers has its own meaning. All three, however, can be visualized graphically as arrows. Analytically, all three can be expressed in either polar or rectangular form. Mathematically, they obey the same rules. The rules for adding and subtracting complex numbers both graphically and analytically were presented in Topic 18-8. It is those rules that are used here as they apply to ac circuits.

The rule for addition is repeated here to reflect the fact that some phasors may be given in polar form. Those must be transformed to rectangular form before adding.

ADDING PHASORS

- Construct a diagram using the tail-to-tip method, Topic 18-8.
- Express all phasors to be added in rectangular form.
- Add the real components of the phasors.
- Add the j-components of the phasors.
- Write the resultant as a phasor in rectangular form.

If desired, the resultant phasor can be transformed to polar form. In many problems, several transformations may be required. To organize the solution to problems, construct a table in which results can be listed. Do the intermediate work in an orderly fashion on separate scratch paper.

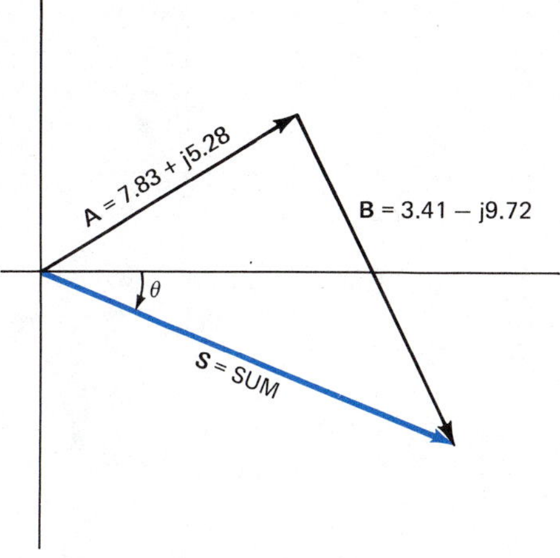

Figure 20-1

Example A Add 7.83 + j5.28 and 3.41 - j9.72. Give the final answer in polar form.

Solution: Construct a freehand tail-to-tip diagram, Figure 20-1. Assign the letters **A** and **B** to represent the two phasors. Plot **A** with its tail at the origin. Phasor **B** is placed with its tail at the tip of **A**. The phasor sum **(S)** has its tail at the origin. Its tip ends on the tip of **B**. An estimate of **S** is $\approx 12\underline{/-20^\circ}$. The phasor sum is **S** = **A** + **B**.

	POLAR		RECTANGULAR
A			7.83 + j5.28
B			3.41 - j9.72
S	$12.1\underline{/-21.6^\circ}$	←	11.24 - j4.44

Note: Horizontal arrows in the middle column indicate a transformation has been made. None of the transformation work is shown.

The sum **(S)** is shown in the last line of the solution table. It is found by adding the two phasors in rectangular form.
Check: Compare the sum with the graphical estimate.

Example B An ac circuit consists of a resistance of $48\underline{/0^\circ}$ Ω, an inductor of 5 + j22 Ω and a capacitor of $65\underline{/-90^\circ}$ Ω in series. Find the total impedance of the circuit in polar form.

Solution: Make a circuit diagram, Figure 20-2(A), and a phasor diagram, Figure 20-2(B). The impedances are labeled $\mathbf{Z_1}$, $\mathbf{Z_2}$ and $\mathbf{Z_3}$, respectively. The total impedance is $\mathbf{Z_T} = \mathbf{Z_1} + \mathbf{Z_2} + \mathbf{Z_3}$.

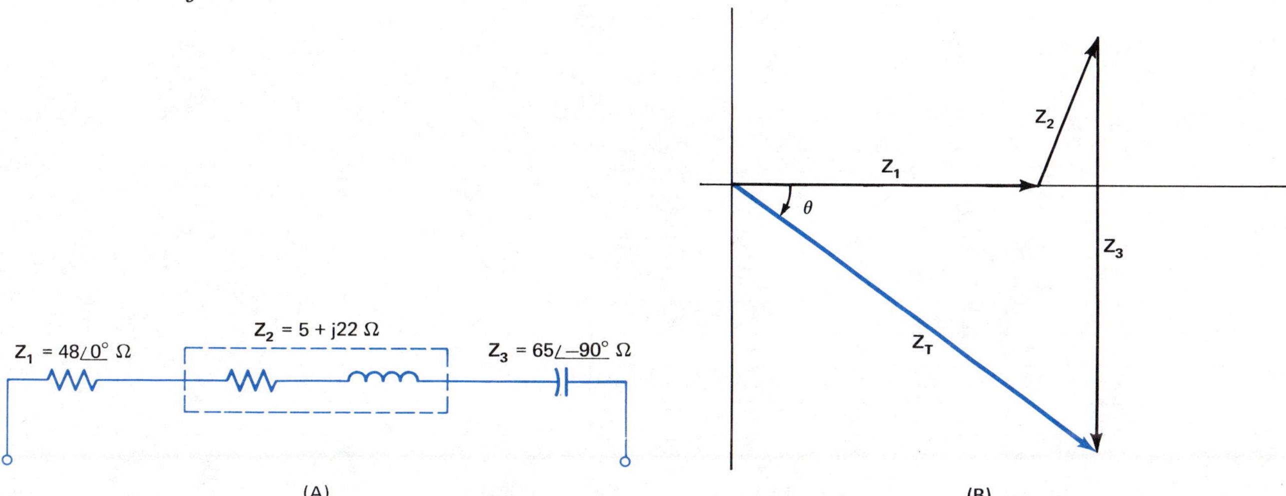

Figure 20-2

	POLAR		RECTANGULAR
Z_1	$48\underline{/0^\circ}$	→	48 + j0
Z_2	–		5 + j22
Z_3	$65\underline{/-90^\circ}$	→	0 – j65
Z_T	$68.2\underline{/-39.1^\circ}$	←	53 – j43

Note: All impedances in the table are in ohms.

Check: Compare with the resultant impedance ($\mathbf{Z_T}$), Figure 20-2(B).

Figure 20-3 is a circuit that is equivalent to that of Figure 20-2(A). It comes from the last line of the solution table, $Z\underline{/\theta} = R + jX$. If the same alternating voltage of the same frequency is applied to both circuits, both will produce the same current. The current will lead the voltage by the same phase angle (39.1°) in both circuits. The power factor is the same and energy dissipation will be the same.

Notice that the total resistance (53 Ω) is the sum of the individual resistances. The total reactance (43 Ω) is the sum of the individual reactances. One kind of reactance tends to nullify the effects of the other kind. In Example B, the capacitive reactance is greater than the inductive reactance. Such a circuit, Figure 20-2(A), is said to be a *capacitive circuit.* If inductive reactance is greater, it would be an *inductive circuit.*

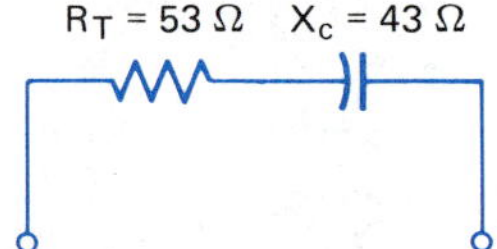

Figure 20-3 Equivalent circuit for the circuit of Figure 20-2(A).

Example C Find the equivalent impedance of a circuit containing the following impedances in series: $96\underline{/-61.9^\circ}$ Ω, $148\underline{/38.2^\circ}$ Ω, $164\underline{/0^\circ}$ Ω and $172\underline{/75.4^\circ}$ Ω.

Solution: Construct circuit and phasor diagrams, Figure 20-4. They are labeled $\mathbf{Z_1}$, $\mathbf{Z_2}$, etc. for convenience. The total impedance is $\mathbf{Z_T} = \mathbf{Z_1} + \mathbf{Z_2} + \mathbf{Z_3} + \mathbf{Z_4}$ added in rectangular form:

	POLAR		RECTANGULAR
Z_1	$96\underline{/-61.9^\circ}$	→	45.2 – j84.7
Z_2	$148\underline{/38.2^\circ}$	→	116.3 + j91.5
Z_3	$164\underline{/0^\circ}$	→	164 + j0
Z_4	$172\underline{/75.4^\circ}$	→	43.4 + j166.4
Z_T	$408\underline{/25.1^\circ}$	←	369 + j173

Note: All impedances in the table are in ohms.

The equivalent circuit is shown in Figure 20-4(C). It is inductive and current lags the voltage by 25.1°

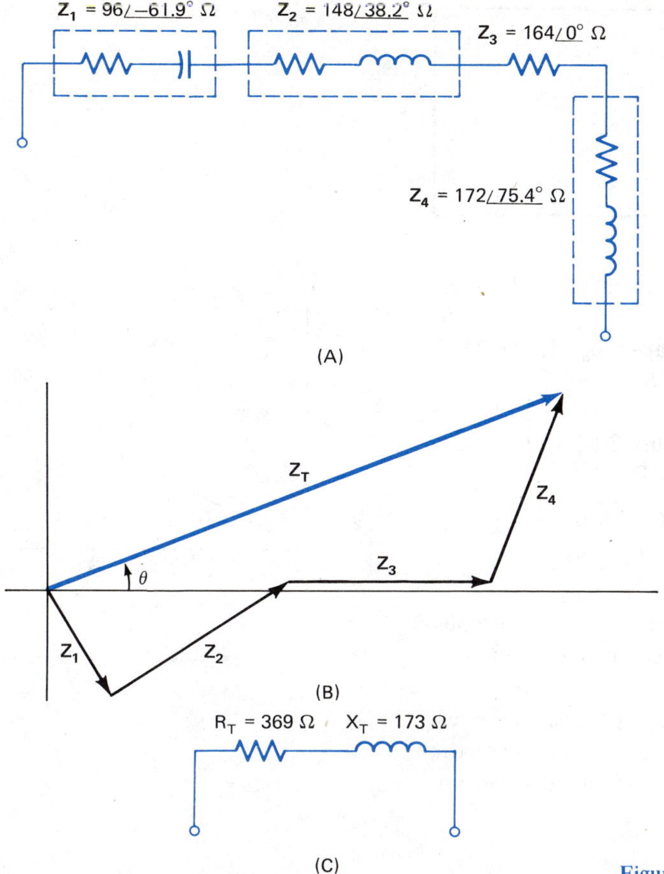

Figure 20-4

Calculator note: If the [P→R] key is used for transformations, some shortcuts can be made. After $\mathbf{Z}_4$ is transformed in the previous example, one of the components is displayed. Add the same components of $\mathbf{Z}_1$, $\mathbf{Z}_2$ and $\mathbf{Z}_3$ to it. Record the sum of that component in the $\mathbf{Z_T}$ line. Now press the [x⇄y] key. The other component of $\mathbf{Z}_4$ should appear on the display. Add the similar components of $\mathbf{Z}_1$, $\mathbf{Z}_2$ and $\mathbf{Z}_3$ to it. Record the sum in the last line. Both components of the sum are now entered in the calculator. Press [x⇄y] if necessary to display the correct one for transformation. Press [INV] [P→R] to find the polar form. Check the Owner's Manual. Not all calculators can perform two-number operations on one component without losing the second component. The use of memory and recall keys will save several steps.

20-2 SUBTRACTING PHASORS

The rule for subtracting phasors is similar to subtracting complex numbers, Topic 18-8:

SUBTRACTING PHASORS

- Rotate the phasor being subtracted (subtrahend) 180°. This is accomplished by changing the sign of both terms when given in rectangular form. If the polar form is given, add 180° to the angle.
- Add the phasors according to the rule for adding phasors.

Example A Subtract $4080\angle -75.2^\circ$ from $6450\angle 26.3^\circ$.

Solution: The first phasor $4080\angle -75.2^\circ$ is the subtrahend. It is assigned the letter **B**. The other one is **A**. The problem is to evaluate **A** – **B** or **A** + (–**B**). The negative sign is the 180° rotation operator. $-75.2^\circ + 180^\circ = 104.8^\circ$

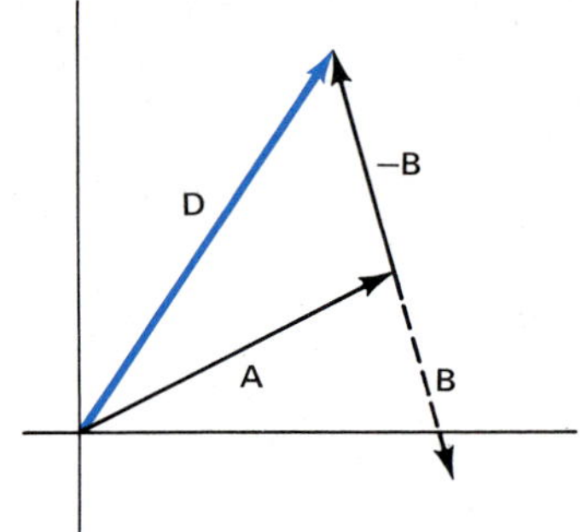

Figure 20-5

The problem is shown graphically in Figure 20-5 with the difference of the phasors assigned the letter **D**.

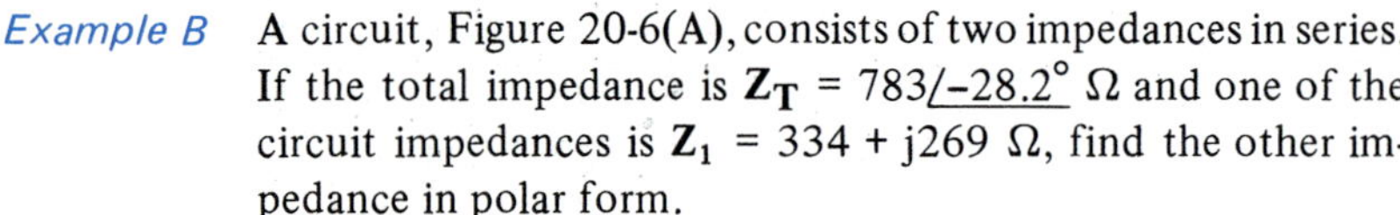

A	$6450\angle 26.3^\circ$	→	3660 + j1810
–B	$4080\angle 104.8^\circ$	→	–1040 + j3940
D	$6320\angle 65.5^\circ$	←	2620 + j5750

The difference **D** is shown in the bottom line of the table. It is found by subtracting like terms in rectangular form.

Example B A circuit, Figure 20-6(A), consists of two impedances in series. If the total impedance is $\mathbf{Z_T} = 783\angle -28.2^\circ\ \Omega$ and one of the circuit impedances is $\mathbf{Z_1} = 334 + j269\ \Omega$, find the other impedance in polar form.

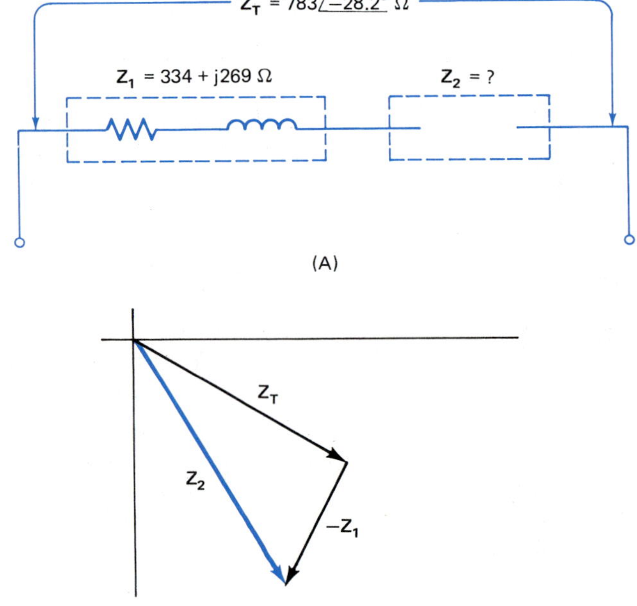

Figure 20-6

Solution: The equation is $\mathbf{Z}_2 = \mathbf{Z}_T - \mathbf{Z}_1$. The phasor $\mathbf{Z}_1$ is the subtrahend so it must be rotated: $-(334 + j269\ \Omega) = -334 - j269\ \Omega$

Z_T	$783\angle -28.2^\circ$	→	$690 - j370$
Z_1	—		$-334 - j269$
Z_2	$731\angle -60.9^\circ$	←	$356 - j639$

The phasor diagram is shown in Figure 20-6(B).

Note: All impedances in the table are in ohms.

EXERCISE 20-1

Construct a circuit diagram and its equivalent, a phasor diagram and a solution table similar to those in the preceding examples. Find the resultant in both rectangular and polar form for each of the following problems. Assume all phasors are circuit impedances (in ohms) connected in series.

1. $(42 + j27) + (18 - j6)$
2. $(624 - j846) + (297 + j528)$
3. $(52.3\angle 31.3^\circ) + (28.3 - j30.1) + (41.9\angle 26.3^\circ)$
4. $(6040\angle -56.2^\circ) + (3280\angle 0^\circ) + (4110\angle 90^\circ)$
5. $(5.43 + j7.18) - (2.76 - j4.18)$
6. $(46 - j39) - (28\angle -24.3^\circ)$
7. $(128\angle 31.3^\circ) - (128\angle -31.3^\circ)$
8. $(8490\angle -16.8^\circ) - (7160\angle -28.3^\circ)$
9. $(78.1 + j64.9) + (37.6\angle -68.3^\circ) + 55.5\angle 39.6^\circ)$
10. Subtract $(0.005\ 43\angle 41.6^\circ)$ from $0.008\ 83\angle 51.6^\circ$

(a) Find the unknown impedance of the following circuits using a solution table. (b) Is the circuit inductive or capacitive? (c) Does the current lead or lag the voltage? (d) What is the power factor? Be sure to construct the vector diagram for each problem.

11.

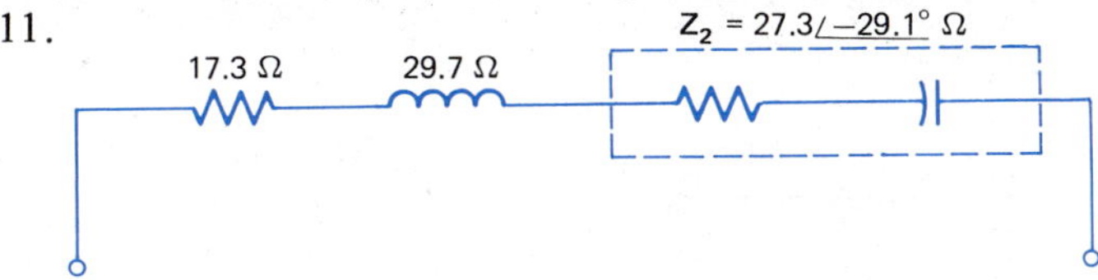

12.

13.

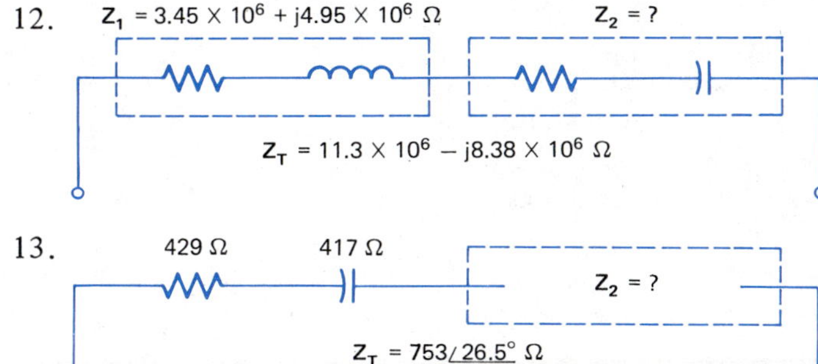

14.

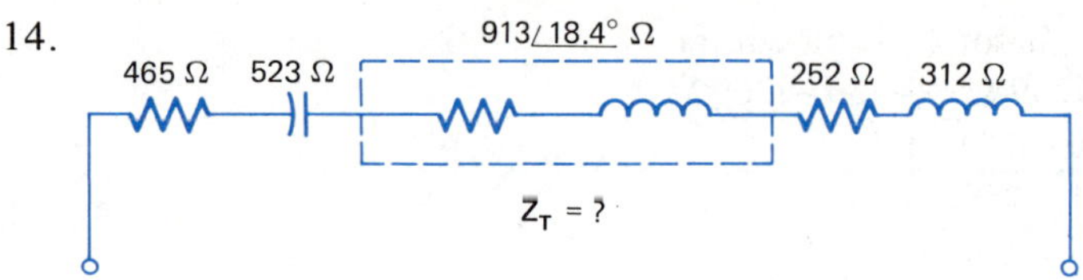

15.

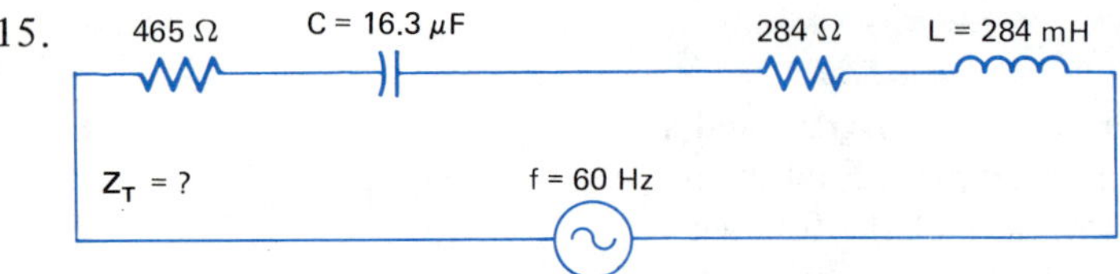

16. Find the impedance of the black box in the following circuit. Determine its resistance and inductance or capacitance. The total impedance of the circuit is $1250\underline{/-16.5^\circ}$ Ω.

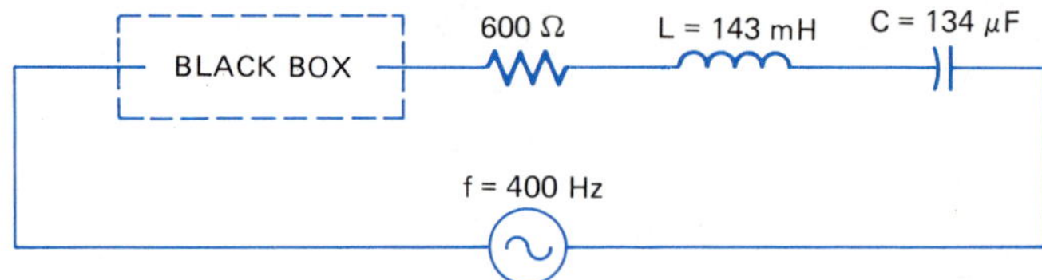

20-3 MULTIPLYING PHASORS

Phasor quantities can be multiplied and divided in either rectangular or polar form. These operations using rectangular form were studied in Topic 18-9 (complex numbers). In rectangular form, the operations are tedious. It is easier to multiply and divide in polar form.

MULTIPLYING PHASORS IN POLAR FORM

- Multiply the magnitudes of the phasors.
- Add the angles.
- Write the product as a phasor in polar form:

$$(a\underline{/\theta})(b\underline{/\phi}) = ab\underline{/\theta + \phi}$$

The product of circuit phasors is not interpreted graphically. If phasors are given in polar form, a table is not required to find the product. If, however, transformation of coordinates is necessary, it is preferable to organize the work in a tabular arrangement.

Example A Find the product of $25\underline{/44^\circ}$ and $10\underline{/22^\circ}$ in both polar and rectangular forms.

Solution: $(25\underline{/44^\circ})(10\underline{/22^\circ})$

$= (25)(10)\underline{/44^\circ + 22^\circ}$

$= 250\underline{/66^\circ} \rightarrow 102 + j228$

Example B Find the product of $-17.3 + j14.8$ and $21.6\angle 204.3^\circ$ in polar and rectangular forms.

Solution:

	POLAR		RECTANGULAR
A	$22.8\angle 139.5^\circ$	←	$-17.3 + j14.8$
B	$21.6\angle 204.3^\circ$		
AB	$492\angle 343.8^\circ$	→	$472 - j138$

Note: $(22.8)(21.6) = 492$,
$139.5^\circ + 204.3^\circ = 343.8^\circ$

Calculator note: Some calculators can be used to simplify the procedure. Transform phasor **A** in Example B using INV P→R. Use x⇄y to display the angle (139.5°). Add the angle of phasor **B** (204.3°) to find the product angle (343.8°). Use x⇄y to display the magnitude of **A** (22.8). Multiply by the magnitude of **B** (21.6) to obtain the product magnitude (492). Use x⇄y to display the correct product component for transformation. Use P→R to find the rectangular form of the product.

The product of several phasors is found by multiplying all of the magnitudes and adding all angles:

$$(a\angle\theta)(b\angle\phi)(c\angle\alpha) = abc\angle\theta + \phi + \alpha$$

Example C Evaluate: $(16 - j43)(21 + j39)(38 - j35)$

Solution:

	POLAR		RECTANGULAR
A	$45.9\angle -69.6^\circ$	←	$16 - j43$
B	$44.3\angle 61.7^\circ$	←	$21 + j39$
C	$51.7\angle -42.6^\circ$	←	$38 - j35$
ABC	$105 \times 10^3\angle -50.5^\circ$	→	$668 \times 10^2 - j81.1 \times 10^3$

Note: $(45.9)(44.3)(51.7) = 105 \times 10^3$
$(-69.6^\circ) + (61.7^\circ) + (-42.6^\circ) = -50.5^\circ$

The square of a phasor is simply the phasor multiplied by itself. Square the magnitude and double the angle:

$$(a\angle\theta)^2 = a^2\angle 2\theta)$$

Example D Square $5\angle 30^\circ$.

Solution: $(5\angle 30^\circ)^2 = 5^2\angle 2(30^\circ) = 25\angle 60^\circ$

20-4 DIVIDING PHASORS

DIVIDING PHASORS IN POLAR FORM

- Divide the magnitude of the numerator by that of the denominator.
- Subtract numerator angle minus the denominator angle.
- Write the quotient as a phasor in polar form:

$$\frac{a\angle\theta}{b\angle\phi} = \frac{a}{b}\angle\theta - \phi$$

Example A Divide $35\angle 65^\circ$ by $7\angle 35^\circ$.

Solution: $\dfrac{35\angle 65^\circ}{7\angle 35^\circ} = \dfrac{35}{7}\angle 65^\circ - 35^\circ = 5\angle 30^\circ$

As with multiplication, the solution should be organized in a table when transformations are required.

Example B Divide $456\angle{-28.3^\circ}$ by 284 + j397. Express the answer in polar and rectangular forms.

Solution: Label the two phasors by letters such as **A** and **B**.

	POLAR		RECTANGULAR
A	$456\angle{-28.3^\circ}$		
B	$488\angle{54.4^\circ}$	←	284 + j397
A/B	$0.934\angle{-82.7^\circ}$	→	0.118 – j0.927

The answer is shown in the final line of the table:

$$\frac{456}{488} = 0.934,\ (-28.3^\circ) - (54.4^\circ) = -82.7^\circ.$$

Since $1 = 1 + j0 = 1\angle{0^\circ}$, the reciprocal of a phasor is simply a division problem. Find the reciprocal of the magnitude and change the sign of the angle.

$$\frac{1}{a\angle{\theta}} = \frac{1\angle{0^\circ}}{a\angle{\theta}} = \frac{1}{a}\angle{-\theta}$$

Example C Find the reciprocal of $17.0\angle{-29.4^\circ}$.

Solution: $\dfrac{1}{17.0\angle{-29.4^\circ}} = \dfrac{1}{17.0}\angle{-(-29.4^\circ)} = 0.0588\angle{29.4^\circ}$

When phasors are substituted in formulas with both multiplication and division operations, the magnitudes are combined according to the formula being used. Angles of multipliers are added. Those of divisions are subtracted:

$$\frac{(a\angle{\theta})(b\angle{\phi})}{c\angle{\alpha}} = \frac{ab}{c}\angle{\theta + \phi - \alpha}$$

Example D Evaluate: $\dfrac{(45.3\angle{-16.2^\circ})(51.9 + j27.8)}{(22.6 - j31.6)}$

Solution: Assign the letters **A, B** and **C** to the phasors:

	POLAR		RECTANGULAR
A	$45.3\angle{-16.2^\circ}$		
B	$58.9\angle{28.2^\circ}$	←	51.9 + j27.8
C	$38.9\angle{-54.4^\circ}$	←	22.6 – j31.6
$\frac{\mathbf{AB}}{\mathbf{C}}$	$68.6\angle{66.4^\circ}$	→	27.3 + j62.5

Note: $\dfrac{(45.3)(58.9)}{38.9} = 68.6$

$(-16.2^\circ) + (28.2^\circ) - (-54.4^\circ) = 66.4^\circ$

EXERCISE 20-2

Evaluate each of the following expressions. Construct a table to display the solution in both polar and rectangular form.

1. $(20\angle{15^\circ})(6\angle{25^\circ})$
2. $(15\angle{75^\circ})(5\angle{-35^\circ})$
3. $(8\angle{26^\circ})^2$
4. $\dfrac{64\angle{18^\circ}}{16\angle{-11^\circ}}$

5. $\dfrac{1}{0.125\underline{/12.5^\circ}}$
6. $(29.6 - j17.4)(21.7\underline{/-48.2^\circ})$
7. $(2.21 + j0.98)^2$
8. $(72.6 - j84.0)^2$
9. $(38.7 + j21.4)(162\underline{/73.2^\circ})$
10. $\dfrac{38.2\underline{/56.9^\circ}}{19.1\underline{/-56.9^\circ}}$
11. $\dfrac{1}{0.008\,43 + j0.001\,14}$
12. $\dfrac{2490\underline{/21.3^\circ}}{3110 - j2940}$
13. $(17.3 - j14.6)(15.6\underline{/67.3^\circ})$
14. $\dfrac{79.2\underline{/28.7^\circ}}{84.1 + j39.4}$
15. $\dfrac{1}{0.0463 - j0.0391}$
16. $\dfrac{1}{16.4 - j17.1}$
17. $\dfrac{-16.4 + j21.6}{11.7 - j13.8}$
18. $\dfrac{4.39 - j5.06}{-2.54 + j4.17}$
19. $(71.4 \times 10^{-3}\underline{/17.6^\circ})^2$
20. $(5.03 - j12.5)(11.6 - j12.9)$
21. $(1.76 + j2.91)(0.88 - j1.47)(1.29\underline{/34.2^\circ})$
22. $(6.35 - j5.12)(35.6 - j7.09)(28.2 + j14.6)$
23. $\dfrac{(12.3 - j9.6)(2.4 + j5.9)}{10.7\underline{/-28.3^\circ}}$
24. $(3.6 - j2.9)^2(5.8 + j4.2)$
25. $\dfrac{(16.1 + j11.5)^2}{3.5\underline{/22.1^\circ}}$
26. $\left(\dfrac{16 - j22}{12 + j13}\right)^2$
27. $\dfrac{(136 + j192) + (87.9 - j70.4)}{35.2\underline{/66.7^\circ}}$
28. $\dfrac{1020\underline{/-18.6^\circ}}{(320 + j540) - (450 - j240)}$

VOLTAGE AND CURRENT IN ac SERIES CIRCUITS

The transformation and arithmetic of phasors have been studied as they apply to impedance. The reader should now be aware of the following facts:

- Impedance is the total opposition to electron flow in an ac circuit.
- Impedance is a combination of reactance and resistance as indicated by the rectangular form of the phasor and is measured in ohms.
- Reactance can be capacitive or inductive and can be calculated knowing the capacitance or inductance at any given frequency.
- The magnitude of the impedance is indicated when it is expressed in polar form.
- The phase angle between voltage and current is also given when impedance is expressed in polar form.
- Current leads voltage in capacitive circuits but lags in inductive circuits.
- The total impedance of a series circuit is found by adding the impedance of all circuit elements in rectangular form. An equivalent circuit consists of one resistance and one reactance having the same total impedance as the series circuit.

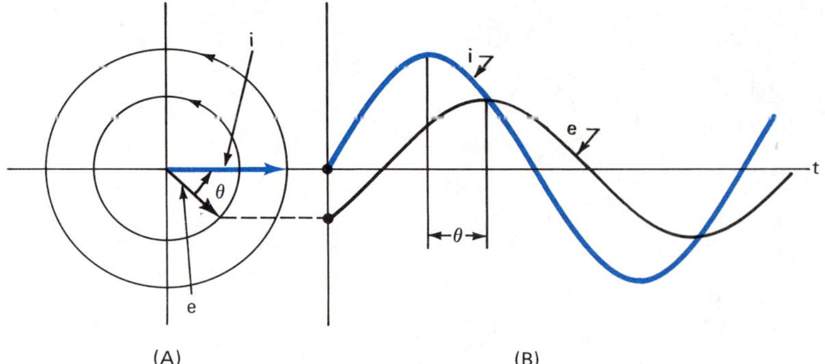

Figure 20-7 Phasor and sine wave diagrams

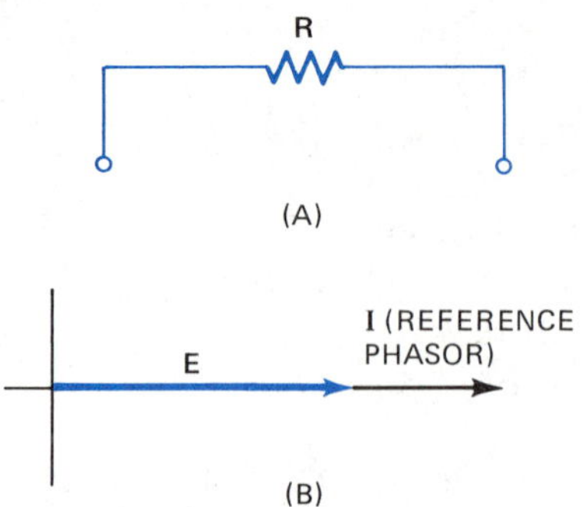

Figure 20-8 Circuit and phasor diagrams for a pure resistive circuit. $Z = R + j0 = R\underline{/0^\circ}$. E and I are in phase.

Effective voltage and effective current, Topic 17-6, are also circuit phasors. They obey the same phasor arithmetic rules described in the first four topics of this chapter. Addition and subtraction operations are performed in rectangular form. To find products, squares, quotients and reciprocals, the easiest method is to use polar form.

20-5 OHM'S LAW IN ac SINGLE-ELEMENT CIRCUITS

Remember that ac currents and voltages are continuously increasing, decreasing and reversing directions, Figure 20-7(B). The phasors, Figure 20-7(A), are those of the maximum voltage and maximum current. They are rotating vectors. They rotate counterclockwise at the same speed and always maintain the same included angle (phase). Think of the figures as a photograph taken with a high-speed camera. The phasors just happened to be in the position shown when the photo was taken.

The effective voltage and effective current are derived from the power used by the circuit. When these values are used, DC circuit relationships apply. From this point on, the words voltage and current should be understood to mean effective or rms values. They should, also automatically, be thought of as phasors.

Since current is the same everywhere in a series circuit, it is sensible to use the current phasor as the reference. It will be plotted in the reference position along the positive x-axis in phasor diagrams.

The voltage and current phasors, the circuit diagram and the rectangular and polar forms of the impedance for a pure resistive circuit are shown in Figure 20-8. Voltage and current are in phase. Figure 20-9 summarizes the same things for a purely inductive circuit. Current lags voltage by 90°. For a purely capacitive circuit, refer to Figure 20-10 in which the current leads the voltage by 90°. In all cases, current is the reference phasor. Current, voltage and impedance are related by Ohm's Law:

$$\mathbf{E} = \mathbf{IZ}$$

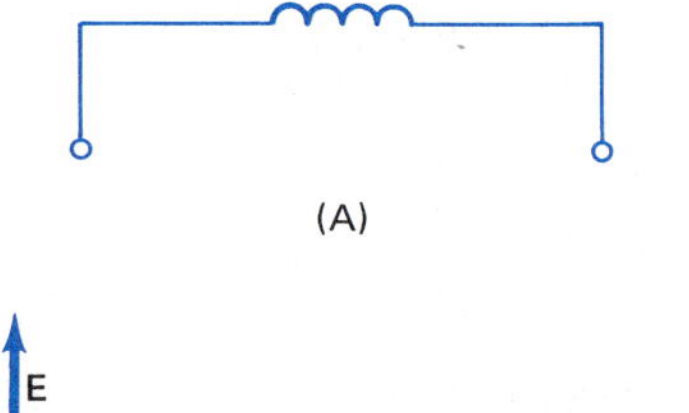

Figure 20-9 Circuit and phasor diagrams for a pure inductive circuit. $Z = 0 + jX_L = X_L\underline{/90^\circ}$. I lags E by 90°.

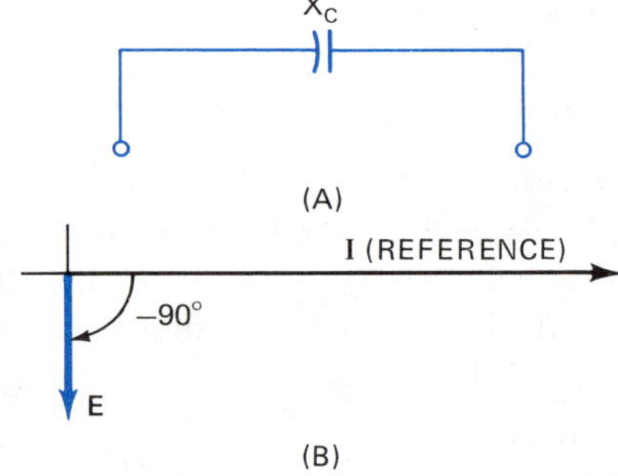

Figure 20-10 Circuit and phasor diagrams for a pure capacitive circuit. $Z = 0 - jX_C = X_C\underline{/-90^\circ}$. I leads E by 90°.

Example A Find the current in a pure resistive circuit with a resistance of 15.0 Ω and a source voltage of 120 V.

Solution: Draw circuit, impedance and voltage-current phasor diagrams, Figure 20-11. In a pure resistive circuit, the impedance equals the resistance: $\mathbf{Z} = R\underline{/0^\circ} = 15.0\underline{/0^\circ}\ \Omega$. The voltage is in phase with the current: $\mathbf{E} = 120\underline{/0^\circ}$ V.

Formula: $\mathbf{I} = \dfrac{\mathbf{E}}{\mathbf{Z}}$

Substitute: $\mathbf{I} = \dfrac{120\underline{/0^\circ}\text{ V}}{15.0\underline{/0^\circ}\ \Omega} = \dfrac{120\text{ V}}{15.0\ \Omega}\underline{/0^\circ - 0^\circ}$

$\mathbf{I} = 8.00\underline{/0^\circ}$ A

Example B Find the source voltage in a pure inductive circuit with an inductance of 934 mH. The current is 0.625 A and the source frequency is 60 Hz.

Solution: Construct diagrams, Figure 20-12. Find the inductive reactance:

Formula: $X_L = 2\pi fL$

Substitute: $X_L = 2\pi(60\text{ Hz})(0.934\text{ H})$

$X_L = 352\ \Omega$

$\mathbf{Z} = \mathbf{X_L} = 352\underline{/90^\circ}\ \Omega$

Formula: $\mathbf{E} = \mathbf{IZ}$

Substitute: $\mathbf{E} = (0.625\underline{/0^\circ}\text{ A})(352\underline{/90^\circ}\ \Omega)$

$\mathbf{E} = (0.625\text{ A})(352\ \Omega)\underline{/0^\circ + 90^\circ}$

$\mathbf{E} = 220\underline{/90^\circ}$ V

Example C Find the capacitive reactance of a pure capacitive circuit if the voltage is $480\underline{/-90^\circ}$ and the current is 10.0 A.

Solution: Construct diagrams, Figure 20-13. The solution is shown in a solution table even though no transformations are necessary.

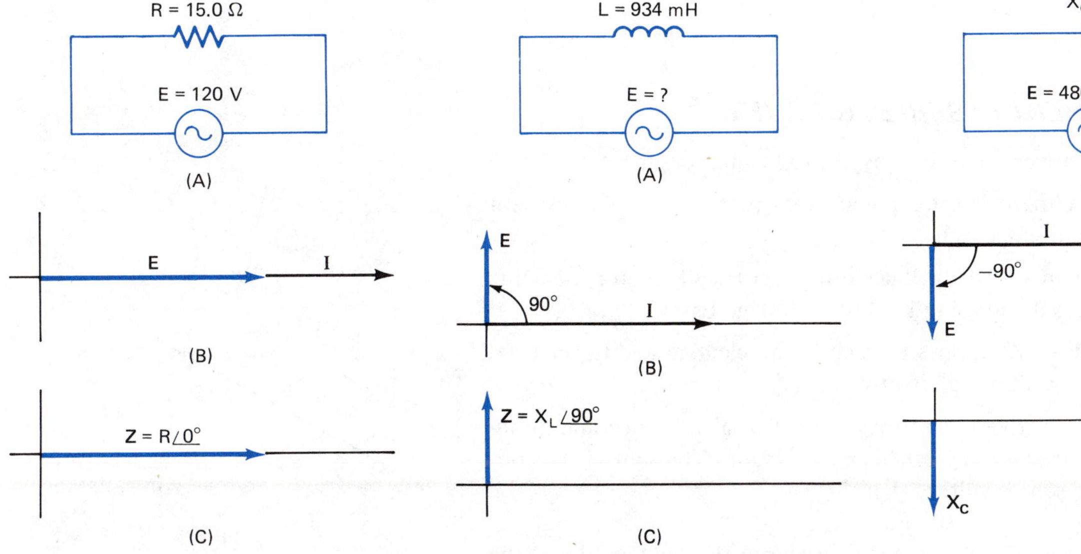

Figure 20-11

Figure 20-12

Figure 20-13

Data:

QUANTITY	POLAR		RECTANGULAR	UNITS
E	$480\angle -90^\circ$			V
I	$10.0\angle 0^\circ$			A
Z = E/I	$48\angle -90^\circ$			Ω
X_C = Z	$48\angle -90^\circ$	→	0 − j48	Ω

Note: $\mathbf{Z} = \frac{\mathbf{E}}{\mathbf{I}} = \frac{480\angle -90^\circ \text{ V}}{10.0\angle 0^\circ \text{ A}} = \frac{480 \text{ V}}{10.0 \text{ A}} \angle -90^\circ - 0^\circ = 48\angle -90^\circ$.

The capacitance (C) could now be found if the frequency was known. Capacitance and inductance (L) are scalar quantities. They would not be included in a solution table. If the rectangular form of the phasors is not used, it may be omitted in the solution table.

EXERCISE 20-3

Construct circuit, voltage-current phasor and impedance phasor diagrams for each of the following problems. Place the data and the results of all phasor calculations in table form. Determine the indicated unknown phasors in polar form.

1. $\mathbf{R} = 60\ \Omega$, $\mathbf{E} = 240$ V, $\mathbf{I} = ?$
2. $\mathbf{R} = 32\ \Omega$, $\mathbf{I} = 500$ mA, $\mathbf{E} = ?$
3. $\mathbf{X_L} = 240\ \Omega$, $\mathbf{E} = 12$ V, $\mathbf{I} = ?$
4. $\mathbf{X_L} = 720\ \Omega$, $\mathbf{I} = 25$ mA, $\mathbf{E} = ?$
5. $\mathbf{E} = 16$ V, $\mathbf{I} = 8.0$ mA, $\mathbf{X_L} = ?$
6. $\mathbf{I} = 40$ mA, $\mathbf{X_C} = 480\ \Omega$, $\mathbf{E} = ?$
7. $\mathbf{E} = 24$ V, $\mathbf{X_C} = 640\ \Omega$, $\mathbf{I} = ?$
8. $\mathbf{I} = 1.25$ A, $\mathbf{E} = 240$ V, $\mathbf{X_C} = ?$
9. $\mathbf{E} = 240$ V, $\mathbf{I} = 2.5$ A, f = 60 Hz, $\mathbf{X_L} = ?$, L = ?
10. $\mathbf{C} = 7.5\ \mu$F, f = 400 Hz, $\mathbf{I} = 6.5$ A, $\mathbf{X_C} = ?$, E = ?

20-6 TWO-ELEMENT ac SERIES CIRCUITS

Series ac circuits are characterized by the following statements:

- The effective current is the same at every point on the circuit path (Kirchhoff's Current Law).
- The phasor sum of the voltage drops across all circuit elements equals the applied voltage (Kirchhoff's Voltage Law).
- Ohm's Law (**E** = **IZ**) applies to each circuit element and to the total or equivalent circuit using effective values.

A two-element inductive circuit may consist of a resistance and an inductance in series. It is usually called an *RL circuit*. The sum of the voltage drops equals the source voltage (E):

$$\mathbf{E} = \mathbf{E_R} + \mathbf{E_X}$$

where $\mathbf{E_R}$ is the voltage drop across the resistance ($E_R \angle 0^\circ$) and $\mathbf{E_X}$ is the reactive voltage drop ($E_X \angle 90^\circ$), Figure 20-14.

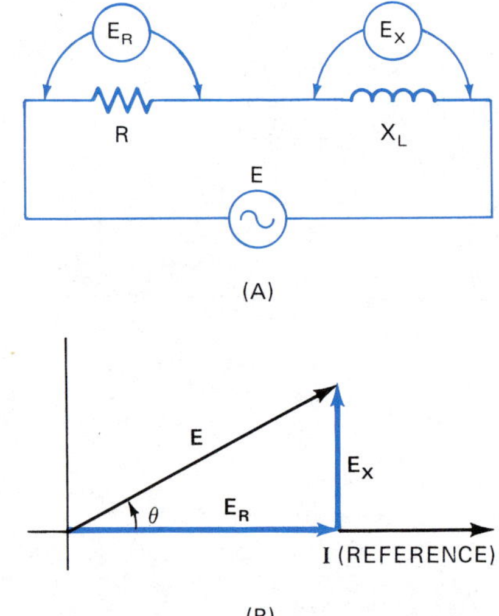

Figure 20-14 Diagrams for an RL circuit

The current in such a circuit is the reference phasor. Current, voltage and impedance are related by Ohm's Law. The voltage drops can also be found using Ohm's Law:

$$\mathbf{E_R} = \mathbf{IR}$$
$$\mathbf{E_X} = \mathbf{IX}$$

Example A A 220-V source is applied to an RL series circuit consisting of a 420-Ω resistor and a 290-Ω inductor. Find (1) the impedance, and (2) the current.

Solution: Draw circuit and phasor diagrams, Figure 20-15. Find the impedance of the circuit, then use Ohm's Law to find the current.

(1) Formula: $Z\angle\theta = R + jX$

Substitute: $Z\angle\theta = 420 + j290 \rightarrow 510\angle 34.6^\circ\ \Omega$

$Z\angle\theta = 510\angle 34.6^\circ\ \Omega$

Since the phase angle is now known, the source voltage can be written as: $\mathbf{E} = 220\angle 34.6^\circ$ V

(2) Formula: $\mathbf{I} = \dfrac{\mathbf{E}}{\mathbf{Z}}$

Substitute: $\mathbf{I} = \dfrac{220\angle 34.6^\circ\ \text{V}}{510\angle 34.6^\circ\ \Omega} = \dfrac{220\ \text{V}}{510\ \Omega}\angle 34.6^\circ - 34.6^\circ$

$\mathbf{I} = 0.431\angle 0^\circ$ A

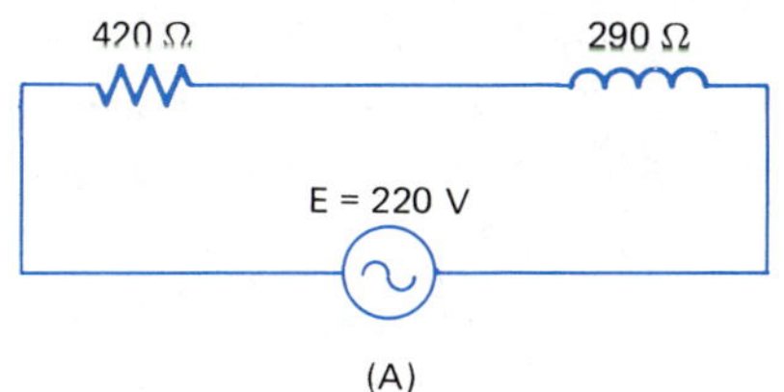

(A)

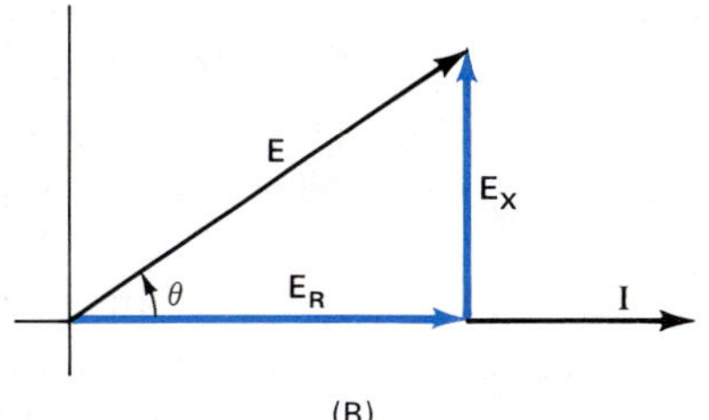

(B)

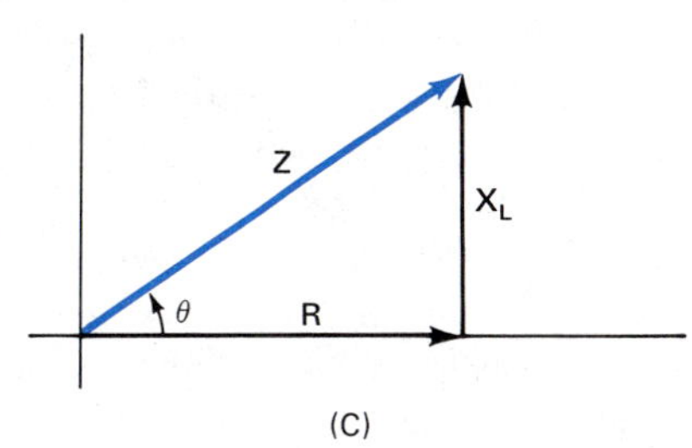

(C)

Figure 20-15

All of the given data and calculated results can be arranged in table form which will organize the solution. The given data is listed first. Calculations are made on the side using the formula given in the first column. Additions can be made within the table in rectangular form. The voltage phase angle is inserted after the polar form of the impedance (**Z**) is determined.

	QUANTITY	POLAR		RECTANGULAR	UNITS
Data:	E	$220\angle 34.6^\circ$			V
	R	$420\angle 0^\circ$	→	420 + j0	Ω
	X	$290\angle 90^\circ$	→	0 + j290	Ω
(1)	Z = R + jX	$510\angle 34.6^\circ$	←	420 + j290	Ω
(2)	I = E/Z	$0.431\angle 0^\circ$			A

Example B Find the voltage drops across each circuit element of Example A. Think of **R** as the phasor $R\angle 0^\circ$ and **X** as $X\angle 90^\circ$.

Solution: Construct the voltage-current diagram, Figure 20-15(B).

Formula: $\mathbf{E_R} = \mathbf{IR}$

Substitute: $\mathbf{E_R} = (0.431\angle 0^\circ\ \text{A})(420\angle 0^\circ\ \Omega)$

$\mathbf{E_R} = 181\angle 0^\circ$ V

Formula: $\mathbf{E_X} = \mathbf{IX}$

Substitute: $\mathbf{E_X} = (0.431\angle 0^\circ\ \text{A})(290\angle 90^\circ\ \Omega)$

$\mathbf{E_X} = (0.431\ \text{A})(290\ \Omega)\angle 0^\circ + 90^\circ$

$\mathbf{E_X} = 125\angle 90^\circ$ V

Check: $\mathbf{E} = \mathbf{E_R} + j\mathbf{E_X}$

$\mathbf{E} = (181\angle 0^\circ) + (125\angle 90^\circ)$

$\mathbf{E} = (181 + j0) + (0 + j125)$

$\mathbf{E} = 181 + j125 \rightarrow 220\angle 34.6^\circ$ V

The following table form of the solution is a continuation of the table from Example A. Given data and calculated answers are not repeated. The two tables form the entire solution of the problem:

	QUANTITY	POLAR		RECTANGULAR	UNITS
	$E_R = IR$	$181\underline{/0^\circ}$	→	$181 + j0$	V
	$E_X = IX$	$125\underline{/90^\circ}$	→	$0 + j125$	V
Check:	$E = E_R + jE_X$	$220\underline{/34.6^\circ}$	←	$181 + j125$	V

Power is dissipated in the resistive elements only in ac circuits and is referred to as true power (P). It is given by the formula:

$$P = EI \cos \theta$$

where $\cos \theta$ is the power factor (Pf) and E and I are the magnitudes of the voltage and current. The product EI is called the *apparent power* (S). The power factor indicates the fraction of the apparent power that is actually consumed. The rest is returned to the source. The power factor is sometimes expressed in percent.

Example C Find the total power dissipation in the circuit shown in Figure 20-15. The source voltage is $220\underline{/34.6^\circ}$ V and the current is $0.431\underline{/0^\circ}$ A. Express the power factor as a percent.

Solution: Data: $E = 220\text{ V}, \theta = 34.6^\circ, I = 0.431\text{ A}$

Formula: $P = EI \cos \theta$

Substitute: $P = (220\text{ V})(0.431\text{ A}) \cos 34.6^\circ$

$P = (94.8\text{ W})(0.823)$

$P = 78.0\text{ W}$

The power factor (0.823) is 82.3%. This means the true power is 82.3% of the apparent power (94.8 W).

A two-element series-capacitive circuit consists of a resistance and a capacitance and is usually called an *RC circuit.* The equations introduced in this chapter also apply to the RC circuit.

Example D A $120\underline{/-51.3^\circ}$ V source is applied to an RC circuit. The current is 315 mA. Find (1) the true power, (2) the resistance, (3) the reactance and (4) the voltage drop across each element.

Solution: Draw circuit and phasor diagrams, Figure 20-16. Use Ohm's Law to find the impedance in polar form. R and X are indicated when Z is transformed to rectangular form. Formulas in the first column of the table show how calculations are made.

(1) Formula: $P = EI \cos \theta$

Substitute: $P = (120\text{ V})(0.315\text{ A}) \cos(-51.3^\circ)$

$P = 23.6\text{ W}$

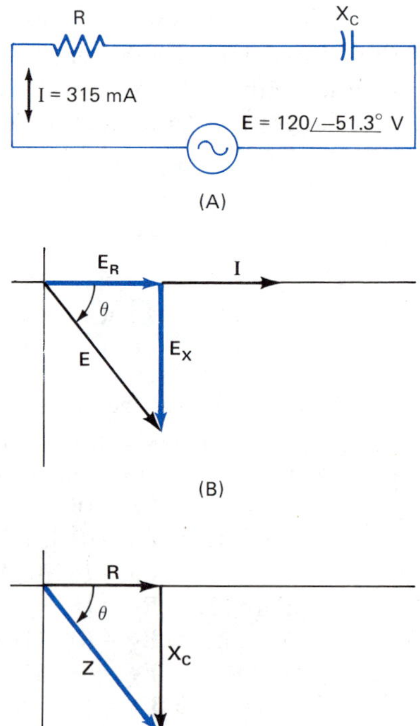

Figure 20-16

	QUANTITY	POLAR		RECTANGULAR	UNITS
Data:	E	$120\angle -51.3°$			V
	I	$0.315\angle 0°$			A
	Z = E/I	$381\angle -51.3°$	→	238 − j297	Ω
(2)	R	$238\angle 0°$	←	238 + j0	Ω
(3)	X	$297\angle -90°$	←	0 − j297	Ω
(4)	E_R = IR	$75.0\angle 0°$	→	75.0 + j0	V
	E_X = IX	$93.6\angle -90°$	→	0 − j93.6	V
Check:	E = E_R + jE_X	$120\angle -51.3°$	←	75.0 − j93.6	V

EXERCISE 20-4

Construct circuit, voltage-current and impedance phasors for each of the following problems. Determine the indicated unknowns in polar form. Construct a solution table for each problem.

1. E = 240 V, R = 34.5 Ω, X_L = 46.8 Ω. Find **Z**, **I**, $\mathbf{E_R}$, $\mathbf{E_X}$ and P.
2. I = 750 mA, R = 425 Ω, X_C = 355 Ω. Find **Z**, **E**, $\mathbf{E_R}$, $\mathbf{E_X}$ and P.
3. **E** = $24\angle 56.2°$, I = 96 mA. Find **Z**, **R**, **X**, $\mathbf{E_R}$, $\mathbf{E_X}$ and P.
4. C = 45 μF, f = 60 Hz, E = 120 V, R = 42.4 Ω. Find $\mathbf{X_C}$, **Z**, **I**, $\mathbf{E_R}$, $\mathbf{E_X}$ and P.
5. L = 46 mH, f = 400 Hz, I = 3.5 A, R = 92 Ω. Find $\mathbf{X_L}$, **Z**, **E**, $\mathbf{E_R}$, $\mathbf{E_X}$ and P.
6. Find $\mathbf{X_C}$, **Z**, **E**, $\mathbf{E_R}$, $\mathbf{E_X}$ and P in the circuit of Figure 20-17.

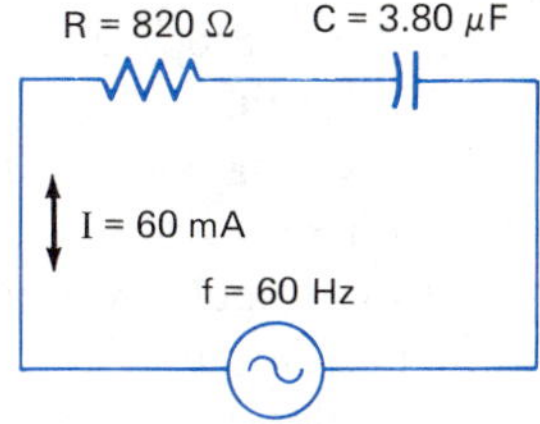

Figure 20-17

7. Find $\mathbf{X_L}$, **Z**, **I**, $\mathbf{E_R}$, $\mathbf{E_X}$ and P in the circuit of Figure 20-18.

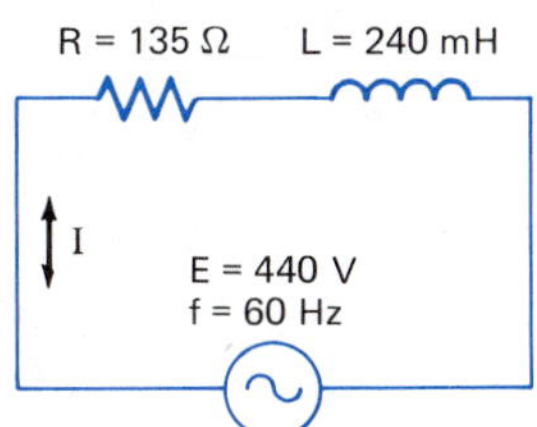

Figure 20-18

8. Find **E**, **R**, $\mathbf{X_C}$, **Z** and P in the circuit of Figure 20-19.
9. A certain circuit has an alternating voltage of 120 V and a leading current of 1.2 A. The power factor is 39.5%. Find the (a) impedance, (b) resistance, (c) reactance and (d) voltage drops across each of the two circuit elements.
10. A certain circuit has a lagging current of 16.0 A and an impedance of 150 Ω. The power factor is 48.8%. Find the (a) resistance, (b) reactance, (c) source voltage and (d) voltage drops across each of the two circuit elements.

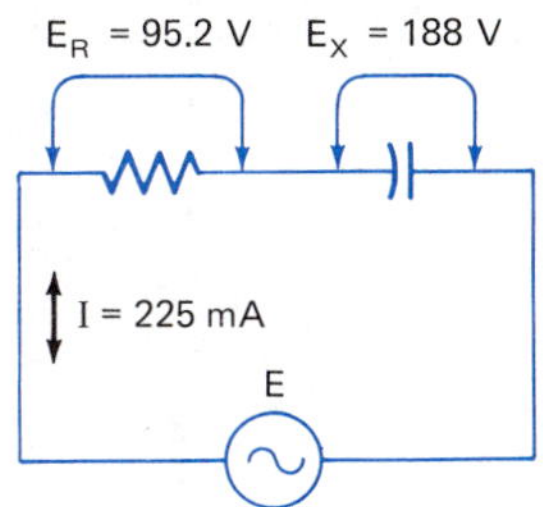

Figure 20-19

20-7 SERIES CIRCUITS WITH THREE OR MORE ELEMENTS

Inductive reactance and capacitive reactance cause opposite effects. The first causes current to lag the voltage; the second makes it lead. Series circuits with resistance and both kinds of reactance will be inductive if X_L is greater than X_C. They will be capacitive if X_C is greater than X_L.

The impedance phasor in rectangular form is the sum of the impedance of each part. In the three-element circuit of Figure 20-20(A), the impedance is

$$Z = R + jX_L - jX_C = R + j(X_L - X_C)$$

The phasor diagram is shown in Figure 20-20(B) with X_C greater than X_L. The circuit is capacitive.

The voltage-current phasor is shown in Figure 20-20(C) with the sum of the voltage drops equal to the source voltage:

$$\mathbf{E} = \mathbf{E_R} + \mathbf{E_L} + \mathbf{E_C}$$

Current is the reference phasor. Voltage drops across the reactance are often greater in magnitude than the source voltage.

Example A Find (1) the total impedance, (2) the current and (3) the voltage drops across each element of the circuit in Figure 20-21(A).

Solution: Draw phasor diagrams, Figure 20-21(B) and Figure 20-21(C). Add resistances and reactances in rectangular form. Transform to polar to find the phase angle.

(1) Formula: $\mathbf{Z_T} = R + jX_L - jX_C$

Substitute: $\mathbf{Z_T} = (85 + j0) + (0 + j120) + (0 - j175)$

$\mathbf{Z_T} = 85 - j55 \rightarrow \underline{101\angle -32.9°}$

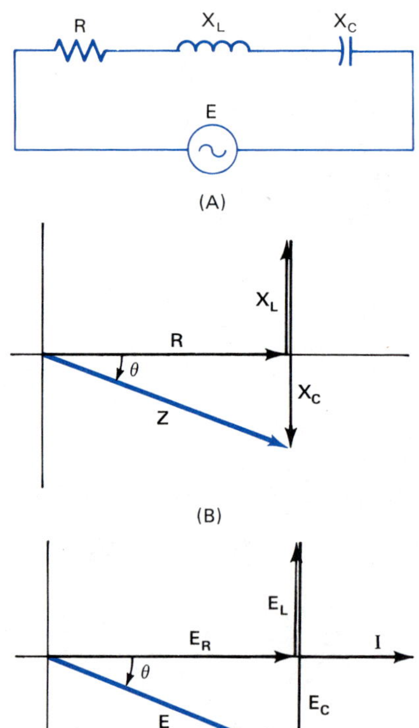

Figure 20-20 Diagrams for an RCL circuit

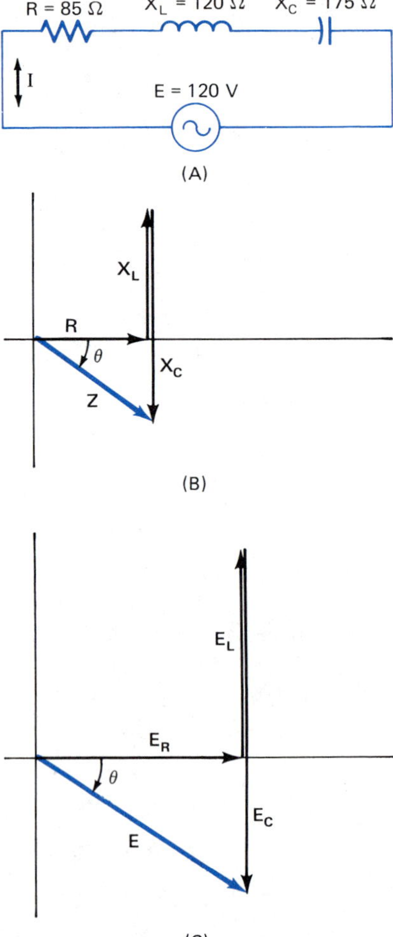

Figure 20-21

In polar form the source voltage is
$\mathbf{E} = 120\angle{-32.9°}$

Use Ohm's Law to find the current.

(2) Formula: $\mathbf{I} = \dfrac{\mathbf{E}}{\mathbf{Z_T}}$

Substitute: $\mathbf{I} = \dfrac{120\angle{-32.9°}\ \text{V}}{101\angle{-32.9°}\ \Omega} = \dfrac{120\ \text{V}}{101\ \Omega}\angle{(-32.9°) - (-32.9°)}$

$\mathbf{I} = 1.19\angle{0°}\ \text{A}$

Use Ohm's Law to find voltage drops:

(3) $\mathbf{E_R} = \mathbf{IR} = (1.19\angle{0°}\ \text{A})(85\angle{0°}\ \Omega) = 101\angle{0°}\ \text{V}$
$\mathbf{E_L} = \mathbf{IX_L} = (1.19\angle{0°}\ \text{A})(120\angle{90°}\ \Omega) = 143\angle{90°}\ \text{V}$
$\mathbf{E_C} = \mathbf{IX_C} = (1.19\angle{0°}\ \text{A})(175\angle{90°}\ \Omega) = 208\angle{-90°}\ \text{V}$

The solution can be checked by adding voltage drops in rectangular form. The sum should equal the applied voltage. This is done in the equivalent solution shown in the table that follows.

	QUANTITY	POLAR		RECTANGULAR	UNITS
Data:	E	$120\angle{-32.9°}$			V
	R	$85\angle{0°}$	→	85 + j0	Ω
	X_L	$120\angle{90°}$	→	0 + j120	Ω
	X_C	$175\angle{-90°}$	→	0 − j175	Ω
(1)	Z_T (Sum)	$101\angle{-32.9°}$	←	85 − j55	Ω
(2)	$I = E/Z_T$	$1.19\angle{0°}$			A
(3)	$E_R = IR$	$101\angle{0°}$	→	101 + j0	V
	$E_L = IX_L$	$143\angle{90°}$	→	0 + j143	V
	$E_C = IX_C$	$208\angle{-90°}$	→	0 − j208	V
Check:	E = (Sum)	$120\angle{-32.8°}$	←	101 − j65	V

When several devices are connected in series, the total impedance is found by adding all impedances in rectangular form, Topic 20-1. This procedure is easily accomplished within the solution table.

Example B Find (1) the total impedance, (2) the source voltage and (3) the voltage drops across each of the four circuit elements, Figure 20-22(A). The current is I = 3.33 A.

Solution: Find the impedance $Z_2 = X_C$

Formula: $X_C = \dfrac{1}{2\pi fC}$

Substitute: $X_C = \dfrac{1}{2\pi(60\ \text{Hz})(90 \times 10^{-6}\ \text{F})} = 29.5\ \Omega$

Construct impedance phasor, Figure 20-22(B). Note that the circuit is inductive.

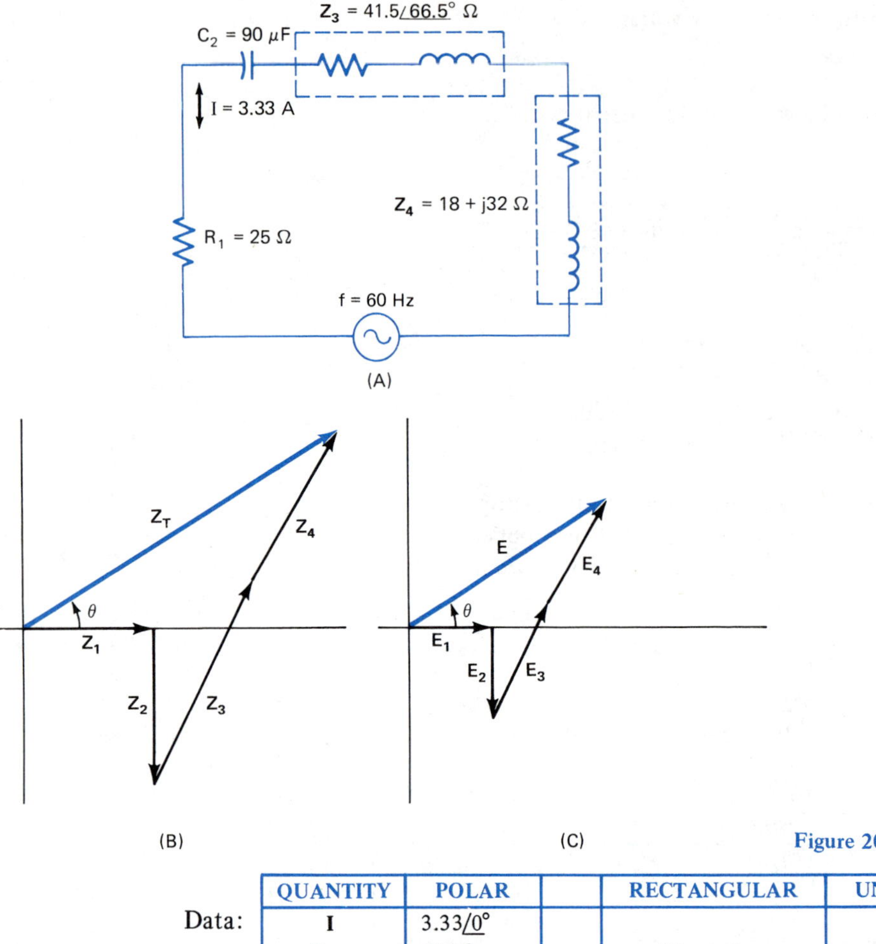

Figure 20-22

	QUANTITY	POLAR		RECTANGULAR	UNITS
Data:	I	$3.33\underline{/0^\circ}$			A
	Z_1	$25\underline{/0^\circ}$	→	$25 + j0$	Ω
	Z_2	$29.5\underline{/-90^\circ}$	→	$0 - j29.5$	Ω
	Z_3	$41.5\underline{/66.5^\circ}$	→	$16.5 + j38.1$	M
	Z_4	$36.7\underline{/60.6^\circ}$	←	$18 + j32$	Ω
(1)	Z_T = (Sum)	$72.0\underline{/34.3^\circ}$	←	$59.5 + j40.6$	Ω
(2)	$E = IZ_T$	$240\underline{/34.3^\circ}$			V
(3)	$E_1 = IZ_1$	$83.3\underline{/0^\circ}$	→	$83.3 + j0$	V
	$E_2 = IZ_2$	$98.2\underline{/-90^\circ}$	→	$0 - j98.2$	V
	$E_3 = IZ_3$	$138\underline{/66.5^\circ}$	→	$55.0 + j127$	V
	$E_4 = IZ_4$	$122\underline{/60.6^\circ}$	→	$59.9 + j107$	V
Check:	E = (Sum)	$240\underline{/34.4^\circ}$	←	$198.2 + j136$	V

Note that the equivalent circuit is given in (1) for total impedance. The equivalent circuit consists of a resistance of 59.5 Ω and an inductive reactance of 40.6 Ω.

Power dissipation in each of the circuit elements can be calculated by multiplying the magnitudes of the voltage drop and the current by the power factor for the element:

$$P_Z = E_Z I \cos \theta_Z$$

Example C Find the true power dissipation in each of the circuit elements and the total true power in the circuit of Figure 20-22.

Solution: Formula: $P = EI \cos \theta$

Substitute: $P_1 = (83.3\text{ V})(3.33\text{ A}) \cos 0° = 277\text{ W}$

$P_2 = (98.2\text{ V})(3.33\text{ A}) \cos (-90°) = 0\text{ W}$

$P_3 = (138\text{ V})(3.33\text{ A}) \cos 66.5° = 183\text{ W}$

$P_4 = (122\text{ V})(3.33\text{ A}) \cos 60.6° = 199\text{ W}$

$P = (240)(3.33\text{ A}) \cos 34.4° = 659\text{ W}$

Check: $P = P_1 + P_2 + P_3 + P_4$

$P = 277\text{ W} + 0\text{ W} + 183\text{ W} + 199\text{ W} = 659\text{ W}$

EXERCISE 20-5

Draw circuit, impedance phasor and voltage-current phasor diagrams for each of the following problems. Make a solution table for each problem showing data and all unknowns. Express all unknowns in polar form.

1. $R = 160\ \Omega$, $X_L = 220\ \Omega$, $X_C = 80\ \Omega$ and $E = 24$ V. Find $\mathbf{Z_T}$, $\mathbf{I}$, $\mathbf{E_R}$, $\mathbf{E_L}$, $\mathbf{E_C}$ and P.
2. $R = 1200\ \Omega$, $X_L = 920\ \Omega$, $X_C = 1600\ \Omega$ and $I = 40$ mA. Find $\mathbf{Z_T}$, $\mathbf{E}$, $\mathbf{E_R}$, $\mathbf{E_L}$, $\mathbf{E_C}$ and P.
3. $R = 85\ \Omega$, $X_L = 265\ \Omega$, $X_C = 300\ \Omega$ and $I = 850$ mA. Find $\mathbf{Z_T}$, $\mathbf{E}$, $\mathbf{E_R}$, $\mathbf{E_L}$, $\mathbf{E_C}$ and P.
4. $R = 95\ \Omega$, $L = 400$ mH, $C = 8.78\ \mu$F, $E = 240$ V and $f = 60$ Hz. Find $\mathbf{Z_T}$, $\mathbf{I}$, $\mathbf{E_R}$, $\mathbf{E_L}$, $\mathbf{E_C}$ and P.
5. $R = 225\ \Omega$, $L = 55$ mH, $C = 4.5\ \mu$F, $I = 80$ mA and $f = 400$ Hz. Find $\mathbf{Z_T}$, $\mathbf{E}$, $\mathbf{E_R}$, $\mathbf{E_L}$, $\mathbf{E_C}$ and P.
6. Solve the circuit in Figure 20-23 for $\mathbf{Z_T}$, $\mathbf{I}$, all voltage drops ($\mathbf{E_1}$, $\mathbf{E_2}$, $\mathbf{E_3}$) and P.
7. Find $\mathbf{Z_T}$, $\mathbf{E}$, $\mathbf{E_1}$, $\mathbf{E_2}$, $\mathbf{E_3}$ and P in the circuit of Figure 20-24.
8. Find $\mathbf{Z_T}$, $\mathbf{I}$, $\mathbf{E_1}$, $\mathbf{E_2}$, $\mathbf{E_3}$ and P in the circuit shown in Figure 20-25.
9. An ac series circuit contains a capacitor of 7.2 μF, an inductor of 975 mH and a resistance of 540 Ω. The circuit is connected to a 60-Hz source of 440 V. Find (a) the total impedance, (b) the current, (c) the voltage drops and (d) the power dissipation.
10. Find the unknowns of the circuit of Problem 9 if it is connected to a 50-Hz source at 440 V.

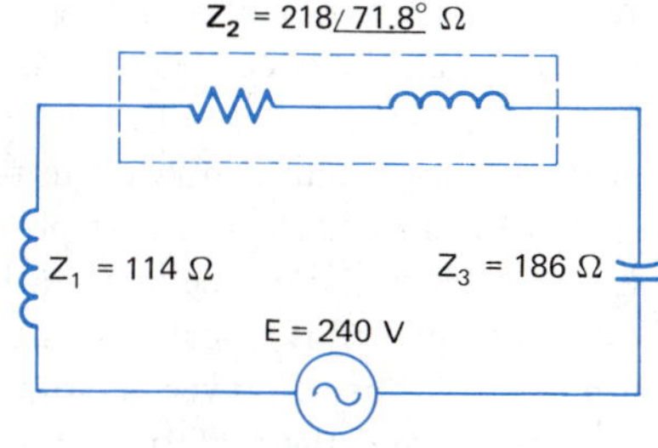

Figure 20-23

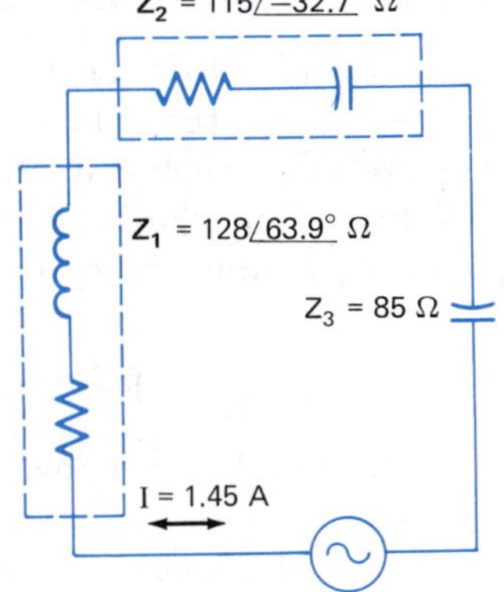

Figure 20-24

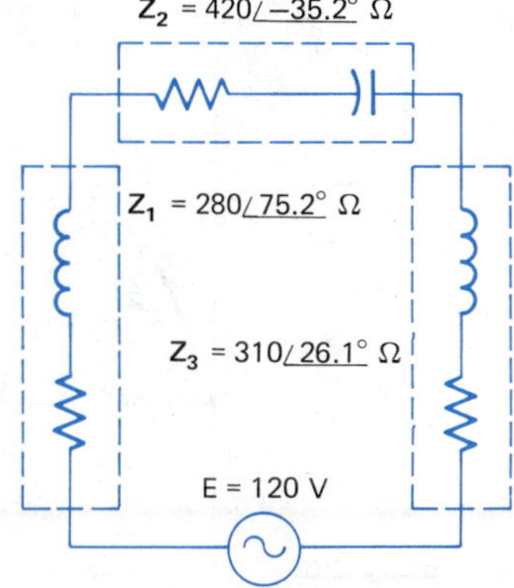

Figure 20-25

POWER IN ac SERIES CIRCUITS

In ac circuits, there are two other measurable "kinds" of power besides true power (P). *Apparent power* (S) is the power delivered by the source. It is the product of the magnitudes of the source voltage and the circuit current:

$$S = EI$$

True power, then, is the product of apparent power and the power factor:

$$P = S \cos \theta$$

Reactive power (Q) is the power that is stored momentarily before being sent back to the source. Storage takes place in the magnetic fields of inductors and in the electrostatic fields of capacitors.

The total true power of a circuit is found by simply adding the power dissipations of all the circuit elements:

$$P = P_1 + P_2 + P_3 + \ldots$$

Power dissipation occurs only in the resistance of the circuit element usually in the form of heat. Reactive power is stored in the reactive elements and is sent back to the source later in the cycle.

While true power is a scalar quantity, the relationship between it and the other two kinds of power is more complex. The meaning of the three kinds of power and their relationship are discussed in the remaining topics of this chapter.

20-8 POWER IN SINGLE-ELEMENT ac CIRCUITS

Power in a pure resistive circuit was discussed in Topic 17-6. At that time, effective voltage (E) and effective current (I) were derived. When an alternating voltage is applied to a pure resistance, the current is in phase with the voltage, Figure 20-26. The true power (P = EI) is equal to the average power (P_{av}) during the cycle. It is equal to one-half the peak power (P_m):

$$P = P_{av} = \frac{1}{2}P_m = EI$$

The energy consumed (transformed to heat) is represented by the shaded area under the curve.

In a pure inductive circuit, current lags the voltage by 90°, Figure 20-27. The product of instantaneous voltage and current yields the double frequency power curve. When voltage and current have the same sign, the instantaneous power is positive. When they have opposite signs, it is negative.

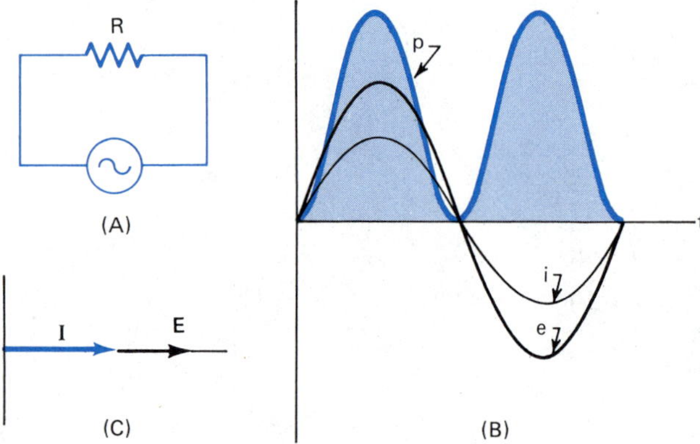

Figure 20-26 Power curve for a pure resistive circuit

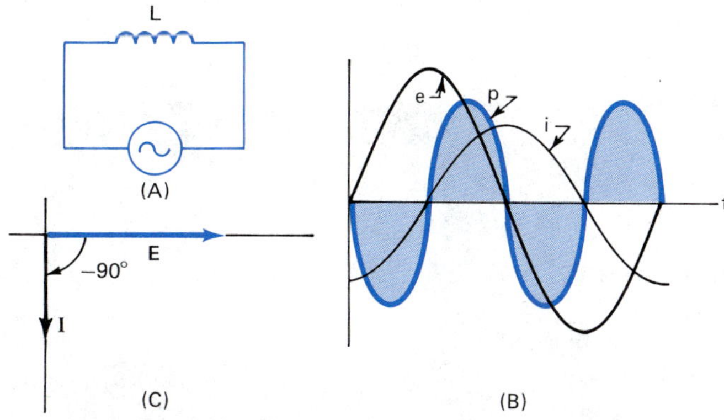

Figure 20-27 Power curve for a pure inductive circuit. The shaded area above the t-axis represents energy being stored in the inductor. That below the t-axis is energy being returned to the source.

From 0° to 90° and from 180° to 270°, current is falling while voltage is increasing. Power is negative, the magnetic field is collapsing and the energy stored in the field is being returned to the source. The shaded area below the axis is called a *negative area* and represents the energy returned.

From 90° to 180° and from 270° to 360°, current is rising as voltage decreases. Both current and voltage have the same sign so power is positive. Energy from the source builds up the magnetic field as it is stored (positive shaded area).

A pure capacitive circuit behaves somewhat like the pure inductive circuit. Unlike the inductive circuit, however, current *leads* the voltage, Figure 20-28. The power curve at any instant is the product of e and i. Again, it has the double frequency feature. Shaded area represents the energy.

In an inductor, energy is stored when current rises and returns to the source when current falls. The capacitor acts differently. Energy is stored in the electrostatic field between capacitor plates when the source voltage rises. It returns to the source when the back emf of the capacitor exceeds the source voltage. This occurs when the source voltage is falling.

Apparent power (S) represents the rate at which energy flows into the circuit from the source. It is the energy absorbed each second by the circuit devices. This can occur two ways. It can be "used," which means it is dissipated as heat, light, etc. in the resistive elements of the circuit. The

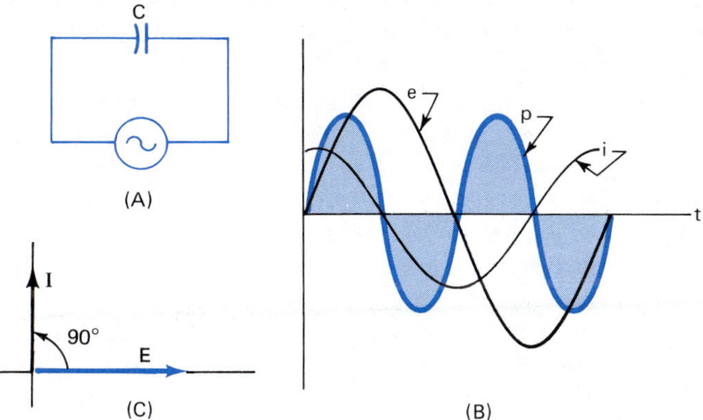

Figure 20-28 Power curve for a pure capacitive circuit

second way a circuit can absorb energy is by storing it in reactive elements.

True power (P) is the rate at which the resistive elements dissipate energy as heat. For the pure resistive circuit, true power equals apparent power. Reactive power (Q) is power that is stored, then returned to the source. The return flow occurs when the circuit acts as a voltage source. The reactive elements become voltage sources (back emf) during parts of the cycle. In the pure reactive circuit, reactive power equals apparent power.

The units of true power (P) is watts (W) and is measured directly on a wattmeter, Figure 20-29. The wattmeter measures the rate at which energy is dissipated as heat, light, work, etc. in the resistive elements. **One watt equals one joule per second.**

In order to better distinguish the different powers, the volt-ampere (VA) is used to indicate apparent power (S). The volt-ampere, the joule per second and the watt are all the same unit. Electricians simply use W as a measure of true power (P) and VA as a unit of apparent power (S). The third quantity, reactive power (Q) is measured in reactive volt-amperes (vars), which is also equivalent to the watt. The letters var are the first letters of the words "volt-ampere-reactive." A varmeter, Figure 20-30, measures reactive power directly.

Figure 20-29 A wattmeter measures true power in an ac circuit (Courtesy of Southeast Nebraska Community College)

Apparent power can always be calculated by the formulas

$$S = EI$$

$$S = \frac{E_2}{Z}$$

$$S = I^2 Z$$

In a pure resistive circuit, true power equals apparent power:

$$P = S \cos \theta$$

Since the phase angle is 0°, the power factor ($\cos \theta$) equals 1 so $P = S$. In a pure reactive circuit, $\theta = \pm 90°$. In these circuits, reactive power is

$$Q = S \sin \theta$$

The *reactive power factor* ($\sin \theta$) equals 1, so $Q = S$.

Figure 20-30 A varmeter measures reactive power in ac circuits. Varmeters are used mainly in power plants and distribution centers.

Example A Find the apparent power in a pure resistive circuit with a 120-V supply. The current is 4.5 A. Find the true power.

Solution: Formula: $S = EI$

Substitute: $S = (120\text{ V})(4.5\text{ A}) = 540\text{ VA}$

Formula: $P = EI \cos \theta$

Substitute: $P = (120\text{ V})(4.5\text{ A}) \cos 0° = 540\text{ W}$

Note that $P = S$.

Example B Determine the apparent power when a 60-Hz voltage of 240 V is applied to an inductance of 235 mH. Find the reactive power.

Solution: Formula: $X_L = 2\pi fL$

Substitute: $X_L = 2\pi(60\text{ Hz})(0.235\text{ H}) = 88.6\ \Omega$

$(Z = X_L)$

Formula: $S = \frac{E_2}{Z}$

Substitute: $S = \frac{(240\ \text{V})^2}{88.6\ \Omega} = 650\ \text{VA}$

Formula: $Q = S \sin\theta$

Substitute: $Q = (650\ \text{VA}) \sin 90^\circ = (650\ \text{VA})(1) = 650$ vars

Note that Q = S.

Example C What is the apparent power in a circuit with a 60-Hz, 3.8-A current? The only circuit element has a capacitance of 3.75 μF.

Solution: Formula: $X_C = \frac{1}{2\pi fC}$

Substitute: $X_C = \frac{1}{2\pi(60\ \text{Hz})(3.75 \times 10^{-6})} = 707\ \Omega$

$(Z = X_C)$

Formula: $S = I^2 Z$

Substitute: $S = (3.8\ \text{A})^2(707\ \Omega) = 10.2\ \text{kVA}$

Formula: $Q = S \sin\theta$

Substitute: $Q = (10\ 200\ \text{VA}) \sin(-90^\circ)$

$Q = (10\ 200\ \text{VA})(1) = 10.2$ kvars

EXERCISE 20-6

Find the magnitudes of the circuit quantities indicated by a question mark. Each circuit contains only a single element. Assume all alternating currents are 60 Hz.

	E	I	R	L	C	S	P	Q
1.	24 V	75 mA	?	–	–	?	?	?
2.	220 V	?	?	–	–	6 kVA	?	?
3.	?	24 A	40 Ω	–	–	?	?	?
4.	120 V	?	40 Ω	–	–	?	?	?
5.	16 V	?	–	0.5 H	–	?	?	?
6.	?	3 A	–	–	5 μF	?	?	?
7.	?	?	–	–	2.1 μF	3 kVA	?	?
8.	?	?	–	850 mH	–	?	?	875 vars
9.	?	6 A	–	1.25 H	–	?	?	?
10.	1760 V	?	–	–	45 μF	?	?	?

20-9 POWER IN RL AND RC SERIES CIRCUITS

Circuits that contain both resistance and reactance cause the current to lead or lag the voltage. The phase angle can be determined by finding the impedance or voltage in polar form. The current in an RL circuit, Figure 20-31, lags the voltage.

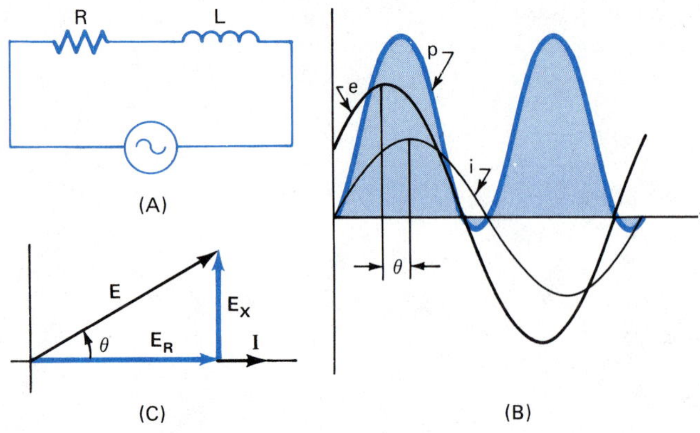

Figure 20-31 Power curve for an RL circuit

The voltage-current phasor diagram, Figure 20-31(C), can be used to derive the relationship between apparent power, true power and reactive power. In the voltage triangle, $E_R = E \cos \theta$ and $E_X = E \sin \theta$. All three sides of the triangle are multiplied by the current (I). This results in the power triangle, Figure 20-32.

The hypotenuse represents the apparent power (S). The horizontal side is the true power (P) where $\cos \theta$ is the power factor. The vertical side is the reactive power (Q). From Figure 20-32, the formulas are as follows:

$$S = EI$$
$$P = E_R I = EI \cos \theta$$
$$Q = E_X I = EI \sin \theta$$

The angle θ is the phase angle between voltage and current. It is the same in the impedance, voltage-current and now the power phasor diagrams.

Apparent, true and reactive power have a phasor relationship and can be expressed in polar and rectangular forms:

$$S\angle\theta = P + jQ$$

When one form is known, the other can be found by transformation of coordinates.

The RC circuit, Figure 20-33, is similar to the RL circuit except current leads instead of lags. The polar and rectangular form are as follows:

$$S\angle-\theta = P - jQ$$

The reactive power (Q) is negative in a capacitive circuit as indicated in the power phasor diagram, polar and rectangular forms. The negative

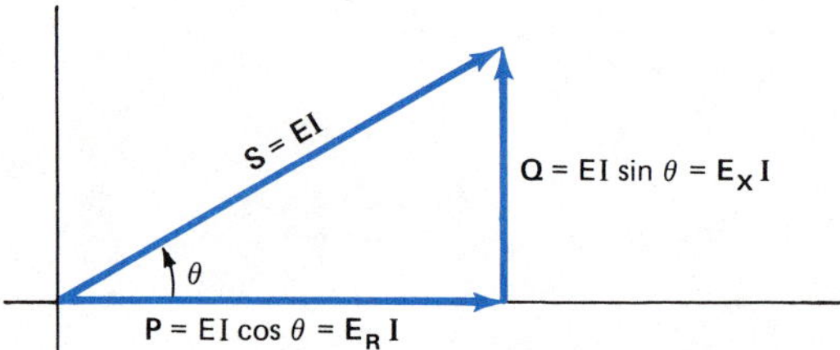

Figure 20-32 Power phasor diagram

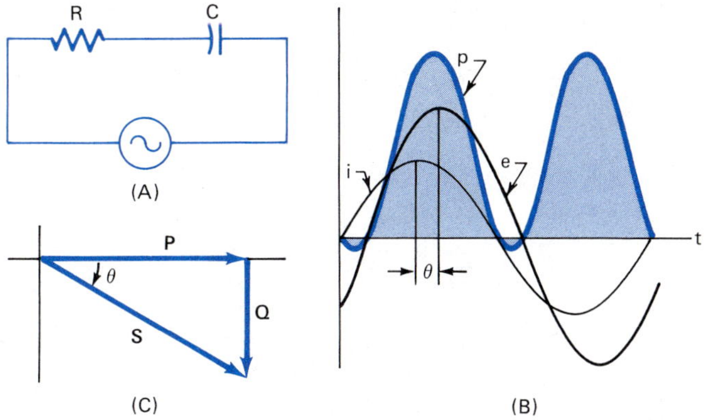

(A) (C) (B)

Figure 20-33 Power curve for an RC circuit

sign simply means the current is leading the voltage. It has the same meaning as a negative sign in voltage and impedance phasor diagrams.

Example A The true power of a circuit is measured with a wattmeter and found to be 45.3 kW. A varmeter indicates the reactive power is 38.2 kvars (kilovolt-amperes reactive) lagging. Express the apparent power in both rectangular and polar form. What is the power factor?

Solution: Construct circuit and phasor diagrams, Figure 20-34. The circuit is inductive because Q lags.

Data: P = 45.3 kW, Q = 38.2 kvars

Together, the given quantities make up the rectangular form of the apparent power:

$P + jQ = 45.3 + j38.2$

45.3 [x⇄y] 38.2 [INV] [P→R] , 40.1°

[x⇄y] , 59.2 VA

[x⇄y] [Cos] , 0.765 = 76.5%

$S\angle\theta = 59.2\angle 40.1°$ kVA

$Pf = \cos\theta = 76.5\%$

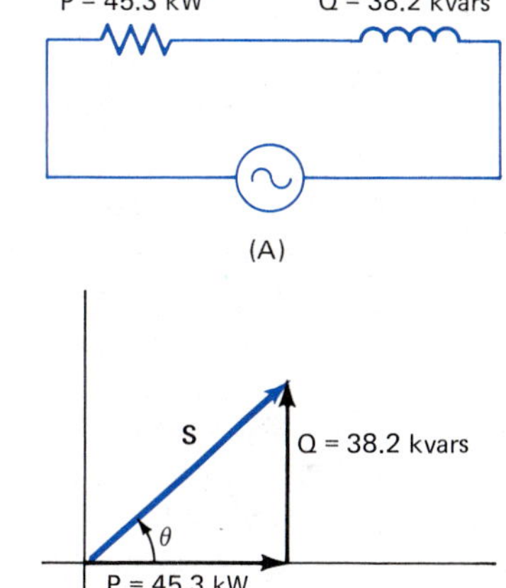

Figure 20-34

Second solution using formulas is

Formula: $\theta = \text{invtan}\dfrac{X}{R}$

Substitute: $\theta = \text{invtan}\dfrac{38.2 \text{ kvars}}{45.3 \text{ kW}} = 40.1°$

Formula: $P = S\cos\theta$

$S = \dfrac{P}{\cos\theta}$

Substitute: $S = \dfrac{45.3 \text{ kW}}{\cos 40.1°} = 59.2 \text{ kVA}$

Example B An ac series circuit is connected across a voltage of $240\angle -68.2°$ V. The current is 10 A. Find the apparent power, the power factor, true power and the reactive power.

Solution: Construct diagrams, Figure 20-35. Note that the circuit is capacitive.

Data: $E = 240\text{ V}, I = 10\text{ A}, \theta = -68.2°$

Formula: $\mathbf{S} = \mathbf{EI}$

Substitute: $\mathbf{S} = (240\underline{/-68.2°}\text{ V})(10\underline{/0°}\text{ A})$

$\mathbf{S} = 2400\underline{/-68.2°}\text{ VA}$

P and Q can now be found by transformation.

$S\underline{/\theta} = P - jQ$

$2400\underline{/-68.2°} \rightarrow 891 - j2230$

Thus: $P = 891\text{ W}$

$Q = 2230\text{ vars (leading)}$

Note that expressing power in polar and rectangular form is the equivalent of listing S, P and Q separately.

$Pf = \cos\theta = 37.1\%$

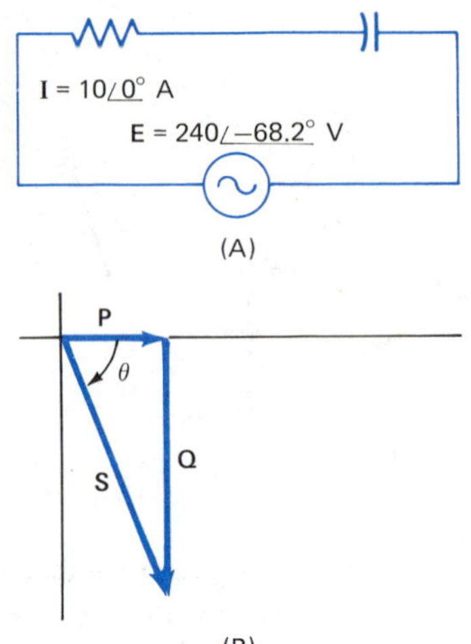

Figure 20-35

Second solution using formulas:

Formula: $P = S\cos\theta$

Substitute: $P = (2400\text{ VA})\cos(-68.2°) = 891\text{ W}$

Formula: $Q = S\sin\theta$

Substitute: $Q = (2400\text{ VA})\sin(-68.2°) = 2230\text{ vars (leading)}$

Example C A circuit has a resistance of 6 Ω in series with an inductive reactance of 8 Ω. It is connected to a 120-V supply line. Find (1) the circuit impedance, (2) the current, (3) the voltage drops, (4) the apparent, true and reactive power and (5) the power factor.

Solution: Construct circuit and phasor diagrams, Figure 20-36, and a solution table. Formulas in the first column indicate the calculations made:

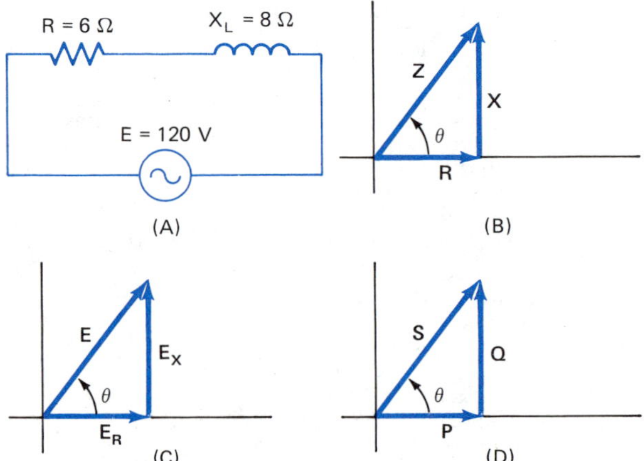

Figure 20-36 Notice the phase angle is indicated in the phasor diagrams of the impedance (Z), the voltage (E) and the apparent power (S).

	QUANTITY	POLAR		RECTANGULAR	UNITS
Data:	E	$120\underline{/53.1^\circ}$			V
	R	$6\underline{/0^\circ}$	→	6 + j0	Ω
	X_L	$8\underline{/90^\circ}$	→	0 + j8	Ω
(1)	Z (Sum)	$10\underline{/53.1^\circ}$	←	6 + j8	Ω
(2)	I = E/Z	$12\underline{/0^\circ}$			A
(3)	E_R = IR	$72\underline{/0^\circ}$			V
	E_X = IX	$96\underline{/90^\circ}$			V
(4)	P = E_RI	$864\underline{/0^\circ}$	→	864 + j0	W
	Q = E_XI	$1152\underline{/90^\circ}$	→	0 + j1152	vars
	S (Sum)	$1440\underline{/53.1^\circ}$	←	864 + j1152	VA, W, vars
Check:	S = EI	$1440\underline{/53.1^\circ}$			VA

(5) Pf = $\cos\theta = \cos 53.1^\circ = 0.6 = 60\%$

The apparent power can also be found using the formulas:

$$\mathbf{S} = \frac{\mathbf{E}^2}{\mathbf{Z}}$$

$$\mathbf{S} = \mathbf{I}^2\mathbf{Z}$$

$\mathbf{E_R}$ and $\mathbf{E_X}$ can also be found by transforming **E** in the first line of the table to rectangular form.

Example D Find the apparent power in both polar and rectangular form for a circuit having $625\underline{/40.7^\circ}$ Ω of impedance. The voltage is 120 V.

Solution: Draw diagrams, Figure 20-37. The voltage has the same phase angle as the impedance which is inductive.

Data: $\mathbf{Z} = 625\underline{/40.7^\circ}\ \Omega$, $\mathbf{E} = 120\underline{/40.7^\circ}$ V

Formula: $\mathbf{S} = \dfrac{\mathbf{E}^2}{\mathbf{Z}}$

Substitute: $\mathbf{S} = \dfrac{(120\underline{/40.7^\circ}\ \mathrm{V})^2}{625\underline{/40.7^\circ}\ \Omega}$

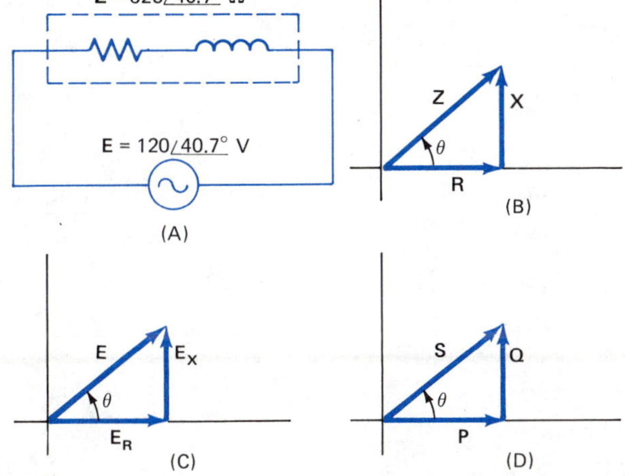

Figure 20-37

$$\mathbf{S} = \frac{(120\angle 40.7^\circ \text{ V})(120\angle 40.7^\circ \text{ V})}{625\angle 40.7^\circ \ \Omega}$$

$$\mathbf{S} = \frac{(120 \text{ V})^2}{625 \ \Omega}\angle 40.7^\circ + 40.7^\circ - 40.7^\circ \text{ VA}$$

$$\mathbf{S} = 23.0\angle 40.7^\circ \text{ VA}$$

Formula: $S\angle\theta = P + jQ$

Substitute: $23.0\angle 40.7^\circ \rightarrow 17.4 + j15.0$

$P = 17.4$ W

$Q = 15.0$ vars

EXERCISE 20-7

Find the apparent power in polar form for each of the following problems. Find the power factor for each.

1. $\mathbf{E} = 120\angle 22.3^\circ$ V, I = 12 A
2. $\mathbf{E} = 240\angle -36.2^\circ$ V, I = 8.6 A
3. $\mathbf{E} = 2400\angle -18.6^\circ$ V, $\mathbf{Z} = 2.8\angle -18.6^\circ \ \Omega$
4. $\mathbf{Z} = 8.20\angle 21.4^\circ \ \Omega$, E = 117 V
5. $\mathbf{Z} = 19.2\angle 37.3^\circ \ \Omega$, I = 12.5 A
6. $\mathbf{I} = 825\angle -57.6^\circ$ mA, Z = 29.1 Ω

Find the true power and the reactive power for each of the following problems. What is the power factor in percent?

7. $S\angle\theta = 45\angle 19.5^\circ$ kVA
8. $S\angle\theta = 7.00\angle -35.9^\circ$ VA
9. $S\angle\theta = 285\angle 68.6^\circ$ kVA
10. $S\angle\theta = 450\angle -73.7^\circ$ VA

Find the apparent power in polar form for the following problems and find the power factor.

11. P = 6500 W, Q = 8800 vars lagging
12. Q = 782 kvars leading, P = 1650 kW
13. Q = 428 kvars lagging, P = 428 kW
14. P = 925 W, Q = 225 vars leading

Data, diagram, formula, substitute, check. Construct solution tables.

15. A refrigerator motor draws 310 W. It draws 3.45 A (lagging) at 120 V. Find (a) the apparent power, (b) the power factor and (c) the reactive power.
16. An electric heater has a resistance of 12 Ω and an inductive reactance of 0.75 Ω. (a) Find the impedance. (b) What is the power factor? (c) If the current is 10 A, find the apparent power in polar and rectangular form. (d) What is the applied voltage?
17. A gasoline pump at a service station uses a motor at $120\angle 35.2^\circ$ V and draws 3.1 A. (a) What is the power factor in percent? (b) Find the apparent power in both polar and rectangular form.
18. A synchronous motor draws 4.5 kVA at a power factor of 72% leading. The impedance is 10,8 Ω. Find (a) the applied voltage and (b) the current. What is (c) the true power and (d) the reactive power?

19. An air-conditioner motor drawn 1.65 kW of true power. The reactive power is 4100 vars lagging. (a) Find the apparent power in polar form. If the line voltage is 230 V, find (b) the current and (c) the impedance.
20. A clock motor draws 20.8 mA of leading current. The resistance is 4.66 kΩ and the reactance is 3.41 kΩ. (a) Find the impedance in polar form. (b) What is the line voltage? (c) Find the power factor in percent. (d) Express the apparent power in polar and in rectangular form. (e) How much energy, in kW•h, does the motor use in one year?

20-10 POWER IN MULTIPLE-ELEMENT SERIES CIRCUITS

When several impedances are connected in series in ac circuits, an equivalent circuit is easily found. Just express each impedance in rectangular form ($R \pm jX$), then add. The resistance components and the reactance components are added separately. This reduces the problem to an RL or an RC circuit problem.

Example A Find (1) the impedance of the equivalent circuit, (2) the apparent power, true power and reactive power and (3) the power factor for a circuit containing a resistance of 65 Ω, a capacitive reactance of 48 Ω and an inductive reactance of 94 Ω. The current is 2.3 A.

Solution: Construct diagrams, Figure 20-38, and a solution table.

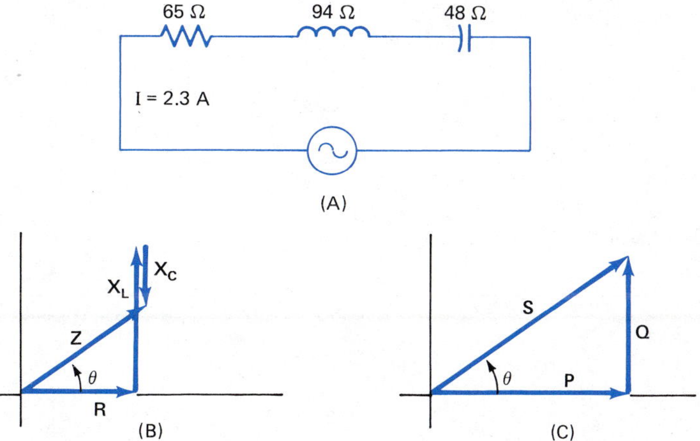

Figure 20-38

Data:

	QUANTITY	POLAR		RECTANGULAR	UNITS
	I	$2.3\angle 0^\circ$			A
	R	$65\angle 0^\circ$	→	65 + j0	Ω
	X_L	$94\angle 90^\circ$	→	0 + j94	Ω
	X_C	$48\angle -90^\circ$	→	0 − j48	Ω
(1)	Z (Sum)	$79.6\angle 35.3^\circ$	←	65 + j46	Ω
(2)	$S = I^2Z$	$421\angle 35.3^\circ$	→	344 + j243	VA, W, vars

The last line is found by multiplying:

$S = (2.3\angle 0^\circ)^2(79.6\angle 35.3^\circ) = 421\angle 35.3^\circ$

The final line yields the answers:

S = 421 VA, P = 344 W and Q = 243 vars lagging.

(3) Pf = cos θ = cos 35.3° = 81.6%

Example B Find (1) the impedance of the equivalent circuit, (2) the apparent power, true power and reactive power and (3) the power factor in the circuit of Figure 20-39.

Solution: Construct phasor diagrams, Figure 20-39 (B) and Figure 20-39 (C) and a solution table:

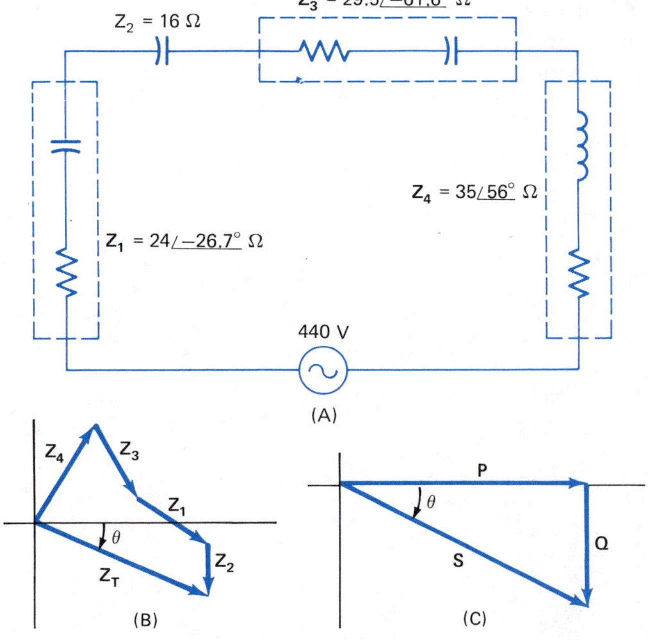

Figure 20-39

Data:

	QUANTITY	POLAR		RECTANGULAR	UNITS
	E	$440\angle{-26.7°}$			V
	Z_1	$24\angle{-35°}$	→	19.7 − j13.8	Ω
	Z_2	$16\angle{-90°}$	→	0 − j16	Ω
	Z_3			14 − j26	Ω
	Z_4	$35\angle{56°}$	→	19.6 + j29.0	Ω
(1)	Z_T (Sum)	$59.7\angle{-26.7°}$	←	53.3 − j26.8	Ω
(2)	$S = \frac{E^2}{Z_T}$	$3250\angle{-26.7°}$	→	2900 − j1460	VA, W vars

The last line is found by multiplying:

$$S = \frac{(440\angle{-26.7})^2}{59.7\angle{-26.7°}} = 3250\angle{-26.7°}$$

From the final line:

S = 3250 VA, P = 2900 W and Q = 1460 vars leading.

(3) Pf = $\cos\theta = \cos(-26.7°)$ = 89.3%

It is usually desirable to have the power factor as high as possible; 100% is ideal. A capacitor or inductor with the correct reactance for offsetting the circuit reactance is often added to the circuit. The process is called *power factor correction.* The reasons and the methods for making such corrections are discussed in the next chapter.

EXERCISE 20-8

Make a circuit diagram and appropriate phasor diagrams for each of the following problems. Construct a solution table and solve for each of the unknowns indicated by a question mark.

	E	Z	I	S	P	Q	cos θ
1.	120 V	60 Ω	?	?	?	?	65% lag
2.	?	1.2 kΩ	0.2 A	?	?	?	82% lead
3.	440 V	?	8 A	?	?	?	33% lead
4.	16 V	64 Ω	?	?	?	?	0% lag
5.	?	16 Ω	4 A	?	?	?	100% –
6.	?	?	16 A	4.4 kVA	?	?	78% lag
7.	1760 V	?	?	?	2.2 kW	?	50% lead
8.	220 V	?	?	?	800 W	500 vars	? lag

Data, diagram, formula, substitute, check. Use a solution table.

9. A circuit contains a resistance of 60 Ω, an inductive reactance of 86 Ω and a capacitive reactance of 94 Ω. It is connected across a 220-V line. Find (a) the total impedance, (b) the current, (c) the apparent, true and reactive powers and (d) the power factor. (e) Is the circuit inductive or capacitive? (f) Does current lead or lag?

10. A circuit with a current of 4.00 A contains a 12-Ω resistance, an 85-Ω inductive reactance and a 76-Ω capacitive reactance. Find (a) the total impedance, (b) the voltage, (c) the apparent, true and reactive powers and (d) the power factor (lag or lead?).
11. A 120-V, 50-Hz voltage is applied to a circuit with a 1.5-kΩ resistance, a 5.8-H coil and a 1.2-μF capacitor. Find (a) the total impedance, (b) the current, (c) the apparent, true and reactive powers and (d) the power factor (lag or lead?).
12. The following 60-Hz impedances are connected in series to a 220-V line: $\mathbf{Z}_1 = 55 + j68\ \Omega$, $\mathbf{Z}_2 = 47 - j23\ \Omega$ and $\mathbf{Z}_3 = 64\underline{/65^\circ}\ \Omega$. Find (a) the total impedance, (b) the current, (c) the apparent, true and reactive powers and (d) the power factor (lag or lead?).
13. A circuit contains: $\mathbf{Z}_1 = 75\underline{/25^\circ}\ \Omega$, $\mathbf{Z}_2 = 40\underline{/-58^\circ}$, $\mathbf{Z}_3 = 105\underline{/75^\circ}$ and $\mathbf{Z}_4 = 65\underline{/-38^\circ}$. The current is 9.89 A at 60 Hz. Find (a) the total impedance, (b) the applied voltage, (c) the apparent, true and reactive powers and (d) the power factor (lag or lead?).
14. Suppose a 103-Ω capacitive reactance is added in series to the circuit of Problem 12. Find (a) the new impedance, (b) the new current, (c) the new apparent, true and reactive powers and (d) the new power factor. Assume the magnitude of the applied voltage stays the same.
15. A 59.2-Ω capacitive reactance is added in series to the circuit of Problem 13. Assume the magnitude of the applied voltage remains the same. Find (a) the new impedance, (b) the new current, (c) the new apparent, true and reactive powers and (d) the new power factor. What is the percentage change of each quantity?

CHAPTER 21

ac PARALLEL CIRCUITS

OBJECTIVES

After satisfactorily completing this chapter, the student should be able to:

- Find the equivalent impedance of parallel circuits using the formula method for two-branch circuits and the admittance method for multiple-branch circuits.
- Determine voltage, current, impedance and power in parallel circuits using Ohm's Law and Kirchhoff's Current Law.
- Find the parallel equivalent when the series equivalent of a parallel circuit is known using the concepts of admittance, conductance and susceptance.
- Apply power factor correction methods to ac circuits.

Most electrical circuits are parallel circuits. In a home, for example, all electrical outlets are connected in parallel. The main characteristic of parallel circuits is that the applied voltage across each element is the same. Many home appliances, operate on 120 V and do not work well or do not work at all at other voltages.

This chapter begins by studying methods of finding the equivalent impedance of parallel circuits and continues with the relationships among voltage, current and power in each circuit branch. The chapter ends with a discussion about power factor correction.

EQUIVALENT CIRCUITS

The total current in a parallel circuit is the phasor sum of the currents in each branch. Branch currents can be found by dividing the applied voltage by the branch impedance. Another method is to find the equivalent impedance of the entire circuit. The applied voltage divided by the equivalent impedance is the total line current.

Methods for finding the equivalent impedance are similar to those for finding the equivalent resistance in DC circuits, Topic 9-3. In that topic, the familiar formula

$$R_T = \frac{R_1 R_2}{R_1 + R_2}$$

was used for finding the total resistance in a two-branch circuit. A similar formula is used for finding the total impedance of two-branch circuits. Such formulas, of course, relate phasors and phasor arithmetic must be used.

21-1 TWO-BRANCH IDEAL CIRCUITS

A two-branch ac circuit can contain one or more elements (resistor, inductor and/or capacitor) in each branch. For one-element branches, the various combinations can be any of those shown in Figure 21-1. The corresponding equivalent circuits are also shown.

Notice that if both branches contain like elements, Figure 21-1(A, B and C), the equivalent circuit contains a single element of the same kind. When the two branches contain unlike elements and one is a resistor, Figure 21-1(D and E), the equivalent circuit contains resistance and a like reactance in series. The final combination, Figure 21-1(F), has two unlike reactances in parallel. The equivalent will be a single element and may be a capacitor or an inductor.

All of the circuits of Figure 21-1 are ideal. Most real circuits will have some resistance and reactance in each branch. To find the equivalent impedance ($\mathbf{Z_T}$) of any two-branch circuits, use the formula

$$\mathbf{Z_T} = \frac{\mathbf{Z_1 Z_2}}{\mathbf{Z_1 + Z_2}}$$

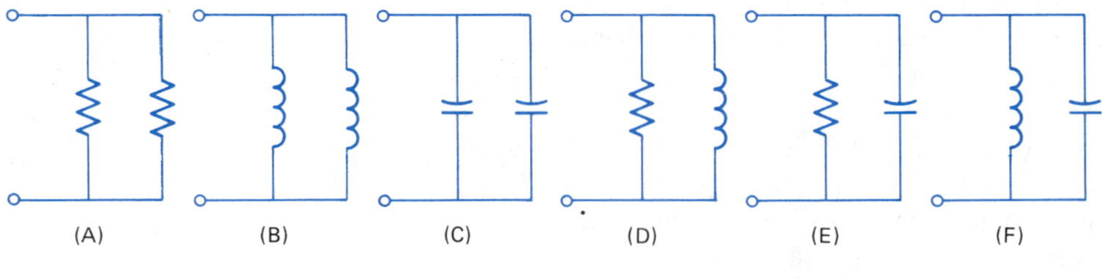

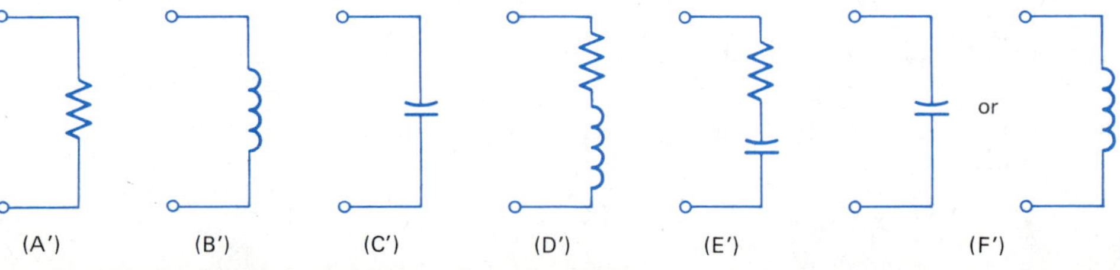

Figure 21-1 Two-branch ideal circuits: (A) Two resistors, (B) Two inductors, (C) Two capacitors, (D) Resistor-inductor, (E) Resistor-capacitor, (F) Inductor-capacitor. The equivalent series circuits are indicated as primes A', B', etc.

Notice the similarity between this formula and the one for equivalent resistance of DC circuits. The main difference is that impedances (ac) are phasor quantities while resistances (DC) are scalars.

The polar form of the impedance displays the phase angle. The phase angle indicates the amount by which the current leads or lags the voltage. This is true in each branch of the circuit and that of the main line with a phase according to the equivalent impedance the source voltage "sees."

The total impedance of a two-branch circuit is the phasor product divided by the phasor sum. This procedure may be called the *two-branch formula method.* The step-by-step procedure is as follows:

FINDING THE EQUIVALENT IMPEDANCE (FORMULA METHOD)

- Draw a circuit diagram.
- Express the given impedances in both polar and rectangular form.
- Add the two impedances in rectangular form and transform the sum to polar form.
- Divide the product of the two phasors by their sum in polar form. Transform the result to rectangular form.
- In rectangular form the circuit components of the equivalent series circuit is indicated: $\mathbf{Z_T} = R_T + jX_T$.

Example A Find the total impedance of a two-branch circuit containing only resistances. One branch has 8 Ω, the other has 12 Ω.

Solution: The impedances of the two branches are $8\angle 0^\circ\ \Omega$ and $12\angle 0^\circ\ \Omega$ respectively. In rectangular form these are $8 + j0\ \Omega$ and $12 + j0\ \Omega$. To divide the product by the sum, first find the polar form of the sum:

$$(8 + j0) + (12 + j0) = 20 + j0 \rightarrow 20\angle 0^\circ\ \Omega$$

Formula: $$\mathbf{Z_T} = \frac{\mathbf{Z_1 Z_2}}{\mathbf{Z_1 + Z_2}}$$

Substitute: $$\mathbf{Z_T} = \frac{(8\angle 0^\circ\ \Omega)(12\angle 0^\circ\ \Omega)}{20\angle 0^\circ\ \Omega}$$

$$\mathbf{Z_T} = \frac{8(12)}{20}\angle 0^\circ + 0^\circ - 0^\circ = 4.8\angle 0^\circ\ \Omega \rightarrow 4.8 + j0\ \Omega$$

Note: 0° indicates a pure resistance, Figure 21-1(A).

Example B A circuit contains 25 Ω and 15 Ω of inductive reactance in parallel. Find the equivalent impedance.

Solution: Find the sum:

$$(0 + j25) + (0 + j15) = 0 + j40 \rightarrow 40\angle 90^\circ\ \Omega$$

Formula: $$\mathbf{Z_T} = \frac{\mathbf{Z_1 Z_2}}{\mathbf{Z_1 + Z_2}}$$

Substitute: $$\mathbf{Z_T} = \frac{(25\angle 90^\circ\ \Omega)(15\angle 90^\circ\ \Omega)}{40\angle 90^\circ\ \Omega}$$

$$\mathbf{Z_T} = \frac{25(15)}{40}\angle 90^\circ + 90^\circ - 90^\circ\ \Omega$$

$$\mathbf{Z_T} = 9.38\angle 90^\circ\ \Omega \rightarrow 0 + j9.38\ \Omega$$

Note: 90° indicates a pure inductor, Figure 21-1(B).

Example C A pure inductance of 14 Ω reactance is in parallel with a pure capacitance of 16 Ω reactance. Find the equivalent impedance.

Solution: Draw the circuit diagram, Figure 21-2. Find the sum:

$(0 + j14) + (0 - j16) = 0 - j2 \rightarrow 2\underline{/-90^\circ}\ \Omega$

Formula: $\mathbf{Z_T} = \dfrac{\mathbf{Z_1 Z_2}}{\mathbf{Z_1 + Z_2}}$

Substitute: $\mathbf{Z_T} = \dfrac{(14\underline{/90^\circ}\ \Omega)(16\underline{/-90^\circ}\ \Omega)}{2\underline{/-90^\circ}\ \Omega}$

$\mathbf{Z_T} = \dfrac{(14)(16)}{2}\underline{/90^\circ - 90^\circ - (-90^\circ)}\ \Omega$

$\mathbf{Z_T} = 112\underline{/90^\circ}\ \Omega \rightarrow 0 + j112\ \Omega$

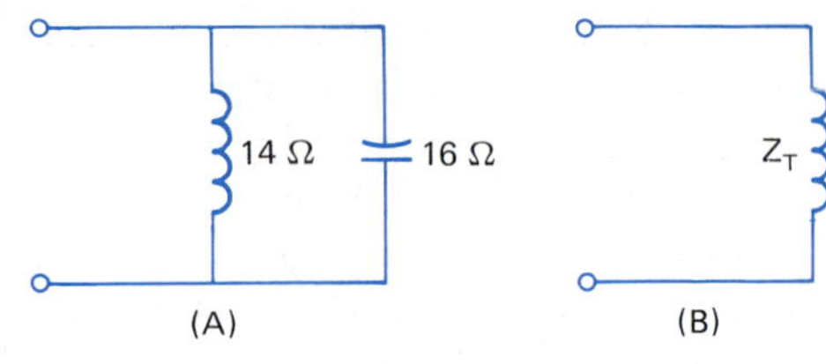

Figure 21-2

Note: The equivalent circuit, Figure 21-2(B), contains an inductance.

For the ideal circuit of Figure 21-1(F), a rule can now be stated. If the circuit contains more capacitive reactance than inductive reactance, the equivalent circuit is inductive. The circuit will be capacitive if the inductive reactance exceeds the capacitive reactance. The total impedance becomes very large when the inductive reactance is approximately equal to the capacitive reactance in the circuit.

When transformations are required that cannot be done mentally, it is probably best to set up the solution in a table.

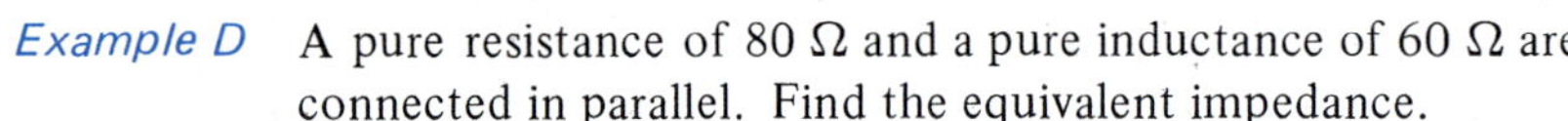

Example D A pure resistance of 80 Ω and a pure inductance of 60 Ω are connected in parallel. Find the equivalent impedance.

Solution: Construct a circuit diagram, Figure 21-3, and a solution table:

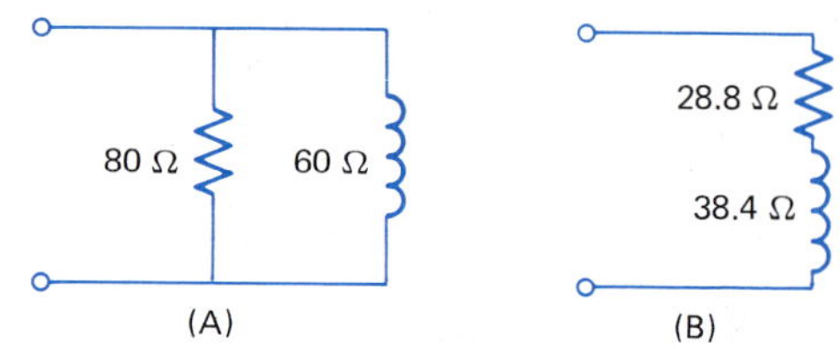

Figure 21-3

	QUANTITY	POLAR		RECTANGULAR	UNITS
Data:	$\mathbf{Z_1}$	$80\underline{/0^\circ}$	→	$80 + j0$	Ω
	$\mathbf{Z_2}$	$60\underline{/90^\circ}$	→	$0 + j60$	Ω
Sum:	$\mathbf{Z_1 + Z_2}$	$100\underline{/36.9^\circ}$	←	$80 + j60$	Ω
	$\mathbf{Z_T}$	$48\underline{/53.1^\circ}$	→	$28.8 + j38.4$	Ω

Note: The given impedances are listed first. The sum (third line) is found by adding in rectangular form. The total impedance (fourth line) is found by multiplying the two given impedances and dividing by the sum, all in polar form.

$\mathbf{Z_T} = \dfrac{\mathbf{Z_1 Z_2}}{\mathbf{Z_1 + Z_2}}$

$\mathbf{Z_T} = \dfrac{(80\underline{/0^\circ}\ \Omega)(60\underline{/90^\circ}\ \Omega)}{100\underline{/36.9^\circ}\ \Omega}$

$\mathbf{Z_T} = \dfrac{80 \times 60}{100}\underline{/0^\circ + 90^\circ - 36.9^\circ}\ \Omega$

$\mathbf{Z_T} = 48\underline{/53.1^\circ}\ \Omega \rightarrow 28.8 + j38.4\ \Omega$

The equivalent circuit, Figure 21-3(B), has 28.8 Ω of resistance and 38.4 Ω of inductance. Both the circuit and its equivalent contain the same kind of reactance. All calculations can be performed with a calculator.

EXERCISE 21-1

Find the equivalent circuit for each of the following two-branch circuits. Each branch contains single-circuit elements as indicated. Draw the circuit diagram and its series equivalent.

1. $R_1 = 28.3\ \Omega$, $R_2 = 44.6\ \Omega$
2. $X_{L1} = 125\ \Omega$, $X_{L2} = 400\ \Omega$
3. $X_{C1} = 93.5\ \Omega$, $X_{C2} = 61.9\ \Omega$
4. $R_1 = 12.6\ \Omega$, $R_2 = 15.3\ \Omega$
5. $L_1 = 320$ mH, $L_2 = 465$ mH, f = 60 Hz
6. $C_1 = 3.5\ \mu F$, $C_2 = 1.7\ \mu F$, f = 60 Hz
7. $X_L = 40\ \Omega$, $X_C = 160\ \Omega$
8. $X_L = 40\ \Omega$, $X_C = 60\ \Omega$
9. $X_L = 40\ \Omega$, $X_C = 50\ \Omega$
10. $X_L = 40\ \Omega$, $X_C = 42\ \Omega$
11. $X_L = 40\ \Omega$, $X_C = 41\ \Omega$
12. $X_L = 45\ \Omega$, $X_C = 40\ \Omega$
13. $X_L = 80\ \Omega$, $X_C = 40\ \Omega$
14. $X_L = 256\ \Omega$, $X_C = 185\ \Omega$
15. L = 715 mH, C = 5.25 μF, f = 60 Hz

Find the equivalent impedance for each of the following two-branch circuits by constructing a solution table. Draw the circuit diagram and its equivalent.

16. R = 65 Ω, $X_L = 80\ \Omega$
17. R = 175 Ω, $X_C = 150\ \Omega$
18. R = 44.3 Ω, $X_C = 31.7\ \Omega$
19. R = 16.5 Ω, $X_L = 12.2\ \Omega$
20. R = 256 Ω, L = 450 mH, f = 60 Hz
21. R = 95.2 Ω, C = 6.75 μF, f = 60 Hz
22. R = 42.5 Ω, C = 7.35 μF, f = 400 Hz

21-2 TWO-BRANCH PRACTICAL CIRCUITS

Real electrical circuits may contain both resistance and reactance in each branch. The equivalent circuit will contain a resistance and a reactance in series. For that reason, the equivalent circuit is called the *series equivalent* of the parallel circuit. The series equivalent is found by the same method used in the previous topic.

Example A Find the series equivalent of a two-branch parallel circuit containing impedances of $456\angle 25.3^\circ$ Ω and 225 – j275 Ω.

Solution: Construct a circuit diagram, Figure 21-4, and a solution table.

QUANTITY	POLAR		RECTANGULAR	UNITS
Z_1	$456\angle 25.3^\circ$	→	412 + j195	Ω
Z_2	$355\angle -50.7^\circ$	←	225 – j275	Ω
$Z_1 + Z_2$	$642\angle -7.2^\circ$	←	637 – j80	Ω
$Z_T = \dfrac{Z_1 Z_2}{Z_1 + Z_2}$	$252\angle -18.2^\circ$	→	239 – j78.9	Ω

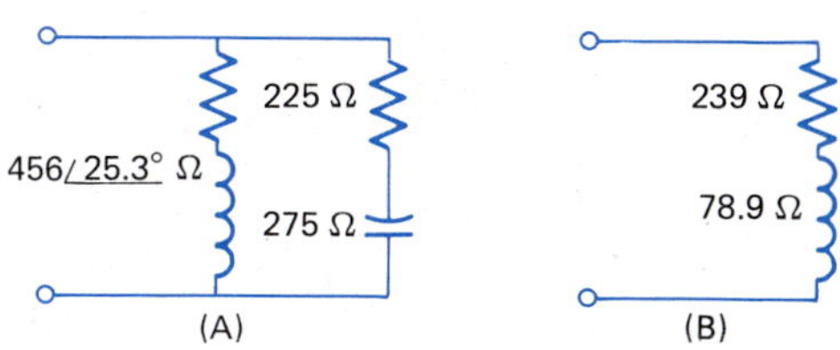

Figure 21-4

Note: Find the sum of Z_1 and Z_2 in rectangular form. Divide the product by the sum in polar form.

$$Z_T = \frac{Z_1 Z_2}{Z_1 + Z_2}$$

$$Z_T = \frac{(456)(355)}{642} \underline{/25.3^\circ + (-50.7^\circ) - (-7.2^\circ)}\ \Omega$$

$$Z_T = 252\underline{/-18.2^\circ}\ \Omega$$

Transform the result to rectangular form to find the components of the series equivalent, Figure 21-4(B).

Calculator note: After transforming $\mathbf{Z_2}$ in the previous example to polar form, some calculators allow the entire calculation to be performed without clearing. See the Owner's Manual. Follow this procedure:

1. Enter $\mathbf{Z_1}$ and transform to rectangular form.
2. Add both components of $\mathbf{Z_1}$ to those of $\mathbf{Z_2}$.
3. Transform the sum to polar form.
4. Subtract the angle from the sum of the angles of $\mathbf{Z_1}$ and $\mathbf{Z_2}$.
5. Divide the product $\mathbf{Z_1 Z_2}$ by the magnitude.
6. Transform the results ($\mathbf{Z_T}$) to rectangular form.

On some occasions, the total impedance may be known. If the impedance of one branch is known, the other can be calculated. The formula for $\mathbf{Z_T}$ must be rearranged to solve for the unknown. This is done according to the rules of algebra, Topic 12-4.

If $\mathbf{Z_2}$ is the unknown, the formula is solved as follows:

$$Z_T = \frac{Z_1 Z_2}{Z_1 + Z_2}$$

$$Z_T(Z_1 + Z_2) = Z_1 Z_2 \qquad \text{(Remove denominator) [MA, } \times (Z_1 + Z_2)]$$

$$Z_T Z_1 + Z_T Z_2 = Z_1 Z_2 \qquad \text{(Remove parentheses)}$$

$$Z_T Z_1 = Z_1 Z_2 - Z_T Z_2 \qquad \text{(Gather like terms) (SA, } - Z_T Z_2)$$

$$Z_T Z_1 = Z_2(Z_1 - Z_T) \qquad \text{(Factor)}$$

$$Z_2 = \frac{Z_T Z_1}{Z_1 - Z_T} \qquad [\text{DA}, \div (Z_1 - Z_T)]$$

The unknown ($\mathbf{Z_2}$) is found by dividing the product ($\mathbf{Z_T Z_1}$) by the difference ($\mathbf{Z_1 - Z_T}$).

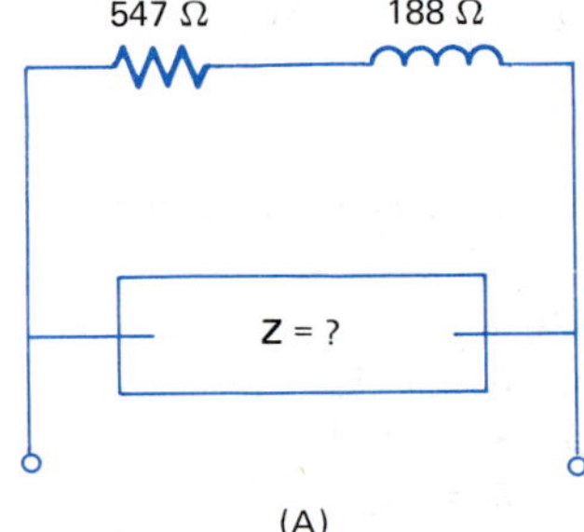

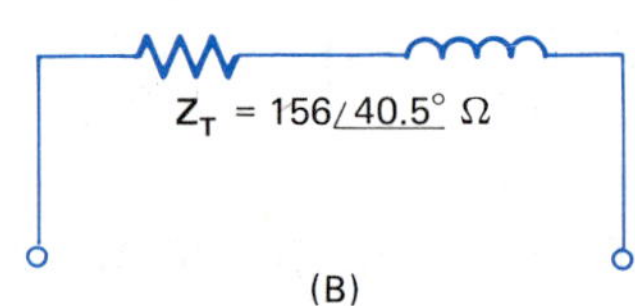

Figure 21-5

Example B The total impedance of a two-branch circuit, Figure 21-5, is $156\underline{/40.5^\circ}\ \Omega$. If the upper branch has an impedance of $547 + j188\ \Omega$, find the impedance of the lower branch.

Solution table:

QUANTITY	POLAR		RECTANGULAR	UNITS
Z_1	$578\underline{/19.0^\circ}$	←	$547 + j188$	Ω
Z_T	$156\underline{/40.5^\circ}$	→	$119 + j101$	Ω
$Z_1 - Z_T$	$437\underline{/11.4^\circ}$	←	$428 + j\ 87$	Ω
$Z_2 = \frac{Z_1 Z_T}{Z_1 - Z_T}$	$206\underline{/48.1^\circ}$	→	$138 + j153$	Ω

Note: Find the difference of $\mathbf{Z_1}$ and $\mathbf{Z_T}$ by subtracting in rectangular form. Divide the product in polar form by the difference.

$$Z_2 = \frac{Z_1 Z_T}{Z_1 - Z_T} = \frac{(578)(156)}{437} \underline{/19.0^\circ + 40.5^\circ - 11.4^\circ}\ \Omega$$

$$Z_2 = 206\underline{/48.1^\circ} \rightarrow 138 + j153\ \Omega$$

The unknown impedance consists of a resistance of 138 Ω an impedance of 153 Ω.

EXERCISE 21-2

Construct parallel circuit and series equivalent diagrams and find the series equivalent impedance (resistance and reactance) in each of the following problems. Use a solution table for each.

1. $\mathbf{Z}_1 = 845\underline{/90^\circ}\ \Omega$, $\mathbf{Z}_2 = 425 - j575\ \Omega$
2. $\mathbf{Z}_1 = 38.4\underline{/-90^\circ}\ \Omega$, $\mathbf{Z}_2 = 66.2\underline{/65.3^\circ}\ \Omega$
3. $\mathbf{Z}_1 = 39.7\underline{/0^\circ}\ \Omega$, $\mathbf{Z}_2 = 65.2 - j72.4\ \Omega$
4. $\mathbf{Z}_1 = 5.17 + j10.3\ \Omega$, $\mathbf{Z}_2 = 14.3\underline{/-31.4^\circ}\ \Omega$
5. $\mathbf{Z}_1 = 175\underline{/71.1^\circ}\ \text{k}\Omega$, $\mathbf{Z}_2 = 250 + j225\ \text{k}\Omega$
6. $\mathbf{Z}_1 = 39.7 - j86.2\ \Omega$, $\mathbf{Z}_2 = 71.4 - j49.7\ \Omega$
7. $\mathbf{Z}_1 = 87.3 + j56.2\ \text{k}\Omega$, $\mathbf{Z}_2 = 38.4\underline{/-17.5^\circ}\ \text{k}\Omega$
8. $\mathbf{Z}_1 = 52.9\underline{/-68.2^\circ}\ \text{k}\Omega$, $\mathbf{Z}_2 = 96.6\underline{/61.9^\circ}\ \Omega$

Find the unknown branch impedance ($\mathbf{Z}_2$).

9. $\mathbf{Z}_T = 24.4\underline{/23.3^\circ}\ \Omega$, $\mathbf{Z}_1 = 54.7 - j35.8\ \Omega$
10. $\mathbf{Z}_T = 22.5 + j11.1\ \Omega$, $\mathbf{Z}_1 = 36.8 - j9.4\ \Omega$

21-3 THE ADMITTANCE METHOD

In DC circuits, the total resistance was found by adding reciprocals, Topic 9-3:

$$\frac{1}{R_T} = \frac{1}{R_1} + \frac{1}{R_2} + \frac{1}{R_3} + \ldots$$

Such reciprocals are called conductance (G) and are measured in units called siemens (S). The equivalent equation in terms of conductance is as follows:

$$G_T = G_1 + G_2 + G_3 + \ldots$$

The total resistance can be found after adding to obtain G_T:

$$R_T = \frac{1}{G_T}$$

This method was used for any number of resistances connected in parallel.

With ac circuits, the procedure is the same. An ac circuit, however, may contain both resistance and reactance in each branch. This is indicated by the impedance, a phasor quantity. Thus, it is the reciprocals of impedances that must be added:

$$\frac{1}{\mathbf{Z}_T} = \frac{1}{\mathbf{Z}_1} + \frac{1}{\mathbf{Z}_2} + \frac{1}{\mathbf{Z}_3} + \ldots$$

The reciprocal of impedance is called *admittance.* It takes the symbol (**Y**) and like conductance, is measured in units of siemens (S). The admittance of a circuit branch is an indication of the ability of the branch to conduct or carry an electric current.

The reciprocal of a phasor, Topic 20-4, is found by dividing in polar form. Find the reciprocal of the magnitude and change the sign of the angle.

Example A Find the admittance of (1) $\mathbf{Z}_1 = 98.4\underline{/36.2^\circ}\ \Omega$ and (2) $\mathbf{Z}_2 = 66.2\underline{/-54.6^\circ}\ \Omega$.

Solution: Formula: $\mathbf{Y} = \dfrac{1}{\mathbf{Z}}$

(1) Substitute: $\mathbf{Y}_1 = \dfrac{1\underline{/0^\circ}}{98.4\underline{/36.2^\circ}\ \Omega}$

$$\mathbf{Y}_1 = \frac{1}{98.4}\underline{/0^\circ - (36.2^\circ)} = 0.0102\underline{/-36.2^\circ}\ \text{S}$$

(2) Substitute: $\mathbf{Y}_2 = \dfrac{1\underline{/0^\circ}}{66.2\underline{/-54.6^\circ}}$

$$\mathbf{Y}_2 = \frac{1}{66.2}\underline{/0^\circ - (-54.6^\circ)} = 0.0151\underline{/54.6^\circ}\ \text{S}$$

In terms of admittance,

$$\mathbf{Y}_\mathrm{T} = \mathbf{Y}_1 + \mathbf{Y}_2 + \mathbf{Y}_3 + \ldots$$

After adding the parallel admittances (in rectangular form), the total or equivalent impedance is calculated with the formula:

$$\mathbf{Z}_\mathrm{T} = \frac{1}{\mathbf{Y}_\mathrm{T}}$$

after transforming to polar form. This method of finding the total impedance is called the *admittance method.*

The procedure is given in the following rule:

FINDING THE EQUIVALENT SERIES CIRCUIT (ADMITTANCE METHOD)

- Draw a circuit diagram.
- Construct a solution table since transformation of coordinates are involved.
- Carry at least three significant figures at every step of the solution.
- Express all branch impedances in polar form.
- Find the admittance of each branch ($\mathbf{Y} = 1/\mathbf{Z}$) and transform to rectangular form.
- Add the components of admittance to obtain the total admittance ($\mathbf{Z}_\mathrm{T}$) in rectangular form. Transform to polar form.
- Find the total impedance ($\mathbf{Z}_\mathrm{T} = 1/\mathbf{Y}_\mathrm{T}$) and transform to rectangular form to find the resistance and reactance ($\mathrm{R} \pm \mathrm{jX}$) of the series equivalent circuit. Draw the equivalent circuit diagram.

Example B Find the equivalent circuit for the following impedances in parallel: $\mathbf{Z}_1 = 0 - \mathrm{j}450\ \Omega$, $\mathbf{Z}_2 = 500 + \mathrm{j}300\ \Omega$.

Solution: Draw the circuit, Figure 21-6, and its equivalent. Construct a solution table.

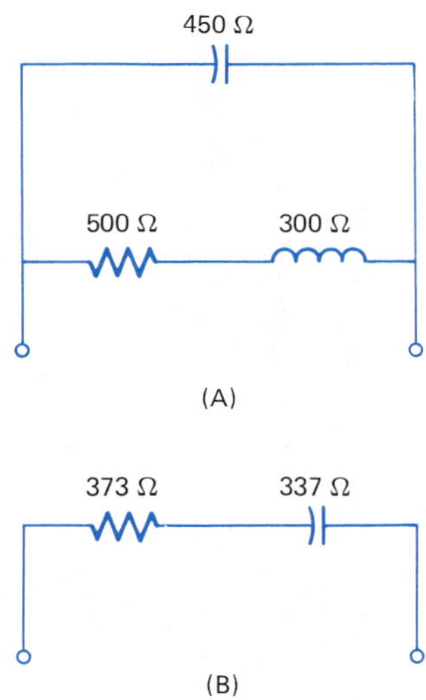

Figure 21-6

Data:

QUANTITY	POLAR		RECTANGULAR	UNITS
$\mathbf{Z}_1$	$450\underline{/-90^\circ}$	←	0 − j450	Ω
$\mathbf{Z}_2$	$583\underline{/31.0^\circ}$	←	500 + j300	Ω
$\mathbf{Y}_1 = 1/\mathbf{Z}_1$	$2.22\underline{/90^\circ}$	→	0 + j2.22	mS
$\mathbf{Y}_2 = 1/\mathbf{Z}_2$	$1.72\underline{/-31.0^\circ}$	→	1.47 − j0.89	mS
$\mathbf{Y}_T$ = (Sum)	$1.99\underline{/42.1^\circ}$	←	1.47 + j1.33	mS
$\mathbf{Z}_T = 1/\mathbf{Y}_T$	$503\underline{/-42.1^\circ}$	→	373 − j337	Ω

Note: $\mathbf{Y}_1$ and $\mathbf{Y}_2$ in the third and fourth lines are reciprocals of $\mathbf{Z}_1$ and $\mathbf{Z}_2$ from the first two lines.

$$\mathbf{Y}_1 = \frac{1}{450\underline{/-90^\circ}} = 0.002\ 22\underline{/90^\circ}\ \text{S}$$

$$\mathbf{Y}_2 = \frac{1}{583\underline{/31.0^\circ}} = 0.001\ 72\underline{/-31.0^\circ}\ \text{S}$$

$\mathbf{Y}_T$ (fifth line) is found by adding in rectangular form. All admittances are expressed in millisiemens in the solution table to save space.

$\mathbf{Z}_T$ is the reciprocal of $\mathbf{Y}_T$:

$$\mathbf{Z}_T = \frac{1}{0.001\ 99\underline{/42.1^\circ}} = 503\underline{/-42.1^\circ}\ \Omega$$

Calculator notes: After $\mathbf{Y}_T$ is found by adding $\mathbf{Y}_1$ and $\mathbf{Y}_2$, transform to polar form: [INV] [P→R]. Use [+/−] to change the sign of the angle. Press [x⇄y] to display the magnitude, press [1/x] to obtain the reciprocal. This yields the impedance in polar form and the rectangular components are found by using [P→R].

For the previous example, assume $\mathbf{Y}_T$ has already been found, (rectangular form) and is entered in the calculator. The procedure is:

[INV] [P→R], 42.1° [+/−], −42.1° (θ)
[x⇄y], 0.001 99 S [1/x], 503 Ω (Z)
[x⇄y] [P→R], −337 Ω (X)
[x⇄y], 373 Ω (R)

The admittance method can be used to find the total impedance of more than two impedances in parallel. The procedure is the same as that of the last example.

Example C Find the equivalent series circuit for the parallel circuit of Figure 21-7(A).

Solution: Label the impedances as $\mathbf{Z}_1$, $\mathbf{Z}_2$, etc. Use a solution table.

Data:

QUANTITY	POLAR		RECTANGULAR	UNITS
$\mathbf{Z}_1$	$25\underline{/53.1^\circ}$			Ω
$\mathbf{Z}_2$	$36.4\underline{/-74.1^\circ}$	←	10 − j35	Ω
$\mathbf{Z}_3$	$15.6\underline{/39.8^\circ}$	←	12 + j10	Ω
$\mathbf{Z}_4$	$20\underline{/-40^\circ}$			Ω

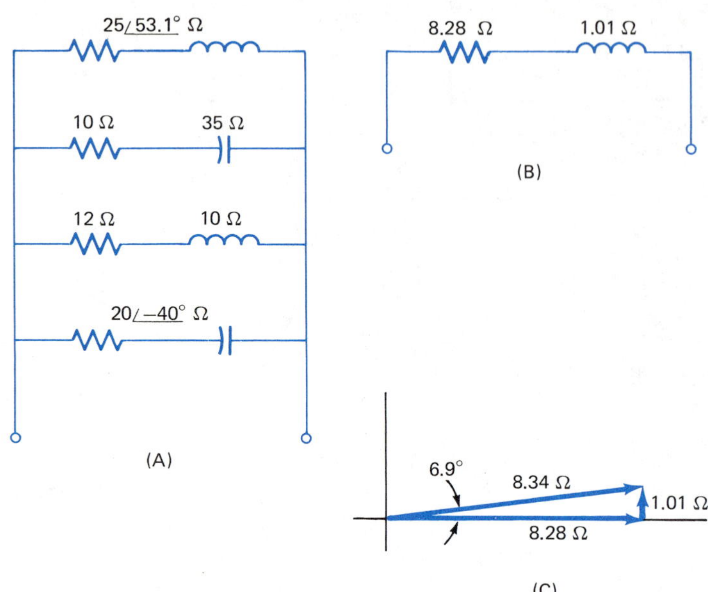

Figure 21-7

QUANTITY	POLAR		RECTANGULAR	UNITS
$\mathbf{Y}_1 = 1/\mathbf{Z}_1$	$0.0400\angle -53.1°$	→	$0.0240 - j0.0320$	S
$\mathbf{Y}_2 = 1/\mathbf{Z}_2$	$0.0275\angle 74.1°$	→	$0.0075 + j0.0264$	S
$\mathbf{Y}_3 = 1/\mathbf{Z}_3$	$0.0640\angle -39.8°$	→	$0.0492 - j0.0410$	S
$\mathbf{Y}_4 = 1/\mathbf{Z}_4$	$0.0500\angle 40°$	→	$0.0383 + j0.0321$	S
$\mathbf{Y}_T$ (Sum)	$0.1199\angle -6.9°$	←	$0.1190 - j0.0145$	S
$\mathbf{Z}_T = 1/\mathbf{Y}_T$	$8.34\angle 6.9°$	→	$8.28 + j1.01$	Ω

The total impedance is given in the last line and indicates the equivalent series circuit shown in Figure 21-7(B). The circuit is inductive with a phase angle of 6.9° lagging, Figure 21-7(C). Notice that $\mathbf{Z}_1$ and $\mathbf{Z}_4$ were given in polar form and do not need to be transformed.

When impedances have large magnitudes, the admittance magnitudes are small fractions. For very small fractions, the polar and rectangular forms of admittance take up a lot of space. This is even true when scientific notation is used. To avoid small fractions, use metric prefixes. If millisiemens (mS) had been used in the previous example, all the decimal fractions could have been avoided. For example, $\mathbf{Y}_1$ would have become $40.0\angle -53.1°$ mS → 24.0 − j32.0 mS. Always carry three significant digits at every step in the solution. Round final answers according to standard procedures.

EXERCISE 21-3

Construct circuit and series equivalent diagrams and find the series equivalent impedance (resistance and reactance) for each of the following parallel circuit problems. Construct a solution table for each using the admittance method.

1. $\mathbf{Z}_1 = 425 + j575\ \Omega$, $\mathbf{Z}_2 = 845\underline{/-90^\circ}\ \Omega$
2. $\mathbf{Z}_1 = j590\ \Omega$, $\mathbf{Z}_2 = 640\underline{/-38.5^\circ}\ \Omega$
3. $\mathbf{Z}_1 = 52.9\underline{/-68.2^\circ}\ k\Omega$, $\mathbf{Z}_2 = 45.5 + j85.2\ k\Omega$
4. $\mathbf{Z}_1 = 3.29 - j5.62\ \Omega$, $\mathbf{Z}_2 = 4.78 + j2.94\ \Omega$
5. $\mathbf{Z}_1 = 5.32\underline{/-29.2^\circ}\ \Omega$, $\mathbf{Z}_2 = 7.78\underline{/79.6^\circ}\ \Omega$
6. $\mathbf{Z}_1 = 39.7 - j86.2\ \Omega$, $\mathbf{Z}_2 = 71.4 + j49.7\ \Omega$

Find the series equivalent for each of the following parallel circuits using the admittance method.

7.

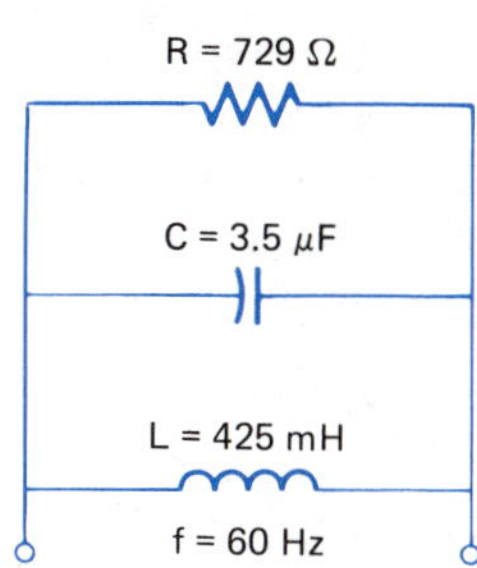

8.

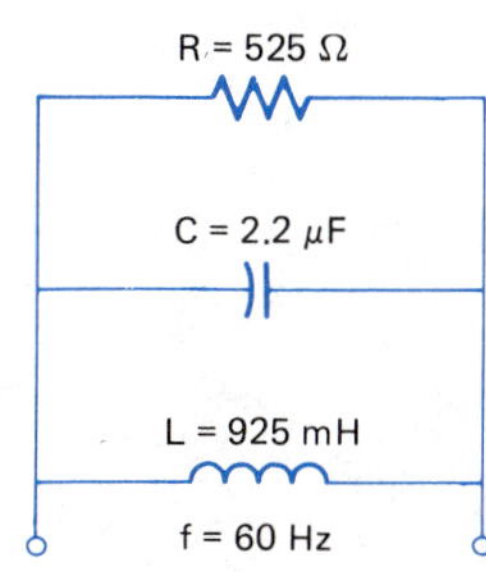

9.

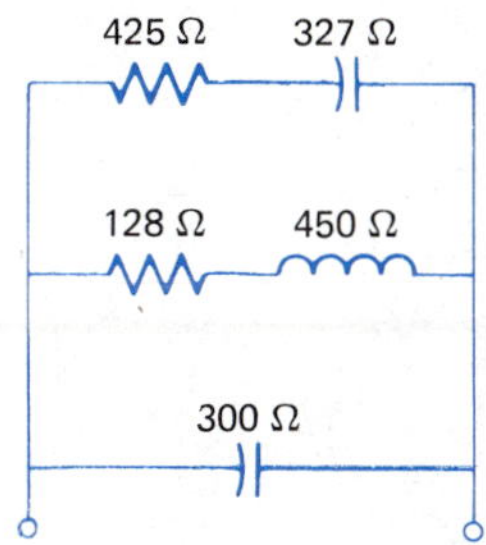

10.

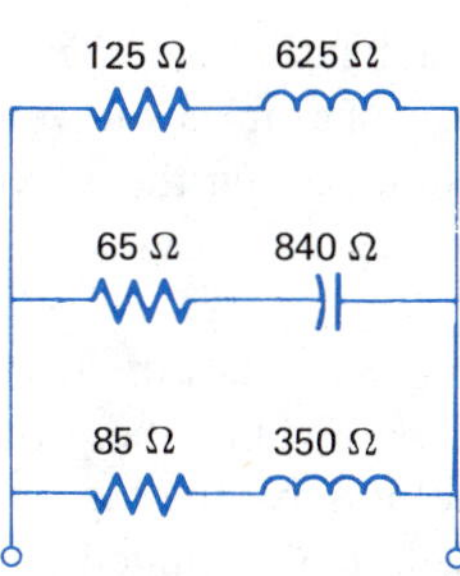

11.

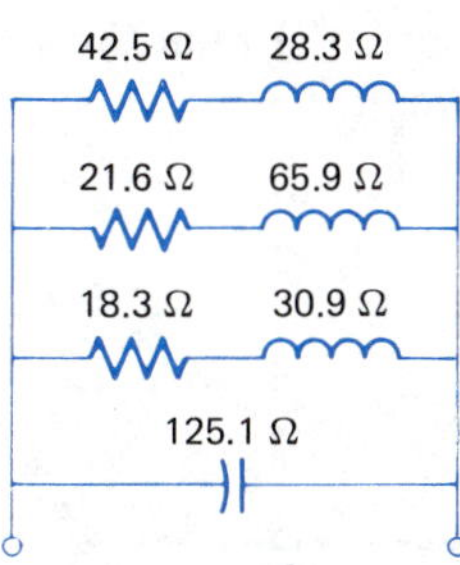

12.

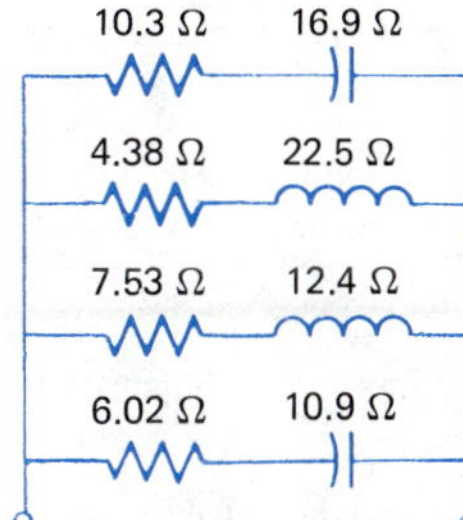

VOLTAGE, CURRENT AND POWER IN PARALLEL CIRCUITS

In the series circuits described in the previous chapter, current is the same in all circuit elements. In a parallel circuit, however, it is the voltage that is the same across each branch. With series circuits, current was taken as the reference phasor. It makes more sense with parallel circuits, though, to use the voltage phasor as the reference.

Voltage, current and impedance are phasor quantities related by Ohm's Law. Ohm's Law applies to each circuit branch and to the overall circuit indicated by the series equivalent. Power (apparent, true and reactive) is calculated using the series equivalent.

21-4 TWO-BRANCH IDEAL CIRCUITS

In a two-branch pure resistive circuit, the phase angle is 0°. The voltage and current are "in phase." Using the voltage phasor as the reference, current phasors are pointed in the same direction along the x-axis. The total current is the simple sum. Branch currents and total current are found by using Ohm's Law (**I** = **EZ**).

There is no reactive power. Apparent power (S) equals the true power (P). The equivalent circuit impedance is found by dividing the product by the sum. In the following examples, the method of finding the total impedance (series equivalent) will not be shown. It is included as part of the data.

Example A (1) Find the current in each branch of the circuit of Figure 21-8. (2) What is the total current? (3) Find the apparent, true and reactive power for the circuit.

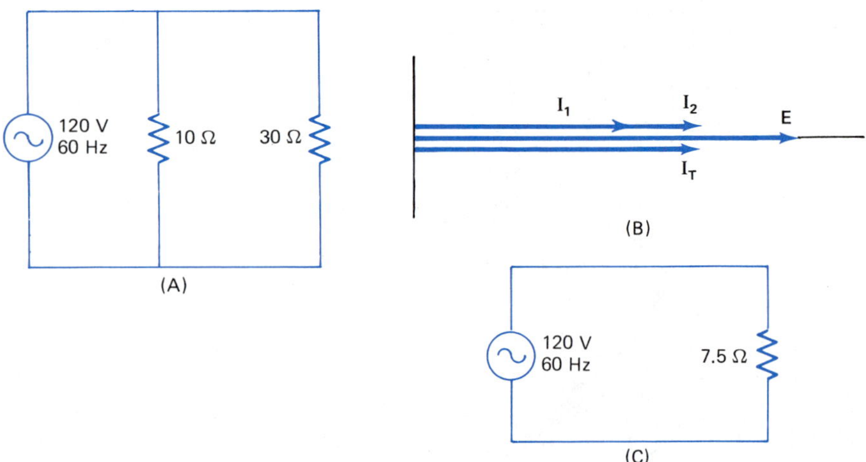

Figure 21-8 A two-branch pure resistive circuit

Solution: Draw the equivalent circuit and phasor diagrams, Figure 21-8, and construct a solution table.

	QUANTITY	POLAR		RECTANGULAR	UNITS
Data:	$\mathbf{E}$	$120\angle 0^\circ$			V
	$\mathbf{Z}_1$	$10\angle 0^\circ$			Ω
	$\mathbf{Z}_2$	$30\angle 0^\circ$			Ω
	$\mathbf{Z}_T$	$7.5\angle 0^\circ$			Ω
(1)	$\mathbf{I}_1 = \mathbf{E}/\mathbf{Z}_1$	$12\angle 0^\circ$	→	$12 + j0$	A
	$\mathbf{I}_2 = \mathbf{E}/\mathbf{Z}_2$	$4\angle 0^\circ$	→	$4 + j0$	A
(2)	$\mathbf{I}_T$ = (Sum)	$16\angle 0^\circ$	←	$16 + j0$	A
Check:	$\mathbf{I}_T = \mathbf{E}/\mathbf{Z}_T$	$16\angle 0^\circ$			A
(3)	$S = P = EI_T$	1920			VA, W
	$Q = S \sin 0^\circ$	0			vars

Note: Currents are found using Ohm's Law in polar form.

$$\mathbf{I}_1 = \frac{\mathbf{E}}{\mathbf{Z}_1} = \frac{120\angle 0^\circ \text{ V}}{10\angle 0^\circ \ \Omega} = \frac{120}{10}\angle 0^\circ - 0^\circ = 12\angle 0^\circ \text{ A}$$

$$\mathbf{I}_2 = \frac{\mathbf{E}}{\mathbf{Z}_2} = \frac{120\angle 0^\circ \text{ V}}{30\angle 0^\circ \ \Omega} = 4\angle 0^\circ \text{A}$$

$$\mathbf{I}_T = \frac{\mathbf{E}}{\mathbf{Z}_T} = \frac{120\angle 0^\circ}{7.5\angle 0^\circ} = 16\angle 0^\circ \text{ A}$$

In pure reactive circuits (two-branch), the current leads or lags the voltage by 90°. The three possible circuits and their corresponding phasor diagrams are shown in Figure 21-9. For the two-inductor circuit Figure

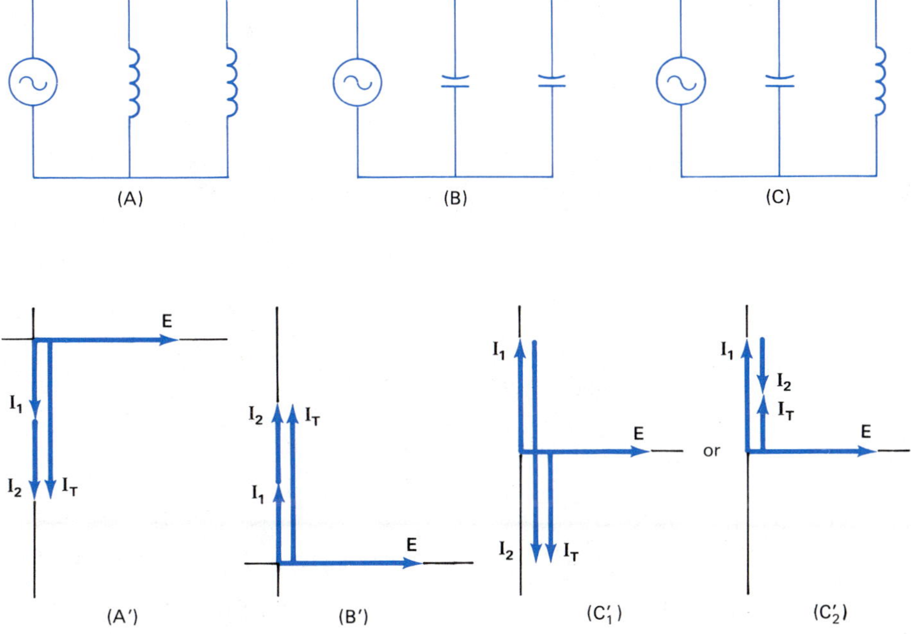

Figure 21-9 Current in ideal parallel circuits

21-9(A), current lags the reference voltage by 90°. Current leads the voltage in the two-capacitor circuit Figure 21-9(B). For the inductor-capacitor circuit, Figure 21-9(C), current may lead or lag depending on which kind of reactance is greater. Power in such circuits is pure reactive: $Q = S = EI_T$.

Example B (1) Find the current in each branch of the circuit in Figure 21-10. (2) Determine the total current and (3) find the values of P, Q and S (power) for the circuit.

Solution: Since inductive reactance is larger than the capacitive reactance, the circuit will be capacitive. Construct the series equivalent and phasor diagrams and show the solution in table form.

	QUANTITY	POLAR		RECTANGULAR	UNITS
Data:	E	$120\angle 0°$			V
	Z_1	$40\angle -90°$			Ω
	Z_2	$60\angle 90°$			Ω
	Z_T	$120\angle -90°$			Ω
(1)	$I_1 = E/Z_1$	$3\angle 90°$	→	$0 + j3$	A
	$I_2 = E/Z_2$	$2\angle -90°$	→	$0 - j2$	A
(2)	$I_T = (\text{Sum})$	$1\angle 90°$	←	$0 + j1$	A
Check:	$I_T = E/Z_T$	$1\angle 90°$			A
(3)	$P = EI_T \cos\theta$	0		0	W
	$S = Q = EI_T$	120			VA, vars

Note:

$$I_1 = \frac{E}{Z_1} = \frac{120\angle 0°\ \text{V}}{40\angle -90°\ \Omega} = 3\angle 90°\ \text{A}$$

$$I_2 = \frac{E}{Z_2} = \frac{120\angle 0°\ \text{V}}{60\angle 90°\ \Omega} = 2\angle -90°\ \text{A}$$

$$I_T = \frac{E}{Z_T} = \frac{120\angle 0°\ \text{V}}{120\angle -90°\ \Omega} = 1\angle 90°\ \text{A}$$

The total current in the feeder lines is less than the branch currents. Such a result happens because electrons flow back and forth between the inductor and the capacitor. True power (3) is zero because cos 90° = 0.

Take special note of the fact that current, like admittance, takes a positive phase angle in parallel capacitive circuits (leading). Both have negative phase angles in inductive circuits (lagging). The current calculations in the note following the solution table of the previous example make the reasons obvious. Current and impedance angles always take opposite signs when voltage is the reference phasor.

In the previous example, the total current is found by adding branch currents in rectangular form:

$$I_T = I_1 + I_2$$

If the total impedance (Z_T) had not been found previously, it could have been found in the next step. Instead of using Ohm's Law to check the

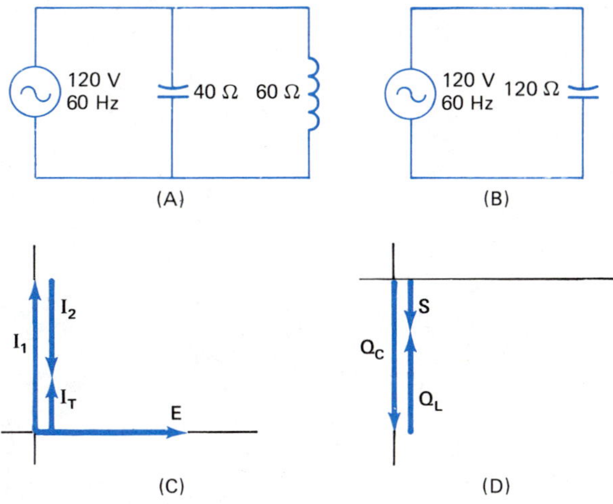

Figure 21-10

total current, it can be used to calculate the total impedance. This procedure is called the *total current method* and can be used in place of one of the methods introduced in previous topics for finding total impedance.

When one branch of a two-branch parallel circuit contains a pure resistance, the phase angle will be more or less than zero degrees. The series equivalent contains the same kind of circuit elements as the parallel circuit, Figure 21-11.

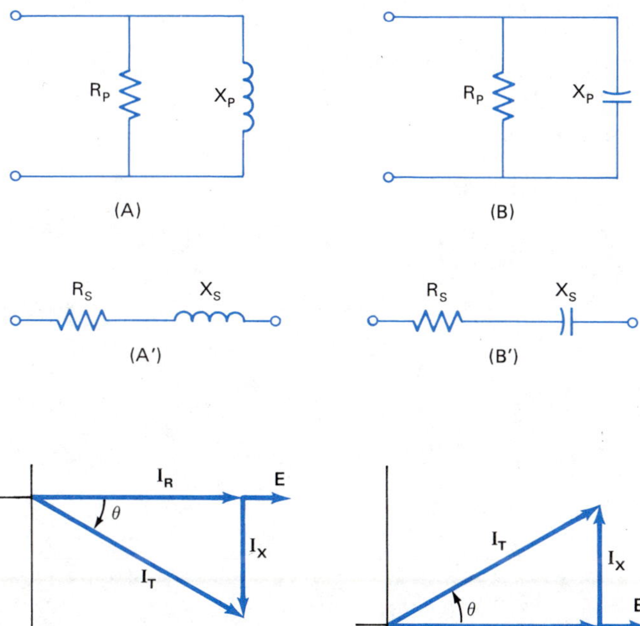

Figure 21-11 Current in two-branch ideal circuits containing resistance in one branch

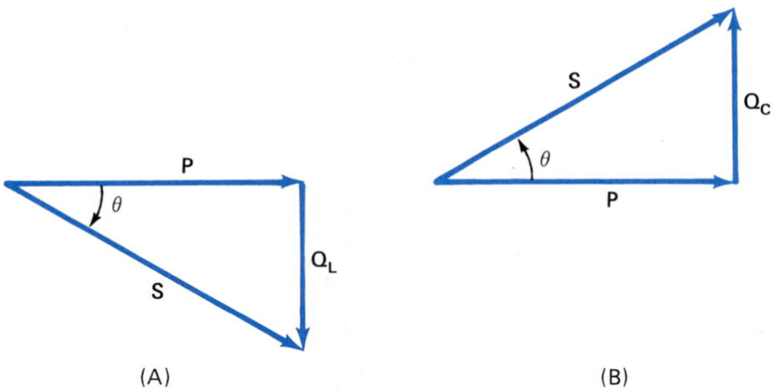

Figure 21-12 Power triangles based on voltage as the reference: (A) Inductive circuits, (B) Capacitive circuits

Power triangles are obtained by multiplying the current phasor by the voltage, Figure 21-12. Note that the direction of θ and that of Q is opposite to that obtained in the previous chapter with series circuits.

Using the series equivalent circuit, Figure 21-11(A′), the voltage-current phasors could be redrawn using the common current as the reference. The result of doing so would be to flip the power triangles upside down, Figure 21-13. The two sets of power triangles both represent power in the same circuit.

The apparent contradiction of Figures 21-12 and 21-13 is a matter of the choice of the reference phasor. The angle simply represents the phase between voltage and current. Either set of triangles is correct. To avoid confusion, it is a matter of choosing one reference or the other. There is no particular trend one way or the other in the latest electrical theory textbooks.

In keeping with the development of the power triangle for series circuits in the previous chapter, the current reference, Figure 21-13, will be used in this text. It is not necessary to construct voltage-current phasors from the series equivalent. Use the following rule:

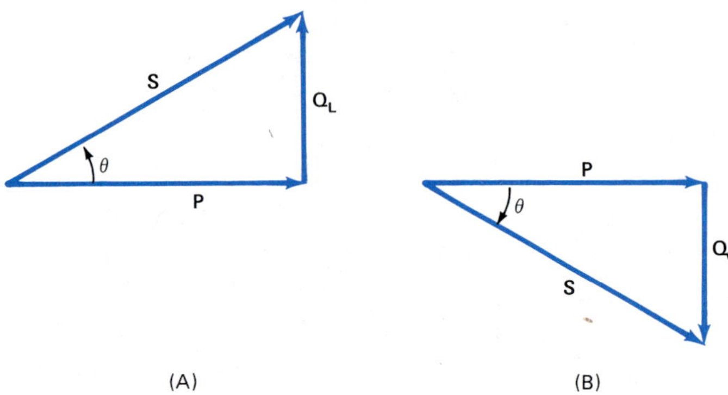

Figure 21-13 Power triangles based on common current as the reference: (A) Inductive circuits, (B) Capacitive circuits

FINDING APPARENT POWER IN ALL ac CIRCUITS

In all power calculations, use the angle measured *from* the current phasor *to* the voltage phasor. The angle is positive if the direction from current to voltage is ccw. It is negative if the direction is cw.

To implement this rule, change the sign of the current angle when voltage is the reference. Apparent power takes positive signs in inductive circuits and negative signs in capacitive circuits:

$$S\underline{/\theta} = P + jQ \quad \text{(Inductive circuits)}$$

$$S\underline{/-\theta} = P - jQ \quad \text{(Capacitive circuits)}$$

All phasors except current and admittance are in reference to the series equivalent circuit. They take the same signs in inductive and capacitive circuits. Current and admittance are parallel circuit phasors. The power factor is the same for a given angle whether it is positive or negative: $\cos\theta = \cos(-\theta)$.

Example C Find (1) the current in each branch of the circuit, Figure 21-14, (2) the total current, (3) the total impedance and (4) the power **S**, P and Q.

Solution: Construct other diagrams, Figure 21-14, and a solution table:

	QUANTITY	POLAR		RECTANGULAR	UNITS
Data:	**E**	$120\underline{/0^\circ}$			V
	$\mathbf{Z_1}$	$40\underline{/0^\circ}$			Ω
	$\mathbf{Z_2}$	$30\underline{/-90^\circ}$			Ω
(1)	$\mathbf{I_1 = E/Z_1}$	$3\underline{/0^\circ}$	→	$3 + j0$	A
	$\mathbf{I_2 = E/Z_2}$	$4\underline{/90^\circ}$	→	$0 + j4$	A
(2)	$\mathbf{I_T}$ (Sum)	$5\underline{/53.1^\circ}$	←	$3 + j4$	A
(3)	$\mathbf{Z_T = E/I_T}$	$24\underline{/-53.1^\circ}$	→	$14.4 - j19.2$	Ω
(4)	$\mathbf{S_T = EI_T}$	$600\underline{/-(53.1^\circ)}$	→	$360 - j480$	VA, W, vars

Note: $\mathbf{Z_T} = \dfrac{\mathbf{E}}{\mathbf{I_T}} = \dfrac{120\underline{/0^\circ}\text{ V}}{5\underline{/53.1^\circ}} = 24\underline{/-53.1^\circ}$

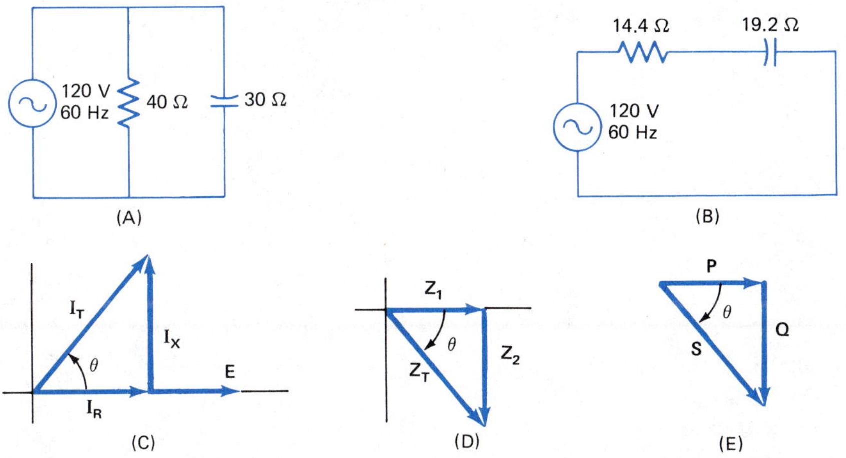

Figure 21-14

To check, divide the product of $\mathbf{Z}_1$ and $\mathbf{Z}_2$ by the sum.

True and reactive power are given in the rectangular form of apparent power in the last line of the table.
$\mathbf{S} = (120\angle 0^\circ \text{ V})(5\angle -53.1^\circ \text{ A}) = 600\angle -53.1^\circ \text{ VA}$

The units of S, P and Q are given in the last column: S = 600 VA, P = 360 W, Q = 480 vars leading. P and Q can also be calculated by multiplying the currents I_1 and I_2 by the applied voltage.

EXERCISE 21-4

In each of the following problems, construct circuit, series-equivalent, voltage-current and impedance diagrams and the power triangle. Use a solution table to solve each problem. For each circuit, find (a) the current in each branch, (b) the total current, (c) the total impedance, (d) the apparent, true and reactive power and (e) the power factor.

1. Two identical lamps with 240 Ω hot resistance are connected in parallel to a 120-V, 60-Hz line.
2. Two capacitors having reactances of 450 Ω and 580 Ω are connected in parallel to a 120-V, 60-Hz source.
3. Two relay coils connected in parallel to a 24-V, 60-Hz source have 42 Ω and 68 Ω of reactance.
4. A clothes dryer with 10-Ω resistance and a built-in oven with a resistance of 8 Ω are connected in parallel to a 220-V, 60-Hz source.
5. Two pure capacitors are connected in parallel to a 220-V, 60-Hz supply. They have capacitive reactances of 39.5 Ω and 62.5 Ω.
6. A circuit contains two pure inductive reactances of 2.35 kΩ and 3.15 kΩ. The circuit is connected to a 12-V, 60-Hz line.
7. A solenoid has an inductive reactance of 225 Ω when supplied by a 12-V, 60-Hz line. The solenoid is in parallel with a lamp having 100 Ω of hot resistance.
8. A 42.3-Ω capacitive reactance and an 86.7-Ω resistance are in parallel with a 115-V, 60-Hz source.
9. A 60-Hz, 120-V line connects a resistance of 3400 Ω and a capacitor of 0.185 μF in parallel.
10. A coil of 1.3 H inductance is in parallel with a 1200-Ω resistor across a 220-V, 60-Hz line.

21-5 PRACTICAL IMPEDANCES IN PARALLEL

Practical impedances are those that exist in actual electrical circuits. Each branch of a parallel circuit may contain both resistance and reactance. While electronic circuits are often capacitive, most *power circuits* (distribution, residential and industrial) are inductive.

The method of solving ac parallel circuits illustrated in the previous topic is a general method. It is used to solve practical as well as ideal circuits. The main difference is in the phase angle of branch currents. They are usually not the –90°, 0° and 90° angles of ideal circuits.

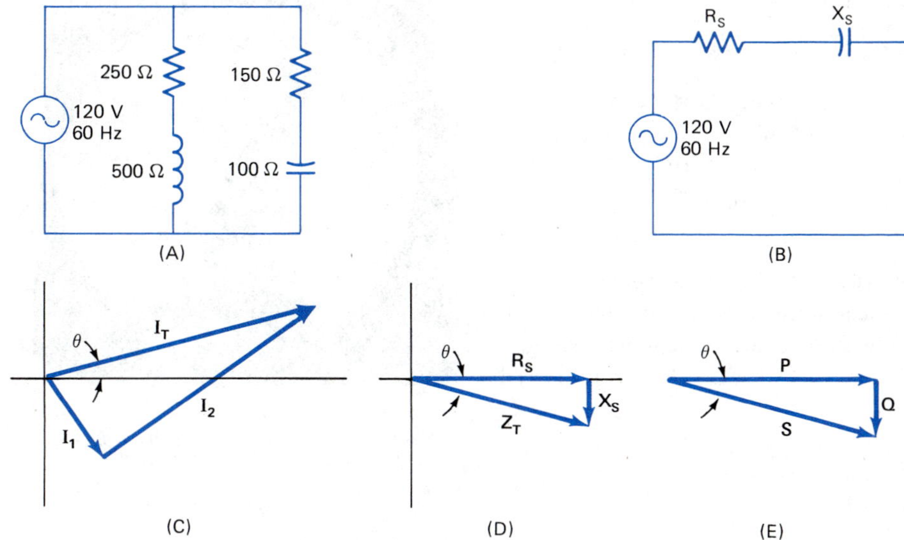

Figure 21-15

Some typical power factors are incandescent lamps and heaters 98%, fractional-horsepower induction motors 60% to 75%, large induction motors 80% to 90% and fluorescent lamps with ballast 50%, all with lagging current. Any of these can be thought of as a resistance and an inductance in series in one branch of the circuit.

Example A A 120-V, 60-Hz source is connected to the parallel circuit of Figure 21-15(A). $\mathbf{Z}_1 = 250 + j500\ \Omega$ and $\mathbf{Z}_2 = 150 - j100\ \Omega$. Find (1) the current in each branch, (2) the total current, (3) the total impedance, (4) the apparent, true and reactive power and (5) the power factor.

Solution: Draw current and impedance phasors, the series equivalent and the power triangle, Figure 21-15. Construct a solution table.

	QUANTITY	POLAR		RECTANGULAR	UNITS
Data:	$\mathbf{E}$	$120\angle 0^\circ$			V
	$\mathbf{Z}_1$	$559\angle 63.4^\circ$	←	$250 + j500$	Ω
	$\mathbf{Z}_2$	$180\angle -33.7^\circ$	←	$150 - j100$	Ω
(1)	$\mathbf{I}_1 = \mathbf{E}/\mathbf{Z}_1$	$215\angle -63.4^\circ$	→	$96 - j192$	mA
	$\mathbf{I}_2 = \mathbf{E}/\mathbf{Z}_2$	$666\angle 33.7^\circ$	→	$554 + j369$	mA
(2)	$\mathbf{I}_T$ = (Sum)	$674\angle 15.3^\circ$	←	$650 + j177$	mA
(3)	$\mathbf{Z}_T = \mathbf{E}/\mathbf{I}_T$	$178\angle -15.3^\circ$	→	$171 - j50$	Ω
(4)	$\mathbf{S} = \mathbf{E}\mathbf{I}_T$	$80.9\angle -15.3^\circ$	→	$78.0 - j21.3$	VA, W, vars

(5) Pf = cos (−15.3°) = 96.5% leading.

Note: $$\mathbf{I}_1 = \frac{120\angle 0^\circ\ \text{V}}{559\angle 63.4^\circ\ \Omega} = \frac{120\ \text{V}}{559\ \Omega}\angle 0^\circ - 63.4^\circ\ \text{A} = 0.096\angle -63.4^\circ\ \text{A}$$

$$\mathbf{I}_2 = \frac{120\angle 0^\circ\ \text{V}}{180\angle -33.7^\circ} = 0.666\angle 33.7^\circ\ \text{A}$$

$$\mathbf{Z_T} = \frac{120\angle 0^\circ \text{ V}}{0.674\angle 15.3^\circ \text{ A}} = 178\angle{-15.3^\circ}\ \Omega$$

$$\mathbf{S} = (120\angle 0^\circ \text{ V})(0.674\angle{-15.3^\circ} \text{ A}) = 80.9\angle{-15.3^\circ}\text{VA}$$

P and Q are found in the rectangular form of **S** in the last line of the table. P = 78.0 W, Q = 21.3 vars leading.

Check: (3) $\mathbf{Z_T} = \dfrac{\mathbf{Z_1 Z_2}}{\mathbf{Z_1} + \mathbf{Z_2}} = \dfrac{(559\angle 63.4^\circ)(180\angle{-33.7^\circ})}{566\angle 45^\circ}$

$$\mathbf{Z_T} = 178\angle{-15.3^\circ}\ \Omega$$

Figure 21-16 A power factor meter (Courtesy of Southeast Nebraska Community College)

In most of the examples and exercises of the previous topics, the applied voltage and the branch impedances were listed as the given data. True power (P), current and voltage measurements are the most commonly measured quantities, however. The amount of lag or lead is measured with a power factor meter, Figure 21-16.

The formulas:

$$\mathbf{E} = \mathbf{IZ}$$

$$\mathbf{S} = \mathbf{EI}$$

$$\mathbf{S} = \frac{\mathbf{E}^2}{\mathbf{Z}}$$

$$\mathbf{S} = \mathbf{I}^2\mathbf{Z}$$

relate the various quantities. These can be used to find the impedance, current, power or voltage phasors in any of the circuit branches as well as the total circuit.

Example B (1) Find the current in a branch of a parallel circuit if the apparent power of the branch is 865 VA. The power factor is 65% lagging and the source is 120 V, 60 Hz. (2) What is the impedance of the branch?

Solution: Find the phase angle.

Data: $\theta = \text{invcos } 0.65 = 49.5^\circ$

$\mathbf{S} = 865\angle 49.5^\circ$ VA, $\mathbf{E} = 120\angle 0^\circ$ V

(1) Formula: $\mathbf{S} = \mathbf{EI}$

$$\mathbf{I} = \frac{\mathbf{S}}{\mathbf{E}}$$

Substitute: $\mathbf{I} = \dfrac{865\angle{-(49.5^\circ)} \text{ VA}}{120\angle 0^\circ \text{ V}}$

$$\mathbf{I} = 7.21\angle{-49.5^\circ} \text{ A}$$

Remember to change the sign of the angle of apparent power before using it in a formula.

(2) Formula: $\mathbf{Z} = \dfrac{\mathbf{E}}{\mathbf{I}}$

Substitute: $\mathbf{Z} = \dfrac{120\angle 0^\circ \text{ V}}{7.21\angle{-49.5^\circ} \text{ A}} = 16.6\angle 49.5^\circ\ \Omega$

A *synchronous motor* can be thought of as a resistance and a capacitance in series in a branch of a parallel circuit. Synchronous motors operate at one speed synchronized to the frequency of the ac source. Current in synchronous motors can be adjusted to lead or lag the voltage. They are often set in the leading mode.

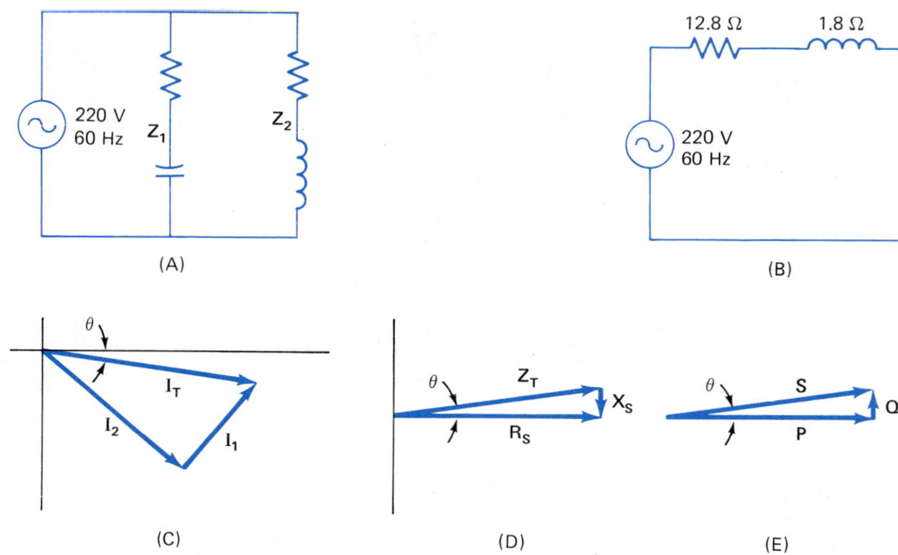

Figure 21-17

Example C The circuit of Figure 21-17 contains a synchronous motor (capacitive) in parallel with an induction motor. The synchronous motor draws an apparent power of 1820 VA with a leading power factor of 65.2%. The induction motor draws a lagging current of 14.3 A at a power factor of 79.8%. The applied voltage is 220 V, 60 Hz. Find (1) the total current, (2) the total impedance, (3) the total apparent, true and reactive power, and (4) the circuit power factor.

Solution: Draw the series equivalent circuit, the current and impedance phasor diagrams and the power triangle for the circuit, Figure 21-17(B), (C), (D) and (E). Determine the phase angles for each branch:

θ_1 = invcos 0.652 = −49.3°

θ_2 = invcos 0.798 = 37.1°

	QUANTITY	POLAR		RECTANGULAR	UNITS
Data:	**E**	$220\underline{/0^\circ}$			V
	$S_1\underline{/\theta_1}$	$1.82\underline{/-49.3^\circ}$	→	1.19 − j1.38	kVA, kW, kvars
	$I_2\underline{/-\theta_2}$	$14.3\underline{/-37.1^\circ}$	→	11.41 − j8.62	A
(1)	$\mathbf{I_1} = \frac{\mathbf{S_1}}{\mathbf{E}}$	$8.27\underline{/49.3^\circ}$	→	5.39 + j6.27	A
	$\mathbf{I_T}$ = (Sum)	$17.0\underline{/-8.0^\circ}$	←	16.8 − j2.35	A
(2)	$\mathbf{Z_T} = \mathbf{E}/\mathbf{I_T}$	$13.0\underline{/8.0^\circ}$	→	12.8 + j1.8	Ω
(3)	$\mathbf{S_T} = \mathbf{EI_T}$	$3.74\underline{/8.0^\circ}$	→	3.70 + j0.52	kVA, kW, kvars
Check:	$\mathbf{S_2} = \mathbf{EI_2}$	$3.15\underline{/37.1^\circ}$	→	2.51 + j1.90	kVA, kW, kvars
	$\mathbf{S_T}$ = (Sum)			3.70 + j0.52	kW, kvars

(4) Pf = cos θ = 99.0%

Note: $\mathbf{I}_1 = \dfrac{1820\angle{-49.3°}\ \text{VA}}{220\angle{0°}\ \text{V}} = 8.27\angle{-(-49.3°)}\ \text{A}$

Current $\mathbf{I}_2$ is listed last in the data so it can readily be added to $\mathbf{I}_1$ in part (1) of the solution table.

The check is done by evaluating the apparent power in the inductive motor. This is added in rectangular form to the apparent power of the synchronous motor given in the data. The sum is the total apparent power, P + jQ, which checks with line (3).

A multiple-branch circuit is also solved using the same procedure. The total current is the sum of all branch currents and is found by adding in rectangular form as usual within the solution table.

Example D (1) Find the total current, (2) total impedance, (3) the total apparent, true and reactive powers and (4) the power factor for the circuit of Figure 21-18.

Solution:

Data:

QUANTITY	POLAR		RECTANGULAR	UNITS
E	$220\angle{0°}$			V
$\mathbf{Z}_1$	$25\angle{53.1°}$			Ω
$\mathbf{Z}_3$	$15.6\angle{39.8°}$	←	12 + j10	Ω
$\mathbf{S}_4$	$2.42\angle{-40.0°}$	←	1.85 − j1.56	kVA, kW, kvars
$\mathbf{I}_2$	$6.04\angle{74.1°}$	→	1.66 + j5.81	A

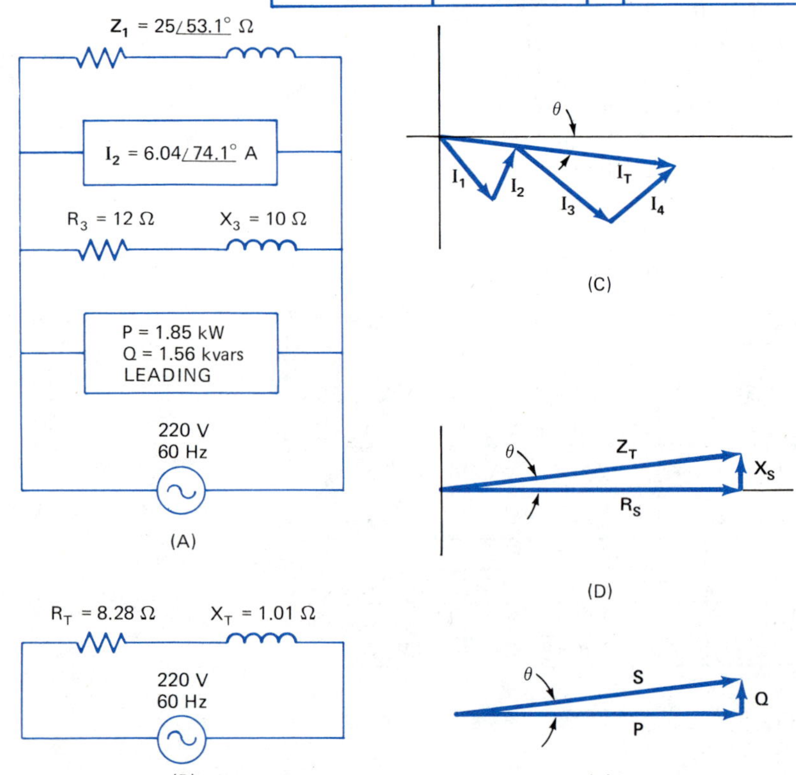

Figure 21-18

	QUANTITY	POLAR		RECTANGULAR	UNITS
(1)	$I_1 = E/E_1$	$8.80\angle -53.1^\circ$	→	$5.28 - j7.04$	A
	$I_3 = E/Z_3$	$14.10\angle -39.8^\circ$	→	$10.83 - j9.03$	A
	$I_4 = S_4/E$	$11.00\angle 40.0^\circ$	→	$8.43 + j7.07$	A
	I_T = (Sum)	$26.4\angle -6.9^\circ$	←	$26.20 - j3.19$	A
(2)	$Z_T = E/I_T$	$8.34\angle 6.9^\circ$	→	$8.28 + j1.01$	Ω
(3)	$S_T = EI_T$	$5.81\angle 6.9^\circ$	→	$5.77 + j0.698$	kVA, kW, kvars

(4) Pf = cos 6.9° = 99.3% lagging

Check: The total impedance can be calculated by the admittance method. Another method is to evaluate the polar form of S_1, S_2 and S_3 by multiplying the voltage by the respective currents. After transforming to rectangular form, they are added to S_4 (data). This results in the P + jQ form of S_T (line 3).

Note: Branch currents are added by the tail-to-tip method, Figure 21-18(C). The series equivalent is shown in Figure 21-18(B), the impedance phasor diagram for the equivalent circuit in Figure 21-18(D) and the power triangle in Figure 21-18(E).

EXERCISE 21-5

Find (a) the total current, (b) the total impedance, (c) the total apparent, true and reactive powers, and (d) the power factor for each of the following circuits. Draw the circuit and series equivalent diagram, the current and impedance phasor diagrams, and the power triangle for each problem. Use a solution table.

1. An induction motor with an impedance of $12.5\angle 65.2^\circ$ Ω is in parallel with another induction motor having an impedance of 8.75 Ω drawing a 31.2° lagging current. The voltage is 220 V, 60 Hz.
2. A 120-V, 60-Hz garbage disposal with an impedance of $14.8\angle 39.2^\circ$ Ω is in parallel with a $10.3\angle 4.5^\circ$ Ω hot plate.
3. A 120-V, 60-Hz refrigerator motor has an impedance of $8.75\angle 49.3^\circ$ Ω. It is in parallel with a washing machine motor with a resistance of 6.25 Ω and an inductive reactance of 5.95 Ω.
4. A $9.75\angle 53.5^\circ$ Ω induction motor is in parallel with a synchronous motor with 6.50 Ω of resistance and a capacitive reactance of 5.25 Ω. They are connected to a 220-V, 60-Hz line.
5. Repeat Problem 1 if the second motor is replaced with a synchronous motor with an impedance of $8.75\angle -31.2^\circ$ Ω.
6. Repeat Problem 4 if a $7.35\angle 44.8^\circ$ Ω induction motor is added in parallel to the original circuit.
7. Repeat Problem 3 if two 100-W lamps are turned on, each in parallel with the other appliances. Assume the lamp draws no reactive power.
8. A bank of fluorescent lamps are in parallel with an induction motor in a 120-V, 60-Hz current. The lamps draw 16 A at a lagging power factor of 55%. The motor has an apparent power of 875 VA at a 65% lagging power factor.
9. An induction motor draws 7.85 A with a lagging power factor of 65%. It is in parallel with another motor drawing a lagging current of 10.5 A at a 45% power factor. The source has a voltage of 480 V at 60 Hz.

10. Two induction motors each drawing 265 W at 60% lagging power factor are both in parallel with a 100-W lamp and a synchronous motor drawing 300 W at an 85% leading power factor. The line voltage is 120 V, 60 Hz.
11. A 220-V, 60-Hz induction motor has a lagging 10.8 kVA apparent power at 83% power factor. A 7.5 kVA synchronous motor with a leading power factor of 72% is also on the same line in parallel.
12. A resistive 120-V, 60-Hz lamp load draws 7.25 A. In parallel with the lamp load is an induction motor having a resistance of 12.0 Ω and an inductive reactance of 15.0 Ω. A synchronous motor with a true power of 750 W and a leading reactive power of 350 vars is also in parallel with the lamp load.

SERIES AND PARALLEL EQUIVALENT CIRCUITS AND POWER FACTOR CORRECTION

Figure 21-19 shows one of the ideal circuits discussed in Topic 21-2. The parallel circuit contains a pure resistance (R_p) and a pure reactance (X_p). The subscript (p) represents the elements in the parallel circuit. The equivalent series circuit has a resistance (R_s) and a reactance (X_s). By defining $\mathbf{Z}_1 = R_p + j0$ and $\mathbf{Z}_2 = 0 + jX_p$, the series equivalent $\mathbf{Z}_T = R_s + jX_s$ is found using the formula

$$\mathbf{Z}_T = \frac{\mathbf{Z}_1\mathbf{Z}_2}{\mathbf{Z}_1 + \mathbf{Z}_2}$$

by dividing the product by the sum in polar form.

Electricians often need to find the *parallel equivalent* of the series circuit, the reverse procedure. Obviously, the two unknowns, $\mathbf{Z}_1$ and $\mathbf{Z}_2$ cannot be determined with just one equation. A different method of finding the series equivalent is developed in the following topic. The reverse procedure of finding the parallel equivalent will be described in a later topic of this chapter.

When the parallel equivalent of an ac circuit has been determined, the process of power factor correction is easy. A new circuit element is added to the existing one to change the power factor to 100%. A 100% power factor has many advantages that will also be described.

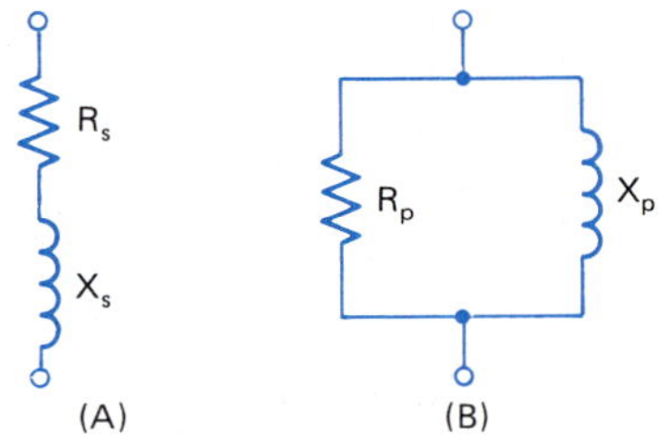

Figure 21-19 Series and parallel equivalent circuits

21-6 ADMITTANCE, CONDUCTANCE AND SUSCEPTANCE

The new method is actually a shortened version of the admittance method, Topic 21-3. For the ideal circuit of Figure 21-19, $\mathbf{Z}_1 = R_p\angle 0^\circ$ and $\mathbf{Z}_2 = X_p\angle 90^\circ$. When the admittances are found ($\mathbf{Y}_1 = \frac{1}{R_p}\angle 0^\circ$ and $\mathbf{Y}_2 = \frac{1}{X_p}\angle -90^\circ$), transformed to rectangular form ($\mathbf{Y}_1 = \frac{1}{R_p} + j0$ and $\mathbf{Y}_2 = 0 - j\frac{1}{X_p}$) and added, the result is $\mathbf{Y}_T = \frac{1}{R_p} - j\frac{1}{X_p}$.

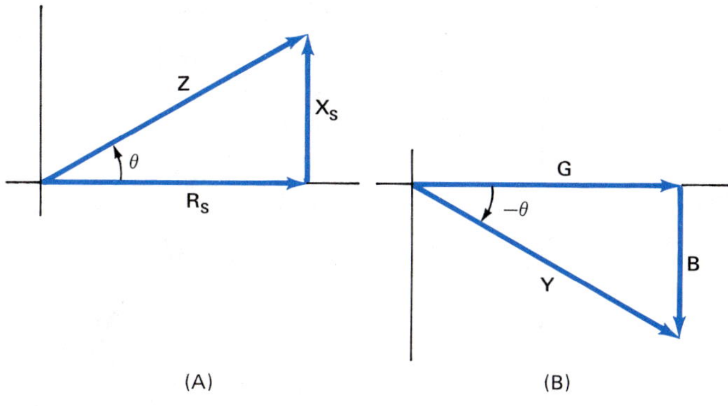

Figure 21-20 (A) Impedance of a series circuit (B) Admittance of the parallel equivalent

The procedure just described can be obtained directly from the circuit diagram, Figure 21-19, simply by taking reciprocals. The reciprocal of resistance is called conductance (G) and is measured in siemens (S), Topic 9-2:

$$G = \frac{1}{R_P}$$

The quantity $\frac{1}{X_P}$ is new and is called *susceptance.* It takes the symbol (B). It, too, is measured in siemens:

$$B = \frac{1}{X_P}$$

The rectangular form of the admittance, then, comes directly from the parallel circuit and has the form

$$Y\angle\pm\theta = G \pm jB$$

The rectangular form is transformed to polar. The reciprocal is taken to obtain the series equivalent impedance (polar form). A quick transformation gives the components of the series equivalent. A comparison of the admittance phasor for the parallel circuit and the impedance of the series equivalent is shown in Figure 21-20. The procedure is outlined in Figure 21-21 and given step-by-step in the following rule.

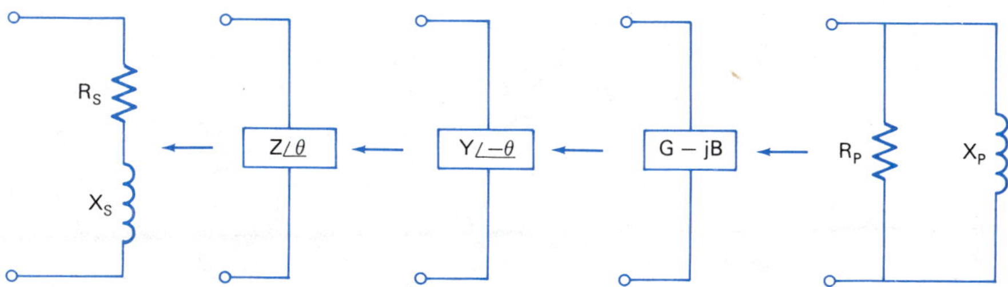

Figure 21-21 Finding the series equivalent for a resistor and a reactance in parallel. Proceed in steps from right to left.

THE CONDUCTANCE-SUSCEPTANCE EQUIVALENT CIRCUIT RULE

- Draw the circuit diagram.
- Find the reciprocals of R_p and X_p and write the rectangular form of the admittance (G – jB). Use the positive sign for capacitance and the negative sign for inductance.
- Transform the admittance to polar form ($Y\angle{-\theta}$).
- Find the reciprocal of the admittance to obtain the impedance of the series equivalent in polar form ($Z\angle{+\theta}$).
- Transform the impedance to rectangular form ($R_s + jX_s$).
- Draw the series equivalent diagram.

Example A Find the series equivalent of a circuit containing 500 Ω of resistance and 250 Ω of inductive reactance in parallel.

Solution: Find the conductance and susceptance and write the admittance in rectangular form.

$$\mathbf{G} = \frac{1}{\mathbf{R_p}} = \frac{1\angle 0^\circ}{500\angle 0^\circ} = 2 \text{ mS}$$

$$\mathbf{B} = \frac{1}{\mathbf{X_p}} = \frac{1\angle 0^\circ}{250\angle 90^\circ} = 4\angle{-90^\circ} \text{ mS} = -j4 \text{ mS}$$

The admittance is:

G – jB = 2 – j4 mS

Transform to polar:

$\rightarrow Y\angle\theta = 4.47\angle{-63.4^\circ}$ mS

Find the impedance:

$$\mathbf{Z} = \frac{1}{\mathbf{Y}} = \frac{1\angle 0^\circ}{4.47\angle{-63.4^\circ}} = 224\angle 63.4^\circ\ \Omega$$

Transform to rectangular form to find the components of the series equivalent circuit:

$\rightarrow 100 + j200\ \Omega$

$R_s = 100\ \Omega$

$X_s = 200\ \Omega$ (Inductance)

The solution can be inserted in a table outlined according to Figure 21-21. The solution proceeds from right to left.

SERIES $R_s + jX_s$	IMPEDANCE $Z\angle\theta$	ADMITTANCE $Y\angle{-\theta} \leftarrow G - jB$	PARALLEL R_p and X_p
100 + j200 Ω	$\leftarrow 224\angle 63.4^\circ$ (Ω)	$4.47\angle{-63.4^\circ} \leftarrow 2 - j4$ (mS)	500 Ω 250 Ω

Calculator note: Enter the parallel resistance (R_p), press [1/x], press [x⇄y]. Then enter the parallel reactance (X_p), press [+/−], press [1/x]. The admittance (G − jB) is now entered in the calculator. From this point, proceed exactly according to the calculator note in Topic 21.3. For the previous example, the entire procedure is as follows:

Note: Numbers in parentheses do not need to be re-entered.

500 [1/x] [x⇄y] 250 [+/−] [1/x] [INV] [P→R] , −63.4°
[x⇄y] , 0.004 47 S
(−63.4) [+/−] [x⇄y] (0.004 47) [1/x] , 224 Ω
[x⇄y] , 63.4°
[P→R] , 200 Ω (X_s)
[x⇄y] , 100 Ω (R_s)

If the circuit contains a capacitor, the admittance takes a positive sign in both polar and rectangular form. The impedance and admittance always have opposite signs since one is the reciprocal of the other.

Example B Find the series equivalence of an 87-Ω resistance and a 240-Ω capacitive reactance in parallel.

Solution: Find reciprocals $\mathbf{G} = \frac{1}{\mathbf{R_p}}$ and $\mathbf{B} = \frac{1}{\mathbf{X_p}}$.

$$\mathbf{G} = \frac{1\angle 0^\circ}{87\angle 0^\circ} = 11.5 \text{ mS}$$

$$\mathbf{B} = \frac{1\angle 0^\circ}{240\angle -90^\circ} = 4.17 \text{ mS}$$

Construct solution table and solve:

SERIES $R_s + jX_s$	IMPEDANCE $Z\angle\theta$	ADMITTANCE $Y\angle-\theta \leftarrow G - jB$	PARALLEL R_p and X_p
76.9 − j27.9 Ω	$81.8\angle -19.9^\circ$ (Ω)	$12.1\angle 19.9^\circ \leftarrow 11.5 + j4.17$ (mS)	87 Ω 240 Ω

EXERCISE 21-6

Construct a matrix for the following problems using the headings of the previous example. Find the series equivalent for each parallel circuit using the conductance-susceptance method. Draw the circuit and equivalent diagrams.

1. R_p = 144 Ω, X_p = 25.3 Ω leading
2. R_p = 639 Ω, X_p = 284 Ω lagging
3. R_p = 10.5 Ω, X_p = 6.73 Ω lagging
4. R_p = 574 kΩ, X_p = 384 kΩ leading
5. R_p = 278 Ω, X_p = 312 Ω leading
6. R_p = 527 kΩ, X_p = 2.99 kΩ lagging
7. R_p = 1.01 kΩ, X_p = 1.25 kΩ lagging
8. R_p = 59.2 Ω, X_p = 74.2 Ω leading
9. R_p = 74.2 Ω, X_p = 59.2 Ω leading
10. R_p = 127 Ω, X_p = 234 Ω lagging

21-7 THE PARALLEL EQUIVALENT CIRCUIT

When a simple two-element series circuit or the series equivalent of a complex circuit is known, the parallel equivalent can be found. The parallel equivalent contains elements that are similar to those in the series circuit. The procedure is the reverse of that described in the previous topic.

THE PARALLEL EQUIVALENT CIRCUIT RULE

- Draw the series equivalent of the given circuit.
- Write the impedance in rectangular form ($R_s + jX_s$).
- Transform the impedance to polar form $Z\angle\theta$.
- Find the reciprocal of the impedance to obtain the admittance of the parallel equivalent in polar form ($Y\angle{-\theta}$).
- Transform the admittance to rectangular form to find the conductance and susceptance ($G - jB$).
- Find the reciprocals of G and B to obtain the resistance (R_p) and the reactance (X_p) of the parallel equivalent circuit.
- Draw the parallel equivalent circuit.

Example A An induction motor has a resistance of 10 Ω and an inductive reactance of 25 Ω. Find the parallel equivalent.

Solution: Construct table working from left to right.

SERIES $R_s + jX_s$	IMPEDANCE $Z\angle\theta$	ADMITTANCE $Y\angle{-\theta} \rightarrow G - jB$	PARALLEL R_p and X_p
10 + j25 Ω	$26.9\angle 68.2^\circ$ (Ω)	$37.1\angle{-68.2^\circ} \rightarrow 13.8 - j34.5$ (mS)	72.5 Ω, 29 Ω

Note: After the impedance is transformed to polar form, calculate the reciprocal:

$$Y\angle{-\theta} = \frac{1\angle 0^\circ}{26.9\angle 68.2^\circ\ \Omega} = 0.0371\angle{-68.2^\circ}$$

Transform the admittance to rectangular form ($G - jB$). Find the reciprocals of the conductance and susceptance to obtain the components of the parallel equivalent:

$$R_p = \frac{1}{0.0138\ \text{S}} = 72.5\ \Omega$$

$$X_p = \frac{1}{0.0345\ \text{S}} = 29.0\ \Omega$$

Caution: Do not write the parallel components as a phasor (complex numbers). Impedance phasors represent resistance and reactance in series. Admittance is the phasor form of the parallel circuit.

Example B A synchronous motor has a resistance of 12.5 Ω and a capacitive reactance of 17.5 Ω. Find the parallel equivalent.

Solution table:

SERIES $R_s + jX_s$	IMPEDANCE $Z\angle\theta$	ADMITTANCE $Y\angle-\theta \to G - jB$	PARALLEL R_p and X_p
12.5 – j17.5 Ω	21.5∠–54.5° (Ω)	46.5∠54.5° → 27.0 + j37.8 (mS)	37 Ω, 26.4 Ω

Note: $R_p = \frac{1}{G} = \frac{1}{0.0270} = 37.0\ \Omega$

$X_p = \frac{1}{B} = \frac{1}{0.0378} = 26.4\ \Omega$

Notice that when the resistance is larger than the reactance in either the series or parallel circuit, it is the exact opposite (reactance is larger than resistance) in the other circuit. This fact can always provide a quick partial check of the calculations. In the previous example, R_p is greater than X_p so R_s should be less than X_s.

EXERCISE 21-7

Construct a matrix for the following problems using the headings of the previous example. Find the parallel equivalent for each series circuit using the conductance-susceptance method. Draw the circuit and equivalent diagrams.

1. 35 + j25 Ω
2. 25 – j35 Ω
3. 604 + j239 Ω
4. 71.5 – j113 Ω
5. 65.1 + j17.3 Ω
6. 13.7 + j21.2 Ω
7. 48.7 – j162 Ω
8. 17.3 + j83.6 Ω
9. 478 + j1160 Ω
10. 385 – j91.4 Ω

Continue the matrix for the following problems for additional practice. Fill in all blanks in the matrix.

11. R_p = 278 Ω, X_p = 312 Ω leading
12. R_p = 527 Ω, X_p = 299 Ω lagging
13. $R_s + jX_s$ = 920 + j980 Ω
14. $R_s + jX_s$ = 277 – j358 Ω
15. Z = 924 Ω, Pf = 44.6% leading
16. Z = 172 Ω, Pf = 50.2% lagging
17. Y = 23.8 μS, Pf = 62.5% leading
18. Y = 68.9 mS, Pf = 78.2% lagging

19. The circuit of Figure 21-7 was found to have a total admittance (G_T - jB_T) of 119 – j14.5 mS. Find (a) the series equivalent and (b) the parallel equivalent for that circuit.
20. Repeat Problem 19 for another circuit if the total admittance is 483 + j969 mS.

21-8 POWER FACTOR CORRECTION

If the switch (A) is closed in the ideal circuit of Figure 21-22, electrons begin to flow producing an alternating current. Current and power in the various parts of the circuit can be found by methods of the previous topics:

$$I_L = \frac{E}{X_L} = 3\angle{-90^\circ}\ \text{A} \qquad Q_L = EI_L = 360\ \text{vars (lag)}$$

$$I_C = \frac{E}{X_C} = 2\angle{90^\circ}\ \text{A} \qquad Q_C = EI_C = 240\ \text{vars (lead)}$$

$$I_T = \frac{E}{X_T} = 1\angle{-90^\circ}\ \text{A} \qquad S = EI_T = 120\ \text{VA (lag)}$$

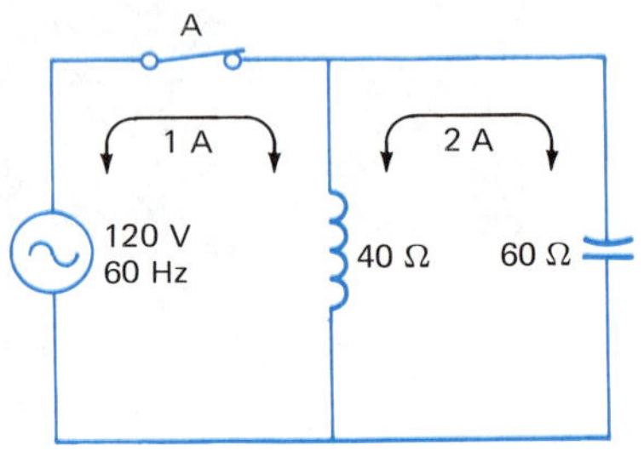

Figure 21-22 An ideal circuit containing an inductor and a capacitor. Energy flows back and forth between the inductor and capacitor without traveling all the way back to the source.

The interesting thing about such a circuit as shown in Figure 21-10 is the fact that there is less current and less power flow in the main lines from the source than in the circuit branches. The inductor and capacitor alternately act as power sources and power absorbers. When the capacitor discharges, the inductor's magnetic field is building. When the inductor's field collapses, the capacitor recharges. A reactive power of 240 vars is driven back and forth between the inductor and capacitor with the 2-A current. An additional 120 vars transfers back and forth between the inductor and the source carried by 1 A of current, Figure 21-22.

Suppose switch A is opened. The 1-A current from the source stops immediately. The capacitor and the inductor continue to act alternately as power sources and power absorbers. Current continues to alternate as electrons flow from one to the other. In the new situation, however, the current must be the same in both circuit elements. This can only happen if the reactances of the two elements have the same value.

Such a change can occur only if the frequency changes.

$$X_L = X_C$$

$$2\pi fL = \frac{1}{2\pi fC}$$

$$f_r = \frac{1}{2\pi\sqrt{LC}}$$

The new frequency (f_r) is known as the *resonant frequency* for the circuit. Although the capacitance and the inductance of the two elements will not change, the capacitive reactance and the inductive reactance will. The current in the isolated circuit loop will continue to alternate at the resonant frequency sending the trapped energy back and forth from capacitor to inductor.

Of course, a practical circuit will contain some resistance in the coil, in the lines and in the capacitor. The trapped energy will soon be dissipated and the current falls to zero. Such a circuit isolated from the source

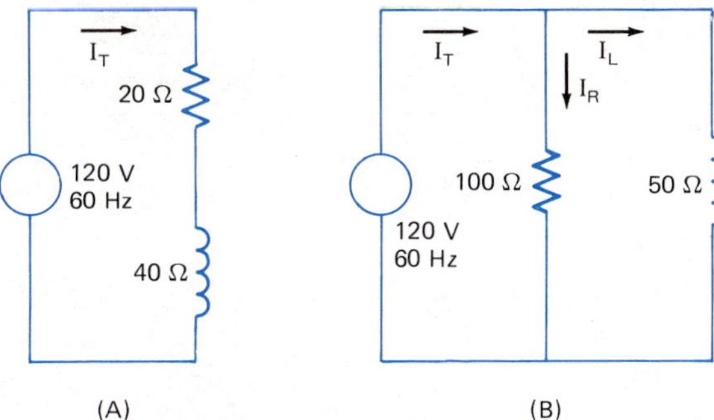

Figure 21-23 (A) A circuit containing a practical device such as a motor is represented by a resistance and impedance in series. (B) The parallel equivalent circuit contains the same circuit elements.

would have little value. Also, the frequency of the source cannot be changed. It is enlightening, however, to study practical circuits for which the 60-Hz source frequency is the resonant frequency.

Suppose the impedance of the ideal circuit, Figure 21-22, is replaced by a practical device such as a motor. The new circuit with the capacitor removed, Figure 21-23(A), contains resistance in series with the inductance. If the motor impedance is replaced with its parallel equivalent, the circuit is that of Figure 21-23(B). The current and power in the various branches of the parallel equivalent are as follows:

$$\begin{aligned} \mathbf{I_L} &= -j2.4 \text{ A} & Q_L &= 288 \text{ vars} \\ \mathbf{I_R} &= \underline{\quad 1.2 \text{ A}} & P &= \underline{144 \text{ W}} \\ \mathbf{I_T} &= 1.2 - j2.4 \text{ A} & \mathbf{S} &= 144 + j288 \text{ VA} \\ &\rightarrow 2.7\underline{/-63.4^\circ} \text{ A} & &\rightarrow 322\underline{/63.4^\circ} \text{ VA} \end{aligned}$$

The power factor is 44.7% lagging. Since the two circuits, Figure 21-23(A) and (B), are equivalent, the total current and the apparent, true and reactive power must be the same. The motor would only need to draw 1.2 A from the source to produce the true power of 144 W. It actually draws more than twice that amount (2.7 A) and transfers a considerable amount of energy back and forth along the lines between the inductor and the source.

To obtain the effect of the ideal circuit, a capacitor is placed in parallel with the motor, Figure 21-24 (A). This will allow energy to circulate in the closed loop containing the motor and the capacitor. If a 60-Ω capacitive reactance is used, the current and power in the various branches of the equivalent circuit are as follows:

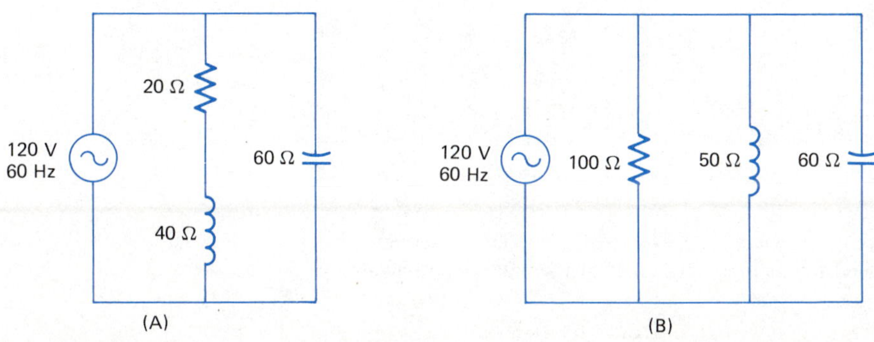

Figure 21-24 (A) The circuit of Figure 21-23 with a capacitor added in parallel, (B) The parallel equivalent

$I_L = -j2.4$ A
$I_C = j2$ A
$I_R = 1.2$ A
$I_T = 1.2 - j0.4$ A
$\rightarrow 1.3\angle{-18.4°}$
Pf = 94.9%

$Q_L = 288$ vars lagging
$Q_C = 240$ vars leading
P = 144 W
S = 144 + j48
$\rightarrow 152\angle{18.4°}$ lagging

The capacitor has had a profound affect. Even though the current and power drawn by the motor is unchanged, the total circuit current and apparent power has dropped. Most of the reactive power required by the inductance is now provided by the capacitor instead of the source. It does not have to travel back and forth all the way to the power source.

The power factor has been improved by the capacitor. It increased from 44.7% to 94.9%. The current in the main lines dropped from 2.7 A to 1.3 A. This fact allows the use of smaller wires from the power source, in the generator and in transformers. Since the motor may be many miles from the power plant, this can result in huge savings by the power company. Some other advantages of power factor improvement are as follows:

- Smaller power ($I^2 R$) losses in the line due to the smaller current.
- Smaller voltage drop (IR) along the power line resulting in better voltage regulation.
- The generator can deliver more true power if reactive power is not circulating in the main lines.
- High power factor increases efficiency and improves the overall operating characteristics of the system.

Large customers usually have to pay a billing penalty if the power factor is low.

By installing a larger capacitor in the circuit, the power factor can be increased to 100%. This is done by matching the inductive reactance of the parallel equivalent circuit. The 60-Hz source will now be the resonant frequency. The capacitive reactance required is 50 Ω, Figure 21-25. In this situation all the reactive power circulates between the capacitor and motor. None travels back and forth to the source:

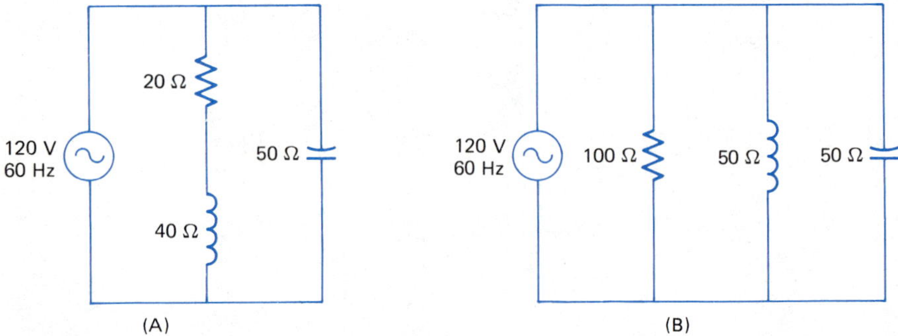

Figure 21-25 (A) The circuit of Figure 21-23 with a capacitor added in parallel, (B) The parallel equivalent. Notice the capacitive reactance matches the inductive reactance of the parallel equivalent.

$\mathbf{I_L} = -j2.4$ A	$Q_L = 288$ vars lagging
$\mathbf{I_C} = +j2.4$ A	$Q_C = 288$ vars leading
$\mathbf{I_R} = \ \ 1.2$ A	$P = 144$ W
$\mathbf{I_T} = \ \ 1.2 - j0$ A	$\mathbf{S} = 144 + j0$ VA
$\rightarrow 1.2\angle 0^\circ$	$\rightarrow 144\angle 0^\circ$ VA
Pf = 100%	

Again, the current and true and reactive power in the motor are unchanged. It is important to realize that the capacitive reactance must match the inductive reactance of the *parallel equivalent circuit* (50 Ω). If a 40-Ω capacitive reactance was used to match the inductive reactance of the motor itself, the capacitor would be too large. The circuit would be a leading one with a power factor of less than 100%.

It is easy to see why. The current circulating between the capacitor and the motor inductance must pass through the motor resistance, also. The motor inductance and resistance cannot actually be isolated from each other.

The resistance and inductive reactance of the motor in the previous discussion can represent the series equivalent of a more complex circuit. The capacitive reactance required to raise the power factor to 100% is found in the same way. Simply, match the inductive reactance of the parallel equivalent. The capacitance required is found by the formula:

$$X_C = \frac{1}{2\pi fC}$$

Example A Find the capacitance of the capacitor, Figure 21-25, used to raise the power factor to 100%. The capacitive reactance is 50 Ω and the source frequency is 60 Hz.

Solution: Formula: $X_C = \dfrac{1}{2\pi fC}$

$$C = \frac{1}{2\pi f X_C}$$

Substitute: $C = \dfrac{1}{2\pi(60\text{ Hz})(50\ \Omega)} = 53.1\ \mu\text{F}$

Example B An induction motor has an impedance of $36.4\angle 74.1^\circ$ Ω. Find the capacitance of a parallel capacitor that will provide a 100% power factor. The source is 208 V, 60 Hz.

Solution: (1) Find the parallel equivalent.

(2) Choose a capacitive reactance (X_C) that matches the parallel inductive reactance (X_L).

(3) Determine the capacitance (C).

	MOTOR	$Z\angle\theta$	$Y\angle-\theta \rightarrow G - jB$	PARALLEL EQUIVALENT
(1)	10 + j35 Ω	$36.4\angle 74.1^\circ$ (Ω)	$27.5\angle -74.1^\circ \rightarrow 75.3 - j26.4$ (mS)	R_p = 133 Ω, X_p = 37.8 Ω

Figure 21-26 An ac industrial analyzer contains a triple-range ammeter, a triple-range voltmeter, a seven-range wattmeter and a power factor meter. (Courtesy of Southeast Nebraska Community College)

(2) Choose a capacitive reactance of 37.8 Ω.

(3) $C = \dfrac{1}{2\pi f X_C}$

$$C = \frac{1}{2\pi(60\ \text{Hz})(37.8\ \Omega)} = 31.7\ \mu\text{F}$$

A different method of finding the capacitance for power factor correction is more commonly used. In Figure 21-25, the inductive reactance of the equivalent circuit was matched by the capacitive reactance. In doing so, it was noted that the reactive power was also matched ($Q_L = Q_C$). In addition, the reactive power has the same value in the series equivalent as in the parallel equivalent.

Usually, an industrial analyzer, Figure 21-26, is used to determine the line voltage and the current. It also measures the true power and the power factor. These measurements could be used to find the impedance of the series equivalent. The procedure then would be the same as that of the previous example.

The same measurements can be used to find the reactive power:

$$\mathbf{S = EI}$$
$$\theta = \cos^{-1} \text{Pf}$$
$$Q = S \sin \theta$$

The reactive power (Q) is almost always lagging (inductive). To correct the power factor to 100%, the electrical industry has rated the power factor correction type capacitors in terms of the reactive power in kvars, Figure 21-27. The voltage is also specified. When the reactive power of a circuit is determined, the capacitor is chosen to match. Usually, standard size capacitors are used so the power factor is "improved" rather than completely corrected to 100%.

Figure 21-27(A) Power factor correction capacitors (Courtesy of Southeast Nebraska Community College)

Figure 21-27(B) A bank of capacitors for power factor correction at a distribution substation

Example C A 208-V, 60-Hz motor draws 7.5 A with a lagging power factor of 58.3%. Find the rating of a capacitor that will improve the power factor to 95%, both in kvars and microfarads.

Solution: $\theta_1 = \cos^{-1}\text{Pf} = \cos^{-1}0.583 = 54.3°$ (Original)

$\theta_2 = \cos^{-1}0.95 = 18.2°$ (Final)

$\mathbf{S}_1 = \mathbf{EI} = (208\text{ V})(7.5\text{ A}) = 1.56\angle 54.3°\text{ kVA}$

$\rightarrow 0.91 + j1.27$ (kW, kvars) (Original)

Since the true power (0.91 kW) is to remain unchanged, Figure 21-28, the new apparent power must be as follows:

$$\mathbf{S}_2 = \frac{P}{\cos\theta} = \frac{0.91\text{ kW}}{0.95} = 0.96\angle 18.2°\text{ kVA}$$

$\rightarrow 0.91 + j0.30$ (kW, kvars) (Final)

For the reactive power to change from 1.27 kvars to 0.30 kvars, a capacitor with a rating of:

$Q_C = 1.27 - 0.30 = 0.97$ kvars

must be added in parallel to the circuit. The capacitive reactance is as follows:

$$X_C = \frac{E^2}{Q_C} = \frac{(208\text{ V})^2}{970\text{ vars}} = 44.6\ \Omega$$

and:

$$C = \frac{1}{2\pi f X_C} = \frac{1}{2\pi(60\text{ Hz})(44.6\ \Omega)} = 59.5\ \mu\text{F}$$

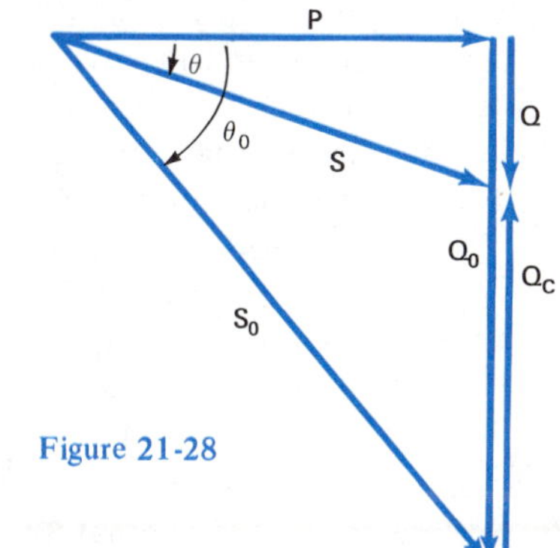

Figure 21-28

EXERCISE 21-8

Data, diagram, formula, substitute, check.

1. Find the resonant frequency of the circuit in Figure 21-22 when the switch is opened.
2. An industrial capacitor is rated at 4800 V, 60 Hz and 5.0 kvars. What is its (a) capacitive reactance and (b) capacitance?
3. An industrial capacitor is rated at 480 V, 60 Hz and 2.5 kvars. Find its (a) capacitive reactance and (b) capacitance.
4. A capacitor used for power factor correction is rated at 240 V, 60 Hz and 2.5 kvars. Find (a) the capacitive reactance and (b) the capacitance.
5. A 60-Hz induction motor has an impedance of 36.7 Ω and a lagging power factor of 80%. Find the capacitive reactance in parallel required to increase the power factor to 100%.
6. A 60-Hz fluorescent lamp and its ballast has an impedance of 120 Ω at a lagging power factor of 55%. What capacitive reactance in parallel will provide a 100% power factor?
7. A factory circuit (60 Hz) has a series equivalent with a total impedance of 1.54 Ω at a lagging power factor of 75%. Determine the capacitive reactance required to boost the power factor to 100% when connected in parallel.
8. Find the capacitance of the capacitor in Problem 5.
9. What is the capacitance of the capacitor in Problem 6?
10. What is the capacitance of the capacitor in Problem 7?

An industrial analyzer is used to make the following measurements. All circuits have lagging currents. Find the indicated unknowns.

11. E = 120 V, I = 18.4 A, Pf = 81.5%. Find **S**, P, Q and X_L.
12. P = 12.5 kW, Pf = 55.3%. Find **S** and Q.
13. E = 220 V, I = 28.4 A, P = 5.00 kW. Find **S**, Q, Pf and X_L.
14. E = 120 V, P = 2.85 kW, Pf = 81.5%. Find **S**, Q, **I** and X_L.
15. The following meter readings were taken on a 60-Hz circuit with an industrial analyzer. P = 2.4 kW, E = 240 V and I = 30 A lagging. Find (a) the reactive power and (b) the power factor. (c) What is the capacitive reactance and (d) the capacitance of a 100% correcting capacitor connected in parallel?
16. An industrial analyzer is used in an industrial plant using a 277-V, 60-Hz alternating current. The true power measures 2.75 kW and the power factor is 78%. Find (a) the apparent and reactive power of the circuit. To correct the power factor to 100%, find (b) the capacitive reactance and (c) the capacitance required.
17. A plant has a lagging power factor of 83% and draws a current of 137 A from a 208-V, 60-Hz line. What is the (a) apparent and reactive power? Find (b) the capacitive reactance and (c) the capacitance required for a power factor correction to 95%.
18. A plant draws 100 kW of true power at a 93% lagging power factor from a 240-V, 60-Hz line. Determine (a) the apparent power, (b) the reactive power and (c) the capacitance of a parallel connected capacitor that will increase the power factor to 98%.

CHAPTER

22

TWO-PHASE AND THREE-PHASE CIRCUITS

OBJECTIVES

After satisfactorily completing this chapter, the student should be able to:

- Define various words and terms associated with the generation of power in a three-phase system and state the advantages of using three-phase circuits.
- Analyze voltage, current and power in two-phase circuits and in delta- and wye-connected three-phase alternators with balanced loads using double subscript notation.

The ac circuits described in previous chapters have all been single-phase. The generated voltage is a sine wave repeating continuously at sixty cycles per second. The energy is delivered in two power surges each cycle. The pulsating nature of single-phase power is often undesirable since the power will be zero twice each cycle also.

If power can be delivered at a more constant rate, the alternator and other electrical equipment operates more smoothly and more quietly. Such a result is possible using a multiple-phase system. This chapter begins with a description of two-phase generation. Two methods of connecting both alternators and loads in three-phase circuits are also explained.

POLYPHASE GENERATION

The single-phase alternator was introduced by describing a single loop coil rotating in a magnetic field, Figure 17-9. A two-phase alternator can be represented by two independent loops mounted rigidly on the rotor. Three-phase alternators have three independent loops. Such descriptions of alternators are over-simplified but help to explain how alternators work.

The advantages of using three-phase generation, distribution and usage are as follows:

- Three-phase alternators operate more efficiently than single-phase alternators.
- Three-phase alternators are less expensive to manufacture.
- Three-phase systems use smaller size copper wire in the distribution system, for alternator and motor windings and in transformers.
- With the same load in each of the three phases, the power output of the alternator is constant. This results in smoother operation and a steady conversion of mechanical energy to electrical energy.
- Three-phase motors are more economical to operate because they use energy at a steady rather than a pulsating rate.

Such advantages will be discussed in more detail as the concepts of two-phase and three-phase circuits are developed.

22-1 DOUBLE-SUBSCRIPT NOTATION AND TWO-PHASE ALTERNATORS

The single-phase alternator, Figure 17-9, is easily changed to a two-phase alternator. A second coil (loop) is added at right angles to the first coil, Figure 22-1. The two coils are mounted on the rotor exactly 90° apart and rotate at the same angular speed.

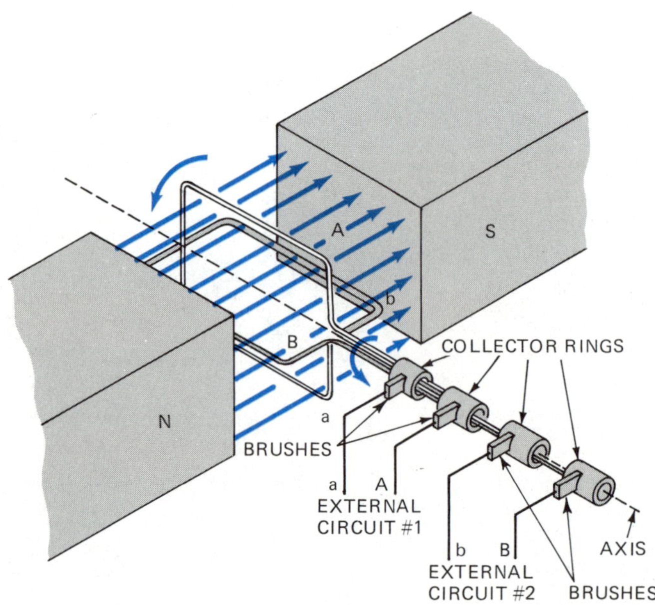

Figure 22-1 The two-phase alternator. The two coils are rigidly mounted on the rotor 90° apart.

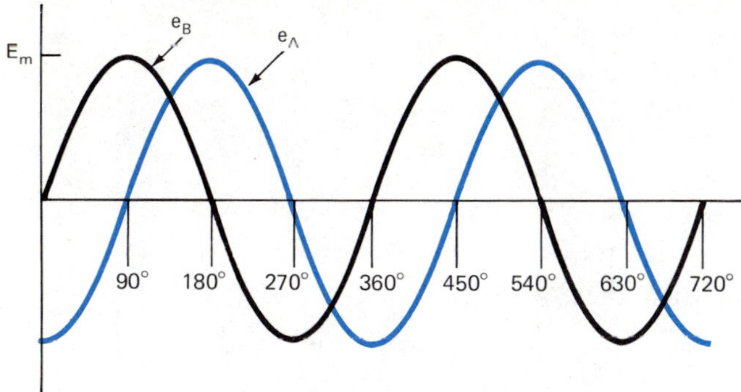

Figure 22-2 The voltage waveforms produced by a two-phase alternator. Voltage e_B leads voltage e_A by 90°.

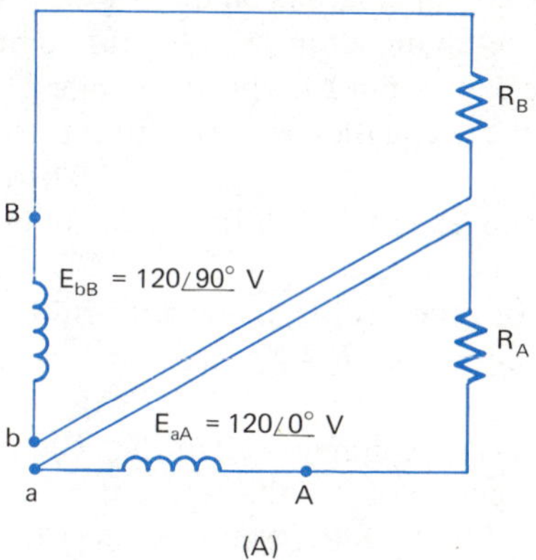

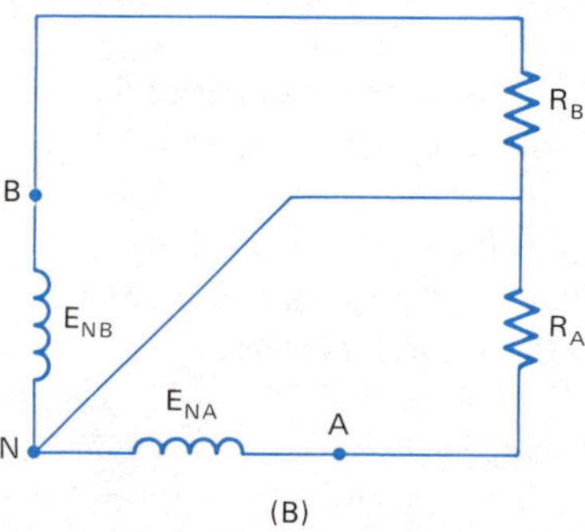

Figure 22-3 Possible circuits for a two-phase alternator. (A) Independent circuits, (B) A circuit containing a common return (neutral) line. The two coils are indicated by the inductor symbol and set at right angles to indicate the 90° phase angle.

If both coils have the same number of turns, they produce equal voltages. Notice that coil B cuts the magnetic field 90° ahead of coil A. The voltage in coil B is always 90° out of phase and leading the voltage produced by coil A, Figure 22-2.

The circuit diagrams for a two-phase circuit are shown in Figure 22-3. The alternators are represented as coils (inductors). The inductors are drawn at right angles to each other to indicate the voltages are 90° out of phase. The two generated voltages can be used as sources for separate independent circuits, Figure 22-3(A). It makes more sense, however, to use a *common line* or *neutral line*. This allows the use of only three wires instead of four, Figure 22-3(B). The abbreviation (2ϕ 3W) is often used to mean "two-phase three-wire systems." Similar abbreviations are used for various three-phase systems.

The voltages measured from the neutral to one of the lines (N to A or N to B) are called the *phase voltages*. They are indicated as E_{NA} and E_{NB}, using a double-subscript notation. The *line-to-line voltage* or simply the *line voltage* (A to B) is indicated as E_{AB}. The first subscript indicates the starting point of the alternator coil. The second subscript is the ending point.

If the leads of one of the coils is reversed, the voltage induced in that coil is changed by 180°. The amplitude and frequency, however, will remain the same. If the leads of coil B are reversed, for example, the voltage will now lag that in coil A by 90°, Figure 22-4. The student should study these figures very carefully.

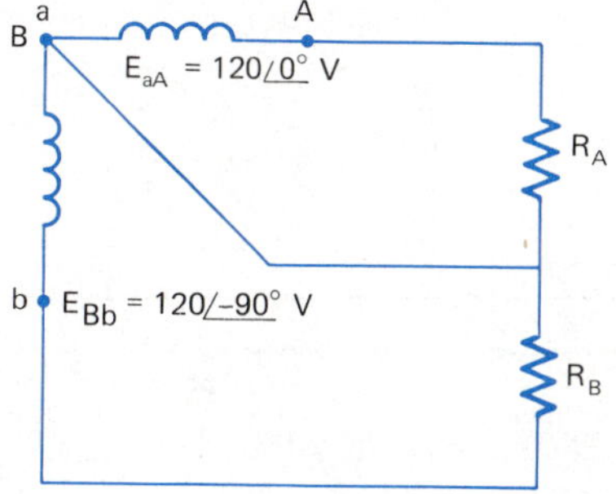

Figure 22-4 A two-phase circuit with coil leads b and B reversed. The voltage in coil B now lags that produced by coil A.

To understand the double-subscript notation, consider the line ABC, Figure 22-5. The line consists of two segments with lengths AB and BC. The total length of the line is AC. Think of the segment AB as an arrow having a direction from A to B. The segment BC is an arrow from B to C. The sum (tail-to-tip method) is AC, an arrow from A to C.

$$AC = AB + BC$$

Figure 22-5 Segments of a line: AC = AB + BC

The common letters on the right side (B, B) are adjacent letters. To obtain the resultant, drop the adjacent common letters (B, B) and the plus sign. The expression AB +BC is combined and becomes AC.

If the equation is solved for AB, it becomes:

$$AB = AC - BC$$

In this equation, the right side contains a common letter (C, C) but they are not adjacent. The vector –BC is the negative of the vector from B to C. In other words, the vector –BC is the vector CB. By substitution, then, the equation for AB becomes:

$$AB = AC + CB$$

Now the common letters (C, C) on the right are adjacent and can be dropped to obtain the resultant AB.

The same ideas can be applied to the subscripts associated with voltage phasors. When two in-phase alternators are connected in series, Figure 22-6(A), their voltages add, Figure 22-6(B):

$$\mathbf{E_{AC}} = \mathbf{E_{AB}} + \mathbf{E_{BC}}$$

Notice the common subscript letters (B, B) are adjacent.

If this equation is solved for $\mathbf{E_{AB}}$ it becomes:

$$\mathbf{E_{AB}} = \mathbf{E_{AC}} - \mathbf{E_{BC}}$$

Voltage $\mathbf{E_{AB}}$ must equal $\mathbf{E_{AC}} + \mathbf{E_{CB}}$ in order to drop the subscript letter C. The negative of a phasor, therefore, must be equal to the phasor itself with the subscripts reversed.

$$\mathbf{E_{CB}} = -\mathbf{E_{BC}}$$

The voltage phasors are shown in Figure 22-6(C). Notice that the negative sign simply means 180° rotation, the same as usual. The double-subscript notation is just a new method of indicating vector or phasor addition and subtraction.

Returning to the two-phase alternator, recall that its two separate armature coils are at right angles. They, too, are connected in series,

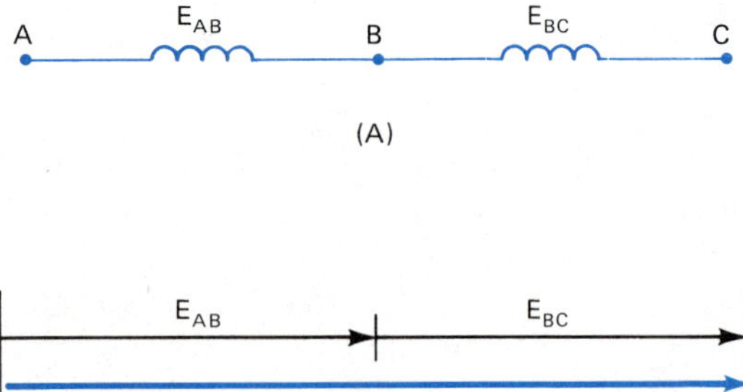

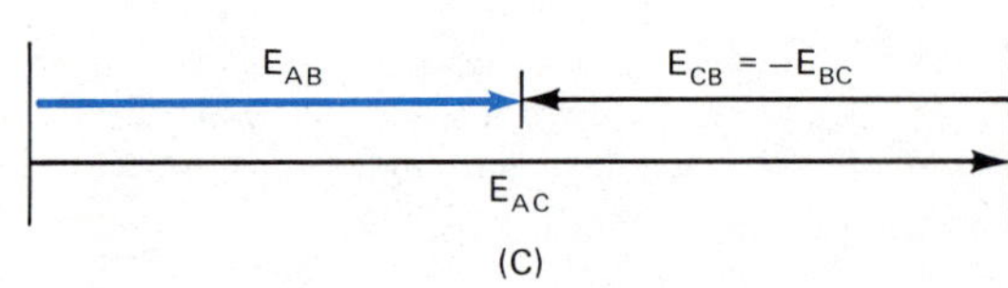

Figure 22-6 Two in-phase alternator coils in series: $E_{AC} = E_{AB} + E_{BC}$

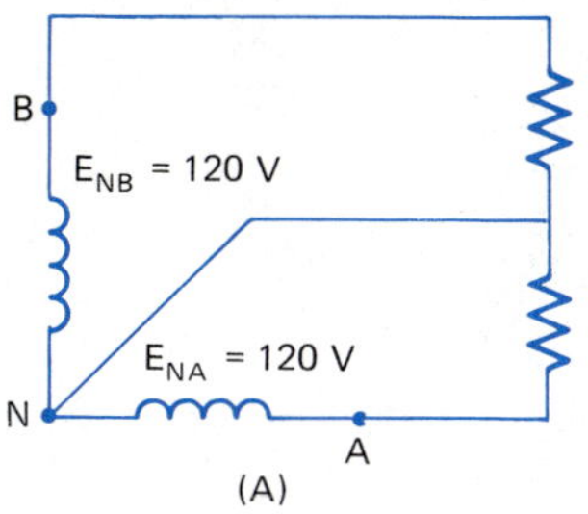

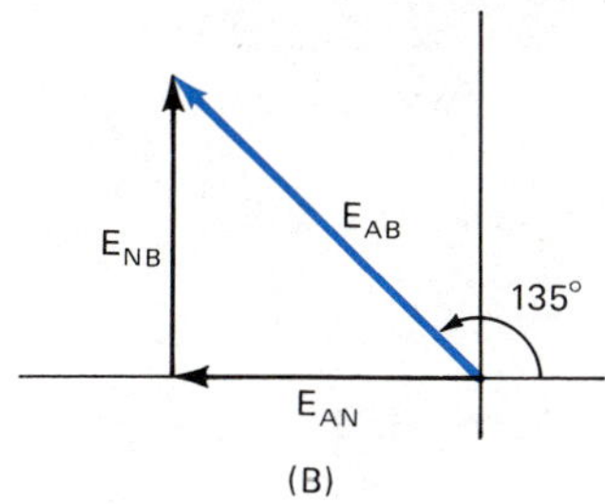

Figure 22-7 The line voltage is the phasor sum of the phase voltages: $\mathbf{E_{AB}} = \mathbf{E_{AN}} + \mathbf{E_{NB}}$

Figure 22-7(A). The phase voltages produced are $\mathbf{E_{NA}}$ and $\mathbf{E_{NB}}$. The phasor $\mathbf{E_{NA}}$ points in the reference direction (0°) and $\mathbf{E_{NB}}$ at 90°. Both phase voltages are measured from the neutral to lines A or B.

The line voltage ($\mathbf{E_{AB}}$) can be found by tracing through the generator from point A to B:

$$\mathbf{E_{AB}} = \mathbf{E_{AN}} + \mathbf{E_{NB}}$$

The first voltage is $\mathbf{E_{AN}}$, the negative of $\mathbf{E_{NA}}$. It has a phase angle of 180°. Since the voltage phasors are at right angles, Figure 22-7(B), they must be combined by phasor addition in rectangular form.

Example A Find the line voltage (A to B) in the two-phase circuit of Figure 22-7(A). The phasor diagram is shown in Figure 22-7(B). The magnitude of each phase voltage is 120 V.

Solution: Transform to rectangular form and add.

Data: $\mathbf{E_{AN}} = 120\angle 180^\circ \text{ V} \rightarrow -120 + j\,0 \text{ V}$

$\mathbf{E_{NB}} = 120\angle 90^\circ \text{ V} \rightarrow 0 + j120 \text{ V}$

Sum: $\mathbf{E_{AB}} = 170\angle 135^\circ \text{ V} \leftarrow -120 + j120 \text{ V}$

The line voltage is greater than either phase voltage. It is not, however, twice as great. The effective value of the line voltage could have been found by using the Pythagorean Theorem. The symbols E_P and E_L are often used to represent the phase voltage and the line voltage respectively when only the magnitude is important. Since the phase voltages have equal effective values, then:

$$E_L = \sqrt{(E_P)^2 + (E_P)^2} = \sqrt{2(E_P)^2}$$

$$\mathbf{E_L = \sqrt{2}E_P}$$

The line voltage is the square root of two times either phase voltage ($\sqrt{2}$ = 1.414).

Example B Find the line voltage if the phase voltage is 440 V.

Solution: Formula: $E_L = \sqrt{2}E_P$

Substitute: $E_L = \sqrt{2}(440 \text{ V})$

$E_L = 622 \text{ V}$

EXERCISE 22-1

Find the magnitude of the line voltage for each of the following phase voltages:

1. 240 V 2. 48 V 3. 4160 V 4. 13.8 kV

Determine the magnitude of the phase voltage for each of the following line voltages:

5. 10.2 kV 6. 3390 V 7. 311 V 8. 488 kV

Data, diagram, formula, substitute, check.

9. A two-phase alternator has a phase voltage of 600 V. (a) Express both phase voltages in polar form. (b) Evaluate the line voltage by adding in rectangular form. (c) Construct a voltage phasor diagram showing the two-phase voltages and the line voltage.
10. A two-phase alternator produces a line voltage of 1000 V. Express the phase voltages and the line voltage in polar form.

22-2 CURRENT AND POWER IN TWO-PHASE CIRCUITS

The current in the lines and in the neutral can be identified by single subscripts, Figure 22-8. Certain currents in three-phase circuits, however, will utilize the double-subscript notation. A two-phase circuit with equal impedances in each branch has a *balanced load,* Figure 22-8. In such circuits, the line currents ($\mathbf{I_A}$ and $\mathbf{I_B}$) are equal but not in phase. The neutral current ($\mathbf{I_N}$) is the phasor sum of the two line currents. Apply Kirchhoff's Current Law at point N:

$$\mathbf{I_N} = \mathbf{I_A} + \mathbf{I_B}$$

Line currents can be found using Ohm's Law. Remember that all quantities are phasors.

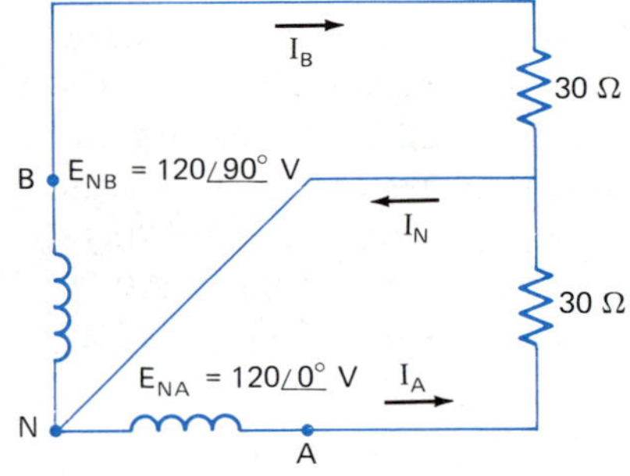

Figure 22-8

Example A Find the line and neutral currents in the circuit of Figure 22-8.

Solution: Data: $\mathbf{E_{NA}} = 120\angle 0^\circ$ V, $\mathbf{E_{NB}} = 120\angle 90^\circ$ V, $\mathbf{Z_A} = \mathbf{Z_B} = 30\angle 0^\circ\ \Omega$

Formula: $\mathbf{I} = \dfrac{\mathbf{E}}{\mathbf{Z}}$

Substitute: $\mathbf{I_A} = \dfrac{120\angle 0^\circ \text{ V}}{30\angle 0^\circ\ \Omega} = 4\angle 0^\circ$ A

Substitute: $\mathbf{I_B} = \dfrac{120\angle 90^\circ \text{ V}}{30\angle 0^\circ\ \Omega} = 4\angle 90^\circ$ A

Construct current phasor diagram, Figure 22-9, transform phase currents to rectangular form and add to find the neutral current.

$\mathbf{I_A} = 4\angle 0^\circ$ A $\rightarrow 4 + j0$ A

$\mathbf{I_B} = 4\angle 90^\circ$ A $\rightarrow 0 + j4$ A

$\mathbf{I_N} = 5.66\angle 45^\circ$ A $\leftarrow 4 + j4$ A

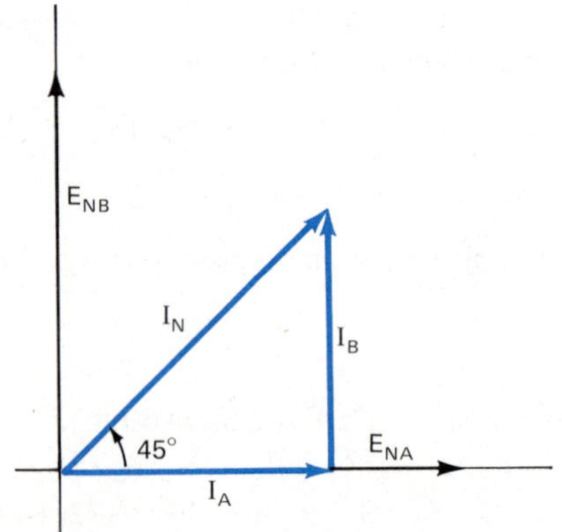

Figure 22-9

Notice the current in phase A is in phase with the voltage E_{NA}. The current in phase B is in phase with the voltage E_{NB}. This is true only because the balanced loads are pure resistive. The circuit, however, may contain reactance as well as resistance. The loads would not be balanced, however, unless both contain equivalent impedance. The two loads need not contain identical circuit elements. In fact, they can be quite different complex combinations. To be balanced, however, the equivalent impedance of phase A must equal the equivalent impedance of phase B.

The current in the neutral line in Figure 22-8 leads the voltage in phase A ($\mathbf{E_{NA}}$) but lags that in phase B ($\mathbf{E_{NB}}$). The magnitude of the neutral current in circuits with balanced loads can be calculated with the formula:

$$\mathbf{I_N = \sqrt{2}I_A}$$

A two-phase circuit with balanced loads has equal currents through the loads. Voltage drops across the loads are equal. The power supplied to the two loads (P_A and P_B) must also be equal. True power in each phase is the product of the phase voltage, phase current and the power factor.

$$P_A = (E_{NA})(I_A)\cos\theta$$

$$P_B = (E_{NB})(I_B)\cos\theta$$

The total power is twice the power supplied to either load:

$$\mathbf{P_T = 2P_A}$$

Example B A balanced two-phase circuit (240 V) has an impedance of 40 Ω at 81.9% lagging power factor in each load. Find (1) the two line currents, (2) the neutral current, and (3) the total power.

Solution: Construct circuit diagram, Figure 22-10(A).

Data: $\mathbf{E_{NA}} = 240\angle 0^\circ$ V, $\mathbf{E_{NB}} = 240\angle 90^\circ$ V,
$\theta = \cos^{-1} 0.819 = 35.0^\circ$, $\mathbf{Z_A} = 40\angle 35.0^\circ\ \Omega$,
$\mathbf{Z_B} = 40\angle 35.0^\circ\ \Omega$

(1) $$\mathbf{I_A} = \frac{\mathbf{E_{NA}}}{\mathbf{Z_A}} = \frac{240\angle 0^\circ \text{ V}}{40\angle 35.0^\circ\ \Omega} = 6\angle -35.0^\circ \text{ A}$$

$$\mathbf{I_B} = \frac{\mathbf{E_{NB}}}{\mathbf{Z_B}} = \frac{240\angle 90^\circ \text{ V}}{40\angle 35.0^\circ\ \Omega} = 6\angle 55.0^\circ \text{ A}$$

(2) $\mathbf{I_A} = 6\angle -35.0^\circ$ A $\rightarrow 4.91 - j3.44$ A
$\mathbf{I_B} = 6\angle 55.0^\circ$ A $\rightarrow 3.44 + j4.91$ A
$\mathbf{I_N} = 8.49\angle 10.0^\circ$ A $\leftarrow 8.35 + j1.47$ A

Check: $I_N = \sqrt{2}(6 \text{ A}) = 8.49$ A

(3) $P_A = (E_{NA})(I_A)\cos\theta = (240 \text{ V})(6 \text{ A}) \cos 35.0^\circ = 1.18$ kW
$P_B = (E_{NB})(I_B)\cos\theta = (240 \text{ V})(6 \text{ A}) \cos 35.0^\circ = 1.18$ kW
$P_T = 2P_A = 2.36$ kW

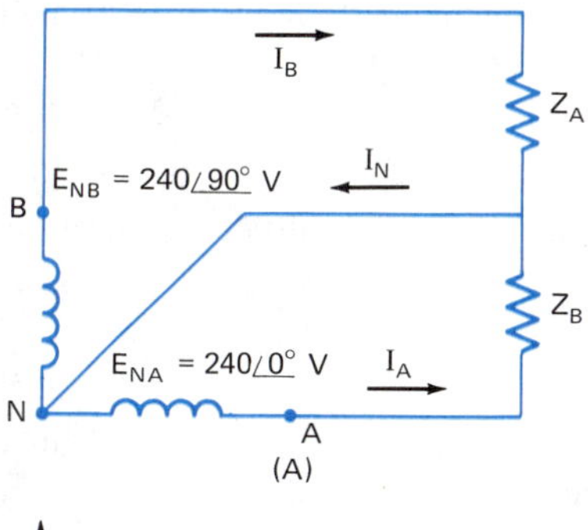

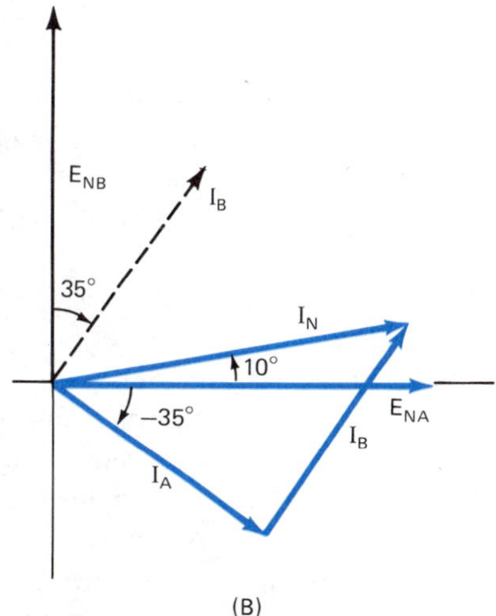

Figure 22-10

Note: The angles indicated by the currents are in reference to the *x*-axis. $\mathbf{I_B}$ lags $\mathbf{E_{NB}}$ by 35°, Figure 22-10(B). The neutral current $\mathbf{I_N}$ bisects the right angle formed by phasors $\mathbf{I_A}$ and $\mathbf{I_B}$.

EXERCISE 22-2

Data, diagram, formula, substitute, check.

1. A two-phase circuit is balanced with pure resistive loads. The line voltage is 240 V and the line current is 45 A. Express (a) the line voltage, (b) both phase voltages, (c) the two line currents and (d) the neutral current in polar form. Assume the phase voltage $\mathbf{E_{NA}}$ is in the reference direction.

Find the magnitude of the neutral current for each of the following line currents.

2. 3.5 A
3. 565 mA
4. 16.8 A
5. 78.9 mA

What is the magnitude of the line current for each of the following neutral currents.

6. 475 mA
7. 2.75 A
8. 9.5 A
9. 27.5 A

In each of the following problems draw the two-phase circuit diagram and the voltage-current phasor diagram. Solve for the indicated unknowns expressing all current phasors in polar form.

10. $E_{NA} = E_{NB} = 220$ V, $Z_A = Z_B = 88\ \Omega$ (pure resistance). Find $\mathbf{I_A}, \mathbf{I_B}, \mathbf{I_N}, P_A, P_B, P_T$.
11. $E_{NA} = E_{NB} = 480$ V, $Z_A = Z_B = 24\ \Omega$ (pure inductance). Find $\mathbf{I_A}, \mathbf{I_B}, \mathbf{I_N}, Q_A, Q_B, Q_T, P_T$.
12. $E_{NA} = E_{NB} = 120$ V, $Z_A = Z_B = 6\ \Omega$ (pure capacitance). Find $\mathbf{I_A}, \mathbf{I_B}, \mathbf{I_N}, Q_A, Q_B, Q_T, P_T$.
13. $E_{NA} = E_{NB} = 240$ V, $\mathbf{Z_A} = \mathbf{Z_B} = 12\underline{/25^\circ}\ \Omega$. Find $\mathbf{I_A}$, $\mathbf{I_B}$, $\mathbf{I_N}$, P_A, P_B and P_T.
14. A two-phase circuit has phase voltages of 120 V. The load in each phase contains an equivalent impedance of 20.8 + 12.0 Ω. (a) Find the phase angle and the power factor. (b) Find the phase and neutral currents. (c) Calculate the total true power.
15. A two-phase circuit has phase voltages of 220 V. The line currents are each 44.0 A with a lagging power factor of 65%. (a) Express the line currents and the neutral current in polar form. (b) What is the impedance of the two loads? (c) What is the true power?
16. If the neutral current is $33.9\underline{/27.5^\circ}$ A, (a) what are the two line currents (polar form)? What is (b) the phase angle and (c) the power factor? Assume the loads are balanced.
17. If the phase voltages in the circuit of Problem 16 are 480 V, (a) find the impedance of each load written in polar form. (b) Find the total true power drawn by the circuit.
18. The phase loads of a balanced circuit are each 9.6 Ω at a lagging power factor of 75%. Find (a) the true power drawn by each load if the line current is 62.5 A. Determine (b) the phase voltages (polar form) and (c) the line voltage.

THREE-PHASE GENERATION

The two-phase generation and distribution system described in the previous topics was used in the early history of the electrical industry. Some industries still use some of the original two-phase equipment, especially motors.

The advantages of two-phase over single-phase are great but it was obvious that three-phase circuits were even better. Many other polyphase systems have been tested and used and some have special applications today. It is, however, the three-phase system that is always used for power generation and distribution.

22-3 THE WYE-CONNECTED ALTERNATOR

The two-phase alternator, Figure 22-1, could easily contain a third coil (loop). With these coils set at exactly 120° intervals, it becomes a three-phase alternator. The three coils each produce independent sine-wave voltages. The voltage of any one coil is one-third of a cycle ahead of one of the other coils and one-third of a cycle behind the third.

An end view of a three-phase alternator is shown in Figure 22-11. The three coils, aA, bB and cC, are wound rigidly onto the rotor. All three have the same number of turns so all three voltages have the same magnitudes. Since rotation is at the same angular speed, all three have the same frequency. The sine-waves are shown in Figure 22-12. The three separate voltages can be brought out by means of brushes on slip rings. The wires lead to terminals at a panel on the alternator.

The six wires can be used to deliver power by means of three separate single-phase circuits. In doing so, however, the advantages of three-phase power would be lost. One common method of linking the terminals is called the *four-wire, wye-connected system,* Figure 22-13. This three-phase

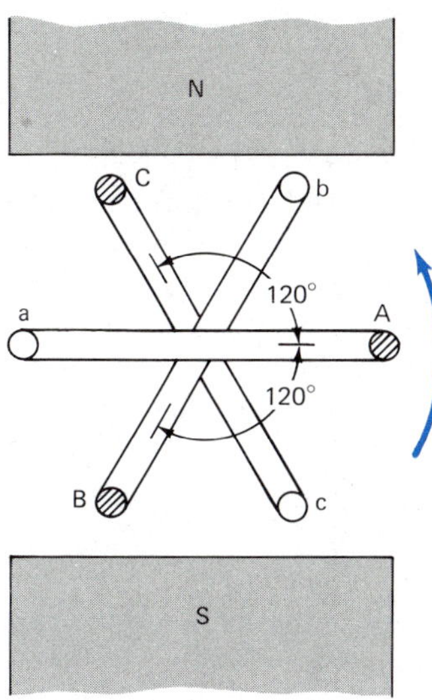

Figure 22-11 The end view of a three-phase alternator. The coils are mounted on the rotor at 120° intervals.

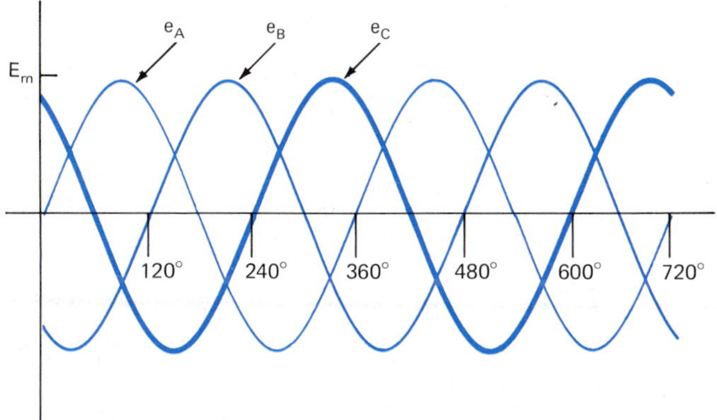

Figure 22-12 The voltage waveforms produced by a three-phase alternator. Voltage e_B lags e_A by 120° but leads e_C by 120°.

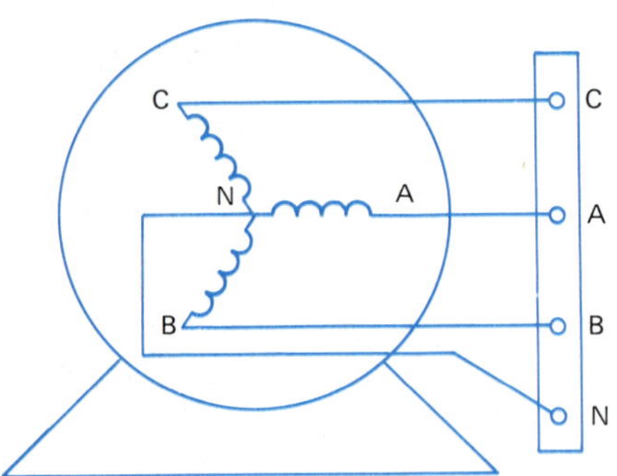

Figure 22-13 The wye-connected alternator

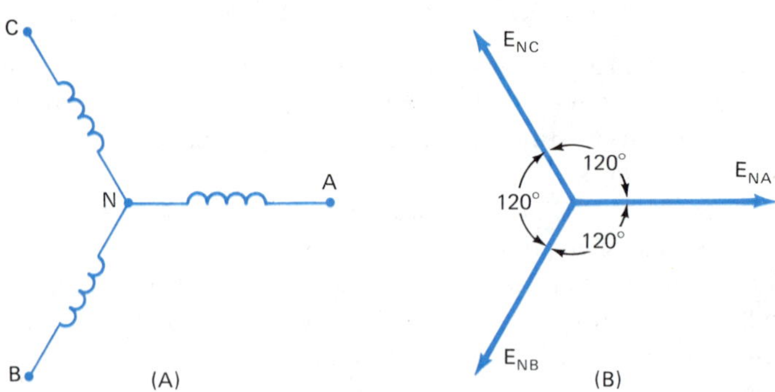

Figure 22-14 Voltage phasors for a wye-connected alternator

connection is shown in Figure 22-14(A). The three lowercase letter terminals are connected together. These letters are replaced by the letter N since the neutral line will connect at this junction.

The coils, represented by the inductor symbol, are at 120° intervals. If the coils were rotated 30° ccw, they would form the letter Y (wye) for which the connection is named. The orientation shown in Figure 22-14 is preferred, however, so phase A is in the reference direction. If the magnitude of the voltage is 220 V, the three-phase voltages are as follows:

$$\mathbf{E_{NA}} = 220\underline{/0^\circ}\ \text{V}$$
$$\mathbf{E_{NB}} = 220\underline{/120^\circ}\ \text{V}$$
$$\mathbf{E_{NC}} = 220\underline{/240^\circ}\ \text{V}$$

The three-phase source does not need to be an actual alternator. The secondary windings of three transformers, Figure 22-15, can represent the coils of a three-phase source just as well. Of course, the primary coils must draw power from a three-phase source.

The loads in the four-wire wye circuit can also be wye-connected. The orientation of the loads is not important since the phase angle indicated by the impedance phasor is the same in all three branches. They are drawn as shown in Figure 22-16 merely to indicate that they are wye-connected.

The four-wire wye circuit is very similar to the two-phase three-wire system, Figure 22-3(B). The circuit is traced from the neutral point (N) in

Figure 22-15 A three-phase source may be the secondary windings of three transformers provided the primary coils of the three transformers are fed by a three-phase system.

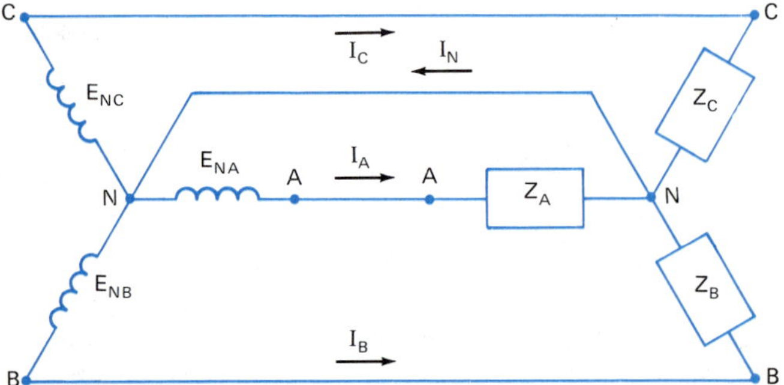

Figure 22-16 A four-wire wye-connected three-phase circuit

the alternator outward through one of the phase coils. It continues along one of the lines to the load and passes through that load. The circuit is completed by tracing back to the starting point along the neutral line.

The *phase voltages* indicated previously are the line-to-neutral voltages. The *line voltages* are the line-to-line voltages: $\mathbf{E}_{AB}$, $\mathbf{E}_{BC}$ and $\mathbf{E}_{CA}$. Line voltages can be determined by tracing through the generator as done with two-phase circuits. Line voltage $\mathbf{E}_{AB}$, for example, is found by tracing from point A to point B:

$$\mathbf{E}_{AB} = \mathbf{E}_{AN} + \mathbf{E}_{NB}$$

Notice the subscripts are correct for phasor addition. Also, recall that:

$$\mathbf{E}_{AN} = -\mathbf{E}_{NA}$$

The analytical addition must be done in rectangular form.

Example A In a four-wire wye system, a three-phase alternator produces a phase voltage of 220 V. (1) Express each phase voltage in polar form, and (2) determine the line voltage $\mathbf{E}_{AB}$.

Solution: Draw circuit and phasor diagrams, Figure 22-17.

(1) Refer to Figure 22-17(A) and (B).

$$\mathbf{E}_{NA} = 220\angle 0^\circ \text{ V}$$
$$\mathbf{E}_{NB} = 220\angle 240^\circ \text{ V}$$
$$\mathbf{E}_{NC} = 220\angle 120^\circ \text{ V}$$

(2) $\mathbf{E}_{AB} = \mathbf{E}_{AN} + \mathbf{E}_{NB}$, Figure 22-17(C). Add in rectangular form:

$$\mathbf{E}_{AN} = 220\angle 180^\circ \text{ V} \rightarrow -220 + j\ \ 0 \text{ V}$$
$$\mathbf{E}_{NB} = 220\angle 240^\circ \text{ V} \rightarrow -110 - j191 \text{ V}$$
$$\mathbf{E}_{AB} = 381\angle 210^\circ \text{ V} \leftarrow -330 - j191 \text{ V}$$

The other line voltages ($\mathbf{E}_{BC}$ and $\mathbf{E}_{CA}$) can be found using the same method. The magnitudes are the same. In the previous example, the line voltage (381 V) is 1.732 times larger than the phase voltage (220 V). The

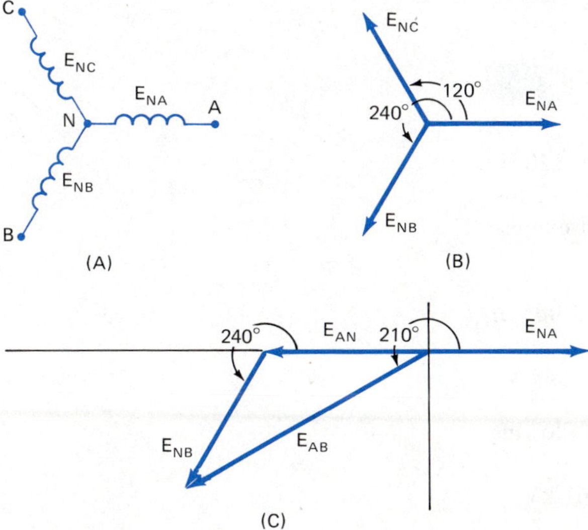

Figure 22-17

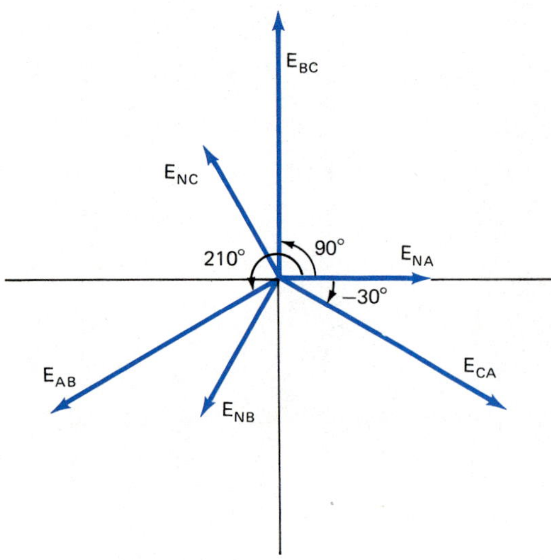

Figure 22-18 Phase voltage and line voltage phasors

number 1.732 is the square root of three ($\sqrt{3}$). For the Y-connected source, the magnitude of the line voltage (E_L) is given by the formula:

$$E_L = \sqrt{3}E_P$$

where E_P is the magnitude of the phase voltage.

Example B Find the line voltage in a four-wire wye circuit if the phase voltage is 440 V.

Solution: Formula: $E_L = \sqrt{3}E_P$

Substitute: $E_L = 1.732(440\ V) = 762\ V$

The phase of line voltage $\mathbf{E_{AB}}$ was found to be 210° in Example A. The phases of $\mathbf{E_{BC}}$ and $\mathbf{E_{CA}}$ are 90° and −30° respectively. The phase and line voltages are all pictured in Figure 22-18. All are taken in reference to phase voltage $\mathbf{E_{NA}}$.

EXERCISE 22-3

Data, diagram, formula, substitute, check.

1. A wye-connected alternator produces a phase voltage of 120 V. (a) Express each of the phase voltages in polar form taking the A phase as reference. (b) Evaluate the line voltage and express each of the three in polar form.

For each of the following phase voltages, determine the magnitude of the line voltage. The circuits are four-wire wye-connected.

2. 19.92 kV
3. 277 V
4. 7.62 kV
5. 347 V

Find the magnitude of the phase voltage for each of the following given line voltages. The circuits are four-wire wye-connected.

6. 345 kV
7. 2400 V
8. 6.93 kV
9. 480 V
10. 161 kV

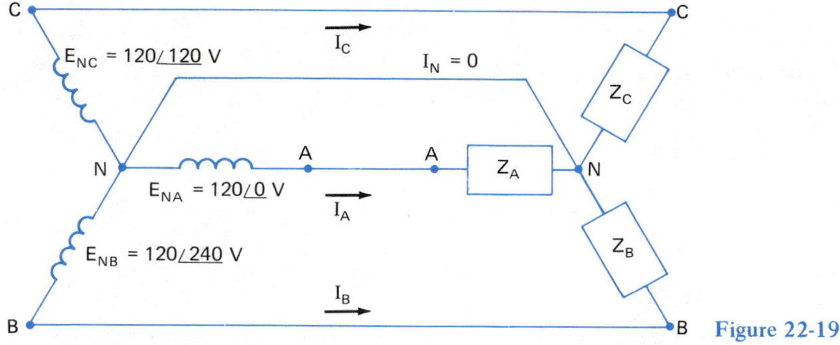

Figure 22-19

22-4 CURRENT AND POWER IN FOUR-WIRE WYE CIRCUITS

In the four-wire wye circuit, Figure 22-19, current in the alternator coil, line and load is the same in any one of the three phases. Trace the circuit from point N in the alternator outward through the coil, along the line and through the load to point N. Phase currents through the load are the same as the line currents. Each phase can be traced in that manner. Currents in all three phases combine and return to point N in the alternator through the neutral line. Currents in the four-wire system can be identified by single subscripts:

$$\mathbf{I_N} = \mathbf{I_A} + \mathbf{I_B} + \mathbf{I_C}$$

The line currents can be calculated by applying Ohm's Law to each phase:

$$\mathbf{I_A} = \frac{\mathbf{E_{NA}}}{\mathbf{Z_A}}$$

$$\mathbf{I_B} = \frac{\mathbf{E_{NB}}}{\mathbf{Z_B}}$$

$$\mathbf{I_C} = \frac{\mathbf{E_{NC}}}{\mathbf{Z_C}}$$

If the load impedances $\mathbf{Z_A}$, $\mathbf{Z_B}$ and $\mathbf{Z_C}$ are equal in magnitude and phase, the load is said to be balanced. With balanced wye loads, the neutral current is zero. This is true regardless of the power-factor angle.

Example A The load in Figure 22-19 is balanced with each phase impedance having a value of $40\angle 35^\circ\ \Omega$ (lagging). Find (1) the line current in each phase in polar form, (2) display the line currents on a voltage-current phasor diagram. Calculate the neutral current by adding in (3) rectangular form and (4) graphically.

Solution: Data: $\mathbf{Z_A} = \mathbf{Z_B} = \mathbf{Z_C} = 40\angle 35^\circ\ \Omega$, $\mathbf{E_{NA}} = 120\angle 0^\circ$ V, $\mathbf{E_{NB}} = 120\angle 120^\circ$ V, $\mathbf{E_{NC}} = 120\angle 240^\circ$ V

(1) $$\mathbf{I_A} = \frac{\mathbf{E_{NA}}}{\mathbf{Z_C}} = \frac{120\angle 0^\circ \text{ V}}{40\angle 35^\circ\ \Omega} = 3\angle -35^\circ \text{ A}$$

$$\mathbf{I_B} = \frac{\mathbf{E_{NB}}}{\mathbf{Z_B}} = \frac{120\angle 120^\circ \text{ V}}{40\angle 35^\circ\ \Omega} = 3\angle 85^\circ \text{ A}$$

$$\mathbf{I_C} = \frac{\mathbf{E_{NC}}}{\mathbf{Z_C}} = \frac{120\underline{/240^\circ}\text{ V}}{40\underline{/35^\circ}\ \Omega} = 3\underline{/205^\circ}\text{ A}$$

(2) See Figure 22-20(A). Notice the currents are 120° apart and lag their respective voltages by the 35° power-factor angle.

(3) $\mathbf{I_N} = \mathbf{I_A} + \mathbf{I_B} + \mathbf{I_C}$

$\mathbf{I_A} = 3\underline{/-35^\circ}\text{ A} \rightarrow 2.46 - j1.72\text{ A}$

$\mathbf{I_B} = 3\underline{/85^\circ}\text{ A} \rightarrow 0.26 + j2.99\text{ A}$

$\mathbf{I_C} = 3\underline{/205^\circ}\text{ A} \rightarrow -2.72 - j1.27\text{ A}$

$\mathbf{I_N} = 0\text{ A} \leftarrow 0 + j0\text{ A}$

(4) See Figure 22-20(B). When adding phasors graphically (tail-to-tip), the resultant is a phasor having a length from the tail of the first phasor plotted to the tip of the last one. In this case, the resultant is zero since the figure begins at the origin and ends at the origin.

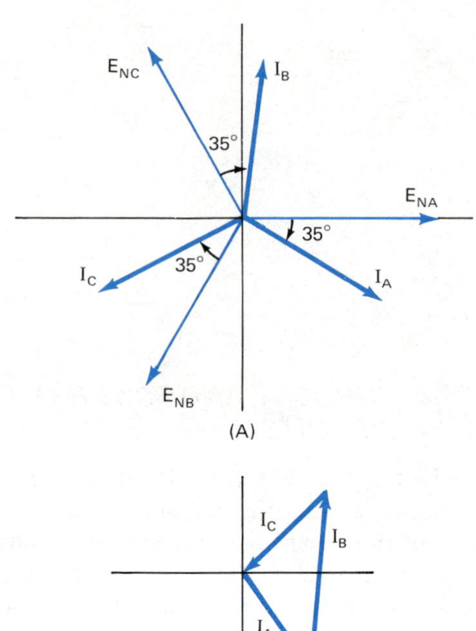

Figure 22-20

In the four-wire circuit, phase voltages have equal magnitudes. They always point in the same (0°, 120° and 240°) directions with $\mathbf{E_{NA}}$ as reference. Line currents lead or lag their respective voltages by the same amount depending on the impedance of the phase loads. If the load is

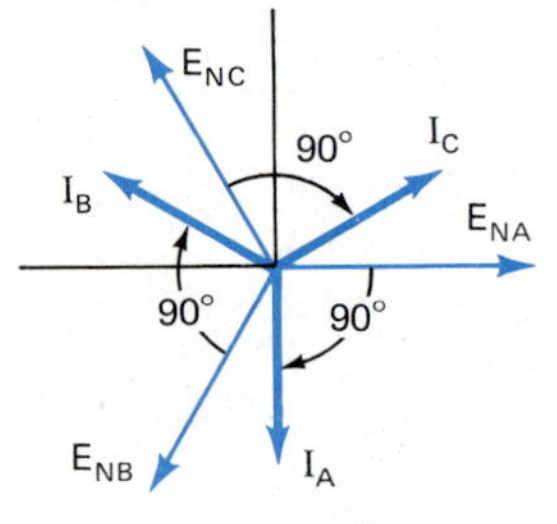

(A) Pf = 0%
$\theta = 90^\circ$ lagging
Pure inductance

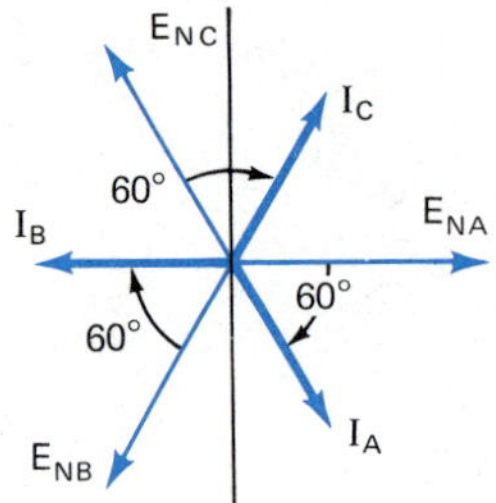

(B) Pf = 50%
$\theta = 60^\circ$ lagging
Inductive

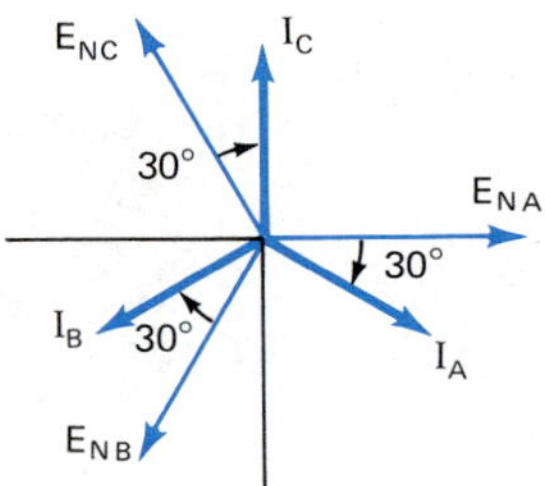

(C) Pf = 86.6%
$\theta = 30^\circ$ lagging
Inductive

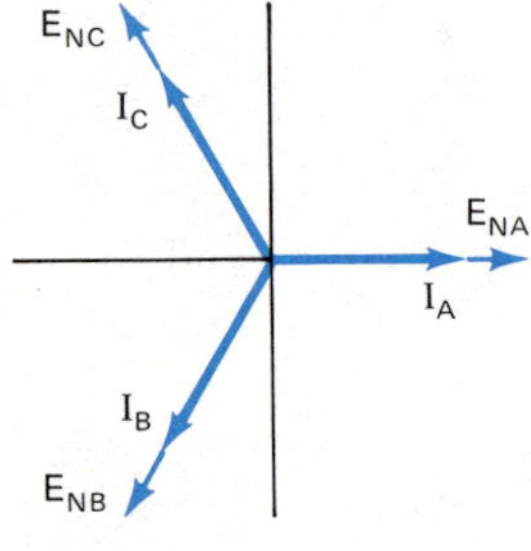

(D) Pf = 100%
$\theta = 0^\circ$
Pure resistance

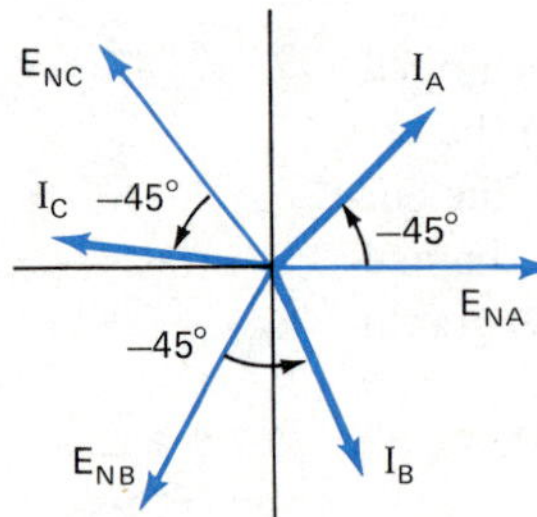

(E) Pf = 70.7%
$\theta = -45^\circ$ leading
Capacitive

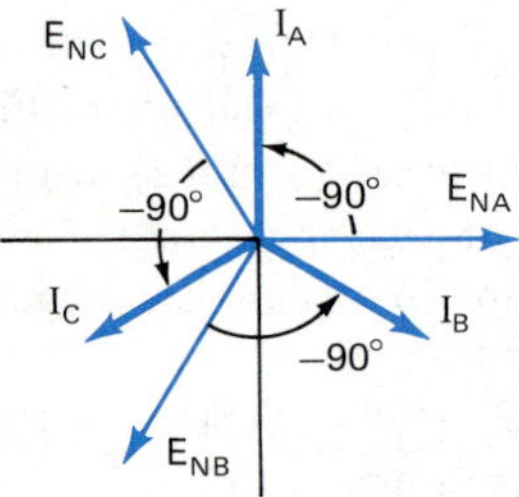

(F) Pf = 0%
$\theta = -90^\circ$ leading
Pure capacitance

Figure 22-21 Voltage-current phasors for various loads

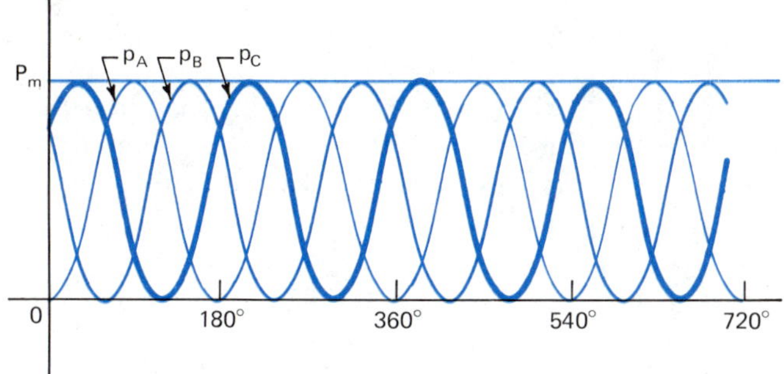

Figure 22-22 Instantaneous power in the four-wire pure resistive circuit. The total power at any instant is constant: $p_A + p_B + p_C = P_m$.

balanced, each pair of current phasors are always separated by 120° and the neutral current is zero. Study the diagrams of Figure 22-21 carefully.

The graphs of the instantaneous power for all three phases are plotted in Figure 22-22 for a pure resistive balanced system. The total instantaneous power at any instant is the sum of the instantaneous power of the three phases at that instant. At any instant, the sum of the three instantaneous powers is constant. This is true even though the instantaneous power in each load is pulsating.

This is one of the important superiorities of three-phase systems. In large equipment such as alternators and motors, the steady conversion of power from mechanical to electrical or electrical to mechanical provides smoother, quieter and more efficient operation.

The total apparent power in a three-phase system is the sum of the powers supplied to each phase load. The power in each phase will be the same if the loads are balanced.

$$S = S_A = S_B = S_C$$

$$S_T = 3S$$

The apparent power in any of the phase loads will be the product of the phase voltage (E_P) and the phase current through the load (I_P).

$$S = E_P I_P$$

$$S_T = 3E_P I_P$$

When the power-factor angle is known, the apparent power can be written as a phasor in polar form. A transformation to rectangular form yields the true power (P) and reactive power (Q):

$$S\angle\theta \rightarrow P + jQ$$

$$S_T\angle\theta \rightarrow P_T + jQ_T$$

Example B In Example A, the phase voltage is 120 V. The current is 3 A with a lagging phase angle (power-factor angle) of 35°. (1) Find the apparent power delivered to each phase, (2) the total apparent power and (3) the total true and reactive powers.

Solution: Data: $E_P = 120$ V, $I_P = 3$ A, $\theta = 35°$

(1) Formula: $S = E_P I_P$

Substitute: $S = (120 \text{ V})(3 \text{ A}) = 360$ VA

(2) Formula: $S_T = 3S$

Substitute: $S_T = 3(360 \text{ VA}) = 1080$ VA

(3) Express the apparent power as a phasor: $1080\angle 35°$ VA

Formula: $S_T\angle\theta \rightarrow P_T + jQ_T$

Substitute: $1080\angle 35°$ VA $\rightarrow 885 + j619$ (W, vars)

Note: Recall that the transformation from polar to rectangular form can be performed with the calculator **P→R** or with the formulas:

$$P_T = S_T \cos\theta$$
$$Q_T = S_T \sin\theta$$

It is not always convenient to measure the phase voltages ($E_P = E_{NA} = E_{NB} = E_{NC}$) at the load. Line voltages ($E_L = E_{AB} = E_{BC} = E_{CA}$) and line currents, ($I_L$), however, are easily measured. Remember that $E_L = \sqrt{3}E_P$ and $I_L = I_P$ in the four-wire wye system. The apparent power in each phase using line measurements is as follows:

$$S = \frac{E_L I_L}{\sqrt{3}}$$

$$S_T = \frac{3E_L I_L}{\sqrt{3}} = \sqrt{3}E_L I_L$$

Example C Each load in a four-wire balanced wye circuit, Figure 22-23, has an impedance of 8 Ω with a 75% lagging power factor. The line voltage is 480 V. Find (1) the phase current, line current and neutral current, (2) the apparent power in each phase, and (3) the total apparent, true and reactive power of the circuit.

Solution: Data: $\theta = \cos^{-1} 0.75 = 41.4°$, $Z_A = Z_B = Z_C = 8\angle 41.4°$ Ω, $E_L = 480$ V

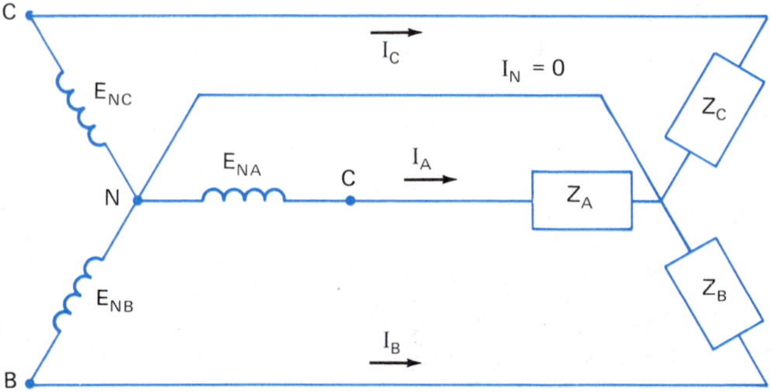

Figure 22-23

(1) $E_P = \frac{E_L}{\sqrt{3}} = \frac{480 \text{ V}}{\sqrt{3}} = 277 \text{ V}$

$I_P = \frac{E_P}{Z} = \frac{277 \text{ V}}{8 \ \Omega} = 34.6 \text{ A}$

$I_L = I_P = 34.6 \text{ A}$

$I_N = 0 \text{ A}$ (Since the circuit has balanced loads)

(2) $S = E_P I_P = (277 \text{ V})(34.6 \text{ A}) = 9.60 \text{ kVA}$

(3) $S_T = 3 \text{ S} = 3(9.60 \text{ kVA}) = 28.8 \text{ kVA}$

Express the total apparent power as a phasor in polar form: $28.8\angle 41.4^\circ$ kVA.

$S_T\angle\theta \rightarrow P_T + jQ_T$

$28.8\angle 41.4^\circ \text{ kVA} \rightarrow 21.6 + j19.0$ (kW, kvars)

Check: $S_T = \sqrt{3}E_L I_L = 1.732(480 \text{ V})(34.6 \text{ A}) = 28.8 \text{ kVA}$

A balanced four-wire wye-connected system is characterized by the following:

- The three loads are equal.
- The phase voltages (line to neutral) have equal magnitudes and are 120° out of phase with each other (0°, 120° and 240°).
- The line voltages (line to line) have magnitudes that are 1.732 times the phase voltage ($E_L = \sqrt{3}E_P$).
- Phase currents (current through each load) are the same as the line current and are equal ($I_L = I_P$).
- The neutral current being the phasor sum of the line currents is zero ($\mathbf{I_N} = \mathbf{I_A} + \mathbf{I_B} + \mathbf{I_C} = 0$).
- The apparent powers ($S = E_P I_P$) delivered to each of the three loads are equal.
- The total apparent power is three times the load power ($S_T = 3S = \sqrt{3}E_L I_L$).

EXERCISE 22-4

Data, diagram, formula, substitute, check.

1. A three-phase, four-wire, wye-connected system has the following phase voltages: $\mathbf{E_{NA}} = 120\angle 0^\circ$ V, $\mathbf{E_{NB}} = 120\angle 120^\circ$ V and $\mathbf{E_{NC}} = 120\angle 240^\circ$ V. The impedances are balanced with a pure resistance of 6 Ω in each arm. (a) Find the current in each phase expressed in polar form. (b) Show that the neutral current is zero by adding in rectangular form.
2. A four-wire wye system has phase voltages with magnitudes of 277 V. The balanced load has impedances of $5.54\angle 30^\circ$ Ω. (a) Find the phase currents expressed in polar form. (b) Construct a voltage-current phasor diagram. (c) What is the magnitude of the line current?

3. A four-wire wye system has a line voltage of 13 200 V. The balanced loads have impedances of $52.8\angle 20^\circ\ \Omega$. (a) Find the three phase currents in polar form. (b) What is the magnitude of the line current?
4. A four-wire wye-connected synchronous motor draws phase currents of 200 A from a 2400-V (line voltage) three-phase supply at an 85% leading power factor. Find the impedance of each of the three-phase windings expressed in polar form.
5. Three loads consist of 40 Ω resistance and 60 Ω inductive reactance. They are connected in wye to a four-wire system across a 480-V (line voltage) supply. Find (a) the power factor angle and (b) the current in each load.
6. Use the circuit of Problem 1. (a) Find the apparent power delivered to each arm of the circuit. (b) Find the total apparent, true and reactive power for the circuit.
7. Use the circuit of Problem 2. (a) Determine the true and reactive power delivered to each phase load. (b) Find the total apparent, true and reactive power for the circuit.
8. Use the circuit of Problem 3. (a) Calculate the apparent, true and reactive power for each arm of the circuit. (b) Find the total true power for the circuit.
9. Use the circuit of Problem 4. What is the total true power drawn by the synchronous motor?
10. Use the circuit of Problem 5. Find the total apparent, true and reactive power for the circuit.
11. Each coil of a three-phase wye-connected alternator produces 347 V. It supplies a load consisting of three wye-connected impedances of $6.94\angle 75^\circ\ \Omega$. Find (a) the line voltage, (b) the current through each load, (c) the neutral current, (d) the line current, (e) the power factor, (f) the total apparent, true and reactive power for the circuit.
12. A four-wire, wye-connected load draws a line current of 50 A at a lagging power factor of 90% from a source that produces a line voltage of 480 V. (a) Express the impedance of each arm in polar form. (b) Find the total apparent, true and reactive power for the circuit.

22-5 THE WYE-DELTA SYSTEM

The wye-connected three-phase alternator has an important characteristic. Two different sets of voltages are available. When the load is also wye-connected, the phase voltages are utilized. The line voltage is also available by connecting the loads in a different configuration. Here, the three loads are connected from line to line, Figure 22-24.

The line-to-line connections form a different arrangement. Instead of the wye orientation, the shape is in the form of a triangle or the Greek capital letter delta (Δ). Thus, the line-to-line connection is called a *delta connection.* Notice that there is no neutral line in the delta connection.

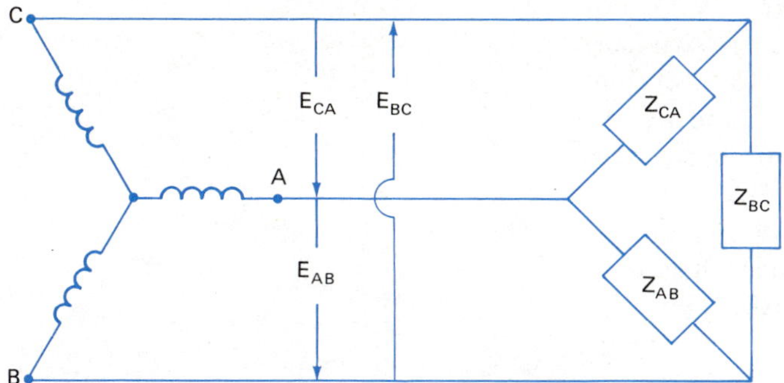

Figure 22-24 The wye-delta circuit

Line voltages in the wye-connected alternator have the following orientation:

$$\mathbf{E_{AB}} = E_L\angle{+210^\circ}$$
$$\mathbf{E_{BC}} = E_L\angle{+90^\circ}$$
$$\mathbf{E_{CA}} = E_L\angle{-30^\circ}$$

where E_L is the line voltage $E_L = \sqrt{3}E_P$. Voltage $\mathbf{E_{AB}}$ was derived in Topic 22-3. The phasor diagram showing the graphical addition of $\mathbf{E_{AN}}$ and $\mathbf{E_{NB}}$ is shown in Figure 22-17(C). The three line voltages are shown in Figure 22-25. Also refer to Figure 22-18 which shows the orientation of the line voltages and the three coil voltages.

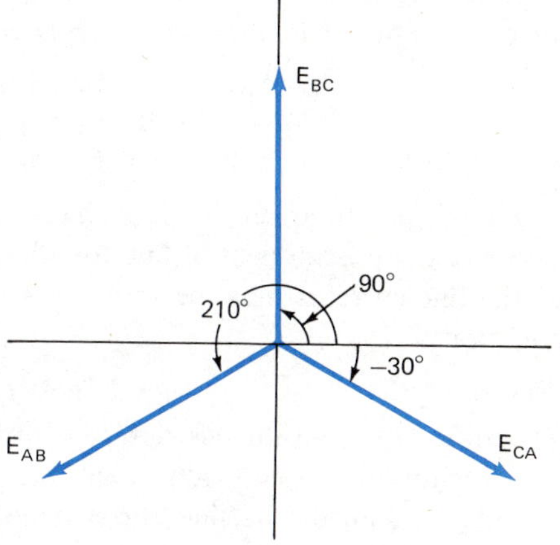

Figure 22-25 Voltage phasors for the wye-delta circuit

One of the most common voltages in power distribution systems in North America is the 120/208 V rating. The 120-V source voltage is used in four-wire wye systems. This is the phase voltage applied to the load. The line voltage in the four-wire system is 208 V. In the wye-delta system, the phase voltage across each arm of the load is the same as the line voltage (208 V):

$$E_P = E_L$$

If the impedance of each arm of the delta load is the same, the load is balanced. The current through each load is called the *phase current*. Phase currents pass from one line to another and must be identified with double subscripts, Figure 22-26. Impedances also require double subscripts.

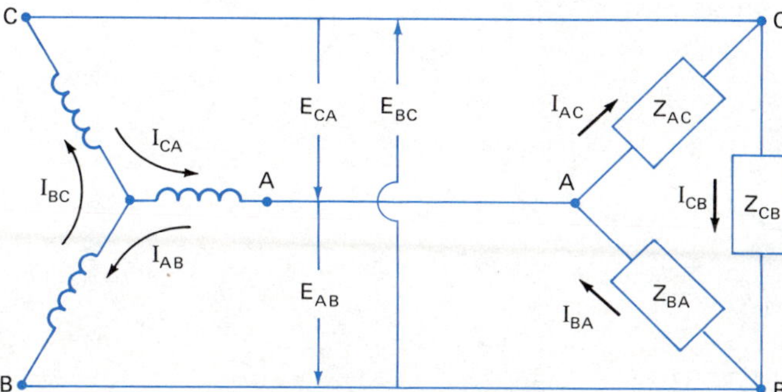

Figure 22-26 Current in the wye-delta circuit

The subscripts are given in the order of tracing in the current loop. For example, $\mathbf{I_{AB}}$ is the current from point A to point B in the generator tracing counterclockwise. Tracing in the same loop through the load, current flows from B to A. The current in the load is labeled $\mathbf{I_{BA}}$. Similarly, the voltage drop across the load is $\mathbf{E_{BA}}$ and the voltage rise in the source is $\mathbf{E_{AB}}$. By Kirchhoff's Voltage Law:

$$\mathbf{E_{AB}} = \mathbf{E_{BA}}$$

The combination of (1) subscript reversal and (2) change to a voltage *drop* has the same effect as two 180° rotations or a double negative.

The phase currents in the loads are given by Ohm's Law:

$$\mathbf{I_{BA}} = \frac{\mathbf{E_{AB}}}{\mathbf{Z_{BA}}} = \frac{\mathbf{E_{BA}}}{\mathbf{Z_{BA}}}$$

$$\mathbf{I_{CB}} = \frac{\mathbf{E_{BC}}}{\mathbf{Z_{CB}}} = \frac{\mathbf{E_{CB}}}{\mathbf{Z_{CB}}}$$

$$\mathbf{I_{AC}} = \frac{\mathbf{E_{CA}}}{\mathbf{Z_{AC}}} = \frac{\mathbf{E_{AC}}}{\mathbf{Z_{AC}}}$$

Line currents are combinations of the phase currents. Refer to Figure 22-26 and apply Kirchhoff's Current Law at points A, B and C in the load.

Point A: $\mathbf{I_A} = \mathbf{I_{AC}} - \mathbf{I_{BA}}$
Point B: $\mathbf{I_B} = \mathbf{I_{BA}} - \mathbf{I_{CB}}$
Point C: $\mathbf{I_C} = \mathbf{I_{CB}} - \mathbf{I_{AC}}$

Of course, phasor arithmetic must be employed.

Since there is no neutral line for the returning current, the phasor sum of the line currents must be zero. This can be proved by adding the three equations:

$$\mathbf{I_A} + \mathbf{I_B} + \mathbf{I_C} = 0$$

This does not mean there is zero current in any one of the lines. The three currents do not actually add in any part of the circuit. The equation for the phasor sum of the line currents does provide a convenient check of current calculations, however.

Example A A 208-V, three-phase, wye source, Figure 22-27(A), is connected to a balanced delta load consisting of three resistances of 10.4 Ω. (1) Express the voltages across each branch load in polar form, (2) calculate the phase currents and (3) the line currents in polar form, (4) construct a voltage-current phasor diagram.

Solution: (1) $\mathbf{E_{BA}} = 208\underline{/210^\circ}$

$\mathbf{E_{CB}} = 208\underline{/90^\circ}$

$\mathbf{E_{AC}} = 208\underline{/-30^\circ}$

(2) $\mathbf{I_{BA}} = \dfrac{\mathbf{E_{BA}}}{\mathbf{Z_{BA}}} = \dfrac{208\underline{/210^\circ}\text{ V}}{10.4\underline{/0^\circ}\ \Omega} = 20\underline{/210^\circ}\text{ A} \rightarrow -17.3 - j10.0\text{ A}$

$\mathbf{I_{CB}} = \dfrac{\mathbf{E_{CB}}}{\mathbf{Z_{CB}}} = \dfrac{208\underline{/90^\circ}\text{ V}}{10.4\underline{/0^\circ}\ \Omega} = 20\underline{/90^\circ}\text{ A} \rightarrow 0 + j20\text{ A}$

$\mathbf{I_{AC}} = \dfrac{\mathbf{E_{AC}}}{\mathbf{Z_{AC}}} = \dfrac{208\underline{/-30^\circ}\text{ V}}{10.4\underline{/0^\circ}\ \Omega} = 20\underline{/-30^\circ}\text{ A} \rightarrow 17.3 - j10.0\text{ A}$

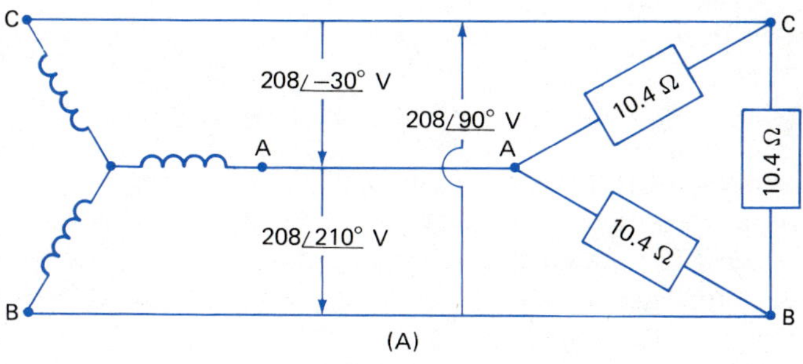

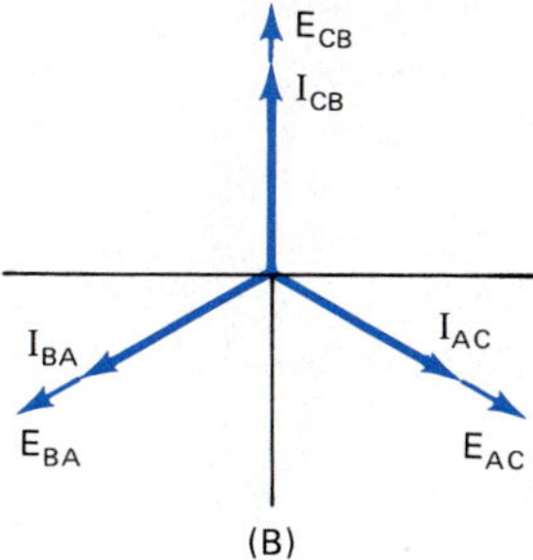

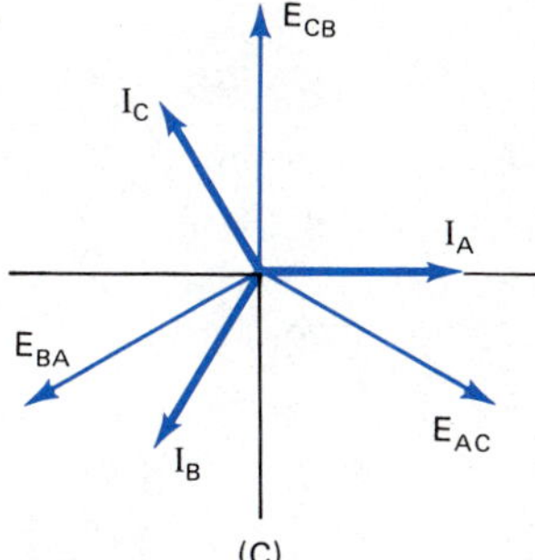

Figure 22-27

(3) $\mathbf{I_A} = \mathbf{I_{AC}} - \mathbf{I_{BA}} = (17.3 - j10.0 \text{ A}) - (-17.3 - j10.0 \text{ A})$
$= 34.6 + j0 \text{ A} \rightarrow 34.6\underline{/0^\circ} \text{ A}$

$\mathbf{I_B} = \mathbf{I_{BA}} - \mathbf{I_{CB}} = (17.3 - j10.0 \text{ A}) - (0 + j20.0 \text{ A})$
$= -17.3 - j30 \text{ A} \rightarrow 34.6\underline{/240^\circ} \text{ A}$

$\mathbf{I_C} = \mathbf{I_{CB}} - \mathbf{I_{AC}} = (0 + j20 \text{ A}) - (17.3 - j10.0 \text{ A}) = -17.3$
$+ j30 \text{ A} \rightarrow 34.6\underline{/120^\circ} \text{ A}$

(4) See Figure 22-27(B). Since the impedance is pure resistive, phase currents are in phase with the applied voltages. Line currents are 30° out of phase with the applied voltage, Figure 22-27(C).

Check: $\mathbf{I_A} + \mathbf{I_B} + \mathbf{I_C} = (34.6 + j0 \text{ A}) + (-17.3 - j30 \text{ A}) +$
$(-17.3 + j30 \text{ A}) = 0$

If the impedance of the balanced load contains reactance, each phase current will lag or lead its respective phase voltage. The amount of lag or lead depends on the power-factor angle. The magnitude of the line current (I_L) is $\sqrt{3}$ times greater than the phase current (I_P):

$$I_L = \sqrt{3} I_P$$

If the line current is measured, the magnitude of the phase current is easily calculated. If the power-factor angle is known, the three current phasors will have simply rotated by that amount. Their new directions are found by adding or subtracting the power-factor angle to the 210°, 90° and –30° reference directions of the three line voltages.

Power in the wye-delta (Y-Δ) circuit is calculated in the usual way. Calculate the apparent power in one branch by multiplying the phase voltage by the phase current:

$$S = E_P I_P$$

Express the voltage in polar form and transform to rectangular form to find the true and reactive power:

$$S\angle\theta \rightarrow P + jQ$$

The total power is three times the power in one branch:

$$S_T = 3S$$

An alternate method is to multiply $\sqrt{3}$ by the product of the line voltage and the line current:

$$S_T = \sqrt{3} E_L I_L$$

Example B In a 208-V wye-delta circuit, Figure 22-28(A), the line current is 45 A at a lagging power factor of 94%. (1) Find the phase current and (2) draw the voltage-current phasor diagrams for the circuit. Find (3) the apparent power in each branch and (4) the total apparent, true and reactive power.

Solution: Data: $E = 208\text{ V}$, $I_L = 45\text{ A}$, $\theta = \cos^{-1} 0.94 = 20°$

(1) $I_P = \dfrac{I_L}{\sqrt{3}} = \dfrac{45\text{ A}}{1.732} = 26.0\text{ A}$

(2) See Figure 22-28(B) and (C). These figures should be compared with Figure 22-27(B) and (C). All current phasors are rotated 20° cw.

(3) $S = E_P I_P = (208\text{ V})(26.0) = 5410\text{ VA}$

(4) $S_T = 3S = 3(5410\text{ VA}) = 16.2\text{ kVA}$

$S_T\angle\theta = 16.2\angle 20°\text{ kVA} \rightarrow 15.2 + j5.54$ (kW, kvars)

Check: $S_T = \sqrt{3} E_L I_L = 1.732(208\text{ V})(45\text{ A}) = 16.2\text{ kVA}$

Besides having equal load impedances, a balanced wye-delta system can be summarized as follows:

- The phase voltages (voltage drop across the load) have equal magnitudes and are 120° out of phase with each other (–30°, 90° and 210°).
- The line voltage phasors are equal to phase voltages ($E_L = E_P$).
- Phase currents (current through the load) are out of phase by 120° with each other.

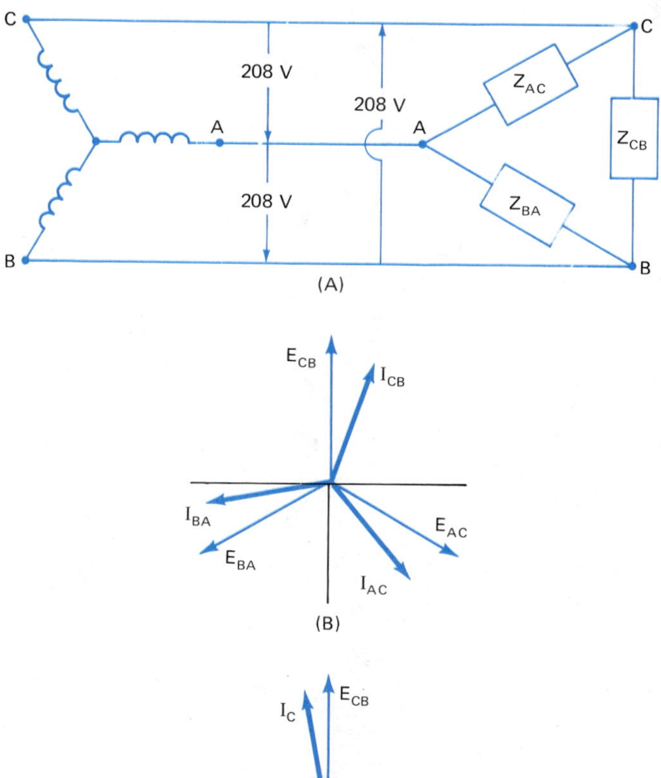

Figure 22-28

- Line currents have magnitudes that are 1.732 times greater than the phase currents ($I_L = \sqrt{3}I_P$).
- The phasor sum of the line currents is zero.
- The apparent powers ($S = E_P I_P$) delivered to each of the three loads are equal.
- The total apparent power is three times the power to any one of the loads ($S_T = 3S = \sqrt{3}E_L I_L$).

EXERCISE 22-5

Find the magnitude of the line current in a wye-delta circuit if the phase current is as follows:

1. 45.3 A
2. 100 A
3. 68.3 A
4. 82.5 A

Determine the magnitude of the phase current in a Y-Δ circuit if the line current is as follows:

5. 100 A 6. 150 A 7. 54.2 A 8. 12.5 A

Data, diagram, formula, substitute, check.

9. Construct phasor diagrams similar to those of Figure 22-28 if the phase current is 10 A and the power factor is 76.6% lagging.
10. Construct phasor diagrams similar to those of Figure 22-28 if the line current is 10 A and the power factor is 86.6% leading.
11. A three-phase, wye-delta system has line voltages of $\mathbf{E}_{\mathbf{AB}} = 208\underline{/210^\circ}$ V, $\mathbf{E}_{\mathbf{BC}} = 208\underline{/90^\circ}$ V and $\mathbf{E}_{\mathbf{CA}} = 208\underline{/-30^\circ}$ V. The impedances of the balanced load are pure resistances of 5.2 Ω each. (a) Find the magnitude of the phase voltage. (b) Find the phase current in each arm expressed in polar form. (c) Determine the three line currents expressed in polar form.
12. A Y-Δ system has phase voltages of 480 V. The balanced load has impedances of $8\underline{/25^\circ}$ Ω. (a) Find the phase currents in polar form. (b) Determine the line currents in polar form. (c) What is the magnitude of the line voltage?
13. A Y-Δ system has balanced loads of $52.8\underline{/20^\circ}$. The line voltage is 13 800 V. (a) Find the three phase currents in polar form. (b) What is the magnitude of the line current?
14. Three loads consist of 60 Ω resistance and 40 Ω inductive reactance. The loads are connected in delta across a 600-V (line voltage) supply. Find (a) the power-factor angle and (b) the current in each load. (c) What is the magnitude of the line current?
15. Use the circuit in Problem 11. (a) Find the apparent, true and reactive power delivered to each arm of the load. (b) What is the total true power?
16. Use the circuit in Problem 12. (a) Calculate the apparent, true and reactive power delivered to each arm of the load. (b) Find the total true power for the circuit.
17. Use the circuit in Problem 13. (a) What is the apparent, true and reactive power delivered to each arm of the load? (b) Find the total true and reactive power for the circuit.
18. Use the circuit in Problem 14. (a) Determine the apparent, true and reactive power delivered to each arm of the load. (b) Find the total true and reactive power for the circuit.
19. Each coil of a three-phase wye-connected alternator produces 347 V. It supplies a load consisting of three delta-connected impedances of $6.85\underline{/38.2^\circ}$ Ω. (a) Find the line voltage and the phase voltage in the load. (b) Find the line current and the phase current in the load. (c) Find the power factor. (d) Find the apparent, true and reactive power delivered to each arm of the load. (e) What is the total true power for the circuit?

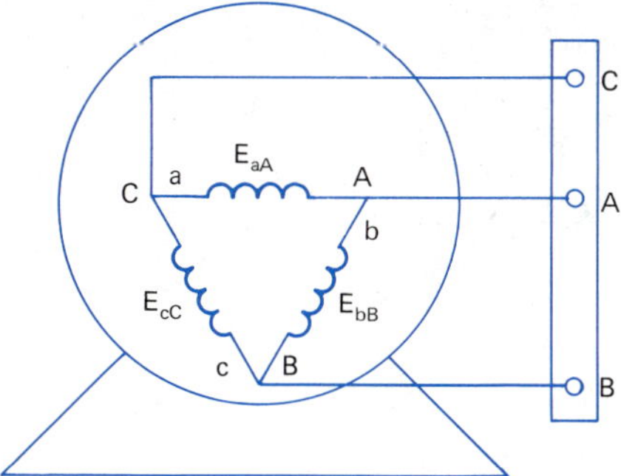

Figure 22-29 The delta-connected alternator

20. A delta-connected load draws a line current of 90 A at a lagging power factor of 70% from a source that produces a line voltage of 600 V. (a) Express the impedance of each arm in polar form. (b) Find the total apparent, true and reactive power for the circuit.

22-6 THE DELTA-CONNECTED ALTERNATOR

Loads can be connected in either a wye or a delta configuration. In the wye-connected load, the applied voltage is the phase (line to neutral) voltage. For delta-connected loads, it is the line voltage (line to line). In the common 120/208 V distribution system, for example, wye loads are driven by the 120 V phase voltage. Delta loads are connected across the 208 V lines.

Delta-connected loads can also be driven by the phase voltages of the alternator. This, however, requires that the alternator coils be connected in delta, also. Figure 22-29 shows the required connections. Leads A and b, B and c and C and a are joined at the terminal board.

The leads must be connected exactly as specified. Otherwise a short circuit will occur. It may appear that the delta connection completes a loop in which current can freely circulate without limit. In tracing the loop using Kirchhoff's Voltage Law, there are no voltage drops. The three voltage rises, however, are not in phase. The resulting sum:

$$\mathbf{E_{aA}} + \mathbf{E_{bB}} + \mathbf{E_{cC}} = 0$$

is equal to zero. The phasor sum is shown in Figure 22-30.

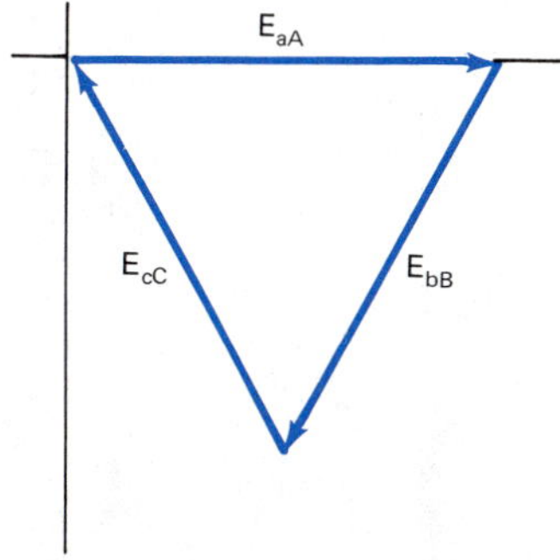

Figure 22-30 Voltage phasors in the delta circuit

The voltages generated across each phase coil are also the voltages across the two corresponding line wires.

$$\mathbf{E_{AB}} = \mathbf{E_{bB}} = E_L \angle 240^\circ \text{ V}$$
$$\mathbf{E_{BC}} = \mathbf{E_{cC}} = E_L \angle 120^\circ \text{ V}$$
$$\mathbf{E_{CA}} = \mathbf{E_{aA}} = E_L \angle 0^\circ \text{ V}$$

The delta-connected alternator has no other voltages available and no neutral point terminal. Each arm of the delta-connected system is connected to two of the three lines. This connects each of them directly across one of the alternator coils, Figure 22-31.

In delta circuits with balanced loads, each line has to carry current for two arms of the load. One current is away from the source and one is

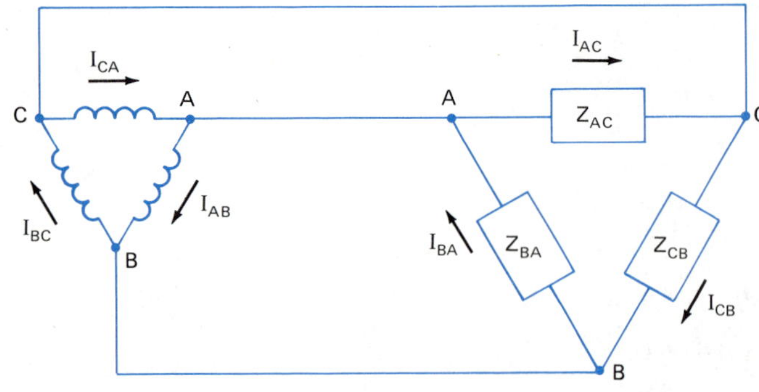

Figure 22-31 Current in a delta-delta circuit

toward the source. The delta-connected load supplied by a delta-connected source is analyzed in exactly the same way as it was with a wye-connected source.

Load phase currents are as follows:

$$\mathbf{I_{BA}} = \frac{\mathbf{E_{BA}}}{\mathbf{Z_{BA}}}$$

$$\mathbf{I_{CB}} = \frac{\mathbf{E_{CB}}}{\mathbf{Z_{CB}}}$$

$$\mathbf{I_{AC}} = \frac{\mathbf{E_{AC}}}{\mathbf{Z_{AC}}}$$

Line currents are combinations of phase currents:

$$\mathbf{I_A} = \mathbf{I_{AC}} - \mathbf{I_{BA}}$$

$$\mathbf{I_B} = \mathbf{I_{BA}} - \mathbf{I_{CB}}$$

$$\mathbf{I_C} = \mathbf{I_{CB}} - \mathbf{I_{AC}}$$

and the sum equals zero:

$$\mathbf{I_A} + \mathbf{I_B} + \mathbf{I_C} = 0$$

Example A The branch impedance of a balanced delta load, Figure 22-32(A), is $40\underline{/50^\circ}$ Ω. If the line voltage is 480 V, find (1) the phase currents in the load in polar form and (2) the line currents in polar form. (3) Construct voltage-current phasor diagrams for both phase and line currents.

Solution: Data: $\mathbf{E_{AC}} = 480\underline{/0^\circ}$ V, $\mathbf{E_{CB}} = 480\underline{/120^\circ}$ V, $\mathbf{E_{BA}} = 480\underline{/240^\circ}$ V, $\mathbf{Z} = 40\underline{/50^\circ}$ V

(1) $\mathbf{I_{AC}} = \frac{\mathbf{E_{AC}}}{\mathbf{Z_{AC}}} = \frac{450\underline{/0^\circ}\text{ V}}{40\underline{/50^\circ}\ \Omega} = 12\underline{/-50^\circ}\text{ A} \rightarrow 7.71 - j9.19\text{ A}$

$\mathbf{I_{CB}} = \frac{\mathbf{E_{CB}}}{\mathbf{Z_{CB}}} = \frac{480\underline{/120^\circ}\text{ V}}{40\underline{/50^\circ}\ \Omega} = 12\underline{/70^\circ}\text{ A} \rightarrow 4.10 + j11.3\text{ A}$

$\mathbf{I_{BA}} = \frac{\mathbf{E_{BA}}}{\mathbf{Z_{BA}}} = \frac{480\underline{/240^\circ}\text{ V}}{40\underline{/50^\circ}\ \Omega} = 12\underline{/190^\circ}\text{ A} \rightarrow -11.8 - j2.08\text{ A}$

(2) $\mathbf{I_A} = \mathbf{I_{AC}} - \mathbf{I_{BA}} = (7.71 - j9.19) - (-11.8 - j2.08) = 19.5 - j7.11\text{ A} \rightarrow 20.8\underline{/-20^\circ}\text{ A}$

$\mathbf{I_B} = \mathbf{I_{BA}} - \mathbf{I_{CB}} = (-11.8 - j2.08) - (4.10 + j11.3) = -15.9 - j13.4\text{ A} \rightarrow 20.8\underline{/-140^\circ}\text{ A}$

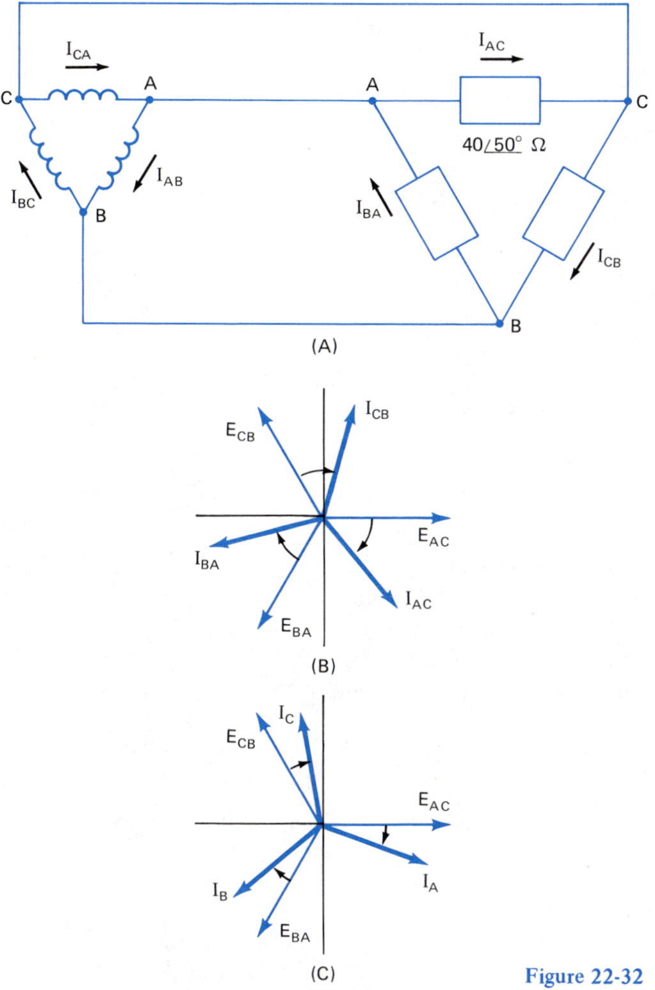

Figure 22-32

$$\mathbf{I_C} = \mathbf{I_{CB}} - \mathbf{I_{AC}} = (4.10 + j11.3) - (7.71 - j9.19)$$
$$= -3.61 + j20.5 \text{ A} \rightarrow 20.8\angle 100^\circ \text{ A}$$

(3) See Figures 22-32(B) and (C).

Check: $\mathbf{I_A} + \mathbf{I_B} + \mathbf{I_C} = (19.5 - j7.11) + (-15.9 - j13.4) + (-3.61 + j20.5) = 0$

The magnitude of the line currents is $\sqrt{3}$ times greater than the phase currents:

$$I_L = \sqrt{3} I_P$$

the same as in the wye-delta circuit. Power is always found in the same way in any circuit. Find the apparent power by multiplying the phase voltage and phase current:

$$S = E_P I_P$$

transform to rectangular form to find the true and reactive power:

$$S\angle\theta \rightarrow P + jQ$$

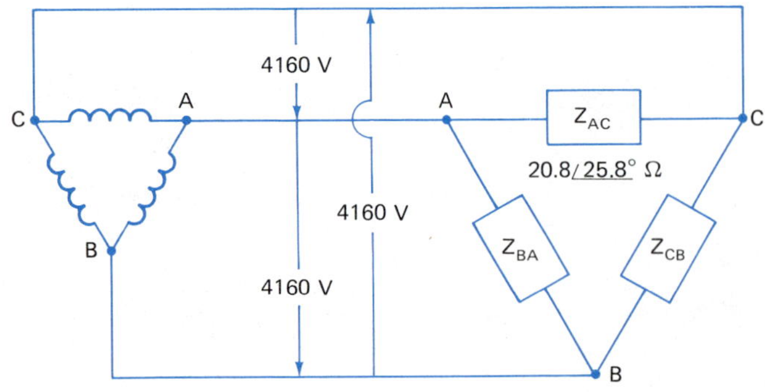

Figure 22-33

The total power is

$$S_T = 3S$$
or
$$S_T = \sqrt{3}E_L I_L$$

Example B The line voltage of a delta (Δ-Δ) system, Figure 22-33, is 4160 V. The impedance of each arm of the delta load is 20.8 Ω with a lagging power factor of 90%. Find the magnitude of (1) the phase voltage, (2) the phase current, (3) the line current, (4) the apparent, true and reactive power in each arm and (5) the total apparent, true and reactive power for the circuit.

Solution: Data: $E_L = 4160\text{ V}, Z = 20.8\ \Omega, \theta = \cos^{-1}0.9 = 25.8^\circ$

(1) $E_P = E_L = 4160\text{ V}$

(2) $I_P = \dfrac{E_P}{Z} = \dfrac{4160\text{ V}}{20.8\ \Omega} = 208\text{ A}$

(3) $I_L = \sqrt{3}I = 1.732(200\text{ A}) = 346\text{ A}$

(4) $S = E_P I_P = (4160\text{ V})(200\text{ A}) = 832\text{ kVA}$

$S\angle\theta = P + jQ$

$832\angle 25.8^\circ\text{ kVA} \rightarrow 749 + j362$ (kW, kvars)

(5) $S_T = 3E_P I_P = 3(4160\text{ V})(200\text{ A}) = 2.5\text{ MVA}$

$2.50\angle 25.8^\circ\text{ MVA} \rightarrow 2.25 + j1.09$ (MW, Mvars)

Check: $S_T = \sqrt{3}E_L I_L = 1.732(4160\text{ V})(346\text{ A}) = 2.50\text{ MVA}$

The delta-connected alternator applied to a balanced delta-connected circuit can be summarized as follows:

- Phase voltage across the load, the line voltage and the alternator coil voltage have equal magnitudes $E_L = E_P$.
- The phase voltages (voltage drop across the load) are 120° out of phase with each other (0°, 120° and 240°).
- Phase currents through the loads are 120° out of phase with each other.
- Line currents have magnitudes that are 1.732 times greater than the phase currents ($I_L = \sqrt{3}I_P$).

- The phasor sum of the line currents is zero.
- The apparent powers ($S = E_P I_P$) delivered to each of the three loads are equal.
- The total apparent power is three times the power to any one of the loads ($S_T = 3S = \sqrt{3}E_L I_L$).

EXERCISE 22-6

For each of the following Δ-Δ circuits, the phase current is given. Find the magnitude of the line current.

1. 100 A
2. 45.3 A
3. 12.6 A
4. 75.6 A

For each of the following Δ-Δ line currents, find the phase currents.

5. 100 A
6. 150 A
7. 75 A
8. 16.9 A

In the following problems, assume all circuit loads are balanced. Find the magnitude of the apparent power delivered to each branch of the following Δ-Δ circuits. Find the total apparent power in each circuit.

9. $E_L = 240$ V, $I_L = 35$ A
10. $E_L = 4160$ V, $I_L = 25$ A
11. $E_L = 480$ V, $I_P = 18.5$ A
12. $E_P = 600$ V, $I_P = 22.3$ A

Find the magnitude of the total true power delivered to the following delta loads if the loads are balanced.

13. $E_L = 480$ V, $I_L = 32.5$ A, Pf = 85%
14. $E_L = 4160$ V, $I_L = 16.9$ A, Pf = 75%
15. $E_L = 600$ V, $I_L = 22.8$ A, Pf = 81.5%
16. $E_L = 13\,800$ V, $I_L = 20.5$ A, Pf = 88.3%

Data, diagram, formula, substitute, check.

17. A three-phase 240-V delta-connected alternator supplies a balanced delta load with pure resistive branches of 12.5 Ω. Determine (a) the line currents and (b) the phase currents in polar form. (c) Find the total true power.
18. A balanced delta-connected load consists of three impedances of $7.5\angle 30^\circ$ Ω. The line voltages are 480 V. Calculate (a) the line current in polar form and (b) the phase current in polar form. (c) Find the total apparent, true and reactive power.
19. Three equal impedances of $6.25 + j7.75$ Ω are connected in delta and to a 600-V delta source. Calculate (a) the line currents in polar form, and (b) the total apparent, true and reactive power.
20. A balanced delta load draws 60 A (line current) from a 480-V delta source at a lagging power factor of 80%. (a) Find the line currents in polar form. (b) Calculate the total apparent, true and reactive power. (c) Find the impedance of each phase.

21. A balanced delta-connected load consists of three impedances of $9.35\angle 25^\circ\ \Omega$. The phase currents have magnitudes of 25.7 A. (a) Express the line currents in polar form. (b) Determine the phase currents in polar form. (c) Calculate the line voltages (Δ source) in polar form. (d) Find the total apparent, true and reactive power.
22. A balanced delta-connected load draws 30 kVA of total apparent power when the line current is 61.9 A. If the power factor is 75% lagging, (a) determine the resistance and the reactance of each phase. (b) Find the magnitude of the line voltage. (c) Find the phase current. (d) Find the total true power drawn by the circuit.

CHAPTER 23

POWER LEVEL AND THE DECIBEL

OBJECTIVES

After satisfactorily completing this chapter, the student should be able to:

- Define various words and terms associated with exponents and logarithms.
- Solve logarithmic equations using the properties of logarithms and the rules of algebra.
- Calculate the values of logarithms and antilogarithms using the calculator when applying logarithms to various electrical concepts.

Logarithms were invented about 400 years ago by John Napier, a Scottish mathematician. By using logarithms, difficult multiplication and division calculations are reduced to simpler addition and subtraction problems. Power and root calculations are reduced to multiplication and division problems.

Calculations that required several hours by longhand methods are solved in a few minutes by using logarithms. However, such calculations can now be evaluated in only a few seconds by using a calculator. Thus, logarithms are no longer necessary to do the things for which they were originally devised.

Logarithms have many other uses. They are used extensively for describing some electrical concepts. These concepts will be described in the following topics. The chapter begins by defining logarithms and logarithmic functions. Methods of solving logarithmic equations are discussed in detail. The chapter ends by describing several electrical applications.

LOGARITHMS

Any rational number can be expressed in many different forms. This is accomplished by using various number notations. A number can be expressed in fraction form:

$$\frac{1501}{4} \qquad \text{(Common fraction)}$$

The number, although improper, is expressed as a common fraction. As a mixed number, the number is written:

$$375\frac{1}{4} \qquad \text{(Mixed number)}$$

In decimal form the number is written:

$$375.25 \qquad \text{(Decimal number)}$$

and can also be expressed in scientific notation:

$$3.7525 \times 10^2 \qquad \text{(Scientific notation)}$$

The number can be expressed as a sum, difference, product, quotient, power or root of other numbers. The power or root form (exponential form) will be discussed in this chapter. The number can be written in exponential form such as 10^x where x is an exponent. The number 375.25 is a number between 100 and 1000. Since $10^2 = 100$ and $10^3 = 1000$, the exponent x must be a number between 2 and 3. The approximate value is 2.574 321.

$$10^{2.574321} = 375.25 \qquad \text{(Exponential notation)}$$

Any positive real number can be expressed in exponential form using the base 10. It is this fact that makes the exponential form so important. The relationship between logarithms and the exponential form will be discussed in the following topic.

23-1 EXPONENTS AND LOGARITHMS

A *logarithm* is an exponent. For the number discussed in the previous paragraphs:

$$10^{2.574321} = 375.25$$

the number 2.574321 is a logarithm. It is simply an exponent and is called the logarithm or log of the number 375.25.

In an expression such as:

$$10^x = N$$

there are three parts: (1) the *number*, N, (2) the *base*, 10, and (3) the *logarithm* or *exponent*, *x*. Logarithms that use the base 10 are called *common logarithms*. Logarithms generally, however, are not restricted to base 10.

Example A Name each of the three parts for each of the following expressions:

(1) $10^{1.39794} = 25$
(2) $2^7 = 128$
(3) $3^{2.5} = 15.59$

Solution:

	Base	Logarithm	Number
(1)	10	1.39794	25
(2)	2	7	128
(3)	3	2.5	15.59

Logarithms (x) can be positive or negative. For a positive base greater than 1:

- If x is positive, N is greater than 1.
- If $x = 0$, N = 1.
- If x is negative, N is less than 1.

The number (N) cannot be negative.

When the base is 10, the exponent is called the *common logarithm.* Another base ($e \approx 2.71828$) is also used extensively. When the base e is used, the exponents are referred to as *natural logarithms.* The number e occurs in nature just as does the number π.

The equation $10^x = N$ can be solved for x: "x is the logarithm (base 10) of the number N." In abbreviated form, this is written:

$$x = \log_{10} N$$

For common logarithms (log), the base is usually omitted.

$$x = \log N$$

(Common log)

The word logarithm is abbreviated with the letters (ln) when the natural logarighms (base 3) are used:

$$x = \ln N$$

(Natural log)

When any base other than 10 or e is used, it must be specified:

Exponential form	Logarithmic form
$b^x = N$	$x = \log_b N$

Example B Rewrite the following exponential expressions in logarithmic form: (1) $2^3 = 8$, (2) $10^4 = 10\ 000$, (3) $e^4 = 54.6$ and (4) $5^{-3} = 0.008$.

Solution:

	Exponential	Logarithmic
(1)	$2^3 = 8$	$\log_2 8 = 3$
(2)	$10^4 = 10\ 000$	$\log 10\ 000 = 4$
(3)	$e^4 = 54.6$	$\ln 54.6 = 4$
(4)	$5^{-3} = 0.008$	$\log_5 0.008 = -3$

Example C Rewrite the following logarithmic expressions in exponential form: (1) $4 = \log_3 81$, (2) $2.176 = \log 150$, (3) $2.75 = \ln 15.64$ and (4) $-3 = \log_4 0.0156$

Solution:

	Logarithmic form	Exponential form
(1)	$\log_3 81 = 4$	$3^4 = 81$
(2)	$\log 150 = 2.176$	$10^{2.176} = 150$
(3)	$\ln 15.64 = 2.75$	$e^{2.75} = 15.64$
(4)	$\log_4 0.0156 = -3$	$4^{-3} = 0.0156$

EXERCISE 23-1

Find the base, logarithm and the number for the following expressions.

1. $6^3 = 216$
2. $\log_4 1024 = 5$
3. $\log 1000 = 3$
4. $e^2 = 7.3891$
5. $\log_{12} 1728 = 3$
6. $10^{4.65} = 44\ 668$
7. $e^{3.75} = 42.52$
8. $\ln 0.1353 = -2$
9. $10^{-2.25} = 0.005\ 62$
10. $\log 1 = 0$

Change the expressions in Problems 11-20 to logarithmic form.

11. $7^2 = 49$
12. $3^5 = 243$
13. $10^{-3} = 0.001$
14. $e^0 = 1$
15. $9^3 = 729$
16. $10^{2.375} = 237.14$
17. $e^2 = 7.3891$
18. $2^{-1} = 0.5$
19. $10^{-1.75} = 0.0178$
20. $e^{-1} = 0.3679$

Change the expressions in Problems 21-30 to exponential form.

21. $\log_2 512 = 9$
22. $\log_5 625 = 4$
23. $\log 421.7 = 2.625$
24. $\ln 20.09 = 3$
25. $\log_7 907.5 = 3.5$
26. $\ln 100 = 4.605$
27. $\log 0.075 = -1.125$
28. $\ln 0.25 = -1.386$
29. $\log 1 = 0$
30. $\log 10 = 1$

23-2 COMMON LOGARITHMS

The common logarithm (x) of a number (N)

$$x = \log N$$

is defined as the exponent to which the base (10) must be raised to yield the number. The common logarithm is easily found using a calculator. It is a one-number function.

FINDING THE COMMON LOGARITHM USING A CALCULATOR

- Enter the number N.
- Press the common logarithm [LOG] key.
- The common logarithm of the number N is now displayed.
- Write the log rounded correctly to four decimal places (four places after the decimal). Four decimal places in the logarithm will return three significant figures in the number.

Example A Find the common logarithms of the following numbers: (1) 150, (2) 2500, (3) 3.14, (4) 525 000, and (5) 0.000 497.

Solution: $x = \log N$

(1) 150 [LOG], 2.1761

(2) 2500 [LOG], 3.3979

(3) 3.14 [LOG], 0.4969

(4) 525 000 [LOG], 5.7202

(5) 0.000 497 [LOG], -3.3036

The log of the number 1 is zero:

$$\log 1 = 0$$

For numbers greater than 1, the log is a positive number. It is negative for numbers less than 1 (fractions). For powers of ten, the log is equal to the exponent:

$$\log 10 = 1$$
$$\log 10^5 = 5$$
$$\log 10^{-3} = -3$$

23-3 ANTILOGARITHMS

When the common logarithm of a number is known, the number can be found by an inverse operation. Just as the arc tangent is the inverse of the tangent function in trigonometry, the antilogarithm is the inverse of the logarithmic function:

$$\text{if } x = \log N, \text{ then } N = \text{antilog } x = 10^x$$

The number is found by raising 10 to the known common logarithm power.

Check the Owner's Manual for the procedure for evaluating the antilogarithms using the calculator. Most calculators use either [10^x] or [INV] [LOG]. The antilogarithm is a one-number function of the calculator:

FINDING THE COMMON ANTILOGARITHM USING A CALCULATOR

- Enter the logarithm (x).
- Press [10^x] or [INV] [LOG] (see Owner's Manual).
- The common antilogarithm (N) is now displayed. Round the answer to three significant figures.
- Check by finding the logarithm of the number.

Example A What is the value of N for each of the following common logarithms: (1) 3.7016, (2) 0.9355, and (3) −1.6126?

Solution: Since log 1 = 0, the answers to (1) and (2) should be greater than 1 while (3) should be less than 1.

(1) $N = \text{antilog } 3.7016 = 10^{3.7016}$

3.7016 [10^x] , 5030

(2) $N = \text{antilog } 0.9355 = 10^{0.9355}$

0.9355 [10^x] , 8.62

(3) $N = \text{antilog } (-1.6126) = 10^{-1.6126}$

1.6126 [+/-] [10^x] , 0.0244

Check: (1) $x = \log 5030$

5030 [LOG] , 3.7016

(2) $x = \log 8.62$

8.62 [LOG] , 0.9355

(3) $x = \log 0.0244$

0.0244 [LOG] , −1.6126

Common logarithms of powers of 10 are integers with the logarithm being equal to the exponent:

$$\log 10^4 = 4$$

The antilogarithm of an integer, therefore, is a power of 10. The exponent is simply the given integer:

(a) antilog $2 = 10^2 = 100$
(b) antilog $(-3) = 10^{-3} = 0.001$
(c) antilog $6 = 10^6 = 1\ 000\ 000$

Such problems can be done manually.

Logarithmic tables and the instructions for using them are given in Appendix C. It is expected, however, that most students will prefer to use the calculator.

EXERCISE 23-2

Find the common logarithm of the following numbers.

1. 2	6. 25	11. 5280	16. 3.57×10^6
2. 5	7. 50	12. 3.57×10^0	17. 0.0983
3. 9	8. 75	13. 3.57×10^1	18. 0.000 481
4. 10	9. 100	14. 3.57×10^2	19. 7.63×10^{-5}
5. 15	10. 500	15. 3.57×10^4	20. 0.174

Find the logarithm of each of the following manually.

21. 0.0001	23. 0.1	25. 10 000
22. 0.001	24. 10	26. 10 000 000

Find the antilogarithm for each of the following common logarithms. Check each answer.

27. 2.1584	32. 7.8971	37. 0.7597
28. 4.2480	33. −3.2434	38. 1.7597
29. 0.5132	34. 2.7033	39. 2.7597
30. 1.5966	35. −1.2907	40. 3.7597
31. 3.9868	36. −5.1765	

Find the antilogarithms of the following manually.

41. 8	43. −4	45. 1
42. 3	44. 0	46. −2

23-4 NATURAL LOGARITHMS

The natural logarithms are based on the number e which has a value of about 2.7183. The symbol for natural logarithms is (ln) which is pronounced "ell-en." Natural logarithms and their antilogarithms can be found with a calculator. Use the [LN], key for finding the natural logarithm. Otherwise, the procedure is the same as for common logs.

Example A Find the natural logarithm of the following numbers: (1) 10, (2) 2, (3) 0.55 and (4) 0.0025.

Solution: $x = \ln N$

(1) 10 [LN], 2.3026
(2) 2 [LN], 0.6931
(3) 0.55 [LN], −0.5978
(4) 0.0025 [LN], −5.9915

Again, record four decimal places to obtain three significant figures in the antilogarithm.

To find antilogarithms (antiln), use the [e^x] or [INV][LN] calculator keys.

Example B Find the antilogarithm of each of the following natural logarithms: (1) 2.9575, (2) −2.6314, (3) 0.2841 and (4) −1.4366.

Solution: $N = \text{antiln } x = e^x$

(1) $N = \text{antiln } 2.9575 = e^{2.9575}$

2.9575 [e^x], 19.2

(2) −2.6314 [e^x], 0.0720
(3) 0.2841 [e^x], 1.33
(4) −1.4366 [e^x], 0.238

23-5 EXPONENTIAL DECAY

A DC circuit containing a resistance and a capacitor in series is known as an *RC circuit,* Figure 23-1. The instantaneous current at the moment the switch is closed is its maximum value (I_0). From that moment on, current decreases as the back emf of the capacitor builds.

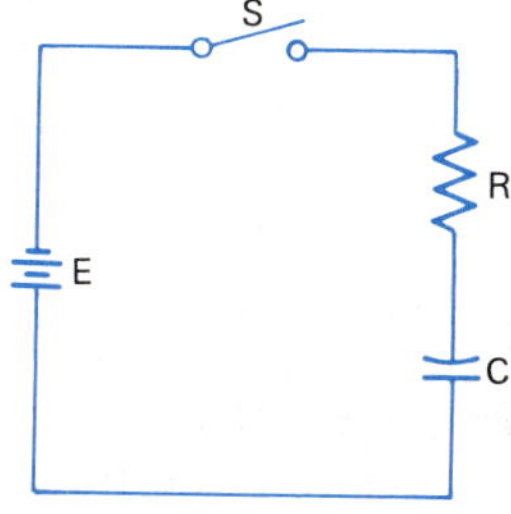

Figure 23-1 An RC circuit

The decrease (*exponential decay*) of current in charging the capacitor is

$$i = I_0 e^{-t/RC}$$

where i is the instantaneous current at time t after the charging has begun. The beginning current I_0 occurs at $t = 0$. R is the resistance and C is the capacitance. The letter e is the base of the natural logarithms.

The graph of the instantaneous current as a function of time is shown in Figure 23-2. If the product RC is large, the time required for the current to stop flowing may be large. If the equation $i = I_0 e^{-t/RC}$ is rearranged, the following equation is obtained:

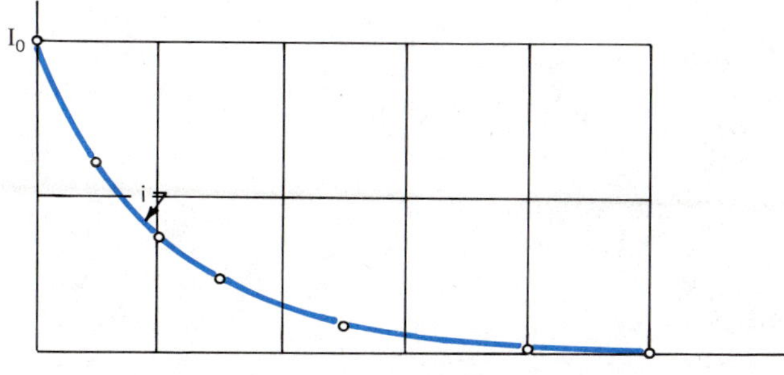

Figure 23-2 The instantaneous current in an RC circuit as a function of time. The current drops (exponential decay) as the capacitor charges.

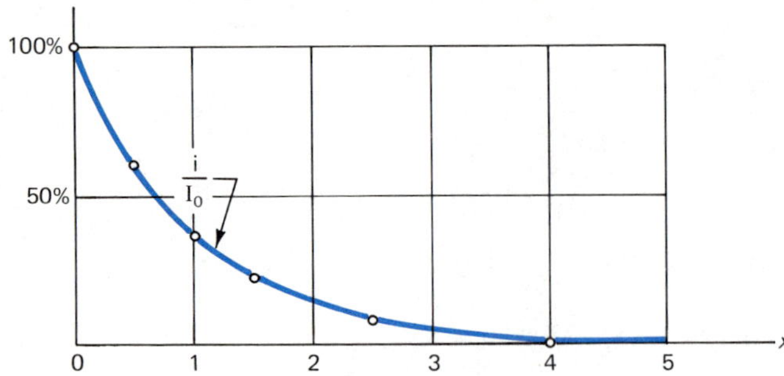

Figure 23-3 The exponential decay as a percent of the original value

$$\frac{i}{I_0} = e^{-x}$$

where $x = \frac{t}{RC}$. The ratio $\frac{i}{I_0}$ can be expressed as a percent of the initial value as indicated in Figure 23-3.

The product RC is called the *time constant* (τ) for the circuit and x is the number of time constants. When t equals the product RC, x has a value of 1 unit and:

$$\frac{i}{I_0} = e^{-1} = 0.3679 \approx 37\%$$

The time constant ($\tau = RC$) represents the time required for the current to decrease by 63%. The product RC has units of seconds when R is expressed in ohms and C is in farads.

Example A Find the time constant for an RC circuit similar to Figure 23-1 if $R = 100\ k\Omega$ and $C = 7.5\ \mu F$.

Solution: Formula: $\tau = RC$

Substitute: $\tau = (100\,000\ \Omega)(7.5 \times 10^{-6}\ F)$

$\tau = 0.75\ s$

A capacitor in an RC circuit is said to be fully charged after 5 time constants. For the circuit of Example A, complete charging requires 3.75 s. In the same period of time, the current falls to zero.

Example B Find the current expressed as a percent of the original current in an RC circuit after (1) 2 time constants and (2) after 5 time constants.

Solution: $N = e^{-x}$ where $N = \frac{i}{I_0}$

(1) $N = e^{-2} = 0.1353 = 13.5\%$

(2) $N = e^{-5} = 0.0067 = 0.67\%$

Example C Find the number of time constants in an RC circuit for the current to drop to (1) 50% and (2) 10% of the initial current.

Solution: Solve the equation $N = e^{-x}$ for x:

$$N = e^{-x} \qquad \left(N = \frac{i}{I_0}\right)$$

$$\ln N = -x$$
$$x = -\ln N$$

(1) $x = -\ln 0.50 = 0.693$ time constants
For the current to fall 50% requires about 0.7 of one time constant.

(2) $x = -\ln 0.10 = 2.30$ time constants
2.3 time constants are required for the current to drop to 10% of the initial current.

Exponential decay also applies to the decreasing voltage across certain circuit elements so the formulas used in the previous examples are important:

$$N = e^{-x}$$
$$x = -\ln N$$

for positive values of x.

EXERCISE 23-3

Find the natural logarithms of the following numbers.

1. 1
2. 3
3. 5
4. 25.0
5. 0.250
6. 0.005 95

Find the antilogarithms of the following natural logarithms.

7. 2.7183
8. −1
9. −2
10. 3.2543
11. 25.6391
12. −5.3844

RC circuits (Figure 23-1) contain the following combinations of resistance and capacitance. (a) Find the time constant for each circuit. (b) How much time is required to charge the capacitor?

13. $R = 50\ k\Omega$, $C = 2.5\ \mu F$
14. $R = 125\ k\Omega$, $C = 5\ \mu F$
15. $R = 10\ k\Omega$, $C = 750\ \mu F$
16. $R = 250\ k\Omega$, $C = 1.5\ \mu F$
17. $R = 500\ k\Omega$, $C = 12.5\ \mu F$
18. $R = 1.00\ M\Omega$, $C = 100\ \mu F$

Find the current in an RC circuit expressed as a percent of the initial current after the following number of time constants (x).

19. 0.1
20. 0.2
21. 0.3
22. 0.4
23. 0.5
24. 0.75
25. 1
26. 1.5
27. 2.5
28. 4
29. 5
30. 6

31. Plot a full page graph of the data obtained in Problems 19 through 30.

How many time constants are required for current in an RC circuit to drop to the following values?

32. 90%
33. 80%
34. 75%
35. 70%
36. 60%
37. 50%
38. 40%
39. 30%
40. 25%
41. 20%
42. 10%

23-6 EXPONENTIAL GROWTH

Another common equation involving natural logarithms is that of *exponential growth.* One example is the growth of current in a circuit containing inductance and resistance. Such a circuit is called an RL circuit, Figure 23-4.

When the switch is closed (t = 0), the current is zero. It increases gradually because the inductor tends to resist the changing current. When the current reaches its final DC value, the inductor no longer produces a counter emf.

The instantaneous current is given by:

$$i = I(1 - e^{-Rt/L})$$

where I is the final DC current, R is the resistance and L is the inductance. The graph of the exponential growth function is shown in Figure 23-5. Rearranging gives:

$$\frac{i}{I} = 1 - e^{-x}$$

where $\frac{i}{I}$ can be expressed in percent and $x = \frac{Rt}{L}$. The ratio $\frac{L}{R}$ is the time constant τ for the RL circuit and x is the number of time constants. The

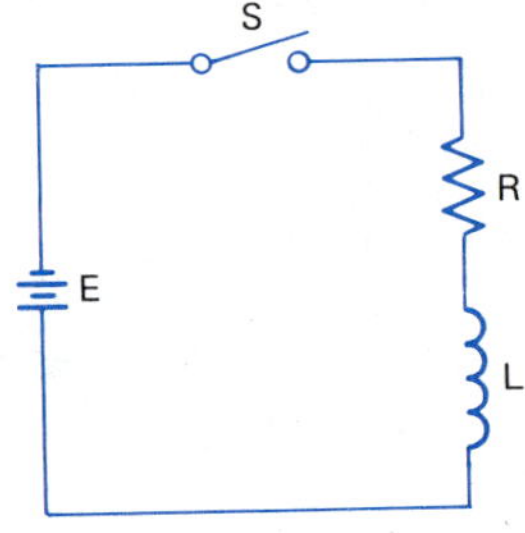

Figure 23-4 An RL circuit

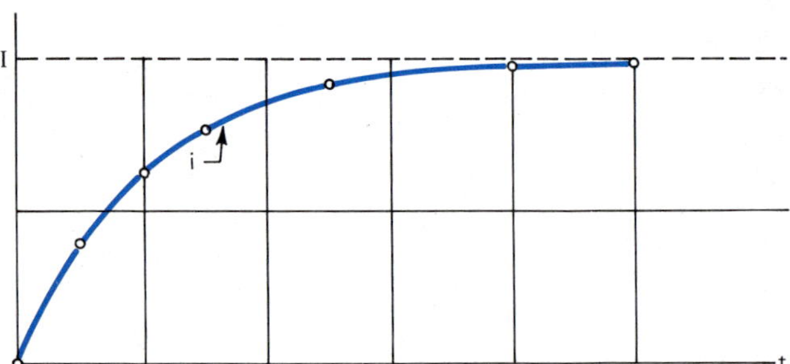

Figure 23-5 The instantaneous current in an RL circuit as a function of time. The current rises (exponential growth) against the resistance of the inductor.

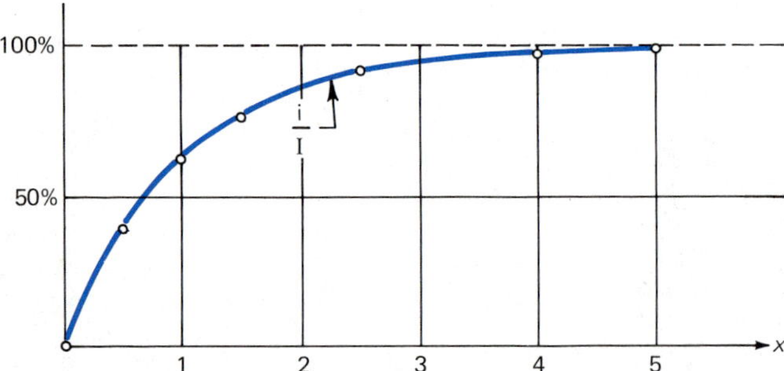

Figure 23-6 The exponential growth as a percent of the final value

time constant is the time required for the current to increase by 63%. The current increases approximately to its DC value after 5 time constants, Figure 23-6.

Example A (1) Find the time constant for an RL circuit containing 4.25 H of inductance and 12.0 Ω of resistance. (2) How much time is required before the current reaches its DC value?

Solution: (1) Formula: $\tau = \frac{L}{R}$

Substitute: $\tau = \frac{4.25\ H}{12.0\ \Omega} = 0.354\ s$

(2) Current builds to its DC value in: 5(0.354 s) = 1.77 s

Example B Find the current in an RL circuit expressed as a percent of its final DC value after 2.5 time constants.

Solution: Formula: $N = 1 - e^{-x}$ where $N = \frac{i}{I}$

Substitute: $N = 1 - e^{-2.5}$

$N = 1 - 0.082$

$N = 0.918 = 91.8\%$

Note: 1 [-] 2.5 [+/-] [e^x] [=], 0.918

Example C How many time constants are required for the current to increase to 90% of its DC value?

Solution: Solve the equation $N = 1 - e^{-x}$ for x:

$N = 1 - e^{-x}$

$N - 1 = -e^{-x}$ (SA, − 1)

$1 - N = e^{-x}$ (MA, − 1)

$\ln(1 - N) = -x$ (Take logarithm)

$x = -\ln(1 - N)$ (MA, − 1)

Substitute: $x = -\ln(1 - 0.90)$

$x = -\ln(0.10)$

$x = 2.3$ time constants

Note: 1 [-] 0.9 [=] [LN] [+/-], 2.30

EXERCISE 23-4

RL circuits (Figure 23-4) contain the following combinations of resistance and inductance. (a) Find the time constant for each circuit and (b) find the time required for the current to build to its DC value.

1. R = 1.5 Ω, L = 4 H
2. R = 6 Ω, L = 500 mH
3. R = 0.35 Ω, L = 8 H
4. R = 250 Ω, L = 7.35 mH

Find the current in an RL circuit expressed as a percent of the final DC value after the following number of time constants (x):

5. 0.1
6. 0.2
7. 0.3
8. 0.4
9. 0.5
10. 0.75
11. 1
12. 1.5
13. 2.5
14. 4
15. 5
16. 6

17. Plot a full page graph of the data obtained in Problems 5 through 16.

How many time constants are required for current in an RL circuit to increase to the following percent of the final DC value?

18. 10%
19. 20%
20. 25%
21. 30%
22. 40%
23. 50%
24. 60%
25. 70%
26. 75%
27. 80%
28. 90%

LOGARITHMIC EQUATIONS

Logarithms have three fundamental properties that are derived from the laws of exponents. These properties along with the rules of algebra are useful in solving equations that contain logarithms. Although these properties apply to any logarithmic system, only common logarithms (base 10) will be discussed from this point on.

The laws of exponents were discussed in Chapter 5. The multiplication law applied to base 10 is

$$10^m \times 10^n = 10^{m+n}$$

and the division law is

$$\frac{10^m}{10^n} = 10^{m-n}$$

These rules can be reviewed in Topic 5-14 and Topic 5-15.

23-7 PROPERTIES OF LOGARITHMS

The rule for multiplying exponents indicates that the product of two powers of 10 is found by adding exponents. Any number can be written in exponential form (base 10) and, of course, logarithms are exponents. Therefore, the rule for finding the logarithm of a product is as follows:

FINDING THE LOGARITHM OF A PRODUCT

- The logarithm of a product of two or more numbers is equal to the sum of the logarithms of the factors.
- $\log xy = \log x + \log y$

The rule for dividing exponents states that the quotient of the two powers of 10 is found by subtracting. The rule for finding the logarithm of a quotient is as follows:

FINDING THE LOGARITHM OF A QUOTIENT

- The logarithm of a quotient of two numbers is equal to the logarithm of the numerator minus the logarithm of the denominator.
- $\log \frac{x}{y} = \log x - \log y$

The third property of logarithms is finding the log of an exponential ($\log x^n$). Since:

$$\log x^n = \log x \cdot x \cdot x \cdot x \ldots \quad \text{(n factors of } x\text{)}$$
$$\log x^n = \log x + \log x + \log x + \ldots \quad \text{(n terms)}$$
$$\log x^n = n \log x$$

FINDING THE LOGARITHM OF A POWER

- The logarithm of a number raised to a power is the exponent times the log of the number.
- $\log x^n = n \log x$

The three properties of logarithms are used mostly as an aid in solving equations, Topic 23-8. Their meaning, however, is best illustrated by showing numerical examples.

Example A Since $24 = 4 \times 6$, show that $\log 24 = \log 4 + \log 6$.

Solution: Left member:

$\log 24 = 1.3802$

24 [LOG], 1.3802

Right member:

$\log 4 + \log 6 = 0.6021 + 0.7782 = 1.3803$

4 [LOG] + 6 [LOG] =, 1.3802

Note: $(10^{0.6021})(10^{0.7782}) = 10^{(0.6021 + 0.7782)} = 10^{1.3803} = 24$

Example B Prove that: $\log \frac{100}{4} = \log 100 - \log 4$.

Solution: $\log \frac{100}{4} = \log 100 - \log 4$

$\log 25 = \log 100 - \log 4$

$1.3979 = 2.0000 - 0.6021$

$1.3979 = 1.3979$

Note: 100 [LOG] – 4 [LOG] =, 1.3979

Example C Show that $\log 7^4 = 4 \log 7$.

Solution: Left member:

$\log 7^4 = \log 2401 = 3.3804$

7 [y^x] 4 [=] [LOG], 3.3804

Right member:

$4 \log 7 = 4(0.8451) = 3.3804$

4 [×] 7 [LOG] [=], 3.3804

The rule for finding logarithms of powers also applies to roots. This is possible since radical signs can be replaced with fractional exponents, Topic 3-16:

$$\sqrt{R} = R^{1/2}$$
$$\sqrt[4]{R} = R^{1/4}$$

Example D Show that $\log \sqrt{144} = \frac{1}{2}(\log 144)$.

Solution:
$$\log \sqrt{144} = \frac{1}{2}(\log 144)$$
$$\log 12 = \frac{1}{2}(2.1584)$$
$$1.0792 = 1.0792$$

For complex mathematical expressions the logarithms can be found by combining the three logarithmic properties.

Example E Find the logarithm of: $\dfrac{24 \times 35^3}{56}$

Solution: Write the logarithm of the expression in expanded form.
$$\log \frac{24 \times 35^3}{56} = \log 24 + 3 \log 35 - \log 56$$
Right member:
$$1.3802 + 3(1.5441) - 1.7482 = 4.2643$$
Left member:
$$\log \frac{24 \times 35^3}{56} = \log 18\,375 = 4.2642$$

Note: 3 [×] 35 [LOG] [=] [+] 24 [LOG] [-] 56 [LOG] [=], 4.2642

EXERCISE 23-5

Rewrite the following expressions in expanded form using the three properties of logarithms. Check all answers by evaluating both forms.

1. $\log 5 \times 17$
2. $\log 25 \times 30$
3. $\log \frac{38}{27}$
4. $\log \frac{17}{11}$
5. $\log 14^3$
6. $\log 27^3$
7. $\log \sqrt{528}$
8. $\log \sqrt[3]{50\,653}$
9. $\log(30^3 \times 7\sqrt{24})$
10. $\log \dfrac{\frac{1}{2}(28)(66)^2}{\sqrt{455}}$

Rewrite the following expressions as the logarithm of a single mathematical expression. Check all answers by evaluating both forms.

11. $\log 22 + \log 16$
12. $\log 9 + \log 32$
13. $\log 75 - \log 5$
14. $\log 26 - \log 13$
15. $\log 144 + \log 35 - \log 12$
16. $\frac{1}{2} \log 121$
17. $\frac{1}{3} \log 1728$
18. $\log 81 + 3 \log 4 - \log 21$
19. $\log 12 - \frac{1}{2} \log 144$
20. $\log 7 + 2 \log 3 + \log 9 - 2 \log 6 - \frac{1}{2} \log 49$

23-8 SOLVING LOGARITHMIC EQUATIONS

A *logarithmic equation* is one that contains the logarithm of the unknown. Some examples of logarithmic equations are

$$25 = 10 \log x$$
$$7 + \log x^2 = 3 \log x$$
$$30 = 20 \log \frac{E}{120}$$

Such equations can be solved using the laws of algebra and the properties of logarithms. The following summarizes the properties of logarithms:

- $\log M \cdot N = \log M + \log N$
- $\log \frac{M}{N} = \log M - \log N$
- $\log M^n = n \log M$
- $\text{antilog } x = 10^x$
- $\text{antilog } (\log x) = x$
- $\log 1 = 0$
- $\log 10 = 1$

To solve logarithmic equations, use the following procedure:

SOLVING LOGARITHMIC EQUATIONS

- Using the properties of logarithms and algebra, solve the equation for the logarithm containing the unknown.
- Evaluate all numerical parts.
- Take the antilogarithm of both sides of the equation.
- If the unknown is not now isolated, solve the resulting equation.

Example A Solve: $25 + 2 \log x = 33$

Solution:

$2 \log x = 8$ (SA, − 25)

$\log x = 4$ (DA, ÷ 2)

$\text{antilog } (\log x) = \text{antilog } 4$ (Apply antilog)

$x = \text{antilog } 4 = 10^4$

$x = 10\,000$

Check: $25 + 2 \log (10\,000) = 33$

$25 + 2(4) = 33$

$25 + 8 = 33$

Example B Solve: $50 = 20 \log \frac{E}{120}$

Solution:

$2.5 = \log \frac{E}{120}$ (DA, ÷ 20)

$\text{antilog } 2.5 = \frac{E}{120}$ (Apply antilog)

$$\frac{E}{120} = 10^{2.5} = 316 \qquad \text{(Evaluate)}$$

$$E = 316(120) \qquad \text{(MA, × 120)}$$

$$E = 37\,900$$

Check: $50 = 20 \log \frac{(37\,900)}{120}$

$$50 = 20 \log 316$$

$$50 = 20(2.5)$$

The previous example can be solved by an alternate procedure. After the first step, apply the rule for finding the logarithm of a quotient. The equation can be solved for the logarithmic term containing the unknown now having no other factors. When the antilog is taken and evaluated, the equation is solved.

Example C Solve: $35 = 10 \log \frac{P_0}{600}$

Solution:

$$3.5 = \log \frac{P_0}{600} \qquad \text{(DA, ÷ 10)}$$

$$3.5 = \log P_0 - \log 600 \qquad \text{(Quotient rule)}$$

$$3.5 + \log 600 = \log P_0 \qquad \text{(AA, + log 600)}$$

$$\log P_0 = 3.5 + 2.7782 \qquad \text{(Evaluate)}$$

$$\log P_0 = 6.2782$$

$$P_0 = \text{antilog } 6.2782 = 10^{6.2782} \qquad \text{(Apply antilog)}$$

$$P_0 = 1\,900\,000$$

Check: $35 = 10 \log \frac{1\,900\,000}{600}$

$$35 = 10 \log 3170$$

$$35 = 10(3.50)$$

Some exceedingly complex equations are reduced to simple ones by applying the properties of logarithms.

Example D Solve: $3 \log 4x + 2 \log x^3 = 5$

Solution:

$$3 \log 4 + 3 \log x + 2 \log x^3 = 5 \qquad \text{(Product rule)}$$

$$3 \log 4 + 3 \log x + 6 \log x = 5 \qquad \text{(Power rule)}$$

$$3 \log 4 + 9 \log x = 5 \qquad \text{(Add)}$$

$$9 \log x = 5 - 3 \log 4 \qquad \text{(SA, − 3 log 4)}$$

$$\log x = \frac{5 - 3 \log 4}{9} \qquad \text{(DA, ÷ 9)}$$

$$\log x = \frac{5 - 1.8062}{9} \qquad \text{(Evaluate)}$$

$$\log x = 0.3549$$

$$x = 2.264 \qquad \text{(Apply antilog)}$$

Check: $3 \log 4(2.264) + 2 \log (2.264)^3 = 5$

$$3 \log 9.056 + 2 \log 11.605 = 5$$

$$3(0.9569) + 2(1.0646) = 5$$

$$2.8708 + 2.1291 = 5$$

EXERCISE 23-6

Solve and check each of the following equations.

1. $3 \log x + 2 \log x = 15$
2. $4 \log x + 7 = 35$
3. $10 \log R - 3 = 47$
4. $8 \log T - 2 \log T = 24$
5. $2 \log 10P = 6$
6. $5 \log 4P - \log P = 3$
7. $2 \log \frac{1}{Q} = -8$
8. $8 \log \sqrt{E} = 4$
9. $10 \log \frac{P_o}{120} = 20$
10. $20 \log \frac{E_o}{120} = 60$
11. $20 \log \frac{I_o}{2.5} = -35$
12. $10 \log \frac{{E_o}^2}{144} = -2.5$
13. $10 \log \frac{{I_o}^2}{0.0625} = 6$
14. $0.2 = 0.62\left(0.161 + 1.48 \log \frac{d}{0.4633}\right)$
15. $20 = \frac{12.1}{\log \frac{d}{0.4633}}$
16. $10 \log \frac{500}{P_i} = 30$
17. $20 \log \frac{900}{E_i} = 45$
18. $\log x^2 - \log x = 2.7497$
19. $5 \log x^3 + 12 \log x^2 = 78$
20. $3 \log \sqrt{x} = 2 \log 2x^2$

ELECTRICAL APPLICATIONS

Power companies often send speech, readout (measurements) and control (switching) signals along existing power lines from one station to another. Such signals are transmitted by superimposing a high frequency signal on the 60-Hz power wave. Telephone lines transmit speech signals in a similar way.

The original signals require power or voltage *amplification* (gain) at the sending end of a circuit using an amplifier. Amplifiers are also used at the receiving end. The transmitted signal always involves power or voltage *attenuation* (loss) along the line. Since these signals often carry speech, amplification and attenuation is usually described in terms of the science of sound.

The human ear responds to sound intensity in a logarithmic fashion. The energy in a sound wave must increase ten times for the ear to interpret it as being twice as loud. Thus, amplified or attenuated signals are described in terms of logarithms.

The calculation of the capacitance and inductance of two parallel (power) lines involves logarithmic equations. These electrical applications are described in the last topic of the chapter.

23-9 POWER RATIO AND POWER LEVEL

Many electrical devices have an input side and an output side. Power is fed into the input side and is called the input power (P_i). Power is obtained on the output side and is referred to as output power (P_o). The ratio of power output to power input is often called efficiency. Efficiency is always less than 100% and applies to such devices as motors, generators, transformers, etc.

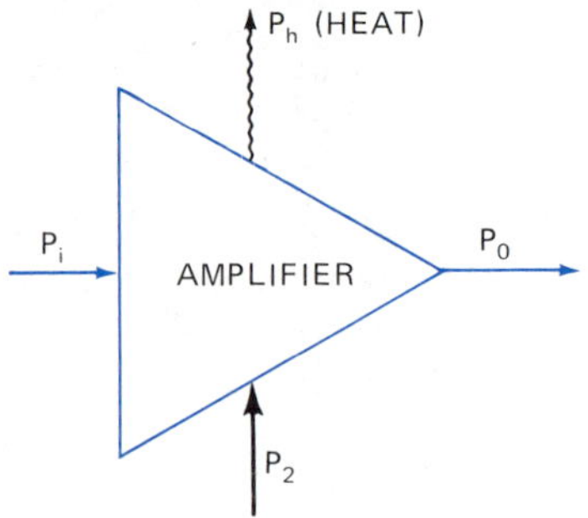

Figure 23-7 The triangular shape of the amplifier circuit symbol. The second power source P_2 is normally not shown.

An amplifier is different. An input signal supplies the input power. A second source also supplies power and some energy is emitted as heat as shown in Figure 23-7. The circuit symbol for the amplifier is the triangular symbol shown. If the second source supplies a large amount of power, then P_o can be much larger than P_i. This, of course, could not be possible without the second source.

The output signal is a reproduction of the input signal. It simply has a larger amplitude (more power) than the input signal. Input and output impedances must be matched to prevent distortion. The efficiency of an amplifier is the ratio of P_o to $P_i + P_2$. The power output to power input ratio is referred to as the *power ratio* (PR):

$$\mathbf{PR = \frac{P_o}{P_i}}$$

where P_o and P_i are measured in the same units.

The logarithm of the power ratio is called the *power level* (PL):

$$\mathbf{PL = \log \frac{P_o}{P_i}}$$

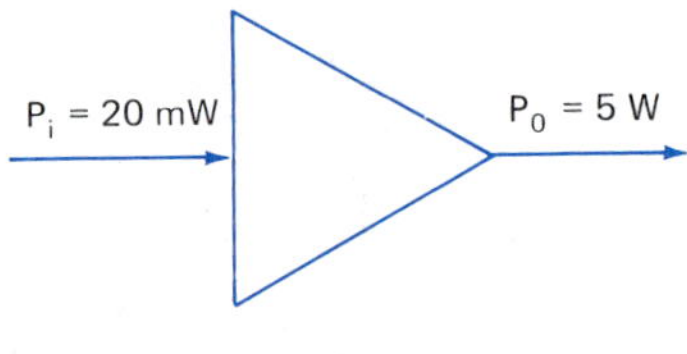

Figure 23-8

Power level is measured in bels (B) to honor Alexander Graham Bell, the inventor of the telephone. The bel unit is a large unit so power level is nearly always measured in decibels (dB). A decibel is one-tenth of a bel.

$$\mathbf{PL = 10 \log \frac{P_o}{P_i}} \qquad \text{(In dB)}$$

Example A An amplifier has an output power of 5 W when the input is 20 mW, Figure 23-8. Find (1) the power ratio and (2) the power level.

Solution: (1) $PR = \frac{P_o}{P_i}$

$PR = \frac{5000 \text{ mW}}{20 \text{ mW}} = 250$

(2) $PL = 10 \log \frac{P_o}{P_i}$

$PL = 10 \log 250 = 24 \text{ dB}$

P_i P_o

Figure 23-9(A) An attenuator circuit symbol.

An attenuator is a device that has a power output which is less than the power input. The circuit symbol is a rectangle, Figure 23-9(A). The power ratio for an attenuator is a fraction. The logarithm of a fraction is negative. Thus, the power level for an attenuator has a negative value but it is still measured in decibels, Figure 23-9(B).

Figure 23-9(B) A decibel meter (Courtesy of Southeast Nebraska Community College)

Example B An attenuator has an output power of 20 mW and an input power of 5 W, Figure 23-10. Find (1) the power ratio and (2) the power level.

Solution: (1) $PR = \frac{P_o}{P_i}$

$$PR = \frac{20 \text{ mW}}{5000 \text{ mW}} = 0.004$$

(2) $PL = 10 \log \frac{P_o}{P_i}$

$$PL = 10 \log (0.004) = -24 \text{ dB}$$

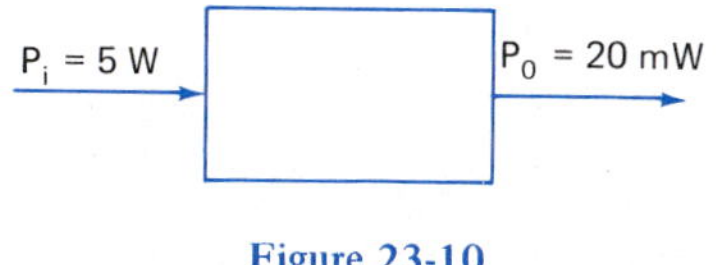

Figure 23-10

Compare Examples A and B. An amplifier produces a *gain*. A gain results in a positive decibel calculation. An attenuating device results in a *loss* and a negative decibel. The power increased by a factor of 250 in the amplifier of Example A which is a gain of 24 dB. The attenuator of Example B reduced the power by a factor of 250 and a loss of 24 dB. The loss is indicated by the negative sign (–24 dB). The negative sign is important. Do not neglect it when making calculations.

Example C A control signal is sent along a transmission line to a substation with an input power of 150 μW. At the substation, the signal is received, having been attenuated by –15 dB, Figure 23-11. What is the power of the signal at the substation?

Solution: The power level of the transmission line for this signal is

$$PL = 10 \log \frac{P_o}{0.000\ 15} = -15 \text{ dB}$$

$$\log \frac{P_o}{0.000\ 15} = -1.5 \qquad \text{(DA, } \div \text{ 10)}$$

$$\log P_o - \log 0.000\ 15 = -1.5 \qquad \text{(Quotient property)}$$

$$\log P_o = -1.5 + \log 0.000\ 15 \qquad \text{(AA, + log 0.000 15)}$$

$$\log P_o = -1.5 - 3.8239$$

$$\log P_o = -5.3239 \qquad \text{(Evaluate)}$$

$$P_o = \text{antilog } (-5.3239) \qquad \text{(Apply antilog)}$$

$$P_o = 4.74\ \mu\text{W}$$

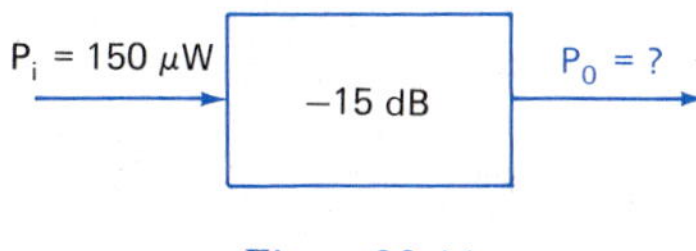

Figure 23-11

23-10 CURRENT AND VOLTAGE RATIOS

The power level in decibels is a measure of the ratio of two powers. Current and voltage ratios can be used, however, to calculate the power level. Since

$$P_o = I_o^2 Z_o$$

and

$$P_i = I_i^2 Z_i$$

the power ratio is as follows:

$$PR = \frac{P_o}{P_i} = \frac{I_o^2 Z_o}{I_i^2 Z_i}$$

If the impedances are matched ($Z_o = Z_i$), they will cancel:

$$PR = \frac{I_o^2}{I_i^2} = \left(\frac{I_o}{I_i}\right)^2$$

$$PL = 10 \log \left(\frac{I_o}{I_i}\right)^2 \qquad \text{(Substitute)}$$

$$PL = 10 \left(2 \log \frac{I_o}{I_i}\right) \qquad \text{(Power property)}$$

$$\mathbf{PL = 20 \log \frac{I_o}{I_i}}$$

It is important to realize that this equation gives the power level. It describes the gain or loss of power.

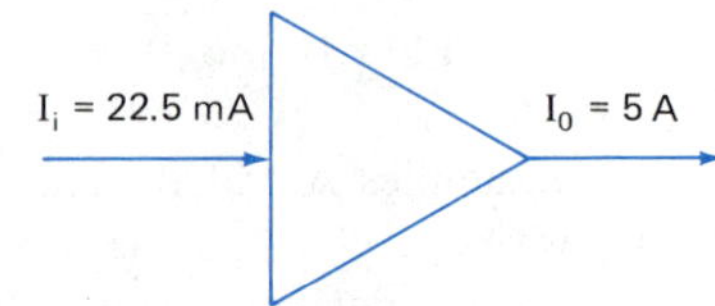

Figure 23-12

Example A (1) Find the power level of an amplifier at the frequency used if the input current is 22.5 mA and the output current is 5 A, Figure 23-12. (2) What is the power ratio?

Solution: (1) $PL = 20 \log \frac{I_o}{I_i}$

$$PL = 20 \log \frac{5000 \text{ mA}}{22.5 \text{ mA}}$$

$$PL = 20 \log 222$$

$$PL = 20(2.3468)$$

$$PL = 46.9 \text{ dB}$$

(2) $PL = 10 \log PR$

$$\log PR = \frac{PL}{10} \qquad (\text{DA}, \div 10)$$

$$PR = \text{antilog} \frac{PL}{10} \qquad \text{(Apply antilog)}$$

$$PR = \text{antilog} \frac{46.9}{10} = \text{antilog } 4.69$$

$$PR = 49\,400$$

Note: This means P_o is 49 400 times greater than P_i. The power ratio also equals the square of the current ratio (222).

A formula using the voltage ratio $\left(\frac{E_o}{E_i}\right)$ is similar:

$$\mathbf{PL = 20 \log \frac{E_o}{E_i}}$$

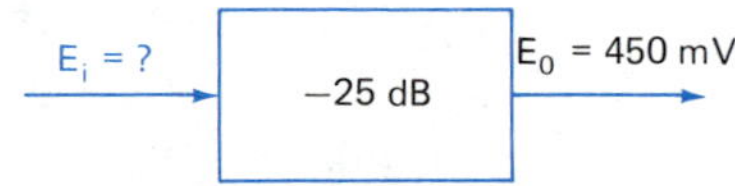

Figure 23-13

Example B The output voltage of an attenuating device, Figure 23-13, is 450 mV at a power level of −25 dB. What is the input voltage?

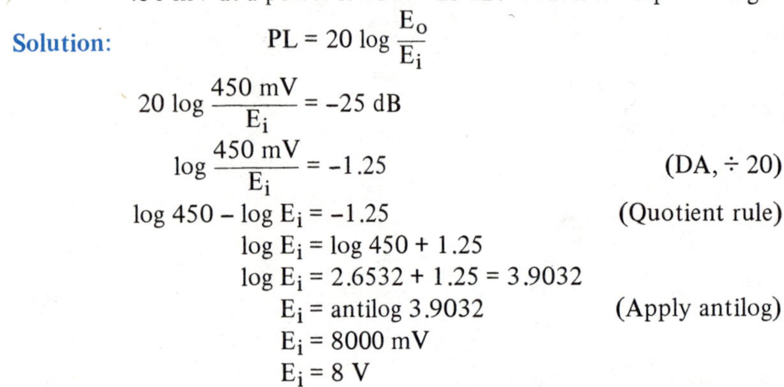

Solution: $PL = 20 \log \frac{E_o}{E_i}$

$$20 \log \frac{450 \text{ mV}}{E_i} = -25 \text{ dB}$$

$$\log \frac{450 \text{ mV}}{E_i} = -1.25 \qquad (\text{DA}, \div 20)$$

$$\log 450 - \log E_i = -1.25 \qquad \text{(Quotient rule)}$$

$$\log E_i = \log 450 + 1.25$$

$$\log E_i = 2.6532 + 1.25 = 3.9032$$

$$E_i = \text{antilog } 3.9032 \qquad \text{(Apply antilog)}$$

$$E_i = 8000 \text{ mV}$$

$$E_i = 8 \text{ V}$$

EXERCISE 23-7

Find the power level in decibels for each of the following.

1. $P_i = 3.15$ W → $P_0 = 47.5$ W

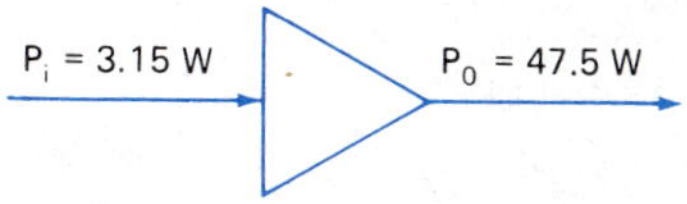

2. $P_i = 1.25$ mW → $P_0 = 67.5$ W

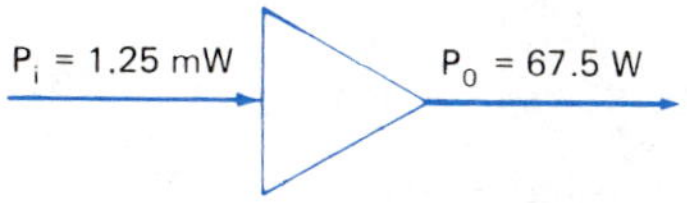

3. $I_i = 650$ mA → $I_0 = 48.5$ mA

4. $P_i = 15.8$ kW → $P_0 = 0.875$ kW

5. $E_i = 3.4$ V → $E_0 = 3650$ V

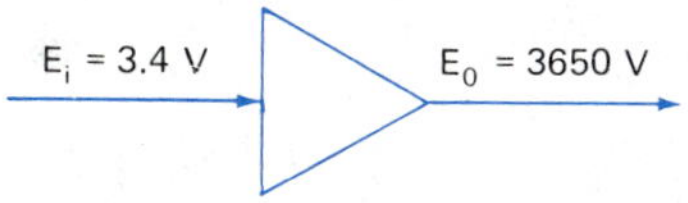

6. $I_i = 875$ mA → $I_0 = 12.5$ A

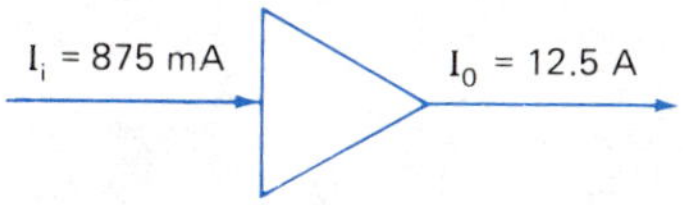

7. $I_i = 650$ mA → $I_0 = 2.46$ mA

8. $E_i = 240$ V → $E_0 = 160$ V

Find the unknown quantity in each of the following problems.

9. $P_i = ?$ → $P_0 = 655$ W; 21.8 dB

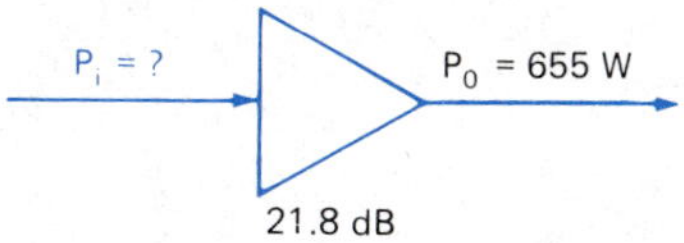

10. $E_i = 35.5$ V → $E_0 = ?$; 36.4 dB

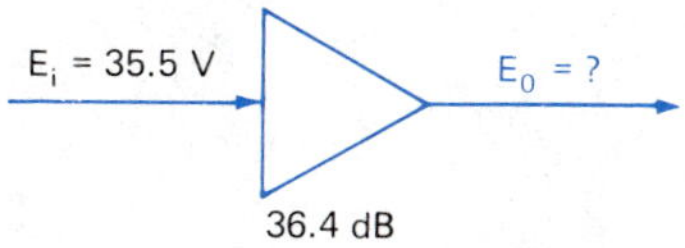

11. $I_i = ?$ → $I_0 = 375$ mA; −28.5 dB

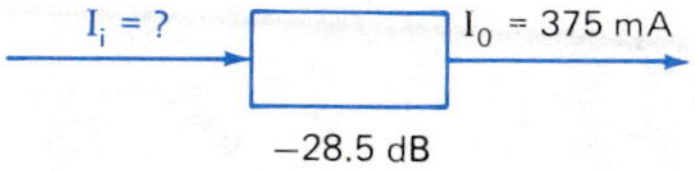

12. $E_i = 120$ V → $E_0 = ?$; −19.6 dB

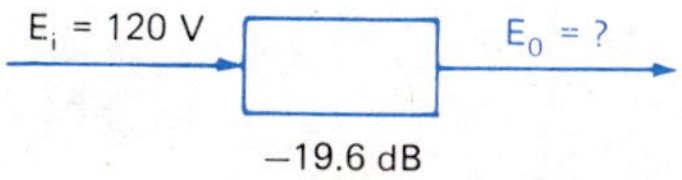

13. A low-pass filter is a capacitor placed in parallel with the load. Its capacitance is such that high frequency waves are directed through it. Only low frequency waves pass on to the load. A low-pass filter attenuates a certain high frequency wave. If the input power of that particular wave is 850 mW and the output power is 2.5 mW, find the power level of the filter in decibels.
14. An amplifier has a gain of 15.8 dB. What is the input power if the output power is 6.5 mW?
15. A 75-km line has an input power of 12.5 mW. If 50 μW is delivered at the end of the line, (a) find the overall attenuation in decibels. (b) What is the attenuation in decibels per kilometre?
16. An amplifier is used to raise the power level at the end of the line in Problem 15 to its original value. If the input voltage is 75 mV, find the output voltage of the amplifier.
17. Find (a) the voltage ratio and (b) the power ratio of the amplifier in Problem 16.
18. Find the output current of an amplifier having a power level of 42 dB. The input current is 680 μA.

23-11 APPROXIMATING DECIBELS

The power level in decibels is a description of the increase or decrease of power. The power level is ten times the logarithm of the power ratio:

$$PL = 10 \log PR$$

A graph of this function is shown in Figure 23-14.

When the power ratio increases from 1 to 2, it doubles. The graph indicates this to be a gain of approximately 3 dB. Another increase of 3 dB is indicated when the power ratio increases from 2 to 4 (doubles again). Changes from 4 to 8 and from 8 to 16 are also increases of 3 dB. Each time the power ratio is doubled, there is an increase of 3 dB. When it increases from 1 to 8, there is an increase of 9 dB. Each doubling of the power ratio adds 3 dB to the increase.

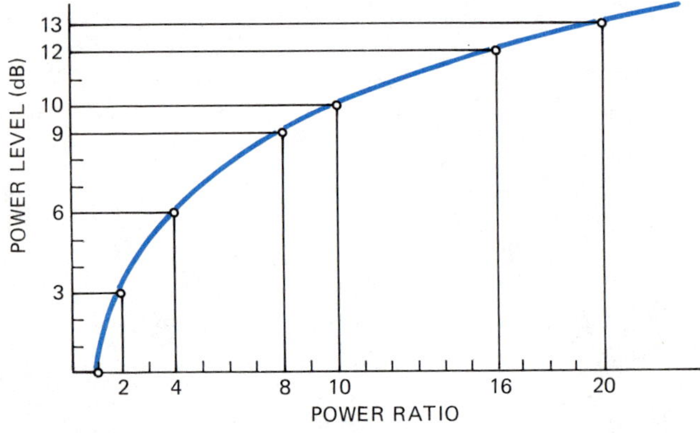

Figure 23-14 Graph of power level vs power ratio; a logarithmic function

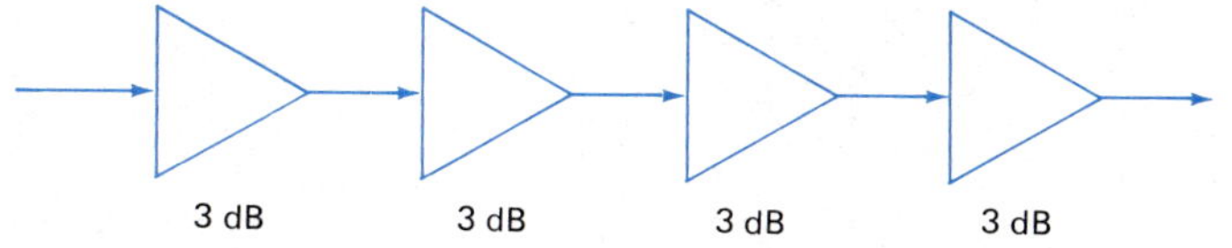

Figure 23-15

Example A Find the decibel increase if the power ratio increases by 16 times.

Solution: An increase of power by a factor of 16 represents four doublings:

$16 = 2 \times 2 \times 2 \times 2$

Since each doubling is 3 dB and decibels add, the power level is:

3 dB + 3 dB + 3 dB + 3 dB = 12 dB

See Figure 23-15.

Figure 23-14 also shows that a power ratio of 10 is indicated by a power level gain of 10 dB. If the graph could be extended, it would indicate that a power ratio of 100 and 1000 correspond to 20 dB and 30 dB, respectively. Thus, each time the power ratio increases by a factor of 10, the power level increases by 10 dB.

Example B Find the increase in decibels if the power ratio is 10 000.

Solution: Write the number 10 000 as factors of ten. Refer to Figure 23-16:

$10\,000 = 10 \times 10 \times 10 \times 10$

Each factor of 10 is 10 dB:

10 dB + 10 dB + 10 dB + 10 dB = 40 dB

If the power ratio can be expressed ad factors of 2s and 10s, the power level can be found mentally.

Example C A power ratio of 40 is how many decibels?

Solution: Write the number 40 as a product of 2s and 10s. Refer to Figure 23-17:

$40 = 10 \times 2 \times 2$

10 dB + 3 dB + 3 dB = 16 dB

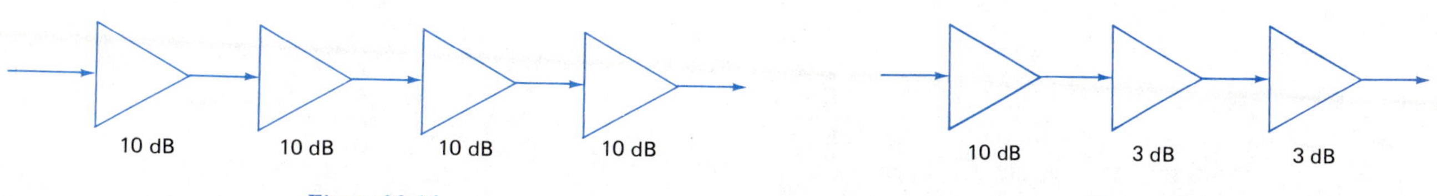

Figure 23-16

Figure 23-17

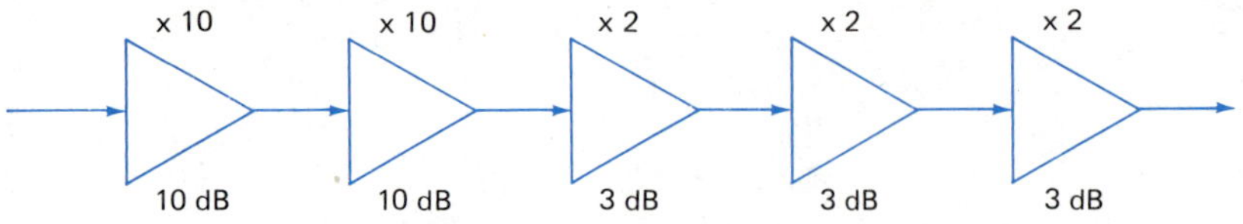

x 10 x 10 x 2 x 2 x 2

10 dB 10 dB 3 dB 3 dB 3 dB

Figure 23-18

Example D Find the output power if 2 W input power undergoes a gain of 29 dB.

Solution: Express the 29 dB as a summation of 10s and 3s. Refer to Figure 23-18:

29 dB = 10 dB + 10 dB + 3 dB + 3 dB + 3 dB

The power ratio is found by multiplying the corresponding 10s and 2s:

$PR = 10 \times 10 \times 2 \times 2 \times 2 = 800$

$PR = \frac{P_o}{P_i}$

$P_o = PR \times P_i$ $(MA, \times P_i)$

$P_o = 800(2\ W) = 1600\ W$

Attenuation decibels can be found in a similar way. Simply find the factor by which the power is decreased. Express the factor as a product of 10s and 2s. Add the corresponding *negative* decibels.

Example E Find the power level of an attenuating device that decreases the power from 800 W to 2 W, Figure 23-19.

Solution: The factor by which the power decreased is 400. (The power level is 1/400.)

$400 = 10 \times 10 \times 2 \times 2$

$PL = -10\ dB - 10\ dB - 3\ dB - 3\ dB = -26\ dB$

EXERCISE 23-8

Find the power level in decibels for each of the following problems. Do the calculations without the aid of a calculator or tables.

1. A change from 5 W to 80 W.
2. A change from 3 W to 3000 W.
3. A gain from 7.5 W to 750 W.
4. A gain from 11 mW to 88 mW.
5. An increase from 3 W to 60 W.
6. An increase from 4 W to 160 W.
7. A change from 10 W to 4000 W.

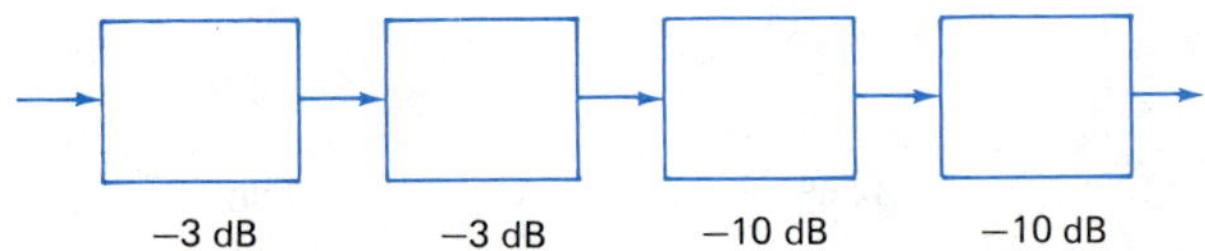

Figure 23-19

8. A change from 6 mW to 2.4 W.
9. A decrease from 40 W to 2 W.
10. A loss from 10 W to 5 mW.
11. A loss from 40 W to 2 mW.
12. A decrease from 250 mW to 0.625 mW.
13. A change from 3 W to 240 W.
14. A loss from 200 mW to 5 mW.
15. A gain with a power ratio of 1600.

Find the power output in each of the following problems. Do not use calculators or tables.

16. 5 W increased by 10 dB.
17. 2 W increased by 20 dB.
18. 3 W increased by 13 dB.
19. 4 W increased by 6 dB.
20. 8 mW increased by 19 dB.
21. 1 mW attenuated by 13 dB.
22. 2 W attenuated by 23 dB.
23. 3 W decreased by 33 dB.
24. 4 W decreased by 39 dB.
25. 150 mW attenuated by 26 dB.
26. 150 mW amplified by 26 dB.
27. 3 W increased by 16 dB.
28. 20 W attenuated by 29 dB.
29. 2 W increased by 17 dB. (Hint: 17 = 20 − 3)
30. 4 W attenuated by 22 dB. (Hint: 22 = 10 + 12)

23-12 TRANSMISSION LINES

Open-air transmission lines, Figure 23-20, consist of wires insulated from each other by an air space. High voltage lines may be separated by several metres. In this topic, the capacitance, inductance and resistance of such lines are discussed.

A two-wire open-air line has a capacitance given by the equation:

Figure 23-20 A long distance transmission line

$$C = \frac{12.08l}{\log \frac{d}{r}}$$

where C is the capacitance of the line in nanofarads (nF).
l is the one way length of the line in kilometres (km).
d is the center-to-center distance between conductors.
r is the radius of each wire expressed in the same units as d and must have a value considerably less than d.

The capacitance of a transmission line is directly proportional to the length of the line. Twenty kilometres of line has twice the capacitance of a similar ten-kilometre line. The only other variable in the equation is the distance separating the lines. (The radius is determined by other requirements.)

The logarithm of larger numbers (large values of d) is a larger number. Since the log is in the denominator, C will have a smaller value. If the distance separating the lines is increased, C will decrease. If d is decreased by mounting the lines closer together, C will increase.

The conductors act as very long, very narrow parallel "plates" separated by a large air gap. Even so, a long line can have a significant capacitance. The capacitance of transmission lines acts the same as a capacitor connected in parallel with the load. It provides a high impedance path for alternating current.

Example A (1) Find the capacitance of a 150-km line, Figure 23-21, consisting of AWG 00 wire spaced 1.25 m apart. (2) What is the capacitive reactance for a 60-Hz source?

Solution: Data: l = 150 km, d = 1.25 m, r = 4.633 mm = 0.004 633 m, f = 60 Hz (find r from wire tables)

(1) Formula: $C = \frac{12.08l}{\log \frac{d}{r}}$

Substitute: $C = \frac{12.08(150 \text{ km})}{\log \frac{1.25 \text{ m}}{0.004\,633 \text{ m}}}$

$$C = \frac{1812}{\log 270} = \frac{1812}{2.43} = 746 \text{ nF}$$

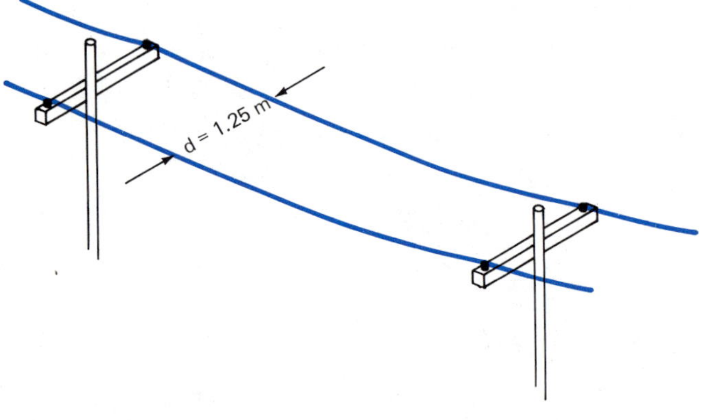

Figure 23-21

(2) Formula: $X_C = \frac{1}{2\pi fC}$

Substitute: $X_C = \frac{1}{2\pi(60\text{ Hz})(746 \times 10^{-9}\text{ F})}$

$X_C = 3550\ \Omega$

A two-wire transmission line also produces its own inductance. It consists of a line carrying energy toward the load and another for the return. The two wires form a closed loop. With alternating current in the loop, the loop itself produces self-inductance. The amount of inductance produced by a two-wire system is

$$L = l(0.921 \log \frac{d}{r} + 0.1)$$

where L is the inductance of the line in millihenries (mH).
l is the one-way length of the line in kilometres (km).
d is the center-to-center distance between conductors.
r is the radius of each wire expressed in the same units as d.

The inductance is directly proportional to the length. A twenty-kilometre line will have twice the inductance of a similar ten-kilometre line. If the spacing between the lines is increased, the inductance increases. It is interesting to note that inductance increases but capacitance decreases when the spacing is increased.

Example B Find (1) the inductance of the line in Example A and (2) the inductive reactance.

Solution: Data: l = 150 km, d = 1.25 m, r = 0.004 633 m, f = 60 Hz

(1) Formula: $L = l(0.921 \log \frac{d}{r} + 0.1)$

Substitute: $L = (150\text{ km})(0.921 \log \frac{1.25\text{ m}}{0.004\ 633\text{ m}} + 0.1)$

$L = (150\text{ km})(0.921 \log 270 + 0.1)$

$L = (150\text{ km})[0.921(2.43) + 0.1]$

$L = (150\text{ km})(2.2393 + 0.1)$

$L = (150\text{ km})(2.3393)$

$L = 351\text{ mH}$

(2) Formula: $X_L = 2\pi fL = 2\pi(60\text{ Hz})(0.351\text{ H}) = 132\ \Omega$

The impedance and capacitance formulas are for open-air, two-wire transmission lines only. Formulas for other kinds of lines such as twisted-pair lines and sheathed cables can be found in advanced textbooks.

The resistance of a line can be determined from tables, Appendix A. The length of the line and the return must be used in the calculation.

Example C Find the resistance of a 50-km solid copper AWG No. 4 line.

Solution: For No. 4 wire, the resistance per kilometre is 0.8155 Ω/km, Appendix A. The total length of line and return is 100 km. The total resistance is the product:

$$\left(0.8155 \frac{\Omega}{\text{km}}\right)(100\text{ km}) = 81.55\ \Omega$$

The following approximate relationships can be found in the AWG wire tables.

- Increasing the gage number by 3 doubles the resistance (per km).
- Increasing the gage number by 10 increases resistance (per km) ten times.

From the experience of estimating decibels, Topic 23-11, you should recognize a similarity in the pattern. Therefore, a simple formula can be written relating AWG gage number (g) and resistance per unit length (R_l):

$$g = 10 \log \frac{R_l}{0.3223}$$

where the constant 0.3223 is the resistance per kilometre of No. 0 wire. The total resistance of the conductor is found by rearranging the equation and multiplying by the total length of the line (l) measured in kilometres *including the return line:*

$$R = 0.3223l\left(\text{antilog } \frac{g}{10}\right)$$

Where l is the length of the line and the return.
g is the AWG gage number.

It should be understood that this formula is an approximation. It gives good results for small gage numbers, about 5% error for No. 26 and 10% error for No. 56. Electricians are seldom concerned with wire having large gage numbers. Such errors are well within the manufacturer's tolerance. For No. 00, 000 and 0000 wires, use the numbers −1, −2 and −3 for g, respectively.

Example D Find the resistance of an AWG No. 2 line that is 60 km long.

Solution: Assume the return line is also 60 km (l = 120 km).

Formula: $R = 0.3223l\left(\text{antilog } \frac{g}{10}\right)$

Substitute: $R = 0.3223(120 \text{ km})\left(\text{antilog } \frac{2}{10}\right)$

$R = 38.7(\text{antilog } 0.2)$

$R = 38.7(1.5849)$

$R = 61.3\ \Omega$

Check: Use values from wire table:
$(120 \text{ km})(0.5128) = 61.5\ \Omega$ (Error is 0.3%)

The inductance and resistance of the lines acts in series with the load. The capacitance acts as a circuit element connected in parallel. The inductive reactance, capacitive reactance and line resistance of the transmission lines act as separate circuit elements connected as shown in Figure 23-22.

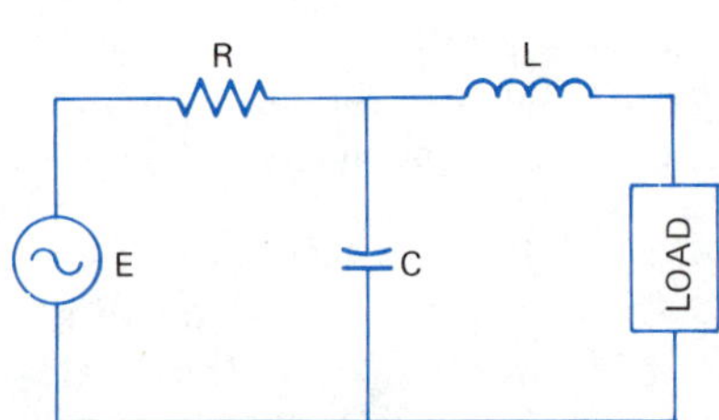

Figure 23-22 Inductance, capacitance and resistance of a transmission line

EXERCISE 23-9

Data, diagram, formula, substitute, check.

1. Find the capacitance of a 100-km line spaced 1 m apart. The lines are constructed of AWG No. 2 wires.
2. Find the capacitance of the line in Problem 1 if the spacing is 2 m.
3. A 50-km transmission line consists of two AWG No. 0 wires spaced 1.5 m apart. Determine the inductance of the line.
4. Find (a) the inductance of the line in Problem 3 if the spacing is 3 m (doubled) and (b) the resistance if the line is copper. Use the equation, $R = 0.3223l\left(\text{antilog } \frac{g}{10}\right)$ and check using tables.
5. Calculate (a) the capacitance and (b) the capacitive reactance of a two-wire (AWG No. 000) transmission line that is 50 km long. The wires are 1.4 m apart and are connected to a 60-Hz source.
6. Find (a) the inductance and (b) the inductive reactance of the line in Problem 5.
7. Find the resistance of the line in Problem 5 assuming the wire is made of copper. Check answer using wire table.
8. Find the resistance of an 85-km line consisting of AWG No. 00 copper wire. Use wire table to check answer.
9. A 150-km transmission line consists of two AWG No. 1 copper wires spaced 1.75 m apart. Determine (a) the inductance, (b) the capacitance and (c) the resistance of the line.
10. In a certain street light circuit, the lights are connected in series. The circuit is 2 km long and returns on an adjacent street with 120 m spacing. The wires are 4 gage. Find (a) the capacitance and (b) the inductance of the line.
11. Increasing the gage number (AWG) of wire by 3 halves the cross-sectional area. An increase of 10 divides the area by 10. The approximate formula relating gage number (g) and cross-sectional area (A) in square millimetres, therefore, is

$$g = -10 \log \frac{A}{53.49}$$

Solve the formula for A and find the cross-sectional area for the following wire sizes. Check with tables and find the percent error:

a. 2　b. 6　c. 10　d. 14　e. 000　f. 0

12. Increasing the gage number (AWG) of wire (g) by 6 halves the diameter (d). An increase of 20 divides the diameter by 10. Therefore:

$$g = -20 \log \frac{d}{8.2525}$$

Solve this approximate equation for d and find the diameter of the following wires. Check all answers with the wire table and find the percent error:

a. 000　b. 1　c. 6　d. 10　e. 16　f. 20

SECTION 4 REVIEW

Chapter 23 concludes Section 4: Analyzing ac Circuits. This review highlights the many topics in Section 4.

- *Graphing.* Sine wave, polar and rectangular coordinate systems, exponential decay, exponential growth, logarithm function.
- *Trigonometry.* Phasors, phasor transformation of coordinates. Adding, subtracting, multiplying, dividing and finding reciprocals of phasors.
- *Algebra.* Solving logarithmic equations.
- *Logarithms.* Common logarithms, natural logarithms, antilogarithms.
- *Electrical concepts.* Phase, lag-lead, inductive and capacitive reactance, impedance, apparent, true and reactive power. Admittance, conductance, susceptance, power factor, power factor correction. Delta-wye connections, three-phase power. Power gain-loss, power ratio, power level, the decibel, resistance, capacitance and inductance of transmission lines.
- *Electric devices.* Varmeter, power factor meter, power factor correction capacitors, three-phase alternators, amplifiers, attenuators.
- *Electric circuits.* ac series circuits, ac parallel circuits, series-parallel equivalent circuits. Two-phase circuits, three-phase circuits, Y-Y, Y-Δ, Δ-Δ.

A-1
AMERICAN WIRE GAGE (AWG) TABLE FOR
SOLID COPPER CONDUCTORS (20°C)

GAGE NO	DIAMETER (mm)	Ω/km	DIAMETER (mil)	Ω/1000 ft
0000	11.68	0.1608	460.0	0.0490
000	10.40	0.2028	409.6	0.0618
00	9.266	0.2557	364.8	0.0779
0	8.252	0.3224	324.9	0.0983
1	7.348	0.4066	289.3	0.1239
2	6.544	0.5129	257.6	0.1563
3	5.827	0.6465	229.4	0.1970
4	5.189	0.8155	204.3	0.2485
5	4.620	1.028	181.9	0.3133
6	4.115	1.297	162.0	0.3951
7	3.665	1.634	144.3	0.4982
8	3.264	2.062	128.5	0.6282
9	2.906	2.599	114.4	0.7921
10	2.588	3.277	101.9	0.9989
11	2.305	4.132	90.74	1.260
12	2.053	5.211	80.81	1.588
13	1.828	6.571	71.96	2.003
14	1.628	8.285	64.08	2.525
15	1.450	10.45	57.07	3.184
16	1.291	13.17	50.82	4.016
17	1.150	16.61	45.26	5.064
18	1.024	20.95	40.30	6.385
19	0.9116	26.42	35.89	8.051
20	0.8118	33.31	31.96	10.15
21	0.7226	42.00	28.45	12.80
22	0.6439	52.96	25.35	16.14
23	0.5733	66.79	22.57	20.36
24	0.5105	84.24	20.10	25.67
25	0.4547	106.2	17.90	32.37
26	0.4049	133.9	15.94	40.81
27	0.3607	168.9	14.20	51.57
28	0.3211	213.0	12.64	64.90
29	0.2860	268.5	11.26	81.83
30	0.2548	338.6	10.03	103.2
32	0.2019	538.5	7.950	164.1
36	0.1270	1361	5.000	414.8
40	0.0790	3540	3.11	1070
44	0.0511	8510	2.01	2582
48	0.0315	22 100	1.24	6705
52	0.0198	55 900	0.780	16 960
56	0.0124	14 200	0.488	43 080

A-2
RESISTIVITY OF VARIOUS MATERIALS TABLE

SUBSTANCE	RESISTIVITIES AT 20°C (Ohm-metres)
Aluminum	28.3×10^{-9}
Carbon	$35\,000 \times 10^{-9}$
Copper	17.2×10^{-9}
Iron	103×10^{-9}
Manganin alloy	447×10^{-9}
Nichrome alloy	1000×10^{-9}
Silver	16.4×10^{-9}
Tungsten	55.2×10^{-9}

A-3
TEMPERATURE COEFFICIENTS OF RESISTANCE TABLE

SUBSTANCE	TEMPERATURE COEFFICIENT PER °C
Aluminum	0.0039
Carbon	-0.0005
Copper	0.003 93
Iron	0.006
Manganin alloy	0.000 01
Nichrome alloy	0.000 17
Silver	0.0038
Tungsten	0.0047

A-4

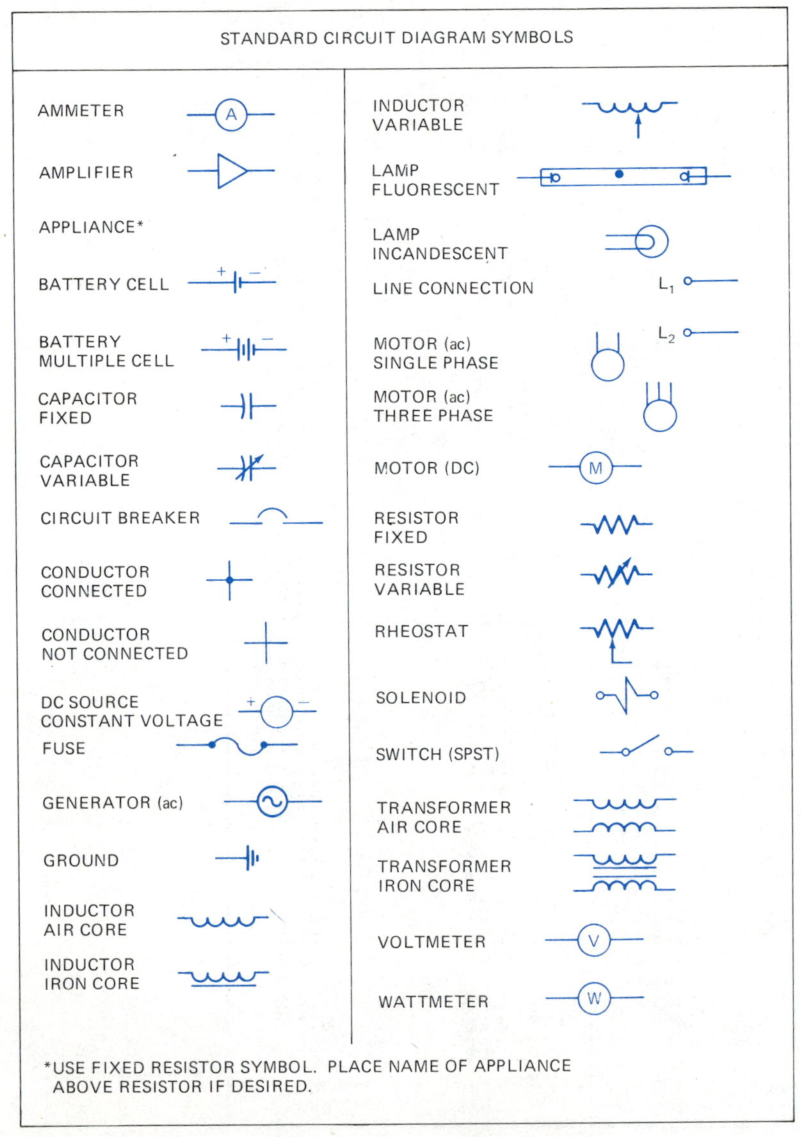

A-5
THE GREEK ALPHABET

GREEK LETTER	GREEK NAME
Α α	Alpha
Β β	Beta
Γ γ	Gamma
Δ δ	Delta
Ε ϵ	Epsilon
Ζ ζ	Zeta
Η η	Eta
Θ θ	Theta
Ι ι	Iota
Κ κ	Kappa
Λ λ	Lambda
Μ μ	Mu
Ν ν	Nu
Ξ ξ	Xi
Ο ο	Omicron
Π π	Pi
Ρ ρ	Rho
Σ σ	Sigma
Τ τ	Tau
Υ υ	Upsilon
Φ φ	Phi
Χ χ	Chi
Ψ ψ	Psi
Ω ω	Omega

A-6
CONVERSION FACTORS

LENGTH

1 m = 3.281 ft = 39.37 in
1 km = 3281 ft = 0.6214 mi
1 in = 2.54 cm
1 ft = 30.48 cm
1 mi = 1.609 km = 5280 ft

AREA

1 m^2 = 10.76 ft^2
1 ft^2 = 144 in^2 = 929 cm^2

VOLUME

1 m^3 = 35.31 ft^3 = 1000 L
1 ft^3 = 1728 in^3 = 7.48 gal = 28.32 L
1 gal = 231 in^3 = 3.7854 L = 4 qt = 8 pt
1 L = 1.06 qt = 61.02 in^3
1 in^3 = 16.39 cm^3

FORCE

1 kg-wt = 2.205 lb = 9.81 N
1 N = 0.102 kg-wt = 0.2248 lb
1 lb = 0.4536 kg-wt = 4.448 N = 16 oz
1 ton = 2000 lb

DENSITY

1 g/cm^3 = 1 kg/L = 62.43 lb/ft^3 = 0.0361 lb/in^3

WORK, ENERGY AND HEAT

1 J = 0.7376 ft•lb
1 kW•h = 3.6 MJ
1 Btu = 778 ft•lb
1 kcal = 4186 J = 3.968 Btu = 3087 ft•lb

POWER

1 W = 0.7376 ft•lb/s = 668 lm
1kW = 1.341 hp
1 hp = 550 ft•lb/s = 33 000 ft•lb/min = 745.7 W

ELECTRIC CHARGE

1 C = 6.242×10^{18} e (electrons)
1 e = 1.602×10^{-19} C

TRIGONOMETRIC TABLE

For angles expressed in degrees, the table at the end of Appendix B can be used. The values of the sine, cosine and tangent for each tenth degree from 0° to 90° are given. To find the value of a particular function, move down the angle column to the given angle. Cross to the right to the desired tenth column and read the value of that function for that angle.

Example B-1 Find the value of the following functions: (a) sin 23°, (b) cos 37.2°, and (c) tan 81.7°.

Solution: (a) Find 23° in the angle column. Move right to the first column to find the answer: sin 23° = 0.3907.

(b) Move down the angle column to 37°. Proceed right to the 0.2° column: cos 37.2° = 0.7965.

(c) Find 81° in the angle column. Cross to the 0.7° column: tan 81.7 = 6.8548.

For vectors and phasors, always draw a diagram. The sign of a function can be determined by inspection. It depends on the signs of the x and y components as described in Topic 16-8. The Trigonometric Table gives the values of functions between 0° and 90° only. To find the functions of angles greater than 90°, use the following procedure:

- Draw a diagram of the radius vector in the given position on an x-y coordinate system.
- Determine the acute angle (smallest) the vector makes with the x-axis.
- Use the tables to determine the value of the desired function of the acute angle.
- Determine the sign of the function by inspection.

Example B-2 Find the sine, cosine and tangent of 150° using the table.

Solution: The acute angle between the vector and the x-axis is $\alpha = 30°$, Figure B-1. Use the table to find: sin 30° = 0.5000, cos 30° = 0.8660 and tan 30° = 0.5774. Since x is negative, y is positive and r is positive, the functions are as follows:

sin 150° = 0.5000
cos 150° = -0.8660
tan 150° = -0.5774

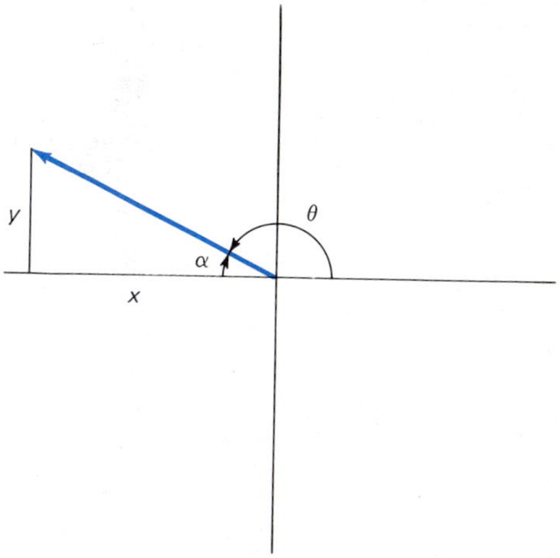

Figure B-1

The acute angle is sometimes called the *working angle.* It is always the smallest angle between the vector and the x-axis. For second and third quadrant vectors it is the angle between the vector and the negative x-axis. The values of the functions of the working angle are the same as those of the given angle. Always construct a diagram to determine the correct signs.

The principal values of the arc functions can also be found using the table. Simply find the arc function of the given value. This will be the working angle:

- For positive functions, the principal value equals the working angle for all three functions.
- For negative sines and tangents, the principal value is the negative of the working angle.
- For negative cosines, the principal value is the supplement of the working angle.

Example B-3 Find the principal value of (a) arcsin 0.5446, (b) $\cos^{-1}$ (-0.9397) and (c) invtan (-6.3138).

Solution:
(a) arcsin 0.5446 = 33°
(b) $\cos^{-1}$ (-0.9397) = 180° - 20° = 160° (Supplement of 20°)
(c) invtan (-6.3138) = -81° (Negative of 81°)

deg	function	0.0°	0.1°	0.2°	0.3°	0.4°	0.5°	0.6°	0.7°	0.8°	0.9°
0	sin	0.0000	0.0017	0.0035	0.0052	0.0070	0.0087	0.0105	0.0122	0.0140	0.0157
	cos	1.0000	1.0000	1.0000	1.0000	1.0000	1.0000	0.9999	0.9999	0.9999	0.9999
	tan	0.0000	0.0017	0.0035	0.0052	0.0070	0.0087	0.0105	0.0122	0.0140	0.0157
1	sin	0.0175	0.0192	0.0209	0.0227	0.0244	0.0262	0.0279	0.0297	0.0314	0.0332
	cos	0.9998	0.9998	0.9998	0.9997	0.9997	0.9997	0.9996	0.9996	0.9995	0.9995
	tan	0.0175	0.0192	0.0209	0.0227	0.0244	0.0262	0.0279	0.0297	0.0314	0.0332
2	sin	0.0349	0.0366	0.0384	0.0401	0.0419	0.0436	0.0454	0.0471	0.0488	0.0506
	cos	0.9994	0.9993	0.9993	0.9992	0.9991	0.9990	0.9990	0.9989	0.9988	0.9987
	tan	0.0349	0.0367	0.0384	0.0402	0.0419	0.0437	0.0454	0.0472	0.0489	0.0507
3	sin	0.0523	0.0541	0.0558	0.0576	0.0593	0.0610	0.0628	0.0645	0.0663	0.0680
	cos	0.9986	0.9985	0.9984	0.9983	0.9982	0.9981	0.9980	0.9979	0.9978	0.9977
	tan	0.0524	0.0542	0.0559	0.0577	0.0594	0.0612	0.0629	0.0647	0.0664	0.0682
4	sin	0.0698	0.0715	0.0732	0.0750	0.0767	0.0785	0.0802	0.0819	0.0837	0.0854
	cos	0.9976	0.9974	0.9973	0.9972	0.9971	0.9969	0.9968	0.9966	0.9965	0.9963
	tan	0.0699	0.0717	0.0734	0.0752	0.0769	0.0787	0.0805	0.0822	0.0840	0.0857
5	sin	0.0872	0.0889	0.0906	0.0924	0.0941	0.0958	0.0976	0.0993	0.1011	0.1028
	cos	0.9962	0.9960	0.9959	0.9957	0.9956	0.9954	0.9952	0.9951	0.9949	0.9947
	tan	0.0875	0.0892	0.0910	0.0928	0.0945	0.0963	0.0981	0.0998	0.1016	0.1033
6	sin	0.1045	0.1063	0.1080	0.1097	0.1115	0.1132	0.1149	0.1167	0.1184	0.1201
	cos	0.9945	0.9943	0.9942	0.9940	0.9938	0.9936	0.9934	0.9932	0.9930	0.9928
	tan	0.1051	0.1069	0.1086	0.1104	0.1122	0.1139	0.1157	0.1175	0.1192	0.1210
7	sin	0.1219	0.1236	0.1253	0.1271	0.1288	0.1305	0.1323	0.1340	0.1357	0.1374
	cos	0.9925	0.9923	0.9921	0.9919	0.9917	0.9914	0.9912	0.9910	0.9907	0.9905
	tan	0.1228	0.1246	0.1263	0.1281	0.1299	0.1317	0.1334	0.1352	0.1370	0.1388
8	sin	0.1392	0.1409	0.1426	0.1444	0.1461	0.1478	0.1495	0.1513	0.1530	0.1547
	cos	0.9903	0.9900	0.9898	0.9895	0.9893	0.9890	0.9888	0.9885	0.9882	0.9880
	tan	0.1405	0.1423	0.1441	0.1459	0.1477	0.1495	0.1512	0.1530	0.1548	0.1566
9	sin	0.1564	0.1582	0.1599	0.1616	0.1633	0.1650	0.1668	0.1685	0.1702	0.1719
	cos	0.9877	0.9874	0.9871	0.9869	0.9866	0.9863	0.9860	0.9857	0.9854	0.9851
	tan	0.1584	0.1602	0.1620	0.1638	0.1655	0.1673	0.1691	0.1709	0.1727	0.1745
10	sin	0.1736	0.1754	0.1771	0.1778	0.1805	0.1822	0.1840	0.1857	0.1874	0.1891
	cos	0.9848	0.9845	0.9842	0.9839	0.9836	0.9833	0.9829	0.9826	0.9823	0.9820
	tan	0.1763	0.1781	0.1799	0.1817	0.1835	0.1853	0.1871	0.1890	0.1908	0.1926
11	sin	0.1908	0.1925	0.1942	0.1959	0.1977	0.1994	0.2011	0.2028	0.2045	0.2062
	cos	0.9816	0.9813	0.9810	0.9806	0.9803	0.9799	0.9796	0.9792	0.9789	0.9785
	tan	0.1944	0.1962	0.1980	0.1998	0.2016	0.2035	0.2053	0.2071	0.2089	0.2107
12	sin	0.2079	0.2096	0.2113	0.2130	0.2147	0.2164	0.2181	0.2198	0.2215	0.2232
	cos	0.9781	0.9778	0.9774	0.9770	0.9767	0.9763	0.9759	0.9755	0.9751	0.9748
	tan	0.2126	0.2144	0.2162	0.2180	0.2199	0.2217	0.2235	0.2254	0.2272	0.2290
13	sin	0.2250	0.2267	0.2284	0.2300	0.2318	0.2334	0.2351	0.2368	0.2385	0.2402
	cos	0.9744	0.9740	0.9736	0.9732	0.9728	0.9724	0.9720	0.9715	0.9711	0.9707
	tan	0.2309	0.2327	0.2345	0.2364	0.2382	0.2401	0.2419	0.2438	0.2456	0.2475
14	sin	0.2419	0.2436	0.2453	0.2470	0.2487	0.2504	0.2521	0.2538	0.2554	0.2571
	cos	0.9703	0.9699	0.9694	0.9690	0.9686	0.9681	0.9677	0.9673	0.9668	0.9664
	tan	0.2493	0.2512	0.2530	0.2549	0.2568	0.2586	0.2605	0.2623	0.2642	0.2661
15	sin	0.2588	0.2605	0.2622	0.2639	0.2656	0.2672	0.2689	0.2706	0.2723	0.2740
	cos	0.9659	0.9655	0.9650	0.9646	0.9641	0.9636	0.9632	0.9627	0.9622	0.9617
	tan	0.2679	0.2698	0.2717	0.2736	0.2754	0.2773	0.2792	0.2811	0.2830	0.2849
16	sin	0.2756	0.2773	0.2790	0.2807	0.2823	0.2840	0.2857	0.2874	0.2890	0.2907
	cos	0.9613	0.9608	0.9603	0.9598	0.9593	0.9588	0.9583	0.9578	0.9573	0.9568
	tan	0.2867	0.2886	0.2905	0.2924	0.2943	0.2962	0.2981	0.3000	0.3019	0.3038
17	sin	0.2924	0.2940	0.2957	0.2974	0.2990	0.3007	0.3024	0.3040	0.3057	0.3074
	cos	0.9563	0.9558	0.9553	0.9548	0.9542	0.9537	0.9532	0.9527	0.9521	0.9516
	tan	0.3057	0.3076	0.3096	0.3115	0.3134	0.3153	0.3172	0.3191	0.3211	0.3230
18	sin	0.3090	0.3107	0.3123	0.3140	0.3156	0.3173	0.3190	0.3206	0.3223	0.3239
	cos	0.9511	0.9505	0.9500	0.9494	0.9489	0.9483	0.9478	0.9472	0.9466	0.9461
	tan	0.3249	0.3269	0.3288	0.3307	0.3327	0.3346	0.3365	0.3385	0.3404	0.3424
19	sin	0.3256	0.3272	0.3289	0.3305	0.3322	0.3338	0.3355	0.3371	0.3387	0.3404
	cos	0.9455	0.9449	0.9444	0.9438	0.9432	0.9426	0.9421	0.9415	0.9409	0.9403
	tan	0.3443	0.3463	0.3482	0.3502	0.3522	0.3541	0.3561	0.3581	0.3600	0.3620
20	sin	0.3420	0.3437	0.3453	0.3469	0.3486	0.3502	0.3518	0.3535	0.3551	0.3567
	cos	0.9397	0.9391	0.9385	0.9379	0.9373	0.9367	0.9361	0.9354	0.9348	0.9342
	tan	0.3640	0.3659	0.3679	0.3699	0.3719	0.3739	0.3759	0.3779	0.3799	0.3819
21	sin	0.3584	0.3600	0.3616	0.3633	0.3649	0.3665	0.3681	0.3697	0.3714	0.3730
	cos	0.9336	0.9330	0.9323	0.9317	0.9311	0.9304	0.9298	0.9291	0.9285	0.9278
	tan	0.3839	0.3859	0.3879	0.3899	0.3919	0.3939	0.3959	0.3979	0.4000	0.4020
22	sin	0.3746	0.3762	0.3778	0.3795	0.3811	0.3827	0.3843	0.3859	0.3875	0.3891
	cos	0.9272	0.9265	0.9259	0.9252	0.9245	0.9239	0.9232	0.9225	0.9219	0.9212
	tan	0.4040	0.4061	0.4081	0.4101	0.4122	0.4142	0.4163	0.4183	0.4204	0.4224
23	sin	0.3907	0.3923	0.3939	0.3955	0.3971	0.3987	0.4003	0.4019	0.4035	0.4051
	cos	0.9205	0.9198	0.9191	0.9184	0.9178	0.9171	0.9164	0.9157	0.9150	0.9143
	tan	0.4245	0.4265	0.4286	0.4307	0.4327	0.4348	0.4369	0.4390	0.4411	0.4431

Figure B-2

deg	*function*	0.0°	0.1°	0.2°	0.3°	0.4°	0.5°	0.6°	0.7°	0.8°	0.9°
24	sin	0.4067	0.4083	0.4099	0.4115	0.4131	0.4147	0.4163	0.4179	0.4195	0.4210
	cos	0.9135	0.9128	0.9121	0.9114	0.9107	0.9100	0.9092	0.9085	0.9078	0.9070
	tan	0.4452	0.4473	0.4494	0.4515	0.4536	0.4557	0.4578	0.4599	0.4621	0.4642
25	sin	0.4226	0.4242	0.4258	0.4274	0.4289	0.4305	0.4321	0.4337	0.4352	0.4368
	cos	0.9063	0.9056	0.9048	0.9041	0.9033	0.9026	0.9018	0.9011	0.9003	0.8996
	tan	0.4663	0.4684	0.4706	0.4727	0.4748	0.4770	0.4791	0.4813	0.4834	0.4856
26	sin	0.4384	0.4399	0.4415	0.4431	0.4446	0.4462	0.4478	0.4493	0.4509	0.4524
	cos	0.8988	0.8980	0.8973	0.8965	0.8957	0.8949	0.8942	0.8934	0.8926	0.8918
	tan	0.4877	0.4899	0.4921	0.4942	0.4964	0.4986	0.5008	0.5029	0.5051	0.5073
27	sin	0.4540	0.4555	0.4571	0.4586	0.4602	0.4617	0.4633	0.4648	0.4664	0.4679
	cos	0.8910	0.8902	0.8894	0.8886	0.8878	0.8870	0.8862	0.8854	0.8846	0.8838
	tan	0.5095	0.5117	0.5139	0.5161	0.5184	0.5206	0.5228	0.5250	0.5272	0.5295
28	sin	0.4695	0.4710	0.4726	0.4741	0.4756	0.4772	0.4787	0.4802	0.4818	0.4833
	cos	0.8829	0.8821	0.8813	0.8805	0.8796	0.8788	0.8780	0.8771	0.8763	0.8755
	tan	0.5317	0.5340	0.5362	0.5384	0.5407	0.5430	0.5452	0.5475	0.5498	0.5520
29.	sin	0.4848	0.4863	0.4879	0.4894	0.4909	0.4924	0.4939	0.4955	0.4970	0.4985
	cos	0.8746	0.8738	0.8729	0.8721	0.8712	0.8704	0.8695	0.8686	0.8678	0.8669
	tan	0.5543	0.5566	0.5589	0.5612	0.5635	0.5658	0.5681	0.5704	0.5727	0.5750
30	sin	0.5000	0.5015	0.5030	0.5045	0.5060	0.5075	0.5090	0.5105	0.5120	0.5135
	cos	0.8660	0.8652	0.8643	0.8634	0.8625	0.8616	0.8607	0.8599	0.8590	0.8581
	tan	0.5774	0.5797	0.5820	0.5844	0.5967	0.5890	0.5914	0.5938	0.5961	0.5985
31	sin	0.5150	0.5165	0.5180	0.5195	0.5210	0.5225	0.5240	0.5255	0.5270	0.5284
	cos	0.8572	0.8563	0.8554	0.8545	0.8536	0.8526	0.8517	0.8508	0.8499	0.8490
	tan	0.6009	0.6032	0.6056	0.6080	0.6104	0.6128	0.6152	0.6176	0.6200	0.6224
32	sin	0.5299	0.5314	0.5329	0.5344	0.5358	0.5373	0.5388	0.5402	0.5417	0.5432
	cos	0.8480	0.8471	0.8462	0.8453	0.8443	0.8434	0.8425	0.8415	0.8406	0.8396
	tan	0.6249	0.6273	0.6297	0.6322	0.6346	0.6371	0.6395	0.6420	0.6445	0.6469
33	sin	0.5446	0.5461	0.5476	0.5490	0.5505	0.5519	0.5534	0.5548	0.5563	0.5577
	cos	0.8387	0.8377	0.8368	0.8358	0.8348	0.8339	0.8329	0.8320	0.8310	0.8300
	tan	0.6494	0.6519	0.6544	0.6569	0.6594	0.6619	0.6644	0.6669	0.6694	0.6720
34	sin	0.5592	0.5606	0.5621	0.5635	0.5650	0.5664	0.5678	0.5693	0.5707	0.5721
	cos	0.8290	0.8281	0.8271	0.8261	0.8251	0.8241	0.8231	0.8221	0.8211	0.8202
	tan	0.6745	0.6771	0.6796	0.6822	0.6847	0.6873	0.6899	0.6924	0.6950	0.6976
35	sin	0.5736	0.5750	0.5764	0.5779	0.5793	0.5807	0.5821	0.5835	0.5850	0.5864
	cos	0.8192	0.8181	0.8171	0.8161	0.8151	0.8141	0.8131	0.8121	0.8111	0.8100
	tan	0.7002	0.7028	0.7054	0.7080	0.7107	0.7133	0.7159	0.7186	0.7212	0.7239
36	sin	0.5878	0.5892	0.5906	0.5920	0.5934	0.5948	0.5962	0.5976	0.5990	0.6004
	cos	0.8090	0.8080	0.8070	0.8059	0.8049	0.8039	0.8028	0.8018	0.8007	0.7997
	tan	0.7265	0.7292	0.7319	0.7346	0.7373	0.7400	0.7427	0.7454	0.7481	0.7508
37	sin	0.6018	0.6032	0.6046	0.6060	0.6074	0.6088	0.6101	0.6115	0.6129	0.6143
	cos	0.7986	0.7976	0.7965	0.7955	0.7944	0.7934	0.7923	0.7912	0.7902	0.7891
	tan	0.7536	0.7563	0.7590	0.7618	0.7646	0.7673	0.7701	0.7729	0.7757	0.7785
38	sin	0.6157	0.6170	0.6184	0.6198	0.6211	0.6225	0.6239	0.6252	0.6266	0.6280
	cos	0.7880	0.7869	0.7859	0.7848	0.7837	0.7826	0.7815	0.7804	0.7793	0.7782
	tan	0.7813	0.7841	0.7869	0.7898	0.7926	0.7954	0.7983	0.8012	0.8040	0.8069
39	sin	0.6293	0.6307	0.6320	0.6334	0.6347	0.6361	0.6374	0.6388	0.6401	0.6414
	cos	0.7771	0.7760	0.7749	0.7738	0.7727	0.7716	0.7705	0.7694	0.7683	0.7672
	tan	0.8098	0.8127	0.8156	0.8185	0.8214	0.8243	0.8273	0.8302	0.8332	0.8361
40	sin	0.6428	0.6441	0.6455	0.6468	0.6481	0.6494	0.6508	0.6521	0.6534	0.6547
	cos	0.7660	0.7649	0.7638	0.7627	0.7615	0.7604	0.7593	0.7581	0.7570	0.7559
	tan	0.8391	0.8421	0.8451	0.8481	0.8511	0.8541	0.8571	0.8601	0.8632	0.8662
41	sin	0.6561	0.6574	0.6587	0.6600	0.6613	0.6626	0.6639	0.6652	0.6665	0.6678
	cos	0.7547	0.7536	0.7524	0.7513	0.7501	0.7490	0.7478	0.7466	0.7455	0.7443
	tan	0.8693	0.8724	0.8754	0.8785	0.8816	0.8847	0.8878	0.8910	0.8941	0.8972
42	sin	0.6691	0.6704	0.6717	0.6730	0.6743	0.6756	0.6769	0.6782	0.6794	0.6807
	cos	0.7431	0.7420	0.7408	0.7396	0.7385	0.7373	0.7361	0.7349	0.7337	0.7325
	tan	0.9004	0.9036	0.9067	0.9099	0.9131	0.9163	0.9195	0.9228	0.9260	0.9293
43	sin	0.6820	0.6833	0.6845	0.6858	0.6871	0.6884	0.6896	0.6909	0.6921	0.6934
	cos	0.7314	0.7302	0.7290	0.7278	0.7266	0.7254	0.7242	0.7230	0.7218	0.7206
	tan	0.9325	0.9358	0.9391	0.9424	0.9457	0.9490	0.9523	0.9556	0.9590	0.9623
44	sin	0.6947	0.6959	0.6972	0.6984	0.6997	0.7009	0.7022	0.7034	0.7046	0.7059
	cos	0.7193	0.7181	0.7169	0.7157	0.7145	0.7133	0.7120	0.7108	0.7096	0.7083
	tan	0.9657	0.9691	0.9725	0.9759	0.9793	0.9827	0.9861	0.9896	0.9930	0.9965
45	sin	0.7071	0.7083	0.7096	0.7108	0.7120	0.7133	0.7145	0.7157	0.7169	0.7181
	cos	0.7071	0.7059	0.7046	0.7034	0.7022	0.7009	0.6997	0.6984	0.6972	0.6959
	tan	1.0000	1.0035	1.0070	1.0105	1.0141	1.0176	1.0212	1.0247	1.0283	1.0319
46	sin	0.7193	0.7206	0.7218	0.7230	0.7242	0.7254	0.7266	0.7278	0.7290	0.7302
	cos	0.6947	0.6934	0.6921	0.6909	0.6896	0.6884	0.6871	0.6858	0.6845	0.6833
	tan	1.0355	1.0392	1.0428	1.0464	1.0501	1.0538	1.0575	1.0612	1.0649	1.0686

Figure B-2 Cont.

deg	*function*	0.0°	0.1°	0.2°	0.3°	0.4°	0.5°	0.6°	0.7°	0.8°	0.9°
47	sin	0.7314	0.7325	0.7337	0.7349	0.7361	0.7373	0.7385	0.7396	0.7408	0.7420
	cos	0.6820	0.6807	0.6794	0.6782	0.6769	0.6756	0.6743	0.6730	0.6717	0.6704
	tan	1.0724	1.0761	1.0799	1.0837	1.0875	1.0913	1.0951	1.0990	1.1028	1.1067
48	sin	0.7431	0.7443	0.7455	0.7466	0.7478	0.7490	0.7501	0.7513	0.7524	0.7536
	cos	0.6691	0.6678	0.6665	0.6652	0.6639	0.6626	0.6613	0.6600	0.6587	0.6574
	tan	1.1106	1.1145	1.1184	1.1224	1.1263	1.1303	1.1343	1.1383	1.1423	1.1463
49	sin	0.7547	0.7559	0.7570	0.7581	0.7593	0.7604	0.7615	0.7627	0.7638	0.7649
	cos	0.6561	0.6547	0.6534	0.6521	0.6508	0.6494	0.6481	0.6468	0.6455	0.6441
	tan	1.1504	1.1544	1.1585	1.1626	1.1667	1.1708	1.1750	1.1792	1.1833	1.1875
50	sin	0.7660	0.7672	0.7683	0.7694	0.7705	0.7716	0.7727	0.7738	0.7749	0.7760
	cos	0.6428	0.6414	0.6401	0.6388	0.6374	0.6361	0.6347	0.6334	0.6320	0.6307
	tan	1.1918	1.1960	1.2002	1.2045	1.2088	1.2131	1.2174	1.2218	1.2261	1.2305
51	sin	0.7771	0.7782	0.7793	0.7804	0.7815	0.7826	0.7837	0.7848	0.7859	0.7869
	cos	0.6293	0.6280	0.6266	0.6252	0.6239	0.6225	0.6211	0.6198	0.6184	0.6170
	tan	1.2349	1.2393	1.2437	1.2482	1.2527	1.2572	1.2617	1.2662	1.2708	1.2753
52	sin	0.7880	0.7891	0.7902	0.7912	0.7923	0.7934	0.7944	0.7955	0.7965	0.7976
	cos	0.6157	0.6143	0.6129	0.6115	0.6101	0.6088	0.6074	0.6060	0.6046	0.6032
	tan	1.2799	1.2846	1.2892	1.2938	1.2985	1.3032	1.3079	1.3127	1.3175	1.3222
53	sin	0.7986	0.7997	0.8007	0.8018	0.8028	0.8039	0.8049	0.8059	0.8070	0.8080
	cos	0.6018	0.6004	0.5990	0.5976	0.5962	0.5948	0.5934	0.5920	0.5906	0.5892
	tan	1.3270	1.3319	1.3367	1.3416	1.3465	1.3514	1.3564	1.3613	1.3663	1.3713
54	sin	0.8090	0.8100	0.8111	0.8121	0.8131	0.8141	0.8151	0.8161	0.8171	0.8181
	cos	0.5878	0.5864	0.5850	0.5835	0.5821	0.5807	0.5793	0.5779	0.5764	0.5750
	tan	1.3764	1.3814	1.3865	1.3916	1.3968	1.4019	1.4071	1.4124	1.4176	1.4229
55	sin	0.8192	0.8202	0.8211	0.8221	0.8231	0.8241	0.8251	0.8261	0.8271	0.8281
	cos	0.5736	0.5721	0.5707	0.5693	0.5678	0.5664	0.5650	0.5635	0.5621	0.5606
	tan	1.4281	1.4335	1.4388	1.4442	1.4496	1.4550	1.4605	1.4659	1.4715	1.4770
56	sin	0.8290	0.8300	0.8310	0.8320	0.8329	0.8339	0.8348	0.8358	0.8368	0.8377
	cos	0.5592	0.5577	0.5563	0.5548	0.5534	0.5519	0.5505	0.5490	0.5476	0.5461
	tan	1.4826	1.4882	1.4938	1.4994	1.5051	1.5108	1.5166	1.5224	1.5282	1.5340
57	sin	0.8387	0.8396	0.8406	0.8415	0.8425	0.8434	0.8443	0.8453	0.8462	0.8471
	cos	0.5446	0.5432	0.5417	0.5402	0.5388	0.5373	0.5358	0.5344	0.5329	0.5314
	tan	1.5399	1.5458	1.5517	1.5577	1.5637	1.5697	1.5757	1.5818	1.5880	1.5941
58	sin	0.8480	0.8490	0.8499	0.8508	0.8517	0.8526	0.8536	0.8545	0.8554	0.8563
	cos	0.5299	0.5284	0.5270	0.5255	0.5240	0.5225	0.5210	0.5195	0.5180	0.5165
	tan	1.6003	1.6066	1.6128	1.6191	1.6255	1.6319	1.6383	1.6447	1.6512	1.6577
59	sin	0.8572	0.8581	0.8590	0.8599	0.8607	0.8616	0.8625	0.8634	0.8643	0.8652
	cos	0.5150	0.5135	0.5120	0.5105	0.5090	0.5075	0.5060	0.5045	0.5030	0.5015
	tan	1.6643	1.6709	1.6775	1.6842	1.6909	1.6977	1.7045	1.7113	1.7182	1.7251
60	sin	0.8660	0.8669	0.8678	0.8686	0.8695	0.8704	0.8712	0.8721	0.8729	0.8738
	cos	0.5000	0.4985	0.4970	0.4955	0.4939	0.4924	0.4909	0.4894	0.4879	0.4863
	tan	1.7321	1.7391	1.7461	1.7532	1.7603	1.7675	1.7747	1.7820	1.7893	1.7966
61	sin	0.8746	0.8755	0.8763	0.8771	0.8780	0.8788	0.8796	0.8805	0.8813	0.8821
	cos	0.4848	0.4833	0.4818	0.4802	0.4787	0.4772	0.4756	0.4741	0.4726	0.4710
	tan	1.8040	1.8115	1.8190	1.8265	1.8341	1.8418	1.8495	1.8572	1.8650	1.8728
62	sin	0.8829	0.8838	0.8846	0.8854	0.8862	0.8870	0.8878	0.8886	0.8894	0.8902
	cos	0.4695	0.4679	0.4664	0.4648	0.4633	0.4617	0.4602	0.4586	0.4571	0.4555
	tan	1.8807	1.8887	1.8967	1.9047	1.9128	1.9210	1.9292	1.9375	1.9458	1.9542
63	sin	0.8910	0.8918	0.8926	0.8934	0.8942	0.8949	0.8957	0.8965	0.8973	0.8980
	cos	0.4540	0.4524	0.4509	0.4493	0.4478	0.4462	0.4446	0.4431	0.4415	0.4399
	tan	1.9626	1.9711	1.9797	1.9883	1.9970	2.0057	2.0145	2.0233	2.0323	2.0413
64	sin	0.8988	0.8996	0.9003	0.9011	0.9018	0.9026	0.9033	0.9041	0.9048	0.9056
	cos	0.4384	0.4368	0.4352	0.4337	0.4321	0.4305	0.4289	0.4274	0.4258	0.4242
	tan	2.0503	2.0594	2.0686	2.0778	2.0872	2.0965	2.1060	2.1155	2.1251	2.1348
65	sin	0.9063	0.9070	0.9078	0.9085	0.9092	0.9100	0.9107	0.9114	0.9121	0.9128
	cos	0.4226	0.4210	0.4195	0.4179	0.4163	0.4147	0.4131	0.4115	0.4099	0.4083
	tan	2.1445	2.1543	2.1642	2.1742	2.1842	2.1943	2.2045	2.2148	2.2251	2.2355
66	sin	0.9135	0.9143	0.9150	0.9157	0.9164	0.9171	0.9178	0.9184	0.9191	0.9198
	cos	0.4067	0.4051	0.4035	0.4019	0.4003	0.3987	0.3971	0.3955	0.3939	0.3923
	tan	2.2460	2.2566	2.2673	2.2781	2.2889	2.2998	2.3109	2.3220	2.3332	2.3445
67	sin	0.9205	0.9212	0.9219	0.9225	0.9232	0.9239	0.9245	0.9252	0.9259	0.9265
	cos	0.3907	0.3891	0.3875	0.3859	0.3843	0.3827	0.3811	0.3795	0.3778	0.3762
	tan	2.3559	2.3673	2.3789	2.3906	2.4023	2.4142	2.4262	2.4383	2.4504	2.4627
68	sin	0.9272	0.9278	0.9285	0.9291	0.9298	0.9304	0.9311	0.9317	0.9323	0.9330
	cos	0.3746	0.3730	0.3714	0.3697	0.3681	0.3665	0.3649	0.3633	0.3616	0.3600
	tan	2.4751	2.4876	2.5002	2.5129	2.5257	2.5386	2.5517	2.5649	2.5782	2.5916
69	sin	0.9336	0.9342	0.9348	0.9354	0.9361	0.9367	0.9373	0.9379	0.9385	0.9391
	cos	0.3584	0.3567	0.3551	0.3535	0.3518	0.3502	0.3486	0.3469	0.3453	0.3437
	tan	2.6051	2.6187	2.6325	2.6464	2.6605	2.6746	2.6889	2.7034	2.7179	2.7326

Figure B-2 Cont.

deg	function	0.0°	0.1°	0.2°	0.3°	0.4°	0.5°	0.6°	0.7°	0.8°	0.9°
70	sin	0.9397	0.9403	0.9409	0.9415	0.9421	0.9426	0.9432	0.9438	0.9444	0.9449
	cos	0.3420	0.3404	0.3387	0.3371	0.3355	0.3338	0.3322	0.3305	0.3289	0.3272
	tan	2.7475	2.7625	2.7776	2.7929	2.8083	2.8239	2.8397	2.8556	2.8716	2.8878
71	sin	0.9455	0.9461	0.9466	0.9472	0.9478	0.9483	0.9489	0.9494	0.9500	0.9505
	cos	0.3256	0.3239	0.3223	0.3206	0.3190	0.3173	0.3156	0.3140	0.3123	0.3107
	tan	2.9042	2.9208	2.9375	2.9544	2.9714	2.9887	3.0061	3.0237	3.0415	3.0595
72	sin	0.9511	0.9516	0.9521	0.9527	0.9532	0.9537	0.9542	0.9548	0.9553	0.9558
	cos	0.3090	0.3074	0.3057	0.3040	0.3024	0.3007	0.2990	0.2974	0.2957	0.2940
	tan	3.0777	3.0961	3.1146	3.1334	3.1524	3.1716	3.1910	3.2106	3.2305	3.2506
73	sin	0.9563	0.9568	0.9573	0.9578	0.9583	0.9588	0.9593	0.9598	0.9603	0.9608
	cos	0.2924	0.2907	0.2890	0.2874	0.2857	0.2840	0.2823	0.2807	0.2790	0.2773
	tan	3.2709	3.2914	3.3122	3.3332	3.3544	3.3759	3.3977	3.4197	3.4420	3.4646
74	sin	0.9613	0.9617	0.9622	0.9627	0.9632	0.9636	0.9641	0.9646	0.9650	0.9655
	cos	0.2756	0.2740	0.2723	0.2706	0.2689	0.2672	0.2656	0.2639	0.2622	0.2605
	tan	3.4874	3.5105	3.5339	3.5576	3.5816	3.6059	3.6305	3.6554	3.6806	3.7062
75	sin	0.9659	0.9664	0.9668	0.9673	0.9677	0.9681	0.9686	0.9690	0.9694	0.9699
	cos	0.2588	0.2571	0.2554	0.2538	0.2521	0.2504	0.2487	0.2470	0.2453	0.2436
	tan	3.7321	3.7583	3.7848	3.8118	3.8391	3.8667	3.8947	3.9232	3.9520	3.9812
76	sin	0.9703	0.9707	0.9711	0.9715	0.9720	0.9724	0.9728	0.9732	0.9736	0.9740
	cos	0.2419	0.2402	0.2385	0.2368	0.2351	0.2334	0.2317	0.2300	0.2284	0.2267
	tan	4.0108	4.0408	4.0713	4.1022	4.1335	4.1653	4.1976	4.2303	4.2635	4.2972
77	sin	0.9744	0.9748	0.9751	0.9755	0.9759	0.9763	0.9767	0.9770	0.9774	0.9778
	cos	0.2250	0.2232	0.2215	0.2198	0.2181	0.2164	0.2147	0.2130	0.2113	0.2096
	tan	4.3315	4.3662	4.4015	4.4374	4.4737	4.5107	4.5483	4.5864	4.6252	4.6646
78	sin	0.9781	0.9785	0.9789	0.9792	0.9796	0.9799	0.9803	0.9806	0.9810	0.9813
	cos	0.2079	0.2062	0.2045	0.2028	0.2011	0.1994	0.1977	0.1959	0.1942	0.1925
	tan	4.7046	4.7453	4.7867	4.8288	4.8716	4.9152	4.9594	5.0045	5.0504	5.0970
79	sin	0.9816	0.9820	0.9823	0.9826	0.9829	0.9833	0.9836	0.9839	0.9842	0.9845
	cos	0.1908	0.1891	0.1874	0.1857	0.1840	0.1822	0.1805	0.1788	0.1771	0.1754
	tan	5.1446	5.1929	5.2422	5.2924	5.3435	5.3955	5.4486	5.5026	5.5578	5.6140
80	sin	0.9848	0.9851	0.9854	0.9857	0.9860	0.9863	0.9866	0.9869	0.9871	0.9874
	cos	0.1736	0.1719	0.1702	0.1685	0.1668	0.1650	0.1633	0.1616	0.1599	0.1582
	tan	5.6713	5.7297	5.7894	5.8502	5.9124	5.9758	6.0405	6.1066	6.1742	6.2432
81	sin	0.9877	0.9880	0.9882	0.9885	0.9888	0.9890	0.9893	0.9895	0.9898	0.9900
	cos	0.1564	0.1547	0.1530	0.1513	0.1495	0.1478	0.1461	0.1444	0.1426	0.1409
	tan	6.3138	6.3859	6.4596	6.5350	6.6122	6.6912	6.7720	6.8548	6.9395	7.0264
82	sin	0.9903	0.9905	0.9907	0.9910	0.9912	0.9914	0.9917	0.9919	0.9921	0.9923
	cos	0.1392	0.1374	0.1357	0.1340	0.1323	0.1305	0.1288	0.1271	0.1253	0.1236
	tan	7.1154	7.2066	7.3002	7.3962	7.4947	7.5958	7.6996	7.8062	7.9158	8.0285
83	sin	0.9925	0.9928	0.9930	0.9932	0.9934	0.9936	0.9938	0.9940	0.9942	0.9943
	cos	0.1219	0.1201	0.1184	0.1167	0.1149	0.1132	0.1115	0.1097	0.1080	0.1063
	tan	8.1443	8.2636	8.3863	8.5126	8.6427	8.7769	8.9152	9.0579	9.2052	9.3572
84	sin	0.9945	0.9947	0.9949	0.9951	0.9952	0.9954	0.9956	0.9957	0.9959	0.9960
	cos	0.1045	0.1028	0.1011	0.0993	0.0976	0.0958	0.0941	0.0924	0.0906	0.0889
	tan	9.5144	9.6768	9.8448	10.02	10.20	10.39	10.58	10.78	10.99	11.20
85	sin	0.9962	0.9963	0.9965	0.9966	0.9968	0.9969	0.9971	0.9972	0.9973	0.9974
	cos	0.0872	0.0854	0.0837	0.0819	0.0802	0.0785	0.0767	0.0750	0.0732	0.0715
	tan	11.43	11.66	11.91	12.16	12.43	12.71	13.00	13.30	13.62	13.95
86	sin	0.9976	0.9977	0.9978	0.9979	0.9980	0.9981	0.9982	0.9983	0.9984	0.9985
	cos	0.0698	0.0680	0.0663	0.0645	0.0628	0.0610	0.0593	0.0576	0.0558	0.0541
	tan	14.30	14.67	15.06	15.46	15.89	16.35	16.83	17.34	17.89	18.46
87	sin	0.9986	0.9987	0.9988	0.9989	0.9990	0.9990	0.9991	0.9992	0.9993	0.9993
	cos	0.0523	0.0506	0.0488	0.0471	0.0454	0.0436	0.0419	0.0401	0.0384	0.0366
	tan	19.08	19.74	20.45	21.20	22.02	22.90	23.86	24.90	26.03	27.27
88	sin	0.9994	0.9995	0.9995	0.9996	0.9996	0.9997	0.9997	0.9997	0.9998	0.9998
	cos	0.0349	0.0332	0.0314	0.0297	0.0279	0.0262	0.0244	0.0227	0.0209	0.0192
	tan	28.64	30.14	31.82	33.69	35.80	38.19	40.92	44.07	47.74	52.08
89	sin	0.9998	0.9999	0.9999	0.9999	0.9999	1.000	1.000	1.000	1.000	1.000
	cos	0.0175	0.0157	0.0140	0.0122	0.0105	0.0087	0.0070	0.0052	0.0035	0.0017
	tan	57.29	63.66	71.62	81.85	95.49	114.6	143.2	191.0	286.5	573.0

Figure B-2 Cont.

APPENDIX C

LOGARITHMIC TABLE

The logarithm of a number greater than one (1) is a positive number. It consists of two parts. One part is called the *characteristic* and the other is the *mantissa* of the number. The reader should review Topic 5-10 regarding significant figures, standard position and the characteristic.

The characteristic is the number of decimal places from the decimal point to standard position. Its value is indicated in the logarithm of the number by the number preceding the decimal point. The mantissa is the part of the logarithm that follows the decimal point:

54 200	Number
4.7340	Logarithm
4	Characteristic
0.7340	Mantissa

The mantissa is *always* a positive number. Its value depends only on the digits of the number. This is readily apparent when the numbers are expressed in standard scientific notation.

Example C-1 Find the common logarithm of the following numbers: (a) 3.75, (b) 375, (c) 375 000 and (d) 375 000 000.

Solution: Express the numbers in standard scientific notation then find the logarithm:

(a) $\log 3.75 \times 10^0 = 0.5740$
(b) $\log 3.75 \times 10^2 = 2.5740$
(c) $\log 3.75 \times 10^5 = 5.5740$
(d) $\log 3.75 \times 10^8 = 8.5740$

The power of ten is the characteristic when the number is expressed in standard scientific notation. The mantissa (0.5740) for each of the four numbers in Example C-1 is the same since the significant digits of the numbers (375) are the same. The logarithm of a number is the sum of the number's characteristic and its mantissa. In the previous example:

(a) 0.5740 = 0 + 0.5740
(b) 2.5740 = 2 + 0.5740
(c) 5.5740 = 5 + 0.5740
(d) 8.5740 = 8 + 0.5740

Thus, when the logarithm of a number is known, the characteristic indicates the position of the decimal point in the original number. The mantissa is an indication of the value of the significant digits of the original number.

Common logarithm tables contain only the mantissa of numbers between 1.00 and 10.0. The characteristic is found by inspection. To find the logarithm of a number, locate the mantissa in the table, determine the characteristic and add.

To locate the mantissa, find the first two significant figures in the left hand column of the table. The mantissa for the desired number is in the column headed by the third significant figure.

Example C-2 Find the mantissa, characteristic and logarithm for (a) 746 000, (b) 628 and (c) 2.74×10^5.

Solution: Use the table to follow along with the discussion.

(a) Find the 74th row in the table. Move right to column 6. The mantissa is 0.8727 with the decimal point understood. The characteristic is 5, by inspection. The logarithm, therefore, is 5.8727 by adding.

(b) Find the 62nd row, column 8. The mantissa is 0.7980 and the characteristic is 2, thus the logarithm is 2.7980.

(c) Proceed to the 27th row, column 4 to find 0.4378. The characteristic is 5 hence the logarithm is 5.4378.

For fractions, the characteristic is negative. This creates a problem. When a negative characteristic is added to the positive mantissa, both characteristic and mantissa are lost. The number 0.001 96 has a characteristic of -3 and a mantissa of 0.2923. When added, the logarithm is -2.7077. To avoid the problem, the -3 is written as 7 - 10. The mantissa is then added to the 7 to obtain 7.2923 - 10 which equals -2.7077.

In the minus-ten notation, the logarithm carries the mantissa and the characteristic. This is important because the tables must also be used to find antilogarithms. The procedure is simply reversed.

Example C-3 Find the number (antilogarithm) whose logarithm is (a) 4.8710, (b) 2.6304, (c) 5.9619 - 10, (d) -1.5952.

Solution: (a) The given logarithm has a mantissa of 0.8710. Locate it in the log table (74th row, column 3). The significant figures of the antilogarithm is 743. Since the characteristic is 4, the decimal is located four places to the right from standard position. The number is 74 300 or in standard scientific notation, it is 7.43×10^4.

(b) The mantissa (0.6304) is in the 42nd row in column 7. The antilogarithm is 427 or 4.27×10^2 since the characteristic is 2. The decimal point is two places from standard position.

(c) The mantissa is in the 91st row, column 6 and the characteristic is -5. The antilogarithm is 0.000 091 6 or 9.16×10^{-5}. The decimal point is placed five places to the left of standard position since the characteristic is negative.

(d) Rewrite the logarithm in minus-ten notation: $-1.5952 = 8.4048 - 10$. The mantissa is 0.4048 and the characteristic is -2. The mantissa is in the 25th row, column 4. The antilogarithm is $0.0254 = 2.54 \times 10^{-2}$.

For mantissas that are not found in the table, use the one with the nearest value to find the antilogarithm. For practice problems, see Exercises 23-2 and 23-3.

To find the natural logarithm of a number, find the common log first, then multiply by the number 2.303:

$$x = \ln N = 2.303 \log N \text{ (where } N = e^x)$$

To find the natural antilogarithm of a number, divide it by 2.303. This gives the common log of the number, then find the common antilogarithm:

$$N = \text{antiln } x = \text{antilog } \frac{x}{2.303}$$

If greater accuracy is required, use the number 2.302 59.

Example C-4 (a) Find the natural logarithm of 100.
(b) Find the natural antilogarithm of 3.

Solution: (a) $\ln 100 = 2.303 \log 100$
$= 2.303(2) = 4{,}606$

(b) $\text{antiln } 3 = \frac{3}{2.303}$

$= \text{antilog } 1.3026 = 20.07$

N	0	1	2	3	4	5	6	7	8	9
10	0000	0043	0086	0128	0170	0212	0253	0294	0334	0374
11	0414	0453	0492	0531	0569	0607	0645	0682	0719	0755
12	0792	0828	0864	0899	0934	0969	1004	1038	1072	1106
13	1139	1173	1206	1239	1271	1303	1335	1367	1399	1430
14	1461	1492	1523	1553	1584	1614	1644	1673	1703	1732
15	1761	1790	1818	1847	1875	1903	1931	1959	1987	2014
16	2041	2068	2095	2122	2148	2175	2201	2227	2253	2279
17	2304	2330	2355	2380	2405	2430	2455	2480	2504	2529
18	2553	2577	2601	2625	2648	2672	2695	2718	2742	2765
19	2788	2810	2833	2856	2878	2900	2923	2945	2967	2989
20	3010	3032	3054	3075	3096	3118	3139	3160	3181	3201
21	3222	3243	3263	3284	3304	3324	3345	3365	3385	3404
22	3424	3444	3464	3483	3502	3522	3541	3560	3579	3598
23	3617	3636	3655	3674	3692	3711	3729	3747	3766	3784
24	3802	3820	3838	3856	3874	3892	3909	3927	3945	3962
25	3979	3997	4014	4031	4048	4065	4082	4099	4116	4133
26	4150	4166	4183	4200	4216	4232	4249	4265	4281	4298
27	4314	4330	4346	4362	4378	4393	4409	4425	4440	4456
28	4472	4487	4502	4518	4533	4548	4564	4579	4594	4609
29	4624	4639	4654	4669	4683	4698	4713	4728	4742	4757
30	4771	4786	4800	4814	4829	4843	4857	4871	4886	4900
31	4914	4928	4942	4955	4969	4983	4997	5011	5024	5038
32	5051	5065	5079	5092	5105	5119	5132	5145	5159	5172
33	5185	5198	5211	5224	5237	5250	5263	5276	5289	5302
34	5315	5328	5340	5353	5366	5378	5391	5403	5416	5428
35	5441	5453	5465	5478	5490	5502	5514	5527	5539	5551
36	5563	5575	5587	5599	5611	5623	5635	5647	5658	5670
37	5682	5694	5705	5717	5729	5740	5752	5763	5775	5786
38	5798	5809	5821	5832	5843	5855	5866	5877	5888	5899
39	5911	5922	5933	5944	5955	5966	5977	5988	5999	6010
40	6021	6031	6042	6053	6064	6075	6085	6096	6107	6117
41	6128	6138	6149	6160	6170	6180	6191	6201	6212	6222
42	6232	6243	6253	6263	6274	6284	6294	6304	6314	6325
43	6335	6345	6355	6365	6375	6385	6395	6405	6415	6425
44	6435	6444	6454	6464	6474	6484	6493	6503	6513	6522
45	6532	6542	6551	6561	6571	6580	6590	6599	6609	6618
46	6628	6637	6646	6656	6665	6675	6684	6693	6702	6712
47	6721	6730	6739	6749	6758	6767	6776	6785	6794	6803
48	6812	6821	6830	6839	6848	6857	6866	6875	6884	6893
49	6902	6911	6920	6928	6937	6946	6955	6964	6972	6981
50	6990	6998	7007	7016	7024	7033	7042	7050	7059	7067
51	7076	7084	7093	7101	7110	7118	7126	7135	7143	7152
52	7160	7168	7177	7185	7193	7202	7210	7218	7226	7235
53	7243	7251	7259	7267	7275	7284	7292	7300	7308	7316
54	7324	7332	7340	7348	7356	7364	7372	7380	7388	7396

N	0	1	2	3	4	5	6	7	8	9
55	7404	7412	7419	7427	7435	7443	7451	7459	7466	7474
56	7482	7490	7497	7505	7513	7520	7528	7536	7543	7551
57	7559	7566	7574	7582	7589	7597	7604	7612	7619	7627
58	7634	7642	7649	7657	7664	7672	7679	7686	7694	7701
59	7709	7716	7723	7731	7738	7745	7752	7760	7767	7774
60	7782	7789	7796	7803	7810	7818	7825	7832	7839	7846
61	7853	7860	7868	7875	7882	7889	7896	7903	7910	7917
62	7924	7931	7938	7945	7952	7959	7966	7973	7980	7987
63	7993	8000	8007	8014	8021	8028	8035	8041	8048	8055
64	8062	8069	8075	8082	8089	8096	8102	8109	8116	8122
65	8129	8136	8142	8149	8156	8162	8169	8176	8182	8189
66	8195	8202	8209	8215	8222	8228	8235	8241	8248	8254
67	8261	8267	8274	8280	8287	8293	8299	8306	8312	8319
68	8325	8331	8338	8344	8351	8357	8363	8370	8376	8382
69	8388	8395	8401	8407	8414	8420	8426	8432	8439	8445
70	8451	8457	8463	8470	8476	8482	8488	8494	8500	8506
71	8513	8519	8525	8531	8537	8543	8549	8555	8561	8567
72	8573	8579	8585	8591	8597	8603	8609	8615	8621	8627
73	8633	8639	8645	8651	8657	8663	8669	8675	8681	8686
74	8692	8698	8704	8710	8716	8722	8727	8733	8739	8745
75	8751	8756	8762	8768	8774	8779	8785	8791	8797	8802
76	8808	8814	8820	8825	8831	8837	8842	8848	8854	8859
77	8865	8871	8876	8882	8887	8893	8899	8904	8910	8915
78	8921	8927	8932	8938	8943	8949	8954	8960	8965	8971
79	8976	8982	8987	8993	8998	9004	9009	9015	9020	9025
80	9031	9036	9042	9047	9053	9058	9063	9069	9074	9079
81	9085	9090	9096	9101	9106	9112	9117	9122	9128	9133
82	9138	9143	9149	9154	9159	9165	9170	9175	9180	9186
83	9191	9196	9201	9206	9212	9217	9222	9227	9232	9238
84	9243	9248	9253	9258	9263	9269	9274	9279	9284	9289
85	9294	9299	9304	9309	9315	9320	9325	9330	9335	9340
86	9345	9350	9355	9360	9365	9370	9375	9380	9385	9390
87	9395	9400	9405	9410	9415	9420	9425	9430	9435	9440
88	9445	9450	9455	9460	9465	9469	9474	9479	9484	9489
89	9494	9499	9504	9509	9513	9518	9523	9528	9533	9538
90	9542	9547	9552	9557	9562	9566	9571	9576	9581	9586
91	9590	9595	9600	9605	9609	9614	9619	9624	9628	9633
92	9638	9643	9647	9652	9657	9661	9666	9671	9675	9680
93	9685	9689	9694	9699	9703	9708	9713	9717	9722	9727
94	9731	9736	9741	9745	9750	9754	9759	9763	9768	9773
95	9777	9782	9786	9791	9795	9800	9805	9809	9814	9818
96	9823	9827	9832	9836	9841	9845	9850	9854	9859	9863
97	9868	9872	9877	9881	9886	9890	9894	9899	9903	9908
98	9912	9917	9921	9926	9930	9934	9939	9943	9948	9952
99	9956	9961	9965	9969	9974	9978	9983	9987	9991	9996

*This table gives the mantissas of numbers with the decimal point omitted in each case. Characteristics are determined by inspection from the numbers.

GLOSSARY

Abscissa. A value of the quantity plotted along the horizontal axis in a rectangular coordinate system.

Accuracy. A word that refers to the reliability of a measuring instrument. More accurate instruments yield more significant figures in the measurements.

Addend. A number to be added.

Addition. The operation of combining numbers to find the total number.

Admittance (Y). The reciprocal of the impedance of an ac circuit. Admittance is the ability of a circuit to conduct electric current and it is expressed in siemens.

Alternator. An alternating-current generator.

Alternator, Polyphase. An alternator containing more than one independent armature winding.

Alternator, Three-phase. An alternator containing three independent armature windings set 120° apart.

Alternator, Two-phase. An alternator containing two independent armature windings set 90° apart.

Ammeter. A measuring device that measures the electric current in a circuit. An ammeter is connected in series with the load.

Ampere (A). A unit of electrical current equal to 6 240 000 000 000 000 000 electrons per second.

Amplifier. An electronic device that increases the power of a signal.

Amplitude. The maximum value of a sine or other periodic wave. It may be a maximum current, maximum voltage, maximum power, etc.

Angle. A figure formed by two lines extended from the same point.

Angle, Acute. An angle that is less than a right angle.

Angle, Generated. The angle made by a rotating radius vector.

Angle, Negative. An angle generated by a radius vector rotating clockwise.

Angle, Obtuse. An angle that is between a right angle and a straight angle.

Angle, Phase. A constant angle between two periodic functions which have the same frequency. The phase angle shifts the entire curve left or right of the reference curve.

Angle, Positive. An angle generated by a radius vector rotating counterclockwise.

Angle, Reflex. An angle that is greater than a straight angle.

Angle, Right. An angle equal to 90°.

Angle, Straight. An angle formed by a straight line and equal to 180°.

Angles, Complementary. Two angles whose sum is 90°.

Angles, Equivalent. Angles having the same measure.

Angles, Supplementary. Two angles whose sum is 180°.

Antilogarithm. The number whose logarithm is the given number.

Approximately Equal to. A relationship between numbers which are *nearly* equal to each other (≈).

Arc, Minute of. A unit of angular measurement equal to one sixtieth of a degree. 1° = 60′

Arc, Second of. A unit of angular measurement equal to one sixtieth of a minute. 1′ = 60″ or 1° = 3600″.

Arc Cosine. An angle of a right triangle in which the corresponding cosine is a known value. Also called inverse cosine.

Arc Sine. An angle of a right triangle in which the corresponding sine is a known value. Also called inverse sine.

Arc Tangent. An angle of a right triangle in which the corresponding tangent is a known value. Also called inverse tangent.

Area (A). A quantity that describes the amount of surface of one face of an object. The SI unit of area is the square metre.

Area, Cross-sectional. The surface on the face of an imaginary cut that is perpendicular to the object's length.

Area, Surface. A measure of the amount of surface on a solid figure. The measure can be internal or external.

Atom. The smallest particle that can take part in a chemical change. Atoms are the building blocks of molecules.

Atom, Balanced. An atom with equal numbers of electrons and protons. A balanced atom is electrically neutral.

Atom, Planetary Model. A model used to explain the behavior of atoms. The positive nucleus is at the center with negative electrons revolving around it.

Atomic Number. The atomic number of an element is the number of protons in the nucleus of atoms of that element.

Attenuator. An electrical or electronic device that reduces the power of a signal.

Axiom. A statement that is assumed to be true without proof.

Axiom, Addition. When equal numbers are added to equal numbers, their sums are equal.

Axiom, Division. When equal numbers are divided by equal numbers, their quotients are equal.

Axiom, Multiplication. When equal numbers are multiplied by equal numbers, their products are equal.

Axiom, Power. When equal numbers are raised to the same power, the powers are equal.

Axiom, Root. When the same root is taken of equal numbers, the roots are equal.

Axiom, Subtraction. When equal numbers are subtracted from equal numbers, their differences are equal.

Axis. One of the reference lines on a coordinate system.

Axis, Horizontal. The *x*-axis or the east-west axis in a rectangular coordinate system.

Axis, Vertical. The *y*-axis or the north-south axis in a rectangular coordinate system.

Base. When a number is raised to a power, that number is the base.

Binomial. A two-term mathematical expression.

Borrowing. Taking one from a digit in the minuend (subtraction) and adding it as ten to the next lower place digit.

Bridge, Wheatstone. A bridge circuit used for precise resistance measurements. Current through the bridge is adjusted to zero.

Cancellation. The process of dividing the two terms of a fraction by the same number.

Candela (cd). An SI unit of luminous intensity.

Capacitance (C). The ratio of the charge stored on the plates of a capacitor to the voltage applied. Capacitance is measured in units of farads.

Capacitive Reactance (X_C). The opposition to the flow of alternating current due to the back emf produced across a capacitor. Capacitive reactance is expressed in ohms.

Capacitor. An electric circuit device consisting of two metal plates separated by a thin layer of dielectric.

Caption, Scale. The designation of the axis scale. It includes the quantity name followed by the symbol for the units.

Carrying. When a column of numbers is added (or multiplied) resulting in a sum greater than 9, there is a transfer of the higher place digits to the next column.

Change. A difference equal to the final value minus the beginning value. A negative change is a decrease; a positive change is an increase.

Characteristic. The number of places in a decimal from standard position to the decimal point.

Charge, Electric (Q). An electrical quantity that obeys the laws of interaction. Electric charge is measured in coulombs. It is defined by the formula $Q = It$.

Circle. A plane figure in which the distance from the center to the edge is the same in all directions.

Circuit, Capacitive. An ac circuit in which the capacitive reactance of the series equivalent circuit is greater than the inductive reactance.

Circuit, Closed. A completed unbroken path around which electric current can flow.

Circuit, Equivalent. A circuit containing one impedance that is equal to the total impedance of the given circuit.

Circuit, Ideal. A circuit containing inductance and/or capacitance with a total disregard for the resistance of the lines, coils, dielectric, etc.

Circuit, Inductive. An ac circuit in which the inductive reactance of the series equivalent circuit is greater than the capacitive reactance.

Circuit, Open. An incomplete circuit due to a loose connection, a broken or disconnected wire or an open switch.

Circuit, Parallel. A circuit, characterized by the fact that current divides and follows separate paths through the loads.

Circuit, Parallel Equivalent. A circuit consisting of a resistance and a reactance in parallel. The total impedance is the same as that of the original circuit.

Circuit, Parallel-series. A circuit in which two loads in series are in parallel with a single load.

Circuit, Practical. An ac circuit in which the resistance of each branch is accounted for.

Circuit, Pure Capacitive. An ac circuit containing a capacitor as the only circuit element.

Circuit, Pure Inductive. An ac circuit containing an inductor as the only circuit element.

Circuit, Pure Resistive. An ac circuit containing a resistance as the only circuit element.

Circuit, RC. An electrical circuit containing a resistance and a capacitance in series.

Circuit, RL. An ac circuit containing a resistance and an inductance.

Circuit, Series. A circuit in which there is only one current path. A series circuit contains two or more loads.

Circuit, Series Equivalent. A simple two-element series circuit that contains the same impedance and phase angle as the complex circuit it replaces.

Circuit, Series-parallel. A circuit with two loads in parallel. The parallel branch is in series with a single load.

Circuit Breaker. A protective circuit element containing a thermal device that opens the circuit when there is excessive current. May be either a manually or automatically reset type.

Circumference (C). The distance around a circle or circular object (see perimeter).

Code, Color. A method of indicating the amount of resistance of a resistor or capacitance of a capacitor.

Coefficient. The numerical factor of a mathematical expression.

Comparison. The equating of two expressions having equal value. A method of eliminating a variable for solving simultaneous equations.

Compounds. A pure substance having molecules made up of two or more kinds of atoms.

Conductance (G). The reciprocal of the resistance in the resistive branch of a parallel equivalent circuit. A quantity that supports the flow of electrical current. Conductance is measured in siemens.

Conduction Band. A band of orbits (normally unoccupied) beyond the valence shell.

Conductor, Electrical. A material such as copper or aluminum that has large numbers of free electrons.

Conjugate, Complex. A pair of complex numbers: $a + jb$ and $a - jb$. The sum and product of a complex number and its conjugate are real numbers.

Constant. A number or literal number in an equation with a value that does not change.

Constant, Time. The time required for a quantity to increase or decrease by 63%. Usually refers to growth or decay of current or voltage in electrical circuits.

Constant, Variation. A constant in a formula that relates two variables associated by direct variation.

Coordinate System, Rectangular. A system by which a plane is divided into quadrants for the purpose of locating points in the plane.

Coordinates. Two numbers used to locate a point in a plane.

Coordinates, Polar. Two ordered pairs of numbers that represent a point in a plane or a complex number (vector). The first number is the distance from the pole; the second indicates the direction (angle).

Coordinates, Transformation of. The process of changing the coordinates of a point or a vector from one form to the other. Rectangular form can be transformed to polar form, for example.

Cosine. A trigonometric function equal to the ratio of the adjacent side to the hypotenuse.

Coulomb (C). A unit of electrical charge equivalent to one ampere-second.

Cross-multiplication. A procedure in which the product of the numerator on the left and the denominator on the right is set equal to the product of the numerator on the right and the denominator on the left.

Cube. (1) A rectangular solid having equal length, width and height. (2) The product of a number and itself taken three times as a factor. The third power of a number.

Current, Alternating (ac). Current that flows first in one direction in a circuit, then the opposite. Reversal can occur many times each second.

Current, Direct (DC). Current that flows continuously in one direction around a circuit.

Current, Effective. The equivalent DC current which determines the heating effect in an ac circuit, sometimes called the rms current.

Current, Electric (I). The rate of flow of electrons in a circuit. Current is measured in units of amperes.

Current, Instantaneous. The value of an alternating current at a particular instant of time.

Current, Line. The current that flows in the main lines of a polyphase circuit.

Current, Maximum. The maximum value of an alternating current.

Current, Phase. The current that flows through a load in polyphase circuits.

Current, Rms. See Current, Effective.

Curve. A smooth line drawn along the general plot of data points.

Curve, Linear. A straight line curve.

Curve, Nonlinear. A curve that is clearly not a straight line.

Cycle. A complete set of events that recur regularly and in the same sequence.

Cylinder. A solid figure with a circular cross section. Its ends are perpendicular to the axis through the centers of the two ends.

D'Arsonval Movement. The stationary permanent-magnet moving-coil meter movement. The coil was originally suspended by a metal ribbon. The use of jeweled bearings was developed by Weston.

Data. A list of various values of the independent variable with the corresponding values of the dependent variable.

Decibel (db). A unit of power level change.

Decimal. A fraction with a denominator of 10, 100, 1000, etc.

Decimal, Repeating. Some common fractions when converted to a decimal do not come out even. They repeat a few digits over and over.

Decimal Point. A period separating the whole number from the decimal fraction in a mixed decimal fraction.

Degree (°). A unit of angular measurement. 1 revolution $= 360^\circ$.

Delta. A Greek letter (Δ). When combined with the symbol for a quantity, the combination represents the change in or the difference in a quantity.

Delta-connection. A three-phase load or source configuration in which the elements are connected along the sides of a triangle. It forms a shape like the capital Greek letter delta (Δ).

Denominator. The lower (second) term of a common fraction. The denominator is the divisor. It indicates the number of equal parts into which an object or quantity has been divided.

Denominator, Common. The denominator of several fractions each having the same denominator.

Denominator, Least Common. The smallest number that can be used as a common denominator for several fractions. It is the smallest number by which all denominators of the fractions can be evenly divided.

Diameter (d). The length of a line across a circle or sphere from edge to edge and passing through the center.

Dielectric. A nonconducting material used to separate the metal plates of a capacitor.

Difference. The number left after subtracting.

Digits, Significant. The digits in a number that are reliable.

Dividend. A number that is divided by another.

Division. A method of finding how many times one number contains another.

Divisor. A number by which the dividend is divided.

Dynamometer. A device used to measure the output power of rotating equipment: motors, engines, etc.

Efficiency. The work (or power) output of a device divided by the work (or power) input. Efficiency is usually expressed in percent.

Efficiency, Luminous. The efficiency of lamps. The ratio of luminous output power to electrical input power. It is expressed in lumens per watt instead of percent.

Electricity. The study of electrical machines, devices and electrical circuits that make up an electrical distribution system.

Electromotive Force (emf). An emf maintains a potential difference between two points. Batteries, generators and other devices are the source of an emf. An emf is expressed in volts.

Electron. A subatomic particle that has a negative charge and a very small mass.

Electron, Free. An electron that has gained energy and occupies an orbit in the conduction band. Free electrons can move easily from one atom to another.

Electron, Valence. An electron occupying the outermost shell of a normal atom.

Electron Shell. An orbit containing electrons revolving about the nucleus of an atom. An electron shell has a definite radius.

Element. A pure substance having molecules made up of atoms of one kind. There are 92 naturally occurring elements.

Element, Active Circuit. A device (voltage source) in which electrons gain energy and flow from plus (+) to minus (–).

Element, Passive Circuit. A device in which electrons give up energy and flow from minus (–) to plus (+).

Energy (*W*). Energy is the ability to do work. It exists in several forms: mechanical, chemical, nuclear, thermal and electrical. Energy is measured in joules.

Energy Level. See electron shell.

Equation. Two mathematical expressions set apart by an equal sign.

Equation, Exponential. An equation that contains the unknown as an exponent.

Equation, Fractional. An equation containing fractions with the unknown in the denominator.

Equation, Linear. An equation that results in a straight line curve when graphed.

Equation, Logarithmic. An equation that contains the logarithm of the unknown.

Equation, Pure Quadratic. A quadratic equation that does not contain a first power term.

Equation, Quadratic. An equation which contains the square of the unknown as the highest power.

Equations, Equivalent. Two or more equations having the same solution. One can be derived from the other by using the multiplication or division axioms.

Equations, Inconsistent. Two or more equations that have no solution. The graphs of inconsistent equations are parallel lines.

Equations, Simultaneous. A system of two or more equations that apply to a given situation at the same time.

Equations, Simultaneous, Second Order. A system of two equations with two unknowns that apply to a situation at the same time.

Equations, Simultaneous, Third Order. A system of three equations with three unknowns that apply to a situation at the same time.

Equations, Unique. A set of simultaneous equations with a unique solution. A graph of the equations involves a single point of intersection.

Exponent. When a number is raised to a power, the number of factors is the exponent.

Exponential Decay. The decrease of a quantity as an exponential function of time.

Exponential Growth. The increase of a quantity as an exponential function of time.

Expression, Algebraic. A collection of numbers and literal numbers combined by one or more basic operations and may include grouping symbols.

Expression, Mathematical. An expression combining two or more numbers and one or more operations.

Expression, Value of. The answer obtained when the expression is evaluated.

Expressions, Equivalent. Two mathematical expressions are said to be equivalent if they have the same value.

Factor. Any number or literal number that evenly divides a mathematical expression.

Factor, Largest Common. The largest number that evenly divides each of two or more numbers.

Factoring. The process of removing common factors from several terms using the distributive law.

Farad (F). A unit of capacitance.

Figure, Plane. An object drawn in one plane. A two-dimensional object.

Figure, Solid Geometrical. A three-dimensional figure.

Figures, Significant. See significant digits.

Flux, Luminous (F). Luminous output power of lamps. Luminous flux is measured in lumens.

Force, Nuclear. A force of attraction between protons and neutrons. A nuclear force is much stronger than electric forces when the particles are close together in the nucleus of atoms.

Formula. A rule or law that describes the relationship between quantities. It is usually expressed in symbols.

Formula, Quadratic. A special formula for finding the roots of a quadratic equation.

Formula, Simple. Formulas with no grouping symbols and no more than two terms.

Formula, Solving a. Transforming a formula into an equivalent formula with the unknown on one side of the equal sign by itself.

Fraction, Common. A two-term number. The upper (first) term is the numerator, the lower (second) term is the divisor.

Fraction, Complex. A fraction whose terms are fractions.

Fraction, Decimal. See decimal.

Fraction, Improper. A common fraction in which the numerator is greater than the denominator.

Fraction, Mixed Decimal. A number composed of a whole number and a decimal fraction.

Fraction, Proper. A common fraction in which the numerator is less than the denominator.

Fractions, Equivalent. Two fractions that reduce to the same fraction.

Fractions, Like. A group of fractions that have the same denominators.

Fractions, Similar. See like fractions.

Frequency (f). The number of complete cycles occurring each second of time. Frequency is measured in hertz.

Frequency, Angular. The number of radians per second in a cyclic sequence. The product of 2π and the frequency.

Frequency, Resonant. The frequency for which the inductive reactance and the capacitive reactance are equal.

Function. If for a value of one quantity, a second quantity assumes only one particular value, the second quantity is said to be a function of the first.

Function, Periodic. A function whereby one quantity is related to another in such a way that a sequence of values of the first quantity repeats over and over.

Function, Trigonometric. Ratios of two sides of a right triangle. Includes sine, cosine and tangent.

Fuse. A protective circuit element. With excessive current, the fuse melts and opens the circuit.

Gain. A description of the change in the power level when it undergoes an increase.

Galvanometer. See microammeter.

Generator. An electrical machine used for producing an emf by electromagnetic induction and generally means a DC device.

Grad. A unit of angular measurement used in some European countries. One right angle equals one hundred grads.

Grade. The ratio of rise (vertical) to the horizontal run. It is usually expressed in percent. A method of measuring nonlevel ground.

Graph. A diagram representing the variation of a variable compared to that of another.

Horsepower (hp). A unit of power. One horsepower equals 746 watts.

Hypotenuse. The longest of the three sides of a right triangle.

Impedance (Z). The collective name used for any combination of resistance, inductive reactance and/or capacitive reactance. Impedance is measured in ohms.

Impedance, Total. The equivalent impedance of a circuit. For a series circuit, the total impedance is found by adding the circuit impedances.

Index of a Radical. A small number to the left of a radical used to indicate higher roots.

Inductance (L). The ability of a coil to produce a back emf in opposition to an ac current. Inductance is measured in henrys.

Induction, Mutual. The effect in which a pair of coils will induce an ac voltage from one to the other.

Induction, Self-. The effect in which a coil will induce a back emf (ac) in itself.

Inductive Reactance (X_L). The opposition to the flow of alternating current due to the back emf produced by self-induction. Inductive reactance is expressed in ohms.

Insulator. A material, such as glass and ceramics, that has few free electrons.

Integer. A set of numbers including the positive whole numbers, negative whole numbers and zero.

Ion. An atom that has gained or lost electrons. An ion is a charged atom.

IR Drop. The voltage drop across a circuit element of resistance (R) due to current flow (I).

j-Numbers. Numbers preceded by the letter j where $j = \sqrt{-1}$. Sometimes called imaginary numbers.

j-Operator. The letter $j = \sqrt{-1}$ used as a sign of operation to indicate a vector rotation of 90°.

Joule (J). A unit of work and energy. The amount of work done when a one-newton load is lifted one metre.

Kelvin (K). An SI unit of temperature. It is the same size unit as the degree Celsius.

Kilogram (kg). An SI unit of mass.

Kilowatt-hour (kW•h). A unit of electrical energy. When energy is used at a rate of one kilowatt for a time of one hour, one kW•h of energy is consumed.

Kilowatt-hour Meter. An electrical measuring device that measures energy consumption.

Lag. When two periodic functions of the same frequency differ in phase, the one *not* shifted to the right lags behind the other (in time) by the given phase angle.

Law, Distributive. A law that allows distributing the multiplication of factors over addition: $a(b + c) = ab + ac$.

Law, Kirchhoff's Current. The sum of all currents flowing into any point in a circuit is equal to the sum of all currents flowing out of that point.

Law, Kirchhoff's Voltage. In any complete conducting path, the sum of the voltage gains is equal to the sum of the voltage drops.

Law, Ohm's. The relationship of the voltage, current and resistance in an electrical circuit.

LCD. Liquid crystal diode.

Lead. When two periodic functions of the same frequency differ in phase, the one shifted to the right leads the other (in time) by the given phase angle.

LED. Light emitting diode.

Line, Neutral. A conductor used in certain polyphase connections which provides a common current path from the load back to the source for all phases of the circuit.

Line, Number. A line with a graduated scale for plotting the real numbers.

Load, Balanced. The load of a polyphase circuit that contains equal impedances in each branch.

Logarithm. The logarithm of a number is the exponent to which a base number is raised to obtain the number.

Logarithm, Common. A logarithmic system using the base 10.

Logarithm, Natural. A logarithmic system using the base e ($e \approx 2.71828$).

Loss. A description of the change in the power level when it undergoes a decrease.

Lumen (lm). A unit of luminous flux (power). One watt = 668.4 lumens.

Magnitude. The value of a quantity expressed as a number and a unit. The size of a vector quantity regardless of its direction.

Mantissa. (1) The numerical factor of a number expressed in scientific notation. (2) The fractional part of a positive logarithm. The fractional part of a negative logarithm when expressed in minus-ten notation.

Matter. Anything that has weight and occupies space.

Mechanical Advantage (MA). The ratio of the output force of a machine to the input force.

Metre (m). The SI unit of distance.

Microammeter. A current-measuring device consisting of a D'Arsonval movement with attached pointer and scale. Measures only very small currents.

Mil. A distance measurement equal to one thousandth of an inch (0.001 in).

Mil, Circular (CM). A unit of area of circles found by squaring the diameter which must be expressed in mils.

Minuend. A number from which the subtrahend is subtracted.

Molecule. The smallest particle of a pure substance having all the physical properties of that substance. The building blocks of matter.

Monomial. A one-term mathematical expression.

Motor, Induction. An electric motor in which the rotor is not connected to any supply but has current induced in it by induction from the current in the stator field coils.

Motor, Synchronous. A constant speed motor in which the rotor is magnetized by a DC source rather than by induced current. The synchronized speed depends on the frequency of the ac current supplied to the stator field coils.

Multiplicand. The number that is to be multiplied by the multiplier.

Multiplication. The operation of "adding" a number to itself a specified number of times.

Multiplier. The number by which the multiplicand is multiplied.

Neutron. A subatomic particle having a neutral charge. It has slightly more mass than the proton.

Newton (N). The SI unit of force or weight.

Nominal value. The value of a specification from which maximum and minimum limits are determined.

Notation, Double-subscript. A subscript notation using two letters. Used for indicating the direction of measurement of various phasor quantities.

Notation, Engineering. (ENG) Scientific notation in which the mantissa has a value between 1 and 1000 and the exponent is a multiple of three.

Notation, Scientific. A number consisting of two factors. One factor is numerical (mantissa), the other a power of ten.

Notation, Standard Scientific. (SSN) Scientific notation in which the mantissa has its decimal point in standard position. The exponent equals the characteristic of the number.

Nucleus. The center of an atom having a positive charge and most of the mass of the atom. The nucleus is made up of protons and neutrons.

Number, Complex. A number consisting of two components; a real number and a j-number.

Number, Composite. Any whole number that can be factored.

Number, Directed. See vector.

Number, Irrational. Numbers such as the square root of prime numbers. Irrationals cannot be written in the form of $\frac{a}{b}$ where a and b are integers.

Number, Literal. A letter symbol that represents the value of a number or a quantity.

Number, Mixed. A number composed of a whole number and a common fraction.

Number, Natural. The numbers used for counting. The positive whole numbers.

Number, Negative. A number less than zero. It is indicated by the minus sign.

Number, Positive. A number greater than zero. It is indicated by the plus sign.

Number, Power of. The product of a number multiplied by itself. The number can be used several times as a factor.

Number, Rational. Any number that can be written in the form of $\frac{a}{b}$ where $b \neq 0$. The numbers a and b are integers.

Number, Real. The set of numbers that consist of all rational and irrational numbers.

Number, Signed. See vector.

Numerals. These are the symbols: 0, 1, 2, 3, 4, 5, 6, 7, 8, 9.

Numerator. The upper (first) term of a common fraction. The numerator is the number of parts.

Ohm (Ω). The unit of electrical resistance. When a current of one ampere flows under the influence of a one-volt emf, the resistance is one ohm.

Ohmmeter. A measuring instrument that measures the resistance of various circuit elements. Never connect an ohmmeter into a live circuit.

Operation, Inverse. An arithmetic operation that "undoes" the original operation.

Operation, Rotation. In the arithmetic of signed numbers (vectors), a minus sign can be interpreted as an operation of rotation of one-half revolution. A j-sign represents a rotation of a quarter revolution.

Orbit. See electron shell.

Ordinate. A value of the quantity plotted along the vertical axis in a rectangular coordinate system.

Origin. The intersection of the x and y axes in a rectangular coordinate system.

Parabola. The curve of an equation such as $y = ax^2 + bx + c$. The cross section of a lamp reflector is parabolic. The path followed by an object thrown upward at an angle is parabolic.

Parentheses, Removing. The removal of parentheses by using (1) the distributive law, or (2) by using the division axiom.

Peak. A term used to indicate the maximum voltage, current, etc.

Percent. A fraction with a denominator of one hundred.

Perimeter (P). The distance around an object (see circumference).

Period (T). The amount of time required for one complete cyclic sequence to occur.

Perpendicular. Two lines or planes are said to be perpendicular if they intersect at right angles (90°).

Phasor. Quantities represented as a vector to indicate phase relationships. The ac circuit quantities, voltage, current and impedance are phasors, for example.

Phasor Reciprocal. A phasor that has a magnitude equal to the reciprocal of the magnitude of the given phasor. Its direction is the negative of the angle of the given phasor.

π-Radian. A unit of angular measurement equal to 3.14 radians. $1\ \pi \cdot \text{rad} = \frac{1}{2}$ revolution = 180°.

Place Value. The value of a numeral depends on its position in a number. (Units, tens, hundreds, etc.).

Polar Form. The polar form of a phasor or vector uses the polar coordinates to indicate the magnitude and direction ($Z\underline{/\theta}$).

Pole. The center or origin of a polar coordinate system.

Polynomial. A mathematical expression containing more than one term.

Position, Initial. The starting position of a radius vector forming an angle. The initial position is the positive x-axis when the angle formed is in standard position.

Position, Standard. (1) The position of the initial side of a generated triangle when it lies on the positive x-axis. (2) The position in a number following the first significant digit.

Position, Terminal. The position of the radius vector in its final position after generating an angle.

Potential Difference (E). A potential difference exists between two points if there are different amounts of electrical charge on the two points. Potential difference is measured in volts.

Prime Numbers. Any whole number that does not have any factors other than 1 and the number itself.

Principal Value of Arc Cosine. The generated angle located in the first or second quadrants for which the cosine has a specific value between −1 and +1.

Principal Value of Arc Sine. The generated angle located in the first or fourth quadrant for which the sine has a specific value between −1 and +1.

Principal Value of Arc Tangent. The generated angle located in the first or fourth quadrant for which the tangent has a specific value.

Power, Apparent (S). The power delivered to the circuit elements by the source. It is indicated by the product of effective voltage and effective current. Apparent power is expressed in volt-amperes.

Power, Electrical (P). The rate at which energy is dissipated or generated. Power is measured in watts.

Power, Reactive (Q). That part of the apparent power that is stored in reactive elements of the circuit and returned to the source during a later part of the cycle. Reactive power is expressed in reactive volt-amperes.

Power, True (P). That part of the apparent power actually consumed by the resistive elements of the circuit. The rate at which energy is converted to heat, light, work, etc. True power is measured in watts.

Power Factor. The cosine of the phase angle between voltage and current. True power equals the product of the apparent power and the power factor.

Power Factor, Reactive. The sine of the phase angle between voltage and current waves. $Q = S \sin \theta$.

Power Factor Correction. Increasing the power factor of a circuit by installing a capacitor in parallel with the load. The capacitive reactance is chosen to match the inductive reactance of the parallel equivalent circuit. Usually, the power factor is improved rather than completely corrected to 100%.

Power Level. The power level of an amplifier or attenuator is the logarithm of the power ratio. Power level is usually expressed in decibels.

Power Loss. The rate at which energy is dissipated in a circuit load.

Power of Ten. A power of ten is any number that has ten as the one and only factor. Can be multiples (10, 100, 1000, etc.) or submultiples (0.1, 0.01, 0.001, etc.).

Power Rating. The power rating of most electrical devices is the wattage drawn (input) under normal conditions. Motors are rated according to output. Resistors are rated at a fraction of the power required for failure.

Power Station. A plant where energy in one form is converted to electrical energy. A station may be powered by coal, diesel fuel, water, wind, nuclear or other kinds of energy.

Prefix, Metric. Prefixes used with metric units representing power of ten multiples.

Product. The total amount obtained as a result of multiplying.

Product, Phasor. A phasor that has a magnitude equal to the product of the magnitudes of the phasors being multiplied. The direction is given by the sum of the angles of the phasors.

Product, Special. Usually refers to binomial products that can be determined without the aid of long multiplication.

Proportion. An equation obtained when two ratios are equal to each other.

Proportion, Direct. An increase in one quantity produces an increase in another quantity. A direct proportion has the form:

$$\frac{A_1}{A_2} = \frac{B_1}{B_2}$$

Proportion, Inverse. An increase in one quantity produces a decrease in another quantity. An inverse proportion has the form:

$$\frac{A_1}{A_2} = \frac{B_2}{B_1}$$

Proton. A positively charged subatomic particle. Its mass is over 1800 times greater than that of an electron.

Pythagorean Theorem. The square of the hypotenuse of a right triangle equals the sum of the squares of the other two sides.

Quadrant. One of four sections into which a plane is divided in a rectangular coordinate system.

Quantity. A property, quality or dimension of an object or phenomenon that is capable of being measured.

Quantity, Basic. A quantity for which fundamental units are chosen.

Quotient. A number that is obtained as a result of dividing.

Quotient, Phasor. A phasor that has a magnitude equal to the ratio of the magnitudes of the phasors being divided. The direction is the difference of the angles.

Radian. A unit of angular measurement defined by the ratio of the arc subtended by the angle to the radius of the arc. $1 \text{ rad} \approx 57.3^\circ$.

Radical. The symbol used to indicate square root ($\sqrt{\ }$).

Radius (r). The distance from the center of a circle or a sphere to the edge in any direction.

Rate. A fraction usually expressed in percent. The ratio of the change or part to the base. Often refers to a time base.

Ratio. The ratio of two numbers is the quotient obtained by dividing the first number by the second.

Ratio, Inverse. The reciprocal of the given ratio.

Ratio, Power. The ratio of the output power of a communication signal to the input power in amplifiers or attenuators.

Ratio, Trigonometric. See Function, Trigonometric.

Reactance (X). The "opposition" to the flow of ac current set up by inductors and capacitors. Reactance is measured in ohms.

Reciprocal. The reciprocal of a number is the number obtained by dividing 1 by the number. A number with an exponent of -1 is the reciprocal of the number. The product of a number and its reciprocal is 1.

Rectangle. A plane figure formed by four sides. All adjacent sides are perpendicular.

Rectangular Form. The rectangular form of a phasor uses the rectangular coordinates to represent the phasor ($x + jy$).

Remainder. (1) See difference. (2) The undivided part left over after a division.

Resistance (R). The opposition to the free flow of electrons. Resistance is measured in ohms.

Resistance, Internal Source. The resistance to the flow of electrons within a voltage source.

Resistance, Temperature Coefficient of (a). A property of matter which determines the change in resistance of a wire when the original resistance and the temperature change is known.

Resistivity (ρ). A property of matter that determines the resistance of wires when the length and cross-sectional area are known.

Resultant. The vector sum of two or more vectors.

Revolution. A unit of angular measurement equal to one complete rotation of the radius vector. 1 revolution = 360°.

Rheostat. An electrical device in which the resistance can be changed.

Root. The root of an equation is any number that will satisfy the equation.

Root, Square. The square root of a number is another number which, when squared, results in the original number.

Root-mean-square. The square root of the average square of the height of a waveform.

Rotor. The rotating component of a generator or motor.

Rounding. A process of dropping digits after a certain significant place in a decimal or a mixed decimal.

Rule, Current Divider. The current in one branch of a two-branch parallel circuit is the product of the total current and the resistance of the other branch divided by the sum of the two resistances.

Rule, Voltage Divider. The voltage drop across a load in a series circuit is the product of the applied voltage and the resistance of the load divided by the total circuit resistance.

Scalar. A quantity that is completely described by a number and a unit.

Scale. A divided or graduated line used to represent units of measure.

Second (s). The SI unit of time.

Semiconductor. A material that is intermediate between those classified as conductors and insulators.

Service Drop. Conductor leading from an outdoor pole to the service entrance of a building.

Shell, Conduction. Orbits that exist beyond the valence shell. They are unoccupied in normal atoms.

Shell, Innermost. The shell or orbit closest to the nucleus. It can contain a maximum of two electrons.

Shell, Outermost. The outermost orbit of electrons in a normal atom. Sometimes called the valence shell. The electrons in this shell are called valence electrons.

Shell, Valence. The outermost occupied shell in a normal atom.

Shunt. A low-resistance conductor placed in parallel with a microammeter to convert it to an ammeter.

Side, Adjacent. The side of a right triangle that is next to (adjacent) and forms one side of the reference angle.

Side, Opposite. The side of a right triangle that is opposite the reference angle.

Siemens (S). A unit of admittance, susceptance and/or conductance.

Signs, Like. Two or more signed numbers having the same sign. All are positive or all are negative.

Signs, Unlike. Two signed numbers having opposite signs. One is positive; one is negative.

Sine. A trigonometric function equal to the ratio of the opposite side to the hypotenuse.

Slope. See grade.

Solid, Rectangular. A six-sided solid figure with all adjacent sides being perpendicular.

Solution, Unique. The solution to a system of equations is unique if there is one and only one set of values for the unknowns.

Speed, Angular (ω). Rotational speed usually expressed in r/min or rad/s.

Speed Ratio (SR). The ratio of input angular speed to output angular speed in rotating machines.

Sphere. A solid figure with all points on the surface being the same distance from the center. A ball.

Square. A rectangle having four equal sides.

Square of a Number. The product found when a number is multiplied by itself. The second power of a number.

Stator. The stationary component of a generator or motor.

Subscript. A small number or letter placed below and to the right of a literal number.

Substitution. The replacement of a variable in an equation with its value or with an expression with an equivalent value.

Subtraction. The operation of deducting the subtrahend from the minuend.

Subtrahend. The number subtracted from the minuend.

Sum. The total amount obtained as a result of adding.

Sum, Vector. See resultant.

Superposition, Principle of. The currents due to each voltage source are superimposed upon one another to yield the total current through any branch of the circuit.

Superscript. A small number placed above and to the right of a number, (an exponent).

Susceptance (B). The reciprocal of the reactance in the reactive branch of a parallel equivalent circuit. Susceptance is expressed in siemens.

Switch. An electrical device used for connecting or disconnecting an electrical circuit.

Symbols, Grouping. Grouping symbols combine terms together and establish priority in the order of operations. (Parentheses, brackets, braces and vinculum.)

System, British Measuring. The measuring system used in the United States. Fundamental units are the foot, pound and second. Sometimes called the fps system, the English system or the British engineering system.

System, mks Measuring. The SI measuring system. The letters m, k and s come from the fundamental units: metre, kilogram and second.

System, Polar Coordinate. A coordinate system consisting of concentric circles which indicate the distance from the pole. Radial lines are marked to indicate the direction (angle) from the reference axis.

System, SI Measuring. The international measuring system, using mks (metric) units. SI is the abbreviation for Le Système International d'Unités.

Tangent. A trigonometric function equal to the ratio of the opposite side to the adjacent side.

Term. An algebraic expression within which numbers and letters are not separated by plus and minus signs. A quantity enclosed in parentheses or under a bar is thought of as a single factor.

Terms, Like. Two or more terms with exactly identical literal factors.

Terms, Unlike. Terms whose literal factors are not exactly the same.
Terms of a Ratio. The numbers a and b that make up the ratio a:b. The letter a is the first term, b is the second.
Thévenin's Theorem. A complex circuit can be reduced to a single source of emf (E_{TH}) and a single resistance (R_{TH}) in series with the circuit load.
Tolerance. A specification that describes by how much a manufactured product may vary in size.
Tolerance, Bilateral. One-half the tolerance.
Tolerance, Relative. The bilateral tolerance expressed as a fraction of the nominal value of a specification, usually expressed in percent.
Transformer. An alternating current device that changes an input at one voltage to an output at a different voltage. It consists of primary and secondary wires coiled around an iron core.
Transformer, Step-down. A transformer in which the output voltage is less than the input.
Transformer, Step-up. A transformer in which the output voltage is greater than the input.
Transmission Lines. A conduit by which electrical energy is transported. Usually constructed of copper or aluminum wires.
Transpose. The process of moving a term from one side of an equal sign to the other by changing its sign.
Triangle. A closed three-sided plane figure.
Triangle, Equilateral. A triangle having three equal angles.
Triangle, First Quadrant. A generated triangle with the radius vector (hypotenuse) in the first quadrant.
Triangle, Fourth Quadrant. A generated triangle formed in the fourth quadrant.
Triangle, Generated. A triangle formed by a rotating radius vector which forms the hypotenuse. A vertical line is drawn through the vector's tip and connected to the x-axis at a right angle.
Triangle, Isosceles. A triangle with two equal angles.
Triangle, Obtuse. A scalene triangle with one obtuse angle.
Triangle, Right. A triangle with one right angle.
Triangle, Scalene. A triangle with three unequal angles.
Triangle, Second Quadrant. A generated triangle formed in the second quadrant.
Triangle, Third Quadrant. A generated triangle formed in the third quadrant.
Triangles, Similar. Two triangles are said to be similar if any two corresponding angles are equal.
Trinomial. A three-term algebraic expression.
Uncertainty. A word that refers to the inexactness of a measurement.
Unit, Common. The common unit for a measurement is formed by writing it as a number between 1 and 1000. The unit then uses a metric prefix for a multiple-of-three exponent. (−9, −6, −3, 0, 3, 6, 9, etc.)
Unit, Compound. A derived unit expressed in terms of fundamental units.
Unit, Derived. Compound units expressed in terms of fundamental units. All nonfundamental units.
Unit, Fundamental. Unit for basic quantities defined independently of all other units.
Unknown. The value of a literal factor in an equation.
Value, Absolute. The numerical part of a number without regard to its sign. Magnitude.
Variable. A quantity that can assume various values by undergoing changes.
Variable, Dependent. A quantity that is affected by changes in the independent variable.
Variable, Independent. A quantity controlled by the technician in an experiment.
Variation. The relationship between one variable and another.
Variation, Direct. A relationship between variables by which a change in one variable changes another variable by the same factor. The curve on a graph of the two variables is a straight line. The two variables are related by the formula: $A = kB$.
Variation, Inverse. If two variables vary inversely, one varies directly with the reciprocal of the other: $A = k/B$.
Variation, Joint. A variation relationship in which one variable depends on the product of two or more others.
Variation, Linear. See direct variation.
Vector. A quantity possessing both magnitude and direction. Vectors are represented graphically as an arrow.
Vector, Radius. A rotating radius of a circle represented by an arrow. Angles are generated as it rotates.
Vector, Resolution of. The process of finding the x and y components of a vector. Polar to rectangular transformation.
Velocity, Angular. Rate or speed of rotation. When the symbol (ω) is used, the units must be measured in radians per second.
Vertex. The intersection of two lines forming an angle.
Vinculum. A bar used as a grouping symbol.
Volt (V). A unit of voltage, potential difference or emf. When one ampere flows through a resistance of one ohm, there is a voltage drop of one volt.
Voltage (E). A voltage is either an emf or a potential difference. Voltage is measured in volts.
Voltage, Effective. The equivalent DC voltage which, when multiplied by the effective current, gives the average power (heating effect) in an ac circuit. Sometimes called rms voltage.
Voltage, Instantaneous. The value of an ac voltage at a particular instant of time.
Voltage, Line. Voltage measured from one line to another in a polyphase system.
Voltage, Maximum. The maximum value of an instantaneous ac voltage.
Voltage, Phase. The voltage drop measured across the load in one arm of a polyphase system. The voltage produced by one winding of the source (coil voltage).

Voltage, Rms. See voltage, effective.

Voltage, Terminal. The voltage across the terminals of a voltage source. The emf minus the voltage drop across the internal resistance.

Volt-ampere. A unit of apparent power which is equivalent to the watt: 1 VA = 1 W.

Volt-ampere, Reactive. A unit of reactive power equivalent to one watt: (1 var = 1 W).

Voltmeter. A voltmeter is a measuring device that measures voltage across a source or a load. It is connected in parallel with the source or load.

Volume (V). A quantity that describes the amount of space an object occupies or contains. The SI unit of volume is the cubic metre.

Watt (W). A unit of power. The power generated when one joule of energy is produced in one second.

Wattmeter. An electrical measuring instrument that measures power.

Whole numbers. Whole numbers or natural numbers are the numbers used for counting.

Work (*W*). The product of force and distance. Work is measured in joules.

Wye-connection. A three-phase load or source configuration in which the elements are connected in the shape of the letter Y (wye).

***x*-Component.** The side of a generated triangle that lies along the x-axis. It is sometimes called the horizontal component or east-west component.

***y*-Component.** The side of a generated triangle that lies parallel to the y-axis. It is sometimes called the vertical component or the north-south component.

ANSWERS TO ODD-NUMBERED PROBLEMS

CHAPTER 1

EXERCISE 1-1

1. Fifty one thousand, eight hundred forty-two
3. Seven billion, nine hundred forty-six million, three hundred sixty-one thousand
5. 39 485
7. 4 208 011 000
9. Five million, thirty-two thousand, six hundred eight

EXERCISE 1-2

1. 17
3. 1837
5. 25 746
7. 3 126 313
9. 22
11. 186
13. 1359
15. 453
17. 4136
19. 308 679
21. 2059 lb
23. 547 000 ohms
25. $17
27. 15 volts

EXERCISE 1-3

1. 846
3. 30 176
5. 4032
7. 328 608
9. 16 325 lb
11. 450 poles
13. 990
15. $378

EXERCISE 1-4

1. 33
3. 22
5. 73
7. 18 kilowatt hours
9. 11 miles

EXERCISE 1-5

1. 2; 3; 5; 7; 11; 13; 17; 19; 23; 29; 31; 37
3. 12 × 2; 8 × 3; 6 × 4
5. 2 × 16; 4 × 8
7. 2 × 9; 3 × 6
9. 50 × 2; 25 × 4; 10 × 10
11. 2; 2; 3; 5
13. 5; 5; 11
15. 3; 7; 11; 17

EXERCISE 1-6

1. 4
3. 2
5. 1
7. 5
9. 3
11. Correct
13. 2
15. Correct
17. Correct
19. Correct
21. 11
23. 12
25. 6
27. 5
29. 12

EXERCISE 1-7

1. 4	5. 49	9. 3	13. 200	17. 16
3. 9	7. 57	11. 116	15. 32	19. 33

CHAPTER 2

EXERCISE 2-1

1. a	3. b	5. d	7. c	9. d	11. c

EXERCISE 2-2

1. b	3. a	5. a	7. a	9. b

EXERCISE 2-3

1. b	5. b	9. d	13. h	17. f	21. g
3. b	7. a	11. d	15. a	19. d	23. e

EXERCISE 2-4

1. (a) Fixed resistor
   (b) Fuse
   (c) Battery
   (d) Incandescent lamp

1. (e) Inductor
   (f) DC motor
   (g) Voltmeter

1. (h) Battery cell
   (i) Constant voltage DC source
   (j) ac generator

EXERCISE 2-5

1.

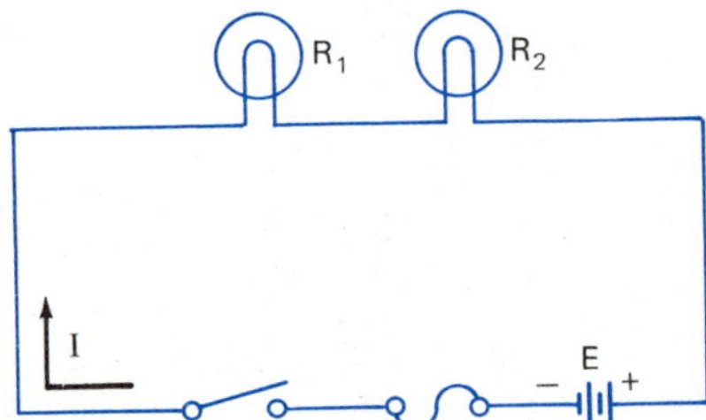

5.

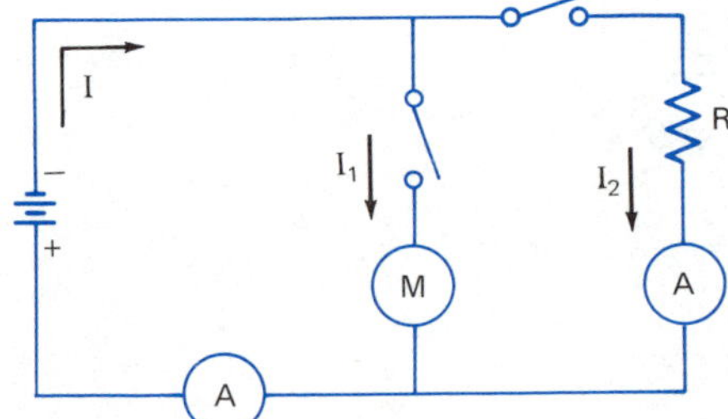

3.

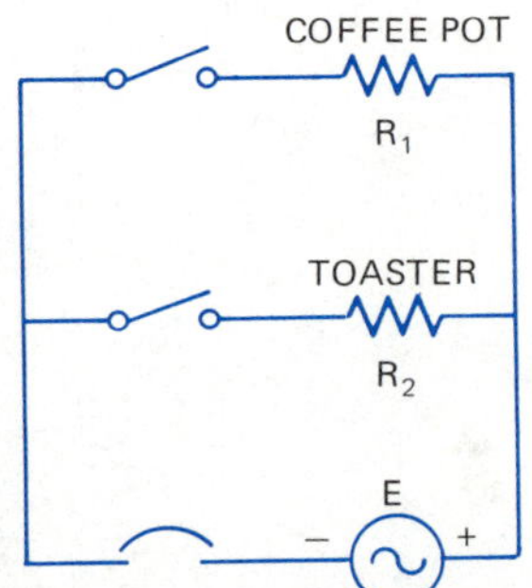

7.

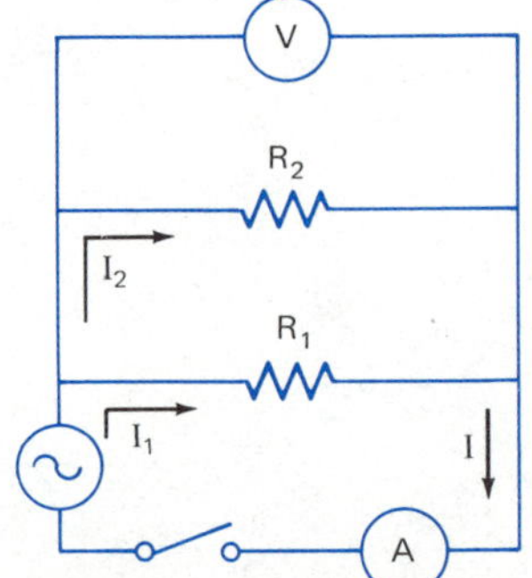

CHAPTER 3

EXERCISE 3-1

1. 1 1/4
3. 2 3/8
5. 7 1/3
7. 3 1/2
9. 7/2
11. 17/3
13. 117/16
15. 3/4
17. 1/3
19. 7/11
21. 12/24; 16/24; 20/24
23. 14/42; 35/42; 24/42
25. 3/5
27. 1/4

EXERCISE 3-2

1. 2 2/3
3. 2 1/6
5. 2
7. 1 1/2
9. 679 7/15
11. 99 ft
13. 34 1/2 L
15. 34 1/8 kg
17. 40 6/7 m
19. 277 830 cubic inches
21. $9467 1/6

EXERCISE 3-3

1. 1/5
3. 1/23
5. 1
7. 3/8
9. 2/25
11. 6/23
13. 6
15. 1 1/2
17. 2/3
19. 1 1/2
21. 3 1/4
23. 13 5/7
25. 13 1/2 rods
27. 300 1/2 ft
29. 1 41/64 kilowatts

EXERCISE 3-4

1. 0.7 (seven tenths)
3. 0.987 (nine hundred eighty-seven thousandths)
5. 5.04 (five and four hundredths)
7. 0.030 25 (three thousand twenty-five hundred-thousandths)
9. 3/4
11. 3/16
13. 601/2000
15. 0.9; 9/10
17. 0.379; 379/1000
19. (a) 0.33
 (b) 1.19
 (c) 2.04
 (d) 2.76
21. (a) 1.08; 5.4; 27
 (b) 4.28; 21.4; 107
 (c) 6.76; 33.8; 169
 (d) 8.6; 43; 215

EXERCISE 3-5

1. 50
3. 6000
5. 20
7. 3.14
9. 0.07
11. 1.00
13. 140
15. 1700
17. 0.56

EXERCISE 3-6

1. 0.125
3. 0.313
5. 0.375
7. 0.016
9. 0.938
11. 0.750
13. 3/8
15. 7/8
17. 5.75
19. 17.28
21. 7.29

EXERCISE 3-7

1. 49.465
3. 0.332 14
5. $260.73
7. 46.611
9. 5.8493
11. 8.8125
13. 74.875 in
15. 3.723 Ω
17. 277 km
19. $1580.30

EXERCISE 3-8

1. 12.566 36
3. 7.488
5. 1.33
7. 872.47
9. 45 300
11. 431
13. 6.568
15. 0.295
17. 0.000 018 2
19. 137.44 ft
21. 34.375 Ω
23. $614.88
25. 47.25 kg tin; 27.75 kg lead

EXERCISE 3-9

1. 18.4
3. 1.3
5. 39.1
7. 12 554 229.2
9. 8.9
11. $5(17 - 8) \div 3$
13. $5(256 - 129) \div 8 \div 17$
15. $\frac{244 \times 18 \times 4}{\pi \times 2}$; 2796.0
17. $\frac{5 - 2}{1728 - 1705}$; 0.13
19. 91.0 V
21. 17.7 Ω

EXERCISE 3-10

1. 0.5
3. 0.25
5. 0.1
7. 0.01
9. 2
11. 0.167
13. 0.125
15. 0.000 001
17. 1.267
19. 0.08
21. 0.595
23. 9.926
25. 24.627
27. 307.7 Ω

EXERCISE 3-11

1. 1
3. 9
5. 25
7. 49
9. 81
11. 121
13. 225
15. 625
17. 324
19. 81
21. 1077.6
23. 0.391
25. 518 400
27. 4.61
29. 1685.6
31. 67 858.4
33. 3.4
35. 904.8 cubic inches

EXERCISE 3-12

1. 3
3. 12
5. 1
7. 4
9. 10
11. 6
13. 25
15. 20
17. 1.47
19. 2.62
21. 0.866
23. 2.51
25. 201.2
27. 11.98
29. 14
31. 14.9 in

EXERCISE 3-13

1. 6.01 A
3. 115 V
5. 16.9 Ω
7. 248 V
9. 178 Ω
11. 26.7 Ω
13. 110 V
15. 2.78 A
17. 0.02 A

EXERCISE 3-14

1. $d = C/\pi$
3. $P = F/A$
5. $C = Q/V$
7. (a) $I = E/R$; $R = E/I$
   (b) $E = P/I$; $I = P/E$
9. 6.03 hp
11. 1 070 000 hp
13. 23 700 watts
15. 0.146 A
17. 12.0 V
19. 2.70 A
21. 0.438 A
23. 6.13 hp

CHAPTER 4

EXERCISE 4-1

1. See Figures 4-1 and 4-2.
3. (a) voltage
   (b) current
5. (a) angular speed
   (b) efficiency
7. (a) temperature
   (b) resistance

EXERCISE 4-2

1.

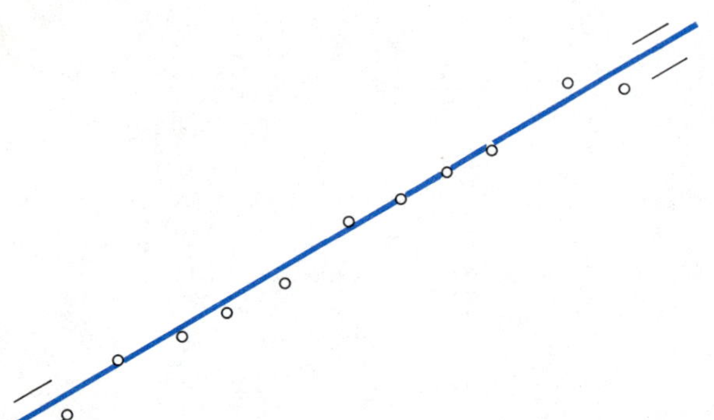

3.

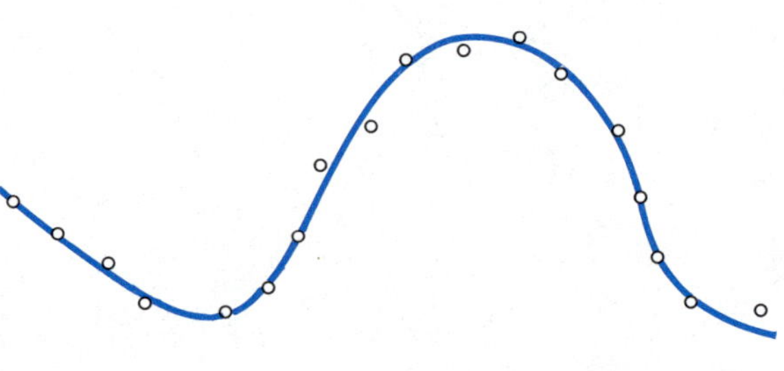

EXERCISE 4-3

1.

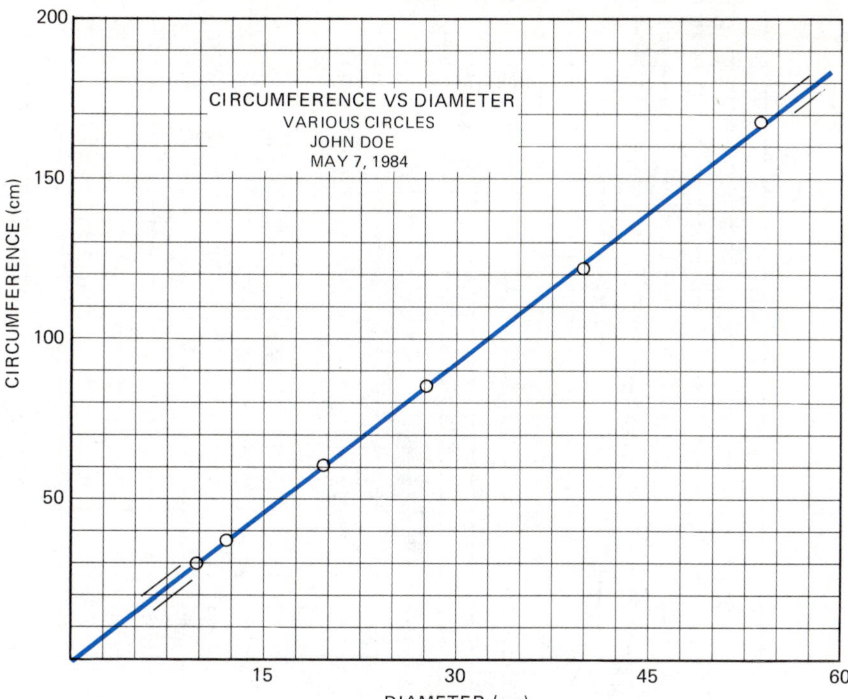

(a) 140 cm
(b) 16.1 cm

3.

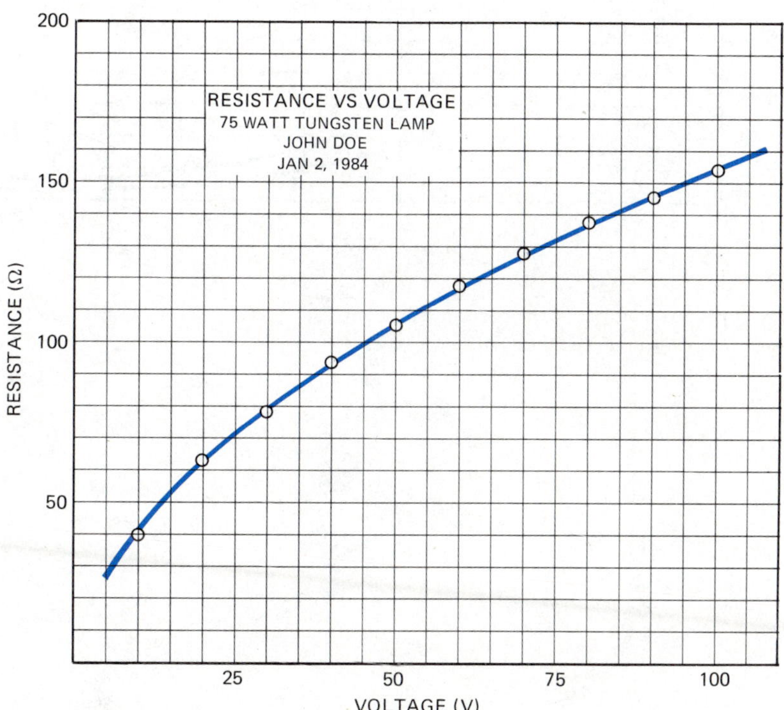

(a) 133 Ω
(b) 46 V

5.

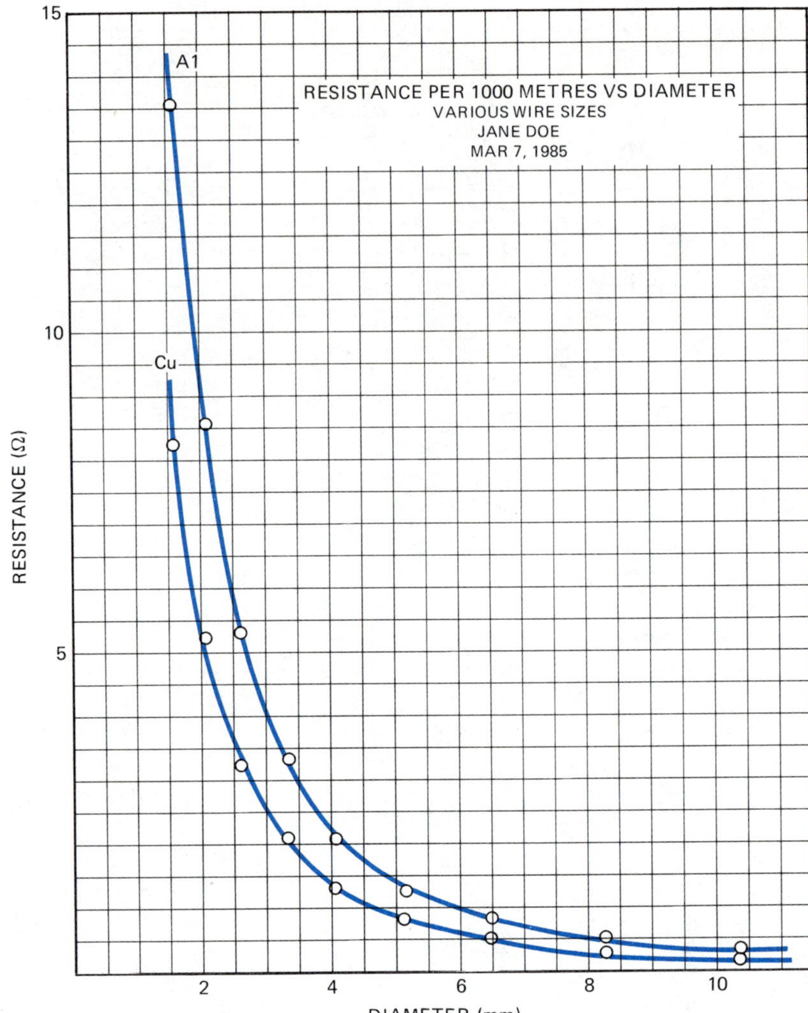

0.4 Ω (Al) and 0.2 Ω (Cu)

7.

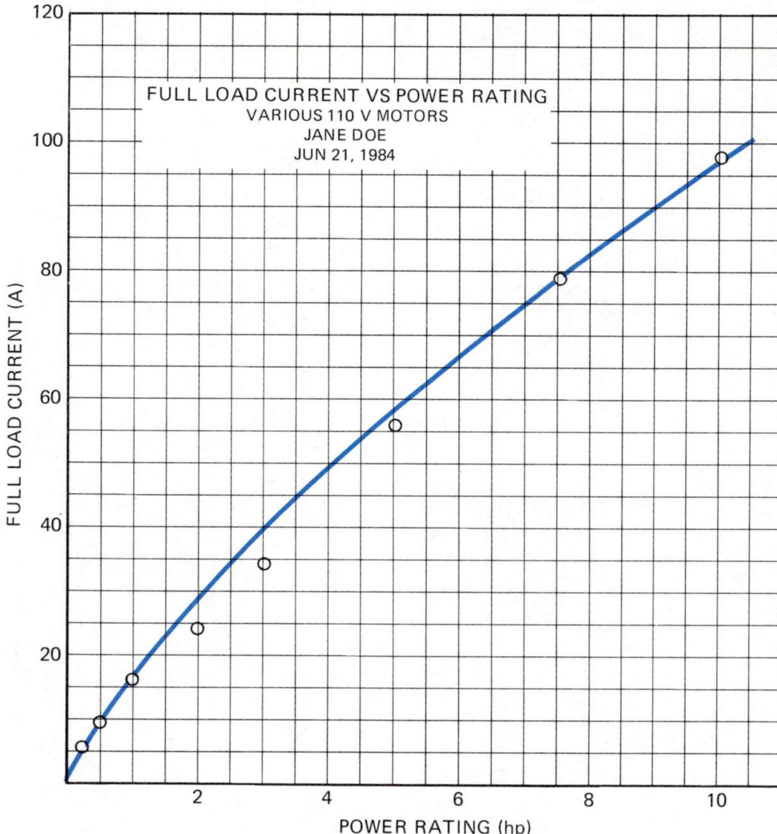

CHAPTER 5

EXERCISE 5-1

1. 7
3. 6
5. 16
7. 3/8
9. +11
11. −3
13. −3
15. +7
17. −1
19. −6
21. −3
23. +12
25. −3
27. −10

EXERCISE 5-2

1. +2
3. −6
5. +4
7. 191
9. 116

EXERCISE 5-3

1. +1
3. +11
5. −5
7. +8
9. −4
11. −1

EXERCISE 5-4

1. +12
3. +24
5. 0
7. −6
9. +36
11. −125
13. +40
15. −6
17. +3
19. −0.125
21. −1
23. −2
25. −6

EXERCISE 5-5

1. 0.000 01
3. 0.000 000 1
5. 0.000 000 000 01
7. 10^{-2}
9. 10^{10}
11. 10^{-8}
13. 10 000; 1 000 000 000 000

EXERCISE 5-6

1. $6_{\uparrow}40$; 2; 2
3. $0.4_{\uparrow}563$; −1; 4
5. $0.000\ 000\ 04_{\uparrow}309$; −8; 4
7. $0.0006_{\uparrow}$; −4; 1
9. $0.000\ 4_{\uparrow}43\ 00$; −4; 5
11. $5_{\uparrow}28\bar{0}$; 3; 4
13. $9_{\uparrow}56\ 00\bar{0}\ 000$; 8; 6
15. 12.0 V
17. 211 000 000 m/s
19. 2.00 s
21. 2480 kW·h
23. $45\bar{0}$ ft
25. 1180 Ω

EXERCISE 5-7

1. 3.80×10^{5}
3. 3.14×10^{-2}
5. 4.43×10^{7}
7. 8.25×10^{11}
9. 5.52×10^{-1}
11. 3.43×10^{4}
13. 592
15. 8 110 000
17. 0.299
19. 0.000 000 000 003 00
21. 33 400
23. 1 400 000 000 000 000 000 000 000
25. 6.24×10^{18}

EXERCISE 5-8

1. (a) $0.000\ 044\ 9 \times 10^{2}$
   (b) 3.48×10^{2}
   (c) 7940×10^{2}
   (d) $0.000\ 001\ 14 \times 10^{2}$
   (e) 0.02×10^{2}
   (f) $0.000\ 998 \times 10^{2}$
   (g) $4\ 280\ 000 \times 10^{2}$
   (h) $48\ 200 \times 10^{2}$
   (i) $0.000\ 661 \times 10^{2}$
   (j) 0.0894×10^{2}
   (k) $0.000\ 000\ 177 \times 10^{2}$
   (*l*) $0.000\ 000\ 297 \times 10^{2}$
3. (a) 4.49×10^{-3}
   (b) 348×10^{0}
   (c) 794×10^{3}
   (d) 114×10^{-6}
   (e) 2×10^{0}
   (f) 99.8×10^{-3}
   (g) 428×10^{6}
   (h) 4.82×10^{6}
   (i) 66.1×10^{-3}
   (j) 8.94×10^{0}
   (k) 17.7×10^{-6}
   (*l*) 29.7×10^{-6}

EXERCISE 5-9

1. 875×10^{6} Ω ± 10%
3. 3.6×10^{3} Ω ± 10%
5. 87.5×10^{6} Ω ± 1%
7. 2.45×10^{6} Ω ± 5%
9. 64×10^{6} Ω ± 10%
11. 930×10^{3} Ω ± 2%
13. 7.5×10^{5} pF ± 10%
15. 2.2×10^{3} pF ± 2%
17. 2.9×10^{1} pF ± 10%
19. 4.0×10^{4} pF ± 20%
21. 2.3×10^{6} pF ± 5%

EXERCISE 5-10

1. 4.3×10^{3}
3. 7.2×10^{-2}
5. 2.02×10^{1}
7. 3.7×10^{3}
9. 2.7×10^{-4}
11. 1.048×10^{6}
13. 10^{5}
15. 10^{-3}
17. 10^{4}
19. 10^{-3}
21. 10^{-9}
23. 1.2×10^{6}
25. 150×10^{3}
27. 3.28×10^{-3}
29. 420×10^{-3}
31. 30×10^{3}

EXERCISE 5-11

1. 10^{-3}
3. 10^{1}
5. 10^{3}
7. 10^{10}
9. 10^{-3}
11. 200×10^{-3}
13. 50×10^{9}
15. 4×10^{3}
17. 4×10^{0}
19. 432×10^{3}
21. 10×10^{-9}
23. 60.0×10^{3}
25. (a) 1.2×10^{2} V
    (b) 2.4×10^{0} W
27. (a) 2.2×10^{1} Ω
    (b) 2.2×10^{3} W
29. (a) 2.087×10^{2} V
    (b) 5.28×10^{-1} W

EXERCISE 5-12

1. 9×10^{4}
3. 8×10^{9}
5. 1.25×10^{-13}
7. 1.6×10^{13}
9. 4×10^{-6}
11. 640×10^{9}
13. 343×10^{-9}
15. 8.1×10^{-15}
17. 2×10^{3}

19. 3×10^4
21. 4×10^{-2}
23. 90×10^0
25. 20×10^{-3}
27. 11×10^{-3}
29. 10^{-5}
31. 5×10^{-3}
33. 4×10^4
35. 125×10^{-12}
37. 204.1×10^6

EXERCISE 5-13

1. 1.00 4
3. 6.30 3
5. −1.53 0
7. −3.47 −24
9. 5.95×10^6
11. 425×10^0
13. 8.01×10^{-9}
15. 235×10^0
17. 3.25×10^3
19. 6.25×10^{-3}
21. 5.57×10^{16}
23. 7.04×10^{-5}
25. 4.12×10^9
27. 2.93×10^8
29. 6.14×10^{12}
31. 6.03×10^{-2}
33. 954×10^0
35. 9.28×10^{-2}
37. 97.7×10^{-3}
39. 1.11×10^3
41. 59.2×10^0

CHAPTER 6

EXERCISE 6-1

1. 1; 6 ± 0.5 cm
3. 3; 24.0 ± 0.05 V
5. 3; 1.61×10^5 ± 500 V
7. 3; 45.5×10^6 ± 50 000 Ω
9. 4; 0.3333 ± 0.000 05 hp
11. 3; 98.6 ± 0.05°F
13. 6.4 ± 0.05 Ω; 2
15. 29.5 ± 0.05 V; 3
17. 1.11 ± 0.005 A; 3
19. 3221 ± 0.5 kW•h; 4
21. 71° ± 0.5°C; 2

EXERCISE 6-2

1. 631.6 V
3. 0.6096 A
5. 0.02 W
7. 0.0221 A
9. 48 W
11. 2300 Ω
13. 1.9 W
15. 480 m
17. 0.0510 Ω

EXERCISE 6-3

1. d
3. h
5. i
7. j
9. b
11. i
13. a
15. b
17. r
19. d
21. p
23. g
25. s
27. h
29. t

EXERCISE 6-4

1. p
3. a
5. g
7. n
9. b
11. ampere; A
13. kΩ
15. kW•h
17. 1 000 000 000; 0.000 000 01
19. (a) microampere
 (b) kilovolt
 (c) megohm
 (d) millisecond
 (e) picofarad
21. (a) mV
 (b) kΩ
 (c) μs
 (d) kg

EXERCISE 6-5

1. (a) 435
 (b) 0.435
3. (a) 38 400
 (b) 38 400 000
5. (a) 75 000 000
 (b) 0.000 075
7. (a) 655
 (b) 0.000 655
9. (a) 75.2
 (b) 75 200 000
11. (a) 4.13×10^2
 (b) 4.13×10^{-1}
13. (a) 1.49×10^0
 (b) 1.49×10^3
15. (a) 8.56×10^0
 (b) 8.56×10^3
17. (a) 3.45×10^0
 (b) 3.45×10^{-3}
19. (a) 7.85×10^2 kΩ
 (b) 7.85×10^5 Ω

EXERCISE 6-6

1. 5.2 kV
3. 776 kΩ
5. 99.6 MHz
7. 229 kΩ
9. 5.54 H
11. 463 kV
13. 285 kW
15. 104.6 MHz

17. 3.50 μF
19. 765 nV
21. 435 kΩ
23. 38.4 A
25. 75 μF
27. 655 ns
29. 75.2 H
31. 413 kΩ
33. 1.49 mA
35. 8.56 mA
37. 3.45 mH
39. 785 kΩ
41. resistance
 (a) 8.10×10^5
 (b) 8.10×10^2
 (c) 810 Ω
43. voltage
 (a) 1.90×10^8
43. (b) 190
 (c) 190 V
45. voltage
 (a) 1340
 (b) 1.34
 (c) 1.34 V
47. time
 (a) 0.204
 (b) 2.04×10^{-7}
 (c) 204 ns
49. inductance
 (a) 4.04×10^5
 (b) 404
 (c) 404 H
51. mass
 (a) 3.29×10^{-4}
 (b) 3.29
 (c) 3.29 μg
53. temperature
 (a) $2.0\bar{0} \times 10^6$ mK
 (b) $20\bar{0}0$ K
 (c) 2.00 kK
55. power
 (a) 7.46×10^6 mW
 (b) 7460 W
 (c) 7.46 kW
57. distance
 (a) 5.50×10^{-4}
 (b) $5.5\bar{0} \times 10^{-7}$
 (c) $55\bar{0}$ J
59. charge
 (a) 0.160
59. (b) 1.60×10^{-19}
 (c) no common prefix
61. velocity (speed)
 (a) 3×10^5
 (b) 3×10^8
 (c) 300 Mm/s
63. power
 (a) 39
 (b) 3.9×10^7
 (c) 39 MW
65. resistance
 (a) 3.4
 (b) 3.4×10^6
 (c) 3.4 MΩ

EXERCISE 6-7

1. 0.087 60 in
3. 9×10^{-31}
5. 0.034 28 s; 34.28 ms
7. 927 L
9. b
11. $21.65
13. 222 000 ft^3/min
15. 610 L/s; 2200 m^3/h

CHAPTER 7

EXERCISE 7-1

1. The have no other factors
3. ω; t
5. 2; 3; 5; R; R; t
7. 2; 2; E_1; I_1
9. $\frac{1}{2}$; C; E; E
11. 2; π; $\sqrt{L/g}$
13. E; $\frac{1}{R}$
15. $\frac{1}{2}$; $\frac{1}{\pi}$; $\frac{1}{f}$; $\frac{1}{C}$

EXERCISE 7-2

1. 15
3. 20
5. 11.2
7. 1.0
9. 16
11. 7
13. −11
15. 1
17. −0.6
19. 403
21. 7
23. −493
25. 4
27. 9
29. 59.5

EXERCISE 7-3

1. 5n = 65; 13
3. n/3 = 17; 51
5. 2E = 628; 314 V
7. $\frac{I}{10} = 0.146$; 1.46 A
9. $\frac{1}{4}L = 1.4$; 5.6 m

EXERCISE 7-4

1. $f = C/\lambda$
 $\lambda = C/f$
3. $h = 2A/b$
 $b = 2A/h$
5. $h = V/\pi R^2$
7. $V = m/d$
 $m = dV$
9. $C = \frac{1}{2\pi f X_C}$
 $f = \frac{1}{2\pi C X_C}$
11. $\Delta T = QL/KAt$
 $L = KA\Delta Tt/Q$
13. $L = X_L/2\pi f$
 $f = X_L/2\pi L$
15. $f = \omega/2\pi$
17. $I_2 = E_1 I_1/E_2$
 $E_2 = E_1 I_1/I_2$
19. $E = Q/C$
 $Q = CE$
21. $I = E/R$
 $R = E/I$
23. $I = Q/t$
 $t = Q/I$
25. $E = P/I$
 $I = P/E$

EXERCISE 7-5

1. 1.5 V
3. 4.2 mC
5. 250 mA
7. 120 V
9. 850 V
11. 1.77 kC; 1.11×10^{22} e
13. 3.60 ks (1h)
15. 130 s
17. 150 mA
19. 120 V
21. 300 s (5 min)
23. 40 kJ
25. 5 kJ
27. 1 kJ
29. 1.0 ks
31. 3 MΩ
33. 9.6 μW
35. 3 s

EXERCISE 7-6

1. 168 cm
3. 44.9 m
5. equilateral
7. obtuse
9. right
11. altitude not given
13. 6.00 ft^2
15. 9.3 cm; 18.6 cm
17. 141.5 ft; 1200 ft^2
19. 34 cm
21. 13 in^2
23. 106 pm
25. c; d; a; b; no
27. 5000 cm^2; 4500 cm^2

EXERCISE 7-7

1. 28.3 in^2
3. 4700 cm^2
5. perimeter
7. area
9. perimeter
11. 14.1 in^3
13. 16 000 cm^3
15. 1.09 m^3
17. 11 000 cm^2
19. 46 L
21. 320 000 mm^3; 320 cm^3
23. 2700 cm^2
25. 0.5890 in^2
27. 38 in
29. 6.5 in
31. 9.2×10^{-38} cm^3
33. (a) 20.65 in^3 (b) 9
35. 5800 m^3; 540 m^2

EXERCISE 7-8

1. 30 μΩ
3. 65.5 m
5. 1.45 Ω
7. 620 mΩ
9. 141 Ω
11. 96 Ω; 1.1 A; 126 W
13. 0.000 17/C°
15. 440 mA
17. 33 A
19. 11 cm; 9.09 A; 1000 W

CHAPTER 8

EXERCISE 8-1

1. 1; mono
3. 2; bi
5. 1; mono
7. 2; bi
9. 4; poly
11. 5
13. $-\frac{1}{2}$
15. 1
17. 26
19. $-\pi$

EXERCISE 8-2

1. 9R
3. −4ac
5. $2P_1 - 2EI_1$
7. $14a^2bc$
9. $3\pi\sqrt{P/R} - 8\pi\sqrt{E^2/R^2}$
11. $9x - 8x^2 - 14y - 1$
13. $29R_1 + 24R_2 + 9$
15. 14 πfC

EXERCISE 8-3

1. 8
3. 6
5. 5.40
7. 16
9. 10
11. 1
13. −16
15. 25
17. −3
19. −39
21. 48
23. 97
25. $25\frac{1}{2}$
27. 7.2
29. 4.60×10^3
31. 0.33
33. −1275
35. 6.7×10^{-6}
37. −16.3
39. 13
41. 2 A
43. 16 V
45. 137°
47. 135 m

EXERCISE 8-4

1. 4
3. 6
5. 17
7. 10
9. 3
11. $3\frac{1}{2}$
13. $-\frac{1}{4}$
15. 18
17. $\frac{1}{2}$
19. 4
21. 25
23. 3
25. 9
27. 1
29. 13
31. 6.67
33. 6
35. 12 A
37. 10 Ω
39. 3.2 m^2

EXERCISE 8-5

1. $E_1 = E_T - E_2$
3. $P_2 = P_T - P_1 - P_3 - P_4$
5. $R = \frac{E - e}{I}$

   $I = \frac{E - e}{R}$
7. $E_3 = \frac{P - (E_1 I + E_2 I)}{I}$
9. $I_1 = \frac{I - 4I_2}{2}$
11. $\phi = 90 - \theta$

EXERCISE 8-6

1. (a) 1840 Ω
   (b) 13 mA
3. (a) 118 Ω
   (b) 1.00 A
5. (a) 41.6 kΩ
   (b) 1.5 mA
7. 885 mA
9. (a) 700 Ω
   (b) 43 mA
   (c) same

EXERCISE 8-7

1. (a) 1250 Ω
   (b) 75 V
   (c) 113 V
   (d) 188 V
3. (a) 79 mA
   (b) 51 V
   (c) 103 V
   (d) 87 V
   (e) 154 V
   (f) 190 V
5. (a) 1.07 A
   (b) 30 V
   (c) open circuit
7. (a) 34.2 V
   (b) 56.1 V
   (c) 216 Ω
   (d) 1420 V
9. (a) 32 100 V
   (b) 45.5 V/km
11. (a) 140 Ω; 35 V
    (b) 160 Ω; 40 V
    (c) 160 Ω; 40 V
    (d) 250 mA
13. 960 mΩ
15. (a) 115 V
    (b) 18 V
    (c) 25 A
    (d) 3.9 A

EXERCISE 8-8

1. (a) 90 mΩ
   (b) 510 mΩ
3. (a) 10.9 V
   (b) 88 mΩ
5. (a) 40 V
   (b) 0.8 Ω
   (c) 4 Ω
7. 0.8 Ω (same as r)

EXERCISE 8-9

1. 25
3. 625
5. 9
7. 230
9. −13
11. 6
13. 3
15. 5×10^{-3}
17. 2
19. 0.75
21. $r = \sqrt{A/\pi}$
23. $R = \sqrt{Z^2 - X^2}$
25. $r = \sqrt[3]{3A/4\pi}$
27. $P = E^2/R$
29. $I = \sqrt{W/Rt}$
31. $R = \sqrt{Z^2 - X_C{}^2}$
33. $X = \sqrt{E^2/I^2 - R^2}$
35. $I = \sqrt{\frac{P}{Rt}}$; 120 mA
37. $E = \sqrt{PR}$; 9.53 V
39. $A = \frac{4\pi r^3}{3}$; 34 ft^3

EXERCISE 8-10

1. 115 V
3. 290 mA
5. (a) 12 kW
   (b) 228 V
7. (a) 10 V; 14 V
   (b) 7.1 W; 9.4 W
9. (a) 7.3 V
   (b) 0.82 A
   (c) 8.9 Ω
   (d) 90 W
11. 2200 V
13. (a) 3.4 W
    (b) 177 W
15. 300 W
17. (a) 1160 V
    (b) 7670 W

19. 52 W
21. (a) 12.5 Ω
(b) 12.5 Ω
(c) 225 Ω
21. (d) 2.9 W
(e) 2.9 W
(f) 52 W
23. (a) 14.4 V
23. (b) 0.56 V
(c) 770 mW
23. (d) 20 W
(e) 19.2 W

CHAPTER 9

EXERCISE 9-1

1. $\frac{6}{12}; \frac{8}{12}; \frac{9}{12}$
3. $\frac{8}{18}; \frac{6}{18}; \frac{3}{18}$
5. $\frac{8}{72}; \frac{54}{72}; \frac{45}{72}$

EXERCISE 9-2

1. $1\frac{1}{5}$
3. 1
5. $\frac{8}{15}$
7. $1\frac{19}{60}$
9. $\frac{1}{33}$
11. $\frac{1}{4}$
13. $\frac{13}{180}$
15. $\$3\frac{7}{50}$
17. $13\frac{7}{8}$ gal
19. $3\frac{1}{120}$ Ω

EXERCISE 9-3

1. 5 Ω
3. 8 Ω
5. 9.9 Ω
7. 1500 Ω
9. 830 Ω
11. (a) 286 μS
(b) 222 μS
(c) 147 μS
11. (d) 655 μS
(e) 1500 Ω
13. 6200 Ω
15. 1200 Ω

EXERCISE 9-4

1. (a) 1.7 A
(b) 4.7 A
(c) 60 V
(d) 12.8 Ω
3. (a) 220 kΩ
(b) 110 kΩ
(c) 73 kΩ
(d) 2 mA
5. 6.4 mA
7. 10 Ω
9. (a) 1.28 Ω
(b) 9.36 A
11. 144 mV
13. (a) 910 mA
(b) 2.09 A
15. (a) 10 A
(b) 6.7 A
(c) 5 A

EXERCISE 9-5

1. 67 W
3. 1180 μW
5. 12.5 A
7. (a) 707 V
(b) 2830 V
9. 15 A
11. (a) 18.5 A
(b) 2220 W

EXERCISE 9-6

1. 9 Ω
3. 26.1 Ω
5. 5.3 kΩ
7. 5.7 kΩ

EXERCISE 9-7

1.
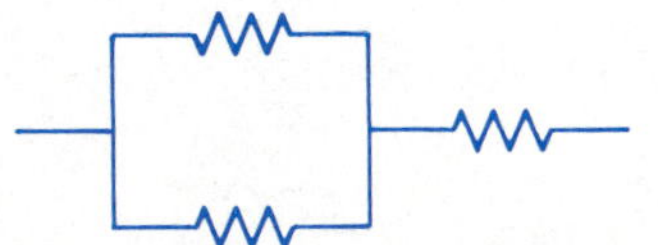

3.
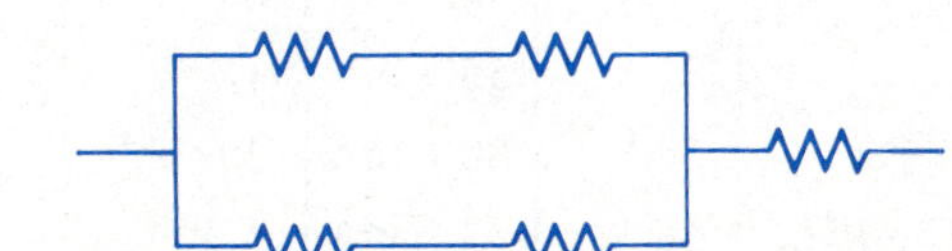

EXERCISE 9-7 (continued)

5.

7.

9.

11.

13.

	E	I	P
R_1	6 V	0.6 A	3.6 W
R_2	6 V	0.4 A	2.4 W
R_3	6 V	1 A	6 W
R_T (12 Ω)	12 V	1 A	12 W

15.

	E	I	P
R_1	35 V	0.70 A	24 W
R_2	49 V	0.70 A	34 W
R_3	21 V	0.52 A	11 W
R_4	63 V	0.52 A	33 W
R_5	37 V	1.22 A	45 W
R_T (100 Ω)	120 V	1.22 A	146 W

17.

	E	I	P
R_1	16 V	80 mA	1.3 W
R_2	16 V	53.3 mA	0.8 W
R_3	32 V	53.3 mA	1.7 W
R_4	32 V	80 mA	2.6 W
R_T (360 Ω)	48 V	133 mA	6.4 W

CHAPTER 10

EXERCISE 10-1

1. (a)

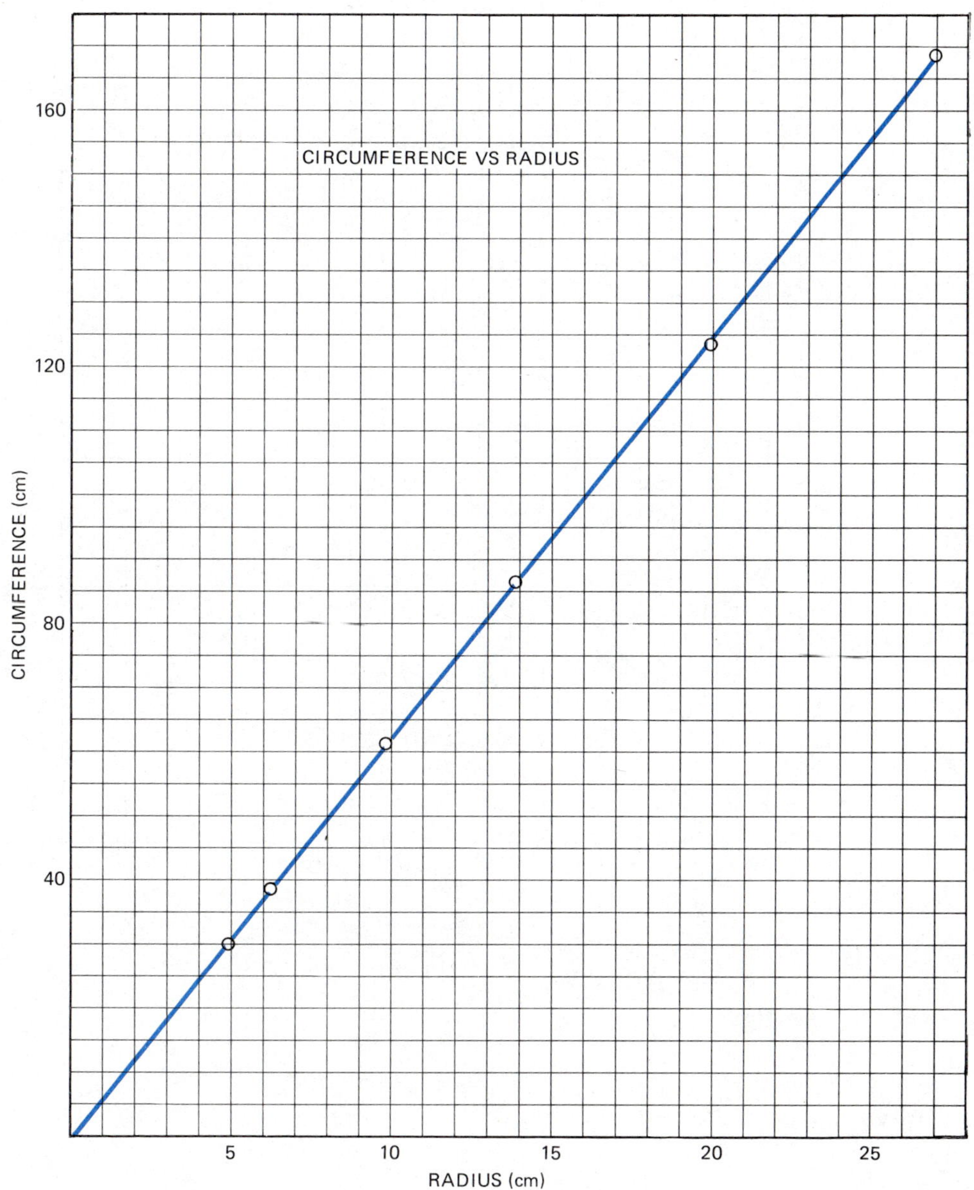

(b) C = 6.29r

(c) 94.4 cm

3. (a)

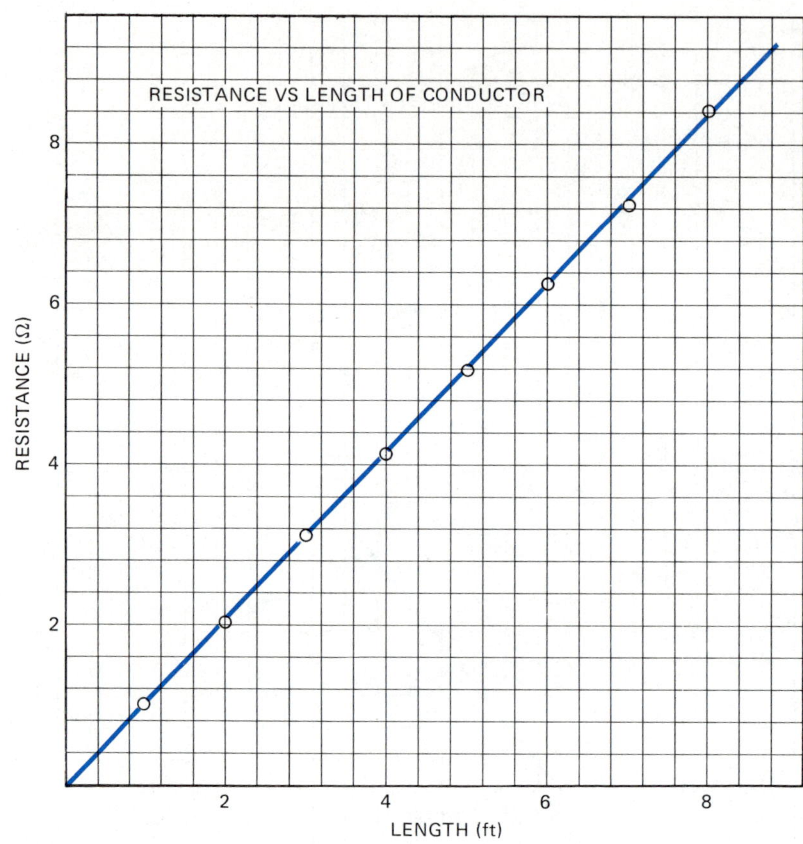

(b) $R = 1.07L$
(c) 5.9 Ω
(d) 9.34

5. $C = 6.54 \times 10^{-6} E$; 3.3 mC
7. (a) $\Delta R = 6\Delta T$
(b) 35 C° decrease

EXERCISE 10-2

1. (a)

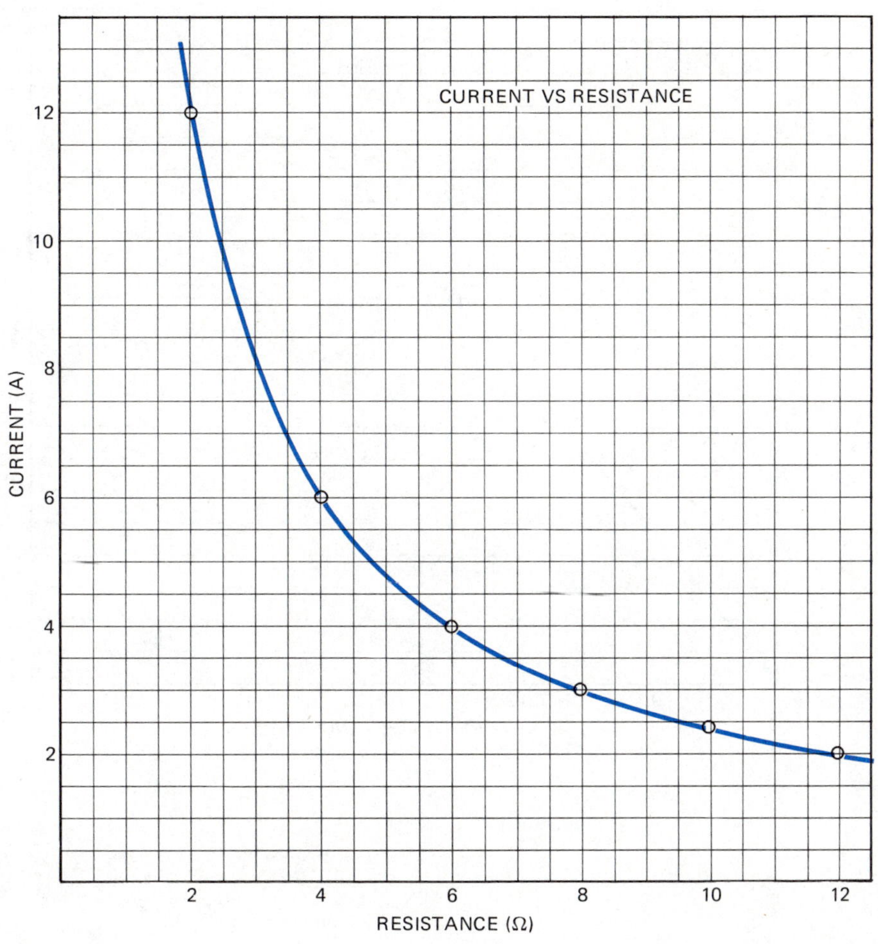

(b)

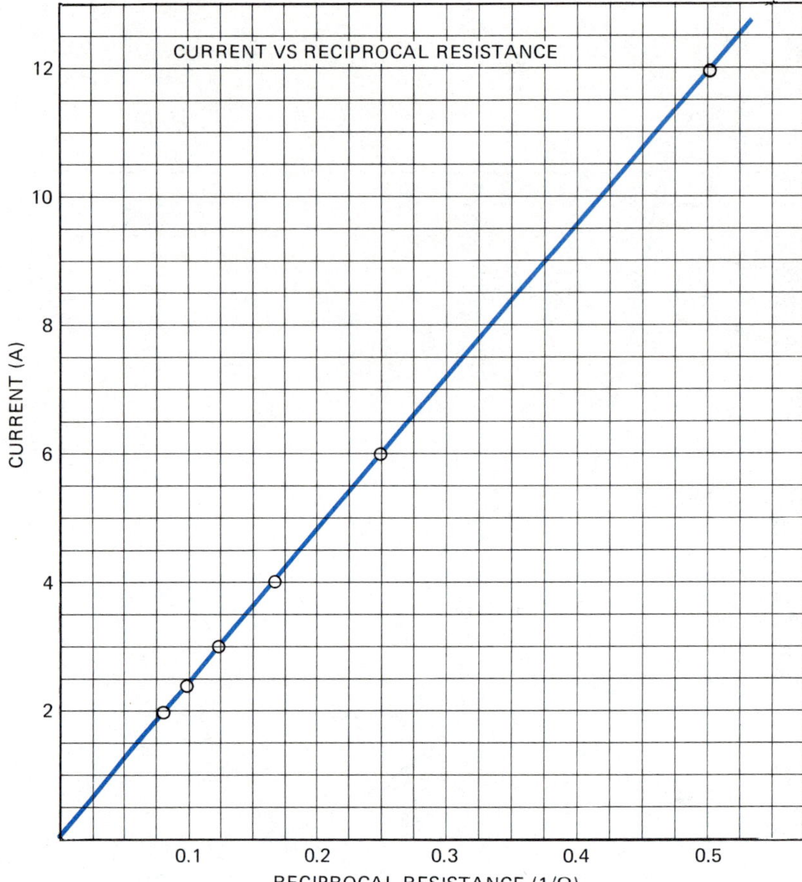

(c) $I = \frac{24}{R}$
(d) 0.5 A

3. (a) $I = \frac{5}{t}$
   (b) 28.6 ms
5. (a) $I = \frac{188}{E}$
   (b) 852 mA

EXERCISE 10-3

1. (a) $W = \frac{1}{2} CE^2$
   (b) 4.15 μJ
3. (a) $W = \frac{1}{2} LI^2$
   (b) 8.2 A
5. (a) $A = 1.39r^2$
   (b) 869 cm^2
   (c) 42.4 cm
7. (a) $E = \frac{120}{r^2}$
   (b) 1.41 m
   (c) 2.83 m
9. (a) 7.26 lx
   (b) 10.4 lx
11. 3.1 m

EXERCISE 10-4

1. 0.1; 10%
3. 1/5; 0.2
5. 3/10; 30%
7. 0.375; 37.5%
9. 1/2; 0.5
11. 5/8; 0.625
13. 0.7; 70%
15. 4/5; 0.8
17. 9/10; 90%
19. 60%
21. 5%

EXERCISE 10-5

1. 45%
3. $873.33
5. 158%
7. 74%
9. 2.5
11. 4%
13. $1716.90
15. (a) 25%
    (b) 75%
17. (a) 51%
    (b) 27%
    (c) 22%
19. 300 m
21. (a) 86.7 kg
    (b) 8.9 kg
21. (c) 5.6 kg
    (d) 111.1 kg
23. 2%

EXERCISE 10-6

1. (a) 78 kΩ
   (b) 15.6 kΩ
   (c) 85.8 kΩ
   (d) 70.2 kΩ
3. 90.0 nF ± 5%
5. Red red orange – black
7. Red white yellow – red
9. (a) 3.6 kΩ
   (b) 0.72 kΩ
   (c) 3.96 kΩ
   (d) 3.24 kΩ
11. (a) 2.45 MΩ
    (b) 0.25 MΩ
    (c) 2.57 MΩ
    (d) 2.32 MΩ

EXERCISE 10-7

1. 85%
3. 25.5 hp
5. (a) 3500 lm
   (b) 5.2 W
   (c) 5.2%
7. (a) 11.9 hp
   (b) 1.4 hp
9. (a) 1/20 hp
   (b) 80%
11. (a) 91.8 kW
    (b) 89.7%
13. (a) 1.03 hp
    (b) 0.84 hp

EXERCISE 10-8

1. (a) 77.8 MJ
   (b) 21.6 kW·h
3. (a) 384 kW·h
   (b) 140 MW·h
5. (a) 485 kW·h
   (b) 4.17×10^5 kcal
7. $5.07
9. $1872.00
11. $4.37
13. $0.75
15. $255.12
17. $133.12

CHAPTER 11

EXERCISE 11-1

1. 1; mono
3. 1; mono
5. 3; poly
7. 2; bi
9. 4; poly
11. 1
13. $4\pi/3$
15. –1
17. $2\pi^2$
19. like
21. unlike
23. like
25. like

EXERCISE 11-2

1. $7a^2$
3. P/8
5. $5\beta(\alpha + 1)$
7. $3.5(\theta - \phi)$
9. 6P + 2EI
11. $\omega_0 + 5\omega$
13. $E^2/R + 4I^2R + 5P$
15. $7(\mu + 1) + 2(\mu - 1)$
17. $6a^3 - 6a^2 + 5a$
19. $42I + 4E_3$
21. $11E^2/8 + 16E$

EXERCISE 11-3

1. $11I^2R$
3. $-37x^3y^2z$
5. $147\mu N$
7. –7R – 4Z + 13X
9. $2\mu + 15\lambda - 1$
11. $5\alpha + 2\beta$
13. 7EIt – 13Pt
15. $4\gamma - 11(90 - \theta)$
17. $40 - 8E_2 - 26I$

EXERCISE 11-4

1. (a) 3^5
   (b) 12^3
   (c) a^7
   (d) a^2b^3
1. (e) I^2R
   (f) $4^6P^3t^3$
3. $15a^2b$
5. $-8a^2b^2$
7. $36r^2s^2t^2$
9. $-24a^5b^3$
11. $6\lambda^8$
13. $-100x^9y^6z^6$
15. $k^3P^6Q^8$

EXERCISE 11-5

1. 25 squares
3. 36 squares
5. $36b + 63$
7. $2\theta^3 + 3\theta^2$
9. $3\theta^2 + 2\theta\phi$
11. $2\pi fR - 10\pi fr$
13. $5IRt + 5irt + 5et - 5Et$
15. $\frac{1}{4}a^2b^2 - \frac{1}{3}a^3b^2 + \frac{1}{5}ab^3$
17. $3ab + ac$
19. $11r^2t - 4rt^2$

EXERCISE 11-6

1. 10^3
3. r^2
5. IR
7. 5R
9. $4a^4$
11. $-2x^3y^{-1}$
13. $1.5I^2R^2t^6$
15. $3\omega f^{-4}\pi^4R^{-2}$
17. $2^3r^8s^{-5}$
19. $2^{-2}\pi k^{-4}m^{-1}$

EXERCISE 11-7

1. $2R + 3$
3. $8\theta + 5$
5. $4\alpha^2 - 3\alpha\beta + 2\beta^2$
7. $3R^2 - 40IR + 5I^2$
9. $\lambda/\pi - 1 + \pi/\lambda$
11. $-Z^{-5}X_C{}^{-3}X_L + 2Z^{-1}X_C{}^{-4}X_L{}^2 - 3X_C{}^3$
13. $P_T = 12I^2 + 36I$
    $E_T = 12I + 36$

EXERCISE 11-8

1. a^2b^2
3. $-\alpha^4\beta^2\gamma^2$
5. $4\pi^2f^2X_C{}^2$
7. $5IP^2$
9. $1/27 i^3R^3$
11. $13\pi f^2C^4$
13. $-27 \times 10^{12}I^6R^3$
15. $-4ab^{-2}c^4$

EXERCISE 11-9

1. $7(R + 3)$
3. $9(1 - 3\mu^2)$
5. $\pi(d + 2r)$
7. $\theta\phi(\theta^2 - \phi^2)$
9. $\frac{1}{2}h(B_1 + B_2)$
11. $7m^3n^3(5n^2 + 10mn - 4m^2)$
13. $4\alpha\beta(\alpha + 2\gamma - 5\beta)$
15. $7ab(6a^2 + 4b^2 - 2ab - 5a^4)$
17. $R - 3$
19. $2t - 5Q^{-1}t^2 + 3Q$
21. $\frac{1}{2}\omega t + \frac{3}{2}t^2 - \omega^2$

CHAPTER 12

EXERCISE 12-1

1. $d = C/\pi$
3. $E = P/I$
5. $W = Pt$
7. $f = \omega/2\pi$
9. $L = RA/\rho$
11. $f = X_L/2\pi L$
13. $f = B^2AL/16\pi\mu$
15. 4
17. 22
19. $\phi = 90 - \theta$
21. 57
23. $r = (E - e)/I$
25. $s = (v^2 - v_0{}^2)/2a$
27. $E_2 = (P - E_1I - E_3I)/I$
29. 11
31. 13
33. 11
35. $I = \sqrt{2W/L}$
37. $R = \sqrt{Z^2 - X^2}$
39. $P = E^2/R$
41. $t = W/RI^2$

EXERCISE 12-2

1. 3
3. −3
5. 4
7. 5
9. 8
11. $h = \dfrac{2A}{B_1 + B_2}$
13. $I = \dfrac{P}{E_1 + E_2}$
15. $I_m = \dfrac{R_sI_T}{R_s + R_m}$
    $R_s = \dfrac{I_mR_m}{I_T - I_m}$
17. $C_1 = \dfrac{C_TC_2}{C_2 - C_T}$
19. $R_2 = \dfrac{R_1R_T}{R_1 - R_T}$

EXERCISE 12-3

1. 1
3. 3
5. 4
7. 0
9. −5
11. $R = \frac{E}{I} - r$
13. $F = \frac{9}{5}C + 32$
15. $R_m = \frac{E}{I} - R_s$
17. $r = \frac{A - P}{Pt}$
19. $T_1 = T_2 - \frac{\Delta L}{\alpha L}$
21. $\frac{\beta}{\beta} + \frac{1}{\beta}$ (intermediate step)
23. $\frac{1}{r} - \frac{r}{r}$ (intermediate step)

EXERCISE 12-4

1. −8
3. 9/4
5. 8
7. 4
9. $t = W/P$
11. $\beta = \frac{\gamma}{\alpha\gamma - 1}$
13. $r = \frac{1}{P - 1}$
15. $R_1 = \frac{R_2 R_T}{R_2 - R_T}$
17. $Z_2 = \frac{Z_1 Z_T}{Z_1 - Z_T}$
19. $R = \frac{I_1 Z}{I_2 - I_1}$

EXERCISE 12-5

1. $x = 2$, $y = 5$
3. R = 14, P = 5
5. s = 5, t = 2
7. $\alpha = 2$, $\beta = 1$

EXERCISE 12-6

1. $x = 7$, $y = 5$
3. $E_1 = -3$, $E_2 = 2$
5. $\theta = 12$, $\phi = -7$
7. $\lambda = 3$, $\mu = -2$
9. $\alpha = 3$, $\beta = -4$

EXERCISE 12-7

1. $x = 5$, $y = -1$
3. $\beta = 1$, $\alpha = -2$
5. i = 2, I = 5
7. inconsistent
9. $\mu = 4$, $\omega = 4$

EXERCISE 12-8

1. $x = 3$, $y = -1$
3. i = −2, I = 3
5. $\theta = 1$, $\phi = 3$
7. $\beta = 3$, $\alpha = 0$
9. r = 2, d = 3

EXERCISE 12-9

1. S = $75 470, L = $9330
3. $I_1 = 1.04$ A, $I_2 = 0.70$ A
5. $I_1 = 15.05$ mA, $I_2 = 7.45$ mA
7. $A = 60°$, $B = 30°$
9. w = 4, L = 10
11. 32 tens, 37 twenty-fives
13. $E_1 = 1.5$ V, $E_2 = 9$ V
15. $P_1 = 40$ W, $P_2 = 75$ W

EXERCISE 12-10

1. $E = \sqrt{PR}$
3. $W = \frac{1}{2}CE^2$
5. $A = 0.7854d^2$
7. $\lambda = c/f$
9. $W = I^2 Rt$

CHAPTER 13

EXERCISE 13-1

1. $x = -1$
   $y = 2$
   $z = 2$

3. $I_1 = -2$
   $I_2 = 2$
   $I_3 = 4$

5. $\alpha = -1$
   $\beta = -2$
   $\gamma = 3$

EXERCISE 13-2

1. (a) $I_1 + I_2 + I_4 = I_3$
   (b) $I_1 + I_4 = I_2 + I_3 + I_5$

3. $14I_2 - 9 + 8I_1 - 12 + 6I_1 = 0$

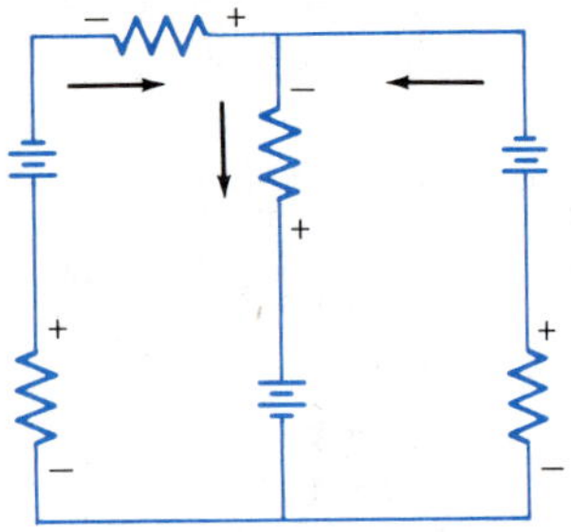

5. $\downarrow 1$, A $\cdot \rightarrow 2$, $\downarrow 3$
   $I_2 = 3$ A
   $I_3 = 6$ A
   $I_1 = 9$ A

7. $\uparrow 2$, $1 \rightarrow \cdot$ A, $\downarrow 3$
   $I_2 = 2$ A
   $I_3 = 0.5$ A
   $I_1 = 2.5$ A

9. A, $1 \leftarrow \cdot \leftarrow 3$, $\downarrow 2$
   $I_2 = 2$ A
   $I_1 = 2$ A
   $I_3 = 4$ A

EXERCISE 13-3

1. $E_1 = 3$ V
   $E_2 = 9$ V

3. $E_1 = 16.8$ V
   $E_2 = 31.2$ V

5. $I_1 = 105$ mA
   $I_2 = 35$ mA

7. $I_1 = 1.47$ A
   $I_2 = 1.03$ A

9. $E_1 = 13.6$ V
   $E_2 = 20.4$ V
   $E_3 = 30.0$ V

EXERCISE 13-4

1. $I_2 = 3$ A
3. $I_1 = 2.5$ A
5. $I_1 = 2$ A
   $I_2 = 2$ A
   $I_3 = 4$ A

EXERCISE 13-5

1. (a) $R_{TH} = 1.25$ kΩ
   (b) $E_{TH} = 250$ V
   (c) 125 mA
   (d) 71 mA
   (e) 12 W
   (f) 11 W

3. (a) 3 A
   (b) 36 W

5. (a) 20 mA
   (b) 800 mW

7. (a) 0.5 A
   (b) 8 V

CHAPTER 14

EXERCISE 14-1

1. 8/3
3. 4/3
5. 15/1
7. 5:3
9. 3:4
11. 2:1
13. 27/32
15. 8/1

EXERCISE 14-2

1. (a) 0.375
   (b) 8/3
3. (a) 4.00
   (b) 1/4
5. (a) 12.5
   (b) 2/25
7. 87.5%
9. 6.25%
11. $40\bar{0}\%$
13. 3/4
15. 3/16
17. 9/20
19. 0.12%
21. 19%
23. 60% tin
    40% lead
25. (a) 51%
    (b) 27%
    (c) 22%

EXERCISE 14-3

1. 65%
3. 25.0 hp
5. 75 lm/W

EXERCISE 14-4

1. 121.5
3. 833 N
5. (a) 1.5:1
   (b) 0.56:1
7. (a) 0.82:1
   (b) 3430 r/min
9. 362 r/min

EXERCISE 14-5

1. 2
3. 84
5. 97.2
7. 2
9. 960

EXERCISE 14-6

1. 11.25 km
3. 55 teeth
5. 4375 kW•h
7. 3000 C

EXERCISE 14-7

1. 1700 turns
3. (a) 220 V
   (b) 455 mA
5. 16 V
7. 2.5 A

EXERCISE 14-8

1. 173 Ω
3. 10.5 V
5. 338 Ω

EXERCISE 14-9

1. 66 600 Ω
3. 55 μΩ
5. (a) 18 800 Ω
   (b) 188 000 Ω
7. 0.064 Ω
9. (a) Circuit current surges through and damages meter coil
   (b) no

EXERCISE 14-10

1. 0.0793 Ω
3. 3.052 Ω
5. $11\bar{0}$ Ω

EXERCISE 14-11

1. 8.75 cm; 4.08 cm
3. 17.6 m
5. 13.0 m
7. a = 260 cm; c = 520 cm

CHAPTER 15

EXERCISE 15-1

1. d
3. e
5. b
7. 45°
9. 25°
11. 17.32°
13. 148°
15. 80°
17. 176°
19. −85°
21. −135°
23. 3
25. 0.25

EXERCISE 15-2

1.

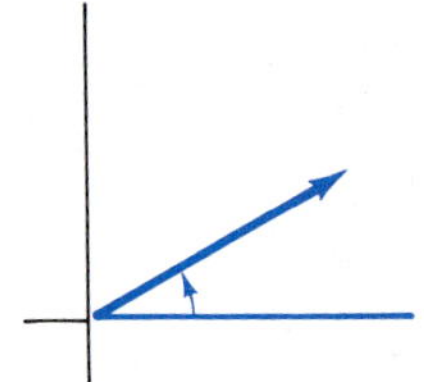

7.

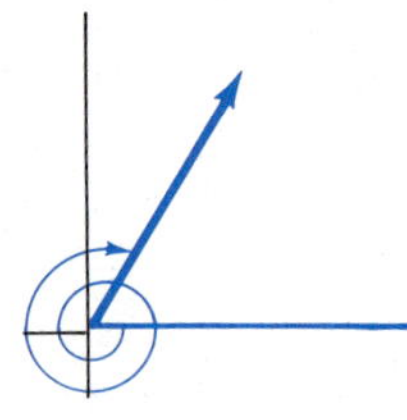

3.

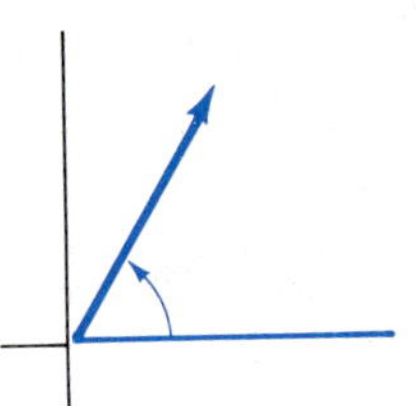

9.

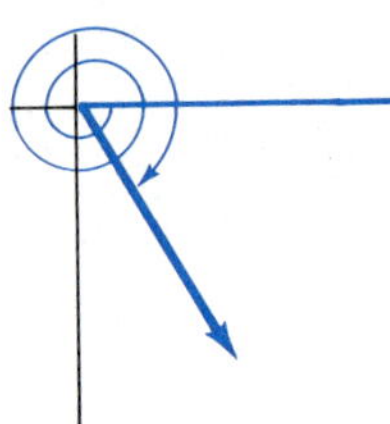

5.

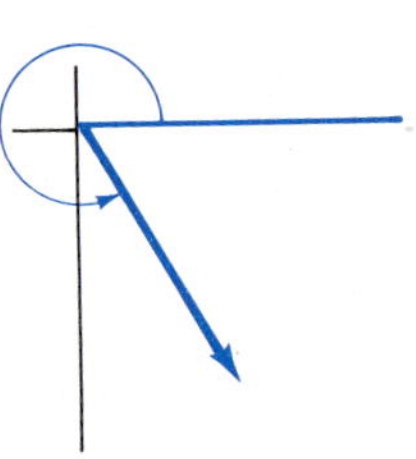

11. IV
13. I
15. I
17. 330°; IV
19. 35°; I
21. 210°; III
23. 325°; IV
25. 0.25 r; 90°; 1 RA
27. 0.708 r; 255°; 2.83 RA
29. 3.58 r; 1290°; 14.3 RA
31. 12.4 r; 4470°; 49.7 RA
33. 3.5 r; 1260°; 14.0 RA
35. 10.5 r; 3780°; 42 RA

EXERCISE 15-3

1. 0.611 rad; 0.194 π•rad
3. 2.48 rad; 0.789 π•rad
5. 3.39 rad; 1.08 π•rad
7. 1.00 rad; 0.318 π•rad
9. 180°; 1 π•rad
11. 3600°; 20 π•rad
13. −57.3°; −0.318 π•rad
15. −90.0°; −0.5 π•rad
17. 180°; 3.14 rad
19. −30.1°; −0.525 rad
21. 3600°; 20π rad = 62.8 rad
23. 1.00°; 0.0175 rad
25. 0.743°
27. 3.12°
29. 88.43°

	Degree	Radian	π-radians	Revolutions
31.	–	0	0	0
33.	45	0.785	–	0.125
35.	–	1.571	0.500	0.250
37.	360	6.283	–	1.000

EXERCISE 15-4

1. 45°; obtuse
3. 90°; right
5. 45°; scalene
7. 1.57 rad; right
9. 42°; isosceles
11. $\frac{1}{2}$ π•rad; right
13. 74.7°
15. 0.57 rad
17. $\frac{5}{12}$ π•rad
19. 0.785 rad
21. 35°
23. 68°

EXERCISE 15-5

1. 19 cm; 5 cm
3. 137 in; 21 in
5. 138.8 cm; 0 cm
7. 5 in
9. 17.0 cm
11. 44.2 m
13. 12.3 mi
15. 22.8 m
17. (a) 8.15′
    (b) 29.75′
19. 10.1 mi
21. 18.9 m

EXERCISE 15-6

1. (a) Z
   (b) R
   (c) X
3. (a) 0.6
   (b) 0.8
   (c) 0.75
5. (a) 0.8
   (b) 0.6
   (c) 1.33
7. (a) 0.707
   (b) 0.707
   (c) 1.00
9. (a) 0.768
   (b) 0.640
   (c) 1.2
11. $x = 43.3$ cm
    $y = 25$ cm

CHAPTER 16

EXERCISE 16-1

1. 3rd side = 5
   sin $\theta = 0.600$
   cos $\theta = 0.800$
   tan $\theta = 0.750$
3. 3rd side = 12
   sin $\theta = 0.800$
   cos $\theta = 0.600$
   tan $\theta = 1.333$
5. 3rd side = 33
   sin $\theta = 0.8615$
   cos $\theta = 0.5077$
   tan $\theta = 1.697$
7. 3rd side = 17
   sin $\theta = 0.4706$
   cos $\theta = 0.8824$
   tan $\theta = 0.5333$
9. 3rd side = 30
   sin $\theta = 0.8824$
   cos $\theta = 0.4706$
   tan $\theta = 1.875$
11. 3rd side = 21
    sin $\theta = 0.6897$
    cos $\theta = 0.7241$
    tan $\theta = 0.9524$

EXERCISE 16-2

1. 0.1392 (sin)
   0.9903 (cos)
   0.1404 (tan)
3. 0.9272 (sin)
   0.3746 (cos)
   2.475 (tan)
5. 0.2474 (sin)
   0.9689 (cos)
   0.2553 (tan)
7. 0.3269 (sin)
   0.9451 (cos)
   0.3459 (tan)
9. 0.5180 (sin)
   0.8554 (cos)
   0.6056 (tan)
11. 0.8368 (sin)
    0.5476 (cos)
    1.5282 (tan)
13. 0.8660 (sin)
    0.5000 (cos)
    1.7321 (tan)
15. 0.3827 (sin)
    0.9239 (cos)
    0.4142 (tan)
17. 0.9239 (sin)
    0.3827 (cos)
    2.4142 (tan)
19. 0.7071 (sin)
    0.7071 (cos)
    1.0000 (tan)

EXERCISE 16-3

1. OPP = 10.6 cm
   ADJ = 27.3 cm
3. OPP = 15.1 in
   HYP = 16.7 in
5. HYP = 416 m
   ADJ = 215 m
7. OPP = 52.6 ft
   ADJ = 4.97 ft
9. HYP = 6.8 cm
   OPP = 6.3 cm

EXERCISE 16-4

The sides are listed alphabetically:

1. 100 m; 109 m; 66.3°
3. 7.7 ft; 11.7 ft; 48.8°
5. 20.3 km; 61.0 km; 71.6°
7. 35.7 m; 48.3 m; 42.4°
9. 82.1 mi; 45.9 mi; 60.8°
11. 9.10×10^4 m; 10.8×10^4 m; 0.57 rad
13. r = 14.8 m
15. 26.0 km east; 20.1 km south
17. 16.6 km of cable
19. (a) 101 mi
    (b) 110 mi
21. 24.9 ft
23. 111 m
25. 83.3 m

EXERCISE 16-5

1. 13.6°
3. 21.1°
5. 58.5°
7. 42.7°
9. 83.3°
11. 10.5°
13. 89.4°
15. 0.7227 rad
17. 0.5300 rad
19. 0.5970 rad
21. 1.3845 rad
23. 0.9030 rad
25. 1.2913 rad
27. 53.1°
29. 0.7141 rad
31. 64.1°
33. 0.4521
35. 4.535
37. 45°

EXERCISE 16-6

1. $\theta = 61.9$; c = 17
3. $\theta = 0.8098$ rad; a = 21
5. $\theta = 28.1^\circ$; b = 30
7. $\theta = 1.0142$ rad; a = 45
9. $\theta = 36.9^\circ$; 20
11. 53.5°; 20.5 m
13. (a) 39.6°
    (b) 1.94 m
15. (a) 16.0°
    (b) 31.4 m
    (c) 33.0 m
17. (a) 48.4°
    (b) 28.2 mi
19. (a) 29.0°
    (b) 15.5 ft
21. 2250 m^2

EXERCISE 16-7

1. (a) 20 m
   (b) 0.800 (sin)
   0.600 (cos)
   1.33 (tan)
3. (a) 32.5 in
   (b) 0.707 (sin)
   −0.707 (cos)
   −1.00 (tan)
5. (a) 50.0 km
   (b) −0.454 (sin)
   0.891 (cos)
   −0.510 (tan)
7. (a) -285°
   (b) $-1.5\ \pi \cdot$rad
   (c) 210°
   (d) 4.71 rad
   (e) -120°
   (f) 315°
   (g) $-1.25\ \pi \cdot$rad
   (h) -40°
9. (a) II
   (b) IV
   (c) III
   (d) I

EXERCISE 16-8

1. 60.0°
3. 60.0°
5. 36.9°
7. -34.5°
9. 102.8°
11. -82.3°
13. -5.3°
15. 154.8°
17. -89.7°

CHAPTER 17

EXERCISE 17-1

1.

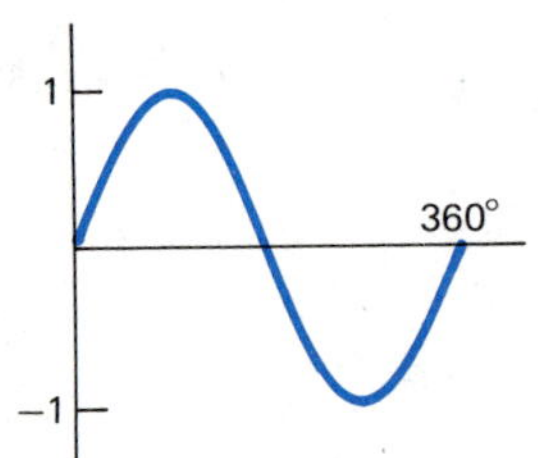

3.

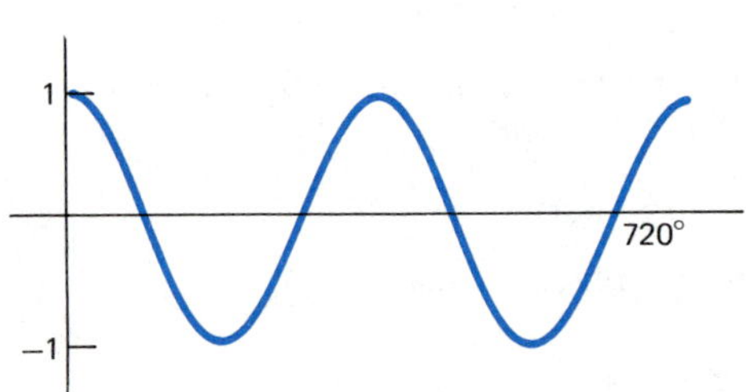

5.

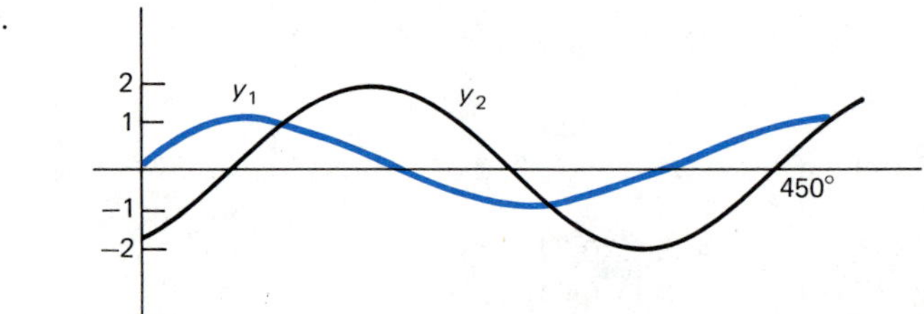

7.

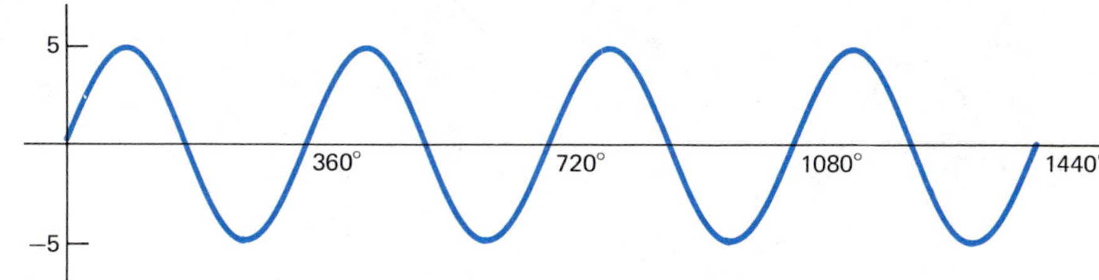

EXERCISE 17-2

1. leftward (←)
3. out of paper (•)
5. upward along wire (↖)
7. leftward (←)
9. zero
11. upward (↑)
13. into paper (×)
15. leftward (←)

EXERCISE 17-3

1. (a) 97.5 V
   (b) 167 V
   (c) 44 V
   (d) −120 V
   (e) −97.5 V
3. 85 V; decreasing
5. 326 mA
7. −4.92 A; decreasing
9. (a) $e = 2.00 \sin \theta$
   (b) 1.97 V
   (c) increasing

EXERCISE 17-4

1. (a) 50 ms
   (b) 3600 s
   (c) 18.2 ms
   (d) 33.3 ms
1. (e) 2.5 ms
3. (a) 60 Hz
   (b) 16.7 ms
5. 20 poles
7. (a) 4770 r/min
   (b) 3000 r/min
   (c) 42 000 r/min
   (d) 6000 r/min
9. 0.141 rad
11. (a) 50
    (b) 10
    (c) 24 000

EXERCISE 17-5

1. (a) 3770 rad/s
   (b) 5030 rad/s
   (c) 25 rad/s
3. 167 V; 377 rad/s; 60 Hz; 16.7 ms
5. 673 V; 1250π rad/s; 625 Hz; 1.6 ms
7. $e = 310 \sin 120\pi t$
9. $e = 157 \sin 200\pi t$
11.

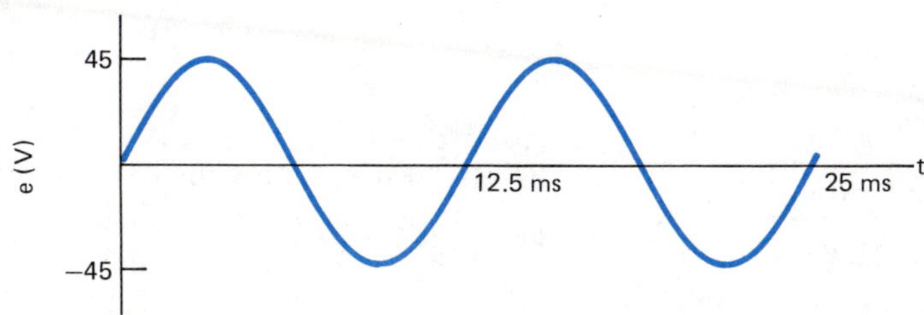

13.

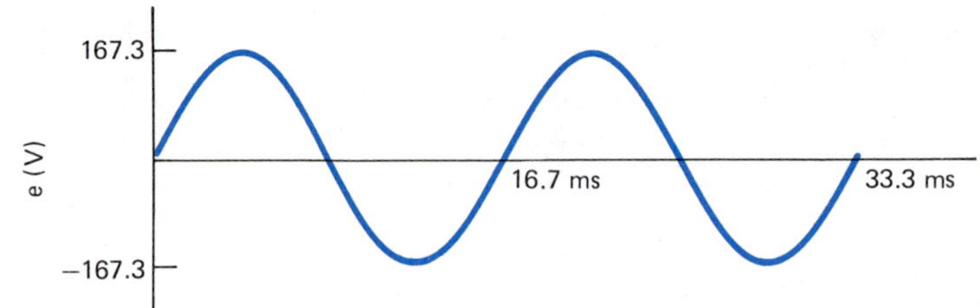

15. (a) 461 V
 (b) 395 V

EXERCISE 17-6

1. 440 V
3. 4.84 mA
5. 880 V
7. 3.75×10^5 V
9. 651 V
11. 70.7 A
13. 1560 kV
15. 25 V
17. (a) 120 V
 (b) 541 mA
 (c) 222 Ω
 (d) $i = 0.76 \sin 120\pi t$
19. (a) 2 kW
 (b) $e = 170 \sin 120\pi t$
 (c) $i = 23.6 \sin 120\pi t$
 (d) 7.2 Ω

EXERCISE 17-7

1. (a) 0.942 Ω
 (b) 13.2 Ω
 (c) 188 Ω
 (d) 942 Ω
3. (a) 6370 Hz
3. (b) 455 Hz
 (c) 31.8 Hz
 (d) 6.37 Hz
5. 94.2 Ω
7. (a) 176 Ω
7. (b) 1.25 A
9. (a) 570 mH
 (b) 215 Ω
 (c) 0.54 A
11. (a) 650 mA
11. (b) 170 Ω
 (c) 110 V
13. $e = 156 \sin 120\pi t$
15. (a) 0.58 A
 (b) 46 V; 69 V

EXERCISE 17-8

1. (a) 3.12 MΩ
 (b) 340 Ω
 (c) 22.3 Ω
 (d) 1.1 Ω
3. (a) 18.7 MHz
3. (b) 2.04 kHz
 (c) 130 Hz
 (d) 6.4 Hz
5. 13.4 Ω
7. (a) 3.28 MΩ
7. (b) 67 μA
9. (a) 65 μF
 (b) 40.8 Ω
 (c) 2.9 A
11. (a) 410 mA
11. (b) 6.24 Ω
 (c) 2.56 V
13. $e = 3.7 \sin 120\pi t$
15. (a) 0.96 A
 (b) 43.1 V; 71.9 V

CHAPTER 18

EXERCISE 18-1

1. $x^2 + 3x + 2$
3. $P + 5P - 24$
5. $3E^2 - 14E - 24$
7. $8\phi^2 - 32\phi + 14$
9. $35E^2 + 29EI + 6I^2$
11. $10 - 9E + 2E^2$
13. $2Q^2 + 10^5 Q + 1.2 \times 10^9$
15. $-18C^4 + 30C^2 - 8$
17. $4E^2 + 20EI + 25I^2$
19. $ac + 2bc - ad - 2bd$

EXERCISE 18-2

1. $x^2 + 10x + 25$
3. $25E^2 - 20E + 4$
5. $P^2 - 14P + 49$
7. $4B^2 - 1$
9. $4\theta^2 - 4\theta\pi + \pi^2$
11. $25\theta^2 + 2.5\theta + \frac{1}{16}$
13. $9L^2 - 12LW + 4W^2$
15. $P^2t^2 - W^2$
17. $10^6 + 6 \times 10^3 E + 9E^2$
19. $\frac{1}{128}E^2 + \frac{3}{8}EI + \frac{9}{2}I^2$

EXERCISE 18-3

1. $(x - 4)^2$
3. $4(R - 3)(R + 3)$
5. $2(R + 3r)^2$
7. none
9. $(2\theta + \pi)(2\theta - \pi)$
11. $9(3t + 4T)(3t - 4T)$
13. $2(2L + 3W)^2$
15. $(7E + 11IR)(7E - 11IR)$

EXERCISE 18-4

1. ± 7
3. ± 9
5. ± 6
7. ± 6
9. ± 1

EXERCISE 18-5

1. $x = \pm 6$
3. $I = -4$ or -1
5. $R = 0$ or 6
7. $t = 2$
9. $E = -4$ or $\frac{1}{2}$
11. $\alpha = 5$ or -1
13. $Q = -4$ or $\frac{1}{2}$
15. $P = 5$ or 1
17. 1 A
19. (a) 1.25 s
    (b) 1.58 s
    (c) 1.94 s
21. 8 s; 1 s
23. r = 18 cm; d = 36 cm

EXERCISE 18-6

1. j9
3. j12
5. −j10
7. j2
9. −j8.66
11. ±j11
13. ±j4
15. ±j2.83
17. 1 ± j

EXERCISE 18-7

1. −9
3. −16
5. j27
7. 16
9. −16

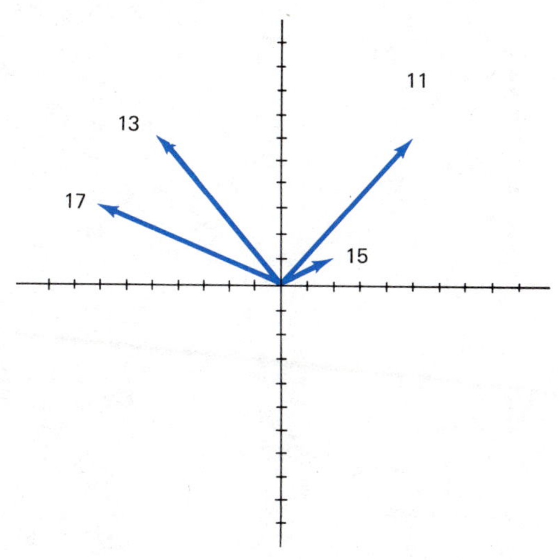

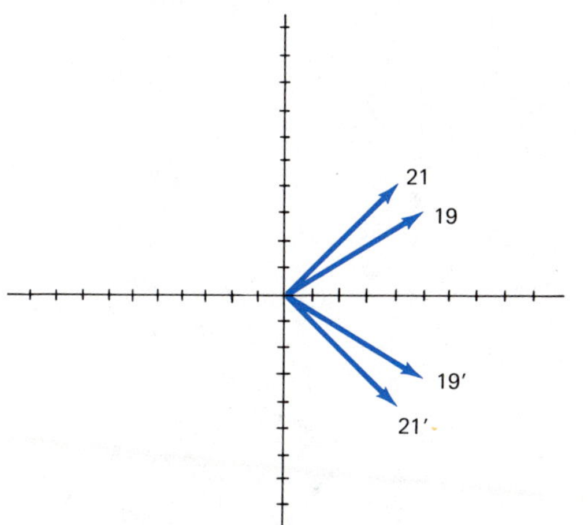

EXERCISE 18-7 (continued)

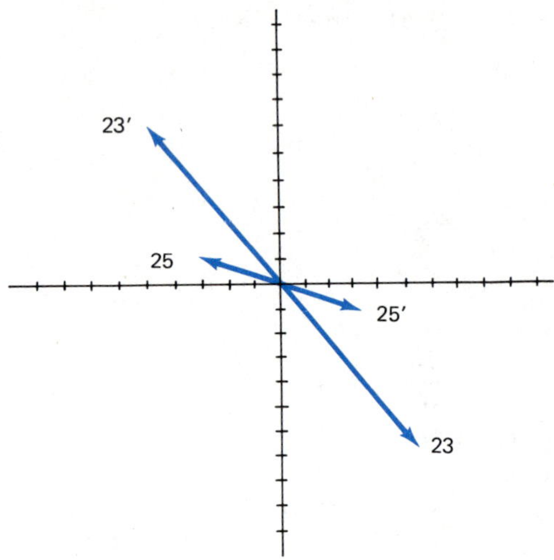

EXERCISE 18-8

1.

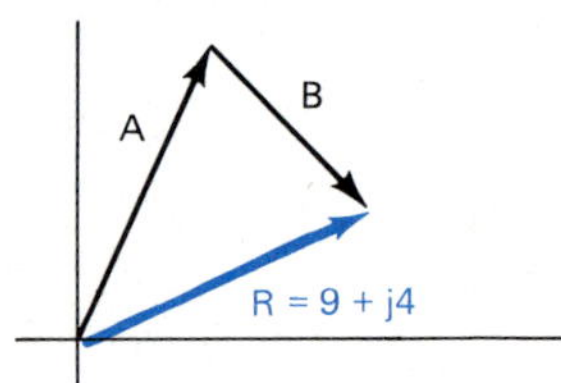

7.

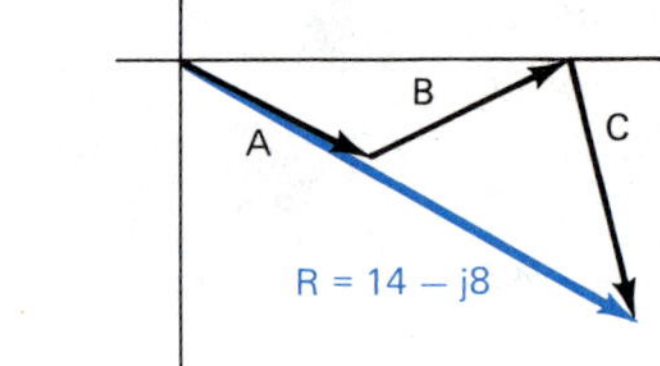

3.

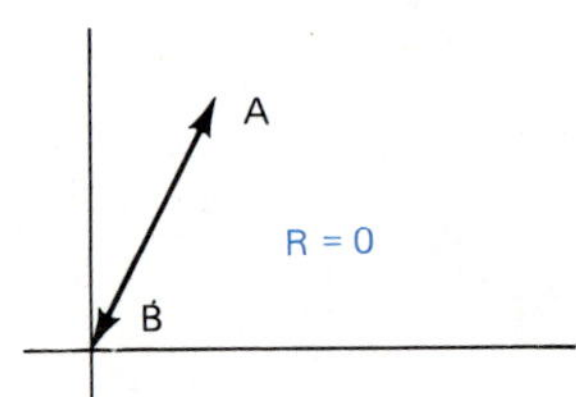

9.

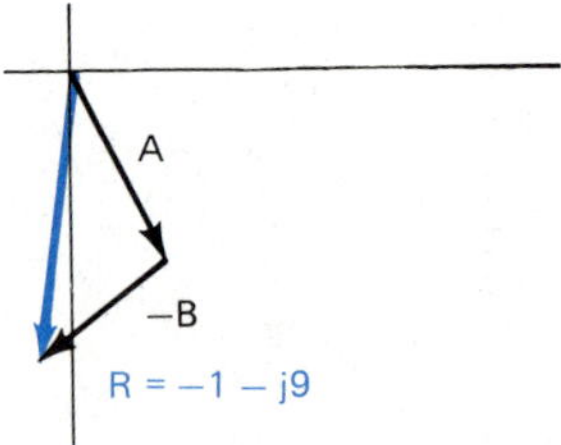

5.

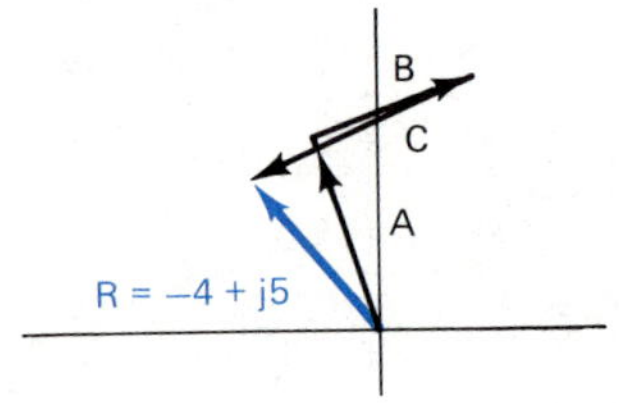

11.

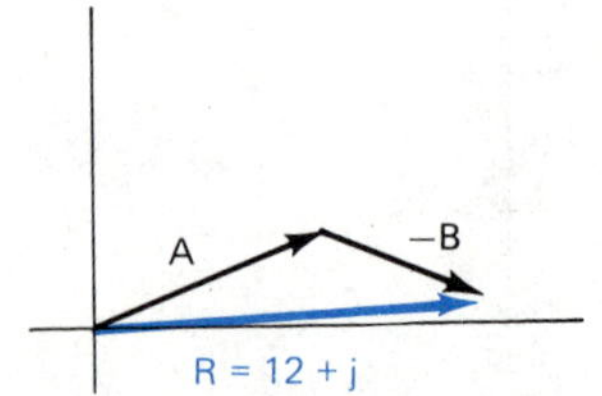

13. 5 − j4
15. 8 + j16
17. 2 − j3
19. −26.7 + j6.9
21. 88
23. j106
25. 3 + j7
27. −7 − j

EXERCISE 18-9

1. 15 + j20
3. 22 + j29
5. 7 + j22
7. −6 − j9
9. 8 + j51
11. 22 + j32
13. 0.7 − j0.4
15. −1.32 + j0.76
17. −5 − j
19. 0.12 − j0.16

CHAPTER 19

EXERCISE 19-1

1. 36.0°; lag
3. 56.3°; lead
5. 54.5°; lead
7. 58.3°; lead
9. 32.6°; lag

EXERCISE 19-2

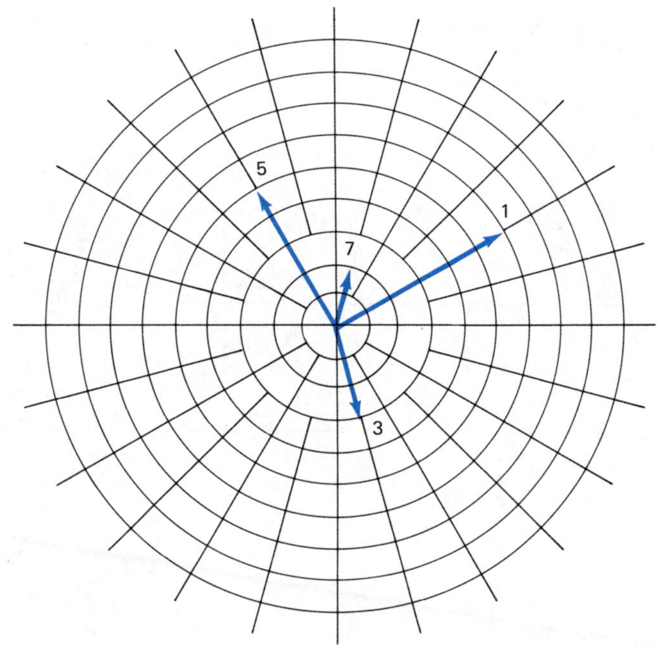

9. $6\underline{/35^\circ}$
11. $5\underline{/215^\circ}$
13. $9\underline{/-65^\circ}$
15. $7.5\underline{/75^\circ}$

EXERCISE 19-3

Note: Answers are calculated values.

1.		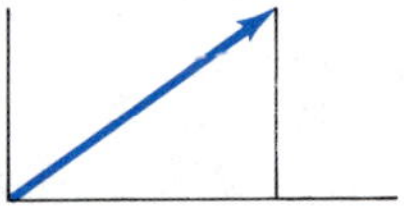		$10\angle 36.9^\circ$
3.		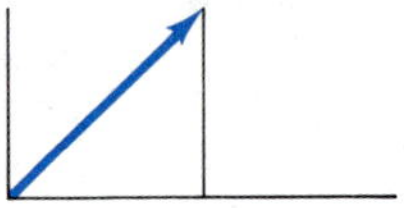	10.6 + j10.6	
5.		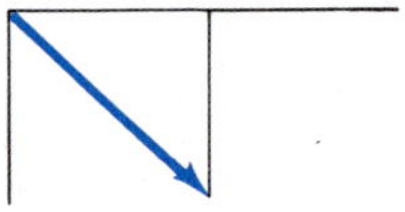		$41.9\angle -40.2^\circ$
7.		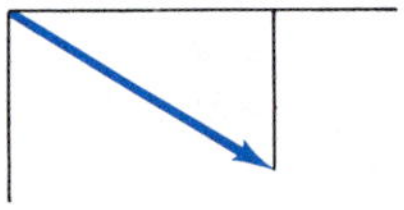	10.6 – j5.67	
9.		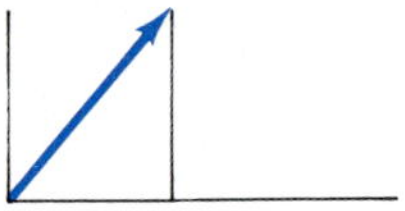		$150\angle 49.1^\circ$
11.			39.3 + j49.2	
13.		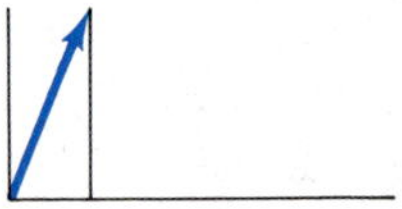	12.2 + j29.9	$32.3\angle 67.8^\circ$
15.		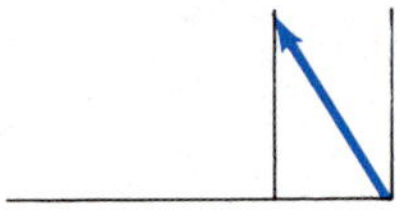		$74.6\angle 122.6^\circ$

EXERCISE 19-4

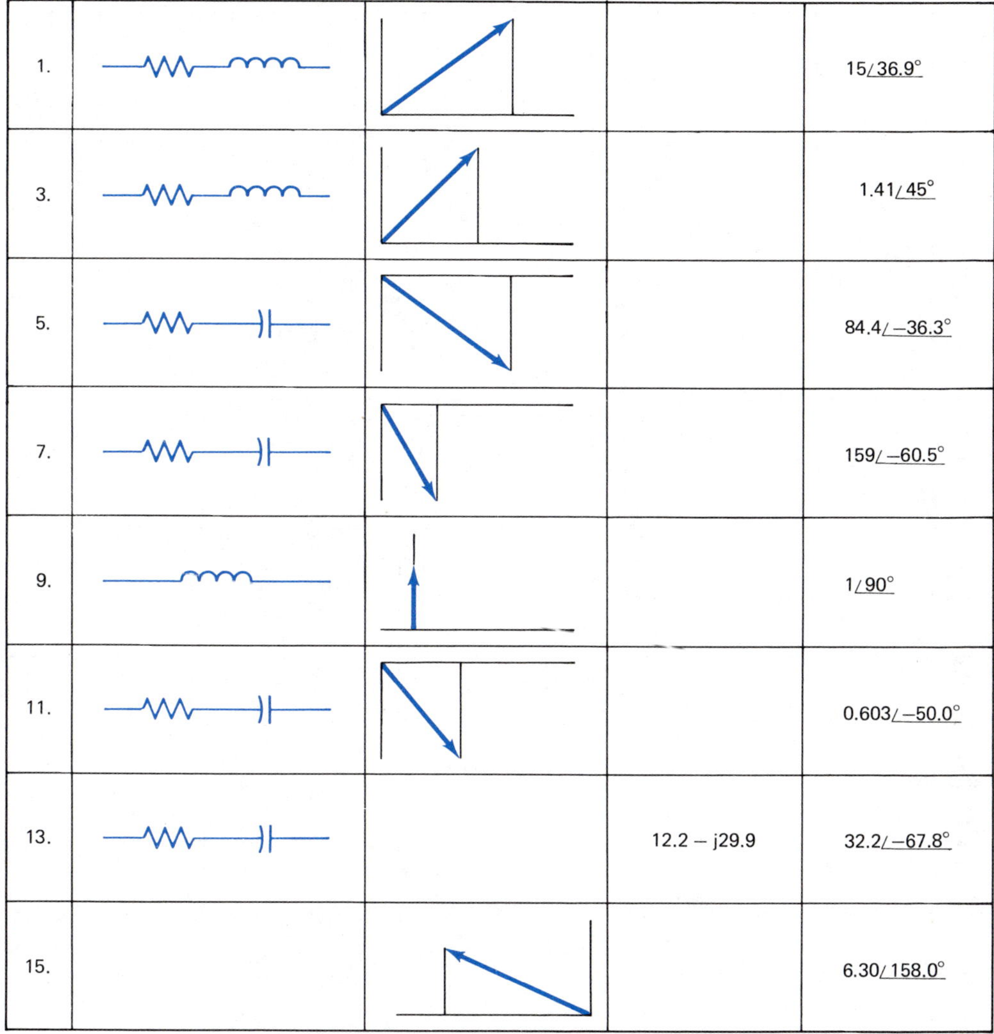

1.				$15\angle 36.9^\circ$
3.				$1.41\angle 45^\circ$
5.				$84.4\angle -36.3^\circ$
7.				$159\angle -60.5^\circ$
9.				$1\angle 90^\circ$
11.				$0.603\angle -50.0^\circ$
13.			$12.2 - j29.9$	$32.2\angle -67.8^\circ$
15.				$6.30\angle 158.0^\circ$

17. $17.9\angle 125.2^\circ$ 19. $4110\angle -56.8^\circ$

EXERCISE 19-5

1.		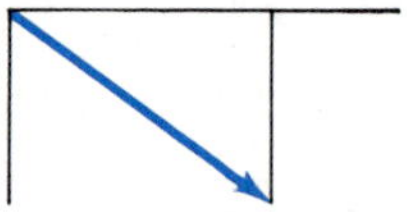	4 – j3	
3.			10.8 + j5.7	
5.		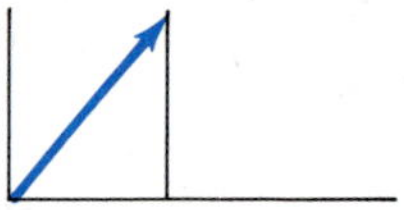	39.6 + j49.0	
7.			137 – j864	
9.		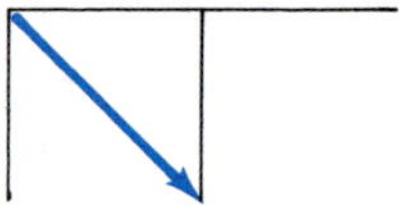	1 – j	
11.		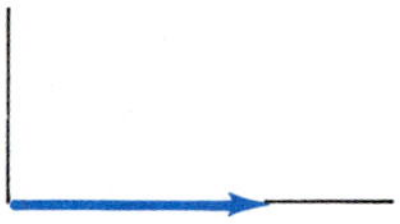	1	
13.			186 – j121	$222\angle -33.1^\circ$
15.			–5.85 + j2.36	$6.31\angle 158^\circ$

17. 9.97 + j22.7 km; –15.3 – j34.3 km

EXERCISE 19-6

1.	Lead Pf = 0.440		1.66 – j3.39	
3.	Lag Pf = 0.346			$0.0899\angle 69.7^\circ$
5.	Lag Pf = 0.943		5.04×10^3 $+ j1.76 \times 10^3$	
7.			16.7 – j24.1	$29.3\angle -55.2^\circ$
9.			4900 + j2340	$5430\angle 25.6^\circ$

EXERCISE 19-7

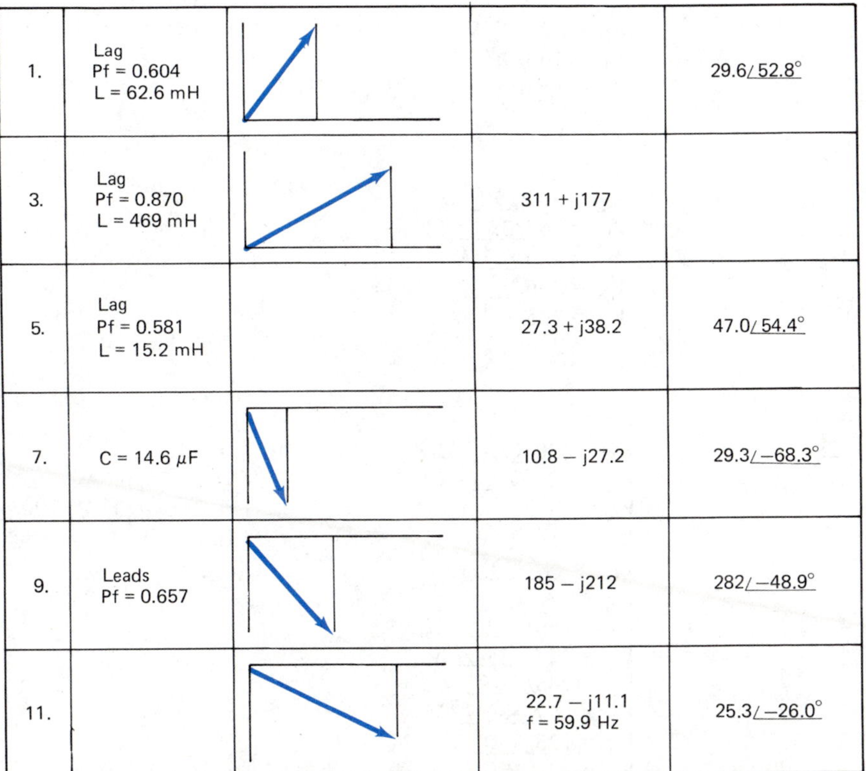

1.	Lag Pf = 0.604 L = 62.6 mH			$29.6\angle 52.8^\circ$
3.	Lag Pf = 0.870 L = 469 mH		311 + j177	
5.	Lag Pf = 0.581 L = 15.2 mH		27.3 + j38.2	$47.0\angle 54.4^\circ$
7.	C = 14.6 μF		10.8 – j27.2	$29.3\angle -68.3^\circ$
9.	Leads Pf = 0.657		185 – j212	$282\angle -48.9^\circ$
11.			22.7 – j11.1 f = 59.9 Hz	$25.3\angle -26.0^\circ$

CHAPTER 20

EXERCISE 20-1

1. $60 + j21\ \Omega \rightarrow 63.6\angle 19.3^\circ\ \Omega$
3. $110.6 + j15.7\ \Omega \rightarrow 112\angle 8.1^\circ\ \Omega$
5. $2.67 + j11.36 \rightarrow 11.7\angle 76.8^\circ$
7. $0 + j133 \rightarrow 133\angle 90^\circ$
9. $135 + j65.4 \rightarrow 150\angle 25.9^\circ$
11. $41.2 + j16.4\ \Omega \rightarrow 44.3\angle 21.7^\circ\ \Omega$
13. $245 - j753 \rightarrow 792\angle -72.0^\circ$
15. $749 - j56 \rightarrow 751\angle -4.3^\circ$

EXERCISE 20-2

1. $120\angle 40^\circ \rightarrow 91.9 + j77.1$
3. $64\angle 52^\circ \rightarrow 39.4 + j50.4$
5. $8\angle -12.5^\circ \rightarrow 7.8 - j1.7$
7. $5.84\angle 47.8^\circ \rightarrow 3.92 + j4.33$
9. $7160\angle 102^\circ \rightarrow -1510 + j7000$
11. $118\angle -7.7^\circ \rightarrow 116 - j16$
13. $353\angle 27.1^\circ \rightarrow 314 + j161$
15. $16.5\angle 40.2^\circ \rightarrow 12.6 + j10.6$
17. $1.50\angle 176.9^\circ \rightarrow -1.50 + j0.08$
19. $5.10 \times 10^{-3}\angle 35.2^\circ \rightarrow 4.17 \times 10^{-3} + j2.94 \times 10^{-3}$
21. $7.51\angle 33.9^\circ \rightarrow 6.24 + j4.19$
23. $9.29\angle 58.2^\circ \rightarrow 4.90 + j7.90$
25. $112\angle 49.0^\circ \rightarrow 73.4 + j84.4$
27. $7.24\angle -38.2^\circ \rightarrow 5.69 - j4.48$

EXERCISE 20-3

1. $4\angle 0^\circ$ A
3. $50\angle 0^\circ$ mA
5. $2000\angle 90^\circ\ \Omega$
7. $37.5\angle 0^\circ$ mA
9. $96\angle 90^\circ\ \Omega$; 255 mH

EXERCISE 20-4

Note: $Z\angle\theta \rightarrow R \pm jX$; $E\angle\theta \rightarrow E_R \pm jE_X$

1. $\mathbf{Z} = 58.1\angle 53.6^\circ\ \Omega$
   $\mathbf{I} = 4.13\angle 0^\circ$ A
   $\mathbf{E_R} = 142\angle 0^\circ$ V
   $\mathbf{E_X} = 193\angle 90^\circ$ V
   P = 588 W
3. $\mathbf{Z} = 250\angle 56.2^\circ\ \Omega$
   $\mathbf{R} = 139\angle 0^\circ\ \Omega$
   $\mathbf{X} = 208\angle 90^\circ\ \Omega$
   $\mathbf{E_R} = 13.3\angle 0^\circ$ V
   $\mathbf{E_X} = 20.0\angle 90^\circ$ V
   P = 1.3 W
5. $\mathbf{X_L} = 116\angle 90^\circ\ \Omega$
   $\mathbf{Z} = 148\angle 51.5^\circ\ \Omega$
   $\mathbf{E} = 517\angle 51.5^\circ$ V
   $\mathbf{E_R} = 322\angle 0^\circ$ V
   $\mathbf{E_X} = 406\angle 90^\circ$ V
   P = 1130 W
7. $\mathbf{X_L} = 90.5\angle 90^\circ\ \Omega$
   $\mathbf{Z} = 163\angle 33.8^\circ\ \Omega$
   $\mathbf{I} = 2.70\angle 0^\circ$ A
   $\mathbf{E_R} = 366\angle 0^\circ$ V
   $\mathbf{E_X} = 245\angle 90^\circ$ V
   P = 987 W
9. (a) $100\angle -66.7\ \Omega \rightarrow 39.5 - j91.9\ \Omega$
   (b) $39.5\angle 0^\circ\ \Omega \rightarrow 39.5 + j0\ \Omega$
   (c) $91.9\angle -90^\circ\ \Omega \rightarrow 0 - j91.9\ \Omega$
   (d) $120\angle -66.7$ V $\rightarrow 475 - j110$ V

EXERCISE 20-5

1. $\mathbf{Z_T} = 213\angle 41.2^\circ\ \Omega$
   $\mathbf{I} = 0.113\angle 0^\circ$ A
   $\mathbf{E_R} = 18.1\angle 0^\circ$ V
   $\mathbf{E_L} = 24.9\angle 90^\circ$ V
   $\mathbf{E_C} = 9.04\angle -90^\circ$ V
   P = 2.04 W
3. $\mathbf{Z_T} = 91.9\angle -22.4^\circ\ \Omega$
   $\mathbf{E} = 78.1\angle -22.4^\circ$ V
   $\mathbf{E_R} = 72.2\angle 0^\circ$ V
   $\mathbf{E_L} = 225\angle 90^\circ$ V
   $\mathbf{E_C} = 255\angle -90^\circ$ V
   P = 61.4 W
5. $\mathbf{Z_T} = 230\angle 12.4^\circ\ \Omega$
   $\mathbf{E} = 18.4\angle 12.4^\circ$ V
   $\mathbf{E_R} = 18\angle 0^\circ$ V
   $\mathbf{E_L} = 11\angle 90^\circ$ V
   $\mathbf{E_C} = 7.1\angle -90^\circ$ V
   P = 1.4 W
7. $\mathbf{Z_T} = 156\angle -11.9^\circ\ \Omega$
   $\mathbf{E} = 227\angle -11.9^\circ$ V
   $\mathbf{E_1} = 185\angle 63.9^\circ$ V
   $\mathbf{E_2} = 167\angle -32.7^\circ$ V
   $\mathbf{E_3} = 123\angle -90^\circ$ V
   P = 322 W
9. (a) $540\angle 0^\circ\ \Omega$
   (b) $0.815\angle 0^\circ$ A
   (c) $300\angle -90^\circ$ V; $300\angle 90^\circ$ V; $440\angle 0^\circ$ V
   (d) 359 W

EXERCISE 20-6

1. 320 Ω; 1.8 VA; 1.8 W; 0 vars
3. 960 V; 23 kVA; 23 kW; 0 vars
5. 84.9 mA; 1.36 VA; 0 W; 1.36 vars
7. 1950 V; 1.55 A; 0 W; 3 kvars
9. 2830 V; 17.0 kVA; 0 W; 17.0 kvars

EXERCISE 20-7

1. 1440 VA; 92.5%
3. 2.06 MVA; 94.8%
5. 3.00 kVA; 79.6%

Hint: $S\angle\theta \rightarrow P \pm jQ$

7. P = 42.4 W
   Q = 15.0 vars
   94.3% lagging
9. P = 104 kW
   Q = 165 kvars
   36.5% lagging
11. $10.9\angle 53.5^\circ$ kVA; 59.4%
13. $605\angle 45^\circ$ kVA; 70.7%
15. (a) 414 VA
    (b) 74.9%
    (c) 274 vars lagging
17. (a) 81.7%
    (b) $372\angle 35.2^\circ$ VA → 304 + j214 (W, vars)
19. (a) $6.20\angle 41.4^\circ$ kVA
    (b) 27.0 A
    (c) $8.53\angle 41.4^\circ$ Ω

EXERCISE 20-8

1. 2 A; 240 VA; 156 W; 182 vars
3. 55 Ω; 3520 VA; 1160 W; 3320 vars
5. 64 V; 256 VA; 256 W; 0 vars
7. 704 Ω; 2.5 A; 4.4 kVA; 3.81 kvars
9. (a) 60 + j86 − j94 Ω → $60.5\angle -7.6^\circ$ Ω
   (b) $3.63\angle 0^\circ$ A
   (c) $800\angle -7.6^\circ$ VA; 793 W; 106 vars
   (d) 99.1%
   (e) capacitive
   (f) leading
11. (a) 1.5 + j1.82 − j2.65 kΩ → $1.71\angle -29.0^\circ$ kΩ
    (b) $70\angle 0^\circ$ mA
    (c) $8.40\angle -29.0^\circ$ VA; 7.35 W; 4.07 vars
    (d) 87.5% leading
13. (a) 168 + j59.2 Ω → $178\angle 19.4^\circ$ Ω
    (b) $1760\angle 19.4^\circ$ V
    (c) $17.4\angle 19.4^\circ$ kVA; 16.4 kW; 5.8 kvars
    (d) 94.3% lagging
15. (a) $168\angle 0^\circ$ Ω
    (b) 10.5 A
    (c) $18.4\angle 0^\circ$ kVA; 18.4 kW; 0 kvars
    (d) 100% in phase
    (e) Inductance: −5.6%
        Current: +6.2%
        Apparent Power: +5.7%
        True Power: +12.2%
        Reactive Power: −100%
        Power Factor: +6.0%

CHAPTER 21

EXERCISE 21-1

1. 17.3 + j0 Ω
3. 0 − j37.2 Ω
5. 0 + j71.5 Ω
7. 0 + j53.3 Ω
9. 0 + j200 Ω
11. 0 + j1680 Ω
13. 0 − j80 Ω
15. 0 + j580 Ω
17. 74.2 − j86.6 Ω
19. 5.84 + j7.89 Ω
21. 89.9 − j21.8 Ω

EXERCISE 21-2

1. 1200 + j86 Ω
3. 29.5 − j7.02 Ω
5. 57 + j104 Ω
7. 30.2 − j2.4 Ω
9. 20.3 + j20.1 Ω

EXERCISE 21-3

1. $1200\angle -4^\circ$ Ω → 1200 − j84 Ω
3. $68.6\angle -35.3^\circ$ kΩ → 56 − j39.6 kΩ
5. $5.25\angle 10.5^\circ$ Ω → 5.16 − j0.96 Ω
7. $196\angle 74.4^\circ$ Ω → 52.6 + j189 Ω
9. $315\angle -49.4^\circ$ Ω → 205 − j239 Ω
11. $18.7\angle 49.2^\circ$ Ω → 12.2 + j14.1 Ω

EXERCISE 21-4

1. (a) 0.5 A; 0.5 A
   (b) 1 A
1. (c) 120 Ω
   (d) 120 VA; 120 W; 0 vars
1. (e) 100%
3. (a) −j571 mA; −j353 mA

3. (b) −j924 mA
   (c) j26 Ω
   (d) 22.2 VA; 0 W; j22.2 vars
   (e) 0%
5. (a) j5.57 A; j3.52 A
   (b) j9.09 A
   (c) −j24.2 Ω
   (d) 2 kVA; 0 W; −j2 kvars
   (e) 0%
7. (a) −j53.3 mA; 120 mA
   (b) $131\angle{-24^\circ}$ mA
   (c) $91.6\angle{24^\circ}$ Ω
   (d) $1.57\angle{24^\circ}$ VA; 1.44 W; j0.64 vars
   (e) 91.3%
9. (a) 35.3 mA; j8.37 mA
   (b) $36.3\angle{13.3^\circ}$ mA
   (c) $3310\angle{-13.3^\circ}$ Ω
   (d) 4.35 VA; 4.24 W; −j1.00 vars
   (e) 97.3%

EXERCISE 21-5

1. (a) $40.9\angle{-45.1^\circ}$ A
   (b) $5.37\angle{45.1^\circ}$ Ω
   (c) $9001\angle{45.1^\circ}$ VA; 6360 W; 6380 vars
   (d) 70.6%
3. (a) $27.6\angle{-46.4^\circ}$ A
   (b) $4.35\angle{46.4^\circ}$ Ω
   (c) $3310\angle{46.4^\circ}$ VA; 2280 W; 2400 vars
   (d) 68.9%
5. (a) $29\angle{-5.8^\circ}$ A
   (b) $7.58\angle{5.8^\circ}$ Ω
   (c) $6.38\angle{5.8^\circ}$ kVA; 6.36 kW; 0.649 kvars
   (d) 99.5%
7. (a) $28.8\angle{-44^\circ}$ A
   (b) $7.65\angle{44^\circ}$ Ω
   (c) $1.88\angle{44^\circ}$ kVa; 1.35 kW; 1.31 kvars
   (d) 71.9%
9. (a) $18.2\angle{-57.4^\circ}$ A
   (b) $26.3\angle{57.4^\circ}$ Ω
   (c) $8.75\angle{57.4^\circ}$ kVA; 4.72 kW; 7.36 kvars
   (d) 53.9%
11. (a) $65.4\angle{-3.3^\circ}$ A
    (b) $3.36\angle{3.3^\circ}$ Ω
    (c) $14.4\angle{3.3^\circ}$ kVA; 14.4 kW; 0.819 kvars
    (d) 99.8%

EXERCISE 21-6

1. 4.31 − j24.5 ← $24.9\angle{-80^\circ}$ Ω; $40.1\angle{80^\circ}$ ← 6.94 + j39.5 mS
3. 3.06 + j4.77 ← $5.67\angle{57.3^\circ}$ Ω; $176\angle{-57.3^\circ}$ ← 95.2 − j148 mS
5. 155 − j138 ← $208\angle{-41.7^\circ}$ Ω; $4.82\angle{41.7^\circ}$ ← 3.60 + j3.21 mS
7. 611 + j494 ← $786\angle{38.9^\circ}$ kΩ; $1270\angle{-38.9^\circ}$ ← 990 − j800 μS
9. 28.9 − j36.2 ← $46.3\angle{-51.4^\circ}$ Ω; $21.6\angle{51.4^\circ}$ ← 13.5 + j16.9 mS

EXERCISE 21-7

1. $43.1\angle{35.5^\circ}$ Ω; $23.2\angle{-35.5^\circ}$ → 18.9 − j13.5 mS; R_p = 52.9 Ω; X_p = 74.2 Ω (lag)
3. $650\angle{21.6^\circ}$ Ω; $1.54\angle{-21.6^\circ}$ → 1.43 − j0.566 mS; R_p = 699 Ω; X_p = 1765 Ω (lag)
5. $67.4\angle{14.9^\circ}$ Ω; $14.8\angle{-14.9^\circ}$ → 14.3 − j3.81 mS; R_p = 69.7 Ω; X_p = 262 Ω (lag)
7. $169\angle{-73.3^\circ}$ Ω; $5.91\angle{73.3^\circ}$ → 1.70 + j5.66 mS; R_p = 588 Ω; X_p = 177 Ω (lead)
9. $12.5\angle{67.6^\circ}$ kΩ; $797\angle{-67.6^\circ}$ → 304 − j737 μS; R_p = 3.29 kΩ; X_p = 1.36 kΩ (lag)
11. 155 − j138 Ω ← $208\angle{-41.7^\circ}$ Ω; $4.82\angle{41.7^\circ}$ ← 3.60 + j3.21 mS
13. $1340\angle{46.8^\circ}$ Ω; $774\angle{-46.8^\circ}$ → 509 − j542 μS; R_p = 1960 Ω; X_p = 1840 Ω (lag)
15. 412 − j827 Ω; $1.08\angle{63.5^\circ}$ → 0.483 + j0.969 mS; R_p = 2070 Ω; X_p = 1030 Ω (lead)
17. 26.3 − j32.8 ← $42.0\angle{-51.3^\circ}$ kΩ; 14.9 + j18.6 μS; R_p = 67.2 kΩ; X_p = 53.8 kΩ (lead)
19. 8.28 + j1.01 ← $8.34\angle{6.9^\circ}$ Ω; $120\angle{-6.9^\circ}$ mS; R_p = 8.40 Ω; X_p = 69.0 Ω (lag)

EXERCISE 21-8

1. f_r = 73.4 Hz
3. (a) 92.2 Ω
   (b) 28.8 μF
5. 61.2 Ω
7. 2.33 Ω
9. 18.4 μF
11. 2.21 kVA; 1.80 kW; 1.28 kvars; 11.3 Ω
13. 6.25 kVA; 3.74 kvars; 80.1%; 12.9 Ω
15. (a) 6.79 kvars
    (b) 33.3%
    (c) 8.49 Ω
    (d) 313 μF
17. (a) 28.5 kVA; 15.9 kvars
    (b) 5.33 Ω
    (c) 498 μF

CHAPTER 22

EXERCISE 22-1

$E_L = \sqrt{2}E_P$

1. 339 V
3. 5880 V
5. 7.21 kV
7. 220 V
9. (a) $600\angle 0^\circ$ V; $600\angle 90^\circ$ V
   (b) $849\angle 135^\circ$ V
   (c)

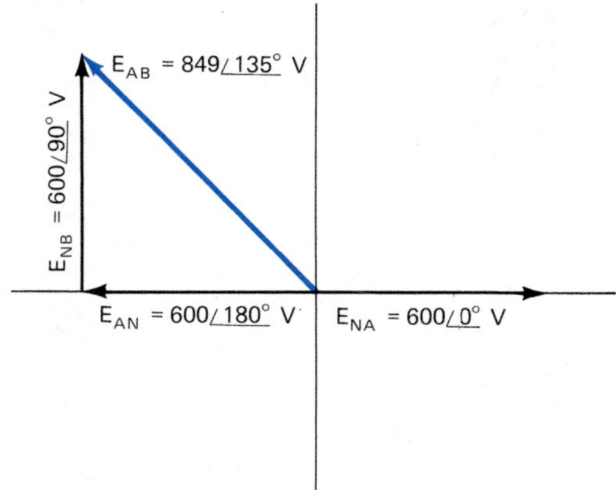

EXERCISE 22-2

1. (a) $240\angle 135^\circ$ V
   (b) $170\angle 0^\circ$ V; $170\angle 90^\circ$ V
   (c) $45\angle 0^\circ$ A; $45\angle 90^\circ$ A
   (d) $63.6\angle 45^\circ$ A
3. 799 mA
5. 112 mA
7. 1.94 A
9. 19.4 A
11. $\mathbf{I_A} = 20\angle -90^\circ$ A
    $\mathbf{I_B} = 20\angle 0^\circ$ A
    $\mathbf{I_N} = 28.3\angle -45^\circ$ A
    $Q_A = 9600$ vars
    $Q_B = 9600$ vars
    $Q_T = 19.2$ kvars
    $P_T = 0$ W
13. $\mathbf{I_A} = 20\angle -25^\circ$ A
    $\mathbf{I_B} = 20\angle 65^\circ$ A
    $\mathbf{I_N} = 28.3\angle 20^\circ$ A
    $P_A = 4350$ W
    $P_B = 4350$ W
    $P_T = 8700$ W
15. (a) $44.0\angle -49.5^\circ$ A; $44.0\angle 40.5^\circ$ A; $62.2\angle -4.5^\circ$ A
    (b) $5.00\angle 49.5^\circ$ Ω
    (c) 12.6 kW (total)
17. (a) $20\angle 17.5^\circ$ Ω
    (b) 22.0 kW

EXERCISE 22-3

1. (a) $E_{NA} = 120\angle 0^\circ$ V; $E_{NB} = 120\angle 120^\circ$ V; $E_{NC} = 120\angle 240^\circ$ V
   (b) $E_{AB} = 208\angle 210^\circ$ V; $E_{BC} = 208\angle 90^\circ$ V; $E_{CA} = 208\angle -30^\circ$ V
3. 480 V
5. 601 V
7. 1390 V
9. 277 V

EXERCISE 22-4

1. (a) $20\angle 0^\circ$ A; $20\angle 120^\circ$ A; $20\angle 240^\circ$ A
   (b) $(20 + j0) + (-10 + j17.3) + (-10 - j17.3) = 0$
3. (a) $144\angle -20^\circ$ A; $144\angle 100^\circ$ A; $144\angle 220^\circ$ A
   (b) 144 A
5. (a) $\theta = 56.3^\circ$
   (b) $3.84\angle -56.3^\circ$ A; $3.84\angle 63.7^\circ$ A; $3.84\angle 183.7^\circ$ A
7. (a) 12.0 W; 6.93 vars
   (b) $41.6\angle 30^\circ$ kVA; 36.0 kW; 20.8 kvars
9. 707 kW
11. (a) 601 V
    (b) $50\angle -75^\circ$ A; $50\angle 45^\circ$ A; $50\angle 165^\circ$ A
    (c) 0 A
    (d) 50 A
    (e) 25.9%
    (f) $52.1\angle 75^\circ$ kVA; 13.5 kW; 50.3 kvars

EXERCISE 22-5

1. 78.5 A
3. 118 A
5. 57.7 A
7. 31.3 A
9. $\theta = 40^\circ$

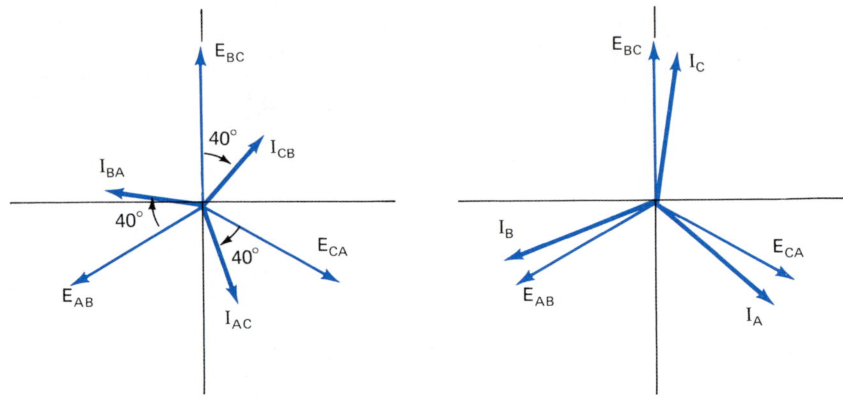

11. (a) 208 V
    (b) $40\angle 210^\circ$ A; $40\angle 90^\circ$ A; $40\angle -30^\circ$ A
    (c) $69.3\angle 240^\circ$ A; $69.3\angle 120^\circ$ A; $69.3\angle 0^\circ$ A
13. (a) $261\angle 190^\circ$ A; $261\angle 70^\circ$ A; $261\angle -50^\circ$ A
    (b) 453 A
15. (a) $8.32\angle 0^\circ$ kVA; 8.32 kW; 0 kvars
    (b) 25.0 kW
17. (a) $3.60\angle 20^\circ$ MVA; 3.38 MW; 1.23 Mvars
    (b) $P_T = 10.2$ MW; $Q_T = 3.70$ Mvars
19. (a) 601 V; 601 V
    (b) 152 A; 87.7 A
    (c) 78.6%
    (d) $52.7\angle 38.2^\circ$ kVA; 41.4 kW; 32.6 kvars
    (e) 124 kW

EXERCISE 22-6

1. 173 A
3. 21.8 A
5. 57.7 A
7. 43.3 A
9. 4.85 kVA; 14.5 kVA
11. 8.88 kVA; 26.6 kVA
13. 23.0 kW
15. 19.3 kW
17. (a) $I_A = 33.3\angle 30^\circ$ A; $I_B = 33.3\angle -90^\circ$ A; $I_C = 33.3\angle 150^\circ$ A
    (b) $I_{BA} = 19.2\angle 240^\circ$ A; $I_{CB} = 19.2\angle 120^\circ$ A; $I_{AC} = 19.2\angle 0^\circ$ A
    (c) $P_T = 13.8$ kW
19. (a) $I_A = 104\angle -21.1^\circ$ A; $I_B = 104\angle -141.1^\circ$ A; $I_C = 104\angle 98.9^\circ$ A
    (b) $108\angle 51.1^\circ$ kVA; 67.9 kW; 84.1 kvars

21. (a) $I_A = 44.5\underline{/5^\circ}$ A; $I_B = 44.5\underline{/-115^\circ}$ A; $I_C = 44.5\underline{/125^\circ}$ A
(b) $I_{BA} = 25.7\underline{/215^\circ}$ A; $I_{CB} = 25.7\underline{/95^\circ}$ A; $I_{AC} = 25.7\underline{/-25^\circ}$ A
(c) $E_{AB} = 240\underline{/240^\circ}$ V; $E_{BC} = 240\underline{/120^\circ}$ V; $E_{CA} = 240\underline{/0^\circ}$ V
(d) $16.8\underline{/25^\circ}$ kVA; 15.2 kW; 7.09 kvars

CHAPTER 23

EXERCISE 23-1

1. 6; 3; 216
3. 10; 3; 1000
5. 12; 3; 1728
7. e; 3.75; 42.52
9. 10; −2.25; 0.005 62
11. $\log_7 49 = 2$
13. $\log 0.001 = -3$
15. $\log_9 729 = 3$
17. $\ln 7.3891 = 2$
19. $\log 0.0178 = -1.75$
21. $2^9 = 512$
23. $10^{2.625} = 421.7$
25. $7^{3.5} = 907.5$
27. $10^{-1.125} = 0.075$
29. $10^0 = 1$

EXERCISE 23-2

1. 0.3010
3. 0.9542
5. 1.1761
7. 1.6990
9. 2.0000
11. 3.7226
13. 1.5527
15. 4.5527
17. −1.0074
19. −4.1175
21. −4
23. −1
25. 4
27. 144
29. 3.26
31. 9700
33. 0.000 571
35. 0.0512
37. 5.75
39. 575
41. 100 000 000
43. 0.0001
45. 10

EXERCISE 23-3

1. 0
3. 1.6094
5. −1.3863
7. 15.2
9. 0.135
11. 1.36×10^{11}
13. (a) 0.125 s
(b) 0.625
15. (a) 7.5 s
(b) 37.5 s
17. (a) 6.25 s
(b) 31.3 s
19. 90.5%
21. 74.1%
23. 60.7%
25. 36.8%
27. 8.2%
29. 0.7%
31.

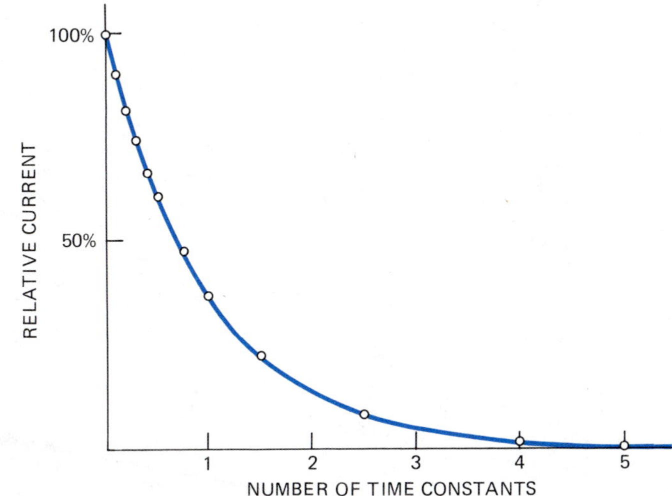

33. 0.223
35. 0.357
37. 0.693
39. 1.20
41. 1.61

EXERCISE 23-4

1. (a) 2.67 s
   (b) 13.3 s
3. (a) 22.9 s
   (b) 114 s
5. 9.5%
7. 25.9%
9. 39.4%
11. 63.2%
13. 91.8%
15. 99.3%
17.

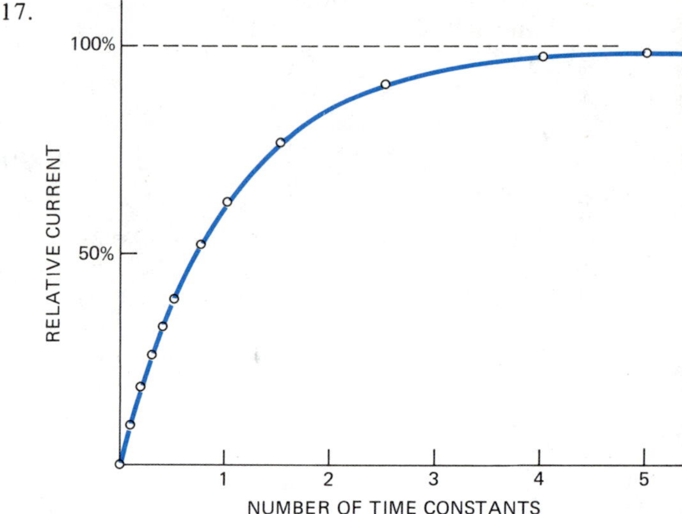

19. 0.223
21. 0.357
23. 0.693
25. 1.20
27. 1.61

EXERCISE 23-5

1. $\log 5 + \log 17 = 1.9294$
3. $\log 38 - \log 27 = 0.1484$
5. $3 \log 14 = 3.4384$
7. $\frac{1}{2} \log 528 = 1.3613$
9. $3 \log 30 + \log 7 + \frac{1}{2} \log 24 = 5.9666$
11. $\log 22(16) = 2.5465$
13. $\log \frac{75}{5} = 1.1761$
15. $\log \frac{144 \times 35}{12} = 2.6232$
17. $\log \sqrt[3]{1728} = 1.0792$
19. $\log \frac{12}{\sqrt{144}} = 0$

EXERCISE 23-6

1. 1000
3. 100 000
5. 100
7. 10 000
9. 12 000
11. 0.0445
13. 0.499
15. 1.87
17. 5.06
19. 100

EXERCISE 23-7

1. 11.78 dB
3. −22.54 dB
5. 60.62 dB
7. −48.44 dB
9. $P_i = 4.33$ W
11. $I_i = 9.98$ A
13. −25.31 dB
15. (a) −23.98 dB
    (b) −0.32 dB
17. (a) 15.8
    (b) 250

EXERCISE 23-8

1. 12 dB
3. 20 dB
5. 13 dB
7. 26 dB
9. −13 dB
11. −43 dB
13. 19 dB
15. 32 dB
17. 200 W
19. 16 W
21. 50 μW
23. 1.5 mW
25. 375 μW
27. 120 W
29. 100 W

EXERCISE 23-9

1. 486 nF
3. 123 mH
5. (a) 249 nF
   (b) 10.7 Ω
7. 20.3 Ω (calculated value)
   20.2 Ω (table value)
9. (a) 385 mH
   (b) 677 nF
   (c) 122 Ω (calculated value)
   122 Ω (table value)
11. A = 53.49 antilog (−g/10)
   (a) 33.75 mm^2 ; 33.62 mm^2 ; 0.4%
   (b) 13.44 mm^2 ; 13.30 mm^2 ; 1.1%
   (c) 5.349 mm^2 ; 5.261 mm^2 ; 1.7%
   (d) 2.13 mm^2 ; 2.08 mm^2 ; 2.4%
   (e) 84.78 mm^2 ; 85.01 mm^2 ; 0.3%
   (f) 53.49 mm^2 ; 53.49 mm^2 ; 0%

INDEX

Abbreviation. *See* Symbols
Abscissa, 72
Absolute value, 87
Accuracy, 115
  in calculations, 119
  of measurements, 115
Active circuit elements, 179
Addend, 7
Addition, 7–8
  axiom, 168
  of common fractions, 194–197
  of complex numbers, 435–439
  of monomials, 248
  of phasors, 470–473
  of polynomials, 248
  in scientific notation, 105–106
  of signed numbers, 88
  of vectors, 91–92, 435–437
  of whole numbers, 7–8
  words meaning, 169
Admittance, 509–512, 526–529
Algebraic expressions. *See* Expressions, mathematical
Alternating current, 392–393. *See also* Circuits, alternating current
Alternators, 392
  polyphase, 539
  single-phase, 395
  three-phase, 547
    delta-connected, 563–567
    wye-connected, 547–550
  two-phase, 540–543
American wire gage tables, 160, 599
Ammeters, 35
  construction of, 329–331
  multirange, 332
  use of, 35
Ampere, 28, 123, 174
Amplifier, 585
Amplitude of sine waves, 391
Analog scales, 48, 116–118
Angle, 339
  acute, 340
  bisecting an, 360
  of depression, 365
  of elevation, 365
  generated, 341–342
  of incline, 365
  negative, 342
  obtuse, 340
  phase, 407
  positive, 342
  reference, 352
  reflex, 340
  right, 340
  symbols for, 339
  trigonometric functions of an, 355–359, 370
  units of, 343–345
    degree, 340
    grad, 345
    minute of arc, 345
    π-radian, 345
    radian, 343
    revolution, 344
    right angle, 340
    second of arc, 345
    sexagesimal system, 345
    straight, 340
  working, 603
Angles, 339–342
  complementary, 340
  equivalent, 342
  supplementary, 340
Angular frequency, 398
Angular measurement. *See* Angle, units of
Angular velocity, 319, 391, 396
Antilogarithm, 573–574
Approximately equal to, 361
Arc functions, 368, 370
Arc functions (Continued)
  of generated angles, 380–381
  from tables, 603
  trigonometric, 368–369
  See also Principal value
Area, 151–153
  cross-sectional, 155
  formulas for, 152, 156
  surface, of solid objects, 155
  units of, 151
Arithmetic, 6–12
  of common fractions, 42–43, 194–197
  of decimal fractions, 53–55
  in scientific notation, 105–108
  of signed numbers, 88–93
  of whole numbers, 8–12
Atomic number, 21
Atoms, 20
  of aluminum, 22–25
  balanced, 23
  normal, 22–23
  planetary model of, 21–23
Attenuator, 585
Axes, 71
  of graphs, 71–72
  of rotation, 386
Axiom, 138
  addition, 168
  commutative, 170
  division, 138
  multiplication, 138
  power, 185
  root, 185
  subtraction, 168

Back emf, 407, 410
Bar. *See* Symbols, grouping
Base, 230
  of common logarithms, 570–571
  of logarithms, defined, 570

Base (Continued)
of natural logarithms, 571
of percents, 230
of powers, 61
of triangles, 152, 349
units of SI system, 121–123
Battery, 26
cells, in parallel, 201–202
cells, in series, 178–180
internal resistance of, 183
Bel, 586
Bell, Alexander Graham, 586
Binomial. *See* Expressions, mathematical
Borrowing, 8
Braces, 17
Brackets, 17
Bridge, Wheatstone, 333–334
British thermal unit, 238

Calculator, electronic, 4–6, 52–64
essential functions, 4–5
operations of, 52–58
antilogarithms, 573
common logarithms, 572
converting angular units, 344–345
engineering notation, 112
inverse trigonometric functions, 368–369
natural logarithms, 574–575
powers, 61–62
reciprocals, 59–61
roots, 63–64
scientific notation, 111–112
signed numbers, 90
transformation of coordinates, 457, 461
trigonometric functions, 358–359
programmable, 4
Cancellation, 42–43, 257
of units, 131–132
Candela, 123
Capacitance, 382, 412
effects of, in ac circuits, 414–415
in parallel circuits, 415
in series circuits, 415
units of, 104, 412
Capacitive reactance, 414–415, 466
units of, 414
Capacitors, 103
charging-discharging, 412–413
color code for, 102–104
construction, 103
power factor correction, 536–537
Carrying, 8, 10
Change, 89
increase-decrease, 89
percent, 231
Characteristic, 98
Charge, electric, 21, 29, 148
Circle, 152
Circuit analysis, 289
equivalent circuit method, 207
Kirchhoff's method, 292–295
Ohm's Law method, 146–148
principle of superposition, 299–304
Circuit analysis (Continued)
Thévenin's Theorem, 305–308
Circuit breaker, 27
Circuit diagrams, 32
Circuit symbols, 30–35
Circuits, electric, 26–36
alternating current, 509
admittance in, 509–512, 526–529
current in, 28, 403
impedance of, 416
parallel, 513
phase relationships, 446
power, 484
reactance in, 406, 414
resistive, 67, 469–502
series, 469–502
three-phase, 547
two-phase, 540–543
voltage in, 403
capacitive, 414–415, 472
combined, 206–213
delta-delta, 563–567
delta-wye, 556–561
direct current, 66–70
equivalent, 175–176, 199
four-wire wye connected, 547
ideal, 469–502
inductive, 408–410
open-closed, 27
parallel, 199–209
parallel equivalent, 199, 526, 530–531
power in, 68–70, 520
practical, 507–508, 520–525
RC, 484
RCL, 486
RL, 482
series, 34, 174–181
series equivalent, 180, 507
single load, 38, 65–70
symbol table, 30
wye-wye, 547–548
Circumference, 151
Code, color, 102–104
Coefficient, 135, 138, 166, 247
defined, 135
of resistivity, 160, 600
Coil, 406
D'Arsonval movement, 329–332
generator, 395–396
induction, 406–408
transformer, 324–327
Color code, 102–104
Commission and sales rate, 230
Common fractions, 39–45, 194–197
Common logarithms, 572–573
Common units, 101, 129
Complementary angles, 340
Complex conjugate, 440
Complex fractions, 45
Complex numbers, 430
Compound, 20
Condensor. *See* Capacitor
Conductance, 200
Conduction band, of electrons, 24
Conductor, electrical, 25
Conjugate, complex, 440
Constant, 72, 135
time, 576
of variation, 216
Conversion factors, table of, 238
Conversion of units, 131
Coordinates, 72
polar, 451
rectangular, 72, 273, 451
transformation of, 453–464
graphical, 453–456
P → R, 461–463
R → P, 457–459
Cosine, 356
Coulomb, 29
Counter-emf, 407, 410
Cross multiplication, 321–322
Cube, 62
of numbers, 62
root, 63–64
solid figure, 155–156
Current, 28–29, 148
alternating. *See* Circuits, alternating current
direct, 28
divider rule, 298
effective, 403
induced, 387
instantaneous, 392–393
line, 551
in delta-delta circuits, 563–564
in delta-wye circuits, 557
in wye-wye circuits, 551–552
maximum, 403
in parallel circuits, 202–203
phase, 551
rms, 404
in series circuits, 176–177
units of, 28
Curve, of graphs, 74–76, 384, 426
Cycle, 394
Cylinder, 155

D'Arsonval movement, 330
Decibels, 586
approximating, 590–592
formulas for, 586, 588
Decimal. *See* Fractions, decimal
Decimal point, 47
Degree, 340, 343
Delta. *See* Symbols
Delta connection, 556
Denominator, 39
common, 194
defined, 39
least common, 194–195, 271
Diameter, 152
of circles, 152
of spheres, 155
Dielectric, of capacitors, 412
Difference, 8
Digits, 7
Digits, significant, 97

Discount rate, 230
Distance, measurement of, 151
Distributive Law, 254, 261. *See also* Factoring
Dividend, 12
Division, 12
- of common fractions, 44–45
- of complex numbers, 440–441
- of decimal fractions, 55
- by monomials, 257–259
- of phasors, 477–478
- of polynomials, 258–259
- in scientific notation, 107–108
- of signed numbers, 92–93
- of vectors, 477–478
- of whole numbers, 12
- words meaning, 141

Divisor, 12
Dynamometer, 74, 81

Effective current and voltage, 402–403
Efficiency, 237–240, 317
Efficiency, luminous, 239–240
Electromotive force (emf), 30, 390
- alternating, 390–393
- counter-, 407
- defined, 30
- units of, 31

Electron, 21
- current, 26
- defined, 21
- drift, 26
- free, 24
- shell, 21–25
- valence, 22

Element, 20
Energy, 68, 148
- consumption, 242–243
- cost, 243
- defined, 68
- levels, 24
- transformation, 238
- units of, 68, 242

Equations, 138
- with combined operations, 171–172, 266–267
- equivalent, 138, 269, 276, 292
- exponential, 186
- fractional, 270–272
- graphing, 273–275
- with grouping symbols, 268–269
- inconsistent, 277, 292
- linear, 273
- logarithmic, 580–585
- quadratic, 424–428
  - complete, 425
  - formula, 425
  - pure, 424
  - root of, 425
  - radical, 185
  - simultaneous, 273
    - addition-subtraction method, 278–280
    - comparison method, 282–283
    - defined, 273

Equations, simultaneous (Continued)
- deriving formulas, 287
- graphical method, 276–277
- substitution method, 281–282
- third order, 290–292
- unique system of, 292
- word problems, 284–285
- solving, 139–140, 168–170, 264–272
- terms of, 166–167
- *See also* Formulas

Exponential decay, 575–577
Exponential growth, 578–579
Exponents, 61, 570
- fractional, 63–64, 260
- negative, 256
- power rule, 109, 581
- power of ten, 94–95
- product rule, 253, 580
- quotient rule, 255–256, 580

Expressions, mathematical, 14–18
- algebraic, 247
- arithmetical, 14
- binomials, 14, 166, 419–422
  - addition-subtraction, 248–251
  - defined, 247
  - division by a monomial, 258–261
  - multiplication by a binomial, 419
  - multiplication by a monomial, 254
  - terms of, 14, 247
- evaluating, 57–61
- monomials, 14, 166, 247, 257
- polynomials, 14, 166, 247
- value of, 14

Factor, 13, 40, 137
- largest common, 40
- literal, 137
- power, 465
- prime, 13, 137

Factoring, 13
- numbers, 13–14
- polynomials, 261–262
- special products, 423–424

Farad, 104
Figure, significant, 97
Flux, luminous, 85, 225
FOIL method, 420
Foot-candle, 226
Formulas, 65, 135, 143
- for area, 152, 156
- circuit, 148
- deriving, 287
- Ohm's Law, 65
- for perimeter, 152
- with plus and minus signs, 173–174
- for power, 68
- quadratic, 425–428
- rule for solving, 143
- simple, 143
- for surface area, 156
- for volume, 156
- *See also* Equations

Fractions, 39–45
- algebraic, 270–272

Fractions (Continued)
- clearing, in equations, 270
- common, 39–45, 194–197
  - adding, 194–197
  - defined, 39
  - dividing, 44–45
  - multiplying, 42–43
  - subtracting, 194–197
- complex, 45
- decimal, 47–51
  - adding, 53
  - dividing, 55
  - exact, 50
  - multiplying, 55
  - repeating, 50
  - subtracting, 53
- equivalent, 40–41
- improper, 39
- like-unlike, 40, 196
- mixed decimal, 47
- proper, 39, 48
- reducing, to lowest terms, 40

Free electron, 24–25
Frequency, 395
- angular, 398
- defined, 395
- resonant, 532

Functions, 72
- defined, 72
- exponential, 575–579
- of generated triangles, 375–377
- logarithmic, 580
- periodic, 394–405, 446–453
- trigonometric. *See* Trigonometric functions
- *See also* Variation

Fundamental units, 120
Fuse, 27

Gage, 159
- American Wire, 159–161
- tables, 160

Gain, 587
Galvanometer, 330
Gears, 319–320
Generation, 383
- of angles, 341–342
- of electricity, 383
- of triangles, 375–379

Generators, 19, 383
- ac. *See* Alternators
- DC, 19–20

Grad, 345
Grade, 352, 365
Graphs, 71–80
- construction of, 77–80
- of exponential functions, 575–576, 578
- of lab data, 80–82
- of linear functions, 74–75, 273–275
- of nonlinear functions, 75–76
- of trigonometric functions, 383–385

Greek letters, 136. *See also* Symbols

Heat, units of, 238

Henry, 408
Hertz, 395
Horsepower, 69, 317
How to study, 2–3
Hypotenuse, 349

Illumination, 225–226
Imaginary numbers. *See* j-Numbers
Impedance, 416
  defined, 416
  in parallel circuits, 503–530
    admittance method, 509–510
    conductance-susceptance method, 526–529
    total current method, 517
    two-branch formula method, 505
  in series circuits, 469–502
Index, of a radical, 63
Inductance, 382, 406–411
  in ac circuits, 408–410
  effect of, 407
  of transmission lines, 595
  units of, 408
  variable, 410
Induction, mutual, 406
Induction, self, 407
Inductive reactance, 408, 466
  calculations of, 408–410
  units of, 408
Insulator, 25
Integer, 431
Interaction, Law of, 21
Interest rate, 230
Inverse logarithms. *See* Antilogarithm
Inverse operations, 171
Inverse trigonometric functions. *See* Arc functions
Inverse variation, 220–223
Ion, 22
IR drop, 180

j-Numbers, 430–431
j-Operator, 432–434
Joule, 68, 242

Kelvin, 123
Kilocalorie, 238
Kilogram, 122
Kilomole, 123
Kilowatt-hour, 242
Kilowatt-hour meter, 73, 237
Kirchhoff's Current Law, 292
  in combined circuits, 292–295
  in parallel circuits, 199
  in series circuits, 176–177, 181
Kirchhoff's Voltage Law, 181, 292–295

Lag, 385, 407
Largest common factor, 40
Law, 65
  distributive, 254, 419
  of exponents, 106–110
  of interaction, 21
  Kirchhoff's Current, 176, 292
  Kirchhoff's Voltage, 181, 292
Law (Continued)
  Ohm's, 65–67, 202
  of proportionality, 322
Lead-lag, 385, 407
Line of sight, 365
Line voltage, 541, 549
Linear equations, 273
Litre, 122
Logarithmic equations, 580–584
Logarithms, 570–580
  common, 571–573
  natural, 571, 574–575
  properties of, 580–582
  tables, 611
Lumen, 225, 239
Luminous efficiency, 239–240
Luminous flux, 85, 225, 239
Luminous intensity, 123
Lux, 226

Machines, 234, 238
Magnetic field, induced, 324
Magnetic flux, 386
Magnitude, 87, 450
Mantissa, 99
  of a logarithm, 608–611
  in scientific notation, 99
  tables of, 611
Mass, 122
Matter, 20
Measuring systems. *See* Systems, measuring
Mechanical advantage, 318
Meter, analog, 36
Meter, digital, 36. *See also* Ammeter, Voltmeter, etc.
Metre, 122
Metric prefixes, 124–125
Metric units, 121–123
  derived, 121
  fundamental, 120
Microammeter, 330
Micrometer, 115
Mil, 161
Mil, circular, 161
Minuend, 8
Molecule, 20
Monomial. *See* Expressions, mathematical
Motors, 521–522
  induction, 521
  synchronous, 522
Multimeters, 36
Multiplicand, 10
Multiplication, 10
  of binomials, 252
  of common fractions, 42–43
  of complex numbers, 440
  of exponentials, 252
  of monomials, 253
  of phasors, 476–477
  in scientific notation, 106
  of signed numbers, 93–94
  signs, 10, 137
  of whole numbers, 10–11
  words meaning, 141
Mutual induction, 406
Napier, John, 569
Natural logarithms, 571, 574–575
Natural numbers, 431
Negative numbers, 86–94
Neutral current, 544–545, 551
  in three-phase circuits, 551
  in two-phase circuits, 544–545
Neutral line, 541, 548
Neutron, 21
Nominal value of measurements, 232
Notation, 95
  decimal, 95
  double-subscript, 540–543
  engineering, 101
  exponential, 95
  logarithmic, 570
  scientific, 96–102
Nucleus, of atoms, 21
Number scale, 87
Number system, 431–432
Numbers, 6
  complex, 430–432
  composite, 13
  directed, 91
  imaginary, 430
  integer, 431
  irrational, 432
  j-, 430
  literal, 135
  mixed, 39–40
  natural, 87, 431
  negative, 86–94, 431
  power of, 61–62
  prime, 13
  rational, 432
  real, 432
  signed, 86–94
  whole, 6–13
Numerals, 6
Numerator, 39

Ohm, 29
Ohm, Georg, 65
Ohmmeter, 36
Ohm's Law, 38, 65–67, 180–181, 202–203, 480–482, 514
Omega, 396
Operation, 14
  inverse, 171
  order of, 14–18
Operator
  j-, 432
  rotation, 92, 432
Orbit, electron, 21
Ordinates, 72
Origin, of coordinate system, 72

Parabola, 426–428
Parallel circuits, 199, 514
  ac, 514–525
  DC, 34, 199–205
Parallel equivalent circuits, 200, 530–531, 535
Parentheses, 17. *See also* Symbols, grouping
Passive circuit elements, 179

Percent, 228–231
changing fraction to, 228
efficiency, 237–240, 317
operations with, 230–231
power factor, 465
Perimeter, 151–153
Period, 394
Periodic cycle, 394
Periodic function, 383
Phase angle, 407, 448
Phase current, 544, 551, 557
Phase relationships in ac circuits, 446–452
Phase voltage, 541, 549, 557
Phasor, 450–452
graphical representations, 435
operations
adding, 435
dividing, 440, 477
multiplying, 440, 476
reciprocal of, 509
subtracting, 438
polar form of, 451
rectangular form of, 451
Pi (π), 135
π-Radian, 345
Place value, 6–7, 47
Planetary model of the atom, 21–23
Polar form of phasors, 451
Polarity of circuit elements, 293, 331
Pole, magnetic, 387, 395–396
Polynomials, 14, 166
adding, 248
defined, 14
dividing by a monomial, 258–259
multiplying by a monomial, 254–255
subtracting, 250
Position, of rotating vectors, 341
initial, 341
standard, 341
terminal, 341
Positive numbers, 87, 431
Potential difference, 30, 178. *See also* Voltage
Power amplitude, 402
Power axiom, 185
Power, electrical, 68, 148
in ac circuits, 402–404
active, 416
amplitude, 402
apparent, 484, 489
average, 404
in DC circuits, 68–70
defined, 68
instantaneous, 402
maximum, 402
in parallel circuits, 205, 519
peak, 402
reactive, 490
in series circuits, 188–190, 489–501
true, 416
units of, 68–69, 188, 492
Power factor, 465, 488
defined, 465
reactive, 492
Power factor (Continued)
typical values, 521
Power factor correction, 501, 532–537
Power level, 586
Power loss, 68
Power rating, 70
Power ratio, 586
Power station, 19
Power triangles, 518
Powers, 61
algebraic, 185
fractional, 260
logarithm of, 581
of a monomial, 259–260
negative, 95, 256
of a number, 61
of a power, 109, 259
of a product, 110
in scientific notation, 94–96, 109–111
of ten, 94
Prefix, metric, 124–125
Prime numbers, 13
Principal value of arc functions, 380
Principle of superposition, 299–304
Problem solving techniques, 66, 135, 141, 145–146, 284
Product, 10
Products, phasor, 476–477
Products, special, 421–422
Profit rate, 230
Proper fractions, 39
Proportion, 321–323
direct, 322
inverse, 323
Proportionality, law of, 322
Proton, 21
Pulleys, 319
Pythagorean Theorem, 349

Quadrants, 71
Quadratic equations, 424–428
Quadratic formula, 425–428
Quantity, 71
Quantity, basic, 120
Quotient, 12
defined, 12
logarithm of, 580
phasor, 440–441, 477

Radian, 343
Radical, 63
Radius, 152
of circles, 152
of spheres, 155
vector, 341
Rate, 230
Ratio, 39, 313–320
defined, 313
efficiency, 317
forms of, 314
inverse, 314
percentage, 315
power, 586
simplifying, 314
Ratio (Continued)
speed, 319
terms of, 39, 313–314
trigonometric, 356–357
Reactance, 406
capacitive, 414–415
inductive, 408
units of, 408
Reciprocals, 59–60
defined, 44
dividing with, 44–45, 60–61
in scientific notation, 109–111
Rectangles, 151
Rectangular form, of phasors, 451
Relationships, variation, 216
Relative tolerance, 232
Remainder, 8, 12
Resistance, 29, 148
in combination circuits, 207–209
defined, 29
internal source, 159, 183
in parallel circuits, 199–201
in series circuits, 175–176
temperature coefficient of, 161, 600
in transmission lines, 595
of wire, 29, 159–161, 600
Resistivity of metals, 160, 600
Resistors, color code for, 102–103
Resonance, 532
Resultant, of vectors, 91, 435
Revolution, 344
Rheostat, 216
Root axiom, 185
Root-mean-square (rms), 404
Roots, 63–64
cube, 63
of a monomial, 260–261
of a quadratic equation, 425
in scientific notation, 109–111
square, 63
Rotating vectors, 341
Rotor, 383
Rounding, 50–51
of calculated values, 119
procedure, 51
Rule
current divider, 298
fundamental, of arithmetic, 196
left-hand motor, 387
for SI usage, 125–126
for solving applied problems, 145–146
for transformation of coordinates, 453, 457, 461
graphical, 453
P → R, 461
R → P, 457
voltage divider, 297

Scalar, 450
Scales, 48
analog, 48, 116–118
graphing, 78, 87
Scientific notation, 96–102
arithmetic of, 105–111

Scientific notation (Continued)
 defined, 96
 engineering, 101–102
 standard, 98
Second, 122
Self-induction, 407
Semiconductor, 25
Series circuits, 34, 176–181
Service drop, 352
Sexagesimal measuring system, 345
Shell, 21
 conduction, 24
 electron, 21–23
 innermost-outermost, 22
 valence, 22
Shunt, ammeter, 330
Siemens, 200
Signed numbers, 86–94
Significant figures, 97
Signs, 10, 137
 of angles, 342
 like-unlike, 88
 of trigonometric functions, 377–379
Simultaneous equations, 273–283
Sine, 356
 calculator calculations, 358
 defined, 356
 tables, 604–607
 using tables, 602–603
 wave, 383
Slope, 352, 365
Solenoid, 407
Solid, rectangular, 155
Solution, 139
 checking the, 141, 146
 of problems, 145–146
Speed, angular, 319
Speed ratio, 319
Sphere, 155
Square, 61
 of algebraic expressions, 259–260
 figures, 151
 of numbers, 61–62
 See also Powers
Square root, 63. *See also* Roots
Standard position, 98, 341
Station, 19
 power, 19
 sub-, 154, 234–235
Stator, 383
Structure of matter, 20–23
Subatomic particles, 21
Subscripts, 136–137
Substitution, 146, 281–282
Subtraction, 8
 axiom, 168
 of common fractions, 194–197
 of complex numbers, 438
 of decimal fractions, 53
 defined, 8
 of monomials, 250
 of phasors, 474
 of polynomials, 250
 in scientific notation, 105–106
Subtraction (Continued)
 of signed numbers, 89
 of vectors, 92–93, 438–439
 of whole numbers, 8–9
 words meaning, 169
Subtrahend, 8
Sum, 7
Sum, vector, 92, 435
Superposition, principle of, 299–304
Superscript, 61
Supplementary angles, 340
Surface area, 155
Susceptance, 527
Switch, electrical, 27
Symbols, 10, 137
 circuit, 33
 delta, 137
 double subscript, 540–543
 greek letters, 136
 grouping, 17, 137
 literal numbers, 135
 metric prefix, 125
 of operation, 7–12, 137
 quantity, 122
 subscripts, 136
Systems
 British measuring, 130–132
 distribution, 234–235
 mks, 121
 polar coordinate, 451
 rectangular coordinate, 71
 SI measuring, 121

Tables
 American wire gage, 160, 599
 appliance usage, 243
 calculator functions, 5
 circuit formulas, 148
 color code, 102–104
 conversion factors, 238, 601
 efficiency of lamps, 240
 Greek letters, 136
 logarithms, 611
 perimeter, area and volume, 152, 156
 resistivity, 160, 600
 SI quantity symbols, 122
 SI unit symbols, 122
 temperature coefficient of resistance, 161, 600
 trigonometric functions, 604–607
Tangent, 356
 calculations of, 358
 defined, 356
 tables, 604–607
 using tables, 602–603
Tax rate, 230
Temperature, units of, 123
Terms, 14
 of algebraic expressions, 166–167, 247, 419
 collecting like, 267
 factors of, 137–138
 of fractions, 39
 like-unlike, 166–167, 247
Terms (Continued)
 of mathematical expressions, 14, 247
 of a ratio, 39
 reducing to lowest, 40
 transposing, 169
Thévenin's Theorem, 305–308
Three-phase alternators, 547–567
Time constant, 576, 578
Time phase, 464–465
Time, units of, 122
Tolerance, 102, 232–233
Transformation of coordinates, 453–464
 graphical, 453–456
 P → R, 461–463
 R → P, 457–459
Transformers, 324–327
 construction of, 324
 current in, 326
 isolation, 325
 step-up, step-down, 325
Transposing, 169
Triangles, 151–152
 angles of, 347
 area of, 152
 construction of, 352–353, 360–362
 estimating angles and sides of, 360–362
 right, 346–347
  defined, 152
  SA, solving, 362–365
  SS, solving, 370–372
 similar, 334–335
 standard position of, 349
 types of, 151–152, 346
Trigonometric functions, 355–359, 370
 defined, 356
 graphs of, 383–385
 inverse, 368–369
 signs of, 377–379
 tables of, 604–607
True power, 416
Two-phase alternators, 540–545

Uncertainty, 115
Units, 31
 British, 130–132
 common, 101, 129
 compound, 126
 conversion factors, 238, 601
 converting, 127–132
 derived, 121
 electrical, 31
 fundamental, 120
 SI, 121–123

Valence electron, 22
Value, absolute, 87
Value, principle, of arcfunctions, 380
Var, 492
Variable, 72, 135
 defined, 72
 dependent-independent, 73, 274
Variation, 215
 constant, 216
 direct, 216–219

Variation (Continued)
inverse, 220–223
inverse square, 224–227
joint, 224–227
linear, 216–219
Varmeter, 492
Vector, rotating, 391
Vectors, 91–92, 432. *See also* Phasors
Vectors, radius, 341, 375
Velocity, angular, 319, 391, 396
Vertex, of an angle, 339
Vinculum, 17
Volt, 31
Voltage, 30, 148, 390
ac, production of, 390–393
alternating, 390–393
divider rule, 297
drop, 30, 179
effective, 402
instantaneous, 391
line, 541, 549
Voltage (Continued)
maximum, 390–392
in parallel circuits, 201
phase, 541, 549
rms, 404
in series circuits, 178–179
sources, 178
terminal, 183
in three-phase circuits, 549
delta-delta, 563
delta-wye, 556
wye-wye, 549
in two-phase circuits, 541
wave equation, 398
Volt-ampere, 188
Voltmeter, 35
connection of, in circuits, 35
construction, 331–332
multirange, 332
Volume, 155–157
formulas for, 156
Volume (Continued)
units of, 155

Watt, 68
Wattmeter, 237
Wheatstone bridge, 333–334
Whole numbers, 6–13
Wire, AWG tables, 160
Wire resistance, 159–161
Word problems, solving, 141, 169–170, 284–285
Work, units of, 68, 236
Wye connection, 547–550

x-Axis, 72

y-Axis, 72
Yoke, of transformers, 324

Zero exponent, 95